W9-BEE-435

MCGRAW-HILL
ONLINE RESOURCES

IMPORTANT:

HERE IS YOUR REGISTRATION CODE TO ACCESS

YOUR PREMIUM McGRAW-HILL ONLINE RESOURCES.

For key premium online resources you need THIS CODE to gain access. Once the code is entered, you will be able to use the Web resources for the length of your course.

If your course is using **WebCT** or **Blackboard**, you'll be able to use this code to access the McGraw-Hill content within your instructor's online course.

Access is provided if you have purchased a new book. If the registration code is missing from this book, the registration screen on our Website, and within your WebCT or Blackboard course, will tell you how to obtain your new code.

Registering for McGraw-Hill Online Resources

To gain access to your McGraw-Hill web resources simply follow the steps below:

http://www.mhhe.com/environmentalscience

1. USE YOUR WEB BROWSER TO GO TO:

2. CLICK ON **FIRST TIME USER**.

3. ENTER THE REGISTRATION CODE* PRINTED ON THE TEAR-OFF BOOKMARK ON THE RIGHT.

4. AFTER YOU HAVE ENTERED YOUR REGISTRATION CODE, CLICK **REGISTER**.

5. FOLLOW THE INSTRUCTIONS TO SET-UP YOUR PERSONAL UserID AND PASSWORD.

6. WRITE YOUR UserID AND PASSWORD DOWN FOR FUTURE REFERENCE.
   KEEP IT IN A SAFE PLACE.

TO GAIN ACCESS to the McGraw-Hill content in your instructor's **WebCT** or **Blackboard** course simply log in to the course with the UserID and Password provided by your instructor. Enter the registration code exactly as it appears in the box to the right when prompted by the system. You will only need to use the code the first time you click on McGraw-Hill content.

Thank you, and welcome to your McGraw-Hill online resources!

 Mc Graw Hill **Higher Education**

pipelining-94836425

REGISTRATION CODE

0-07-250931-7 CUNNINGHAM: PRINCIPLES OF ENVIRONMENTAL SCIENCE, 2E

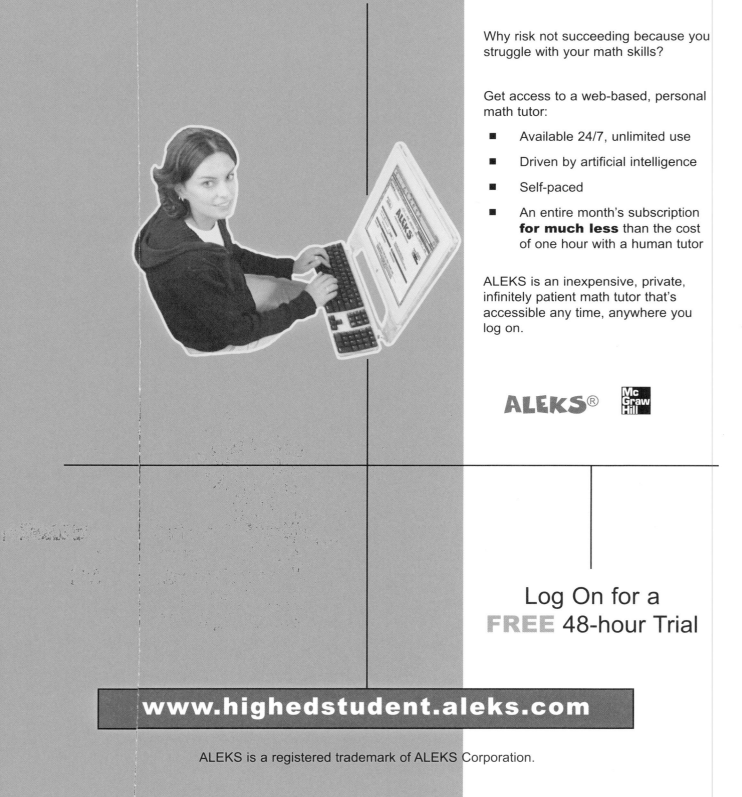

William P.
CUNNINGHAM
University of Minnesota

Mary Ann
CUNNINGHAM
Vassar College

Second Edition

Principles of ENVIRONMENTAL
Science

Inquiry and Applications

Wisconsin Center for Environmental Education
403 Learning Resources Center
University of Wisconsin-Stevens Point
900 Reserve Street
Stevens Point, WI 54481-4419

Higher Education

Boston Burr Ridge, IL Dubuque, IA Madison, WI New York San Francisco St. Louis
Bangkok Bogotá Caracas Kuala Lumpur Lisbon London Madrid Mexico City
Milan Montreal New Delhi Santiago Seoul Singapore Sydney Taipei Toronto

PRINCIPLES OF ENVIRONMENTAL SCIENCE: INQUIRY AND APPLICATIONS
SECOND EDITION

♻ This book is printed on recycled, acid-free paper containing 10% postconsumer waste.

International   1 2 3 4 5 6 7 8 9 0 QPD/QPD 0 9 8 7 6 5 4 3 2
Domestic   1 2 3 4 5 6 7 8 9 0 QPD/QPD 0 9 8 7 6 5 4 3 2

ISBN 0–07–250931–7
ISBN 0–07–121449–6 (ISE)

Publisher: *Margaret J. Kemp*
Senior developmental editor: *Kathleen R. Loewenberg*
Marketing manager: *Heather K. Wagner*
Project managers: *Jane Mohr / Rose Koos*
Lead production supervisor: *Sandy Ludovissy*
Senior media project manager: *Stacy A. Patch*
Media technology producer: *Renee Russian*
Coordinator of freelance design: *Michelle D. Whitaker*
Cover/interior designer: *Jamie E. O'Neal*
Cover image: *© David Woodfall/Stone;* back cover image: *Visions of Nature/Corbis*
Senior photo research coordinator: *Lori Hancock*
Photo research: *Connie Mueller*
Compositor: *Precision Graphics*
Typeface: *10/12 Times Roman*
Printer: *Quebecor World Dubuque, IA*

The credits section for this book begins on page 393 and is considered an extension
of the copyright page.

Library of Congress Cataloging-in-Publication Data

Cunningham, William P.
      Principles of environmental science : inquiry and applications /
William P. Cunningham, Mary Ann Cunningham. — 2nd ed.
          p. cm.
      Includes index.
      ISBN 0–07–250931–7 (hard copy : alk. paper)
      1. Environmental sciences. I. Cunningham, Mary Ann. II. Title.

GE105 .C865   2004
363.7—dc21
                                            2002151069
                                            CIP

www.mhhe.com

CONTENTS IN BRIEF

GE
105
C865
2004

CONTENTS

Chapter 4 Human Populations

Chapter 5 Biomes and Biodiversity

PREFACE

NEEDED: A SCIENCE OF SUSTAINABILITY

When world leaders and delegates from 190 nations met in Johannesburg, South Africa in 2002 for the World Summit on Sustainable Development, a host of urgent issues filled their agenda. Among the pressing problems they debated were global climate change, freshwater shortages, dwindling forests, biodiversity losses, food insecurity, over-exploited fish stocks, emergent ecological diseases, dependence on fossil fuels, and the role of social, political and economic factors in environmental degradation.

Underlying these issues was a division between the world's rich citizens, who want to protect nature, and the 80 percent of the world's population living in the poorest countries who insist that they can't be concerned with safeguarding natural resources until they achieve simple survival. This division was highlighted by South Africa's President Thabo Mbeki, who said in his opening comments: "A global human society based on poverty for many and prosperity for a few, characterized by islands of wealth, surrounded by a sea of poverty is unsustainable. The tragic result of this is the avoidable increase in human misery and ecological degradation."

One answer to this dilemma is sustainable development that addresses immediate human and social needs while preserving the earth's fragile life support systems. But good science is needed to provide answers about how we can achieve these goals. The Paris-based International Council for Science, which was invited to prepare briefing documents as background for the World Summit, said "the transition towards sustainable development is inconceivable without science, engineering, and technology. We need rigorous scientific information that addresses issues people care about."

To reach the goals of sustainability, we also need an educated public that understands some of the scientific basis of the complex problems and difficult trade-offs we face. A broad-based environmental science course is an excellent way to teach a wide range of students about both scientific and social issues. It can show how valid information is gathered and analyzed as well as how to think critically about complex issues. Many instructors have asked for a textbook that gives students a strong foundation in the basic principles of environmental science. We understand and agree with that goal.

This book provides a solid foundation in scientific approaches to environmental problems and solutions. We integrate information from a wide range of disciplines from both the natural and social sciences. And we attempt to present a **balanced, objective perspective** that presents both sides of controversial issues.

"I appreciate the way the authors have treated the subject matter. They are not judgmental or try to overtly express one idea over another. I find students want to be provided balanced information to allow for good in-class discussions and to make informed decisions."

Ron Sundell
Northern Michigan University

Although this book is suitable for non-science majors, you will find that it isn't simplistic or condescending. The presentation, while condensed, remains sophisticated and discerning.

WHY A SMALLER BOOK?

In recent years, environmental science textbooks have gotten bigger, more encyclopedic, and increasingly expensive. Both instructors and students call for a **concise and affordable text.** This book was written to fill that need. Rather than the 25 to 30 chapters found in most environmental science textbooks, we've limited this book to 15 chapters, or about one chapter per week for a typical semester course. The more concise presentation focuses on key principles, on scientific methods and ideas, and on life-long learning skills for students. We have also included enough case studies and current events to provide the real-world context for the themes discussed here. Additional case studies and current issues are available on the book's web site to help enrich your course's content. At the same time, the moderate size and price of this book should allow you to add supplementary materials to meet your individual teaching and learning objectives.

"I chose this text primarily based on size and its emphasis on scientific principles. Past textbooks were somewhat overwhelming for my students. For many of them, this is their first college-level science course."

Pamela Pape-Lindstrom
Everett Community College

The outline of this book follows a topic sequence widely used in many environmental science textbooks and courses, but we know that many instructors choose to organize their courses around their own outlines. We've written each chapter in a way that doesn't assume that students have already read other chapters in any particular order. If instructors prefer a different organization of course topics, chapters can be presented in any order that suits their course's needs.

ACTIVE LEARNING AND CRITICAL THINKING

Learning how scientists approach problems can help students develop habits of independent, orderly, and objective thought. But it takes active involvement to master these skills. Throughout this book, we encourage students to practice thinking for themselves. Data and interpretations aren't presented as immutable truths, but rather as evidence to be examined and tested.

> *"I found the Application Boxes to be another strong feature in this book because they require the student to utilize a diversity of critical thinking skills. Specifically, they require the student to make observations, collect and analyze data, and reach conclusions. These are vital skills and I would assign many of these as homework assignments or quizzes in class."*
>
> Dan Gleason
> *Georgia Southern University*

We try to give a balanced view of controversial topics. Orderly, critical assessment of complex problems is a key part of **scientific literacy,** which is essential for understanding current environmental science. In every chapter, students are invited to practice **critical thinking** and to apply new ideas. We also include case studies demonstrating how scientists have thought about important environmental questions.

Because we think a discouraged student is unlikely to take positive action toward sustainability, we also strive to avoid "gloom and doom" or "shame and blame" attitudes. Instead, we adopt a thoughtful but cautiously optimistic view that will encourage readers to look for ways that we can solve problems and make our world a better place in which to live. In nearly every chapter, we include "What can you do?" boxes that give practical suggestions for things individuals can do to make a difference. Most chapters also have short applications boxes that invite the reader to stop for a moment and practice using the principles they've learned.

An introductory story at the beginning of each chapter illustrates an important current issue and shows how it relates to practical environmental concerns. These stories also start the process of exploring how scientists study complex issues. In addition to these introductory stories, case studies and examples of how scientists investigate our environment appear periodically throughout the book to remind readers about the practical importance of these issues.

INTEGRATION AND SUSTAINABILITY

Environmental problems and their solutions occur at the intersection of natural systems and the human systems that manipulate the natural world. In this book we present an **integrated approach** to physical sciences—biology, ecology, geology, air and water resources—and to human systems that affect nature—food and agriculture, population growth, urbanization, environmental health, resource economics, and policy. Although it is tempting to emphasize purely natural systems, we feel that students can never understand why coral reefs are threatened or why tropical forests are being cut down if they don't know something about the cultural, economic, and political forces that shape our decisions.

This integrated approach is essential if we are to work toward sustainable solutions in our environment. Throughout the book, we present **sustainability** as an ultimate goal for both preserving nature and improving the lives of people everywhere. Sustainability implies that human well-being and environmental health need to be complementary, not contradictory, efforts. The goal of sustainability also requires a global view. We take a **global perspective** in this book because we believe that the most important and difficult environmental problems we now face involve worldwide resources and international institutions. To help students gain geographical literacy, we have included many maps and international case studies in this book, including a valuable set of world maps in the appendices.

CURRENCY

Throughout this book, we present up-to-date tables and graphs with the most current available data. We hope these data will give students an appreciation of the kinds of information available in environmental science. Among the sources we have called upon here are geographic information systems (GIS) data and maps, current census and population data, international news and data sources, and federal data collection agencies.

This text has had the benefit of input from more than 400 researchers, professionals, and instructors who have reviewed this book or our larger text, *Environmental Science: A Global Concern.* These reviewers have helped us keep the text current and focused. We deeply appreciate their many helpful suggestions and comments.

WEB-ENHANCEMENT

The world wide web has become a vast and valuable resource for students. You can find a wealth of information there to update or supplement topics in environmental science. We incorporate this resource more fully in this text than in any other environmental science book currently available. Every chapter opens with a list of **web-based resources** that relate to and enrich the chapter contents. We have also placed further readings and extra case studies on the web, and referenced these readings in the text. We encourage instructors to make use of these additional resources. At the end of each chapter a **web-based exercise** invites students to visit specific web pages and use the data found there to create graphs, make comparisons, or do some other practical analysis of real data. These aren't simplistic exercises that just ask the reader to look at a site and report on what it contains. Rather, we ask the student to really explore these resources and use the information they find in pragmatic ways. Note that these exercises are just a beginning: instructors and students can modify or add to these exercises if they wish. One of our objectives in these exercises is simply to expose students to important data sources. Another objective in these exercises is to make students work with data, create graphs, map data, and experience some of the ways that scientists create and share information today.

NEW FEATURES IN THIS EDITION

While retaining the chapter sequence and basic structure of the original version, this edition has been thoroughly updated with the most recently available data. We use the latest current events to show how environmental science principles can be used to analyze problems and find solutions to pressing problems. Among the most important changes in this edition are:

Chapter 1 A major revision enhances this chapter with a discussion of science and its practice, including the scientific worldview, scientific design, the importance of statistics, reproducibility and significance, sample size, blind and double blind experiments, cooperation and insight in science, and additional examples of inductive reasoning. A brief overview of conservation history has been added to this chapter to give students a sense of how the field has developed.

Following a section on pseudoscience and baloney detection, we offer a critical thinking box on the controversy engendered by Danish political scientist, Bjorn Lomborg's book *The Skeptical Environmentalist*. We present both pro and con sides of this debate so students can form their own opinions about this issue.

 Chapter 2 Following reviewers' suggestions, we have rewritten the entire section in this chapter on environmental chemistry to eliminate errors and improve coverage. The hydrological cycle has been moved into this chapter so it can be discussed together with other biogeochemical cycles. An updated and corrected periodic table is presented in figure 2.2.

Chapter 3 Again, following reviewers' comments, we have strengthened and expanded our discussion of evolution to include effects of positive and negative mutations, geographic and behavioral isolation, and to tie competition to tolerance range and critical ecological factors. The classic snowshoe hare/Canadian lynx

story of predator/prey relations is updated to include modern questions about how data was collected.

 Chapter 4 Using the most recent information from the Population Reference Bureau and the United Nations Population Division, graphs, charts, and tables in this chapter have been revised with up-to-the-minute data. The chapter opens with a vignette about population growth in India and conflicts about the best way to reach stability.

Chapter 5 In response to requests from reviewers, we've added a section on marine ecosystems to the biome discussion, and improved the explanation of climate graphs in figure 5.4.

 Chapter 6 Drawing on the 2002 FAO report, State of the World's Forests, we have updated information on tropical and temperate forest area and rates of deforestation. New tables and graphs show rates of loss by region and by source.

The hot, dry summer of 2002 resulted in large forest fires in the western United States. Competing ideas about how forests should be managed and fires could be suppressed or prevented provide an opportunity to examine resource policies and practices. The highly controversial issue of logging in roadless, defacto wilderness areas also is discussed. Ecosystem management has become the operative paradigm of several U.S. resource agencies, so we've added a section on this topic. Contentious issues of park and wildlife refuge area management also are discussed.

Marine refuges and reserves are being declared by a number of nations. Research shows that a ban on fishing in refuges can increase fish populations significantly in adjacent areas. We review some of the important developments in this subject.

 Chapter 7 Reports prepared for the 2002 World Summit on Sustainable Development in South Africa provide data for a major revision of this chapter on world food supplies and food security issues.

Controversial topics such as livestock confinement operations, availability of irrigation water, and the pros and cons of pesticides and organic farming are also included. A new case study illustrates the ways in which shade-grown coffee and cocoa can help protect biodiversity.

A major new section on genetic engineering and the debate over ecological and human health safety of genetically modified organisms is added to this chapter. Coverage of this controversial topic is balanced and objective, giving students information with which to form their own opinions. Without going too deeply into either genetics or biochemistry, students are taught some basic biology about genetic modifications that will help them understand this complex but important topic.

Chapter 8 Following the tragic events of September 11, 2001, we all must become better informed about the risks of terrorist attacks. This chapter opens with a vignette about the potential for bioterrorism. This leads into a newly revised discussion of infectious diseases and environmental health hazards that help us understand how diseases can spread and what factors present the greatest risk. Newly published data on leading causes of death and disability put these risks in perspective.

Another important environmental health topic is emergent diseases—those that were previously unknown or that have been dormant for many years but suddenly reappear. We expand this important coverage to include emergent ecological diseases such as chronic wasting disease in deer and elk, black band disease of corals, and the widespread deaths of North Atlantic marine mammals.

A boxed reading on the statistical significance of disease clusters—using the recently announced contraction of Parkinson's disease by actor Michael J. Fox and several of his coworkers as an example—will interest students looking for patterns in environmental diseases.

 Chapter 9 Many environmental scientists regard global climate change as the most important issue facing us today. Drawing on recently published reports of the Intergovernmental Panel on Climate Change, this chapter has been extensively revised and updated to reflect current information on this important topic. Opening with a compelling vignette of recent evidence of global warming, the chapter goes on to a thorough and detailed discussion of climate models and projections for possible effects of climate change. Some of the basic science underlying this debate is revealed in a new box on satellite imaging and how it is used in climate analysis. Policy choices for reducing climate change are explored, including reducing emissions, changing to new energy sources, and storing carbon dioxide. Critical thinking on this issue is encouraged by a new boxed reading on scientific uncertainty and general circulation models.

 Chapter 10 Again drawing on preparatory material for the Johannesburg World Summit on Sustainable Development, this chapter presents a wealth of information on world water supplies and impending shortages. The opening vignette on the Three Gorges Dam illustrates both the promise and the dangers of intrabasin water transfers. A fascinating case study of the controversy around water diversion projects on the Klamath River in California shows how these principles apply at home as well as abroad.

A greatly expanded box on water quality testing will help students understand both the major types of water pollutants and how they are analyzed. This information compliments new data on water quality assessments in U.S. rivers and lakes by the Environmental Protection Agency.

 Chapter 11 This chapter opens with a vignette concerning another highly controversial current issue, the storage of high-level radioactive waste at Yucca Mountain, Nevada. An unusual aerial photograph of Yucca Mountain gives an interesting perspective on why it was chosen as a nuclear waste repository. Using questions about the geological stability of Yucca Mountain illustrates how geology is important in environmental science. After reviewing some basic geology, the chapter discusses current issues in economic mineralogy and resource extraction such as mountain top removal in Appalachia and heap-leach mineral extraction techniques.

 Chapter 12 Opening with a discussion of proposed oil and gas drilling in the Alaska National Wildlife Refuge, this chapter presents a detailed examination of the pros and cons of both traditional (fossil fuels and nuclear power) and renewable energy sources such as wind, solar, fuel cells, and geothermal energy. Many students are unaware that unconventional fossil fuel reserves such as tar sands and oil shales could potentially contain far more energy than liquid oil supplies. The environmental effects of extracting and using these sources, however, require careful consideration.

This chapter ends with energy scenarios just published in 2002 by the World Energy Council as a tool for envisioning future developments. The chapter also reviews some of the controversy surrounding Vice President Dick Cheney's Energy Policy Task Force and its vision of our energy future.

Chapter 13 New data on waste production and recycling rates bring this chapter up to date. Recycling successes and challenges are discussed including recent bans on single use beverage containers and requirements for recycling of computers and automobiles in Europe. A new boxed reading on bioremediation explores alternative methods for cleaning up toxic wastes.

 Chapter 14 Sustainable development is a major international focus and a key theme in this book. To understand this topic, students need some background in ecological economics, which we present in this chapter. By 2050, two-thirds of the world population is expected to live in urban areas. Thus, most economic development and progress towards sustainability will have

to occur in cities. To appreciate both the promise and problems of urban areas, we include a brief overview of city planning. Urban blight, sprawl, and smart growth receive a more thoughtful and contemporary treatment in this chapter than in any other environmental science textbook on the market. A new boxed reading on urban ecology illustrates the environmental quality of these issues; while a web exercise in calculating ecological footprints at the end of the chapter helps students apply urban issues to their own lives.

 Chapter 15 A unique feature of this book is its systematic and meaningful presentation of environmental policy. This section has been strengthened and made even more thorough with the addition of information on science and policy in local communities. It also has updated information on international bodies and conventions such as the World Trade Organization, the World Bank, and the North American Free Trade Agreement.

New photographs and art Approximately 60 new full color photographs help maintain student interest by showing real life examples of environmental issues. In addition, 34 new or revised graphs, drawings, and other graphics provide valuable information as well as making the text attractive and highly readable. Ten new tables allow readers to compare and understand data in greater depth than just what is provided by straight text.

New case studies, opening stories, and other special readings Every chapter starts with an opening vignette that presents a fascinating current environmental problem illustrating the principles to be presented in the text. Interspersed throughout each chapter are other special features that show concrete application of theoretical knowledge. Ranging from controversy over dredging PCBs from the Hudson River to the threat of bioterrorism, these special readings also provide an opportunity to bring up current events and to make connections to real-life issues. Of the 30 special readings in the book, 16 are new to this edition, making the text thoroughly up-to-date and relevant.

LET US KNOW WHAT YOU THINK

We'd appreciate hearing from both students and instructors about where—and how—we could improve this text. You, the users, are the real test of whether we have accomplished our goal of presenting the principles of environmental science in an engaging and understandable way. Please let us know what you think. We value your comments and suggestions. Please send your recommendations to the Integrative Biology team at McGraw-Hill Publishers, 2460 Kerper Boulevard, Dubuque, IA, 52001.

USEFUL SUPPLEMENTS

- **Instructor's Testing and Resource CD-ROM** This cross-platform CD-ROM provides a wealth of resources for the instructor. Supplements featured on this CD-ROM include lab activities, and a computerized test bank utilizing Brownstone Diploma® testing software to quickly create customized exams. This user-friendly program allows instructors to search for questions by topic, format, or difficulty level; edit existing questions or add new ones; and scramble questions and answer keys for multiple versions of the same test. Other assets on the Instructor's Testing and Resource CD-ROM are grouped within easy-to-use folders.

- **Transparencies** A set of 100 transparencies is available to users of the text. These acetates include key figures from the text, including new art from this edition.

- **Environmental Science Essential Study Partner CD-ROM** A complete, interactive student study tool, this CD features animations, videos, and learning activities. From quizzes to interactive diagrams, you'll find that there has never been a better study partner to ensure the mastery of core concepts. Best of all, it's available FREE with a new textbook purchase in an optional package.

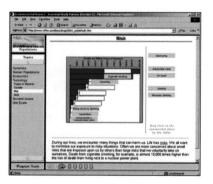

- **Digital Content Manager (DCM) CD-ROM** This multimedia collection of visual resources allows instructors to utilize artwork from the text in multiple formats to create customized classroom presentations, visually based tests and/or quizzes, dynamic course website content, or attractive printed support materials. The digital assets on this cross-platform CD-ROM are grouped within the following easy-to-use folders:

  - *Illustrations and Photos* All of the line drawings from the text and hundreds of photos in ready-to-use digital files.

- *PowerPoint Lecture Outline* Ready-made presentations combine art from the text with customized, instructor-written lecture notes, covering all 15 chapters.
- *Tables* Every table that appears in the text is provided in electronic form.
- *Active Art* These special art pieces consist of key images from the text that are converted to a format which allows instructors to break the art down into core elements and then group the various pieces and create customized images. This is especially helpful with difficult concepts; they can be presented step-by-step.
- *Animations* Numerous full-color animations illustrating many different concepts covered in the study of environmental science are provided. The visual impact of motion will enhance classroom presentations and increase comprehension.
- **Online Learning Center** (http://www.mhhe.com/environmentalscience/). This comprehensive website offers numerous resources for both students and instructors.

Student Resources—Everything you need in one place:
- Practice quizzing
- How-to study tips
- Hyperlinks on chapter topics
- Web exercises
- Guide to electronic research
- Regional perspectives (case studies)
- Environmental issues world map
- Key term flashcards
- How to write a paper
- How to contact your elected officials
- Further readings
- Metric equivalents and conversion tables
- Career information
- PowerWeb's hundreds of current articles and daily news items integrated into each chapter
- Access Science offering the advantage of an online, interactive encyclopedia

Instructor Resources—In addition to <u>all of the above</u>, you'll receive:
- Supplements resource chart for each chapter
- Answers to web exercises
- Additional case studies
- Answers to critical thinking questions
- PageOut (create your own course website)

RELATED TITLES OF INTEREST

1. *Field and Laboratory Activities Manual, 7ed.* (0-07-290913-7) by Enger and Smith.

2. *Interactive World Issues: Of Place and Planet CD-ROM* (0-07-255648-X), Cambridge Studios.

3. *Annual Editions: Environment 03/04* (0-07-283851-5) Editor: John L. Allen.

4. *Taking Sides: Clashing Views on Controversial Environmental Issues, 10ed.* (0-07-285531-2) Editor: Theodore D. Goldfarb.

5. *Sources: Notable Selections in Environmental Studies, 2ed.* (0-07-303186-0) Editor: Theodore D. Goldfarb.

6. The Dushkin *Student Atlas of Environmental Issues* (0-697-36520-4) Editor: John Allen. University of Connecticut.

7. *Life Science Living Lexicon* (CD = 0-697-37993-0; Print = 0-697-12133-X) by William Marchuk.

8. *You Can Make a Difference: Be Environmentally Responsible* (0-07-292416-0) by Judy Getis.

9. *Environmental Ethics: Divergence and Convergence* (0-07-006180-7) by Botzler and Armstrong.

10. *Environmental Problem-Solving: A Case Study Approach* (0-07-027686-2) by Isobel W. Heathcote.

11. *Eyewitness World Atlas CD-ROM* (0-07-233220-4), published by Dorling-Kindersley.

ACKNOWLEDGMENTS

We're indebted to all the instructors who have reviewed the manuscript and made helpful suggestions, corrections, and recommendations for improving this book. Space does not permit inclusion of all the excellent ideas that were provided, but we will continue to do our best to incorporate the insights reviewers have given us. In addition, all of us owe a big debt to the many scholars whose work forms the basis of our understanding of environmental science. We stand on the shoulders of giants. If errors persist in spite of our best efforts to root them out, it is our responsibility.

We express our gratitude to the entire McGraw-Hill book team for their wonderful work in putting together this edition. Kathy Loewenberg oversaw the developmental stages and has made many creative contributions to this book. Cathy Conroy has done a superb job of copyediting, correcting errors, and improving our prose. Jane Mohr and Rose Koos were the production project managers and kept everything running smoothly. Connie Mueller found exceptional photographs. The folks at Precision Graphics did an excellent job of composition and page layout. Heather Wagner and Marge Kemp have supported this project with their enthusiasm and creative ideas. We would also like to thank the attendees of the McGraw-Hill Environmental Science Symposium in Bellingham, Washington. Matthew Laposata, Allan Matthias, Kenneth Rhinehart, Royal Berglee, Nicole Welch, Paul Rowland, Lynnette Danzl-Tauer, Catherine Christie, Linda Fitzhugh, David Charlet, David Hassenzahl, and Stacy Smith contributed to many lively discussions and their suggestions were gratefully received and incorporated wherever possible in this edition.

The following individuals provided reviews for this book. We appreciated their input and thank them for their many valuable comments and suggestions.

Shannon A. Bliss
Cerro Coso Community College

Catherine W. Carter
Georgia Perimeter College

Laurie S. Eberhardt
Valparaiso University

Brad C. Fiero
Pima Community College

Steven P. Frysinger
James Madison University, Harrisonburg

Zofia E. Gagnon
Marist College

Carey A. Gazis
Central Washington University

Marcia L. Gillette
Indiana University, Kokomo

Daniel F. Gleason
Georgia Southern University

Lawrence J. Gray
Utah Valley State College

Mark F. Hammer
Wayne State College

Charles Kaminski
Middlesex Community College

Paul Kimball
Northeast Iowa Community College

Ned J. Knight
Linfield College

Jeff Kushner
James Madison University

Matthew Laposata
Kennesaw State University

Kurt M. Leuschner
College of the Desert

Edward M. Lignowski
Holy Family College

Les M. Lynn
Bergen Community College

Timothy F. Lyon
Ball State University

Janet S. MacFall
Elon University

Nancy Jean Mann
Cuesta College

Ken R. Marion
University of Alabama, Birmingham

Thomas C. Moon
California University of Pennsylvania

Michele Morek
Brescia University

Michael J. Neilson
The University of Alabama, Birmingham

Pamela Pape-Lindstrom
Everett Community College

David R. Perault
Lynchburg College

Ervand M. Peterson
Sonoma State University

Julie Phillips
De Anza College

Mark D. Plunkett
Bellevue Community College

John M. Pratte
Kennesaw State University

Robert J. Sager
Pierce College

Ronald L. Sass
Rice University

Bruce A. Schulte
Georgia Southern University

Julie A. Seiter
Oakland Community College

Jill Singer
Buffalo State College

Ronald C. Sundell
Northern Michigan University

John A. Tiedemann
Monmouth University

Edward R. Wells
Wilson College

Thomas B. Wilson
University of Arizona

Bruce C. Wyman
McNeese State University

GUIDED TOUR

FIGURE 5.1

"Investigating Our Environment" Essays

These examples give students some experience with tools and approaches scientists use to solve contemporary environmental problems. Each boxed essay shows how scientists work to solve complex environmental questions.

Unique Features

This book is intentionally written in a **lively, accessible style** and illustrated with nearly 400 full color photographs and figures that help explain important concepts. We have also integrated a number of learning tools to help students gain an informed, thoughtful view of our environment.

INVESTIGATING Our Environment

When Are Disease Clusters Significant?

When actor Michael J. Fox revealed that he suffers from Parkinson's disease, he opened some doors that may help us understand the cause of this mysterious illness. Parkinson's occurs when neurons degenerate in the substantia nigra, a part of the midbrain that produces dopamine, a chemical messenger essential for normal muscle control. The classic symptoms of the disease are tremors, muscle rigidity and weakness, and a shuffling gait. These symptoms usually progress slowly, but eventually the escalating paralysis is fatal.

Mr. Fox's illness is unusual in two ways. First, he began showing symptoms at age 30, much earlier than most Parkinson's victims. Moreover, three others who worked with him at a television studio in Vancouver, British Columbia, in the late 1970s also have contracted the disease. Typically, Parkinson's afflicts one in 300 people. To have four victims out of the 125 people on the studio staff makes the incidence in this group more than eight times above average. Furthermore, one of the victims also discovered she had the illness at age 38. Was there something in the air or water—a virus or toxic chemi-

cal perhaps—that triggered this disease cluster, or are the similarities just a coincidence?

There are good reasons to suspect environmental involvement in Parkinson's. Other clusters of the disease have been identified in heavily industrialized areas or where heavy use of herbicides and pesticides has occurred. In the 1980s, 65 people in California died after taking a synthetic heroin contaminated with methyl phenyl tetrahydropyridine (MPTP). This drug caused symptoms indistinguishable from Parkinson's (they could be temporarily reversed by administering l-dopa, which is converted to dopamine in the brain), and quickly led to paralysis, coma, and death. MPTP is very similar in its chemical structure and characteristics to paraquat, a widely used herbicide.

Evidence also suggests infectious agents may be involved in Parkinson's. After World War I, people who had contracted sleeping sickness later developed what came to be known as postencephalitic parkinsonism. Parkinson's is also more prevalent among people such as teachers, doctors, nurses, and others who have relatively high exposure to infections. In one notable study, several members in each of six Canadian families were shown to have developed Parkinson's within a few years of each other even though

they were very different ages. In this study, the common thread seemed to be the health food diet eaten by those who contracted the disease.

Suppose you notice what seems like an abnormally large number of people in your family, neighborhood, or workplace have a particular illness. How can you know whether this is a significant connection or simply a statistical fluke? It may take a great deal of work to establish the meaning of a disease cluster, but this is an important way to study how illnesses start and spread through populations. The Centers for Disease Control and Prevention (CDC) in Atlanta searches for patterns in disease distribution to identify epidemics and to zero in on their causes. Rapid analysis of statistical and geographical data is particularly important in recognizing and responding to terrorist attacks as described at the beginning of this chapter.

Perhaps, as in Mr. Fox's case, celebrity status will help educate the public about Parkinson's. He might also illuminate environmental factors that will help us understand the causes of this puzzling disease. Finally, sharing knowledge of his condition with former co-workers calls our attention to the importance of cluster studies in understanding environmental health issues.

---

skin cancer. It also is an endocrine hormone disrupter. Nine countries, including Sweden, Germany, Vietnam and Indonesia, have banned CCA-treated wood. Nonarsenic alternatives exist, but the $4 billion/year U.S. wood-treating industry has resisted change. Nevertheless, two major U.S. manufacturers agreed, in 2002, to begin a phase-out of CCA.

- Perflurooctane sulfonate (PFOS), the major ingredient in a popular stain and spill repellent made by the 3M Company, was found to be extremely widely distributed in the global environment, from remote Pacific Islands to the Arctic. No serious adverse effect of this compound is known, but the fact that it is long lasting and is accumulating in the environment creates worries that it may become a problem. The manufacturer agreed in 2000 to a phase-out.

- Atrazine, one of the most widely used herbicides in the Midwestern corn belt, was shown to cause abnormal development and sexual dysfunction in frogs. Farmers who use Atrazine heavily have high rates of certain lymphomas, but these farmers generally are exposed to other toxic compounds as well. A study of farm families in northwestern Minnesota found considerably higher rates of birth defects

than urban families. "The data is associative," said Dr. Vince Gary, who directed the research. No definitive cause and effect can be shown with any particular pesticide, but the sensitivity of other species raises concerns (see Investigating Our Environment, above).

Chemical Interactions

Some materials produce *antagonistic* reactions—that is, they interfere with the effects or stimulate the breakdown of other chemicals. For instance, vitamins E and A can reduce the response to some carcinogens. Other materials are *additive* when they occur together in exposures. Rats exposed to both lead and arsenic show twice the toxicity of only one of these elements. Perhaps the greatest concern is synergistic effects. **Synergism** is an interaction in which one substance multiplies the effects of another. For example, occupational asbestos exposure increases lung cancer rates 20-fold. Smoking increases lung cancer rates by the same amount. Asbestos workers who also smoke, however, have a 400-fold increase in cancer rates. How many other toxic chemicals are we exposed to that are below threshold limits individually but combine to give toxic results?

Biomes and Biodiversity

The first rule of intelligent tinkering is to save all the pieces.
—Aldo Leopold

Biodiversity is essential for the ecological services on which we all depend.

OBJECTIVES

After studying this chapter, you should be able to

- recognize the characteristics and general distribution of major biomes, and understand the most important factors that determine biome distributions.
- evaluate the degree of disturbance of major biomes, and identify the biomes that are most important for human activities and for biodiversity protection.
- define biodiversity and explain its importance.
- report on the total number and relative distribution of living species on the earth.
- describe how human activities cause biodiversity losses.
- identify regions and ecosystems of high biodiversity.
- evaluate the effectiveness of efforts to protect endangered species.

LEARNING ONLINE

Visit our webpage at www.mhhe.com/environmentalscience for data sources, further readings, additional case studies, current environmental news, and regional examples within the Online Learning Center to help you understand the material in this chapter. You'll also find active links to information pertaining to this chapter including:

Biomes
Center for Restoration Ecology
Landscape Ecology and Birds
Restoration and Reclamation Review
Biodiversity and WORLDMAP
Ecology and Biodiversity
Biodiversity Profile of India

Study Aids

Each chapter opens with a list of objectives that summarizes the main points of the chapter. These objectives are written in active terms that suggest to the student that it requires active involvement on their part to learn environmental science. Notice that while some objectives call for simple, concrete thinking skills, others are deliberately aimed at higher cognitive levels to encourage students to think reflectively, analytically, and critically. We ask students to analyze, understand, explore, or question concepts that don't have simple, clear endpoints that can be measured objectively, because we believe it's important to point out the need for higher-level thinking about complex issues. Every chapter ends with a summary of main points, a list of key terms, and review questions that help the student review material and prepare for tests.

Scientific Thinking

A more challenging, open-ended set of questions titled "Thinking Scientifically" encourages students to think deeply and independently about issues and principles presented in the chapter. These questions make excellent starting points for discussion sections. They also could be used to practice for essay exams, or might even serve as an essay exam themselves.

element cycles that occur in ecosystems. Unlike matter, energy is not cycled. Energy always flows through systems in a one-way process in which some energy is converted from a high-quality, concentrated form to a lower-quality, less useful, dispersed form. We describe this increase in disorder as entropy. Primary producers, which synthesize organic compounds from solar energy and carbon dioxide, capture nearly all energy in ecosystems. Resulting compounds provide chemical energy and materials to consumers.

A species is all the organisms of the same kind; often this means organisms genetically similar enough to breed in nature and produce live, fertile offspring. The populations of different species that live and interact within a particular area at a given time make up a biological community. An ecosystem is composed of a biological community together with all the biotic and abiotic factors that make up the environment in a defined area.

Matter and energy are processed through the trophic levels of an ecosystem via food chains and food webs, and biogeochemical cycling. At each energy transfer point, less energy is available for work because of the laws of thermodynamics, so ecosystems require a continuous supply of energy. The relationships between producers and consumers in an ecosystem, often depicted as pyramids, demonstrate this principle. Most of the energy that enters an ecosystem comes, ultimately, from the sun. Living organisms constantly use and reuse elements. Carbon, nitrogen, sulfur, and phosphorus, for instance, recycle through complex biogeochemical paths.

QUESTIONS FOR REVIEW

1. Define the terms *atom* and *element*. Are these terms interchangeable?
2. Your body contains vast numbers of carbon atoms. How is it possible that some of these carbon atoms may have been part of the body of a prehistoric creature?
3. In the biosphere, matter follows a circular pathway while energy follows a linear pathway. Explain.
4. The oceans store a vast amount of heat, but this huge reservoir of energy is of little use to humans (except for climate moderation). Explain the difference between high-quality and low-quality energy.
5. Ecosystems require energy to function. From where does this energy come? Where does it go? How does the flow of energy conform to the laws of thermodynamics?
6. Heat is released during metabolism. How is this heat useful to a cell and to a multicellular organism? How might it be detrimental, especially in a large, complex organism?
7. Photosynthesis and cellular respiration are complementary processes. Explain how they exemplify the laws of conservation of matter and thermodynamics.
8. What do we mean by carbon-fixation or nitrogen-fixation? Why is it important to humans that carbon and nitrogen be "fixed"?
9. The population density of large carnivores is always very small compared to the population density of herbivores occupying the same ecosystem. Explain this in relation to the concept of an ecological pyramid.
10. A species is a specific kind of organism. What general characteristics do individuals of a particular species share? Why is it important for ecologists to differentiate among the various species in a biological community?

THINKING SCIENTIFICALLY

1. How does the principle of entropy (or the second law of thermodynamics) explain the idea of trophic pyramids?
2. Which wavelengths do our eyes respond to, and why (refer to fig. 2.10)? What is the ratio of short ultraviolet wavelengths to microwave wavelengths (used in a microwave oven)?
3. Ecosystems are often defined as a matter of convenience, because we can't study everything at once. How would you describe the character and extent of the natural ecosystem in which you live? What characteristics make your ecosystem an open system?
4. How would you delimit the human ecosystem (including food and energy production) in which you live?
5. If you had to define a research project to evaluate the relative biomass of producers and consumers in an ecosystem, what would you measure? (Note: This could be a natural ecosystem or an enclosed system such as a terrarium.)
6. Understanding storage compartments is essential to understanding material cycles such as the carbon cycle. If you look around your back yard, how many carbon storage compartments are there? Which ones are biggest? Which ones are longest-lasting?

Applications

Because few of us learn effectively without an opportunity to actively apply new ideas, we have included application boxes. These boxes provide a break in reading the text and invite students to practice or apply skills they have just learned. In addition, "What can you do?" boxes in nearly every chapter suggest practical things that individuals can do to help improve environmental quality and resource conservation.

APPLICATION: Calculating Probabilities

You can calculate the statistical danger of a risky activity by multiplying the probability of danger by the frequency of the activity. For example, in the United States, one person in three will be injured in a car accident in their lifetime (so the probability of injury is 1 per 3 persons, or $\frac{1}{3}$). In a population of 30 car-riding people, the cumulative risk of injury is: 30 people × (1 injury/3 people) = 10 injuries over 30 lifetimes.

1. If the average person takes 50,000 trips in a lifetime, and the accident risk is $\frac{1}{3}$ per lifetime, what is the probability of an accident per trip?

2. If you have been riding safely for 20 years, what is the probability of an accident during your next trip?

Answer: 1. Probability of injury per trip = (1 injury/3 lifetimes) × (1 lifetime/50,000 trips) = 1 injury/150,000 trips. 2. 1 in 150,000. Statistically, you have the same chance each time.

---

dying from lung cancer if you smoke one pack of cigarettes per day is about 1 in 1,000. By comparison, the risk from drinking water with the EPA limit of trichloroethylene is about 2 in 1 billion. Strangely, many people demand water with zero levels of trichloroethylene, while continuing to smoke cigarettes.

More than 1 million Americans are diagnosed with skin cancer each year. Some of these cancers are lethal, and most are disfiguring, yet only one-third of teenagers routinely use sunscreen. Tanning beds more than double your chances of cancer, especially if you're young, but about 10 percent of all teenagers admit regularly using these devices.

Table 8.6 lists some activities estimated to increase your chances of dying in any given year by 1 in 1 million. These are statistical averages, of course, and there clearly are differences in where one lives or how one rides a bicycle that affect the danger level of these activities. Still, it is interesting how we readily accept some risks, while shunning others.

Our perception of relative risks is strongly affected by whether risks are known or unknown, whether we feel in control of the outcome, and how dreadful the results are. Risks that are unknown or unpredictable and results that are particularly gruesome or disgusting seem far worse than those that are familiar and socially acceptable.

Studies of public risk perception show that most people react more to emotion than to statistics. We go to great lengths to avoid some dangers, while gladly accepting others. Factors that are involuntary, unfamiliar, undetectable to those exposed, catastrophic, or that have delayed effects or are a threat to future generations are especially feared, while those that are voluntary, familiar, detectable, or immediate cause less anxiety. Even though the actual number of deaths from automobile accidents, smoking, or alcohol, for instance, are thousands of times greater than those from pesticides, nuclear energy, or genetic engineering, the latter preoccupy us far more than the former.

TABLE 8.6 Activities Estimated to Increase Your Chances of Dying in Any Given Year by 1 in 1 Million

ACTIVITY	RESULTING DEATH RISK
Smoking 1.4 cigarettes	Cancer, heart disease
Drinking 0.5 liter of wine	Cirrhosis of the liver
Spending 1 hour in a coal mine	Black lung disease
Living 2 days in New York or Boston	Air pollution
Traveling 6 minutes by canoe	Accident
Traveling 10 mi by bicycle	Accident
Traveling 150 mi by car	Accident
Flying 1,000 mi by jet	Accident
Flying 6,000 mi by jet	Cancer caused by cosmic radiation
Living 2 months in Denver	Cancer caused by cosmic radiation
Living 2 months in a stone or brick building	Cancer caused by natural radioactivity
Having one chest X ray	Cancer caused by radiation
Living 2 months with a cigarette smoker	Cancer, heart disease
Eating 40 tablespoons of peanut butter	Cancer from aflatoxin
Living 5 years at the site boundary of a typical nuclear power plant	Cancer caused by radiation from routine leaks
Living 50 years 5 mi from a nuclear power plant	Cancer caused by accidental radiation release
Eating 100 charcoal-broiled steaks	Cancer from benzopyrene

Source: From William Allman, "Staying Alive in the Twentieth Century," *Science 85*, 5(6): 31, October 1985. Reprinted by permission.

ESTABLISHING PUBLIC POLICY

Risk management combines principles of environmental health and toxicology with regulatory decisions based on socioeconomic, technical, and political considerations (fig. 8.17). The biggest problem in making regulatory decisions is that we are usually exposed to many sources of harm, often unknowingly. It is difficult to separate the effects of all these different hazards and to evaluate their risks accurately, especially when the exposures are near the threshold of measurement and response. In spite of often vague and contradictory data, public policymakers must make decisions.

The case of the sweetener saccharin is a good example of the complexities and uncertainties of risk assessment in public health. Studies in the 1970s at the University of Wisconsin and the Canadian Health Protection Branch suggested a link between saccharin and bladder cancer in male rats. Critics of these studies pointed out that humans would have to drink 800 cans of diet soda per day to get a saccharin dose equivalent to that given to the rats. Further-

---

Welcome to McGraw-Hill's Online Learning Center

Location: http://www.mhhe.com/environmentalscience

WEB EXERCISE

Calculating Your Ecological Footprint

One way of assessing the resource consumption patterns of communities and individuals is to calculate an ecological footprint. To learn about this concept and how it is done, go to www.rprogress.org/progsum/nip/ef/ef_main.html, the web page of Redefining Progress. You can find information here about individual consumption levels and those of whole nations as well as an interesting discussion of genuine progress indicators and real measures of satisfaction. For a quick calculation of your ecological footprint, go to www.lead.org/leadnet/footprint/intro.htm. By answering 13 simple questions, you can find out how much land it takes to support your present lifestyle. Suggestions are given for average consumption rates of residents of Canada and the United States. At the end, you are asked how much space should be reserved for other species. This shows you how many earths it would take to support the whole human population at your level of consumption. For a more detailed calculation of your footprint, go to www.rprogress.org/resources/nip/ef/ef_household_calculator.html. You can download a detailed household footprint spreadsheet in either a zipped or stuffed version (both small files) or as an Excel 5.0 file.

Land Use and Sustainability

Land is a resource that is at the center of many of our sustainability debates. How land should be used is a perennial policy and development problem. In the United States the Natural Resources Conservation Service (NRCS) conducts periodic surveys of land use and agricultural production. The NRCS website at www.nhq.nrcs.usda.gov/land/index/ includes a large collection of mapped data.

1. Click on several themes in the index at the address above. Which topics would be most useful for providing information about sustainability of land and water use in the United States? What information could

you use from the different topics to support arguments about wise or foolish resource use decisions? Which of the resources mapped are renewable? Which are nonrenewable?

2. Urban growth into agricultural lands is a problem in many cities. Look at the maps of acreage conversion to urban land use at www.nhq.nrcs.usda.gov/land/meta/m5123.html, the annual rates of development at www.nhq.nrcs.usda.gov/land/meta/m5095.html and the acres of cropland developed at www.nhq.nrcs.usda.gov/land/meta/m5252.html. For each map you can see an enlarged version by clicking on "Large GIF" just below the map image. Note the dates on the maps and the meaning of the colors. What do these maps tell about development rates in your area? How does growth impact sustainability in your area? Which regions are experiencing the most rapid urban expansion?

Cities and Sustainability

The New Colonist website at 64.70.131.22/ is a website devoted to making cities more sustainable and more livable. Look at the list of current featured stories on the web page. Discuss the following questions with a group of your colleagues: What are the main themes? (transportation? food? housing? energy?) What important topics are missing? What are some of the recommendations for change in the stories on the main topics? Based on your understanding of sustainability, economics, and resource use, do you think the recommendations in these articles are practical? If so, what could you do to help pursue recommended changes? If not, how could conditions be changed to make the recommendations more practical?

Web Exercises

The World Wide Web has become an extremely important source of current data, but many students know little about the kinds of information available or the agencies and organizations that produce it. Our web exercises make use of current data and ask students to perform activities such as graphing data, comparing maps, and using live GIS sources to learn about environmental issues and information sources.

Maps and Appendices

The appendices include conversion factors for weights and measures in an easy-to-locate position on the inside back cover. A new and expanded set of maps serve as references for the whole book and should help students expand their geographical knowledge and global perspective. The glossary defines all key terms as well as other important vocabulary words. A list of further readings is posted on the web page, where it can be updated periodically.

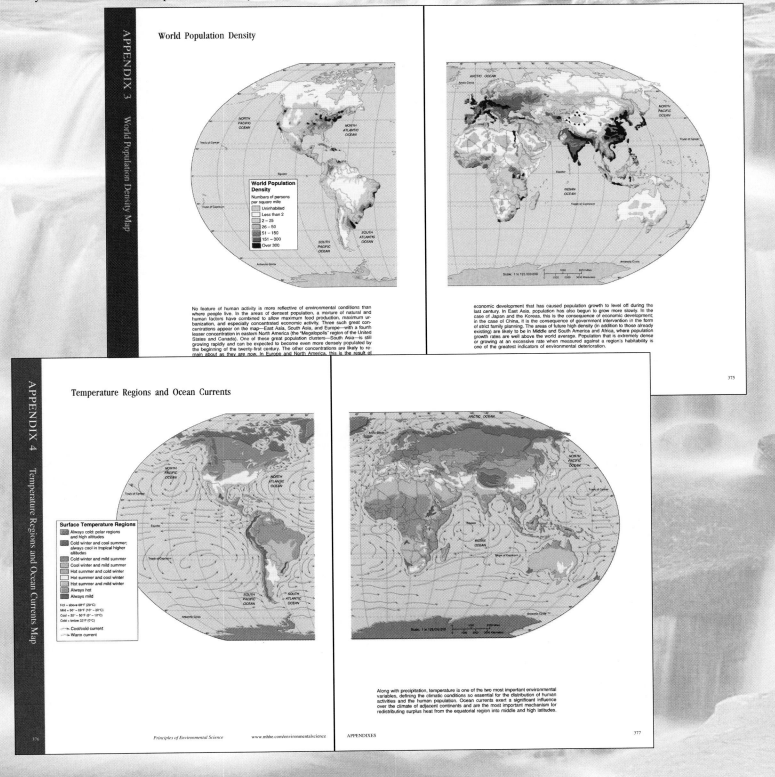

World Population Density

World Population Density

Numbers of persons
per square mile

- Uninhabited
- Less than 2
- 2 – 25
- 26 – 50
- 51 – 150
- 151 – 300
- Over 300

No feature of human activity is more reflective of environmental conditions than where people live. In the areas of densest population, a mixture of natural and human factors have combined to allow maximum food production, maximum urbanization, and especially concentrated economic activity. Three such great concentrations appear on the map—East Asia, South Asia, and Europe—with a fourth lesser concentration in eastern North America (the "Megalopolis" region of the United States and Canada). One of these great population clusters—South Asia—is still growing rapidly and can be expected to become even more densely populated by the beginning of the twenty-first century. The other concentrations are likely to remain about as they are now. In Europe and North America, this is the result of

economic development that has caused population growth to level off during the last century. In East Asia, population has also begun to grow more slowly. In the case of Japan and the Koreas, this is the consequence of economic development; in the case of China, it is the consequence of government intervention in the form of strict family planning. The areas of future high density (in addition to those already existing) are likely to be in Middle and South America and Africa, where population growth rates are well above the world average. Population that is extremely dense or growing at an excessive rate when measured against a region's habitability is one of the greatest indicators of environmental deterioration.

Temperature Regions and Ocean Currents

Surface Temperature Regions

- Always cold: polar regions and high altitudes
- Cold winter and cool summer; always cool in tropical higher altitudes
- Cold winter and mild summer
- Cool winter and mild summer
- Hot summer and cold winter
- Hot summer and cool winter
- Hot summer and mild winter
- Always hot
- Always mild

Hot = above 68°F (20°C)
Mild = 50° – 68°F (10° – 20°C)
Cool = 32° – 50°F (0° – 10°C)
Cold = below 32°F (0°C)

- Cool/cold current
- Warm current

Along with precipitation, temperature is one of the two most important environmental variables, defining the climatic conditions so essential for the distribution of human activities and the human population. Ocean currents exert a significant influence over the climate of adjacent continents and are the most important mechanism for redistributing surplus heat from the equatorial region into middle and high latitudes.

The Online Learning Center with PowerWeb

www.mhhe.com/environmentalscience

Imagine the advantages to having so many learning and teaching tools all in one place—at your fingertips.

Students, you'll appreciate extensive study aids including practice quizzes, interactive animations/quizzing, case studies, career information, research search engines, help on writing papers, and access to PowerWeb, which offers hundreds of pre-sorted, quality articles and daily news items relating to environmental science. An additional new feature, Access Science, allows you to tap into the equivalent of an online, interactive encyclopedia. The Online Learning Center is a one-stop site for all of your research and study needs.

Instructors, you'll want to take advantage of all of the above features offered to students PLUS classroom activities, lab exercises, answers to the end-of-chapter questions, a Resource Locator Chart for every chapter, and access to PageOut: Course Website Development Center.

Digital Content Manager | Version 2.0

This multimedia collection of visual resources allows instructors to utilize artwork from the text in multiple formats to create customized classroom presentations, visually based tests and/or quizzes, dynamic course website content, or attractive printed support materials. The digital assets on this cross-platform CD-ROM are grouped within the following easy-to-use folders:

- **Art Library.** Full-color digital files of all illustrations in the book.

- **Photo Library.** Hundreds of discipline appropriate photos in digital files.

- **PowerPoint Lecture Outline.** Ready-made presentations combine art from the text with customized lecture notes, covering all 15 chapters.

- **Table Library.** Every table that appears in the text is provided in electronic form.

- **Active Art.** These special art pieces consist of key images from the text that are converted to a format that allows instructors to break the art down into core elements and then group the various pieces and create customized images. This is especially helpful with difficult concepts; they can be presented step-by-step.

- **Animations Library.** Numerous full-color animations illustrating many different concepts covered in the study of environmental science are provided. The visual impact of motion will enhance classroom presentations and increase comprehension.

Call the McGraw-Hill Customer Service Department at 800-338-3987, or contact your local sales representative to obtain this valuable teaching aid.

We travel together, passengers on a little spaceship, dependent upon its vulnerable reserves of air and soil, all committed for our safety to its security and peace; preserved from annihilation only by the care, the work, and I will say, the love that we give to our fragile craft.

—Adlai Stevenson

Students monitor water quality in a stream.

OBJECTIVES

After studying this chapter, you should be able to

- define the term *environment* and identify some important environmental concerns that we face today.
- explain the scientific method and why it refutes or supports theories but never proves them beyond any doubt.
- apply the scientific method to problem solving.
- explain how statistics can help evaluate the accuracy and significance of results.
- summarize four stages in the history of conservation.
- distinguish among analytical, creative, logical, critical, and reflective thinking.
- summarize some major environmental dilemmas and issues that shape our current environmental agenda.
- discuss the implications of sustainability and sustainable development.

Welcome to McGraw-Hill's Online Learning Center

Location: http://www.mhhe.com/environmentalscience

McGraw Hill

LEARNING ONLINE

Visit our webpage at www.mhhe.com/environmentalscience for data sources, further readings, additional case studies, current environmental news, and regional examples within the Online Learning Center to help you understand the material in this chapter. You'll also find active links to information pertaining to this chapter including:

Frogweb: Amphibian Declines and Deformities
John Muir
Rachel Carson
Gifford Pinchot
Environmental News Network
Mount Graham International Observatory
Forest Service Employees for Environmental Ethics
Environmental Justice

Dredging the Hudson River

In February 2002, the U.S. Environmental Protection Agency (EPA) signed a decision to begin the most expensive river cleanup in U.S. history. Over the next decade, a $460 million dredging project will remove some 2 million m^3 of contaminated sediment from a 40-mile stretch of the Hudson River north of Albany, New York. The main target of this expensive operation is polychlorinated biphenyls (PCBs)—oily, persistent, organic compounds once widely used as electrical insulators in capacitors and transformers.

Between 1946 and 1977, General Electric Company (GE) legally discharged an estimated 500,000 kg (1.1 million lbs) of PCBs into the river from its factories in Hudson Falls and Fort Edward. Production, processing, and distribution of PCBs were banned in the United States in 1976 after these chemicals were shown to cause cancer in laboratory animals. GE stopped releasing PCBs in 1977, but seepage from the river bottom still continues. In addition to being classified as potential carcinogens in humans, PCBs also are suspected of triggering other serious health problems, including low birth weight, thyroid disease, and mental, reproductive, and immunological abnormalities. Because of these health concerns, a $40 million per year Hudson River fishing industry was shut down in 1976. Although PCB concentrations in the Hudson have dropped significantly in recent years, women of childbearing age and children under age 15 still are warned not to eat fish from the river, and no one should eat fish caught between Troy and Hudson Falls. In 1983, a 200-mile stretch of the Hudson south of Hudson Falls was declared the largest Superfund site in United States.

Despite decades of research and debate, what to do about polluted sediment from the river bottom remains highly controversial. Although environmentalists and most residents further south support dredging, many who live along affected sections of the river remain strongly opposed. Local residents fear the noise, lights, and commotion of the 19-hour per day operation will disrupt fish and wildlife populations, drive away tourists, and spoil the quiet beauty of their neighborhoods. They argue that dredging the river will stir up sediments and contaminate the water even more.

These fears were magnified by GE's multimillion-dollar advertising campaign showing giant clamshell dredges dumping dripping loads of muck into barges. A much better plan, according to GE—which will have to pay most of the dredging costs—would be to leave the PCBs in the sediment where bacteria would eventually metabolize and deactivate them. The EPA replies that the environmental dredges to be used on the Hudson are not much larger than the average recreational fishing boat (fig. 1.1). Operating like large vacuum cleaners, they suck up

FIGURE 1.1 *An environmental dredge removes contaminants from the river bottom.*

sediment inside silt curtains that keep contaminants from escaping into nearby water. The EPA claims residents will hardly notice that the operation is underway and that it will be completely open and transparent in reporting water quality data and cooperating with affected communities.

EPA scientists agree that bacterial action would eventually destroy PCBs in the river sediment, but they point out that this process would take centuries. Meanwhile, according to EPA estimates, about 500 lbs (227 kg) of PCBs seep from the river mud and from fractured bedrock beneath the Hudson Falls factory every year. The total amount of PCBs released in six years of dredging, according to the EPA, will be about 200 lbs (91 kg). We can't be sure if these estimates are correct until the dredging has been carried out, but EPA scientists argue that the benefits of dredging far outweigh its potential risks. Many local residents distrust these calculations and doubt the benefits are as large as promised. Meanwhile, GE has challenged the entire Superfund program as an unconstitutional constraint on business.

The dilemma over what to do about contamination of the Hudson River shares aspects of many issues we will discuss in this book. We are changing our environment in ways that we suspect are hazardous both to us and to other living things. It often is uncertain, however, exactly how dangerous these threats are or what we ought to do about them. How can science help us understand these complex problems? What do we need to know to make reasoned, responsible judgments? Studying environmental science will help you find answers to some of these questions and to become informed environmental citizens. Welcome to an important and, we hope, engaging voyage of discovery.

UNDERSTANDING OUR ENVIRONMENT

The debate about dredging the Hudson River illustrates some of the complexity and importance of contemporary environmental issues. Human actions are having widespread impacts on our world and the other organisms with which we share it. Science and technology have become pervasive forces, both to explain how things work and to reveal how we can make our environment safer, more comfortable, and more enduring. The knowledge being gained by scientists is fundamental to our ability to manage the earth's resources in a sustainable manner and to improve the quality of our lives and those of our children. Environmental scientists work on many problems that critically affect our well-being in many ways. Because of the significance of its findings, an understanding of environmental science is becoming increasingly necessary for any educated person.

As you study environmental science, you will learn about many serious problems. But environmental science can also be exciting and highly gratifying. Environmental scientists explore coral reefs, live with great apes, collect ice samples from deep within glaciers, study exotic plant species, and listen to whales as they study the world around us. They also examine the social institutions and built environment that we create for ourselves using science, technology, and political organization. There is room for many different kinds of interests and abilities within this broad discipline. Whether you are a professional scientist or a concerned citizen, you can apply your knowledge of environmental science in enjoyable and useful ways.

A Marvelous Planet

Before proceeding in our discussion of current dilemmas and how scientists are trying to understand them, we should pause for a moment to consider the extraordinary natural world that we inherited and that we hope to pass on to future generations in as good—or perhaps even better—condition than we found it.

Imagine that you are an astronaut returning to the earth after a long trip to the moon or Mars. What a relief it would be, after experiencing the hostile environment of outer space, to come back to this beautiful, bountiful planet (fig. 1.2). Although there are dangers and difficulties here, we live in a remarkably prolific and hospitable world that is, as far as we know, unique in the universe. Compared to the conditions on other planets in our solar system, temperatures on the earth are mild and relatively constant. Plentiful supplies of clean air, fresh water, and fertile soil are regenerated endlessly and spontaneously by biogeochemical cycles (discussed in chapters 2 and 3).

Perhaps the most amazing feature of our planet is its rich diversity of life. Millions of beautiful and intriguing species populate the earth and help sustain a habitable environment (fig. 1.3). This vast multitude of life creates complex, interrelated communities where towering trees and huge animals live together with, and depend upon, such tiny life-forms as viruses, bacteria, and fungi. Together, all these organisms make up delightfully diverse, self-

FIGURE 1.2 *The life-sustaining ecosystems on which we all depend are unique in the universe, as far as we know.*

sustaining ecosystems, including dense, moist forests; vast, sunny savannas; and richly colorful coral reefs.

From time to time, we should pause to remember that, in spite of the challenges and complications of life on earth, we are incredibly lucky to be here. We should ask ourselves: what is our proper place in nature? What *ought* we do and what *can* we do to protect the irreplaceable habitat that produced and supports us? These are some of the central questions of environmental science.

What Is Environmental Science?

We inhabit two worlds. One is the natural world of plants, animals, soils, air, and water that preceded us by billions of years and of which we are a part. The other is the world of social institutions and artifacts that we create for ourselves using science, technology, and political organization. Both worlds are essential to our lives, but integrating them successfully causes enduring tensions.

Environment (from the French *environner:* to encircle or surround) can be defined as (1) the circumstances and conditions that surround an organism or group of organisms, or (2) the social and cultural conditions that affect an individual or community. Since humans inhabit the natural world as well as the "built" or technological, social, and cultural world, all constitute important parts of our environment (fig. 1.4).

Environmental science is the systematic study of our environment and our place in it. A relatively new field, environmental science is highly interdisciplinary. It integrates information from biology, chemistry, geography, agriculture, and many other fields. To apply this information to improve the ways we treat our world,

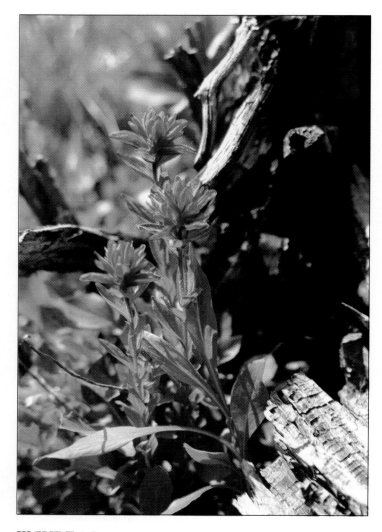

FIGURE 1.3   *Perhaps the most amazing feature of our planet is its rich diversity of life.*

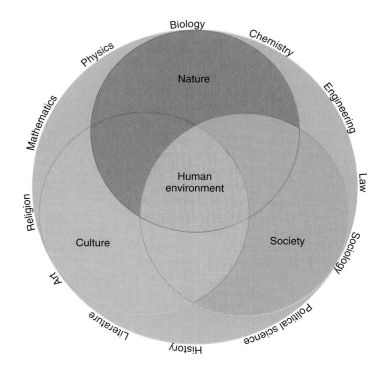

FIGURE 1.4   *The intersections of the natural world with the social and cultural worlds constitute our environment. Many disciplines contribute to environmental science and help us understand how our worlds intertwine as well as our proper role in them.*

environmental scientists also incorporate knowledge of social organization, politics, and the humanities. In other words, environmental science is inclusive and holistic. Environmental science is also mission-oriented: it implies that we all have a responsibility to get involved and try to do something about the problems we have created.

As distinguished economist Barbara Ward pointed out, for an increasing number of environmental issues, the difficulty is not to identify remedies. Remedies are now well understood; the problem is to make them socially, economically, and politically acceptable. Foresters know how to plant trees, but not how to establish conditions under which villagers in developing countries can manage plantations for themselves. Engineers know how to control pollution, but not how to persuade factories to install the necessary equipment. City planners know how to design urban areas, but not how to make them affordable for the poorest members of society. The solutions to these problems increasingly involve human social systems as well as natural science.

Criteria for environmental literacy have been suggested by the National Environmental Education Advancement Project in Wisconsin. These criteria include awareness and appreciation of the natural and built environment, knowledge of natural systems and ecological concepts, understanding of current environmental issues, and the ability to use analytical and problem-solving skills on environmental issues. These are good goals to keep in mind as you study this book.

SCIENCE AS A WAY OF KNOWING

Science, derived from "knowing" in Latin, is a process for producing knowledge. It depends on making precise observations of natural phenomena and on formulating reasonable theories to make sense out of those observations. Science rests on the assumptions that the world is knowable and that we can learn about how things work through careful empirical study and logical analysis. Moreover, because science provides information about both materials and mechanisms in the world around us, it can help us find practical solutions for many problems (table 1.1).

An important value of scientific thinking is that it reduces our tendency to rely on emotional reactions and unexamined assumptions. In the Middle Ages, the ultimate source of knowledge about matters such as how crops grow or how diseases spread was religious authorities or cultural traditions. While these sources may have provided useful insights in many cases, there was no way to test their explanations independently and objectively. They were

TABLE 1.1 Some Scientific Assumptions

1. The world is knowable. With careful, impartial observation and logical analysis, we can make sense of the fundamental processes and laws that shape our environment.

2. Basic patterns that describe events in the natural world are uniform throughout time and space. The forces at work now are the same as those that shaped the world in the past and will continue to do so in the future.

3. Where two equally plausible explanations for a phenomenon are possible, we should choose the simpler one. (Also known as the law of **parsimony,** or Ockham's razor after the English philosopher who first proposed this rule.)

4. Change in knowledge is inevitable because new evidence may challenge prevailing theories. No matter how well one theory explains a set of observations, it is possible that another theory may fit just as well or better, or may fit a still wider range of data.

5. Although new facts can disprove existing theories, science can never provide absolute proof that a theory is correct. Every theory should be considered only conditionally or provisionally correct until contrary evidence is found.

6. Even if there is no way to secure complete and absolute truth, increasingly accurate approximations can be made to account for the world and how it works.

7. Because science provides information about both mechanisms and processes in the world around us, it can help find practical solutions for many problems.

FIGURE 1.5 *Plato and Aristotle debate moral philosophy in a painting by Raphael. Plato* (left) *motions upward, indicating a transcendent, universal moral truth, while Aristotle* (right) *motions downward to suggest grounded, contingent knowledge.*

right because custom, politics, or theology said so. The benefit of scientific thinking is that it searches for testable evidence: if you suspect that a disease spreads through contaminated water, you can close off access to that water source and see if the disease stops spreading.

Ideally, scientists are skeptical. They don't accept proposed explanations until there is substantial evidence to support them. Even then, every explanation is considered only provisionally true because there is always a possibility that some additional evidence may appear to disprove it. Scientists also try to be methodical, rigorous, and unbiased. Because bias and methodical errors are hard to avoid, scientific tests are subject to review by informed peers, who can help evaluate results and conclusions.

Modern science has its roots in antiquity. Greek philosophers such as Pythagoras (about 520 B.C.), Socrates (464–399 B.C.), and Aristotle (380–320 B.C.) laid the foundations for inquiry and logic (fig. 1.5). Arabic mathematicians and astronomers added to this store of knowledge and reintroduced it to the Western world at the end of the Middle Ages. Much of what we now know as the scientific worldview and scientific method, however, originated in seventeenth- and eighteenth-century Europe. Among the scholars who pioneered empirical, experimental approaches to science were Galileo Galilei (1564–1642), a pioneer in physics and astronomy; philosophers such as René Descartes (1596–1650) and Francis Bacon (1561–1626), who emphasized the use of objective

observation and inductive reasoning; and Isaac Newton (1642–1727), who invented calculus and studied optics and gravitation. The insights and methods introduced by these scientific pioneers laid the foundation for much of the material progress that we now enjoy.

Cooperation and Insight in Science

Good science rarely is carried out by a single individual working in isolation. Instead, a community of scientists collaborates in a cumulative, self-correcting process. Ideas and information are freely exchanged, debated, and tested to find the most correct ones. Ideally, scientists compare results and challenge interpretations in a democratic, impartial way. You often hear about big breakthroughs and dramatic discoveries that change our understanding overnight, but these are rare in ordinary science. Instead, many people work on different aspects of a common problem, each adding small insights into how a system works (fig. 1.6).

Confusing or erroneous results often misdirect researchers temporarily until mistakes are recognized and the search for truth gets back on the right track. One of the main benefits of

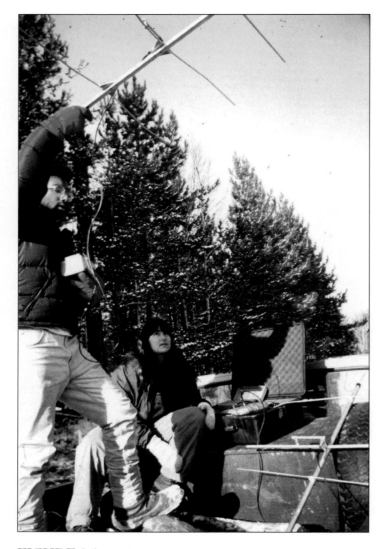

FIGURE 1.6 *Researchers use a directional antenna to track radio-collared wolves.*

research work is the realization it gives you of how contradictory science often is. If you're working at the cutting edge of either your theories or your instruments, you may well get wrong answers as often as you get right ones. Making sense of a baffling mass of information is one of the greatest challenges that scientists face.

Many people—even scientists—fail to recognize the role that creativity, insight, aesthetics, and luck play in research. Often our most important discoveries were made not through superior scientific method or equipment, but because the investigators were passionate and tenacious in their pursuit of knowledge. Some great physicists such as Albert Einstein, Richard Feynman, and Murray Gell-Mann have been attracted to certain theories as much by their elegance and beauty as by their scientific importance.

Scientific Design

Scientists demand lots of evidence before they are willing to accept the accuracy of any data or the usefulness of a particular interpretation. One of the most important tests of any data is its **reproducibility.** Making an observation or obtaining a particular result just once doesn't count for much. You have to produce the same result consistently to be sure that your first outcome wasn't a fluke. Even more importantly, you ought to be able to describe in sufficient detail the conditions under which you made the observation or obtained the result so that someone else can reproduce your findings.

It's often difficult to study natural systems because so many important factors vary simultaneously. To avoid this problem, scientists design **controlled studies** in which comparisons are made between experimental and control populations that are identical (as far as possible) in every factor except the one being studied. Studies must be designed very carefully so that experimenters don't inadvertently treat the experimental and control groups differently or that data coders don't unconsciously let bias creep into their scoring. To avoid investigator bias, **blind experiments** are designed in which those carrying out the experiment don't know until after data has been gathered and analyzed. In medical research involving human subjects you have to be careful that expectations of the subjects and the experimenters don't bias results (the placebo effect). In a **double-blind design,** neither the subject (participant) nor the experimenter knows which participants are receiving the experimental or the control treatments. Analysis of the results is done by a third party who does know which participants are allocated to which group.

Researchers try very hard to be accurate, but it's important to avoid unnecessary or meaningless precision. Suppose you've had a fresh snowfall in your area, and you want to know its depth. You have a ruler marked in millimeters. The first measurement you make is 24 mm, but the surface of the snow isn't even. By moving your ruler slightly, you can get measurements of 21 mm and 25 mm. What's the true depth of the snowfall? If you average these three measurements, you get 23.3333 mm, but this gives a false notion of precision. The last three digits aren't **significant numbers.** Your ruler isn't accurate to tenths of a millimeter, let alone thousandths or ten-thousandths, nor is the surface of the snow even enough to make such a precise measurement. The most accurate answer is an approximation of about 23 mm.

Deductive and Inductive Reasoning

In 1919, Sir Arthur Eddington led an expedition to Principe Island off the coast of West Africa to photograph stars during a solar eclipse. These photographs were a landmark in modern science because they were the first successful empirical test of Einstein's theory of general relativity. Einstein's calculations predicted that photons of light should be affected by gravity. Thus, light coming from distant stars should be deflected if it passes close to our sun's surface. These stars should appear shifted slightly in position during the day, when the sun is present, compared to the night, when

Principles of Environmental Science

it is not. Normally, you can't make such an observation because the stars close to the sun are invisible in the daytime. During an eclipse, however, it is possible to photograph stars and compare their apparent day and night locations. This is what Eddington did, and his corroboration of Einstein's prediction was a triumphant moment in modern physics, one that helped Einstein's theory of relativity become widely accepted.

This series of logical steps is **deductive reasoning.** Starting with a general principle (the theory of general relativity), a testable prediction is derived about a specific case: that certain stars should appear closer together in the daytime than at night. You might think of this as "top-down" reasoning—proceeding from the general to the specific. Although deduction is often regarded as a scientific ideal, in many cases we don't have a general theory to explain certain phenomena in nature. Instead, we depend on "bottom-up" **inductive reasoning** in which we study specific examples and try to discover patterns and derive general explanations from collected observations.

An example of inductive reasoning is seen in the brilliant work of Austrian zoologist Karl Von Frisch, who discovered how honeybees communicate. Although people had noticed for many years that bees seem to have an uncanny ability to find their way to nectar-bearing flowers, no theoretical explanation gave any insight into this phenomenon. Through meticulous, methodical observations, Von Frisch discovered and described the dances by which foraging honeybees (*Apis mellifera*) indicate the direction and distance of food sources to their hive mates. In 1973, Von Frisch shared the Nobel Prize in Medicine with Konrad Lorenz and Nikolaas Tinbergen who made similar discoveries in animal behavior through equally detailed and painstaking observations. While we still don't have many widely accepted laws of animal behavior or communication, these studies have given us fascinating insights into the lives of other organisms.

Hypotheses and Theories

You may already be using scientific technique without being aware of it. Suppose your flashlight doesn't work. The problem could be in the batteries, the bulb, or the switch—or all of them could be faulty. How can you distinguish among these possibilities? If you change all the components at once, you may have a working flashlight, but you won't know which was the faulty part. A series of methodical steps to test each component can be helpful.

Starting with the observation that your flashlight doesn't work, you make inferences about the component you think most likely to be defective. Perhaps that might be the batteries. You propose a **hypothesis,** or a conditional explanation that can be tested by further observation or experiment. In this case, the hypothesis might be, "The batteries are dead." From this tentative assumption, you make a prediction: "If I replace the dead batteries with fresh ones, the flashlight should work." If your hypothesis is correct, then when you perform the experiment (replacing the batteries), you will see the predicted result. If not, you reject your first hypothesis and form a new one: for example, "The bulb is burned out." By formulating testable hypotheses and evaluating each

component in turn, you should eventually be able to isolate the problem and find a way to solve it.

What you have just employed in this example is the scientific method of observations, systematic testing, and interpretation of results shown in figure 1.7. As mentioned earlier, the results of one experiment may give us information that leads to further hypotheses and additional experiments. In each case, prior knowledge and experience help us design experiments and interpret results. Eventually, with evidence from a group of related investigations, we may be able to formulate a theory to explain a set of general principles.

Logically, you can show a hypothesis based on inductive reasoning to be wrong, but you can almost never show it to be unquestionably true. The philosopher Ludwig Wittgenstein gave the following example to illustrate this principle: Suppose all the swans you have ever seen are white. You might assume that *all* swans are white. This hypothesis could be tested by examining a large number of swans. If you never find a black one, you might tentatively conclude that your hypothesis is right. But even if you look at a million white swans, there could still be a black one somewhere (actually, there are black swans in Australia) that would refute your conjecture. Thus, you could show your hypothesis is wrong, but you could never be absolutely sure that it is correct, no matter how much evidence you collect.

As a result of this uncertainty, we should always regard evidence in science as provisional. Additional information might come along to undermine the observations we have made thus far.

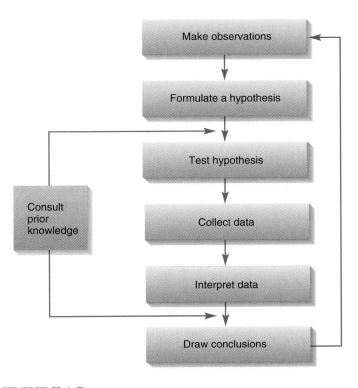

FIGURE 1.7 *Ideally, scientific investigation follows a series of logical, orderly steps to formulate and test hypotheses.*

What Are Statistics, and Why Are They Important?

Statistics are numbers that let you evaluate and compare things. "Statistics" is also a field of study that has developed meaningful methods of comparing those numbers. By both definitions, statistics are widely used in environmental sciences, partly because they can give us a useful way to assess patterns in a large population, and partly because the numbers can give us a measure of confidence in our research or observations. Understanding the details of statistical tests can take years of study, but a few basic ideas will give you a good start toward interpreting statistics.

1. *Descriptive statistics help you assess the general state of a group.* In many towns and cities, the air contains a dust, or particulate matter, as well as other pollutants. From personal experience you might know your air isn't as clean as you'd like, but you may not know how clean or dirty it is. You could start by collecting daily particulate measurements to find average levels. An averaged value is more useful than a single day's values, because daily values may vary a great deal, but general, long-term conditions affect your general health. Collect a sample every day for a year, then divide the sum by the number of days, to get a **mean** (average) dust level. Suppose you found a mean particulate level of 30 micrograms per cubic meter ($\mu g/m^3$) of air. Is this level high or low? In 1997 the EPA set a standard of 50 $\mu g/m^3$ as a limit for allowable levels of coarse particulates (2.5–10 micrometers in diameter). Higher levels tend to be associated with elevated rates of asthma and other respiratory diseases. Now you know that your town, with an annual average of 30 $\mu g/m^3$, has relatively clean air, after all.

2. *Statistical samples.* Although your town is clean by EPA standards, how does it compare to the rest of the cities in the country? Testing the air in *every* city is probably not possible. You could compare your town's air quality to a **sample,** or subset of cities, however. A large, random sample of cities should represent the general "population" of cities reasonably well. Taking a large sample reduces the effects of outliers (unusually high or low values) that might be included. A random sample minimizes the chance that you're getting only the worst sites, or only a collection of sites that are close together, which might all have similar conditions. Suppose you get average annual particulate levels from a sample of 50 randomly selected cities. You can draw a frequency distribution, or histogram, to display your results (fig. 1). The mean value of this group is 36.8 $\mu g/m^3$, so by comparison, your town (at 30 $\mu g/m^3$) is relatively clean.

FIGURE 1 *Average annual airborne dust levels for 50 hypothetical cities in 2001.*
Source: U.S. Environmental Protection Agency.

Many statistical tests assume that the sample has a normal or Gaussian frequency distribution, often described as a bell-shaped curve (fig. 2). In this distribution, the mean is near the center of the range of values, and most values are fairly close to the mean. Large and random samples are more likely to fit this shape than small or nonrandom samples.

3. *Confidence.* How do you know that the 50 cities you sampled really represent all the cities in the country? You can't ever be completely certain, but you can use estimates, such as confidence limits, to express the reliability of your mean statistic. Depending on the size of your sample (not 10, not 100, but 50) and the amount of variability in the

---

When a large number of tests supports an explanation, however, and a majority of experts in a given field have reached a general consensus that it is the best description or explanation available, we call it a **scientific theory.** Note that scientists' use of this term is very different from that of the general public. To many people, a theory is speculative and unsupported by facts. To a scientist, it means just the opposite; while all explanations are tentative and open to revision and correction, one that counts as a scientific theory is supported by an overwhelming body of data and experience, and is generally accepted in the scientific community, at least for the present.

Modeling and Natural Experiments

Not every area that we might want to study scientifically is open to simple, direct experiments like those described above. A geologist, for instance, might want to study mountain building, or an ecologist might like to know how species co-evolve, but neither scientist would be able to re-create the millions of years and complex environmental conditions required for these processes to occur. Similarly, a toxicologist might want to know about the effects of toxins but couldn't deliberately poison people, no matter how useful that information might be.

One way to gather information in fields such as these is to look for historic evidence, or what might be called "natural experiments," for support or contradiction of an idea. Can you find examples that would support or refute a given hypothesis from what has already happened? The geologist, for example, might propose the hypothesis that mountain building always involves folding and uplifting of surface strata. If she can find counter examples, this hypothesis could be discarded.

Another way to gather information about environmental systems is to use models. A model could be a substitute organism or a physical mock-up that simulates a real system, or it could be a set of mathematical equations, often calculated in a computer program, that represents the phenomenon you want to study. For

FIGURE 2 *A normal distribution.*

FIGURE 3 *A plot showing relationships between variables.*

sample data, you can calculate a confidence interval that the mean represents the whole population (all cities). Confidence levels, or confidence intervals, represent the likelihood that your statistics represent the entire population correctly. For the mean of your sample, a confidence interval tells you the probability that your sample is similar to other random samples of the population. A common convention is to compare values to a 95 percent confidence level, or a probability of 5 percent or less that your conclusions are misleading. Using statistical software, we can calculate that for our 50 cities, the mean is 36.8 $\mu g/m^3$, and the confidence interval is 35.0 to 38.6. This suggests that if you took 1,000 samples from the entire population of cities, 95 percent of those samples

ought to be within 2 $\mu g/m^3$ of your mean. This indicates that your mean is reliable and representative.

4. *Is your group unusual?* Once you have described your group of cities, you can compare it to other groups. For example, you might believe that Canadian cities have cleaner air than U.S. cities. You can compare mean air quality levels for the two groups. Then you can calculate confidence intervals for the difference between the means, to see if the difference is meaningful.

5. *Evaluating relationships between variables.* Are respiratory diseases correlated with air pollution? For each city in your sample, you could graph pollution and

asthma rates (fig. 3). If the graph looks like a loose cloud of dots, there is no clear relationship. A tight, linear pattern of dots trending upward to the right indicates a strong and positive relationship. You can also use a statistical package to calculate an equation to describe the relationship and, again, confidence intervals for the equation. This is known as a regression equation.

6. *Lies, Damned Lies, and Statistics.* Can you trust a number to represent a complex or large phenomenon? One of the devilish details of representing the world with numbers is that those numbers can be tabulated in many ways. If we want to assess the greatest change in air quality statistics, do we report rates of change, the total amount of change? Do we look at change over five years? Twenty-five years? Do we accept numbers selected by the EPA, by the cities themselves, by industries, or by environmental groups? Do we trust that all the data were collected with a level of accuracy and precision that we would accept if we knew the hidden details in the data-gathering process? Like all information, statistics need to be interpreted in terms of who produced them, when, and why. Awareness of some of the standard assumptions behind statistics, such as sampling, confidence, and probability, will help you interpret statistics that you see and hear.

instance, the toxicologist might feed the toxin in question to rats and then attempt to extrapolate or infer how the rats' reaction is related to humans. The geologist might build a scale model of a mountain range to see what forces affect it. Mathematical models are especially useful when there are simultaneous variables. Input conditions (initial population size, population growth rate, and so on) can be manipulated to test their effect on resulting conditions.

Models, however, represent researchers' assumptions about how a system works, so models run by different people can produce contradictory results. Global warming models, for example, disagree on the amount of climate change expected, and the rate of change. However, models do provide heuristic information—suggestions of how things might be—and they allow us to test our understanding of the relationships in a system. Even though climate models disagree on the degree of change, nearly all agree that temperatures will increase. Models have become an important scientific tool for exploring complex

systems. In chapter 3 we will look at a few simple population models and at how these models help us understand population dynamics.

Statistics and Probability

You will encounter a lot of numbers in this book. It is important to think carefully about what they mean and what they don't mean. Quantitative data can be precise and easily compared, and provide good benchmarks for measuring change. Suppose that you read— as you will in chapter 6—that 2.7 million ha (6.7 million acres) of forest burned in the United States in 2002. Is that a lot or a little? Is it an ominous change from previous years, or is it within the range that we might expect? It happens that 2000 was an exceptionally hot, dry year and that the number and sizes of forest fires were much greater than anytime in the previous 90 years. The total area of the fires was slightly larger than Maryland or Vermont, but some of the forests were probably only lightly burned, while other sections were more severely affected. You have to look carefully at what numbers mean, not just take them at face value. It also helps to try to develop some benchmarks to which you can compare statistics. How big is a hectare? How many people live in Canada? What is a gigaton? A useful conversion table can be found inside the back cover of this book.

Probability is another important concept in science. Probability, risk, and chance are all ideas that try to measure how likely something is. They never tell what *will* happen, only what might happen. If you have a 20 percent chance of catching a cold this month, that means that about twenty out of every 100 people will probably get a cold. You are relatively unlikely to get one, but you might. We rarely can be 100 percent sure of anything, but scientists consider a 95 percent probability—or confidence level—that an event is not random a high degree of certainty (see Investigating Our Environment, p. 8).

A related question concerns appropriate sample size. How many individuals would you need to examine to have a reliable representation of a population? Is it better to sample small areas in many different places or one larger area (fig. 1.8)? How many times do you need to replicate your investigation to ensure a reliable result? The answers to these questions depend on the variables that you intend to study, as well as the accuracy level you hope to attain. These aspects of research design are the domain of statistics. Statistics is an important tool in both planning and evaluating scientific studies.

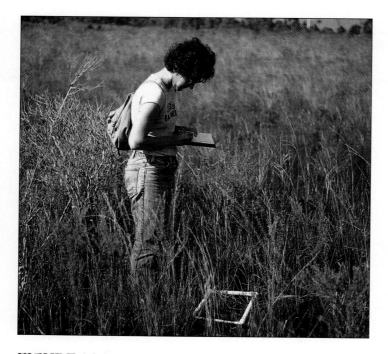

FIGURE 1.8 *What is an appropriate sample size for an investigation? How many individuals would you need to examine to have a representative sample of a population? These are statistical questions of variability, probability, and accuracy.*

APPLICATION: Calculating Probability

An understanding of statistical analysis and probability is fundamental in most areas of modern science. Working with these concepts is critical to your ability to comprehend scientific information.

Every time you flip a coin, the chance that heads will end up on top is 1 in 2 (50 percent) (assuming you have a normal coin). The odds of getting heads two times in a row is $1/2 \times 1/2$, or $1/4$.

1. What are the odds of getting heads five times in a row?
2. As you start the fifth flip, what are the odds of getting heads?
3. If there are 100 students in your class and everybody flips a coin five times, how many people are likely to get five heads in a row?

Answers: 1. $(1 \text{ in } 2 \text{ to the } 5\text{th}) = 1/32$; *2.* $1 \text{ in } 2$; *3.* $100 \text{ students} \times 1/32 = \text{about } 3$.

Paradigms and Scientific Consensus

The notion of paradigm shifts has been important since science historian Thomas Kuhn introduced it in the 1960s. **Paradigms,** according to Kuhn, are overarching models of the world that guide our interpretation of events. Two centuries ago, Noah's flood was considered one of the principal events that shaped our world. Geological evidence was interpreted in terms of the flood and its impacts. Today geologists interpret the world's history in terms of tectonic plates that have rearranged themselves repeatedly over billions of years. Biogeographers and ecologists also now have entirely different explanations for the same physical evidence (such as distribution of fossils on different continents). Paradigms are also important because they guide the sorts of questions we ask. Today few people would search for evidence that Noah's flood redistributed fish species, but they might be interested in how mountain building separated species into different watersheds.

The shift from one guiding paradigm to another occurs when a majority of scientists accept that the old explanation no longer explains new observations very well. This shift can be contentious and political because a new model can undermine whole careers that are based on one sort of research and explanation. Sometimes this revolution is accomplished rather quickly. Quantum mechanics and Einstein's theory of relativity, for example, overturned classical physics in only about 30 years. In other cases, it may take a century or more for new ideas to supplant old ones. Often a paradigm shifts as a new generation replaces an older generation.

As you study this book and think about environmental science, try to identify some of the paradigms that guide our investigations and explanations today. This is one of the skills involved in critical thinking, which we will discuss in further detail later in the chapter.

Pseudoscience and Baloney Detection

In every controversial topic, a few contrarian scientists always champion a view diametrically opposed to the majority position of the scientific community. There are several reasons to adopt a contradictory view. One might be sincerely convinced, for instance, that the majority opinion is wrong. Often the most brilliant and original thinkers are regarded as hopelessly off track by conventional scientists. Others may be motivated simply by a desire to consider all possibilities in case unexpected evidence comes up.

Some people also may be tempted to take an opposing viewpoint because they know it will generate publicity and even, perhaps, generous payments from those who benefit from counter testimony. In many court cases, for example, opposing sides are able to produce paid "experts" who are willing to present a partisan agenda rather than objective science. A common tactic is to use scientific uncertainty as an excuse to postpone or reverse an action or policy that a vast majority of the scientific community believes to be prudent. Opponents will claim that the evidence doesn't constitute conclusive proof that the action needs to be taken, or that the policy is based only on a scientific "theory" and therefore has no validity. As we've discussed earlier, theory describes an entirely different level of certainty in science than it does in common usage. In addition, while scientists usually avoid claims that a theory is unquestionably true, an overwhelming body of evidence is usually required before a consensus is reached on the provisional validity of a particular position.

Harvard's Edward O. Wilson writes, "We will always have contrarians whose sallies are characterized by willful ignorance, selective quotations, disregard for communications with genuine experts, and destructive campaigns to attract the attention of the media rather than scientists. They are the parasite load on scholars who earn success through the slow process of peer review and approval." How can we identify junk science that presents bogus analysis dressed up in quasi-scientific jargon but that really represents a perversion of objective inquiry? The astronomer Carl Sagan proposed a "Baloney Detection Kit" containing the questions in table 1.2. An example of a current scientific controversy is presented in Investigating Our Environment (p. 12).

THINKING ABOUT THINKING

Perhaps the most valuable skill you can learn in any of your classes is the ability to think clearly, creatively, and purposefully. Much of the most important information in environmental science is highly contested. As previous sections of this chapter show, facts can vary, depending on when and by whom they were gathered. How can you make sense out of a welter of conflicting information? Developing your ability to learn new skills, examine new facts, evaluate new theories, and formulate your own interpretations is essential to keep up in a changing world.

A flood of information and misinformation inundates us everyday. Competing explanations and contradictory ideas battle for our attention. The rapidly growing complexity of our world and our lives intensifies the difficulties of knowing what to believe

TABLE 1.2	Questions for Baloney Detection

1. How reliable are the sources of this claim? Is there reason to believe that they might have an agenda to pursue in this case?
2. Have the claims been verified by other sources? What data are presented in support of this opinion?
3. What position does the majority of the scientific community hold in this issue?
4. How does this claim fit with what we know about how the world works? Is this a reasonable assertion or does it contradict established theories?
5. Are the arguments balanced and logical? Have proponents of a particular position considered alternate points of view or only selected supportive evidence for their particular beliefs?
6. What do you know about the sources of funding for a particular position? Are they financed by groups with partisan goals?
7. Where was evidence for competing theories published? Has it undergone impartial peer review or is it only in proprietary publication?

or how to act. Consider how the communications revolution has brought us computers, email, cell phones, mobile faxes, pagers, the World Wide Web, hundreds of channels of satellite TV, and direct mail or electronic marketing that overwhelm us with conflicting information. We have more choices than we can possibly manage, and we know more about the world around us than ever before but, perhaps, understand less.

An additional complication is that distinguished authorities vehemently disagree about many important topics. Some cynics say that, for any expert, there is always an equal and opposite expert. How can you decide what is true and meaningful? Is it simply a matter of what feels good at the moment or what supports our preconceived notions? Or can we use logical, orderly, creative thinking procedures to reach decisions?

Approaches to Knowledge and Meaning

A number of skills, attitudes, and approaches can help us evaluate information and make decisions. **Analytical thinking** asks, "How can I break this problem down into its constituent parts?" **Creative thinking** asks, "How might I approach this problem in new and inventive ways?" **Logical thinking** asks, "How can orderly, deductive reasoning help me think clearly?" **Critical thinking** asks, "What am I trying to accomplish here, and how will I know when I've succeeded?" **Reflective thinking** asks, "What does it all mean?" (See fig. 1.9.)

While much of science is based on analytical and logical thinking, critical thinking adds elements of contextual sensitivity and empathy that can be very helpful in your study of environmental science. It challenges us to examine theories, facts, and options in a systematic, purposeful, and responsible manner. It shares many methods and approaches with other methods of reasoning but adds some important skills, attitudes, and dispositions.

INVESTIGATING Our Environment

A Skeptical Environmentalist?

In 2001, Cambridge University Press published *The Skeptical Environmentalist: Measuring the Real State of the World* by the Danish statistician Bjørn Lomborg. The book triggered a firestorm of criticism as natural scientists argued that Lomborg's book was serving to "confuse legislators and regulators, and poison the well of public environmental information."[1] In January 2002, *Scientific American* published a series of articles by five distinguished environmental scientists contesting Lomborg's claims. To some observers, the ferocity of the attack was surprising. Why was *The Skeptical Environmentalist* so controversial, and how can we judge the relative correctness of the claims and counterclaims in this debate?

Lomborg begins by saying that although he is an "old left-wing Greenpeace member," he is worried about the unrelenting "doom and gloom" of mainstream environmentalism. He describes what he regards as an all-pervasive litany that says, among other things, "Our resources are running out. The population is ever growing, leaving less and less to eat. The air and water are becoming ever more polluted. The planet's species are becoming extinct in vast numbers . . . The forests are disappearing, fish stocks are collapsing and coral reefs are dying."

Lomborg, however, says these claims about the collapse of ecosystems are "simply not in keeping with reality. We are not running out of energy or natural resources. There will be more and more food per head of the world's population. Fewer and fewer people are starving. In 1900 we lived for an average of 30 years; today we live for 67. According to the UN we have

reduced poverty more in the last 50 years than we did in the preceding 500, and it has been reduced in practically every country." He goes on to challenge conventional scientific assessment of global warming, forest losses, fresh water scarcity, energy shortages, and a host of other environmental problems. Is Lomborg being deliberately (and some would say, hypocritically) optimistic, or are others being unreasonably pessimistic? Is this simply a case of regarding the glass as half full or half empty?

Part of the discrepancy is that Lomborg is unabashedly anthropocentric. He dismisses the value of biodiversity, for example. As long as there are plants and animals to supply human needs, what does it matter if a few nonessential species go extinct? Furthermore, he is intentionally positive. He cheerfully predicts that progress in population control, use of renewable energy, and extension of water supplies through desalination technology will spread to the whole world, thus avoiding crises in resource supplies and human impacts on our environment. Others, particularly Lester Brown of the Worldwatch Institute, according to Lomborg, seem to deliberately adopt worst-case scenarios. Are they merely being realistic or might they be attempting to scare us into supporting their agenda?

Perhaps the most effective criticism of Lomborg comes from his reporting of statistics and research results. Stephen Schneider, a distinguished climate scientist from Stanford University, for instance, writes in *Scientific American*, "most of [Lomborg's] nearly 3,000 citations are to secondary literature and media articles. Moreover, even when cited, the peer-reviewed articles come elliptically from those

studies that support his rosy view that only the low end of the uncertainty ranges [of climate change] will be plausible. IPCC authors, in contrast, were subjected to three rounds of review by hundreds of outside experts. They didn't have the luxury of reporting primarily from the part of the community that agrees with their individual views."[2]

Lomborg also criticizes extinction rate estimates as much too large, citing evidence from places like Brazil's Atlantic forest, where about 90 percent of the trees have been cleared without large numbers of recorded extinctions. Thomas Lovejoy, chief biodiversity adviser to the World Bank, responds, "First, this is a region with very few field biologists to record either species or their extinction. Second, there is abundant evidence that if the Atlantic forest remains as reduced and fragmented as it is, it will lose a sizable fraction of the species that at the moment are able to hang on."

How would you evaluate the claims and counterclaims in this debate? What questions from the Baloney Detection Kit in table 1.2 might be most appropriate here? Based on these limited quotes from Lomborg and his critics, what hidden agendas might you look for in the original articles, and what further evidence would you want to gather by examining the primary sources?

[1]Richard C. Bell. 2002. Media sheep: How did *The Skeptical Environmentalist* pull the wool over the eyes of so many editors? *Worldwatch* 15(2):11–13.

[2]Stephen Schneider. 2002. Global warming: Neglecting the complexities. *Scientific American* 286(1):62–65.

Furthermore, it challenges us to plan methodically and to assess the process of thinking as well as the implications of our decisions. Thinking critically can help us discover hidden ideas and means, develop strategies for evaluating reasons and conclusions in arguments, recognize the differences between facts and values, and avoid jumping to conclusions. Richard Paul, chair of the National Council for Critical Thinking, identifies ten steps in critical thinking (table 1.3).

Notice that many critical thinking processes are self-reflective and self-correcting. This form of thinking is sometimes called "thinking about thinking." It is an attempt to plan rationally how to analyze a problem, to monitor your progress while you are doing it, and to evaluate how your strategy worked and what you have learned when you are finished. It is not critical in the sense of finding fault, but it makes a conscious, active, disciplined effort to be aware of hidden motives and assumptions, to uncover bias, and to recognize the reliability or unreliability of sources.

While critical thinking shares many of the orderly, systematic approaches of formal logic, it also invokes traits like empathy, sensitivity, courage, and humility. Formulating intelligent opinions about some of the complex issues you will encounter in environmental science requires more than simple logic. Developing these attitudes and skills is not easy or simple. It takes practice. You have to develop your mental faculties just as you need to train for a sport. Intellectual integrity, modesty, fairness, compassion, and fortitude are not traits you use only occasionally. They must be cultivated until they become a part of your normal way of thinking (fig. 1.10).

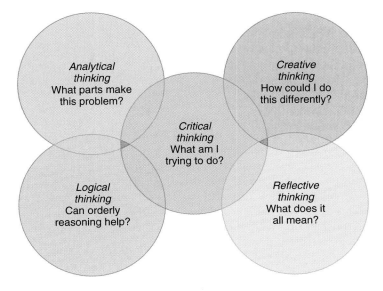

FIGURE 1.9 *All these ways of thinking can be useful in scientific studies. Critical thinking is central because it draws on aspects of each of the other four.*

| **TABLE 1.3** | **Steps in Critical Thinking** |

1. What is the purpose of my thinking?
2. What precise question am I trying to answer?
3. Within what point of view am I thinking?
4. What information am I using?
5. How am I interpreting that information?
6. What concepts or ideas are central to my thinking?
7. What conclusions am I aiming toward?
8. What am I taking for granted; what assumptions am I making?
9. If I accept the conclusions, what are the implications?
10. What would the consequences be, if I put my thoughts into action?

Applying Critical Thinking

We all use critical or reflective thinking at times. Suppose a television commercial tells you that a new breakfast cereal is tasty and good for you. You may be suspicious and ask yourself a few questions: What do they mean by good? Good for whom or what? Does "tasty" simply mean more sugar and salt? Might the sources of this information have other motives besides your health and happiness? Although you may not have been aware of it, you already have used some of the techniques of critical analysis. Working to expand these skills helps you recognize how information and analysis can be distorted, misleading, prejudiced, superficial, unfair, or otherwise defective.

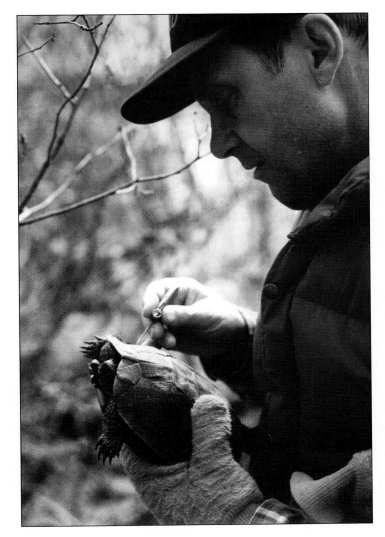

FIGURE 1.10 *A wildlife biologist examines an endangered Blandings turtle. Critical thinking is essential in environmental research.*

Here are some steps in critical thinking:

1. *Identify and evaluate premises and conclusions in an argument.* What is the basis for the claims made here? What evidence is presented to support these claims and what conclusions are drawn from this evidence? If the premises and evidence are correct, does it follow that the conclusions are necessarily true?

2. *Acknowledge and clarify uncertainties, vagueness, equivocation, and contradictions.* Do the terms used have more than one meaning? If so, are all participants in the argument using the same meanings? Are ambiguity or equivocation deliberate? Can all the claims be true simultaneously?

3. *Distinguish between facts and values.* Can claims be tested? (If so, these are statements of fact and should be verifiable by gathering evidence.) Are claims made about the worth or

lack of worth of something? (If so, these are value statements or opinions and probably cannot be verified objectively.) For example, claims of what we ought to do to be moral or righteous or to respect nature are generally value statements.

4. *Recognize and assess assumptions.* Given the backgrounds and views of the protagonists, what underlying reasons might there be for the premises, evidence, or conclusions presented? Does anyone have an "axe to grind" or a personal agenda in this issue? What do they think you know, need, want, or believe? Is there a subtext based on race, gender, ethnicity, economics, or some belief system that distorts this discussion?

5. *Distinguish source reliability or unreliability.* What qualifies the experts on this issue? What special knowledge or information do they have? What evidence do they present? How can we determine whether the information offered is accurate, true, or even plausible?

6. *Recognize and understand conceptual frameworks.* What are the basic beliefs, attitudes, and values that this person, group, or society holds? What dominating philosophy or ethics control their outlooks and actions? How do these beliefs and values affect the way people view themselves and the world around them? If there are conflicting or contradictory beliefs and values, how can these differences be resolved?

Some Clues for "Unpacking" an Argument

A logical argument is made up of one or more introductory statements (called premises) and a conclusion that supposedly follows logically from the premises. In ordinary conversation, different kinds of statements often are mixed together, so distinguishing among them or deciphering hidden or implied meanings is difficult. Social theorists call the process of separating and analyzing textual components "unpacking." Applying this type of analysis to an argument can be useful.

An argument's premises are usually claimed to be based on facts; conclusions are usually opinions and values drawn from, or used to interpret, those facts. Words that often introduce a premise include *as, because, assume that, given that, since, whereas,* and *we all know that* . . . . Words that generally indicate a conclusion or statement of opinion or values include *and, so, thus, therefore, it follows that, consequently, the evidence shows,* and *we can conclude that.* . . .

For instance, in the example we used earlier, the television cereal ad might have said: "Since we all need vitamins, and since this cereal contains vitamins, the cereal must be good for you." Which are the premises, and which is the conclusion? Does one necessarily follow from the other? Remember that, even if the facts in a premise are correct, the conclusions drawn from those facts may not be. Information may be withheld from the argument such as the fact that the cereal is also loaded with unhealthy amounts of sugar.

Using Critical Thinking

In this book, you will have many opportunities to practice critical thinking skills. Every chapter includes many facts, figures, opinions, and theories. Are all of them true? No, probably not. They were the best information available when this text was written, but much in environmental science is in a state of flux. Data change constantly, as does our interpretation of data. Do the ideas presented here give a complete picture of the state of our environment? Unfortunately, they probably don't. No matter how comprehensive our discussion is of this complex, diverse subject, it can never capture everything worth knowing, nor can it reveal all possible points of view.

When reading this text, try to distinguish between statements of fact and opinion. Ask yourself if the premises support the conclusions drawn from them. Although we have tried to be fair and even-handed in presenting controversies, our personal biases and values—some of which we may not even recognize—affect how we see issues and present arguments. Watch for cases in which you need to think for yourself, and use your critical and reflective thinking skills to uncover the truth.

A BRIEF HISTORY OF CONSERVATION AND ENVIRONMENTALISM

Although some early societies had negative impacts on their surroundings, others lived in relative harmony with nature. In modern times, however, growing human populations and the power of our technology have heightened concern about what we are doing to our environment. We can divide conservation history and environmental activism into at least four distinct stages: (1) pragmatic resource conservation, (2) moral and aesthetic nature preservation, (3) a growing concern about health and ecological damage caused by pollution, and (4) global environmental citizenship. These eras aren't necessarily mutually exclusive, however; parts of each persist today in the environmental movement and one person may embrace them all simultaneously.

Historic Roots of Nature Protection

Recognizing human misuse of nature is not unique to modern times. Plato complained in the fourth century B.C. that Greece once was blessed with fertile soil and clothed with abundant forests of fine trees. After the trees were cut to build houses and ships, however, heavy rains washed the soil into the sea, leaving only a rocky "skeleton of a body wasted by disease." Springs and rivers dried up while farming became all but impossible. Many classical authors regarded Earth as a living being, vulnerable to aging, illness, and even mortality. Periodic threats about the impending death of nature from human misuse have persisted to our own times. As Mostafa K. Tolba, former Director of the United Nations Environment Programme said, "The problems that overwhelm us today are precisely those we failed to solve decades ago."

Some of the earliest scientific studies of environmental damage were carried out in the eighteenth century by French or British colonial administrators who often were trained scientists and who considered responsible environmental stewardship as an aesthetic and moral priority, as well as an economic necessity. These early conservationists observed and understood the connection between deforestation, soil erosion, and local climate change. The pioneering British plant physiologist Stephen Hales, for instance, suggested that conserving green plants preserved rainfall. His ideas were put into practice in 1764 on the Caribbean island of Tobago, where about 20 percent of the land was marked as "reserved in wood for rains."

Pierre Poivre, an early French governor of Mauritius, an island in the Indian Ocean, was appalled at the environmental and social devastation caused by destruction of wildlife (such as the flightless dodo) and the felling of ebony forests on the island by early European settlers. In 1769, Poivre ordered that one-quarter of the island was to be preserved in forests, particularly on steep mountain slopes and along waterways. Mauritius remains a model for balancing nature and human needs. Its forest reserves shelter a larger percentage of its original flora and fauna than most other human-occupied islands.

Pragmatic Resource Conservation

Many historians consider the publication of *Man and Nature* in 1864 by geographer George Perkins Marsh as the wellspring of environmental protection in North America. Marsh, who also was a lawyer, politician, and diplomat, traveled widely around the Mediterranean as part of his diplomatic duties in Turkey and Italy. He read widely in the classics (including Plato) and personally observed the damage caused by excessive grazing by goats and sheep and by the deforestation of steep hillsides. Alarmed by the wanton destruction and profligate waste of resources still occurring on the American frontier in his lifetime, he warned of its ecological consequences. Largely because of his book, national forest reserves were established in the United States in 1873 to protect dwindling timber supplies and endangered watersheds.

Among those influenced by Marsh's warnings were U.S. President Theodore Roosevelt and his chief conservation advisor, Gifford Pinchot. In 1905, Roosevelt, who was the leader of the populist, progressive movement, moved forest management out of the corruption-filled Interior Department into the Department of Agriculture. Pinchot, who was the first American-born professional forester, became the first chief of the new Forest Service (fig. 1.11). He put resource management on an honest, rational, and scientific basis for the first time in American history. Together with naturalists and activists such as John Muir, William Brewster, and George Bird Grinnell, Roosevelt and Pinchot established the framework of the national forest, park, and wildlife refuge system. They passed game protection laws and tried to stop some of the most flagrant abuses of the public domain. In 1908, Pinchot organized and chaired the White House Conference on Natural Resources, perhaps the most prestigious and influential environmental meeting ever held in the United States. Pinchot also was governor of Pennsylvania and founding head of the Tennessee Valley Authority, which provided inexpensive power to the Southeastern United States.

The basis of Roosevelt's and Pinchot's policies was pragmatic **utilitarian conservation.** They argued that the forests should be saved "not because they are beautiful or because they

FIGURE 1.11   *Gifford Pinchot, first chief of the U.S. Forest Service and founder of the utilitarian conservation movement.*

shelter wild creatures of the wilderness, but only to provide homes and jobs for people." Resources should be used "for the greatest good, for the greatest number, for the longest time." "There has been a fundamental misconception," Pinchot wrote, "that conservation means nothing but husbanding of resources for future generations. Nothing could be further from the truth. The first principle of conservation is development and use of the natural resources now existing on this continent for the benefit of the people who live here now. There may be just as much waste in neglecting the development and use of certain natural resources as there is in their destruction." This pragmatic approach still can be seen in the multiple use policies of the U.S. Forest Service.

Moral and Aesthetic Nature Preservation

John Muir (fig. 1.12), amateur geologist, popular author, and first president of the Sierra Club, strenuously opposed Pinchot's utilitarian policies. Muir argued that nature deserves to exist for its own sake, regardless of its usefulness to us. Aesthetic and spiritual values formed the core of his philosophy of nature protection. This outlook has been called **biocentric preservation** because it emphasizes the fundamental right of other organisms to exist and to pursue their own interests. Muir wrote: "The world, we are told, was made for man. A presumption that is totally unsupported by the facts. . . . Nature's object in making animals and plans might possibly be first of all the happiness of each one of them. . . . Why ought man to value himself as more than an infinitely small unit of the one great unit of creation?"

FIGURE 1.12   *John Muir and Teddy Roosevelt.*

FIGURE 1.13   *Rachel Carson's book* Silent Spring *was a landmark in modern environmental history. She alerted readers to the dangers of indiscriminate pesticide use.*

Muir, who was an early explorer and interpreter of California's Sierra Nevada range, fought long and hard for establishment of Yosemite and Kings Canyon National Parks. The National Park Service, established in 1916, was first headed by Muir's disciple, Stephen Mather, and has always been oriented toward preservation of nature in its purest state. It has often been at odds with Pinchot's utilitarian Forest Service. One of Muir and Pinchot's biggest battles was over the damming of Hetch Hetchy Valley in Yosemite. Muir regarded flooding the valley a sacrilege against nature. Pinchot, who championed publicly owned utilities, viewed the dam as a way to free San Francisco residents from the clutches of greedy water and power monopolies.

Modern Environmentalism

The undesirable effects of pollution probably have been recognized as long as people have been building smoky fires. In 1273, King Edward I of England threatened to hang anyone burning coal in London because of the acrid smoke it produced. In 1661, the English diarist John Evelyn complained about the noxious air pollution caused by coal fires and factories, and suggested that sweet-smelling trees be planted to purify city air. Increasingly dangerous smog attacks in Britain led, in 1880, to formation of a national Fog and Smoke committee to combat this problem.

The tremendous expansion of chemical industries during and after the World War II added a new set of concerns to the environmental agenda. *Silent Spring,* written by Rachel Carson (fig. 1.13) and published in 1962, awakened the public to the threats of pollution and toxic chemicals to humans as well as other species. The movement she engendered might be called **modern environmentalism** because its concerns extended to include both natural resources and environmental pollution. Among some other pioneers of this movement were activist David Brower and scientist Barry Commoner. Brower, as executive director of the Sierra Club, Friends of the Earth, and Earth Island Institute, introduced many of the techniques of environmental lobbying and activism, including litigation, intervention in regulatory hearings, book and calendar publishing, and use of mass media for publicity campaigns. Commoner, who was trained as a molecular biologist, has been a leader in analyzing the links between science, technology, and society. Both activism and research remain hallmarks of the modern environmental movement.

Under the leadership of a number of other brilliant and dedicated activists and scientists, the environmental agenda was expanded in the 1960s and 1970s to most of the issues addressed in this textbook, such as human population growth, atomic

Principles of Environmental Science

weapons testing and atomic power, fossil fuel extraction and use, recycling, air and water pollution, wilderness protection. Environmentalism has become well established in the public agenda since the first national Earth Day in 1970. A majority of Americans now consider themselves environmentalists, although there is considerable variation in what that term means.

Global Concerns

Increased opportunities to travel and expanded international communications now enable us to know about daily events in places unknown to our parents or grandparents. We have become, as Marshal McLuhan famously announced in the 1960s, a global village. As in a village, we are all interconnected in various ways. Events that occur on the other side of the globe have profound and immediate effects on our lives.

Photographs of the earth from space (see fig. 1.2) provide a powerful icon for the fourth wave of ecological concern that might be called **global environmentalism.** These photos remind us how small, fragile, beautiful, and rare our home planet is. We all share a common environment at this global scale. As the ambassador Adlai Stevenson noted, in his 1965 farewell address to the United Nations (quoted at the beginning of this chapter), we now need to worry about the life-support systems of the planet as a whole. He went on to say in this speech, "We cannot maintain it half fortunate, half miserable, half confident, half despairing, half slave to the ancient enemies of mankind and half free in a liberation of resources undreamed of until this day. No craft, no crew, can travel with such vast contradictions. On their resolution depends the security of us all."

Among the leaders of the worldwide movement combining environmental protection with social justice have been British economist Barbara Ward, French/American scientist René Dubois, Norwegian Prime Minister Gro Harlem Brundtland, and Canadian diplomat Maurice Strong. All have been central in major international environmental conventions, such as the 1972 UN Conference on the Human Environment in Stockholm or the 1992 UN Earth Summit in Rio de Janeiro. Once again, new issues have become part of the agenda as our field of vision widens. We have begun to appreciate the links between poverty, injustice, oppression, and exploitation of humans and our environment. A concern about the human dimensions of environmental science is an important focus of this textbook.

CURRENT ENVIRONMENTAL CONDITIONS

While there are many things to appreciate and celebrate about the world in which we live, many pressing environmental problems cry out for our attention. In a study published by the World Resources Institute, the United Nations, and the World Bank in September 2000, more than 175 scientists assessed the health of global ecosystems. They discovered a widespread decline in the ability of ecosystems to produce the goods and services on which we currently depend. They found, for instance, that half the world's wetlands were lost in the last 100 years, while logging and land conversion have shrunk the world's

APPLICATION: Making Sense of Data

You will see many numbers in this book. Instead of simply glossing over them, think about what they mean relative to some benchmark numbers.

1. For example, you just read that we are adding 85 million people to the world's population each year. How does that compare to the number of people in the United States? in Canada? in Mexico? (If you don't know, where can you find those numbers?)

2. What proportion of the current global population of about 6 billion is the 85 million people added each year?

Answers: 1. The U.S. has about 270 million people; Canada has about 25 million, and Mexico has about 100 million. 2. About 1.4 percent.

forests by as much as half. Nearly three-quarters of the world's major marine fish stocks are overfished or are being harvested beyond a sustainable rate, and soil degradation has affected two-thirds of the world's agricultural lands in the last 50 years.

Causes of Environmental Degradation

Human populations grew at alarming rates in the last century. More than 6 billion people now occupy the earth, and we are adding about 85 million more each year. In the next decade, our numbers will increase by almost as many people as are now alive in India. Most of that growth will be in the poorer countries where present populations already strain resources and services.

In most places, however, population growth rates slowed dramatically in the last decades of the twentieth century. Some demographers predict that our population will stabilize in about 50 years, and may eventually drop back below its present size. Others warn, however, that we have already surpassed the carrying capacity of our environment, and that unless dramatic steps are taken, we may double our present population by 2050.

Food shortages and famines are already too familiar in many places and may increase in frequency and severity if population growth, soil erosion, and nutrient depletion continue at the same rate as they have in the past (fig. 1.14). We are coming to realize, however, that food security often has more to do with poverty, democracy, and equitable distribution than it does with the amount of food available. Water deficits and water contamination threaten to be critical environmental issues in the future for agricultural production, as well as for domestic and industrial uses. Many countries already have serious water shortages, and more than one billion people lack access to clean water or adequate sanitation. Violent conflicts over control of natural resources may flare up in many places if we don't learn to live within nature's budget.

How we obtain and use energy is likely to play a crucial role in our environmental future. Fossil fuels (oil, coal, and natural gas)

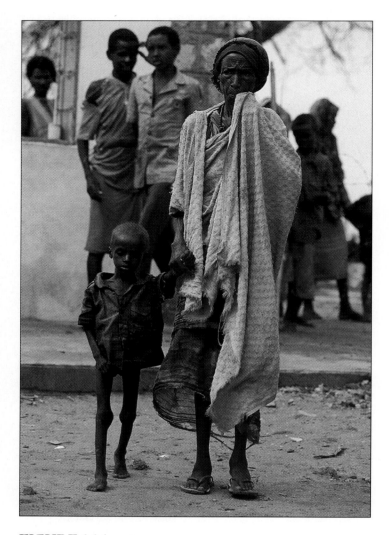

FIGURE 1.14   *Three-quarters of the world's poorest nations are in Africa. Millions of people lack adequate food, housing, medical care, clean water, and safety. The human suffering engendered by this poverty is tragic.*

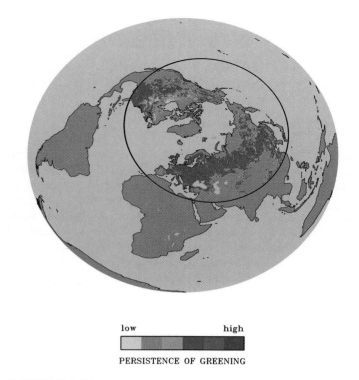

low   high

PERSISTENCE OF GREENING

FIGURE 1.15   *Satellite images and surface temperature data show that polar regions, especially in Eurasia, are becoming green earlier and staying green longer than ever in recorded history. This appears to be evidence of a changing global climate.*
Source: NASA, 2002.

presently supply about 80 percent of the energy that industrialized countries use. Supplies of these fuels are finite, and problems associated with their acquisition and use—air and water pollution, mining damage, shipping accidents, and political insecurity—may limit where and how we use remaining reserves. Cleaner, renewable energy resources (solar power, wind, and biomass), together with conservation, may replace environmentally destructive energy sources if we invest in appropriate technology in the next few years.

As we burn fossil fuels, we release carbon dioxide and other heat-absorbing gases that cause global warming and may bring about sea-level rises and catastrophic climate changes. Some of the most evident climate changes have occurred in the Arctic, where growing seasons are at least three weeks longer than normal and average annual temperatures have risen as much as 7.5°C (13.5°F) over the past 30 years (fig. 1.15). Burning of fossil fuels also releases toxic compounds. At least 100 million Americans breathe air pollutants that cause unacceptably high cancer risks.

Chlorinated compounds, such as the chlorofluorocarbons used in refrigeration and air conditioning, also contribute to global warming, as well as damaging the stratospheric ozone that protects us from cancer-causing ultraviolet radiation in sunlight.

Destruction of tropical forests, coral reefs, wetlands, and other biologically rich landscapes is causing an alarming loss of species and a reduction in biological variety and abundance that could severely limit our future options. Many rare and endangered species are threatened directly or indirectly by human activities (fig. 1.16). In addition to practical values, aesthetic and ethical considerations suggest that we should protect these species and the habitat necessary for their survival. On the other hand, cultural or economic interests sometimes compete with biodiversity conservation.

Toxic air and water pollutants, along with mountains of solid and hazardous wastes, are becoming overwhelming problems in industrialized countries. We produce hundreds of millions of tons of these dangerous materials annually, and methods of disposal are often dangerous and irresponsible. No one wants this noxious stuff dumped in their own backyard, and too often, the solution is to export it to someone else's backyard. We may reach a political impasse where our failure to decide where to put wastes or how to dispose of them safely will close down industries and result in wastes being spread everywhere. The health effects of pollution, toxic wastes, stress, and the other environmental ills of modern society have become a greater threat than infectious diseases for many of us in industrialized countries.

These and other similarly serious problems illustrate the importance of environmental science and environmental education for everyone. What we are doing to our world and what that may mean for our future and that of our children is of paramount concern as we begin the twenty-first century. We hope that's why you're reading this book.

Signs of Hope

The dismal litany of environmental problems we have just reviewed seems pretty overwhelming, doesn't it? But is there hope that we may find solutions to these problems? As you will see in subsequent chapters, progress has been made in many areas in controlling air and water pollution and reducing wasteful resource uses. Many cities in North America and Europe are cleaner and less polluted than they were a generation or so ago. Population has stabilized in most industrialized countries and even in some very poor countries where social security and democracy have been established. Over the last twenty years, the average number of children born per woman worldwide has decreased from 6.1 to 3.4. This is still above the zero population growth rate of 2.1 children per couple, but it is an encouraging improvement. If this rate of progress continues in the next twenty years as it has in the past twenty, the world population could stabilize early in the twenty-first century.

The incidence of life-threatening infectious diseases fell sharply in most countries during the last century, while the average life expectancy nearly doubled. We have discovered many new resources and invented more efficient ways of using existing supplies, which allows us to enjoy luxuries and conveniences that would have seemed miraculous only a few generations ago. Although modern life has many stresses and strains, few people would willingly return to conditions that existed 10,000 or even 100 years ago. Would you?

Still, we can do much more, both individually and collectively, to protect and restore our environment. Being aware of the problems we face is the first step toward finding their solutions. Increased media coverage has brought environmental issues to public attention. More than 80 percent of the Americans polled in public opinion surveys agree that "protecting the environment is so important that requirements and standards cannot be too high and continuing environmental improvements must be made regardless of cost." This growing understanding and concern is itself a hopeful sign. Young people today may be in a unique position to address these issues because, for the first time in history, we now have the resources, motivation, and knowledge to do something about our environmental problems. Unfortunately, if we don't act now, we may not have another chance to do so.

HUMAN DIMENSIONS OF ENVIRONMENTAL SCIENCE

Because we live in both the natural and social world, and because we and our technology have become such dominant forces on the planet, environmental science must take human institutions and the human condition into account. We live in a world of haves and

FIGURE 1.16   *This emperor tamarin is a marmoset from South America. More than 99 percent of the once-extensive dry tropical forests of Latin America, where these animals live, have been destroyed, threatening a host of interesting and important species with extinction.*

have-nots; a few of us live in increasing luxury, while many others lack the basic necessities for a decent, healthy, productive life. The World Bank estimates that more than 1.3 billion people—about one-fifth of the world's population—live in acute poverty with an income of less than $1 (U.S.) per day. These poorest of the poor generally lack access to an adequate diet, decent housing, basic sanitation, clean water, education, medical care, and other essentials for a humane existence. Seventy percent of those people are women and children. In fact, four out of five people in the world live in what would be considered poverty in the United States or Canada.

Policymakers are becoming aware that eliminating poverty and protecting our common environment are inextricably interlinked because the world's poorest people are both the victims and the agents of environmental degradation. The poorest people are often forced to meet short-term survival needs at the cost of long-term sustainability. Desperate for croplands to feed themselves and their families, many move into virgin forests or cultivate steep, erosion-prone hillsides, where soil nutrients are exhausted after only a few years. Others migrate to the grimy, crowded slums and ramshackle shanty-towns that now surround most major cities in the developing world. With no way to dispose of wastes, the residents often foul their

FIGURE 1.17   *While many of us live in luxury, more than a billion people lack access to resources for a healthy, productive life. Often the poorest people are both the victims and agents of environmental degradation as they struggle to survive.*

FIGURE 1.18   *About four-fifths of the world's population lives in countries where nearly everyone is poor by American standards.*

environment further and contaminate the air they breathe and the water on which they depend for washing and drinking (fig. 1.17).

The cycle of poverty, illness, and limited opportunities can become a self-sustaining process that passes from one generation to another. People who are malnourished and ill can't work productively to obtain food, shelter, or medicine for themselves or their children, who also are malnourished and ill. About 250 million children—mostly in Asia and Africa and some as young as 4 years old—are forced to work under appalling conditions weaving carpets, making ceramics and jewelry, or in the sex trade. Growing up in these conditions leads to educational, psychological, and developmental deficits that condemn these children to perpetuate this cycle.

Faced with immediate survival needs and few options, these unfortunate people often have no choice but to overharvest resources; in doing so, however, they diminish not only their own options, but also those of future generations. And in an increasingly interconnected world, the environments and resource bases damaged by poverty and ignorance are directly linked to those on which we depend.

Rich and Poor Countries

Where do the rich and poor live? About one-fifth of the world's population lives in the twenty richest countries, where the average per capita income is above $25,000 (U.S.) per year. Most of these countries are in North America or Western Europe, but Japan, Singapore, Australia, New Zealand, the United Arab Emirates, and Israel also fall into this group. Almost every country, however, even the richest, such as the United States and Canada, has poor people. No doubt everyone reading this book knows about homeless people or other individuals who lack resources for a safe, productive life. The *Journal of Public Health* reported in 1999 that 10 million Americans—including 4 million children—do not have enough to eat.

The other four-fifths of the world's population lives in middle- or low-income countries, where nearly everyone is poor

by North American standards (fig. 1.18). More than 3 billion people live in the poorest nations, where the average per capita income is below $620 (U.S.) per year. China and India are the largest of these countries, with a combined population of about 2.3 billion people. Among the 41 other nations in this category, 33 are in sub-Saharan Africa. All the other lowest-income nations, except Haiti, are in Asia. Although poverty levels in countries such as China and Indonesia have fallen in recent years, most countries in sub-Saharan Africa and much of Latin America have made little progress. The destabilizing and impoverishing effects of earlier colonialism continue to play important roles in the ongoing problems of these unfortunate countries. Meanwhile, the relative gap between rich and poor has increased dramatically (fig. 1.19).

The ten poorest countries in the world in 1999 were (in ascending order from very poorest): Mozambique, Ethiopia, Tanzania,

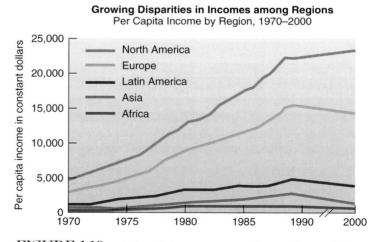

FIGURE 1.19   *Although the percentage of the world's population living in poverty has decreased slightly over the past 30 years, the relative gap between the rich and poor nations has increased sharply.*
Source: World Resources Institute, 2000–2001.

TABLE 1.4 Average Indicators of Quality of Life for the Ten Richest and Poorest Countries[1]

INDICATOR	POOR COUNTRIES	RICH COUNTRIES
Per capita GNP	$175	$35,000
Life expectancy	46 years	77 years
Infant mortality[2]	102	6
Child deaths[3]	194	7.5
Safe drinking water	42%	NA[4]
Female literacy	38%	97%
Birthrate[5]	42	12

[1]Averaged as a group
[2]per 1,000 live births
[3]per 1,000 children before age 5
[4]not available, but close to 100 percent
[5]per 1,000 people
Source: Population Reference Bureau, 2000–2002 Data Sheet.

TABLE 1.5 The United States, with 4.5 Percent of the World's Population . . .

CONSUMES	PRODUCES
26 percent of all oil	50 percent of all toxic wastes
24 percent of aluminum	26 percent of nitrogen oxides
20 percent of copper	25 percent of sulfur oxides
19 percent of nickel	22 percent of chlorofluorocarbons
13 percent of steel	26 percent of carbon dioxide

Source: Data from World Resources Institute, 1998–99.

Congo, Malawi, Rwanda, Chad, Sierra Leone, Nepal, and Niger. Each of these countries has annual per capita gross national product (GNP) of less than $200 (U.S.) per year. They also have low levels of food security, social welfare, and quality of life, as table 1.4 indicates.

By contrast, each of the ten richest countries in the world—Switzerland, Japan, Norway, Denmark, Germany, the United States, Austria, Singapore, France, and Iceland (in descending order from richest)—has an annual per capita GNP more than a hundred times that of the poorest countries. As you can see in table 1.4, other conditions in the rich countries reflect this wide disparity in wealth.

The gulf between rich and poor is even greater at the individual level. The richest 200 people in the world have a combined wealth of $1 trillion. This is more than the total owned by the 3 billion people who make up the poorest half of the world's population.

A Fair Share of Resources?

The affluent lifestyle that many of us in the richer countries enjoy consumes an inordinate share of the world's natural resources and produces a shockingly high proportion of pollutants and wastes. The United States, for instance, with less than 5 percent of the total population, consumes about one-quarter of most commercially traded commodities and produces a quarter to half of most industrial wastes (table 1.5).

To get an average American through the day takes about 450 kg (nearly 1,000 lbs) of raw materials, including 18 kg (40 lbs) of fossil fuels, 13 kg (29 lbs) of other minerals, 12 kg (26 lbs) of farm products, 10 kg (22 lbs) of wood and paper, and 450 liters (119 gal) of water. Every year, Americans throw away some 160 million tons of garbage, including 50 million tons of paper, 67 billion cans and bottles, 25 billion styrofoam cups, 18 billion disposable diapers, and two billion disposable razors (fig. 1.20).

This profligate resource consumption and waste disposal strains the planet's life-support systems. If everyone in the world tried to live at consumption levels approaching ours, the results would be disastrous. Unless we find ways to curb our desires and produce the things we truly need in less destructive ways, the sustainability of human life on our planet is questionable.

Sustainability

An overarching theme of this book is **sustainability:** a search for ecological stability and human progress that can last over the long term. Of course, neither ecological systems nor human institutions can continue forever. We can work, however, to protect the best aspects of both realms, and to encourage resiliency and adaptability in both of them. World Health Organization Director Gro Harlem Brundtland has defined **sustainable development** as "meeting the needs of the present without compromising the ability of future generations to meet their own needs." In these terms, development means bettering people's lives. Sustainable development, then,

FIGURE 1.20 *"And may we continue to be worthy of consuming a disproportionate share of this planet's resources."*
© The New Yorker Collection 1992 Lee Lorenz from cartoonbank.com. All Rights Reserved.

means progress in human well-being that we can extend or prolong over many generations, rather than just a few years. To be truly enduring, the benefits of sustainable development must be available to all humans and not just to the members of a privileged group. We will discuss this topic further in chapter 14.

Indigenous Peoples

In both rich and poor countries, indigenous or native peoples are generally the least powerful, most neglected groups in the world. Typically descendants of the original inhabitants of an area taken over by more powerful outsiders, they are distinct from their country's dominant language, culture, religion, and racial communities. Of the world's nearly 6,000 recognized cultures, 5,000 are indigenous ones that account for only about 10 percent of the total world population. In many countries, traditional caste systems, discriminatory laws, economics, or prejudice repress indigenous peoples. Their unique cultures are disappearing, along with biological diversity as natural habitats are destroyed to satisfy industrialized world appetites for resources. Traditional ways of life are disrupted further by dominant Western culture sweeping around the globe.

At least half of the world's 6,000 distinct languages are dying because they are no longer taught to children. When the last few elders who still speak the language die, so will the culture that was its origin. Lost with those cultures will be a rich repertoire of knowledge about nature and a keen understanding of a particular environment and way of life (fig. 1.21).

Nonetheless, in many places, the 500 million indigenous people who remain in traditional homelands still possess valuable ecological wisdom and remain the guardians of little-disturbed habitats that are refuges for rare and endangered species and undamaged ecosystems. In his recent book, *The Future of Life,* the eminent ecologist E. O. Wilson argues that the cheapest and most effective way to preserve species is to protect the natural ecosystems in which they now live. Interestingly, just twelve countries account for 60 percent of all human languages (fig. 1.22). Seven of these are also among the "megadiversity" countries that contain more than half of all unique plant and animal species. Conditions that support evolution of many unique species seem to favor development of equally diverse human cultures as well.

Recognizing native land rights and promoting political pluralism can be among the best ways to safeguard ecological processes and endangered species. As the Kuna Indians of Panama say, "Where there are forests, there are native people, and where there are native people, there are forests." A few countries, such as Papua New Guinea, Fiji, Ecuador, Canada, and Australia, acknowledge indigenous title to extensive land areas.

Other countries, unfortunately, ignore the rights of native people. Indonesia, for instance, claims ownership of nearly three-quarters of its forest lands and all waters and offshore fishing rights, ignoring the interests of indigenous inhabitants. Similarly, the Philippine government claims possession of all uncultivated land in its territory, while Cameroon and Tanzania recognize no rights at all for forest-dwelling pygmies who represent one of the world's oldest cultures.

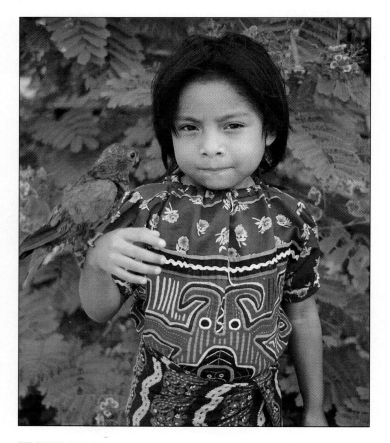

FIGURE 1.21   *Indigenous cultures often represent a valuable source of knowledge about nature. At least half of the world's 6,000 distinct languages are dying because they no longer are taught to children.*

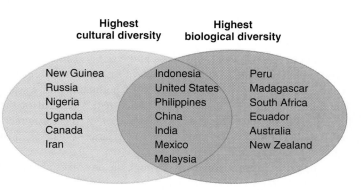

FIGURE 1.22   *Cultural diversity and biological diversity often go hand in hand. Seven of the countries with the highest cultural diversity in the world are also on the list of "megadiversity" countries with the highest number of unique biological organisms. (Listed in decreasing order of importance.)*

Source: Norman Myers, Conservation International and Cultural Survival Inc., 1999.

SUMMARY

Humans inhabit two worlds: one of nature and another of human society and technology. Environmental science is the systematic study of the intersection of these worlds. An interdisciplinary field, environmental science draws from many areas of inquiry to help us understand the worlds in which we live and how we might improve both of them.

The most amazing features of our planet may be the self-sustaining ecological systems that make life possible and the rich diversity of life that is part of, and dependent upon, those ecological processes. In spite of the many problems that beset us, the earth is wonderfully bountiful and beautiful.

Science is a way of exploring and understanding the world around us. It depends on making careful observations of natural phenomena and on formulating reasonable theories to explain those observations. Science rests on certain assumptions such as that the world is understandable and that we can learn about how things work through careful empirical study and rational analysis. Scientists strive to be methodical, rigorous, and unbiased. They also are skeptical, reserve judgment, and generally consider conclusions and interpretations to be conditional because new information can always overturn what we think to be true. Scientists can use either deductive reasoning (arguing from laws or general theories to specific detail) or inductive reasoning (using individual observations to derive general rules). In either case, scientists formulate testable hypotheses (provisional explanations) based on their inferences, and attempt to test them through experimentation or looking for evidence in nature. Once a substantial body of evidence has been accumulated and a consensus emerges within the scientific community, an explanation is considered a theory.

Statistics and probability are important concepts in environmental science. Descriptive statistics help us assess the general state of a group. Using a large, random sample helps prevent errors that arise from variations within the study group. Confidence levels represent the likelihood that your results are in error. Scientists usually compare values to a 95 percent confidence level, or a probability of 5 percent or less that the conclusions are misleading.

Critical thinking shares many of the aspects of logical, analytical thinking but also includes empathy, context, and justice in its search for truth and meaning.

Unprecedented population growth, food shortages, scarce energy supplies, air and water pollution, and destruction of habitats and biological resources are all serious threats to our environment and our way of life. As international travel and communication become easier, we realize that these problems encompass our whole planet and require global cooperation to find solutions. Still, there is good news. Pollution has been reduced and population growth has slowed in many places. Perhaps we can extend these advances to other areas as well.

The 20 percent of us in the world's richest countries consume an inordinate amount of resources and produce a shocking amount of waste and pollution. Meanwhile, at least 1.3 billion people live in acute poverty and lack access to an adequate diet, decent housing, basic sanitation, clean water, education, medical care, and other essentials for a humane existence. Faced with immediate survival needs, the poorest poor often have little choice but to overharvest resources and reduce long-term sustainability for themselves and their children. Development means a real increase in standard of living for the average person. Sustainable development attempts to meet the needs of present generations without reducing the ability of future generations to meet their own needs.

Indigenous or native peoples are generally among the poorest and most oppressed of any group. Nevertheless, they possess valuable ecological knowledge and remain the guardians of nature in many places. Recognizing the rights of indigenous people and minority communities is an important way to protect natural resources and environmental quality.

QUESTIONS FOR REVIEW

1. Define the terms *environment* and *environmental science.*

2. What is parsimony and why is it important to scientists?

3. Describe how deductive and inductive reasoning differ.

4. Diagram the scientific method, and give one example of how you might apply it.

5. What are statistics and probability? How are they useful in science?

6. How do analytical, creative, critical, logical, and reflective thinking differ?

7. List six environmental dilemmas that we now face, and describe how each concerns us.

8. Compare some indicators of quality of life between the richest and poorest nations.

9. Define the terms *sustainability* and *sustainable development.*

10. What role might indigenous or tribal land rights have in protecting biodiversity?

THINKING SCIENTIFICALLY

1. If you were voting on whether to dredge PCBs out of the Hudson River, what information would you need and what sources would you trust to supply it?

2. Do you believe the world is ultimately knowable? Why or why not? How does this affect the way you live?

3. Why can we falsify a hypothesis but never prove it to be true?

4. Suppose you wanted to investigate the prevalence of some trait among your classmates (say, the percentage who are left-handed). How many out of a class of 100 would you need to question to get a sufficiently accurate result? How would you define "sufficiently accurate"?

5. Some field biologists claim that mathematical models and laboratory simulations can never capture the complexity and variability of the real world. Experimental scientists retort that complex systems have too many simultaneous variables to allow meaningful analysis. Where would you stand in this debate?

6. Identify and explain a paradigm that you and your friends use to help explain how things work.

7. Many social theorists argue that there is no such thing as objective truth or an impartial observer. If you were researching some controversial topic—perhaps whether grizzly bears can be safely restored to the Northern Rockies—what steps would you take to try to maintain objectivity and impartiality?

8. Does the world have enough resources for 8 or 10 billion people to live decent, secure, happy, fulfilling lives? What do those terms mean to you? Try to imagine what they mean to others in our global village.

9. Suppose you wanted to study the environmental impacts of a rich versus poor country. What factors would you examine, and how would you compare them?

KEY TERMS

analytical thinking 11
biocentric preservation 15
blind experiments 6
controlled studies 6
creative thinking 11
critical thinking 11
deductive reasoning 7
double-blind design 6
environment 3
environmental science 3
global environmentalism 17
hypothesis 7
inductive reasoning 7
logical thinking 11

mean 8
modern environmentalism 16
paradigms 10
parsimony 5
probability 10
reflective thinking 11
reproducibility 6
sample 8
scientific theory 8
significant numbers 6
statistics 8
sustainability 21
sustainable development 21
utilitarian conservation 15

Welcome to McGraw-Hill's Online Learning Center

Location: http://www.mhhe.com/environmentalscience

WEB EXERCISE

Satellites in Environmental Science

Satellite data are an important source of environmental information and monitoring. Satellite imagery is available from the National Atmospheric and Space Agency (NASA) at http://seawifs.gsfc.nasa.gov/SEAWIFS.html

What are some of the current topics shown in images on this page? Have you heard or read about these topics in the news?

Go to the big maps if you have a fast Internet connection: http://seawifs.gsfc.nasa.gov/cgi/level3.pl

Click on the NDVI (normalized difference vegetation index) option in the table of themes. NDVI is a measure of how green the vegetation is.

Then click on the current month in the table of months and years. (Click on a map to see a larger version.) Compare the view of the current month with the views at several different times of year. Find your location on the map. Can you identify local features that should (or should not) be green in the various months? How does the amount of vegetation in your area affect local environmental conditions? If satellites had existed 100 years ago, how might the view they would have shown differed from what we see today?

> To see things in the seed, that is the genius.
> –*Lao-tzu*

Sunlight, water, and living things are all part of the energy flow and material cycles that make up our environment.

OBJECTIVES

After studying this chapter, you should be able to

- describe matter, atoms, and molecules, and give simple examples of the roles of four major kinds of organic compounds in living cells.
- define energy, and explain the difference between kinetic and potential energy.
- understand the principles of conservation of matter and energy, and appreciate how the laws of thermodynamics affect living systems.
- explain how photosynthesis captures energy for life and how cellular respiration releases that energy to do useful work.
- define species, populations, biological communities, and ecosystems, and understand the ecological significance of these levels of organization.
- discuss food chains, food webs, and trophic levels in biological communities, and explain why there are pyramids of energy, biomass, and numbers of individuals in the trophic levels of an ecosystem.
- explain the importance of material cycles, such as carbon and nitrogen cycles, in ecosystems.

Welcome to McGraw-Hill's Online Learning Center

Location: http://www.mhhe.com/environmentalscience

LEARNING ONLINE

Visit our webpage at www.mhhe.com/environmentalscience for data sources, further readings, additional case studies, current environmental news, and regional examples within the Online Learning Center to help you understand the material in this chapter. You'll also find active links to information pertaining to this chapter including:

The Biology Place
Bioweb: Resources for Cell Biologists
What is Photosynthesis?
Photosynthesis Center

Measuring Energy Flows in Cedar Bog Lake

In 1936, a young graduate student named Ray Lindeman began his Ph.D. research on a small, marshy pond in Minnesota called Cedar Bog Lake. His pioneering work helped reshape the way ecologists think about the systems they study. At the time, most ecologists were concerned primarily with descriptive histories and classifications of biological communities. A typical lake study might classify the taxonomy and life histories of resident species and describe the lake's stage in development from open water to marsh and then to forest. Blind in one eye, Lindeman couldn't do the microscopy necessary to identify the many species of algae, protozoans, and other aquatic organisms in the lake. Instead, following the ideas of two contemporary English ecologists, Charles Elton and A. G. Tansley, he concentrated on biological communities as systems and looked at broad categories of feeding relationships, for which he coined the term *trophic levels* (from the Greek word for eating). Aided by his wife, Eleanor, Lindeman spent many hours collecting samples of aquatic plants and algae, grazing and predatory zooplankton and fish, and the benthic (bottom-dwelling) worms, insect larvae, crustaceans, and sediment. Back in the laboratory, he measured the plants' photosynthetic rates, the animals' respiration rates, and the total energy content of organic compounds in each of the different trophic levels.

Describing the system in terms of energy flows was a radical departure from ecological methods at the time. Lindeman made a careful balance sheet of the total energy content in the biomass at each trophic level, the energy used in respiration, and the energy content of organic matter deposited in the sediment. To his surprise, he found that each successive feeding level contained only about 10 percent of the energy captured by the level below it. The remainder is lost as heat or deposited in sediments, he argued, because of the work that organisms perform and the inefficiency of biological energy transformations. In his dissertation, Lindeman showed that energy represents a common denominator that allows us to sum up all the processes of production and consumption by the myriad organisms in a biological community.

Lindeman also broke from standard procedure by representing the relationships in his study lake as a mathematical model: He used a series of equations to describe thermodynamic relationships and the efficiency of energy capture and transfer. Ironically, Lindeman's most important paper was rejected by the journal *Ecology* as being too theoretical and too quantitative. It was only after the intercession of G. Evelyn Hutchinson from Yale, with whom Lindeman had a postdoctoral fellowship after finishing his studies at Minnesota, that his mathematical model and energy analysis of Cedar Bog Lake was finally published. Unfortunately, Ray Lindeman died of liver failure before his article appeared. It has since become a landmark in ecological history.

In the years since Lindeman's work, the idea of taking a systemic view of a biological community together with its physical and inorganic environment has become standard in ecology. Energy flows and nutrient cycles are central to the way we understand the workings of ecological systems. Constructing quantitative models to describe, explain, and explore ecological processes has become routine. In this chapter, we will investigate the ways living things use energy and matter and the ways these flows create relationships in ecosystems. (See Lindeman, R. L. 1942. The trophic-dynamic aspect of ecology. *Ecology* 23:399–418.)

PRINCIPLES OF MATTER AND ENERGY

How and why materials are cycled between the living and nonliving parts of our environment are the domain of **ecology,** the scientific study of relationships between organisms and their environment. Modern biology covers a wide range of scales and themes, from molecules to ecosystems to global systems. Ecology examines the life histories, distribution, and behavior of individual species, as well as the structure and function of natural systems at the level of populations, communities, ecosystems, and landscapes. The systems approach of ecology encourages us to think holistically about interconnections that make whole systems more than just the sum of their individual parts.

In a sense, every organism is a chemical factory that captures matter and energy from its environment and transforms them into structures and processes that make life possible. Therefore, to understand how ecosystems function it is important first to know something of how energy and matter behave—both in the universe and in living things. In this chapter, we will survey some fundamental aspects of energy flow and material recycling within ecosystems.

What Is Matter?

Everything that takes up space and has mass is matter. All matter has three interchangeable physical forms, or phases: gas, liquid, and solid. Water, for example, can exist as a gas (water vapor), a liquid (water), or a solid (ice). Under ordinary circumstances, matter is neither created nor destroyed but is recycled over and over again. The elements in your body have been recycled through many other organisms, over millions of years. Matter is transformed and combined in different ways, but it doesn't disappear; everything goes somewhere. These statements paraphrase the physical principle of **conservation of matter.**

How does this principle apply to human relationships with the biosphere? It implies that, as we use more resources to produce more "disposable" goods, we should pay attention to where our waste products go. What happens to the garbage the truck hauls away? Where does the exhaust from your car go? Ultimately, we need to answer these questions because there is no "away" where we can throw things we don't want any more.

What Is Energy?

Energy and matter are essential constituents of both the universe and living organisms. If **matter** is the material of which things are made, **energy** is the capacity to do work such as moving matter over a distance. Energy can take many different forms. Heat, light, electricity, and chemical energy are common forms. The energy contained in moving objects is called **kinetic energy.** A rock rolling down a hill, the wind blowing through the trees, water flowing over a dam (fig. 2.1), or electrons speeding around the nucleus of an atom are all examples of kinetic energy. **Potential energy** is stored energy that is latent but available for use. A rock poised at the top of a hill contains potential energy, which is converted to kinetic energy when the rock starts rolling down the hill. Chemical energy, stored in the food that you eat and the gasoline that you put into your car, is also potential energy that can be released to do useful work. Energy is often measured in units of heat (calories) or work (joules). One joule (J) is the work done when one kg is accelerated 1 m per second per second ($1 \text{ J} = 1 \text{ kg} \cdot \text{m}^2/\text{s}^2$). One calorie is the amount of energy needed to heat one gram of pure water one degree Celsius. A calorie can also be measured as 4.184 J.

Heat describes the energy that can be transferred between objects of different temperature. When two objects of different temperature are placed in contact, heat transfers to the cooler one until the two reach the same temperature. When a substance absorbs heat, its internal energy increases; the kinetic energy (motion) of its molecules increases, or it may change state: a solid may become liquid, or a liquid become a gas. When you heat a tea kettle on a stove, for example, energy is transferred to the water. We sense the change in heat content as a change in temperature. Some of the water may change state from liquid to vapor. After the

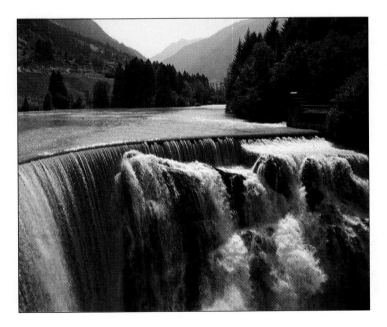

FIGURE 2.1 *Water stored behind this dam represents potential energy. Water flowing over the dam has kinetic energy, some of which is converted to heat.*

stove is turned off, heat gradually dissipates from the water into the surrounding air. Water requires a relatively large exchange of heat in order to warm or change to steam. One kilogram of iron, for example would warm much faster than one kilogram of water on the same stove. We measure this difference in terms of *specific heat,* the amount of heat required to warm one gram of a substance one degree C. Water requires 4.18 J to warm 1 degree C; a gram of wood takes less than half as much energy (1.7 J), and iron just over one tenth as much energy (0.45 J) to warm 1 degree C.

A substance can have a low temperature but a high heat content, as in the case of a lake that freezes slowly in the fall. Other objects, such as a burning match, have a high temperature but little heat content. Low-quality energy is diffused, dispersed, or low in temperature, so it is difficult to gather and use for productive purposes. Oceans store vast amounts of heat, but converting that heat to useful purposes is difficult. High-quality energy is intense, concentrated, or high in temperature, and it is useful in carrying out work. The intense flames of a very hot fire or high-voltage electrical energy are useful for many purposes. This distinction is important because many of our most common energy sources are low-quality and must be concentrated or transformed into high-quality before they are useful to us.

Why is understanding heat and energy important to understanding environmental science? Because physical characteristics of substances, and their ability to absorb and release energy, control environmental systems. Consider the properties of water, for example. Water's high specific heat keeps lakeshores and seashores relatively cool in the summer and warm in the winter (see "The Miracle of Water," p. 32). Because water absorbs so much heat as it evaporates, atmospheric water vapor redistributes heat around the globe, as well as contributing to the formation of thunderstorms and hurricanes (chapter 9). In addition, the concepts that energy can be converted from work to heat, or from potential to kinetic energy, help explain the ways we store and use energy, both in our bodies and in our electrical utility systems. Further, a common problem in alternative energy production is that alternative energy sources are often more diffuse and difficult to capture than conventional sources such as oil or coal (chapter 12).

Thermodynamics and Energy Transfers

Matter is recycled endlessly through living things, but this recycling is made possible by something that cannot be recycled: energy. Most energy used in ecosystems originates as sunlight. Green plants capture and convert some of this energy to chemical energy, which can be used or stored. Animals consume plants and convert some of the stored chemical energy to kinetic energy and heat. Eventually the energy dissipates and becomes no longer useful. Thus, energy is reused, but it is degraded from higher quality to lower quality forms as it moves through living systems.

Thermodynamics is the study of how energy is transferred, its rates of flow and transformation from one form or quality to another. Thermodynamics is a complex, quantitative discipline, but you don't need a great deal of math to understand some of the broad principles that shape our world and our lives.

The **first law of thermodynamics** states that energy is conserved: it is neither created nor destroyed under normal conditions. It may be transferred or transformed, but the total amount of energy remains the same.

The **second law of thermodynamics** states that, with each successive energy transfer or transformation in a system, less energy is available to do work. Even though the total amount of energy remains the same, its intensity and usefulness deteriorates. The second law recognizes the principle known as *entropy,* the tendency of all natural systems to go from a state of order (for example, high-quality energy, such as chemical energy) toward a state of increasing disorder (for example, low-quality energy, such as heat or kinetic energy).

How does the second law of thermodynamics apply to organisms and biological systems? Organisms are highly organized, both structurally and metabolically. Constant care and maintenance is required to keep up this organization, and a constant supply of energy is required to maintain these processes. Every time some energy is used by a cell to do work, some of that energy is dissipated or lost as heat and movement. If cellular energy supplies are interrupted or depleted, the result—sooner or later—is death.

THE BUILDING BLOCKS OF EARTH AND LIFE

Matter consists of elements, which are combined to form molecules and compounds. Each of the 112 known elements has distinct chemical characteristics. Among the more common elements in biology are carbon (C), hydrogen (H), oxygen (O), nitrogen (N), and phosphorus (P) (fig. 2.2, table 2.1).

Atoms, Molecules, and Compounds

An **atom** is the smallest particle that exhibits the characteristics of an element. Atoms are tiny units of matter composed of positively charged *protons,* negatively charged *electrons,* and electrically neutral *neutrons.* Protons and neutrons, which have approximately the same mass, are clustered in the nucleus at the center of the atom (fig. 2.3). Electrons, which are tiny in comparison to protons and neutrons, move at high speed around the nucleus. Atoms that have equal numbers of electrons and protons are electrically neutral. Atoms frequently lose or gain electrons, acquiring a positive or negative electrical charge. Charged atoms are called **ions.** A *cation* has a positive charge (the atom has lost one or more electrons); an

FIGURE 2.2   *The periodic table arranges elements by atomic number (number of protons). The rows and columns are organized to show groups of similar chemical characteristics. For example, the right-most column contains "noble" gases that do not react readily with other elements, and the next column to the left (F, Cl, Br, I, At) includes highly reactive elements known as halogens.*

TABLE 2.1	Important Elements in Environmental Science
Living things	C, H, O, N
Atmosphere	N, O, Ar, C, Ne, He
Earth, rocks	Fe, O, Si, Mg, Ni, Ca, Al, Na
Economic Metals	Al, Cr, Cu, Fe, Pb, Mn, Ni
Primary toxins	Hg, Pb, Se, Br, Cd, Be, Rn, Ni, As

6 protons

6 neutrons

6 electrons

FIGURE 2.3 *As difficult as it may be to imagine when you look at a solid object, all matter is composed of tiny, moving particles, separated by space and held together by energy. This model represents a carbon-12 atom, with a nucleus containing six protons and six neutrons; the six electrons are represented as a fuzzy cloud of potential locations, rather than as individual particles.*

anion has a negative charge (the original atom gained one or more extra electrons). Some elements tend to lose electrons and form cations readily. Chemically, these elements are known as *metals.* Others tend to gain electrons; these are known as *nonmetals.* A bold line separates the two groups in the periodic table (fig. 2.2), but elements known as *metalloids,* which share some traits of both groups, occur in the table along this line.

We identify atoms by their atomic number, the number of protons in their nuclei. A hydrogen (H) atom has one proton in its nucleus, while a carbon (C) atom has 6. The number of neutrons in atoms of the same element can vary, producing slight variations in atomic mass (the sum of protons and neutrons). Atoms of a single element that differ in atomic mass are *isotopes.* For example, the nuclei of most hydrogen atoms contain only one proton. A small percentage of hydrogen nuclei in nature contain one proton *and* one neutron. We call this isotope deuterium (^{2}H). An even smaller percentage of hydrogen atoms have one proton and two neutrons in their nuclei. This isotope is known as tritium (^{3}H).

Tritium is an example of a radioactive isotope, an unstable form that spontaneously decays by emitting high-energy electromagnetic radiation or subatomic particles, or both. The rate of radioactive decay is indicated by the half-life of an isotope, or the time it takes for half of the atoms to decay. Tritium decays to helium-3 (^{3}He), with a half-life of 12.5 years. If we started with 1,000 ^{3}H nuclei, 500 of them would have decayed to ^{3}He after 12.5 years. Some isotopes of iodine have half-lives measured in seconds, while plutonium, a waste product of nuclear power reactions, has a half-life of 24,000 years. The radioactive emissions from these atoms may be neutrons, beta particles (high-energy electrons), alpha particles (helium nuclei consisting of two protons and two neutrons), positrons (high-energy, positively charged particles), or gamma rays (very short-wavelength radiation). These high-energy emissions can badly damage living cells and tissues. The decades-long debate about storing nuclear waste at Yucca Mountain, Nevada (chapter 12) reflects the danger of radioactive emissions and the long half-life of plutonium.

Because most atoms have some tendency to lose or gain electrons, most are not electrically stable as individuals. They gain stability by joining to form molecules. A **molecule** is a group of atoms, such as O_2 (two oxygen atoms) or H_2O (two hydrogen atoms and one oxygen), that can exist as an individual unit and that has unique properties. We often refer to molecular oxygen (O_2) or molecular nitrogen (N_2), meaning a molecule composed of

identical atoms, also called the elemental form of these atoms. A molecule containing different kinds of atoms is called a **compound.** Water (H_2O, fig. 2.4) is a familiar, simple chemical compound. Some compounds are extremely complex. The genetic information in your cells is encoded in molecules called deoxyribonucleic acid (DNA), which contain millions of atoms.

When metals, which tend to give up electrons, combine with nonmetals, which tend to gain electrons, an *ionic compound*

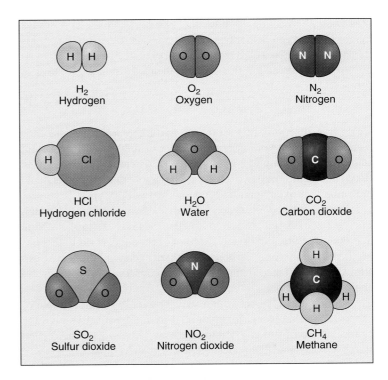

FIGURE 2.4 *These common molecules, with atoms held together by covalent bonds, are important components of the atmosphere or important pollutants.*

results. A common example is NaCl (table salt). The metal (sodium) releases one or more electrons, becoming a positively charged cation. The nonmetal (chlorine) gains one or more electrons, becoming an anion. When two or more nonmetal atoms join, they do so by sharing electrons in what is called a *covalent bond.* Look at the molecules in figure 2.4. Are the atoms metals or nonmetals? Would you expect the bonds to be ionic or covalent?

Because atoms are more stable when bonded ionically or covalently than when they stand alone, energy is needed to break chemical bonds, and energy is released when bonds are formed. Some bonds are stronger than others, so that more energy is needed to break them, and more energy is released when they are broken. For example, plants require energy from the sun to disrupt the very strong covalent bonds in carbon dioxide (CO_2) and water. Plants use atoms from these compounds to form complex molecules, such as sugars and cellulose. An animal that eats a plant breaks the relatively weak bonds of carbohydrates (sugars and starches), and the resulting atoms recombine to form much stronger bonds of CO_2 and H_2O. The net effect is a release of energy, which powers muscles and allows cells to perform functions such as transferring nutrients and synthesizing proteins.

Chemical Reactions

Chemical reactions occur when bonds are broken and re-formed among atoms and compounds. These reactions form the basic compounds on which life depends, and they underlie many of the processes (and problems) in environmental science. Some reactions, such as the breakdown of sugar molecules, can be very complex, but all reactions follow the basic principles of physics: matter (atoms) can be neither created nor destroyed, and energy is transformed or dissipated as reactions occur.

Here is a reaction that occurs when you light a gas stove or furnace:

$$CH_4 + 2\ O_2 \rightarrow CO_2 + 2\ H_2O$$

This reaction shows how atoms are rearranged into new compounds when you burn natural gas, or methane. The reaction starts with two input components, or *reactants:* one methane (CH_4) molecule and two oxygen (O_2) molecules. The methane molecule is broken apart and recombined with oxygen. Two new compounds, or *products,* are formed: carbon dioxide and water. Energy (heat) is released as the methane bonds are broken. Note that the same number of each kind of atom appears on both sides of the reaction (one carbon, four hydrogen, and four oxygen atoms). This balance demonstrates the conservation of matter in chemical reactions.

Some reactions occur in a sequence, as in the breakdown of ozone (O_3) to molecular oxygen (O_2) by highly reactive chlorine atoms released from chlorofluorocarbons (CFCs) (see the related story, "Ozone Hole Continues to Grow," at www.mhhe.com/apps.):

Step 1: CF_2Cl_2 + high energy sunlight $\rightarrow CF_2Cl + Cl$

Step 2: $Cl + O_3 \rightarrow ClO + O_2$

Step 3: $ClO + O_3 \rightarrow Cl + 2\ O_2$

When a molecule or atom loses electrons it is said to be *oxidized.* For example, when iron in the steel body of your car is exposed to oxygen in the air, each iron atom gives up three electrons to the oxygen (O_2) molecules. The formerly blue-black iron becomes a red, iron oxide compound commonly known as rust. In this process, we say that the iron, which lost electrons, was oxidized, while the oxygen, which gained electrons, was *reduced.* Not all oxidation-reduction reactions involve oxygen, but many do. Can you think of any other such reactions? What happens when a candle burns?

Acids and Bases

You are probably familiar with **acids,** such as vinegar or battery acid. Mild acids add a sour flavor to foods; strong acids can burn your skin. Chemically, acids are compounds that readily release hydrogen ions (H^+) in water. Familiar alkaline substances, or **bases,** are also common, especially as cleaning agents: these include baking soda, ammonia, bleach, and drain cleaners. Strong bases are extremely caustic to your skin. Chemically, bases are substances that readily take up H^+ ions and release hydroxide ions (OH^-) in solution. The strength of an acid or base solution can be described by its concentration of H^+ ions, or **pH** (fig. 2.5). Acids have pH values less than 7; they have more H^+ than OH^- ions. Bases have a pH greater than 7; they have more OH^- than H^+ ions. Substances with a pH of exactly 7 are neutral. The pH scale is logarithmic, which means it increases by factors of 10. A substance with pH 5, therefore, has ten times as many hydrogen ions as one with pH 6. In spring 2002, a train car filled with ammonia (NH_3), the most common nitrogen fertilizer on midwestern farms, overturned in North Dakota. Nearby residents had to be evacuated, because in contact with moisture, ammonia is a strong basic compound that is extremely caustic to living tissue.

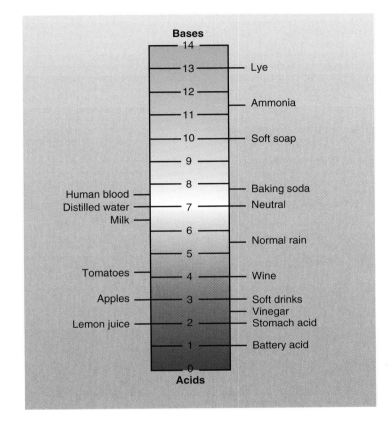

FIGURE 2.5 *The pH scale. The numbers represent the negative logarithm of the hydrogen ion concentration in water. Alkaline (basic) solutions have a pH greater than 7. Acids (pH less than 7) have high concentrations of reactive H^+ ions.*

Organic Compounds

Organic compounds, the compounds that make up living things, are large, often complex molecules built on structures of carbon atoms. The carbon atoms can be arranged in chains or rings, or in complex arrangements of many rings and chains. Aside from carbon, some of the most common elements in organic compounds are hydrogen, oxygen, nitrogen, phosphorus, and potassium. There are many types of bioorganic compounds, but they can be grouped into four major categories: lipids, carbohydrates, proteins, and nucleic acids (fig. 2.6, table 2.2). Together these form the structural and functional characteristics of cells.

Cells: The Fundamental Units of Life

All living organisms are composed of cells, minute compartments within which the processes of life are carried out (fig. 2.7). Microscopic organisms, such as bacteria, some algae, and protozoa, are composed of single cells. By contrast, your body contains several trillion cells of about two hundred distinct types. Surrounding every cell is a thin membrane of lipid and protein that receives information and regulates the flow of materials between the cell and its environment. Inside, cells are subdivided into tiny organelles and subcellular particles that provide the machinery for life. Some of these organelles store and release energy. Others manage and distribute information. Still others create the internal structure that gives the cell its shape and allows it to fulfill its functions.

A special class of proteins called enzymes facilitate all the chemical reactions in cells—providing energy, disposing of wastes, building proteins, and creating new cells. Enzymes are molecular catalysts: they initiate chemical reactions without being used up or inactivated in the process. Think of them as tools. Like hammers or wrenches, they do their jobs without being consumed or damaged as they work. There are generally thousands of different kinds of enzymes in every cell, all specially designed to carry out the many individual processes on which life depends. Most enzymes work by temporarily binding molecules in a unique cavity or slot, much as a key fits a lock. The multitude of enzymatic reactions an organism performs is called its **metabolism.**

TABLE 2.2 Major Classes of Organic Compounds

SUBSTANCE	EXAMPLES	SOME USES	BASIC STRUCTURES
Lipids and hydrocarbons	Fats, oils	Cell membranes, internal cell organelles	Chains of C with attached H
Carbohydrates	Sugars, starches, cellulose	Cell structure, energy storage	Rings or chains of C with attached H and O
Proteins	Muscles, enzymes	Cell structure, function	Chains of amino acids
Nucleic acids	DNA, RNA	Genetic information, protein synthesis	Chains of nucleotides linked by phosphate-sugar bonds

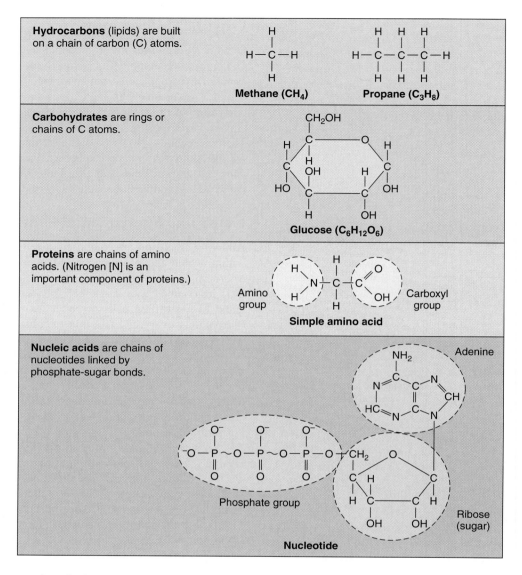

FIGURE 2.6 *Compounds in the four major classes of biologically important molecules are built on a skeleton of carbon atoms and provide the structure and energy for living cells (or are by-products of organic processes). Some are as simple as methane or glucose, but most are large, complex molecules. A DNA molecule contains millions of atoms.*

The Miracle of Water

If travelers from other solar systems were to visit our lovely, cool, blue planet, they might call it Aqua rather than Terra because of its most outstanding feature: an abundance of streams, rivers, lakes, and oceans of liquid water. Without liquid water we would not exist. Water is the (nearly) universal solvent that carries nutrients and wastes to and from our cells. It participates in many organic functions and reactions that allow life as we know it to exist. The earth is the only place that we know of where water exists in liquid form in any appreciable quantity. Liquid water covers nearly three-fourths of the earth's surface, and during the winter, snow and ice cover a good deal of the rest. Not only is water essential for cell structure and metabolism, but water's unique physical and chemical properties directly affect the earth's surface temperatures, its atmosphere, and the interactions of life-forms with their environments. Water has many unique, almost magical, qualities, including the following:

1. Water makes up 60 to 70 percent (on average) of the weight of living organisms. It fills cells, thereby giving form and support to many tissues. Water is the medium in which all of life's chemical reactions occur, and it is an active participant in many of these reactions. Water is a solvent that dissolves the nutrients that cells need for life, as well as the wastes that cells produce. Thus, water is essential in delivering materials to and from cells.

Chloroplast

Nucleus

Cell wall

FIGURE 2.7 *In the center of this high-magnification photograph, a heavy, cellulose cell wall encloses a single, spherical plant cell. Parts of five other cells surround the center cell. The bean-shaped structures with multiple layers of internal membranes are chloroplasts. Chlorophyll inside the chloroplasts captures light energy and uses it to carry out photosynthesis, discussed later in this chapter.*

2. Water dissolves salts and other compounds, producing solutions that conduct electricity. These solutions, called electrolytes, also provide the energy that drives photosynthesis.

3. Water molecules are cohesive, tending to stick together tenaciously. You have experienced this property if you have ever done a belly flop off a diving board. Water has the highest surface tension of any common, natural liquid (fig. 2.8). As a result, water is subject to *capillary action;* that is, it can be drawn into small channels. Without capillary action, movement of water and nutrients into groundwater reservoirs and through living organisms might not be possible.

4. Water exists as a liquid over a wide temperature range that, for most of the world (at least during summer months), corresponds to the ambient temperature range. Most substances exist as either a solid or a gas, with only a very narrow liquid temperature range. Organisms synthesize important organic compounds, such as oils and alcohols, that remain liquid at ambient temperatures, but the original and predominant liquid in nature is water.

5. Water is unique in that it expands when it crystallizes. Most substances shrink as they change from liquid to solid. Ice floats because it is less dense than liquid water. When temperatures fall below freezing, the surface layers of lakes, rivers, and oceans cool faster and freeze before deeper water. Floating ice then insulates underlying layers, keeping most water bodies liquid (and aquatic organisms alive) throughout the winter. Without this feature, lakes, rivers,

FIGURE 2.8 *Surface tension (cohesiveness of water molecules) prevents this water strider from breaking through as it walks on water.*

and even oceans in high latitudes would freeze solid and never melt.

6. Water has a high heat of vaporization: a great deal of energy is needed to convert it from liquid to vapor. Consequently, evaporating water is an effective way for organisms to shed

excess heat. Many animals pant or sweat to moisten evaporative cooling surfaces. Why do you feel less comfortable on a hot, humid day than on a hot, dry day? Because the water vapor-laden air inhibits the rate of evaporation from your skin, thereby impairing your ability to shed heat.

7. Water also has a high specific heat; that is, water absorbs or loses a great deal of energy as it changes temperature. Water bodies such as lakes and oceans warm and cool slowly as seasons change. This temperature lag helps moderate global temperatures, keeping the environment warm in winter and cool in summer. This effect is especially noticeable near the ocean, but it is important globally.

Altogether, these unique properties of water not only shape life at the molecular and cellular level, they also determine many of the features of both the biotic (living) and abiotic (nonliving) components of our world.

SUNLIGHT: ENERGY FOR LIFE

Ultimately, most organisms depend on the sun for the energy needed to create structures and carry out life processes. Our sun is a star, a fiery ball of exploding hydrogen gas (fig. 2.9). Its thermonuclear reactions emit powerful forms of radiation, including potentially deadly ultraviolet and nuclear radiation (fig. 2.10). Radiant energy is classified by its wavelengths. Intense energy, such as X rays or gamma rays, has short wavelengths. Lower-energy, longer wavelengths include heat (infrared) and radio wavelengths. By the time solar energy reaches the earth's surface, our atmosphere has filtered out most of the dangerous shortwave radiation. The remaining energy provides two critical factors for life: warmth and light.

Warmth is essential because most organisms survive within a relatively narrow temperature range. At very high temperatures, biomolecules break down or become distorted and nonfunctional. At very low temperatures, the chemical reactions of metabolism occur too slowly to enable organisms to grow and reproduce. Other planets in our solar system are either too hot or too cold to support life as we know it.

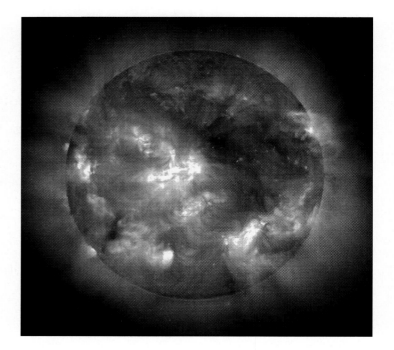

FIGURE 2.9 *Thermonuclear reactions in the sun provide the energy for life on earth.*

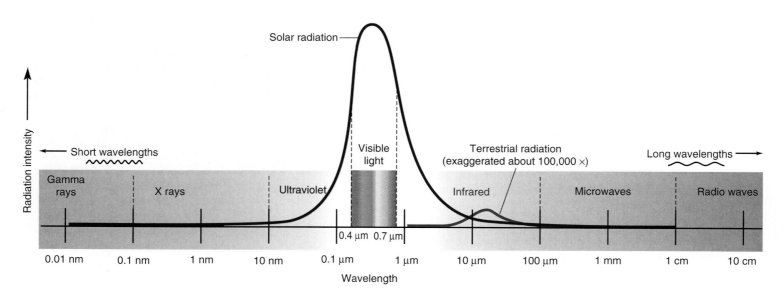

FIGURE 2.10 *The electromagnetic spectrum. Our eyes are sensitive to light wavelengths, which make up nearly half the energy that reaches the earth's surface (represented by the area under the "solar radiation" curve). Photosynthesizing plants use the most abundant solar wavelengths (light and infrared). The earth re-emits lower-energy, longer wavelengths (shown by the "terrestrial radiation" curve), mainly the infrared part of the spectrum.*

The dominant form of solar energy that reaches the earth's surface is in and near the wavelengths of visible light. Not coincidentally, these are the wavelengths that drive **photosynthesis** in green plants. Photosynthesis converts radiant energy into useful, high-quality chemical energy in the bonds that hold together organic molecules. Energy captured by photosynthesis supports nearly all life on earth.

How much incoming solar energy do organisms actually use? The amount of incoming solar radiation is enormous, about 1,372 watts/m^2 at the top of the atmosphere (1 watt = 1 J per second). However, more than half of the incoming sunlight may be reflected or absorbed by atmospheric clouds, dust, and gases. In particular, harmful, short wavelengths are filtered out by gases (especially ozone) in the upper atmosphere. Of the solar radiation that reaches the earth's surface, about 10 percent is ultraviolet, 45 percent is visible, and 45 percent is infrared. Photosynthesizing plants use light energy, especially the blue and red wavelengths, to power photosynthesis. Because most plants reflect, rather than absorb, green wavelengths, they appear green to our eyes.

How Does Photosynthesis Capture Energy?

Photosynthesis occurs in tiny, membranous organelles called chloroplasts that reside within plant cells (see fig. 2.7). The key to this process is chlorophyll, a unique green molecule that captures light energy and uses it to do useful work. Chlorophyll doesn't do this important job alone, however. Many molecules within the chloroplast membranes participate in a series of oxidation-reduction steps called the light-dependent reactions (because they occur only when light strikes the chloroplast) (fig. 2.11). In these reactions, high-energy electrons pass from one intermediate to another. The electrons from this process are drawn from water. The oxygen released when water is split is the source of all the atmospheric oxygen on which all higher animals, including us, depend for life.

Ultimately, some of the energy captured in the light-dependent reactions is stored in two more stable compounds called adenosine triphosphate (ATP) and reduced nicotinamide adenine dinucleoside phosphate (NADPH). ATP and NADPH provide the energy for a series of steps called the light-independent reactions (they can take place after light is no longer present) that fix carbon; that is, they create complex organic molecules such as sugars out of atmospheric carbon dioxide (CO_2). The following equation summarizes the overall set of reactions in photosynthesis:

$$6\,H_2O + 6\,CO_2 + \text{solar energy} \xrightarrow[\text{chlorophyll}]{} C_6H_{12}O_6\ (\text{sugar}) + 6\,O_2$$

We read this as "water plus carbon dioxide plus energy (in the presence of chlorophyll) produces sugar plus oxygen." The sugar shown here is glucose, a simple six-carbon molecule that is one of the main products of photosynthesis in most plants. Glucose and other simple sugars are the building block for all the organic molecules that make up cells and also provides the energy that drives cell physiology. The process of releasing chemical energy, called

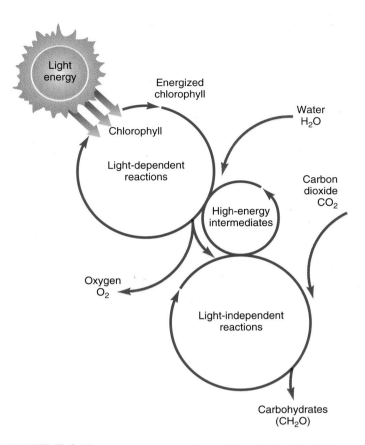

FIGURE 2.11 *During photosynthesis, chlorophyll molecules capture light energy and use it to synthesize energy-rich chemical compounds that cells can use to drive life processes. These reactions consume carbon dioxide and water, and produce oxygen and carbohydrates.*

cellular respiration, essentially reverses photosynthesis by splitting carbon and hydrogen atoms from sugar molecules and recombining them with oxygen to re-create carbon dioxide and water. The net chemical reaction, then, is:

$$C_6H_{12}O_6 + 6\,O_2 \rightarrow 6\,H_2O + 6\,CO_2 + \text{released energy}$$

Note that in photosynthesis energy is captured, while in respiration energy is released. Similarly, photosynthesis consumes water and carbon dioxide to produce sugar and oxygen, while respiration does the opposite. In both sets of reactions, energy is stored temporarily in chemical bonds, which constitute a kind of energy currency for the cell.

Organisms that cannot photosynthesize, such as ourselves, must get energy for cellular respiration by consuming plants, or animals that eat plants (fig. 2.12). In the process, we also consume oxygen and release carbon dioxide, thus completing the cycle of photosynthesis and respiration. Later in this chapter we will see more of how these feeding relationships work.

FIGURE 2.12 *Energy exchange in an ecosystem. Plants take water and carbon dioxide from the environment and use the energy from sunlight to convert them into energy-rich sugars and other organic chemicals, releasing oxygen in the process. Consumers and decomposers take up oxygen and break down sugars during cellular respiration to release useful energy for living. In this process, water, carbon dioxide, and low-quality heat are released to the environment.*

ENERGY AND MATTER IN THE ENVIRONMENT

Living systems are maintained by the use and recyling of energy and matter. Conceptually, we organize living systems in terms of species, populations, biological communities, and ecosystems. A **species** ("kind" in Latin) is most often defined as all organisms that are genetically similar enough to breed and produce live, fertile offspring in nature. A **population** consists of all members of a species that live in the same area at the same time. We might refer to the population of elk in Yellowstone National Park, for example: this group is genetically similar to an elk population in Oregon, but the two are considered separate populations because they are geographically separate. A **biological community** consists of all the populations living and interacting in an area. A biological community and its physical environment (water, mineral resources, air, sunlight) make up an ecological system, or **ecosystem.** Much of ecology is concerned with understanding the ways energy and matter move through ecosystems.

For the sake of simplicity, we often think of ecosystems and communities as fixed ecological units with distinct boundaries. A patch of woods surrounded by farm fields, for instance, has relatively sharp boundaries that separate the two environments. Contrasting levels of light, moisture, and wind occur in the fields and the woods, as well as different kinds of birds, rodents, and insects. Such contrasts differentiate ecosystems. On the other hand, the fields and forest are also interconnected: some animals cross the boundary, carrying nutrients and water from one to the other. Permeable boundaries, across which resources move, make most ecosystems *open systems.* In contrast, a *closed system* would be one which does not exchange energy or matter with its surroundings. In fact, few such closed systems exist in nature.

Ecosystems can also be delimited in terms of their dominant, or most visible, components. The Greater Yellowstone Ecosystem, which has been defined as an area reaching from southwestern Wyoming to southern Montana and Idaho, is often outlined as the area that would support a viable population of grizzly bears. A coral reef ecosystem is defined in terms of the extent and composition of the community of coral polyps (organisms) that give it structure and the myriad organisms that depend on the primary productivity of coral for sustenance.

Regardless of how we define the extent of an ecosystem, we understand its structure and organization largely in terms of the storage and movement of energy through organisms and their environment.

Food Chains, Food Webs, and Trophic Levels

Photosynthesis provides all the energy for nearly all ecosystems. (The chief exception is deep-sea ecosystems supported by geothermal heat and minerals.) One of major properties of ecosystems is **productivity,** the amount of **biomass** (biological material) produced in a given area during a given period of time. Photosynthesis is often called primary productivity because it is the basis for almost all other growth in an ecosystem. Photosynthesizing organisms are called **primary producers.** Organisms that do not photosynthesize are **consumers,** and get their nutrients and energy by eating other things. Net productivity is the amount of primary production that accumulates in a system. An ecosystem may have very high total productivity, but if decomposers break down organic material as rapidly as it is formed, the net productivity will be low.

Think about what you have eaten today, and trace it back to its photosynthetic source. If you have eaten an egg, you can trace it back to a chicken, which ate corn. This is an example of a food chain, a linked feeding series. Now think about a more complex food chain involving you, a chicken, a corn plant, and a grasshopper. The chicken could eat grasshoppers that had eaten leaves of the corn plant. You also could eat the grasshopper directly—some humans do. Or you could eat corn yourself, making the shortest possible food chain. Humans have several options of where we fit into food chains.

In ecosystems, some consumers feed on a single species, but most consumers have multiple food sources. Similarly, some species are prey to a single kind of predator, but many species in an ecosystem are beset by several types of predators and parasites.

FIGURE 2.13 *Each time an organism feeds, it becomes a link in a food chain. In an ecosystem, food chains become cross-connected when predators feed on more than one kind of prey, thus forming a food web. How many food chains make up the food web shown here? The arrows in this diagram indicate the direction in which matter and energy are transferred through feeding relationships.*

In this way, individual food chains interconnect to form a **food web.** Figure 2.13 shows feeding relationships among some of the larger organisms in a woodland and lake community. If we were to add all the insects, worms, and microscopic organisms that belong in this picture, however, it would be overwhelmingly complex.

An organism's feeding position in an ecosystem is its **trophic level.** A corn plant, for example, is at the primary producer level: it produces chemical energy from sunlight, water, and carbon dioxide. A grasshopper that eats the corn is a primary consumer. A chicken that eats the grasshopper is a secondary consumer, and if you eat the chicken, you are a tertiary consumer (sometimes called a "top carnivore").

Most terrestrial food chains are relatively short (seeds → mouse → owl), but aquatic food chains can be quite long (microscopic algae → copepod → minnow → crayfish → bass → osprey). The length of a food chain also may reflect the physical characteristics of a particular ecosystem. A harsh arctic landscape may have a much shorter food chain than a temperate or tropical one.

Organisms also can be identified by the kinds of food they eat (fig. 2.14). **Herbivores** eat plants, **carnivores** eat flesh, and **omnivores** eat both plant and animal matter. Humans are natural omnivores, something you can tell by looking at our teeth. Our teeth are suited for an omnivorous diet, with a combination of cutting and crushing surfaces that are not highly adapted for one specific kind of food, as are the teeth of a wolf (carnivore) or a horse (herbivore).

In the end everything is consumed by the many organisms that recycle the dead bodies and waste products of others. Scavengers, such as crows, jackals, and vultures, clean up dead carcasses of larger animals. Detritivores, such as ants and beetles, consume detritus (litter, debris, and dung), while **decomposer** organisms, such as fungi and bacteria, complete the final breakdown and return nutrients to the soil to fertilize the primary producers. Without the activity of these microorganisms, nutrients would remain locked up in the organic compounds of dead organisms and discarded body wastes, rather than being made available to successive generations of organisms. Think for a moment about the food web that supports you. How many organisms, in how many places, does it include?

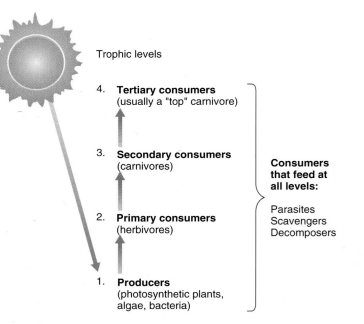

FIGURE 2.14 *Organisms in an ecosystem may be identified by how they obtain food for their life processes (producer, herbivore, carnivore, omnivore, scavenger, decomposer, reducer) or by consumer level (producer; primary, secondary, or tertiary consumer) or by trophic level (first, second, third, fourth).*

FIGURE 2.15 *A classic example of an energy pyramid from Silver Springs, Florida. The numbers in each bar show the percentage of the energy captured at the primary producer level that is incorporated in the biomass at each succeeding level. Detritivores and decomposers feed at every level but are shown attached to the producer bar because this level provides most of their energy.*

Source: Data from Howard T. Odum, "Trophic Structure and Productivity of Silver Springs, Florida" in *Ecological Monographs,* 27:55–112, 1957, Ecological Society of America.

Ecological Pyramids

Most ecosystems have a huge number of primary producers supporting a smaller number of herbivores, which in turn support an even smaller number of secondary consumers. This relationship can be described in terms of a trophic pyramid, which represents the amount of energy stored in the bodies of living things. Most energy in most ecosystems is stored in the bodies of primary producers (fig. 2.15). The second law of thermodynamics (that energy dissipates as it is used), explains this relationship.

This loss of energy occurs in several ways. First, some consumed energy is not used efficiently: a prairie dog that eats grass does not digest all the plant efficiently. Second, a great deal of energy dissipates as kinetic energy and heat. Prairie dogs spend a lot of energy running, playing, hiding, and simply living. When a coyote catches a prairie dog, only a small part of the energy the prairie dog has eaten is available to the coyote.

Furthermore, ecosystems don't operate at 100 percent efficiency. Some surplus usually exists at the lower levels. If there were enough foxes to catch all the rabbits available in the summer, when the rabbit supply is abundant, there would be too many foxes in the middle of the winter, when rabbits are scarce. A popular rule of thumb is that only about 10 percent of the energy at one consumer level is represented at the next higher level. The amount of energy available is often expressed in terms of biomass. For example, we could say that it takes about 100 kg of clover to make 10 kg of rabbit, and 10 kg of rabbit to make 1 kg

of fox (fig. 2.16). The energy efficiency of real ecosystems varies a great deal, however. In most natural ecosystems a more reliable estimate is that about one percent of energy passes to the next trophic level, because predators expend a great deal of energy chasing prey.

The amount of biomass at each trophic level may also form a pyramid. A large population of rabbits, for example, is needed to support one fox. The total mass of those rabbits is much greater than the mass of the fox. Rapid fluctuations in producer and consumer populations can invert this pyramid, however: in lakes, winter populations of producers (plants and algae) can fall well below that of consumers. In some cases the number of individuals in a community can also be described as a pyramid (fig. 2.17). However, the numbers pyramid is highly variable and can be inverted. For example, one fox can support numerous tapeworms.

BIOGEOCHEMICAL CYCLES AND LIFE PROCESSES

The elements and compounds that sustain us are cycled endlessly through living things and through the environment. On a global scale, this movement is referred to as biogeochemical cycling. Substances can move quickly or slowly: you might store carbon for hours or days, while carbon is stored in the earth for millions of years. When human activity alters flow rates or storage times in these natural cycles, overwhelming the environment's ability to process them, these materials can become pollutants. Sulfur, nitrogen, carbon dioxide, and phosphorus are some of the more serious examples of these. Global-scale movement of energy is discussed in detail in chapter 9. Here we will explore some of the paths involved in cycling several important elements: water, carbon, nitrogen, sulfur, and phosphorus.

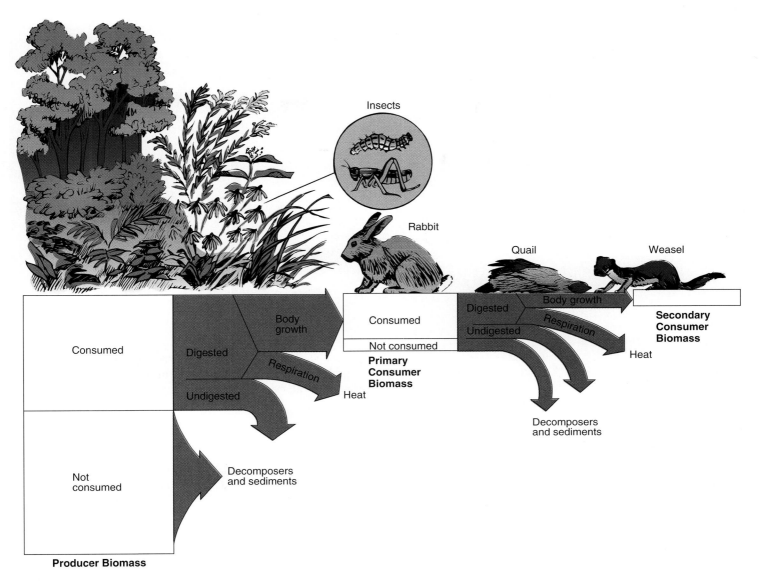

Producer Biomass

FIGURE 2.16 *The energy pyramid is more clearly understood if it is related to a biomass period, which represents the amount of biomass at each trophic level in a food chain. This figure illustrates how nutrients and energy become increasingly less available to successive consumers.*

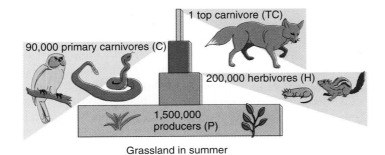

Grassland in summer

FIGURE 2.17 *Usually, larger organisms eat smaller organisms, and it takes numerous small organisms to feed one large organism. The classic study represented in this pyramid shows numbers of individuals at each trophic level per 1,000 m² of grassland. To support one individual at the top carnivore level, there were 90,000 primary carnivores feeding upon 200,000 herbivores that, in turn, fed upon 1.5 million producers.*

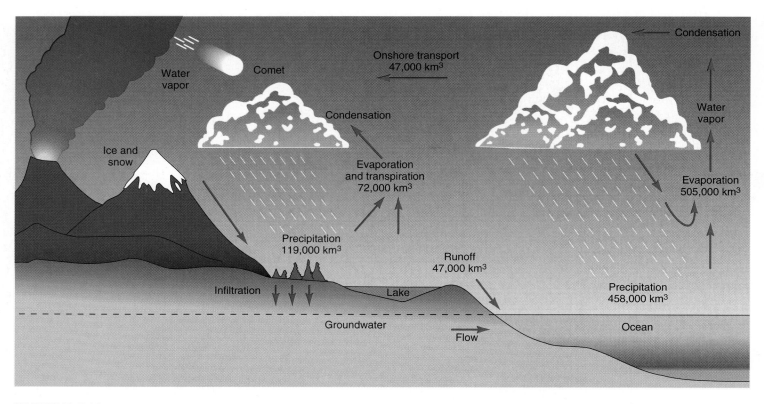

FIGURE 2.18 *In the hydrologic cycle, water moves constantly between aquatic, atmospheric, and terrestrial compartments driven by solar energy and gravity. The total annual runoff from land to the oceans is about 12.2 quadrillion gallons, or 46 quadrillion liters.*

The Hydrologic Cycle

The path of water through our environment is perhaps the most familiar material cycle, and it is discussed in greater detail in chapter 10 (fig. 2.18). Most of the earth's water is stored in the oceans, but solar energy continually evaporates this water, and winds distribute water vapor around the globe. Water that condenses over land surfaces, in the form of rain, snow, or fog, supports all terrestrial (land-based) ecosystems. Living organisms emit the moisture they have consumed through respiration and perspiration. Eventually this moisture reenters the atmosphere or enters lakes and streams, from which it ultimately returns to the ocean again.

As it moves through living things and through the atmosphere, water is responsible for metabolic processes within cells, for maintaining the flows of key nutrients through ecosystems, and for global-scale distribution of heat and energy (chapter 9). Water performs countless services because of its unusual properties (see previous section, "The Miracle of Water"). Water is so important that when astronomers look for signs of life on distant planets, traces of water are the key evidence they seek.

The Carbon Cycle

Carbon serves a dual purpose for organisms: (1) it is a structural component of organic molecules, and (2) chemical bonds in carbon compounds provide metabolic energy. The **carbon cycle** begins with photosynthetic organisms taking up carbon dioxide (CO_2) (fig. 2.19). Once a carbon atom is incorporated into organic compounds, its path to recycling may be very quick or extremely slow. Imagine for a moment what happens to a simple sugar molecule you swallow in a glass of fruit juice. The sugar molecule is absorbed into your bloodstream, where it is made available to your cells for cellular respiration or for making more complex biomolecules. If it is used in respiration, you may exhale the same carbon atom as CO_2 in an hour or less, and a plant could take up that exhaled CO_2 the same afternoon.

Alternatively, your body may use that sugar molecule to make larger organic molecules that become part of your cellular structure. The carbon atoms in the sugar molecule could remain a part of your body until it decays after death. Similarly, carbon in the wood of a thousand-year-old tree will be released only when fungi and bacteria digest the wood and release carbon dioxide as a by-product of their respiration.

Sometimes, recycling takes a very long time. Coal and oil are the compressed, chemically altered remains of plants or microorganisms that lived millions of years ago. Their carbon atoms (and hydrogen, oxygen, nitrogen, sulfur, etc.) are not released until the coal and oil are burned. Enormous amounts of carbon also are locked up as calcium carbonate ($CaCO_3$), used to build shells and skeletons of marine organisms from tiny protozoans to corals. The world's extensive surface limestone deposits

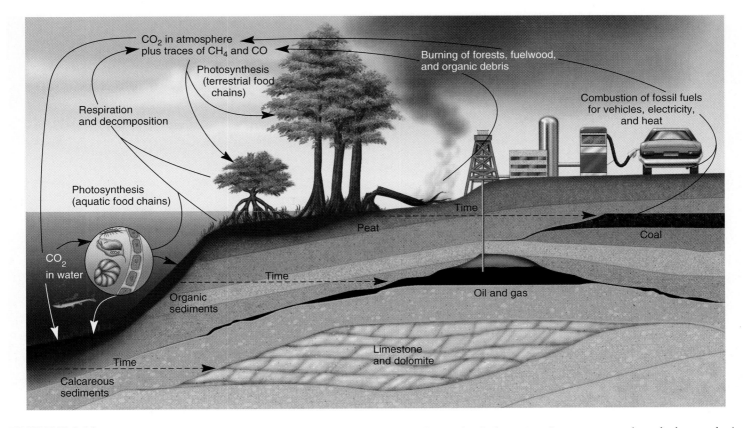

FIGURE 2.19 *Atmospheric carbon dioxide is the "source" of carbon in the carbon cycle. Carbon passes into ecosystems through photosynthesis and is captured in the bodies and products of living organisms. Weathering, respiration, and combustion release carbon to the atmosphere. Carbon may be locked up for long periods in both organic (coal, oil, gas) and inorganic (limestone, dolomite) geological formations, called carbon "sinks."*

are biologically formed calcium carbonate from ancient oceans, exposed by geological events. The carbon in limestone has been locked away for millennia, which is probably the fate of carbon currently being deposited in ocean sediments. Eventually, even the deep ocean deposits are recycled as they are drawn into deep molten layers and released via volcanic activity. Geologists estimate that every carbon atom on the earth has made about 30 such round trips over the last 4 billion years.

Materials that store carbon, including geologic formations and standing forests, are known as carbon sinks. When carbon is released from these sinks, as when we burn fossil fuels and inject CO_2 into the atmosphere, or when we clear extensive forests, natural recycling systems may not be able to keep up. This is the root of the global warming problem discussed in chapter 9. Alternatively, extra atmospheric CO_2 could support faster plant growth, speeding some of the recycling processes.

The Nitrogen Cycle

Organisms cannot exist without amino acids, peptides, and proteins, all of which are organic molecules that contain nitrogen. Nitrogen is therefore an extremely important nutrient for living things. (Nitrogen is a primary component of many household and agricultural fertilizers.) Even though nitrogen makes up about 78

percent of the air around us, plants cannot use N_2, the stable diatomic (two-atom) molecule in the air.

Plants acquire nitrogen through an extremely complex **nitrogen cycle** (fig. 2.20). The key to this cycle is nitrogen-fixing bacteria (including some blue-green algae or cyanobacteria). These organisms have a highly specialized ability to "fix" nitrogen, or combine gaseous N_2 with hydrogen to make ammonia (NH_3).

Other bacteria then combine ammonia with oxygen to form nitrites (NO_2^-). Another group of bacteria converts nitrites to nitrates (NO_3^-), which green plants can absorb and use. After plant cells absorb nitrates, the nitrates are reduced to ammonium (NH_4^+), which cells use to build amino acids that become the building blocks for peptides and proteins.

Members of the bean family (legumes) and a few other kinds of plants are especially useful in agriculture because nitrogen-fixing bacteria actually live in their root tissues (fig. 2.21). Legumes and their associated bacteria add nitrogen to the soil, so interplanting and rotating legumes with crops such as corn that use but cannot replace soil nitrates are beneficial farming practices that take practical advantage of this relationship.

Nitrogen reenters the environment in several ways. The most obvious path is through the death of organisms. Fungi and bacteria decompose dead organisms, releasing ammonia and ammonium ions, which then are available for nitrate formation. Organisms don't

FIGURE 2.20 *Nitrogen is incorporated into ecosystems when plants and bacteria use it to build their own amino acids. Bacterial decomposition releases nitrogen from ecosystems. Both natural and human interactions with the nitrogen cycle are depicted here.*

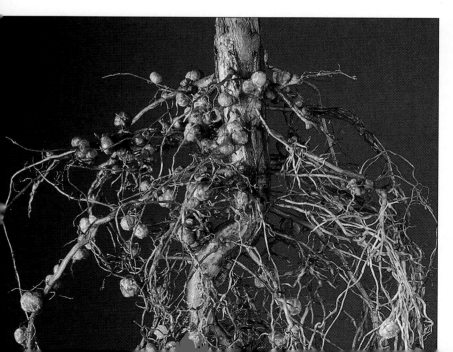

FIGURE 2.21 *Bumps called nodules cover the roots of this adzuki bean plant. Each nodule is a mass of root tissue containing many bacteria that help to convert nitrogen in the soil to a form that the bean plant can assimilate and use to manufacture amino acids.*

have to die to donate proteins to the environment, however. Plants shed their leaves, needles, flowers, fruits, and cones; animals shed hair, feathers, skin, exoskeletons, pupal cases, and silk. Animals also produce excrement and urinary wastes that contain nitrogenous compounds. Urine is especially high in nitrogen because it contains the detoxified wastes of protein metabolism. All of these by-products of living organisms decompose, replenishing soil fertility.

How does nitrogen reenter the atmosphere, completing the cycle? Denitrifying bacteria break down nitrates into N_2 and nitrous oxide (N_2O), gases that return to the atmosphere; thus, denitrifying bacteria compete with plant roots for available nitrates. However, denitrification occurs mainly in waterlogged soils that have low oxygen availability and a large amount of decomposable organic matter. These are suitable growing conditions for many wild plant species in swamps and marshes, but not for most cultivated crop species, except for rice, a domesticated wetland grass.

In recent years, humans have profoundly altered the nitrogen cycle. By using synthetic fertilizers, cultivating nitrogen-fixing crops, and burning fossil fuels, we now convert more nitrogen to ammonia and nitrates than all natural land processes combined. This excess nitrogen input is causing serious loss of soil nutrients such as calcium and potassium, acidification of rivers and lakes, and rising atmospheric concentrations of nitrous oxide, a greenhouse gas. It also encourages the spread of weeds into areas such as prairies, where native plants are adapted to nitrogen-poor environments. Blooms of toxic algae and dinoflagellates in coastal areas also may be linked to excess nitrogen carried by rivers.

The Phosphorus Cycle

Minerals become available to organisms after they are released from rocks. Two mineral cycles of particular significance to organisms are phosphorus and sulfur. Why do you suppose phosphorus is a primary ingredient in fertilizers? At the cellular level, energy-rich phosphorus-containing compounds are primary participants in energy-transfer reactions, as discussed earlier.

The amount of available phosphorus in an environment, therefore, can dramatically affect productivity. Abundant phosphorus stimulates lush plant and algal growth, making phosphorus a major contributor to water pollution.

The phosphorus cycle begins when phosphorus compounds leach from rocks and minerals over long periods of time (fig. 2.22).

FIGURE 2.22 *The phosphorus cycle carries this essential mineral from its source in rocks and soil, through plants and animals, and back to sediments in wastes and carrion.*

Because phosphorous has no atmospheric form, it is usually transported in water. Producer organisms take in inorganic phosphorus, incorporate it into organic molecules, and then pass it on to consumers. Phosphorus returns to the environment by decomposition. An important aspect of the phosphorus cycle is the very long time it takes for phosphorus atoms to pass through it. Deep ocean sediments are significant phosphorus sinks of extreme longevity. Phosphate ores that now are mined to make detergents and inorganic fertilizers represent exposed ocean sediments that are millennia old. You could think of our present use of phosphates, which are washed out into the river systems and eventually the oceans, as an accelerated mobilization of phosphorus from source to sink. Aquatic ecosystems often are dramatically affected in the process because excess phosphates can stimulate explosive growth of algae and photosynthetic bacteria populations ("algae blooms"), upsetting ecosystem stability (see Investigating Our

Environment: Environmental Chemistry of Phosphorus). Can you think of ways we could reduce the amount of phosphorus we put into our environment?

The Sulfur Cycle

Sulfur plays a vital role in organisms, especially as a minor but essential component of proteins. Sulfur compounds are important determinants of the acidity of rainfall, surface water, and soil. In addition, sulfur in particles and tiny airborne droplets may act as critical regulators of global climate. Most of the earth's sulfur is tied up underground in rocks and minerals such as iron disulfide (pyrite) or calcium sulfate (gypsum). Weathering, emissions from deep seafloor vents, and volcanic eruptions release this inorganic sulfur into the air and water (fig. 2.23).

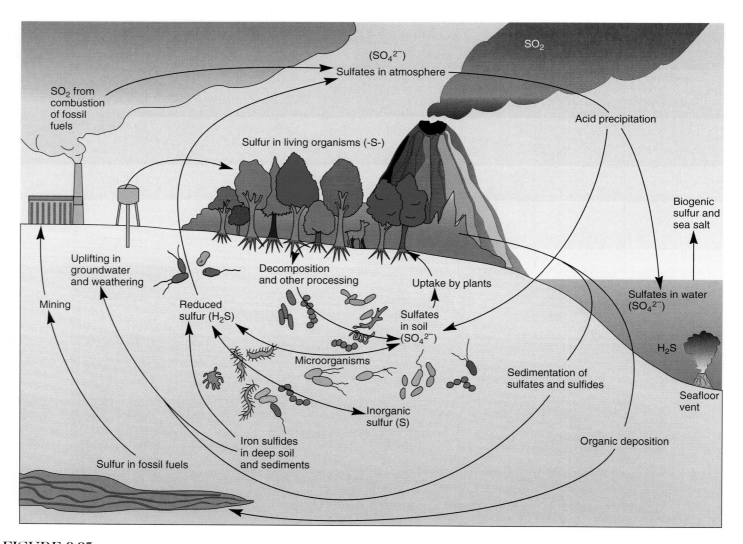

FIGURE 2.23 *Sulfur is present mainly in rocks, soil, and water. It cycles through ecosystems when organisms take it in. Combustion of fossil fuels increases the levels of atmospheric sulfur compounds, which create problems related to acid precipitation.*

INVESTIGATING Our Environment

Environmental Chemistry of Phosphorus

In the 1950s and 1960s people across Canada, the United States, and Europe began to notice increasing rates of smelly algae growth in lakes and wetlands. Increased nutrient loading was accepted as the cause. But which nutrient? Nitrogen, phosphorus, potassium, calcium, and other elements are all known to increase plant growth. To find an answer, ecologist David Schindler decided to do a large-scale experiment. Schindler and his colleagues divided an Ontario lake into two parts. They enriched one half of it with nitrogen and the other half with phosphorus. The phosphorus-enriched half of the lake rapidly became eutrophic: algae and plants grew in such profusion that they choked out the sunlight, died, and decayed, producing a warm, smelly, chemically and biologically altered ecosystem. The other half showed little change. With this classic study, Schindler and his colleagues demonstrated that phosphorus was the key limiting factor for plant growth in aquatic ecosystems. Add abundant phosphorus, and eutrophication results. This discovery led to drastic reduction in phosphates in household detergents and other products.

Phosphorus is a natural, and essential component of all living systems. Released by weathered rocks into soil and water, it is taken up by bacteria and algae, then incorporated into the food web. Our cells use it to produce countless essential compounds, including DNA and the energy-releasing molecule ATP. By weight, your bones are 7 percent phosphorus, and your body contains about 700 grams of the element.

But since the 1950s the use of phosphorus fertilizers on farms has increased dramatically, and their distribution has expanded globally. At the same time, agricultural intensification has increased erosion rates, sending fertilizer-laden soils into streams and lakes at an accelerating rate. Limnologists Elena Bennett and Steve Carpenter estimate that recent expansion in phosphorus mining and trading have more than quadrupled natural rates of phosphorus inputs into freshwater ecosystems around the world, from 3.5 terragrams (trillion grams) per year normally to 13 terragrams now. Eutrophication is increasingly evident in developing countries, where phosphorus-rich fertilizers have become common in recent years. Further, our past activities will affect our water bodies for decades or

centuries to come. Fertilizers will continue to wash off of farm fields, and the phosphorus currently in aquatic ecosystems may be recycled for centuries.

As this case shows, understanding the chemistry of ecosystems is essential to minimizing damage to our environment, or to restoring already-damaged ecosystems. Environmental chemistry is therefore a growing discipline that needs dedicated students to ensure that our environment stays healthy—or returns to health.

This case also demonstrates the broad, multidisciplinary nature of environmental science. Understanding global phosphorus imbalances requires the expertise of hydrologists (who study water sources and the geography of water movement), climatologists (who study atmospheric distribution of moisture and airborne compounds), ecologists (who study ecosystem structure and functions), and toxicologists (who study organically harmful compounds), in addition to environmental chemists. As you read this book, we hope you will discover the ways different scientific disciplines help us identify, and hopefully solve, emerging threats to our environment and our health.

The sulfur cycle is complicated by the large number of oxidation states the element can assume, producing hydrogen sulfide (H_2S), sulfur dioxide (SO_2), sulfate ion (SO_4^{2-}), among others. Inorganic processes are responsible for many of these transformations, but living organisms, especially bacteria, also sequester sulfur in biogenic deposits or release it into the environment. Which of the several kinds of sulfur bacteria prevails in any given situation depends on oxygen concentrations, pH, and light levels.

Human activities also release large quantities of sulfur, primarily through burning fossil fuels. Total yearly anthropogenic sulfur emissions rival those of natural processes, and acid rain (caused by sulfuric acid produced as a result of fossil fuel use) is a serious problem in many areas (see chapter 9). Sulfur dioxide and sulfate aerosols cause human health problems, damage buildings and vegetation, and reduce visibility. They

also absorb ultraviolet (UV) radiation and create cloud cover that cools cities and may be offsetting greenhouse effects of rising CO_2 concentrations.

Interestingly, the biogenic sulfur emissions of oceanic phytoplankton may play a role in global climate regulation. When ocean water is warm, tiny, single-celled organisms release dimethylsulfide (DMS), which is oxidized to SO_2 and then SO_4^{2-} in the atmosphere. Acting as cloud droplet condensation nuclei, these sulfate aerosols increase the earth's albedo (reflectivity) and cool the earth (fig. 2.24). As ocean temperatures drop because less sunlight gets through, phytoplankton activity decreases, DMS production falls, and clouds disappear. Thus, DMS, which may account for half of all biogenic sulfur emissions, could be a feedback mechanism that keeps temperature within a suitable range for all life.

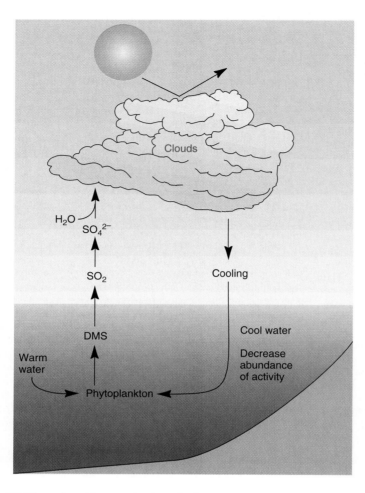

FIGURE 2.24 *Dimethylsulfide (DMS), produced by oceanic phytoplankton (microscopic, single-celled plants), may act as a governor, or feedback control, for global climate. As oceans warm, more DMS is produced, causing clouds to block sunlight. Less solar energy means a cooler earth, less active phytoplankton, and thus, less DMS.*

SUMMARY

Certain conditions are essential for life on earth, including the availability of required chemical elements, a steady influx of solar energy, mild surface temperatures, the presence of liquid water, and a suitable atmosphere. Energy is the capacity to do work. Matter is the substance of which the universe is composed. It exists in three interchangeable phases: gas, liquid, and solid. Matter is made up of atoms, which are composed of particles called protons, neutrons, and electrons.

The energy that holds atoms together forms the basis for energy transfers in the bodies of living organisms and, therefore, in the biosphere. Chemical reactions occur when bonds joining atoms break and new compounds or molecules form. Energy is released when bonds are formed. Ionic compounds form between charged metals and nonmetals. Covalent bonds form when nonmetal atoms share electrons.

Solar radiation provides the heat and light energy needed to support life in the biosphere. Water, which covers approximately three-fourths of the earth's surface, is a remarkable substance. Because of its unique characteristics, it stabilizes the biosphere's temperature and provides the medium in which life processes occur. The earth's atmosphere provides gases necessary for life, helps maintain surface temperatures, and filters out dangerous radiation.

Physical laws, including the principle of conservation of matter and the first and second laws of thermodynamics, govern ecosystem dynamics. The recycling of matter is the basis for the element cycles that occur in ecosystems. Unlike matter, energy is not cycled. Energy always flows through systems in a one-way process in which some energy is converted from a high-quality, concentrated form to a lower-quality, less useful, dispersed form. We describe this increase in disorder as entropy. Primary producers, which synthesize organic compounds from solar energy and carbon dioxide, capture nearly all energy in ecosystems. Resulting compounds provide chemical energy and materials to consumers.

A species is all the organisms of the same kind; often this means organisms genetically similar enough to breed in nature and produce live, fertile offspring. The populations of different species that live and interact within a particular area at a given time make up a biological community. An ecosystem is composed of a biological community together with all the biotic and abiotic factors that make up the environment in a defined area.

Matter and energy are processed through the trophic levels of an ecosystem via food chains and food webs, and biogeochemical cycling. At each energy transfer point, less energy is available for work because of the laws of thermodynamics, so ecosystems require a continuous supply of energy. The relationships between producers and consumers in an ecosystem, often depicted as pyramids, demonstrate this principle. Most of the energy that enters an ecosystem comes, ultimately, from the sun. Living organisms constantly use and reuse elements. Carbon, nitrogen, sulfur, and phosphorus, for instance, recycle through complex biogeochemical paths.

QUESTIONS FOR REVIEW

1. Define the terms *atom* and *element*. Are these terms interchangeable?

2. Your body contains vast numbers of carbon atoms. How is it possible that some of these carbon atoms may have been part of the body of a prehistoric creature?

3. In the biosphere, matter follows a circular pathway while energy follows a linear pathway. Explain.

4. The oceans store a vast amount of heat, but this huge reservoir of energy is of little use to humans (except for climate moderation). Explain the difference between high-quality and low-quality energy.

5. Ecosystems require energy to function. From where does this energy come? Where does it go? How does the flow of energy conform to the laws of thermodynamics?

6. Heat is released during metabolism. How is this heat useful to a cell and to a multicellular organism? How might it be detrimental, especially in a large, complex organism?

7. Photosynthesis and cellular respiration are complementary processes. Explain how they exemplify the laws of conservation of matter and thermodynamics.

8. What do we mean by carbon-fixation or nitrogen-fixation? Why is it important to humans that carbon and nitrogen be "fixed"?

9. The population density of large carnivores is always very small compared to the population density of herbivores occupying the same ecosystem. Explain this in relation to the concept of an ecological pyramid.

10. A species is a specific kind of organism. What general characteristics do individuals of a particular species share? Why is it important for ecologists to differentiate among the various species in a biological community?

THINKING SCIENTIFICALLY

1. How does the principle of entropy (or the second law of thermodynamics) explain the idea of trophic pyramids?

2. Which wavelengths do our eyes respond to, and why (refer to fig. 2.10)? What is the ratio of short ultraviolet wavelengths to microwave wavelengths (used in a microwave oven)?

3. Ecosystems are often defined as a matter of convenience, because we can't study everything at once. How would you describe the character and extent of the natural ecosystem in which you live? What characteristics make your ecosystem an open system?

4. How would you delimit the human ecosystem (including food and energy production) in which you live?

5. If you had to define a research project to evaluate the relative biomass of producers and consumers in an ecosystem, what would you measure? (Note: This could be a natural ecosystem or an enclosed system such as a terrarium.)

6. Understanding storage compartments is essential to understanding material cycles such as the carbon cycle. If you look around your back yard, how many carbon storage compartments are there? Which ones are biggest? Which ones are longest-lasting?

KEY TERMS

acids 30
atom 28
bases 30
biological community 36
biomass 36
carbon cycle 40
carnivores 37
cellular respiration 35
compound 29

conservation of matter 26
consumers 36
decomposer 37
ecology 26
ecosystem 36
energy 27
first law of thermodynamics 28
food web 37

herbivores 37
ions 28
kinetic energy 27
matter 27
metabolism 31
molecule 29
nitrogen cycle 41
omnivores 37
organic compounds 31
pH 30

photosynthesis 35
population 36
potential energy 27
primary producers 36
productivity 36
second law of thermodynamics 28
species 36
trophic level 37

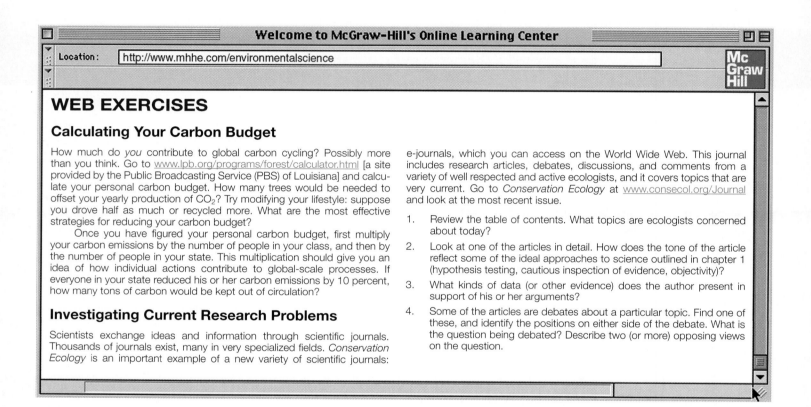

Welcome to McGraw-Hill's Online Learning Center

Location: http://www.mhhe.com/environmentalscience

WEB EXERCISES

Calculating Your Carbon Budget

How much do *you* contribute to global carbon cycling? Possibly more than you think. Go to www.lpb.org/programs/forest/calculator.html [a site provided by the Public Broadcasting Service (PBS) of Louisiana] and calculate your personal carbon budget. How many trees would be needed to offset your yearly production of CO_2? Try modifying your lifestyle: suppose you drove half as much or recycled more. What are the most effective strategies for reducing your carbon budget?

Once you have figured your personal carbon budget, first multiply your carbon emissions by the number of people in your class, and then by the number of people in your state. This multiplication should give you an idea of how individual actions contribute to global-scale processes. If everyone in your state reduced his or her carbon emissions by 10 percent, how many tons of carbon would be kept out of circulation?

Investigating Current Research Problems

Scientists exchange ideas and information through scientific journals. Thousands of journals exist, many in very specialized fields. *Conservation Ecology* is an important example of a new variety of scientific journals:

e-journals, which you can access on the World Wide Web. This journal includes research articles, debates, discussions, and comments from a variety of well respected and active ecologists, and it covers topics that are very current. Go to *Conservation Ecology* at www.consecol.org/Journal and look at the most recent issue.

1. Review the table of contents. What topics are ecologists concerned about today?

2. Look at one of the articles in detail. How does the tone of the article reflect some of the ideal approaches to science outlined in chapter 1 (hypothesis testing, cautious inspection of evidence, objectivity)?

3. What kinds of data (or other evidence) does the author present in support of his or her arguments?

4. Some of the articles are debates about a particular topic. Find one of these, and identify the positions on either side of the debate. What is the question being debated? Describe two (or more) opposing views on the question.

Populations, Communities, and Species Interaction

> Any species of bug is an irreplaceable marvel, equal to the works of art which we religiously preserve in our museums.
>
> –*Claude Levi-Strauss*

Arctic hares battle over territory. Competition for resources is an important factor in regulating populations and in species interactions.

OBJECTIVES

After studying this chapter, you should be able to

- describe how environmental factors determine which species live in a given ecosystem and where or how they live.
- understand how random genetic variation and natural selection lead to evolution, adaptation, niche specialization, and partitioning of resources in biological communities.
- compare and contrast interspecific predation, competition, symbiosis, commensalism, mutualism, and coevolution.
- explain population growth rates, carrying capacity, and factors that limit population growth.
- discuss productivity, diversity, complexity, and structure of biological communities and how these characteristics might be connected to resilience and stability.
- explain how ecological succession results in ecosystem development and allows one species to replace another.
- list some examples of exotic species introduced into biological communities, and describe the effects such introductions can have on indigenous species.

Welcome to McGraw-Hill's Online Learning Center

Location: http://www.mhhe.com/environmentalscience

LEARNING ONLINE

Visit our webpage at www.mhhe.com/environmentalscience for data sources, further readings, additional case studies, current environmental news, and regional examples within the Online Learning Center to help you understand the material in this chapter. You'll also find active links to information pertaining to this chapter including:

USGS Biological Resources Division
National Marine Mammal Laboratory
Cornell Lab of Ornithology
U.S. Long Term Ecological Research Network
Quantitative Population Ecology
Biodiversity and Protected Areas
Population Dynamics, Interactive Models

Darwin and the Theory of Evolution

Why do living things vary so much, and how is it that they fit so neatly into their environment and their biological communities? These questions, at the core of environmental science today, inspired Charles Darwin to explore the origin of species a century and a half ago. Darwin's work demonstrates the importance of careful observation, cautious explanation, and a willingness to consider new ideas, even controversial ones.

Darwin was only 22 years old when he set out on his epic five-year voyage aboard the ship *Beagle*, which left England in 1831 to map sailing routes around South America (fig. 3.1). Initially an indifferent student, Darwin had found inspiring professors in his last years in college. One of these helped him get a position as an unpaid naturalist on board the *Beagle*. Darwin turned out to be an inquisitive thinker, an avid collector of specimens, and a careful reader of explorers, natural historians, and geologists.

After four years of exploring and collecting, Darwin reached the Galápagos Islands. Six hundred miles from the coast of Ecuador, these islands are isolated from the mainland and from each other by strong, cold currents and high winds. By then, Darwin had read and seen enough to recognize that numerous species on each island occurred nowhere else in the world. Yet, their features suggested common ancestors on the South American mainland. The finches were especially interesting: each island had its own species, marked by distinct bill shapes, which graded from large and parrot-like to small and warbler-like. Each bird's beak type was suited to an available food source on its island. It seemed obvious that, somehow, these birds had been modified to survive in their distinct environments.

To convince others that species could adapt to their environments, and to make a useful contribution to scientific understanding, Darwin had to come up with an *explanation* of *how* they did so. Shortly after his return to England, he read an essay by the Reverend Thomas Malthus, proposing that humans invariably reproduce beyond the carrying capacity of their resources, after which famine, disease, and competition eliminate the surplus population. Here Darwin found a plausible explanation for evolution: species produce many offspring, but only the fittest survive to reproduce again; poorer competitors do not reproduce and pass on their traits.

It's important to note that Darwin wasn't the first to think about biological evolution. Many people in his time understood that organisms could share common ancestors but drift apart over time. Darwin spent a great deal of time, for example, interviewing animal and plant breeders about how traits changed from generation to generation. What was missing—and what Darwin supplied with his brilliant insights about natural selection—was how this process works in nature. Darwin didn't know, however, how char-

FIGURE 3.1   *Charles Darwin, in a portrait painted shortly after the voyage on the* Beagle.

acteristics were passed from generation to generation. Only after many years of further research were Ernst Mayer and others able to put together a modern synthesis that explains the genetic basis of evolution.

An overwhelming majority of biologists now consider the theory of evolution through natural selection to be the cornerstone of their science. A huge body of evidence from both laboratory and field studies supports this theory. Neither Darwin nor his ideas are above scrutiny, however. Darwin believed, for example, that evolution probably proceeds very slowly and gradually. In 1972, the Harvard paleontologist Stephen Jay Gould challenged this idea of gradualism, suggesting instead that species and biological communities go through long periods of relatively little change, but then have brief episodes of very rapid evolutionary development. Gould called this punctuated equilibrium. Darwin's work remains one of our foremost examples of bold, creative thinking and meticulous observation. A good theory is able to explain new evidence as it accumulates; it is also a springboard for further exploration and debate. How species multiply and how they interact, the subjects of this chapter, are aspects of evolution that scientists continue to explore and debate almost 150 years after Darwin's theory was first published.

For further reading, see Charles Darwin, *The Voyage of the Beagle* (1837) and *On the Origin of Species by Means of Natural Selection* (1859).

WHO LIVES WHERE, AND WHY?

"Why" questions often stimulate scientific research, but the research itself centers on "how" questions. Why, we wonder, does a particular species live where it does? More to the point, how does it deal with the physical resources and stresses of its environment? How does it interact with the other species present? And what gives one species an edge over another species in a particular habitat?

In this chapter, we will examine some specific ways organisms respond to physical aspects of their environment. We then will discuss how members of a biological community interact, pointing out a few of the difficulties ecologists encounter when they attempt to discern patterns and make generalizations about community interactions and organization.

Critical Factors and Tolerance Limits

Every living organism has limits to the environmental conditions it can endure. Temperatures, moisture levels, nutrient supply, soil and water chemistry, living space, and other environmental factors must be within appropriate levels for organisms to persist. In 1840, Justus von Liebig proposed that the single factor in shortest supply relative to demand is the critical determinant in species distribution. Ecologist Victor Shelford later expanded this principle of limiting factors by stating that each environmental factor has both minimum and maximum levels, called **tolerance limits,** beyond which a particular species cannot survive or is unable to reproduce (fig. 3.2). The single factor closest to these survival limits, Shelford postulated, is the critical limiting factor that determines where a particular organism can live.

At one time, ecologists accepted the critical limiting factor concept so completely that they called it Liebig's or Shelford's Law and tried to identify unique factors limiting the growth of every plant and animal population. For many species, however, we find that the interaction of several factors working together, rather than a single limiting factor, determines biogeographical distribution. If you have ever explored the rocky coasts of New England or the Pacific Northwest, for instance, you have probably noticed that mussels and barnacles endure extremely harsh conditions but generally are sharply limited to an intertidal zone, where they grow so thickly that they often completely cover the substrate. No single factor determines this distribution. Instead, a combination of temperature extremes, drying time between tides, salt concentrations, competitors, and food availability limits the number and location of these animals.

For other organisms, there may be a specific critical factor that, more than any other, determines the abundance and distribution of that species in a given area. A striking example of cold intolerance as a critical factor is found in the giant saguaro cactus (*Carnegiea gigantea*), which grows in the dry, hot Sonoran desert of southern Arizona and northern Mexico (fig. 3.3). Saguaros are extremely sensitive to low temperatures. A single exceptionally cold winter night with temperatures below freezing for 12 hours or more kills growing tips on the branches.

Animal species, too, exhibit tolerance limits that often are more critical for the young than for the adults. The desert pupfish (*Cyprinodon*), for instance, occurs in small, isolated populations in warm springs in the northern Sonoran desert. Adult pupfish can survive temperatures between 0°C and 42°C (a remarkably high temperature for a fish) and are tolerant to an equally wide range of salt concentrations. Eggs and juvenile fish, however, can only live between 20°C and 36°C, and are killed by high salt levels. Reproduction, therefore, is limited to a small part of the range of adult fish.

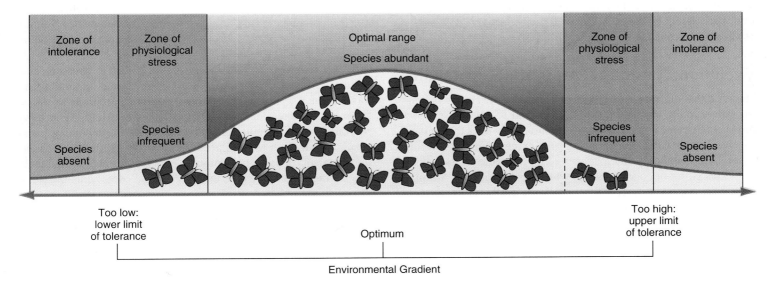

FIGURE 3.2 *The principle of tolerance limits states that for every environmental factor, an organism has both maximum and minimum levels beyond which it cannot survive. The greatest abundance of any species along an environmental gradient is around the optimum level of the critical factor most important for that species. Near the tolerance limits, abundance decreases because fewer individuals are able to survive the stresses imposed by limiting factors.*

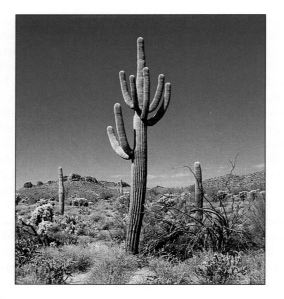

FIGURE 3.3 *Saguaro cacti, symbolic of the Sonoran desert, are an excellent example of distribution controlled by a critical environmental factor. Extremely sensitive to low temperatures, saguaros grow only where minimum temperatures never dip below freezing for more than a few hours at a time.*

Sometimes, the requirements and tolerances of species are useful indicators of specific environmental characteristics. The presence or absence of such species can tell us something about the community and the ecosystem as a whole. Lichens and eastern white pine, for example, are indicators of air pollution because they are extremely sensitive to sulfur dioxide and acid precipitation. Bull thistle is a weed that grows on disturbed soil but is not eaten by cattle; therefore, an abundant population of bull thistle in a pasture is a good indicator of overgrazing. Similarly, anglers know that trout species require clean, well-oxygenated water, so the presence or absence of trout can be an indicator of water quality.

Evolution: Natural Selection and Adaptation

How is it that mussels have developed the ability to endure pounding waves, daily exposure to drying sun and wind, and seasonal threats of extreme cold or hot temperatures? How does the saguaro survive in the harsh temperatures and extreme dryness of the desert? We commonly say that each of these species is "adapted" to its special set of conditions, but what does that mean? **Adaptation,** when species acquire traits that allow them to survive in their environments, is one of the most important concepts in biology.

We use the term *adapt* in two ways. One is a limited range of physiological modifications (called acclimation) available to individual organisms. If you keep houseplants inside all winter, for example, and then put them out in full sunlight in the spring, they get sunburned. If the damage isn't too severe, your plants will probably grow new leaves with a thicker cuticle and denser pigments that protect them from the sun. They can adapt to some degree, but the change isn't permanent. Another winter inside will

make them just as sensitive to the sun as before. Furthermore, the changes they acquire are not passed on to their offspring. Although the potential to acclimate is inherited, each generation must develop its own protective epidermis.

Another type of adaptation operates at the population level and is brought about by inheritance of specific genetic traits that allow a species to live in a particular environment. This process is explained by the theory of **evolution**. This theory is mainly attributed to Charles Darwin (see "Darwin and the Theory of Evolution" at the beginning of this chapter), but it was also developed independently and simultaneously by Alfred Russel Wallace, who documented species differentiation in Indonesia. According to this theory, species change gradually through competition for scarce resources. **Natural selection** describes the process in which better competitors survive and reproduce more successfully. Poorer competitors are more likely to die or to fail to reproduce. Note that "better competitors" could have any trait that provides individuals with some advantage over their peers. Bigger bills, better camouflage, or more attractive plumage or songs each could provide advantages in different circumstances. Individuals carrying advantageous traits are most likely to reproduce successfully, thus passing on those traits.

Natural selection occurs because small, random mutations (changes in genetic material) occur spontaneously in every population. These random differences create genetic diversity (natural variability in traits) in a population. Depending on environmental conditions, some traits may be more advantageous than others. Limited resources or environmental conditions may exert **selective pressure** on a population, reducing the chance that less fit individuals will reproduce successfully. In such cases, individuals with advantageous traits become relatively more abundant in the population.

It is important to remember that mutations can be negative as well as positive. That is, the characteristics created by a particular genetic change can be either harmful or beneficial. Natural selection tends to discard the negative traits and preserve the helpful ones.

What environmental factors cause selective pressure and influence fertility or survivorship in nature? Some important factors include (1) physiological stress due to inappropriate levels of some critical environmental factor, such as moisture, light, temperature, pH, or specific nutrients; (2) predation, including parasitism and disease; (3) competition; and (4) luck. In some cases, the organisms that survive environmental catastrophes or find their way to a new habitat where they start a new population may simply be lucky rather than more fit or better suited to subsequent environmental conditions than their less fortunate contemporaries.

Speciation

Given enough time, many small changes (mutations) may collectively allow a species to become better suited to new environmental conditions. Sometimes, evolution creates entirely new species, physically distinct from their ancestors. The development of new species is known as speciation. Speciation can result from the

Principles of Environmental Science

development of new opportunities (the appearance of a new food source or other resource) or risks (climate change or new predators or competitors). Isolation can also produce new species: if a small group is separated from the rest of the population (called geographic isolation), its genetic characteristics can diverge from those of the main group. Similarly, if a small population colonizes a new environment, island, or habitat, it might encounter new environmental conditions that produce selective pressure. The barriers that divide subpopulations are not always physical. In some cases, behaviors such as when and where population members feed, sleep, or mate, or even how they communicate, may separate them sufficiently for **divergent evolution** (separation into new types) to occur even though they occupy the same territory. This is called behavioral isolation. Although each genetic change may be very slight, many mutations over a long time have produced the incredible variety of different life-forms that we observe in nature.

Cheetahs are remarkable examples of evolutionary adaptation (fig. 3.4). The fastest running of any mammal, these big cats have a flexible skeleton, a sleek shape, and musculature that allows them to catch even the fleetest prey. Think about the adaptations of house cats. Which features might have been useful to their ancestors but are no longer so important?

The variety of finches observed by Charles Darwin on the Galápagos Islands is another classic example of speciation driven by availability of different environmental opportunities (fig. 3.5). Originally derived from a single seed-eating species that was perhaps carried over 1,000 km from the mainland in a storm, the finches have evolved into 13 distinct species that differ markedly in appearance, food preferences, and habitat. Fruit eaters have thick, parrot-like bills; seed eaters have heavy, crushing bills; insect eaters have thin, probing beaks to catch their prey. One of the most unusual species is the woodpecker finch, which pecks at tree bark for hidden insects. Lacking the woodpecker's long tongue, however, the finch uses a cactus spine as a tool to extract bugs.

FIGURE 3.4 *Cheetahs are adapted by natural selection to run extremely fast.*

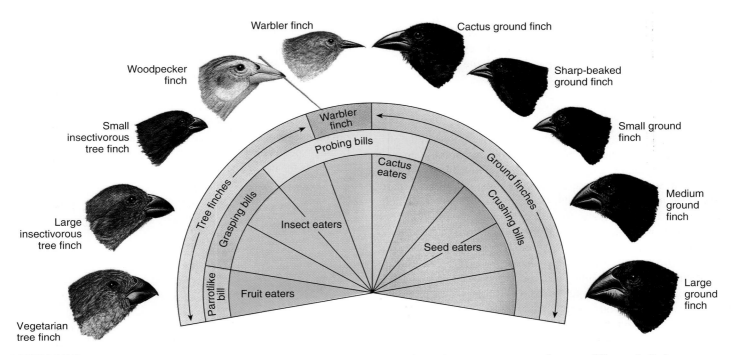

FIGURE 3.5 *Some species of Galápagos Island finches. Although all are descendents of a common ancestor, they now differ markedly in appearance, habitat, and feeding behavior. Ground finches (lower right) eat cactus leaves; warbler finches (upper left) eat insects; others eat seeds or have mixed diets. The woodpecker finch (upper left) pecks tree bark as do woodpeckers, but lacks a long tongue. Instead, it uses cactus spines as tools to extract insects.*

The amazing variety of colors, shapes, and sizes of dogs, cats, rabbits, fish, flowers, vegetables, and other domestic species shows the effects of deliberate selective breeding. The various characteristics of these organisms arose through mutations. We have simply kept the ones we liked.

Natural selection and adaptation can produce very different organisms from a common origin, but it can also cause unrelated organisms to look and act very much alike. We call this latter process **convergent evolution**. The fruit-eating Galápagos finches, for example, look and act very much like parrots even though they are genetically very dissimilar. The features that enable parrots to eat fruit successfully work well for these finches also.

Be sure you understand that while selection affects individuals, evolution and adaptation work at the population level. Individuals don't evolve; species do. Each individual is locked in by genetics to a particular way of life. Most plants, animals, or microbes have relatively limited ability to modify their physical makeup or behavior to better suit a particular environment. Over time, however, random genetic changes and natural selection can change an entire population.

Another common mistake is to believe that organisms develop certain characteristics because they want or need them. This is incorrect. A duck doesn't have webbed feet because it *wants* to swim or *needs* to swim in order to eat; it has webbed feet because some ancestor happened to have a gene for webbed feet that gave it some advantage over other ducks in acquiring food or evading predators. That successful duck produced more offspring than other ducks, and its webbed feet were passed on. A variety of different genetic types is always present in any population, and natural selection simply favors those types best suited for particular conditions. Whether this process has a purpose or direction is a philosophical question rather than a scientific one and is beyond the scope of this book.

The Taxonomic Naming System

Taxonomy, or the study of types of organisms and their relationships, traces how organisms have evolved from common ancestors. Like a family tree, taxonomic relationships can be displayed in a tree-like arrangement. The most widely used part of this tree are the final levels, the genus and species, which make up the binomial name (or scientific name or Latin name) that identifies a species. These names follow a Latin-like naming style, although species are often named with non-Latin words or with names of people. These names help scientists communicate with each other, because they provide an international language to refer to organisms. Common names like buttercup or bluebird might refer to a wide variety of different organisms in different locations, and multiple names are often used for a single species. A scientific name such as *Pinus resinosa,* on the other hand, always means the same kind of pine tree, regardless of whether you think of it as a red pine or a Norway pine.

In general, taxonomic organization uses the following levels:

Kingdom
  Phylum (or division)
    Class
      Order
        Family
          Genus
            Species

For example, you are a *Homo sapiens,* and corn (maize) is *Zea mays* (table 3.1). These two species represent the two best-known kingdoms: animals and plants. Scientists recognize six different kingdoms, however: animals, plants, fungi (molds and mushrooms), protists (algae, protozoans), eubacteria (ordinary bacteria) and archaebacteria (ancient, single-celled organisms that live in harsh environments like hot springs). Within these kingdoms are millions of different species (see chapter 5).

The Ecological Niche

Habitat describes the place or set of environmental conditions in which a particular organism lives. A more functional term, the **ecological niche**, is a description of either the role played by a species in a biological community or the total set of environmental factors that determine species distribution. The concept of niches, as community roles, describing how a species obtains food, what relationships it has with other species, and the services it provides its community, was first defined by the British ecologist, Charles Elton in 1927. Thirty years later, the American limnologist G. E. Hutchinson proposed a more biophysical definition of this concept. Every species, he pointed out, has a range of physical and chemical conditions (temperature, light levels, acidity, humidity, salinity, etc.) as well as biological interactions (predators and prey present, defenses, nutritional resources available, etc.) within which it can exist. Figure 3.2, for example, shows the abundance of a hypothetical species along a single factor gradient. If it were possible to graph simultaneously all of the factors that affect a particular species, a multidimensional space would result that describes the ecological niche available to that species.

	TABLE 3.1	Taxonomy of Two Common Species	
TAXONOMIC LEVEL	HUMANS	CORN
Kingdom	Animalia	Plantae
Phylum	Chordata	Anthophyta
Class	Mammalia	Monocotyledons
Order	Primates	Commenales
Family	Hominidae	Poaceae
Genus	*Homo*	*Zea*
Species	*Homo sapiens*	*Zea mays*

Some species, like raccoons or coyotes, are generalists that eat a wide variety of food and live in a broad range of habitats (including urban areas). Others, such as the panda, are specialists that occupy a very narrow niche (fig. 3.6). Specialists often tend to be rarer than generalists and less resilient to disturbance or change.

A few species, such as elephants, chimpanzees, and baboons, learn how to behave from their social group and can invent new ways of doing things when presented with new opportunities or challenges. Most organisms, however, are limited by genetically determined physical structure and instinctive behavior to established niches.

Over time, niches can change as species develop new strategies to exploit resources. The law of competitive exclusion states that no two species will occupy the same niche and compete for exactly the same resources in the same habitat for very long. Eventually, one group will gain a larger share of resources, while

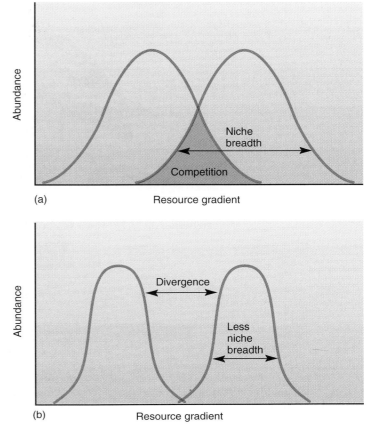

FIGURE 3.7 *Resource partitioning and niche specialization caused by competition. (a) Where niches of two species overlap along a resource gradient, competition occurs* (shaded area). (b) *Individuals occupying this part of the niche are less successful in reproduction. Characteristics of the population then diverge to produce more specialization, narrower niche breadth, and less competition between species.*

FIGURE 3.6 *The giant panda feeds exclusively on bamboo. Although its teeth and digestive system are those of a carnivore, it is not a good hunter and has adapted to a vegetarian diet. In the 1970s, huge acreages of bamboo flowered and died, and many pandas starved.*

the other will either migrate to a new area, become extinct, or change its behavior or physiology in ways that minimize competition. We call this latter process of niche evolution **resource partitioning** (fig. 3.7). Partitioning can allow several species to utilize different parts of the same resource and coexist within a single habitat (fig. 3.8). Species can specialize in time, too. Swallows and insectivorous bats both catch insects, but some insect species are active during the day and others at night, providing noncompetitive feeding opportunities for day-active swallows and night-active bats.

What Are Weedy Species?

Most of us have spent time trying to control weeds, but have you ever thought about what weeds are, ecologically? In your yard or garden, weeds are species that grow rapidly where you don't want them. They are *opportunistic*, quickly appearing wherever an opportunity arises. Weeds such as dandelions, crabgrass, and creeping Charlie can reproduce quickly, easily spreading many

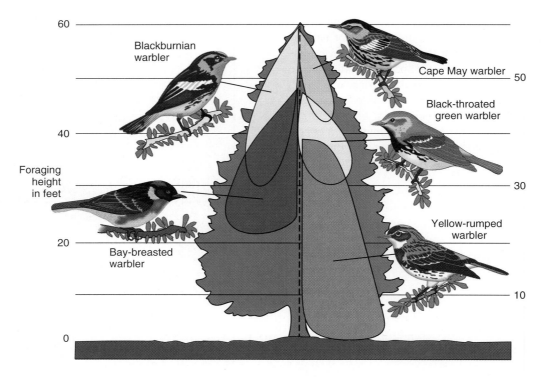

FIGURE 3.8 *Several species of wood warblers use different strata of the same forest, thereby demonstrating resource partitioning and the concept of the ecological niche. This is a classic example of the principle of competitive exclusion.*
Source: Original observation by R. H. MacArthur.

seeds long distances (fig. 3.9). Many weeds are generalist species, able to tolerate a wide variety of environmental conditions (hot, cold, sunny, shady, dry, wet) and to endure environmental change. Ecologically, these attributes make such species excellent competitors in the environmental conditions of your yard. Weeds are also frequently *pioneer species,* able to quickly colonize open, disturbed, or bare ground. Such species can play important ecological roles, covering bare soil and reducing erosion, for example. If they have no natural predators, they can also quickly overwhelm other vegetation, causing significant environmental change.

Many animals can be considered "weedy" species as well. Generalist, opportunistic species, such as starlings and English sparrows, raccoons, and rats, all spread readily and can tolerate a wide range of climate and environmental conditions. Ecologically, these traits are advantageous to these species. Their aggressiveness can quickly alter environments in which they arrive, however.

SPECIES INTERACTIONS

Predation and competition for scarce resources are major factors in evolution and adaptation. Not all biological interactions are competitive, however. Organisms also cooperate with, or at least tolerate, members of their own species as well as individuals of other species in order to survive and reproduce. In some cases, different organisms depend on each other to help acquire resources. In this section, we will look more closely at the different interactions within and between species that shape biological communities.

FIGURE 3.9 *Dandelions and other opportunistic species generally produce many highly mobile offspring.*

Predation

All organisms need food to live. Producers make their own food, and consumers eat organic matter created by other organisms. In most communities, as we saw in chapter 2, photosynthetic plants and algae are the producers. Consumers include herbivores, carnivores, omnivores, scavengers, detritivores, and decomposers. With which of these categories do you associate the term predator? Ecologically, a **predator** is any organism that feeds directly upon another living organism, whether or not it kills its prey to do so (fig. 3.10). By this definition, herbivores, carnivores, and omnivores that feed on live prey are predators, but scavengers, detritivores, and decomposers that feed on dead things are not. In this broad sense, parasites (organisms that feed on a host organism or steal resources from it without necessarily killing it) and even pathogens (disease-causing organisms) might also be considered predator organisms.

Predation is a potent and complex influence on the population balance of communities involving (1) all stages of the life cycles of predator and prey species, (2) many specialized food-obtaining mechanisms, and (3) specific prey-predator adaptations that either resist or encourage predation.

Predatory relationships can change dramatically with life stages. In marine intertidal ecosystems, many crustaceans, mollusks, and worms release eggs directly into the water, and the eggs and free-living larval and juvenile stages are part of the floating community, or plankton (fig. 3.11). Planktonic animals feed upon each other and are food for successively larger carnivores, including small fish. As prey species mature, their predators change. Barnacle larvae are planktonic and are eaten by fish. Adult barnacles, on the other hand, build hard shells that protect them from fish but can be crushed by limpets and other mollusks. Predators also may change their feeding targets. Adult frogs, for instance, are carnivores, but the tadpoles of most species are grazing herbivores. Sorting out the trophic levels in these communities can be very difficult.

Predation can be an important factor in controlling populations of both prey organisms and their consumers. The cyclamen mite (*Tarsonemus sp.*), for example, is a pest of strawberry crops in California. It is kept in check by a predatory mite in the genus *Typhlodromus*. When predators are removed, cyclamen mite pop-

FIGURE 3.11 *Microscopic plants and animals form the basic levels of many aquatic food chains and account for a large percentage of total world biomass. Many oceanic plankton are larval forms that have very different habitats and feeding relationships than their adult forms.*

ulations explode and can severely damage a crop. If predators are re-introduced, they quickly reduce the cyclamen mite populations and bring them under control.

Given enough time, these trophic relationships can exert selective pressure that favors evolutionary adaptation. The predator becomes more efficient at finding and feeding, and the prey becomes more effective at escape or avoidance. In plants, predator avoidance is often accomplished with thick bark, spines, thorns, or chemical defenses. Animal prey may become very adept at hiding, fleeing, or fighting back against predators. Predators, in turn, evolve mechanisms to overcome the defenses of their prey. This process in which species exert selective pressure on each other is called **coevolution**. Coevolution can also be mutually beneficial: many plants and pollinators have evolved together, each aiding the other.

Competition

Competition is another kind of antagonistic relationship within a community. For what do organisms compete? To answer this question, think again about what all organisms need to survive: energy and matter in usable forms, space, and specific sites for life activities. Plants compete for growing space for root and shoot systems so they can absorb and process sunlight, water, and nutrients. Animals compete for living, nesting, and feeding sites, as well as for food, water, and mates. Competition among members of the same species is called intraspecific competition, whereas competition between members of different species is called interspecific competition.

FIGURE 3.10 *Insect herbivores are predators as much as lions and tigers are. In fact, insects consume the vast majority of biomass in the world. Complex patterns of predation and defense have often evolved between insect predators and their plant prey.*

FIGURE 3.12 *Plants compete for light and growing space in this Indonesian rainforest. Epiphytes, such as the ferns and bromeliads shown here, find a place to grow in the forest canopy by perching on the limbs of large trees. This may be a commensal relationship if the epiphytes don't hurt their hosts. Sometimes, however, the weight of the epiphytes breaks off branches and even topples whole trees.*

If you look closely at a patch of weeds growing on good soil early in the summer, you likely will see several types of interspecific competition. First of all, many weedy species attempt to crowd out their rivals by producing prodigious numbers of seeds. After the seeds germinate, the plants race to grow the tallest, cover the most ground, and get the most sun. You may observe several strategies to do this. For example, vines don't build heavy stems of their own; they simply climb up over their neighbors to get to the light.

We often think of competition among animals as a bloody battle for resources, "nature red in tooth and claw." In fact, many animals tend to avoid fighting if possible. It's not worth getting injured. Most confrontations are more noise and show than actual fighting. Often competition is a matter of getting to food or habitat first or being able to use it more efficiently. As we discussed earlier, each species has a tolerance range for abiotic factors. Repeated studies have shown that when two species compete, the one closest to its optimum range of abiotic conditions will have an advantage and, more often than not, will prevail.

Intraspecific competition (within species) can be just as intense as interspecific competition (between species). Members of the same species have the same space and nutritional requirements; therefore, they compete directly for these environmental resources. How do organisms cope with intraspecific competition? One way is by dispersing: plants use wind, water, and passing animals to disperse seeds away from the parent plant. Young animals are forced to leave the parents' territory as soon as they are independent. Territoriality, vigorously defending territory, is a way of minimizing competition within a species. From grizzly bears to song birds, many species vigorously defend territories—and food and water sources—from others of their own kind. Another way to reduce intraspecific competition is resource partitioning. Often the young of a species use resources differently than mature individuals. A leaf-munching caterpillar uses different food sources than a nectar-sipping adult butterfly. Crabs start out as floating planktonic larvae, which do not compete with bottom-crawling adult forms. In these examples, the adults and juveniles of each species do not compete because they occupy different ecological niches.

Symbiosis

In contrast to predation and competition, symbiotic interactions between organisms can be nonantagonistic. **Symbiosis** is the intimate living together of members of two or more species. Symbiotic relationships often enhance the survival of one or both partners. **Commensalism** is a type of symbiosis in which one member clearly benefits and the other apparently is neither benefited nor harmed. Cattle, for example, are often accompanied by cattle egrets and cowbirds, both of which catch insects kicked up as the cattle graze through a field. The birds benefit, while the cattle seem indifferent. Many of the mosses, bromeliads, and other plants growing on trees in the moist tropics are also considered to be commensals (fig. 3.12). These epiphytes get water from rain and nutrients from leaf litter and dust fall, and often neither help nor hurt the trees on which they grow. In a way, the robins and sparrows that inhabit suburban yards are commensals with humans.

Lichens are a combination of a fungus and a photosynthetic partner, either an alga or a cyanobacteria. Their association is a type of symbiosis called **mutualism**, in which both members of the partnership benefit (fig. 3.13). Some ecologists believe that cooperative, mutualistic relationships may be more important in evolution than we have commonly thought. Aggressive interactions often are dangerous and destructive, while cooperation and compromise may have advantages that we tend to overlook. Survival of the fittest often may mean survival of those organisms that can live best with one another.

Parasitism, described earlier as a form of predation, also could be considered a type of symbiosis, where one species benefits and the other is harmed. All of these relationships have a bearing on such ecological issues as resource utilization, niche specialization, diversity, predation, and competition.

FIGURE 3.13 *Lichens, such as the various species growing on this log, are a combination of algae and fungi in a classic example of mutualistic symbiosis.*

FIGURE 3.14 *Poison arrow frogs of the family* Dendrobatidae *use brilliant colors to warn potential predators of the extremely toxic secretions from their skin. Native people in Latin America use the toxin on blowgun darts.*

Symbiotic relationships often entail some degree of coadaptation or coevolution of the partners, shaping—at least in part—their structural and behavioral characteristics. An interesting case of mutualistic coadaptation is seen in Central and South American swollen thorn acacias and their symbiotic ants. Acacia ant colonies live within the swollen thorns on the acacia tree branches and feed on two kinds of food provided by the trees: nectar produced in glands at the leaf bases and special protein-rich structures produced on leaflet tips. The acacias thus provide shelter and food for the ants. Although they spend energy to provide these services, the trees are not physically harmed by ant feeding.

What do the acacias get in return? Ants defend their colonies and territories aggressively, and they drive off herbivorous insects that attempt to feed on their home acacia, thus reducing predation on the tree. Ants also trim away vegetation that grows around the tree, thereby reducing competition for water and nutrients. This is a fascinating example of how a symbiotic relationship fits into community interactions. It is also an example of coevolution based on mutualism rather than competition or predation.

Defensive Mechanisms

Evolution has produced a bewildering array of ingenious defensive adaptations. Toxic chemicals, body armor, extraordinary speed, and the ability to hide are a few strategies organisms use to protect themselves from competitors or predators. Arthropods, amphibians, snakes, and some mammals, for instance, produce noxious odors or poisonous secretions to induce other species to leave them alone. Poison ivy and stinging nettles use chemicals to discourage you from disturbing them. Often, species possessing these chemical defenses evolve distinctive colors or patterns to warn potential enemies (fig. 3.14).

Sometimes, species that actually are harmless evolve colors, patterns, or body shapes that mimic unpalatable or poisonous species. This is called **Batesian mimicry** after the English naturalist H. W. Bates, who described the strategy in 1857.

Wasps, for example, often have bold patterns of black and yellow stripes to warn off potential predators (fig. 3.15a). The rarer longhorn beetle has no stinger, but looks and acts much like a wasp, tricking predators into avoiding it (fig. 3.15b). Another form of mimicry, called **Müllerian mimicry,** named for the German biologist Fritz Müller, who described it in 1878, involves two species, both of which are unpalatable or dangerous and have evolved to look alike. When predators learn to avoid either species, both benefit.

Species also evolve amazing abilities to avoid being discovered. You may have seen examples of insects that look exactly like dead leaves or twigs to hide from predators. Predators also use camouflage to hide as they lie in wait for their prey. Not all cases of mimicry are to avoid or carry out predation, however. Some tropical orchids have evolved flower structures that look exactly like female flies. Males attempting to mate unwittingly carry away pollen.

Keystone Species

A **keystone species** is a species or group of species whose impact on its community or ecosystem is much larger and more influential than would be expected from mere abundance. Originally, keystone species were thought to be top predators, such as wolves, whose presence limits the abundance of herbivores and thereby reduces their grazing or browsing on plants. Recently, scientists have recognized that less conspicuous species also play essential community roles. Certain tropical figs, for example, bear during seasons when no other fruit is available for frugivores (fruit-eating animals). If these figs were removed, many animals would starve to death during periods of fruit scarcity. With those animals gone, many other plant species that depend on them at

(a)

(b)

FIGURE 3.15 *An example of Batesian mimicry. (a) This dangerous wasp has bold yellow and black bands to warn away predators. (b) The much rarer longhorn beetle has no poisonous stinger, but looks and acts like a wasp and thus avoids predators as well.*

other times of the year for pollination and seed-dispersal would disappear as well.

Even microorganisms can play vital roles. In some forest ecosystems, mycorrhizae (fungi associated with tree roots) are essential for mineral mobilization and absorption. If the fungi die, so do the trees and many other species that depend on a healthy forest community. Rather than being a single species, mycorrhizae are actually a group of species that together fulfill a keystone function.

Often a number of species are intricately interconnected in biological communities so that it is difficult to tell which is the essential key. In the kelp "forests" off the California coast, the giant kelp (a kind of algae) provides shelter for many fish and shellfish species and so could be regarded as the key to community structure (fig. 3.16). However, kelp depends on sea otters, which eat the sea urchins that graze on the kelp. Are kelp or otters most important? Each depends on and affects the other. Perhaps we should think in terms of a "keystone set" of organisms in some ecosystems. Some ecological communities are functionally redundant in the sense that if one important species disappears, another will replace it, and essential ecological functions will continue without much change. Such a community might be said to have no keystone species. We'll discuss the role of keystone species in preserving biodiversity in chapter 5.

FIGURE 3.16 *Giant kelp is a massive alga that forms dense "forests" off the Pacific coast of California. It is a keystone species in that it provides food, shelter, and structure essential for a whole community. Removal of sea otters allows sea urchin populations to explode. When the urchins destroy the kelp, many other species suffer as well.*

POPULATION DYNAMICS

Many biological organisms can produce amazing numbers of offspring, given optimum environmental resources (fig. 3.17). Consider the example of the common fruitfly (*Drosophila melanogaster*). Under ideal conditions, 24 fly generations can be produced in a year. Typically, each female lays 50 to 100 eggs per generation. If one female fly were to lay 100 eggs and if all her offspring lived long enough to reproduce at the same rate with the same survival success, in a single year she would have about 6×10^{40} offspring. That's 60 billion trillion quadrillion insects! If this rate of reproduction continued for a decade, the whole earth would be covered several meters deep in fruitflies. Fortunately for us, fruitfly reproduction, like that of most organisms, is limited by a variety of environmental factors. This example demonstrates, however, the remarkable potential amplification of biological reproduction. Population dynamics describes the changes in number of organisms in a population.

Population Growth

We call the unrestricted increase in populations **exponential growth** because its rate can be expressed as a constant fraction, or exponent, by which the existing population is multiplied. The mathematical formula for exponential growth is:

$$\frac{dN}{dt} = rN$$

FIGURE 3.17 *High reproductive rates give many organisms the potential to expand populations explosively. The cockroaches in this kitchen could have been produced in only a few generations. A single female cockroach can produce up to 80 eggs every six months. This exhibit is in the Smithsonian Institute's National Museum of Natural History.*

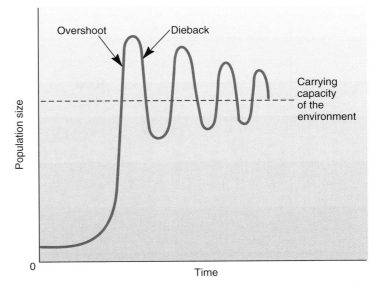

FIGURE 3.18 *Population oscillations. Some species demonstrate a pattern of cyclic overshoot and dieback.*

That is, the change in numbers of individuals (*dN*) per change in time (*dt*) equals the rate of growth (*r*) times the number of individuals in the population (*N*). The *r* term is a fraction representing the average individual contribution to population growth, perhaps 1.2 if each individual produces 1.2 times its own number in each time step. If *r* is less than 1, then the population is shrinking. If *r* is exactly 1, then there is no change, and *dN/dt* = 0.

This equation for population growth is also referred to as **biotic potential**, the potential of a population to grow if nothing were limiting its expansion. Note that the equation is a very simple *model* (an idealized, simple description of the real world). In reality, many factors prevent most populations from growing at their biotic potential. (The same equation is used to calculate growth in your bank account due to interest rates; unfortunately, many factors also may keep your savings from growing at their potential rate.)

Boom and Bust Population Cycles

A graph of exponential population growth is described as a **J curve** (left portion of fig. 3.18) because of its shape. As you can see in this graph, the number of individuals added to a population at the beginning of an exponential growth curve can be rather small. But within a very short time, the numbers begin to increase quickly because a fixed percentage becomes a much larger amount as the population increases.

In the real world, however, there are limits to growth. We call the maximum number of individuals of any species that can be supported by a particular ecosystem on a sustainable basis the **car-**

rying capacity. When a population **overshoots**, or surpasses the carrying capacity of its environment, death rates will begin to surpass birth rates. The growth curve becomes negative rather than positive, and the population may decrease as fast as, or faster, than it grew. We call this dieback a population crash. Populations may go through repeated oscillating cycles of population growth and decline as shown in fig. 3.18. These cycles may be very regular if they depend on a few simple factors, such as the seasonal light- and temperature-dependent bloom of algae in a lake. They also may be very irregular if they depend on complex environmental and biotic relationships that control cycles, such as the outbreaks of migratory locusts in the desert or tent caterpillars in northern forests. We call long periods of low population size followed by a sudden population explosion irruptive growth.

Sometimes, predator and prey populations oscillate in a sort of synchrony with each other as is shown in fig. 3.19. This is a classic study of the number of furs brought into Hudson Bay Company trading posts in Canada between 1840 and 1930. As you can see, the numbers of Canada lynx fluctuate on about a ten-year cycle that is similar to, but slightly out of phase with, the population peaks of snowshoe hares. When the hare population is high and food is plentiful, lynx reproduction is very successful, and lynx populations grow rapidly.

Eventually, declining food supplies limit hare populations. For awhile lynx populations continue to grow because starving hares are easier to catch than healthy ones. As hares become more scarce, however, so do the lynx. When hares are at their lowest levels, food supplies recover and population growth of both prey and predator begins again. This predator-prey oscillation is known as the Lotka-Volterra model, after the scientists who first described it mathematically.

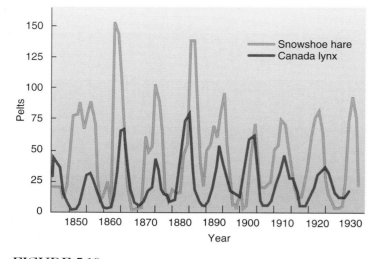

FIGURE 3.19 *Oscillations in the population of snowshoe hare and lynx in Canada suggest the close interdependency of this prey-predator relationship. These data are based on the number of pelts received by the Hudson Bay Company. Both predator and prey show a ten-year cycle in population growth and decline.*

Source: Data from D. A. MacLulich. *Fluctuations in the Numbers of the Varying Hare (Lepus americus).* Toronto: University of Toronto Press, 1937, reprinted 1974.

Growth to a Stable Population

Not all biological populations go through these cycles of exponential overshoot and catastrophic diebacks. Many species are regulated by both internal and external factors so that they come into equilibrium with their environmental resources and maintain relatively stable population sizes. These species may grow exponentially when resources are unlimited, but their growth slows as they approach the carrying capacity of the environment. This pattern is called **logistic growth** because of its constantly changing rate.

Mathematically, this growth pattern is described by the following equation, which adds a term for carrying capacity (K) to the biotic potential growth equation:

$$\frac{dN}{dt} = rN\left(1 - \frac{N}{K}\right)$$

This equation says that the change in numbers over time (dN/dt) equals the exponential growth rate (r times N) *times the portion of the carrying capacity (K) represented by the population size (N).* The term (1 – N/K) represents the relationship between N at any given time step and K, the number of individuals the environment can support. If N is less than K, say 100 compared to 120, then (1 – N/K) is a positive number (1 – 100/120 = 0.17), and population growth, dN/dt, is slow but positive. If N is greater than K, that is, if the population is greater than the environment can support, then (1 – N/K) is a negative number. For example, if N is 150 and K is 120, then (1 – N/K) is (1 – 150/120), which is equal to (1 – 1.25), or –0.25. In this case, the growth rate is negative. The logistic growth model, then, describes a population that decreases if its numbers exceed carrying capacity.

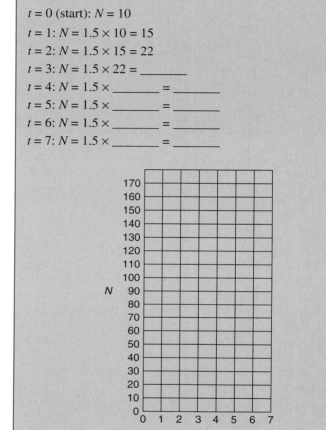

APPLICATION: Calculate Population Growth

Think of exponential growth as occurring in time steps. Start with ten cockroaches that can produce enough young to increase at a rate of 150 percent per month (r = 1.5). Calculate the number of roaches after seven time steps: for each t, multiply r times the N of the previous time step (round N to the nearest whole number). Then graph the results.

t = 0 (start): N = 10

t = 1: N = 1.5 × 10 = 15

t = 2: N = 1.5 × 15 = 22

t = 3: N = 1.5 × 22 = _____

t = 4: N = 1.5 × _____ = _____

t = 5: N = 1.5 × _____ = _____

t = 6: N = 1.5 × _____ = _____

t = 7: N = 1.5 × _____ = _____

Answer: The final graph should be a J-shaped curve.

How does the growth curve of a stable population differ from the J curve of an exploding population? Figure 3.20 shows an idealized comparison between exponential and logistic growth. The J curve on the left in this figure represents the growth without restraint toward the biotic potential, or the maximum number a species might possibly attain. The curve to the right represents logistic growth. We call this later pattern an **S curve** or sigmoidal curve (for the Greek letter sigma).

Limiting Factors

In many species, population growth is regulated both by internal and external factors. Maturity, body size, and hormonal status are

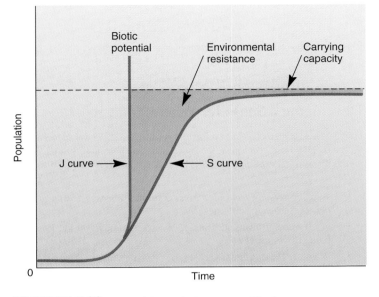

FIGURE 3.20 *J and S population curves. The J curve represents theoretical unlimited growth. The S curve represents population growth and stabilization in response to environmental resistance.*

examples of internal factors. Habitat and food availability, or interactions with other organisms, are examples of external factors. Some of these limits are dependent on population density. Food and water, for example, become more limited as populations grow. Disease, stress, and exposure to predators or parasites can all increase mortality rates as populations increase. These factors are called *density-dependent.*

Other limits to growth are *density-independent.* Many of these factors are abiotic: drought or early frost can reduce populations of mosquitoes drastically, regardless of how many mosquitoes there were to start with. Habitat destruction—because of floods, landslides, or human activities—can also limit population growth.

Together, factors that tend to reduce population growth rates are called **environmental resistance**. The area between the two curves in fig. 3.20 is the cumulative effect of environmental resistance. Note that the resistance becomes larger and the rate of logistic growth becomes smaller as the population approaches the carrying capacity of the environment.

K-adapted and r-adapted Species

Some organisms, such as dandelions, persist by depending on a high rate of reproduction and growth (*rN*). These organisms are described as **r-adapted.** They tend to have rapid reproduction and high mortality of offspring, and they may frequently overshoot carrying capacity and die back. Other organisms tend to reproduce more slowly as they approach the carrying capacity (*K*) of their environment. These species are referred to as **K-adapted.**

Many species don't fit neatly into either exponential (r-adapted) or logistic (K-adapted) growth patterns. Still, it's useful to contrast the advantages and disadvantages of some organ-

isms at the extremes of a continuum of growth patterns. Although we should avoid implying intention in natural systems, it sometimes helps us see these differences in terms of "strategies" of adaptation and "logic" in different modes of reproduction.

Organisms with *r*-adapted or exponential growth patterns often tend to occupy low trophic levels in their ecosystems (see chapter 2) or to be successional pioneers. As generalists or opportunists, they move quickly into disturbed environments, grow rapidly, mature early, and produce many offspring. They usually do little to care for their offspring or protect them from predation. They depend on sheer numbers and dispersal mechanisms to ensure that some offspring survive to adulthood. They have little investment in individual offspring, using their energy to produce vast numbers instead (table 3.2).

A female clam, for example, can release up to one million eggs over her lifetime. As the eggs drift away on water currents, they are entirely on their own. The mother clam can neither protect them from predators nor help them find food or a place to live. The vast majority of all young clams die before reaching maturity, but if even a few survive, the species will continue. Many marine invertebrates, parasites, insects, rodents, and annual plants follow this reproductive strategy. Predators or other external factors generally limit their numbers. Also included in this group are the weeds, pests, or other species we consider nuisances that reproduce profusely, adapt quickly to environmental change, and survive under a broad range of conditions.

So-called *K*-adapted organisms are usually larger, live longer, mature more slowly, produce fewer offspring in each generation, and have fewer natural predators than the species below them in the ecological hierarchy. Elephants, for example, are not reproductively mature until they are 18 to 20 years old. During

TABLE 3.2	**Reproductive Strategies**
R ADAPTED SPECIES	*K* ADAPTED SPECIES
1. Short life	1. Long life
2. Rapid growth	2. Slower growth
3. Early maturity	3. Late maturity
4. Many, small offspring	4. Few, large offspring
5. Little parental care and protection	5. High parental care or protection
6. Little investment in individual offspring	6. High investment in individual offspring
7. Adapted to unstable environment	7. Adapted to stable environment
8. Pioneers, colonizers	8. Later stages of succession
9. Niche generalists	9. Niche specialists
10. Prey	10. Predators
11. Regulated mainly by intrinsic factors	11. Regulated mainly by extrinsic factors
12. Low trophic level	12. High trophic level

youth and adolescence, a young elephant is part of a complex extended family that cares for it, protects it, and teaches it how to behave. A female elephant normally conceives only once every four or five years after she matures. The gestation period is about 18 months; thus, an elephant herd doesn't produce many babies in a given year. Since they have few enemies and live a long life (often 60 or 70 years), however, this low reproductive rate usually produces enough elephants to keep the population stable, given appropriate environmental conditions.

An important underlying question to much of the discussion in this book is which of these strategies do humans follow. Do we more closely resemble wolves and elephants in our population growth, or does our population growth pattern more closely resemble that of moose and rabbits? Will we overshoot our environment's carrying capacity (or are we already doing so), or will our population growth come into balance with our resources?

COMMUNITY PROPERTIES

The processes and principles that we have studied thus far in this chapter—tolerance limits, species interactions, resource partitioning, evolution, and adaptation—play important roles in determining the characteristics of populations and species. In this section, we will look at some fundamental properties of biological communities and ecosystems—productivity, diversity, complexity, resilience, stability, and structure—to learn how they are affected by these factors.

Productivity

A community's **primary productivity** is the rate of biomass production, or the conversion of solar energy into chemical energy stored in living (or once-living) organisms. Since much energy is

what can you do?

Developing a Sense for Where You Live

One of the first steps toward conserving biological diversity is to educate yourself. The more you know, the more you can share your knowledge—and skills—to help the natural world. Look for answers to questions like these:

- What ecosystems and biological communities existed in your area before European settlement?
- What impact, if any, did indigenous people have on the flora and fauna of your area?
- What are the dominant species (besides humans) in your neighborhood? Where did they originate?
- How much rain falls in your region each year? Is precipitation seasonal? Is water a limiting factor for biological communities?
- What are the seasonal high and low temperatures where you live? How do native plants and animals adapt to seasonal variations?
- Is there a keystone species or group of species especially important in determining the structure and functions of your local ecosystems? What factors might threaten those keystone components?
- Where do your drinking water, food, and energy come from? What local and regional environmental impacts are caused by production, use, and disposal of those resources? Could you lessen those impacts by changing your sources or use patterns of resources?
- Is there a park or wildlife refuge near where you live? Does it contain any rare, threatened, or endangered species? What makes them rare, threatened, or endangered?
- Are there opportunities for volunteer work to improve your local environment, such as planting native species, cleaning up a river or lake, restoring a wetland, recycling trash, or helping to maintain a refuge or park?

used in respiration, a more useful term is often *net* primary productivity, or the amount of biomass stored after respiration. Productivity depends on light levels, temperature, moisture, and nutrient availability. Figure 3.21 shows approximate productivity levels for some major ecosystems. As you can see, tropical forests, coral reefs, and estuaries (bays or inundated river valleys where rivers meet the ocean) have high levels of productivity because they have abundant supplies of all the required resources. In deserts, lack of water limits photosynthesis. On the arctic tundra or in high mountains, low temperatures inhibit plant growth. In the open ocean, a lack of nutrients reduces the ability of algae to make use of plentiful sunshine and water.

Even the most photosynthetically active ecosystems capture only a small percentage of the available sunlight and use it to make energy-rich compounds. In a temperate-climate oak forest, leaves absorb only about half the available light on a midsummer day. Of this absorbed energy, 99 percent is used to evaporate water

APPLICATION:	*K*-Adapted and *r*-Adapted Species

Do *K*-adapted or *r*-adapted characteristics best describe the following species (or are some attributes of each appropriate)?

ant	_____
housefly	_____
bald eagle	_____
bison	_____
wildebeest	_____
antelope	_____
cheetah	_____
shark	_____

Answer: The ant, housefly, bison, wildebeest, and antelope are generally considered *r*-adapted, while the bald eagle, cheetah, and shark are best described as *K*-adapted.

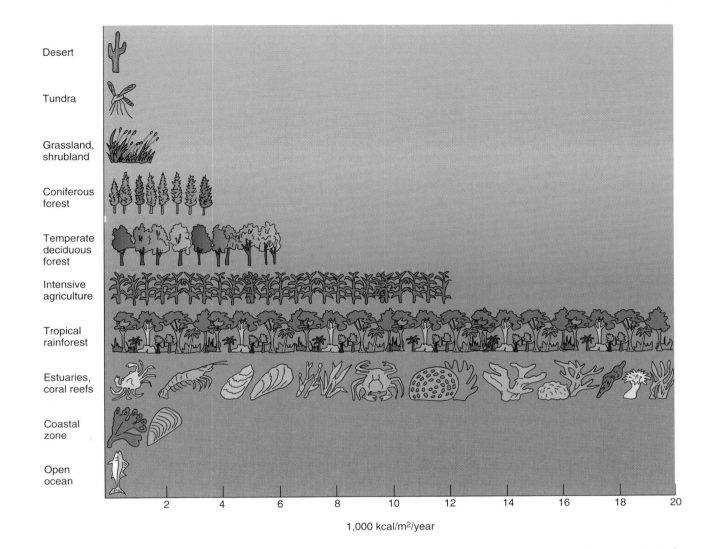

Desert

Tundra

Grassland,
shrubland

Coniferous
forest

Temperate
deciduous
forest

Intensive
agriculture

Tropical
rainforest

Estuaries,
coral reefs

Coastal
zone

Open
ocean

2   4   6   8   10   12   14   16   18   20

1,000 kcal/m2/year

FIGURE 3.21   *Relative biomass accumulation of major world ecosystems. Only plants and some bacteria capture solar energy. Animals consume biomass to build their own bodies.*

in respiration and cooling. A large oak tree can transpire (evaporate) several thousand liters of water on a warm, dry, sunny day, while making only a few kilograms of sugars and other energy-rich organic compounds.

Abundance and Diversity

Abundance is an expression of the number of individuals of a species in an area. **Diversity** is the number of different species in an area. Diversity is also a useful measure of the variety of ecological niches or genetic variation in a community. Communities with high diversity often have low abundance of most species. As a general rule, diversity decreases as we go from the equator toward the poles, but abundance of a smaller number of species increases. The Arctic has a vast abundance of mosquitoes, for example, but relatively few other insect species. The tropics, on the other hand, have vast numbers of insect species—some of

which have incredibly bizarre forms and habits—but often only a few individuals of any particular species in a given area.

Bird populations also vary dramatically with latitude. Greenland is home to 56 species of breeding birds, while Colombia, which is only one-fifth the size of Greenland, has 1,395. Why so many species in Colombia and so few in Greenland?

Climate and history are important factors. Greenland has such a harsh climate that the need to survive through the winter or escape to milder climates becomes the single most important critical factor that overwhelms all other considerations and severely limits the ability of species to specialize or differentiate into new forms. Furthermore, because glaciers covered Greenland until about 10,000 years ago, new species have had little time to develop.

Many areas in the tropics, by contrast, have relatively abundant rainfall and warm temperatures year-round so that ecosystems there are highly productive. The year-round availability of

CASE STUDY

I s a more diverse community more stable and resilient in the face of environmental stress? This has been an ongoing debate among ecologists. One of the important contributors to the debate has been experimental ecologist David Tilman. Since 1982, Tilman and his colleagues have been studying the effects of nitrogen fertilization on grassland plots. Initially, Tilman's goal was to study the effect of fertilizer on regenerating grass on abandoned farm fields and natural oak savannas in central Minnesota. The experimental design was to randomly assign different fertilizer treatments to a set of sample plots on old fields and natural savannas. Study plots varied considerably in species diversity: some contained only one species, some had as many as 26. Using the same set of sample plots for nearly two decades, Tilman and his colleagues—and platoons of student assistants—have carefully gathered, counted, and weighed all the plants growing on the plots. The result has been a long record of highly detailed data.

In a fortuitous turn of events, an extreme case of environmental stress occurred several years into the experiment. The summer of 1988 was the hottest, driest summer in 50 years. When researchers tallied up the total plant productivity on the study plots that year, they found that the plots with the most species suffered much less than those with few species. On species-rich plots, drought-tolerant plants still grew, while more sensitive species languished. By contrast, the most species-poor plots had only about one-eighth their pre-drought productivity.

In subsequent years, the species-poor plots also took longer to recover from the drought. Ability to recover from stress—or resilience—is an important factor in ecosystem stability.

The idea that biodiversity increases ecosystem stability is one that ecologists have long believed was true, but that they have had difficulty proving. Many field researchers have concluded that complex communities are more stable, but their evidence has been ambiguous because natural environments have so many simultaneously changing variables. Other field ecologists have found simple systems to be very stable and resilient, depending on the types of organisms present. Mathematical models have suggested that simple communities of a few generalist species can be more stable than more complex assemblages of specialists. Experiments like Tilman's, with long-term data gathering on a set of controlled sample plots, provide an important type of evidence to the debate. Experimental data are relatively controlled—reducing the number of simultaneously changing variables of a natural system, but they are more realistic (and often more convincing) than computer models.

Ecologists continue to dispute whether it is the variety of organisms in a community, or the particular type of organisms (or both) that controls ecosystem stability and resilience, but experiments like Tilman's provide important contributions to the discussion. This work also shows the value of careful, long-term record keeping in science. This work was carried out on one of 18 Long-Term Ecological Research sites, funded by the National Science Foundation precisely to allow long-term monitoring of ecological systems. Tilman's work also shows the importance of attention to unexpected implications of your research. As Louis Pasteur said, "Chance favors the prepared mind." In this case, the 1988 drought provided an unexpected opportunity to explore critical questions about stability and resilience.

---

food, moisture, and warmth supports a great exuberance of life and allows a high degree of specialization in physical shape and behavior. Coral reefs are similarly stable, productive, and conducive to proliferation of diverse and exotic life-forms. The enormous abundance of brightly colored and fantastically shaped fish, corals, sponges, and arthropods in the reef community is one of the best examples we have of community diversity.

Productivity is related to abundance and diversity (both of which depend on total resource availability in an ecosystem), as well as the reliability of resources, the adaptations of the member species, and the interactions between species. You shouldn't assume that all communities are perfectly suited to their environment. A relatively new community that hasn't had time for niche specialization, or a disturbed one where roles such as top predators are missing, may not achieve maximum efficiency of resource use or reach its maximum level of either abundance or diversity.

We will discuss the importance of biodiversity and abundance further in chapter 5.

Complexity, Resilience, and Stability

Community complexity involves diversity and community functions. **Complexity** in ecological terms refers to the number of species at each trophic level and the number of trophic levels in a community. A diverse community may not be very complex if all of its species are clustered in only a few trophic levels and form a relatively simple food chain.

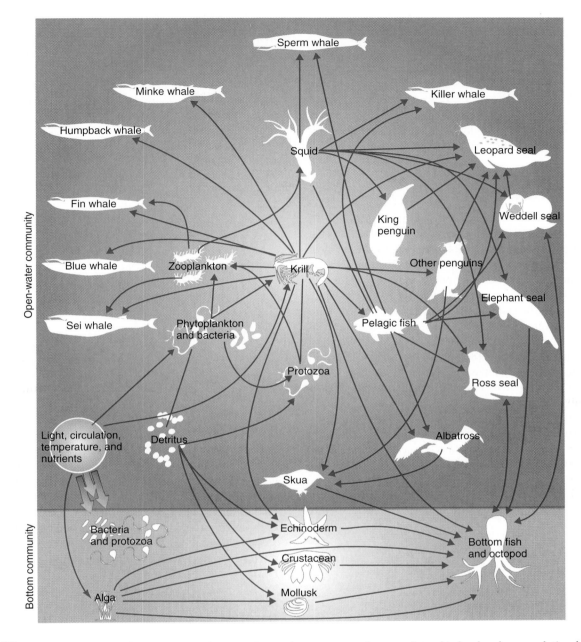

FIGURE 3.22 *A complex and highly interconnected community can have many species at each trophic level and many relationships, as this Antarctic marine food web illustrates.*

A complex, highly interconnected community might have many trophic levels, some of which can be compartmentalized into subdivisions (fig. 3.22). In tropical rainforests, for instance, herbivores can be grouped into "guilds," based on the specialized ways they feed on plants. There may be fruit-eaters, leaf-nibblers, root-borers, seed-gnawers, and sap-suckers—each composed of species of very different size, shape, and even biological kingdom, but that feed in related ways. A highly interconnected community such as this can form a very elaborate food web.

How is complexity related to stability in an ecosystem and to resilience (the ability to recover from disturbance)? Ecologists have debated this question for many years. We can identify three kinds of stability or resiliency in ecosystems: (1) constancy (lack of fluctuations in composition or functions), (2) inertia (resistance to perturbations), and (3) renewal (ability to repair damage after disturbance).

In 1955, Robert MacArthur, who was then a graduate student at Yale, proposed that the more complex and interconnected a community is, the more stable and resilient it will be in the face of disturbance. If many different species occupy each trophic level, some can fill in if others are stressed or eliminated by external forces, making the whole community resistant to perturbations and able to recover relatively easily from disruptions. On the other hand, in a diverse and highly specialized ecosystem, removal of a few keystone members can eliminate many other associated species. Eliminating a major tree species from a tropical forest, for example, may destroy pollinators and fruit distributors as well. We might replant the trees, but could we replace the whole web of relationships on which they depend? In this case, diversity makes the forest less resilient, rather than more. This relationship between diversity and stability remains controversial (see Case Study, p. 66).

Community Structure

Ecological structure refers to patterns of spatial distribution of individuals and populations within a community, as well as the relation of a particular community to its surroundings. At the local level, even in a relatively homogeneous environment, individuals in a single population can be distributed randomly, clumped together, or in highly regular patterns (fig. 3.23). In randomly arranged populations, individuals live wherever resources are available. Ordered patterns may be determined by the physical environment but are more often the result of biological competition. For example, competition for nesting space in a penguin colony is often fierce. Each nest tends to be just out of reach of the neighbors sitting on their own nests. Constant squabbling produces a highly regular pattern. Similarly, sagebrush releases toxins from roots and fallen leaves that inhibit the growth of competitors and create a circle of bare ground around each bush. As neighbors fill in empty spaces up to the limit of this chemical barrier, a regular spacing results.

Some other species cluster together for protection, mutual assistance, reproduction, or to gain access to a particular environmental resource (fig. 3.23c). Dense schools of fish, for instance, cluster closely together in the ocean, increasing their chances of detecting and escaping predators (fig. 3.24). Similarly, predators, whether sharks, wolves, or humans, often hunt in packs to catch their prey. A flock of blackbirds descending on a cornfield or a troop of baboons traveling across the African savanna band together both to avoid predators and to find food more efficiently.

Plants can cluster for protection, as well. A grove of wind-sheared evergreen trees is often found packed tightly together at the crest of a high mountain or along the seashore. They offer mutual protection from the wind not only to each other but also to other creatures that find shelter in or under their branches.

Most environments are patchy at some scale. Organisms cluster or disperse according to patchy availability of water, nutrients, or other resources. Distribution in a community can be vertical as well as horizontal. The tropical forest, for instance, has many layers, each with different environmental conditions and combinations of species. Distinct communities of smaller plants, animals, and microbes live at different levels. Similarly, aquatic communities are often stratified into layers based on light penetration in the water, temperature, salinity, pressure, or other factors.

Edges and Boundaries

An important aspect of community structure is the boundary between one habitat and its neighbors. We call these relationships edge effects. Sometimes, the edge of a patch of habitat is relatively sharp and distinct. In moving from a woodland patch into a grassland or cultivated field, you sense a dramatic change from the cool, dark, quiet forest interior to the windy, sunny, warmer, open space of the field or pasture. In other cases, one habitat type may intergrade very gradually into another so that there is no distinct border.

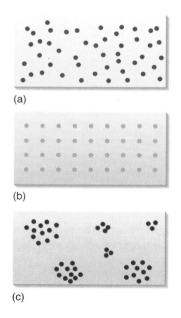

(a)

(b)

(c)

FIGURE 3.23 *Distribution of members of a population in a given space can be* (a) *random,* (b) *ordered, or* (c) *clustered. The physical environment and biological interactions determine these patterns. The patterns may produce a graininess or patchiness in community structure.*

FIGURE 3.24 *Fish and birds often flock in dense bands for protection or mutual feeding.*

Ecologists call the boundaries between adjacent communities **ecotones** (fig. 3.25). Ecotones are often rich in species diversity because individuals from both environments occupy the boundary area. In addition many species actively occupy an ecotone, taking advantage of resources in both environments. Whitetail deer, for example, browse in open fields but hide in forest cover.

Other species avoid edges and ecotones and prefer interior environments. Spotted owls of the Pacific Northwest, for example, nest only in forest "core," relatively deep, cool rainforest. **Edge effects**, the environmental and biotic conditions on the edges, may extend hundreds of meters into a forest fragment. In a rainforest, edges are sunnier, drier, hotter, and more susceptible to storm damage than the interior. Edges also have a different set of predators and competitors than in the forest core. These conditions discourage spotted owls from nesting near edges.

Depending on how far edge effects extend from the boundary, differently shaped habitat patches may have very dissimilar amounts of interior area (fig. 3.26). If edge effects extended 200 m into a forest, a 40-acre square block (400 m^2) surrounded by clearcut would have no true core habitat at all.

Many popular game animals, such as white-tailed deer and pheasants that are adapted to human disturbance, often are most plentiful in boundary zones between different types of habitat. Game managers once were urged to develop as much edge as possible to promote large game populations. Today, however, most wildlife conservationists recognize that the edge effects associated with habitat fragmentation are generally detrimental to biodiversity. Preserving large habitat blocks and linking smaller blocks with migration corridors may be a good way to protect some rare and endangered species. (See chapter 6; see also related story, "Where have all the songbirds gone?" at www.mhhe.com/apps.)

COMMUNITIES IN TRANSITION

So far, our view of communities has focused on the day-to-day interactions of organisms with their environments, set in a context of survival and selection. In this section, we'll step back to look at some transitional aspects of communities, including where communities meet and how communities change over time.

Ecological Succession

Biological communities have a history in a given landscape. The process by which organisms occupy a site and gradually change environmental conditions by creating soil, shade, shelter, or increasing humidity is called ecological succession or development. **Primary succession** occurs when a community begins to develop on a site previously unoccupied by living organisms, such as an island, a sand or silt bed, a body of water, or a new volcanic flow (fig. 3.27). **Secondary succession** occurs when an existing community is disrupted and a new one subsequently develops at the site (fig. 3.28). The disruption may be caused by some natural catastrophe, such as fire or flooding, or by a human activity, such as deforestation, plowing, or mining. In each case, organisms modify the environment in ways that allow one species to replace another.

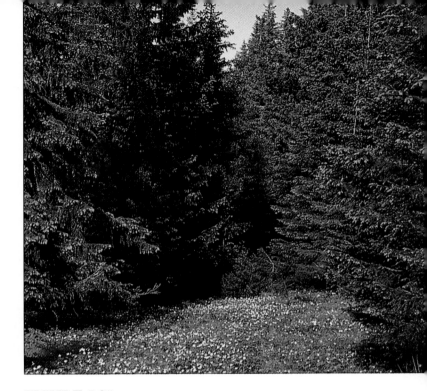

FIGURE 3.25 *Ecologists call the sharp edge, or boundary, between the woods and sunny glade in this picture an ecotone. Edge effects, such as increased sunshine, wind, and temperature, as well as decreased humidity, may extend a considerable distance into the forest. When fragmented into many small patches, the forest may become all edge, with no true interior core at all.*

Total area: 39 ha   Core area: 0 ha

Total area: 47 ha   Core area: 20 ha

FIGURE 3.26 *Shape can be as important as size in small preserves. While these areas are similar in size, no place in the top figure is far enough from the edge to have characteristics of core habitat, while the bottom patch has a significant core.*

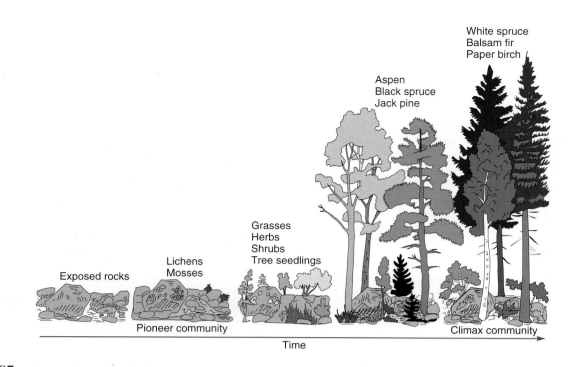

White spruce
Balsam fir
Paper birch

Aspen
Black spruce
Jack pine

Grasses
Herbs
Shrubs
Tree seedlings

Lichens
Mosses

Exposed rocks

Pioneer community

Climax community

Time

FIGURE 3.27   *A generalized example of primary succession on a terrestrial site is shown in five stages* (left to right), *beginning with rocks that are initially colonized by a pioneer community of lichens and mosses and ending with a climax forest community.*

FIGURE 3.28   *Clear-cutting and burning have turned this cool, shady, black spruce bog into a dry, sunny, barren opening in which few of the former residents can survive. Secondary succession may eventually restore conditions, if the climate doesn't change and further disturbance is prevented.*

In primary succession on a terrestrial site, the new site first is colonized by a few hardy **pioneer species**, often microbes, mosses, and lichens that can withstand harsh conditions and lack of resources. Their bodies create patches of organic matter. Debris accumulates in pockets and crevices, providing soil in which seeds can become lodged and grow. We call this process of environmental modification by organisms **ecological development** or facilita-

tion. The community of organisms often becomes more diverse and increasingly competitive as development continues and new niche opportunities appear. The pioneer species gradually disappear as the environment changes and new species combinations replace the preceding community.

Examples of secondary succession are easy to find. Observe an abandoned farm field or clear-cut forest (fig. 3.28) in a temperate climate. The bare soil first is colonized by rapidly growing annual plants (those that grow, flower, and die the same year) that have light, wind-blown seeds and can tolerate full sunlight and exposed soil. They are followed and replaced by perennial plants (those that live for several to many years), including grasses, flowering plants, shrubs, and trees. As in primary succession, plant species progressively change the environmental conditions. Biomass accumulates and the site becomes richer, better able to capture and store moisture, more sheltered from wind and climate change, and biologically more complex. Species that cannot survive in a bare, dry, sunny, open area find shelter and food as the field turns to prairie or forest.

Climax Communities or Individualistic Succession?

Eventually, in either primary or secondary succession, many communities reach a state that resists further change. Fifty years ago, ecologists called this a mature or climax community because it appeared to be the culmination of the successional process. An analogy was made between community succession and organism maturation. Beginning with a primitive or juvenile state and going

Principles of Environmental Science

www.mhhe.com/environmentalscience

through a complex developmental process, both individual organisms and communities progress until complex, stable, and mature forms are reached. It's dangerous to carry this analogy too far, however, because no mechanism is known to regulate communities in the same way that genetics and physiology regulate development of the body.

The concept of succession to a climax community was first championed by the pioneer biogeographer F. E. Clements. He viewed this process as being like a parade or relay, in which species replace each other in predictable groups and in a fixed, regular order. Clements argued that every landscape has a characteristic climax community, determined mainly by climate conditions. If left undisturbed, this community would mature to a characteristic set of organisms, each performing its optimal functions. The climax community would represent the maximum possible state of complexity and stability.

This community-unit theory was opposed by Clements' contemporary, H. A. Gleason, who saw community history as a much more unpredictable process. He argued that species are individualistic, each getting established according to its ability to colonize and reproduce in an area. Myriad temporary associations of plants and animals might form, according to the conditions prevailing at a particular time and the species available to colonize a given area. You might think of the Gleasonian model as a time-lapse movie of a busy railroad station. Passengers come and go; groups form and then dissipate. Patterns and assemblages that seem significant to us may not mean much in the long run. Gleason suggested that we see ecosystems as stable and uniform only because our lifetimes are so short and our view so limited.

The process of succession may not be as deterministic as we once thought, yet steady-state or highly developed ecological communities do often tend to be resilient and stable over long periods. Complex, mature communities, characterized by high species diversity, narrow niche specialization, well-organized community structure, good nutrient conservation and recycling, can be highly resistant to disturbance. Community functions, such as productivity and nutrient cycling, tend to be self-stabilizing or self-perpetuating. What we once regarded as "final" climax communities, however, may still be changing. It's probably more accurate to say that the rate of succession is so slow in some mature communities that, from the perspective of a single human lifetime, it appears to be unchanging.

Some landscapes never reach a stable climax in the traditional sense because they are characterized by, and adapted to, periodic disruption. Grasslands, the chaparral shrubland of California, and some kinds of coniferous forests, for instance, are shaped and maintained by periodic fires that have long been a part of their history (fig. 3.29). Plants in these communities are adapted to resist fires, reseed quickly after fires, or both. In fact, many of the plant species we recognize as dominants in these communities require fire to eliminate competition, to prepare seedbeds for germination of seedlings, or to open cones or thick seed coats. Without fire, community structure may be quite different.

The differences in these points of view influence how we manage ecosystems. Disturbance in ecosystems was long seen as

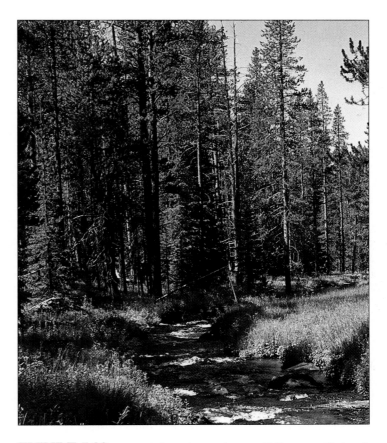

FIGURE 3.29 *This lodgepole pine forest in Yellowstone National Park was once thought to be a climax forest, but we now know that this forest must be constantly renewed by periodic fire.*

disrupting climax conditions, and thus, inherently bad. Accordingly, fire suppression and flood control have been central policies in the United States and elsewhere for most of the twentieth century. In recent years, more conservationists have come to see disturbance as a natural part of ecosystems. Grasslands and some forests are now described as "fire-adapted," for example, and many fires are allowed to burn naturally. Similarly, floods are increasingly considered healthy for floodplains and river corridors, and policymakers are more cautious about funding dams and levees than they once were. (See further discussion in chapter 6 and related stories "Prescribed Fires in New Mexico" and "Fires in the Boundary Waters Canoe Area Wilderness" at www.mhhe.com/apps.)

Introduced Species and Community Change

Succession requires the continual introduction of new community members and the disappearance of previously existing species. Sometimes communities can be completely altered, however, by the introduction of exotic species, organisms originating in completely different locations. Usually we think of exotic species as introduced by humans from different continents. Because these organisms arrive without their natural predators, they can invade aggressively, drastically altering local environments. European rats in Polynesia, for example, have decimated breeding birds on many islands, and kudzu, a Japanese vine, has run rampant in much of the southern United States. Many ecologists consider exotic species invasions the most pressing hazard for biological communities in the coming century (see chapter 5).

Successful exotics tend to be prolific, opportunistic species. Sailing ships visiting oceanic islands, for example, brought rats, goats, cats, and pigs. All these animals are prolific, quickly developing large populations. Goats are efficient, nonspecific herbivores; they eat nearly any vegetation, from grasses and herbs to seedlings and shrubs. In addition, their sharp hooves are hard on plants rooted in thin island soils. Rats and pigs are opportunistic omnivores, eating the eggs and nestlings of seabirds that tend to nest in large, densely packed colonies, and digging up sea turtle eggs. Cats prey upon nestlings of both ground- and tree-nesting birds. Native species, those that originate in an area or have occupied the area for a very long time, are particularly vulnerable to exotics because they have not evolved under circumstances that required them to have defensive adaptations to these predators.

Sometimes we introduce new species in an attempt to solve problems created by previous introductions but end up making the situation worse. In Hawaii and on several Caribbean islands, for instance, mongooses were imported to help control rats that had escaped from ships and were destroying indigenous birds and devastating pineapple plantations (fig. 3.30). Since the mongooses were diurnal (active in the day), however, and rats are nocturnal, they tended to ignore each other. Instead, the mongooses killed native birds and further threatened endangered species. Our lessons from this and similar introductions have a

FIGURE 3.30 *Mongooses were released in Hawaii in an effort to control rats. The mongooses are active during the day, however, while the rats are night creatures, and so they ignored each other. Instead, the mongooses attacked defenseless native birds and became as great a problem as the rats.*

new technological twist. Some of the ethical questions currently surrounding the release of genetically engineered organisms are based on concerns that they are novel organisms, and we might not be able to predict how they will interact with other species in natural ecosystems—let alone how they might respond to natural selective forces. It is argued that we can't predict either their behavior or their evolution.

SUMMARY

A vast majority of scientists regard the theory of evolution via natural selection to be the cornerstone of biology. Random, spontaneous mutations (changes in genetic material) create diversity in populations. Advantageous traits allow some individuals to survive and reproduce more successfully than less fit members of the species. Eventually, the traits conveying fitness will come to dominate the population. Evolution allows organisms to adapt to a particular set of environmental conditions. Both geographic and behavior isolation can lead to speciation: the creation of new species.

The principle of limiting factors states that for every physical factor in the environment, there are both maximum and minimum tolerable limits beyond which a given species cannot survive. Often the factor in shortest supply or closest to the tolerance limit for a particular species at a particular time is the critical factor that will determine the abundance and distribution of that species in that ecosystem.

Habitat describes the place in which an organism lives; niche describes either the role an organism plays or the total set of conditions that control its distribution. Natural selection often leads to niche specialization and resource partitioning that reduce competition between species. Organisms interact within communities in many ways. Predation, competition, and symbiosis are principal types of interaction among species. These interactions influence adaptation in species.

Theoretically, populations can grow exponentially or logistically. In either case, increase depends on the growth rate times the size of the population. In logistic growth, populations decline if they exceed their environmental carrying capacity. A variety of intrinsic and extrinsic factors limit population growth. Some of these are density dependent (including competition for resources), while others are density-independent (such as birth rates or climate hazards).

Some fundamental properties of biological communities are productivity, diversity, complexity, resilience, stability, and structure. Productivity is a measure of the rate at which photosynthesis produces biomass made of energy-rich compounds. Tropical rainforests are generally the most productive of all terrestrial communities; coral reefs and estuaries are generally the most productive aquatic communities. Diversity is a measure of the number of different species in a community, while abundance is the total number of individuals. Often the most productive and stable communities are highly diverse and profusely populated, but sometimes a high degree of specialization makes an ecosystem more, rather than less, susceptible to disturbance. Ecological complexity refers to the number of species at each trophic level as well as the total number of trophic levels in a community. Structure concerns the patterns of organization, both spatial and functional, in a community. Often a keystone species, or group of species, plays an unusually important role in determining community structure, composition, or function. All of these characteristics are affected both by physical and chemical factors as well as biological interactions between the organisms that make up the community. Edge effects at the boundaries between different habitat types can be important where landscapes are fragmented into isolated patches.

Ecological succession and development are processes by which organisms alter the environment in ways that allow some species to replace others. Primary succession starts with a previously unoccupied site. Secondary succession occurs on a site that has been disturbed by external forces. Often succession proceeds until a mature, diverse, relatively stable community is established. These mature communities may have self-perpetuating processes that make them resistant to change and resilient to disturbance. Whether diversity always leads to stability, however, is controversial. Communities that are disrupted regularly by fires or other natural disasters sometimes establish dynamic equilibrium or disclimax communities dependent on constant renewal. Introduction of new species by natural processes, such as opening of a land bridge, or through human intervention can upset the natural relationships in a community and cause catastrophic changes for indigenous species.

QUESTIONS FOR REVIEW

1. Explain how tolerance limits to environmental factors determine distribution of a highly specialized species, such as the desert pupfish. Compare this to the distribution of a generalist species, such as cowbirds or starlings.

2. Productivity, diversity, complexity, resilience, and structure are exhibited to some extent by all communities and ecosystems. Describe how these characteristics apply to the ecosystem in which you live.

3. Describe the general niche occupied by a bird of prey, such as a hawk or an owl. How can hawks and owls exist in the same ecosystem and not adversely affect each other?

4. Define keystone species and explain their importance in community structure and function.

5. Relationships between predators and prey play an important role in the energy transfers that occur in ecosystems. They also influence the process of natural selection. Explain how predators affect the adaptations of their prey. How do prey species affect the adaptations of their predators?

6. Competition for a limited quantity of resources occurs in all ecosystems. This competition can be interspecific or intraspecific. Explain some of the ways an organism might deal with these different types of competition.

7. Explain the difference between exponential and logistic population growth. In the exponential growth model, $dN/dt = rN$, what do r, N, and dN represent?

8. What are some factors that prevent exponential growth in a population? What are some density-dependent and density-independent limiting factors?

9. Explain the concept of climax community. Why do mature, late-stage communities often exhibit a higher level of stability than that found in other successional stages?

10. Discuss the dangers posed to existing community members when new species are introduced into ecosystems. What type of organism would be most likely to survive and cause problems in a new habitat?

THINKING SCIENTIFICALLY

1. List all the physical attributes of an American bison that you can think of. The bison is a social animal that lives on open grasslands. How are the traits you listed suitable to this lifestyle? What traits might make the bison unsuited to live in a forested environment?

2. A scientist suspects that fruit flies do better living on bananas than on oranges. What attributes of two populations (one living on bananas and the other on oranges) would you measure to determine whether the scientist's hunch is right? How would you state a simple, testable hypothesis for this problem?

3. In the fruit fly experiment in question 2, both populations are living and reproducing in glass-encased aquaria. How can you be sure that you are controlling all variables except food source? What environmental (or population) variables might be hard to control?

4. Scientific experiments like the fruit-fly test in question 2 should be replicable. That is, another researcher should be able to conduct the same experiment and come up with comparable results. Why might it be worthwhile to conduct the fruit fly experiment more than once?

5. Suppose you were interested in whether a population of frogs was mainly influenced by intrinsic or extrinsic population controls. What types of intrinsic and extrinsic factors would you put on your list to test for? State how you would test for at least one of these.

6. Taxonomists identify relationships among species by identifying common physical traits. Which of the following do you think might be most reliable in identifying similarity in two rodent species: length of legs; size of skull; digestive system; number of teeth? Why?

KEY TERMS

adaptation 52	*K*-adapted species 63
Batesian mimicry 59	keystone species 59
biotic potential 61	logistic growth 62
carrying capacity 61	Müllerian mimicry 59
coevolution 57	mutualism 58
commensalism 58	natural selection 52
complexity 66	overshoots 61
convergent evolution 54	pioneer species 70
divergent evolution 53	predator 57
diversity 65	primary productivity 64
ecological development 70	primary succession 69
ecological niche 54	*r*-adapted species 63
ecotones 69	resource partitioning 55
edge effects 69	S curve 62
environmental resistance 63	secondary succession 69
evolution 52	selective pressure 52
exponential growth 60	symbiosis 58
habitat 54	tolerance limits 51
J curve 61	

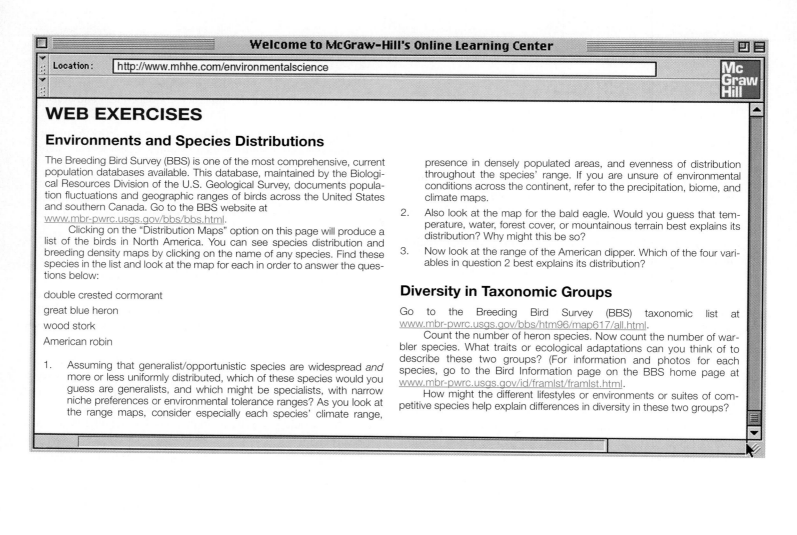

WEB EXERCISES

Environments and Species Distributions

The Breeding Bird Survey (BBS) is one of the most comprehensive, current population databases available. This database, maintained by the Biological Resources Division of the U.S. Geological Survey, documents population fluctuations and geographic ranges of birds across the United States and southern Canada. Go to the BBS website at www.mbr-pwrc.usgs.gov/bbs/bbs.html.

Clicking on the "Distribution Maps" option on this page will produce a list of the birds in North America. You can see species distribution and breeding density maps by clicking on the name of any species. Find these species in the list and look at the map for each in order to answer the questions below:

double crested cormorant

great blue heron

wood stork

American robin

1. Assuming that generalist/opportunistic species are widespread *and* more or less uniformly distributed, which of these species would you guess are generalists, and which might be specialists, with narrow niche preferences or environmental tolerance ranges? As you look at the range maps, consider especially each species' climate range, presence in densely populated areas, and evenness of distribution throughout the species' range. If you are unsure of environmental conditions across the continent, refer to the precipitation, biome, and climate maps.

2. Also look at the map for the bald eagle. Would you guess that temperature, water, forest cover, or mountainous terrain best explains its distribution? Why might this be so?

3. Now look at the range of the American dipper. Which of the four variables in question 2 best explains its distribution?

Diversity in Taxonomic Groups

Go to the Breeding Bird Survey (BBS) taxonomic list at www.mbr-pwrc.usgs.gov/bbs/htm96/map617/all.html.

Count the number of heron species. Now count the number of warbler species. What traits or ecological adaptations can you think of to describe these two groups? (For information and photos for each species, go to the Bird Information page on the BBS home page at www.mbr-pwrc.usgs.gov/id/framlst/framlst.html.

How might the different lifestyles or environments or suites of competitive species help explain differences in diversity in these two groups?

> The salvation of this human world lies nowhere else than in human responsibility.
>
> –*Václav Havel*

More than a billion people now live in India.

OBJECTIVES

After studying this chapter, you should be able to

- summarize historic factors that have contributed to human population growth.
- calculate doubling times for different annual growth rates.
- describe Malthusian and Marxian theories of limits to population growth, and explain why technological optimists and supporters of social justice oppose these theories.
- explain the process of demographic transition and why it produces a temporary population surge.
- understand how changes in life expectancy, infant mortality, women's literacy, standards of living, and democracy affect population changes.
- evaluate pressures for and against family planning in traditional and modern societies.
- compare modern birth control methods and think about a personal family planning agenda.

Welcome to McGraw-Hill's Online Learning Center

Location: http://www.mhhe.com/environmentalscience

LEARNING ONLINE

Visit our webpage at www.mhhe.com/environmentalscience for data sources, further readings, additional case studies, current environmental news, and regional examples within the Online Learning Center to help you understand the material in this chapter. You'll also find active links to information pertaining to this chapter including:

World Population Information
International Demographic Data
World Population
Population Research Institute

A Billion People and Growing

In 1999, having added more than 180 million people in just a decade, India reached a population of 1 billion humans. If current growth rates persist, India will have 1.63 billion residents in 2050, and will surpass China as the world's most populous country. How will the country, which already has more than a quarter of its population living in abject poverty, feed, house, educate, and employ all those being added each year? And what's the best way to slow this rapid growth? The fierce debate now taking place about how to control India's population has ramifications for the rest of the world as well.

On one side of this issue are those who believe that the best way to reduce the number of children born is poverty eradication and progress for women. Drawing on social justice principles established at the 1994 UN Conference on Population and Development in Cairo, some argue that responsible economic development, a broad-based social welfare system, education and empowerment of women, and high-quality health care—including family planning services—are essential components of population control. Without progress in these areas, they believe, efforts to provide contraceptives or encourage sterilization are futile.

On the other side of this debate are those who contend that, while social progress is an admirable goal, India doesn't have the time or resources to wait for an indirect approach to population control. The government must push aggressively, they argue, to reduce births now or the population will be so huge and its use of resources so great that only mass starvation, class war, crime, and disease will be able to bring it down to a manageable size.

Unable to reach a consensus on population policy, the Indian government decided in 2000 to let each state approach the problem in its own way. Some states have chosen to focus on social justice, while others have adopted more direct, interventionist policies.

The model for the social justice approach is the southern state of Kerala, which achieved population stabilization in the mid-1980s, the first Indian state to do so. Although still one of the poorest places in the world, economically, Kerala's fertility rate is comparable to that of many industrialized nations, including the United States. Both women and men have a nearly 100 percent literacy rate and share affordable and accessible health care, family planning, and educational opportunities; therefore, women have only the number of children they want, usually two. The Kerala experience suggests that increased wealth isn't a prerequisite for zero population growth.

Taking a far different path to birth reduction is the neighboring state of Andra Pradesh, which just reached a stable growth rate in 2001. Boasting the most dramatic fertility decline of any large Indian state, Andra Pradesh has focused on targeted, strongly enforced sterilization programs. The poor are encouraged—some would say, compelled—to be sterilized after having only one or two children. The incentives include cash payments. You might receive 500 rupees—equivalent to $11 (U.S.) or a month's wages for an illiterate farm worker—if you agree to have "the operation." In addition, participants are eligible for better housing, land, wells, and subsidized loans.

The pressure to be sterilized is overwhelmingly directed at women, for whom the procedure is major abdominal surgery. Sterilizations often are done by animal husbandry staff and carried out in government sterilization camps. This practice raises troubling memories of the 1970s for many people, when then-Prime Minister Indira Gandhi suspended democracy and instituted a program of forced sterilization of poor people. There were reports at the time of people being rounded up like livestock and castrated or neutered against their will.

While many feminists and academics regard Andra Pradesh's policies as appallingly intrusive and coercive to women and the poor, the state has successfully reduced population growth. By contrast, the hugely populous northern states of Uttar Pradesh and Bihar have increased their growth rates slightly over the past two decades to a current rate above 2.5 percent per year. How will they slow this exponential growth, and what might be the social and environmental costs of not doing so?

India's population problems introduce several of the important themes of this chapter. What are the trends in human populations around the world, what do those trends mean for our resources and environment, and what is the best way to approach population planning? Keep in mind, as you read this chapter, that resource limits aren't simply a matter of total number of people, they also depend on consumption levels and the types of technology used to produce the things we use and consume.

POPULATION GROWTH

Every second, on average, four or five children are born somewhere on the earth. In that same second, two other people die. This difference between births and deaths means a net gain of nearly 2.5 more humans per second in the world population. This means we are growing at a little less than 9,000 per hour, 214,000 per day, or almost 78 million more people per year. In 1999, the world population passed 6 billion, making us the most numerous vertebrate species on the planet. For the families to whom these children are born, this may well be a joyous and long-awaited event (fig. 4.1). But is a continuing increase in humans good for the planet in the long run?

Many people worry that overpopulation will cause—or perhaps already is causing—resource depletion and environmental degradation that threaten the ecological life-support systems on which we all depend. These fears often lead to demands for immediate, worldwide birth-control programs to reduce fertility rates and to eventually stabilize or even shrink the total number of humans.

Principles of Environmental Science   www.mhhe.com/environmentalscience

Others believe that human ingenuity, technology, and enterprise can extend the world's carrying capacity and allow us to overcome any problems we encounter. From this perspective, more people may be beneficial, rather than disastrous. A larger population means a larger workforce, more geniuses, more ideas about what to do. Along with every new mouth comes a pair of hands. Proponents of this worldview argue that continued economic and technological growth can both feed the world's billions and enrich everyone enough to end the population explosion voluntarily.

Yet another perspective on this subject derives from social justice concerns. According to this worldview, resources are sufficient for everyone. Current shortages are only signs of greed, waste, and oppression. The root cause of environmental degradation, in this view, is inequitable distribution of wealth and power rather than population size. Fostering democracy, empowering women and minorities, and improving the standard of living of the world's poorest people are what are really needed. A narrow focus on population growth only fosters racism and an attitude that blames the poor for their problems, while ignoring the deeper social and economic forces at work.

Whether human populations will continue to grow at present rates and what that growth would imply for environmental quality and human life are among the most central and pressing questions in environmental science. In this chapter, we will look at some causes of population growth, as well as at how populations are measured and described. Family planning and birth control are essential for stabilizing populations. The number of children a couple decides to have and the methods they use to regulate fertility, however, are strongly influenced by culture, religion, politics, and economics, as well as basic biological and medical considerations. We will examine how some of these factors influence human demographics.

Human Population History

For most of our history, humans have not been very numerous compared to other species. Studies of hunting and gathering societies suggest that the total world population was probably only a few million people before the invention of agriculture and the domestication of animals around 10,000 years ago. The agricultural revolution produced a larger and more secure food supply and allowed the human population to grow, reaching perhaps 50 million people by 5000 B.C. For thousands of years, the number of humans increased very slowly. Archaeological evidence and historical descriptions suggest that only about 300 million people were living at the time of Christ (table 4.1).

As you can see in figure 4.2, human populations began to increase rapidly after about A.D. 1600. Many factors contributed to this rapid growth. Increased sailing and navigating skills stimulated commerce and communication among nations. Agricultural developments, better sources of power, and better health care and hygiene also played a role. We are now in an exponential or J curve pattern of growth described in chapter 3.

It took all of human history to reach 1 billion people in 1804, but little more than 150 years to reach 3 billion in 1960. It took us only 12 years—from 1987 to 1999—to add the sixth billion. Another way

FIGURE 4.1 *A Mayan family in Guatemala with four of their six living children. Decisions on how many children to have are influenced by many factors, including culture, religion, need for old-age security for parents, immediate family finances, household help, child survival rates, and power relationships within the family.*

to look at population growth is that the number of humans tripled during the twentieth century. Will it do so again in the twenty-first century? If it does, will we overshoot our environment's carrying capacity and experience a catastrophic dieback similar to those described in chapter 3? As you will see later in this chapter, there is some evidence that population growth already is slowing, but whether we will reach equilibrium soon enough and at a size that can be sustained over the long term remains a difficult but important question.

Population Doubling Times

If the current world population is growing at approximately 1.3 percent per year, how long will it take to double the population if that rate persists? A useful rule of thumb is that if you divide 70 by the annual percentage growth, you get the approximate doubling time in years for anything growing exponentially. For example, a savings account or a biological population growing at a compound interest rate of 1 percent per year will double in about 70 years. By

TABLE 4.1 World Population Growth and Doubling Times

DATE	POPULATION	DOUBLING TIME
5000 B.C.	50 million	?
800 B.C.	100 million	4,200 years
200 B.C.	200 million	600 years
A.D. 1200	400 million	1,400 years
A.D. 1700	800 million	500 years
A.D. 1900	1,600 million	200 years
A.D. 1965	3,200 million	65 years
A.D. 1990	5,300 million	38 years
A.D. 2020 (estimate)	8,230 million	55 years

Source: Data from Population Reference Bureau, Inc., Washington, D.C.

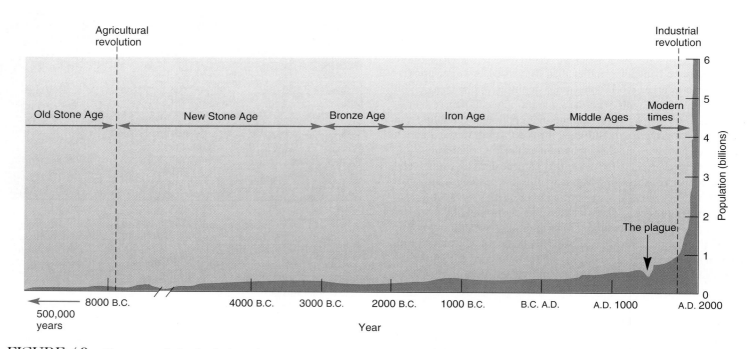

FIGURE 4.2 *Human population levels throughout history. Since about A.D. 1000, our population curve has assumed a J shape. Are we on the upward slope of a population overshoot? Will we be able to adjust our population growth to an S curve? Or can we just continue the present trend indefinitely?*

APPLICATION: Population Doubling Times

Using the "rule of 70," calculate the doubling time of populations growing at the following annual percentage growth rates: 0.1 percent, 1.0 percent, 1.3 percent, 5 percent.

Example: $\dfrac{70 \text{ years}}{(\text{percent growth})} = $ doubling time in years

Answers: 0.1 = 700 years; 1.0 = 70 years; 1.3 = 54 years; 5.0 = 14 years.

the same measure, a population growing at 35 percent doubles every 2 years. Countries growing at 4 percent per year will double their populations in 17.5 years. A country growing at a rate of 0.1 percent annually will double its population in 700 years.

LIMITS TO GROWTH: SOME OPPOSING VIEWS

As with many topics in environmental science, people have widely differing opinions about population and resources. Some believe that population growth is the ultimate cause of poverty and environmental degradation. Others argue that poverty, environmental degradation, and overpopulation are all merely symptoms of deeper social and political factors. Which of these opposing worldviews we accept as most correct will have a great impact on our population policies. In this section, we will examine some of the major figures and their arguments in this debate.

Malthusian Checks on Population

In 1798, the Reverend Thomas Malthus wrote *An Essay on the Principle of Population* to refute the views of progressives and optimists—including his father—who, inspired by the egalitarian principles of the French Revolution, predicted a coming utopia. The younger Malthus argued that human populations tend to increase at an exponential or compound rate, while food production either remains stable or increases only slowly. The result, according to Malthus, is that human populations inevitably outstrip their food supply and eventually collapse into starvation, crime, and misery. Figure 4.3a summarizes his theory.

According to Malthus, the only ways to stabilize human populations are "positive checks," such as diseases or famines that kill people, or "preventative checks," including all the factors that prevent human birth. Among the preventative checks he advocated were "moral restraint," including late marriage and celibacy until a couple could afford to support children. Malthus himself, however, had several illegitimate children and didn't have much faith in moral restraint. Many social scientists and biologists have been influenced by Malthus. Charles Darwin, for instance, derived his theories about the struggle for scarce resources and survival of the fittest after reading Malthus's essay.

If Malthus's views of the consequences of exponential population growth were dismal, the corollary he drew was even more bleak. He believed that most people are too lazy and immoral to regulate birth rates voluntarily. Consequently, he opposed efforts to feed and assist the poor in England because he feared that more food would simply increase their fertility and thereby perpetuate the problems of starvation and misery. (See the related story "The Saga of Easter Island" at <u>www.mhhe.com/apps</u>.)

Principles of Environmental Science

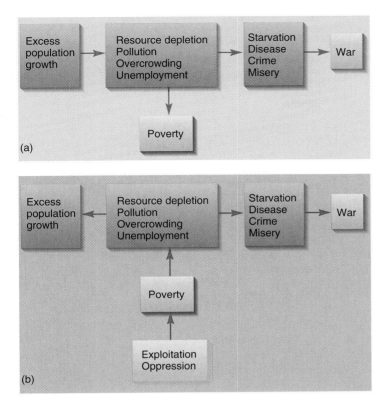

FIGURE 4.3 (a) *Thomas Malthus argued that excess population growth is the ultimate cause of many other social and environmental problems.* (b) *Karl Marx argued that oppression and exploitation are the real causes of poverty and environmental degradation. In his view, population growth is a symptom or result of other problems, not the source.*

Not surprisingly, Malthus's ideas provoked a great social and economic debate. Karl Marx was one of his most vehement critics, claiming that Malthus was a "shameless sycophant of the ruling classes." According to Marx, population growth is a symptom rather than a root cause of poverty, resource depletion, pollution, and other social ills. The real causes of these problems, he believed, are exploitation and oppression (fig. 4.3b). Marx argued that workers always provide for their own sustenance, given access to means of production and a fair share of the fruits of their labor. According to Marxians, the way to slow population growth and to alleviate crime, disease, starvation, misery, and environmental degradation is through social justice.

Malthus and Marx Today

Both Marx and Malthus developed their theories about human population growth in the nineteenth century, when understanding of the world, technology, and society were much different than they are now. Still, as the opening story in this chapter shows, the questions they raised are relevant today. While the evils of racism, classism, and other forms of exploitation that Marx denounced still beset us, it is also true that at some point available resources must limit the numbers of humans that the earth can sustain.

Those who agree with Malthus that we are approaching—or may already have surpassed—the earth's carrying capacity are called **neo-Malthusians.** In their view, we should address the issue of surplus population directly by making birth control our highest priority. Cornell University entomologist David Pimentel expressed a version of this view when he said, "By 2100, if current trends continue, twelve billion miserable humans will suffer a difficult life on Earth." The optimum population, he argues, would be about 2 billion, or about the number on the planet at the beginning of World War II. He believes that this population would allow everyone to enjoy a standard of living equal to the average European today.

Neo-Marxians, on the other hand, believe that only eliminating oppression and poverty through technological development and social justice will solve population problems. Claims of resource scarcity, they argue, are only an excuse for inequity and exclusion. They believe that, if distribution of wealth and resources were more fair, there would be plenty for everyone. As Mohandas Gandhi said, "There is enough for everyone's need, but not enough for anyone's greed."

Perhaps a compromise position between these opposing views is that population growth, poverty, and environmental degradation are all interrelated. No factor exclusively causes any other, but each influences and, in turn, is influenced by, the others.

Can Technology Make the World More Habitable?

Technological optimists argue that Malthus was wrong in his predictions of famine and disaster 200 years ago because he failed to account for scientific progress. In fact, food supplies have increased faster than population growth since Malthus's time. There have been terrible famines in the past two centuries, but they were caused more by politics and economics than lack of resources or sheer population size. Whether this progress will continue remains to be seen, but technological advances have vastly increased human carrying capacity so far.

The burst of world population growth that began 200 years ago was stimulated by scientific and industrial revolutions. Progress in agricultural productivity, engineering, information technology, commerce, medicine, sanitation, and other achievements of modern life have made it possible to support approximately 1,000 times as many people per unit area as was possible 10,000 years ago. Economist Stephen Moore of the Cato Institute in Washington, D.C. regards this achievement as "a real tribute to human ingenuity and our ability to innovate." There is no reason, he argues, to think that our ability to find technological solutions to our problems will diminish in the future.

Much of our growth in the past 200 years, however, has been based on availability of easily acquired natural resources, especially cheap, abundant fossil fuels. Whether we can develop alternative, renewable energy sources in time to avert disaster when current fossil fuels run out is a matter of great concern.

Could More People Be Beneficial?

Larger populations can have benefits as well as disadvantages. More people mean larger markets, more workers, and efficiencies of scale in mass production of goods. Greater numbers also provide more intelligence and enterprise to overcome such problems as underdevelopment, pollution, and resource limitations. Human ingenuity and intelligence can create new resources through substitution of new materials and can discover new ways of doing things for old materials and old ways. For instance, utility companies are finding it cheaper and more environmentally sound to finance insulation and energy-efficient appliances for their customers rather than build new power plants. The effect of saving energy that was formerly wasted is comparable to creating a new fuel supply.

Economist Julian Simon was one of the most outspoken champions of this rosy view of human history. He argued that people are the "ultimate resource" and that no evidence suggests that pollution, crime, unemployment, crowding, the loss of species, or any other resource limitations will worsen with population growth. Leaders of many developing countries share this outlook and insist that, instead of being obsessed with population growth, we should focus on the inordinate consumption of the world's resources by people in richer countries.

HUMAN DEMOGRAPHY

Demography (derived from the Greek words *demos* [people] and *graphein* [to write or to measure]) encompasses vital statistics about people, such as births, deaths, and where they live, as well as total population size. In this section, we will survey ways to measure and describe human populations and discuss demographic factors that contribute to population growth.

How Many of Us Are There?

On October 12, 1999, the United Nations officially declared that the human population had reached 6 billion. The U.S. Census Bureau, however, had put the date for this landmark three months earlier on July 19. Even in this age of information technology and communication, counting the number of people in the world is like shooting at a moving target. Some countries have never even taken a census, and those that have been done may not be accurate. Governments may overstate or understate their populations to make their countries appear larger and more important or smaller and more stable than they really are. Individuals, especially if they are homeless, refugees, or illegal aliens, may not want to be counted or identified.

We really live in two very different demographic worlds. One of these worlds is poor, young, and growing rapidly. It is occupied by the vast majority of people who live in the less-developed countries of Africa, Asia, and Latin America. These countries represent 80 percent of the world population but more than 90 percent of all projected growth (fig. 4.4). (See Investigating Our Environment, p. 81.)

The highest population growth rates occur in a few "hot spots" such as sub-Saharan Africa and the Middle East, where economics, politics, religion, and civil unrest keep birth rates high and contraceptive use low. In Chad and the Democratic Republic

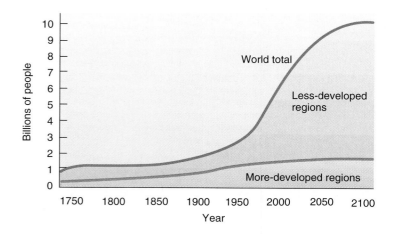

FIGURE 4.4 *Estimated human population growth, 1750–2100, in less-developed and more-developed regions. More than 90 percent of all growth in the twentieth century and projected for the twenty-first century is in the less-developed countries.*

of Congo, for example, annual population growth is above 3.2 percent. Less than 10 percent of all couples use any form of birth control, women average more than seven children each, and nearly half the population is less than 15 years old. Even faster growth rates occur in Oman and Palestine, where the population doubling time is only 18 years.

Some countries in the developing world have experienced amazing growth rates and are expected to reach extraordinary population sizes by the middle of the twenty-first century. Table 4.2 shows the ten largest countries in the world, arranged by their projected size in 2050. Note that, while China was the most populous

TABLE 4.2   Ten Most Populous Countries in 2000 and 2050

MOST POPULOUS IN 2000		MOST POPULOUS IN 2050	
COUNTRY	POPULATION (MILLIONS)	COUNTRY	POPULATION (MILLIONS)
China	1,265	India	1,628
India	1,002	China	1,369
United States	281	United States	404
Indonesia	212	Indonesia	312
Brazil	170	Nigeria	304
Pakistan	151	Pakistan	285
Russia	145	Brazil	244
Bangladesh	128	Bangladesh	211
Japan	127	Ethiopia	188
Nigeria	123	Congo	182

Source: Data from Population Reference Bureau, 2002.

Principles of Environmental Science   www.mhhe.com/environmentalscience

INVESTIGATING Our Environment

Interpreting Graphs

An important part of any scientific investigation is presenting the data. Graphs and other pictorial representations of information are particularly good for helping others to understand what we have to say. It is one thing to read that "the human population is rising explosively," but it is considerably more compelling to see the J curve along with concrete numbers as evidence. The combination of picture and numbers often conveys powerful impressions beyond the numbers themselves. But it is important to recognize that graphs can be as truthful or misleading as the person who creates them intends to be.

Central questions addressed in this chapter are: How serious is the worldwide population problem? Does it demand action? The graphs in figures 1 and 2 present information relevant to human population studies. Yet, the impressions these figures convey vary considerably, depending on the type of data used and the design of the graph.

Reexamine figure 4.2, which presents an historical perspective on the growth in human numbers. What conclusions do you draw from it? Does it suggest that we are experiencing an explosive rise in human numbers? Does it indicate that the current increase is unprecedented in our species history? Does it also imply, perhaps more indirectly, that the earth's population problem is serious and urgent? In light of the concepts of environmental resistance, carrying capacity, and overshoot and dieback, many people would conclude that the current population growth, highlighted by the graph, is unsustainable.

Now examine figure 1. It plots the same variables as figure 4.2 but suggests that population is rising at a modest rate. It does not produce the same sense of explosiveness and urgency as figure 4.2. How can the impressions be so different?

Notice that the graph in figure 1 covers a much shorter time period. This greatly changes the time interval lengths on the horizontal scale. In figure 4.2, 1 mm represents about 50 years, but in figure 1, it represents less than 1 year. This changes the line's slope from nearly vertical in figure 4.2 to a modest incline in figure 1. Slope impacts the visual impression created by a graph, and therefore, its interpretation. How could you change the time axis in figure 1 to flatten the slope even more?

Next examine figure 2. This graph plots the stabilization ratio, rather than population size over time. The graph reveals that the growth rate for most regions except Africa has been declining since about 1970. Do the graph's downward-trending lines suggest that there really isn't much of a population problem, or at least not much reason for concern?

All three graphs present information relevant to a discussion of worldwide population growth, yet each gives a different impression of the seriousness of the problem. So, how can we analyze graphs and their messages in a thoughtful way? Certainly, no single formula is applicable to all situations, but the following questions are worth keeping in mind:

1. Is the graph's time frame of reference appropriate, or is it too restricted to allow a valid, comprehensive assessment of the issue?
2. Are the unit intervals on the graph appropriately sized?
3. Is the impression the graph creates real or simply an artifact of the graph's format?
4. Do the data the graph presents provide only a partial, perhaps misleading, view of the whole?

Because graphs can create powerful impressions, we need to use and interpret them with care.

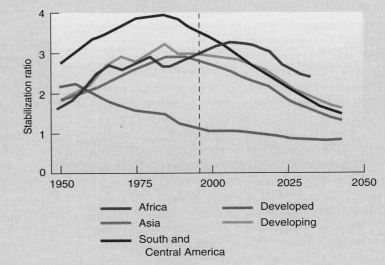

FIGURE 1 *The UN Population Division projects continued population growth, although at a gradually slowing rate, over the next 50 years.*
Source: World Resources Institute, *World Resources 1998–99.*

FIGURE 2 *The stabilization ratio is measured by dividing crude birth rate by crude death rate. A ratio of 1 indicates zero population growth.*
Source: United Nations (UN) Population Division, *World Population Prospects, 1950–2050 (The 1996 Revision).* The United Nations, New York, 1996.

INVESTIGATING Our Environment

Getting to Know the Neighbors

Imagine the world as a village of a thousand people: who are our neighbors?

- Ethnically, this global village has 385 East Asians, 209 South Asians, 129 Africans, 91 Europeans, 78 Latin Americans, 52 from the former Soviet States, 51 from the United States and Canada, and 5 from Oceania (Australia, New Zealand, and the Pacific Islands).

- About half the residents speak one or more of the six major world languages (Mandarin, English, Hindi/Urdu, Spanish, Russian, and Arabic, in that order), while the rest speak some of the other 6,000 known languages.

- Children under age 15 make up 32 percent of the village, while those over age 65 make up 6 percent. The number of senior citizens is increasing rapidly, while the number of children is falling. At some point in the future, retirees may outnumber the workers in the village, bankrupting the social security system.

- Every year, 26 children are born in the village, and nine people die. One-third of yearly deaths are children under age five, mostly from infectious diseases aggravated by malnutrition. One person dies from heart disease, one from cancer, two from chronic or acute lung diseases, and two from violence, accidents, and various infectious diseases.

- The average life expectancy is 70 years for women and 65.6 for men, but longevity varies considerably. In the richest families, women live to an average age of 80 and men to an average of 78 years, while in the poorest families the life expectancy is less than 48 years for both men and women.

- The average annual income for the village is $4,880 (U.S.), but this average also obscures great discrepancies. More than half the village live in households where the annual income is less than $620 per person, while the richest 100 citizens (mostly Americans, Canadians, Europeans, and Japanese) enjoy annual incomes over $25,000 each.

- Half of the 620 adults in the village are illiterate. Lacking an education, most of these people work as day laborers or seasonal farm workers. Among the poorer families, girls are half as likely as boys to attend school.

- Women and girls make up slightly more than half the village population. They do two-thirds of all manual labor, receive one-tenth of the wages, and own less than one-hundredth of the property. Seventy percent of the poorest members of the population are women and children.

- About 400 villagers suffer from ill health at any given time. Much of that illness is related to lack of clean water, sanitation, and food. Some 250 of our neighbors don't have clean water to drink or adequate sanitation. About 150 are chronically hungry, lacking the calories and nutrients needed for normal growth and development in children or a healthy, productive life for adults.

- Only 452 people actually live in the village itself; the other 548 live in the surrounding countryside. Sixty percent of the rural families are landless or have too little land to subsist. They make up the majority of those who lack clean water, sanitation, food, housing, and health care.

- Generally, the worst pollution problems are borne by those who live in the poorest parts of the village where air and water pollution, noise, congestion, and toxic wastes are most common. Those who live in the better neighborhoods actually enjoy a cleaner, safer environment than their parents did only a few decades ago.

Source: Data from UN Development Programme and Population Reference Bureau, 2002.

---

country throughout the twentieth century, as the opening of this chapter shows, India is expected to pass China in the twenty-first century. Nigeria, which had only 33 million residents in 1950, is forecast to have more than 300 million in 2050. Ethiopia, with about 18 million people 50 years ago, is likely to grow at least ten-fold over a century. In many of these countries, rapid population growth is a serious problem. Bangladesh, about the size of Iowa, is already overcrowded at 128 million people. Another 83 million people by 2050 will only add to current problems.

The other demographic world is made up of the richer countries of North America, Western Europe, Japan, Australia, and New Zealand. This world is wealthy, old, and shrinking. Italy, Germany, Hungary, and Japan, for example, all have negative growth rates. The average age in these countries is now 40, and life expectancy of their residents is expected to exceed 90 by 2050. With many couples choosing to have either one or no children, the populations of these countries are expected to decline significantly over the next century. Japan, which has 126 million residents now, is expected to shrink to about 100 million by 2050. Europe, which now makes up about 12 percent of the world popu-lation, will constitute less than 7 percent in 50 years, if current trends continue. Even the United States and Canada would have nearly stable populations if immigration were stopped.

It isn't only wealthy countries that have declining populations. Russia, for instance, is now declining by nearly 1 million people per year as death rates have soared and birth rates have plummeted. A collapsing economy, hyperinflation, crime, corruption, and despair have demoralized the population. Horrific pollution levels left from the Soviet era, coupled with poor nutrition and health care, have resulted in high levels of genetic abnormalities, infertility, and infant mortality. Abortions are twice as common as live births, and the average number of children per woman is now 1.3, one of the lowest in the world. Death rates, especially among adult men, have risen dramatically. According to some medical experts, four out of five Russian men are drunk when they die, and male life expectancy dropped from 68 years in 1990 to 58 years in 2000. After having been the fourth largest country in the world in 1950, Russia is expected to have a smaller population than Vietnam, the Philippines, or the Democratic Republic of Congo by 2050.

Principles of Environmental Science

The situation is even worse in many African countries, where AIDS and other communicable diseases are killing people at a terrible rate. In Zimbabwe, Botswana, Zambia, and Namibia, for example, up to 30 percent of the adult population have AIDS or are HIV positive. Health officials predict that more than two-thirds of the 15-year-olds now living in Botswana will die of AIDS before age 50. Many of these countries are soon expected to have declining populations. Overall, however, Africa is expected to grow by at least 1.5 billion over the next century.

The human density map in Appendix 3 on p. 374 shows human population distribution around the world. Notice the high densities supported by fertile river valleys of the Nile, Ganges, Yellow, Yangtze, and Rhine Rivers and the well-watered coastal plains of India, China, and Europe. Historic factors, such as technology diffusion and geopolitical power, also play a role in geographic distribution.

Fertility and Birth Rates

Fecundity is the physical ability to reproduce, while fertility describes the actual production of offspring. Those without children may be fecund but not fertile. The most accessible demographic statistic of fertility is usually the **crude birth rate,** the number of births in a year per thousand persons. It is statistically "crude" in the sense that it is not adjusted for population characteristics, such as the number of women of reproductive age.

The **total fertility rate** is the number of children born to an average woman in a population during her entire reproductive life. Upper-class women in seventeenth- and eighteenth-century Europe, whose babies were given to wet nurses immediately after birth and who were expected to produce as many children as possible, often had 25 or 30 pregnancies. The highest recorded total fertility rates for working-class people are among some Anabaptist agricultural groups in North America, who have averaged up to 12 children per woman. In most tribal or traditional societies, food shortages, health problems, and cultural practices limit total fertility to about six or seven children per woman, even without modern methods of birth control.

Fertility is usually calculated as births per woman because, in many cases, establishing paternity is difficult. Nevertheless, a few demographers argue that we should pay more attention to birth rates per male because, in some cultures men have far more children, on average, than do women. In Cameroon, for instance, due to multiple marriages, extramarital affairs, and a high rate of female mortality, men are estimated to have 8.1 children in their lifetime, while women average only 4.8.

Zero population growth (ZPG) occurs when births plus immigration in a population just equal deaths plus emigration. It takes several generations of replacement level fertility (where people just replace themselves) to reach ZPG. Where infant mortality rates are high, the replacement level may be five or more children per couple. In the more highly developed countries, however, this rate is usually about 2.1 children per couple because some people are infertile, have children who do not survive, or choose not to have children.

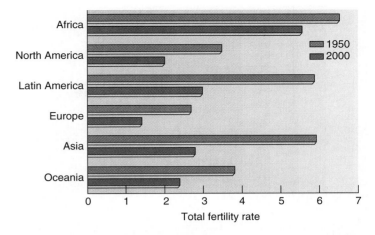

FIGURE 4.5   *Declines in fertility rates by region, 1950 and 2000.*
Source: Data from Population Reference Bureau, 2002.

Fertility rates have declined dramatically in every region of the world except Africa over the past 50 years (fig. 4.5). In the 1960s, total fertility rates above six were common in many countries. The average family in Mexico in 1975, for instance, had seven children. By 2000, however, the average Mexican woman had only 2.5 children. According to the World Health Organization, 61 of the world's 190 countries are now at or below a replacement rate of 2.1 children per couple. The greatest fertility reduction has been in Southeast Asia, where rates have fallen by more than half, primarily in the past few decades. Contrary to what many demographers expected, some of the poorest countries in the world have been remarkably successful in lowering growth rates. Bangladesh, for instance, reduced its fertility rate from 6.9 in 1980 to only 3.1 children per woman in 1998.

China's one-child-per-family policy decreased the fertility rate from six in 1970 to 1.8 in 1990. This policy, however, has sometimes resulted in abortions, forced sterilizations, and even infanticide. Another adverse result is that the only children (especially boys) allowed to families may grow up to be spoiled "little emperors" who have an inflated impression of their own importance (fig. 4.6). Furthermore, there may not be enough workers to maintain the army, sustain the economy, or support retirees when their parents reach old age.

Although the number of boys and girls normally should be fairly balanced in a population, 20 years of female infanticides and sex-based abortions in China have created ratios as high as 140 boys to 100 girls in some regions. This shortage of women has created a flourishing trade in abducted brides. A crackdown in 2002 released 110,000 women who had been kidnapped and sold into marriage, but it's thought that the total number abducted is much higher.

While the world as a whole still has an average fertility rate of 2.7, growth rates are now lower than at any time since World War II. If fertility declines like those in Bangladesh and China were to occur everywhere in the world, our total population could begin to decline by the end of the twenty-first century. Interestingly,

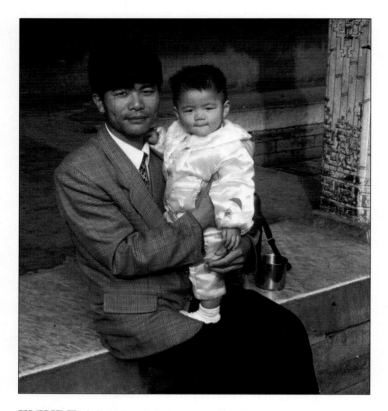

FIGURE 4.6 *China's one-child-per-family policy has been remarkably successful in reducing birth rates. It may, however, have created a generation of "little emperors," since parents and grandparents focus all their attention on an only child.*

Life Span and Life Expectancy

Life span is the oldest age to which a species is known to survive. Although there are many claims in ancient literature of kings living for a thousand years or more, the oldest age that can be certified by written records was that of Jeanne Louise Calment of Arles, France, who was 122 years old at her death in 1997. While modern medicine has made it possible for many of us to survive much longer than our ancestors, it doesn't appear that the maximum life span has increased much at all. Apparently, cells in our bodies have a limited ability to repair damage and produce new components. Sooner or later they simply wear out, and we fall victim to disease, degeneration, accidents, or senility.

Life expectancy is the average age that a newborn infant can expect to attain in any given society. It is another way of expressing the average age at death. For most of human history, life expectancy in most societies probably has been 35 to 40 years. This doesn't mean that no one lived past age 40, but rather that many people died at earlier ages (mostly early childhood), which balanced out those who managed to live longer.

The twentieth century has seen a global transformation in human health unmatched in history. This revolution can be seen in the dramatic increases in life expectancy in most places (table 4.3). Worldwide, the average life expectancy has risen from about 40 to 65.5 years over the past century. The greatest progress has been in developing countries. For example, in 1900, the average Indian man could expect to live less than 23 years, while the average woman would reach just over 23 years. By 2000, although India had an annual per capita income of less than $440 (U.S.), the average life expectancy for both men and women had nearly tripled and was very close to that of countries with ten times its income level. Longer lives were due primarily to better nutrition, improved sanitation, clean water, and education, rather than to miracle drugs or high-tech medicine. While the gains were not as great for the already industrialized countries, residents of the United States, Sweden, and Japan, for example, now live about half-again as long as they did at the beginning of the twentieth century, and they can expect to enjoy much of that life in relatively

Spain and Italy, although predominately Roman Catholic, have the lowest reported fertility rates (1.2 children per woman) of any countries.

Mortality and Death Rates

A traveler to a foreign country once asked a local resident, "What's the death rate around here?" "Oh, the same as anywhere," was the reply, "about one per person." In demographics, however, **crude death rates** (or crude mortality rates) are expressed in terms of the number of deaths per thousand persons in any given year. Countries in Africa where health care and sanitation are limited may have mortality rates of 20 or more per 1,000 people. Wealthier countries generally have mortality rates around 10 per 1,000. The number of deaths in a population is sensitive to the population's age structure. Rapidly growing, developing countries, such as Belize or Costa Rica, have lower crude death rates (four per 1,000) than do the more-developed, slowly growing countries, such as Denmark (12 per 1,000). This is because a rapidly growing country has proportionately more youths and fewer elderly than a more slowly growing country. Declining mortality, not rising fertility, is the primary cause of most population growth in the past 300 years. Crude death rates began falling in Western Europe during the late 1700s.

TABLE 4.3 Life Expectancy at Birth for Selected Countries in 1900 and 2000

COUNTRY	1900		2000	
	MALES	FEMALES	MALES	FEMALES
India	22.6	23.3	60.3	60.5
Japan	42.4	43.7	77.4	84.2
Russia	30.9	33.0	61.7	73.6
Sweden	56.6	59.5	77.0	82.1
United States	45.6	48.3	74.7	79.3

Source: Data from Population Reference Bureau, 2002.

good health. The Disability Adjusted Life Years (DALYs) that someone living in Japan can expect is now 74.5 years, compared to only 64.5 DALYs two decades ago.

As figure 4.7 shows, annual income and life expectancy are strongly correlated up to about $4,000 (U.S.) per person. Beyond that level—which is generally enough for adequate food, shelter, and sanitation for most people—life expectancies level out at about 75 years for men and 80 for women. Russia is a striking exception to this pattern. With a gross national product per person of near $5,000 (U.S.), Russian life expectancy is only 58 years for men and 71 for women. Russian men now live about 14 years less, on average, than they did before the breakup of the USSR. As mentioned earlier, a disastrous economy, alcoholism, poor nutrition, and substandard medical care have all contributed to this decline.

Large discrepancies in how the benefits of modernization and social investment are distributed within countries are revealed in differential longevities of various groups. The greatest life expectancy reported anywhere in the United States is for women in Stearns County, Minnesota, who live to an average age of 86. By contrast, Native American men on Pine Ridge Indian Reservation in neighboring South Dakota, live, on average, only to age 45. Only a few countries in Africa have a lower life expectancy. The Pine Ridge Reservation is the poorest area in America, with an unemployment rate near 75 percent and high rates of poverty, alcoholism, drug use, and alienation. Similarly, African-American men in Washington, D.C. live, on average, only 57.9 years, which is less than life expectancies in Lesotho or Swaziland.

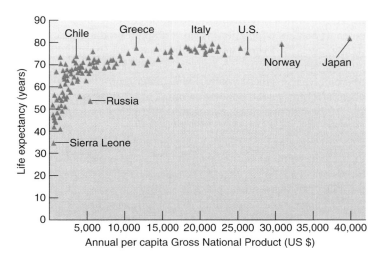

FIGURE 4.7 *As incomes rise, so does life expectancy up to about $4,000 (U.S.). Russia is an exception, with a life expectancy nearly 20 years less than that of Chile, even though their gross national products are about the same.*

Sources: The World Bank, *World Development Indicators 1997,* the World Bank, Washington, D.C., 1997; United Nations (UN) Population division; *World Population Prospects,* 1950–2050 (The 1996 Revision) UN, New York. 1996.

Living Longer: Demographic Implications

A population growing rapidly by natural increase has more young people than does a stationary population. One way to show these differences is to graph age classes in a histogram, as figure 4.8 shows.

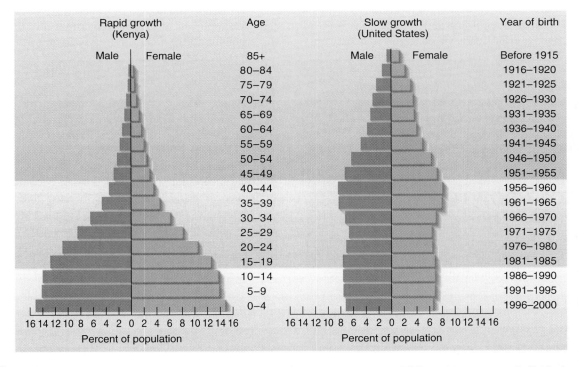

FIGURE 4.8 *Population by age and sex in rapidly growing and slowly growing countries. Middle regions represent individuals of reproductive ages. Note the high proportion of children in the rapidly growing population and the high proportion of elderly in the slowly growing population.*

Sources: Data from the U.S. Bureau of the Census, the United Nations, and the Population Reference Bureau.

In Kenya, which is growing at a rate of 2.5 percent per year, 42 percent of the population is in the prereproductive category (below age 15). Even if total fertility rates fell abruptly, the total number of births, and the population size, would continue to grow for some years as these young people enter reproductive age. This phenomenon is called population momentum.

A population that has recently entered a slower growth rate pattern, such as the United States, will have a bulge in the age classes for the last high-birth rate generation. Notice in figure 4.8 that females outnumber males in the older age group in the United States because of differences in longevity between the sexes. After a country has had a stable population for many years, it will have approximately the same numbers in all age classes, and the columns will be nearly uniform from top to bottom.

Both rapidly growing countries and slowly growing countries can have a problem with their **dependency ratio,** or the number of nonworking compared to working individuals in a population. In Mexico, for example, each working person supports a high number of children. In the United States, by contrast, a declining working population is now supporting an ever larger number of retired persons. In 1970, the median age in the United States was 30, and there were four workers for every retired person. By 2100, if current trends continue, the median age could be over 60. If workers continue to retire at age 65, half of the population could be unemployed, and retirement could last 35 or 40 years.

This changing age structure and shifting dependency ratio are occurring worldwide (fig. 4.9). In 1950, there were only 130 million people in the world over 65 years old. In 2000, more than 420 million had reached this age. By 2150, those over 65 might make up 25 percent of the world population. Countries such as Japan, Singapore, and Taiwan already are concerned that they may not have enough young people to fill jobs and support their retirement system. They are encouraging couples to have more children and are recruiting immigrants who might bring down the average age of the population. A group of business leaders, including Federal Reserve chairman, Alan Greenspan, recently called for increased immigration into the United States to fill jobs, prevent inflation, and boost economic productivity.

POPULATION GROWTH: OPPOSING FACTORS

A number of social and economic pressures affect decisions about family size, which, in turn, affects the population at large. In this section, we will examine both positive and negative pressures on reproduction.

Pronatalist Pressures

Factors that increase people's desires to have babies are called **pronatalist pressures.** Raising a family may be the most enjoyable and rewarding part of many people's lives. Children can be a source of pleasure, pride, and comfort. They may be the only

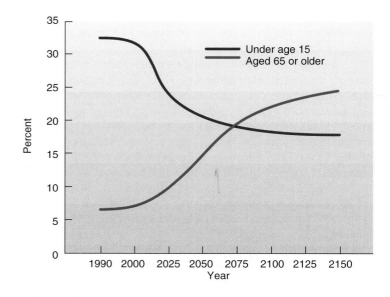

FIGURE 4.9 *Changing age structure of world population. In the twenty-first century, children under age 15 will make up a smaller percentage of world population, while people over age 65 will make up a rapidly rising share of the population.*

source of support for elderly parents in countries without a social security system. Where infant mortality rates are high, couples may need to have many children to ensure that at least a few will survive to take care of them when they are old. Where there is little opportunity for upward mobility, children give status in society, express parental creativity, and provide a sense of continuity and accomplishment otherwise missing from life. Often children are valuable to the family not only for future income, but even more as a source of current income and help with household chores. In much of the developing world, small children tend domestic animals and younger siblings, fetch water, gather firewood, help grow crops, or sell things in the marketplace (fig. 4.10). Parental desire for children rather than an unmet need for contraceptives may be the most important factor in population growth in many cases.

Society also has a need to replace members who die or become incapacitated. This need often is codified in cultural or religious values that encourage bearing and raising children. Some societies look upon families with few or no children with pity or contempt. The idea of deliberately controlling fertility may be shocking, even taboo. Women who are pregnant or have small children have special status and protection. Boys frequently are more valued than girls because they carry on the family name and are expected to support their parents in old age. Couples may have more children than they really want in an attempt to produce a son.

Male pride often is linked to having as many children as possible. In Niger and Cameroon, for example, men, on average, want 12.6 and 11.2 children, respectively. Women in these countries consider the ideal family size to be only about half of what their husbands desire. Even though a woman might desire fewer

Principles of Environmental Science

children, however, she may have few choices and little control over her own fertility. In many societies, a woman has no status outside of her role as wife and mother. Without children, she may have no source of support in her old age.

Birth Reduction Pressures

In more highly developed countries, many pressures tend to reduce fertility. Higher education and personal freedom for women often result in decisions to limit childbearing. A desire to spend time and money on other goods and activities offsets the desire to have children. When women have opportunities to earn a salary, they are less likely to stay home and have many children. Not only do many women find the challenge and variety of a career attractive, but the money that they earn outside the home becomes an important part of the family budget. Thus, education and socioeconomic status are usually inversely related to fertility in richer countries. In some developing countries, however, fertility initially increases as educational levels and socioeconomic status rise. With higher income, families are better able to afford the children they want. More money also means that women are healthier and therefore better able to conceive and carry a child to term. It may be a generation before this unmet desire for children abates.

In less-developed countries, where feeding and clothing children can be a minimal expense, adding one more child to a family usually doesn't cost much. By contrast, raising a child in a developed country can cost hundreds of thousands of dollars by the time the child finishes school and is independent. Under these circumstances, parents are more likely to choose to have one or two children on whom they can concentrate their time, energy, and financial resources.

Figure 4.11 shows U.S. birth rates between 1910 and 2000. As you can see, birth rates have fallen and risen in an irregular pattern. The period between 1910 and 1930 was a time of industrialization and urbanization. Women were getting more education than ever before and entering the workforce in large numbers. The

FIGURE 4.10 *Children in rural areas can help with many household chores, such as tending livestock or caring for younger children.*

Great Depression in the 1930s made it economically difficult for families to have children, and birth rates were low. The birth rate increased at the beginning of World War II (as it often does in wartime). For reasons that are unclear, a higher percentage of boys are usually born during war years.

A "baby boom" followed World War II, as couples were reunited and new families started. During this time, the government encouraged women to leave their wartime jobs and stay home. A high birth rate persisted through the times of prosperity and optimism of the 1950s, but began to fall in the 1960s. Part of this decline was caused by the small number of babies born in the

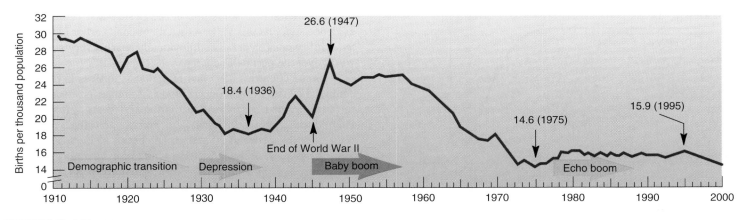

FIGURE 4.11 *Birth rates in the United States, 1910–2001. The falling birth rate from 1910 to 1929 represents a demographic transition from an agricultural to an industrial society. Note that this decline occurred before the start of the Great Depression. The baby boom following World War II lasted from 1945 to 1957. A much smaller "echo boom" occurred around 1980 when the baby boomers started to reproduce, but it produced far fewer births than anticipated.*

Sources: Data from Population Reference Bureau, Inc., and U.S. Bureau of the Census.

1930s, which resulted in fewer young adults to give birth in the 1960s. Part was due to changed perceptions of the ideal family size. Whereas in the 1950s women typically wanted four children or more, the norm dropped to one or two (or no) children in the 1970s. A small "echo boom" occurred in the 1980s as baby boomers began to have babies, but changing economics and attitudes seem to have permanently altered our view of ideal family size in the United States.

DEMOGRAPHIC TRANSITION

In 1945, demographer Frank Notestein pointed out that a typical pattern of falling death rates and birth rates due to improved living conditions usually accompanies economic development. He called this pattern the **demographic transition** from high birth and death rates to lower birth and death rates. Figure 4.12 shows an idealized model of a demographic transition. This model is often used to explain connections between population growth and economic development.

Development and Population

The left side of figure 4.12 represents the conditions in a premodern society. Food shortages, malnutrition, lack of sanitation and medicine, accidents, and other hazards generally keep death rates in such a society around 30 per 1,000 people. Birth rates are correspondingly high to keep population densities relatively constant. As economic development brings better jobs, medical care, sanitation, and a generally improved standard of living, death rates often fall very rapidly. Birth rates may actually rise at first as more money and better nutrition allow people to have the children they always wanted. Eventually, however, birth rates fall as people see that all their children are more likely to survive and that the whole family benefits from concentrating more resources on fewer chil-

dren. Note that populations grow rapidly during the time that death rates have already fallen but birth rates remain high. Depending on how long it takes to complete the transition, the population may go through one or more rounds of doubling before coming into balance again.

The right-hand side of each curve in figure 4.12 represents conditions in many developed countries, where the transition is complete and both birth rates and death rates are low, often, a third or less than those in the predevelopment era. The population comes into a new equilibrium in this phase, but at a much larger size than before. Most of the countries of northern and western Europe went through a demographic transition in the nineteenth or early twentieth century similar to the curves shown in figure 4.12. In countries like Italy, where fertility levels have fallen below replacement rates, there are now fewer births than deaths, and the total population curve has started to decline.

Many of the most rapidly growing countries in the world, such as Kenya, Yemen, Libya, and Jordan, now are in the middle phase of this demographic transition. Their death rates have fallen close to the rates of the fully developed countries, but birth rates have not fallen correspondingly. In fact, both their birth rates and total population are higher than those in most European countries when industrialization began 300 years ago. The large disparity between birth and death rates means that many developing countries now are growing at 3 to 4 percent per year. Such high growth rates in the Third World could boost total world population to 9 billion or more before the end of the twenty-first century. This raises what may be the two most important questions in this entire chapter: why are birth rates not yet falling in these countries, and what can be done about it?

An Optimistic View

Some demographers claim that a demographic transition already is in progress in most developing nations. They believe that problems in taking censuses and a normal lag between falling death and birth rates may hide this for a time, but that the world population should stabilize sometime in the twenty-first century. Some evidence supports this view. As we mentioned earlier in this chapter, fertility rates have fallen dramatically nearly everywhere in the world over the past half century.

Some countries have had remarkable success in population control. In Thailand, Indonesia, and Colombia, for instance, total fertility dropped by more than half in 20 years. Morocco, the Dominican Republic, Jamaica, Peru, and Mexico all have seen fertility rates fall between 30 and 40 percent in a single generation. Surprisingly, one of the most successful family planning programs in recent years has been in Iran (see Case Study, p. 89).

The following factors help to stabilize populations:

- Growing prosperity and social reforms that accompany development reduce the need and desire for large families in most countries.

- Technology is available to bring advances to the developing world much more rapidly than was the case a century ago,

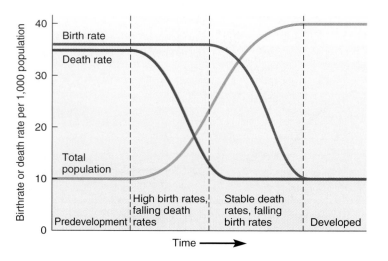

FIGURE 4.12 *Theoretical birth, death, and population growth rates in a demographic transition accompanying economic and social development.*

Principles of Environmental Science

FAMILY PLANNING IN IRAN

After the Islamic Revolution in 1979, Iran had one of the world's highest population growth rates. In spite of civil war, large-scale emigration, and economic austerity, the country surged from 34 million to 63 million in just 20 years. A crude birth rate of 43.4/1,000 people and a total fertility rate of 5.1 per woman during this time resulted in annual population growth of 3.9 percent and a doubling time of less than 18 years. Religious authorities exhorted couples to have as many children as Allah would give them. Any mention of birth control or family planning (other than to have as many children as possible) was forbidden, and the marriage age for girls was dropped to nine years old. When a devastating war with Iraq in the 1980s killed at least 1 million young soldiers, producing more children to rebuild the army became a civic as well as religious duty.

In the late 1990s, however, the Iranian government became aware of the costs of such rapid population growth. With religious moderates gaining greater political power, public policy changed abruptly. Now, the Iranian government is spending millions of dollars to lower birth rates. Couples must pass a national family planning course before they are allowed to marry. While it took a few years to convince people that this change will be long-lasting, most Iranian citizens are now eager for access to birth-control information. Family planning classes are sought out both by engaged couples and those already married. A wide range of birth control methods are available. Implantable or injectable slow-release hormones, condoms, intrauterine devices (IUDs), pills, and male or female sterilization are free to all. Billboards, newspapers, television, and even water towers advertise this national program. Religious leaders have issued a *fatwah,* or command, that all faithful Muslims participate in family planning.

As a consequence, Iran has been remarkably successful in stemming its population growth. Between 1986 and 1996, the fertility rates for urban residents dropped almost by half, to less than three children per women, and the crude birth rate dropped from 43 to 18 per 1,000 people. By 2000, the average annual growth rate had fallen to 1.4 percent. While the population is still increasing, another decade of such progress would bring the country to a stable or even declining rate of growth.

Several societal changes have contributed to this rapid birth reduction. While the minimum marriage age has been returned to 15, couples are encouraged to wait until at least age 20 to begin their families. The educational benefits of concentrating the family resources on just one or two children are being promoted. Although women's roles are still highly restricted in the Islamic Republic, greater gender equity has given women more control over their reproductive lives. Access to modern, information-age jobs gives people an incentive to seek out education both for themselves and their children.

The demographic transition hasn't spread to all levels of Iranian society. Rural families, ethnic minorities, and some urban poor still tend to have many children. Still, this example of how quickly both ideals of the perfect family size and information about modern birth control can spread through a society—even a highly religious, fundamentalist one—is encouraging for what might be accomplished worldwide in a surprisingly short time.

and the rate of technology exchange is much faster than it was when Europe and North America were developing.

- Less-developed countries have historic patterns to follow. They can benefit from the mistakes of more-developed countries and chart a course to stability relatively quickly.

- Modern communications (especially television) have caused a revolution of rising expectations that act as stimuli to spur change and development.

A Pessimistic View

Economist Lester Brown of the Worldwatch Institute takes a more pessimistic view of world populations. He warns that many of the poorer countries of the world appear to be caught in a "demographic trap" that prevents them from escaping from the middle phase of the demographic transition. Their populations are now growing so rapidly that human demands exceed the sustainable yield of local forests, grasslands, croplands, and water resources. The resulting resource shortages, environmental deterioration, economic decline, and political instability may prevent these countries from ever completing modernization. Their populations may continue to grow until catastrophe intervenes.

Many people believe that the only way to break out of the demographic trap is to immediately and drastically reduce population growth by whatever means are necessary. They argue strongly for birth control education and bold national policies to encourage lower birth rates. Some agree with Malthus that helping the poor will simply increase their reproductive success and further threaten the resources on which we all depend. Author Garret Hardin described this view as lifeboat ethics:

> Each rich nation amounts to a lifeboat full of comparatively rich people. The poor of the world are in other much more

crowded lifeboats. Continuously, so to speak, the poor fall out of their lifeboats and swim for a while, hoping to be admitted to a rich lifeboat, or in some other way to benefit from the goodies on board. . . . We cannot risk the safety of all the passengers by helping others in need. What happens if you share space in a lifeboat? The boat is swamped and everyone drowns. Complete justice, complete catastrophe.

A Social Justice View

A third view is that social justice (a fair share of social benefits for everyone) is the real key to successful demographic transitions. The world has enough resources for everyone, in this view, but inequitable social and economic systems cause maldistributions of those resources. Hunger, poverty, violence, environmental degradation, and overpopulation are symptoms of a lack of justice, rather than a lack of resources. Although overpopulation exacerbates other problems, a narrow focus on overpopulation alone encourages racism and hatred of the poor. Proponents of the social justice view argue that people in the richer nations should recognize our excess consumption levels and the impact that this consumption has on others. Figure 4.13 expresses the opinion of many people in less-developed countries about the relationship between resources and population.

Infant Mortality and Women's Rights

The 1994 International Conference on Population and Development in Cairo, Egypt, signaled a new approach to population issues. A broad consensus reached by the 180 participating countries agreed that responsible economic development, education and empowerment of women, and high-quality health care (including family-planning services) must be accessible to everyone if population growth is to be slowed. Child survival is one of the most critical factors in stabilizing population. When infant and child mortality rates are high, as they are in much of the developing world, parents tend to have high numbers of children to ensure that some will survive to adulthood. There has never been a sustained drop in birth rates that was not first preceded by a sustained drop in infant and child mortality.

One of the most important distinctions in our demographically divided world is the high infant mortality rate in the less developed countries. Better nutrition, improved health care, simple oral rehydration therapy, and immunization against infectious diseases (see chapter 8) have dramatically reduced child mortality rates, which have been accompanied in most regions by falling birth rates. Estimates indicate that saving 5 million children each year from easily preventable communicable diseases would avoid 20 or 30 million extra births.

FIGURE 4.13 *Controlling our population and resources—there may be more than one side to the issue.*
Used with permission of the Asian Cultural Forum on Development.

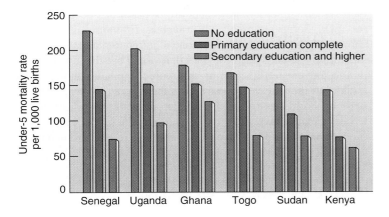

FIGURE 4.14 *Childhood deaths drop as female education rises. The under-5 mortality rate for selected African countries is compared to level of female education.*
Source: World Resources Institute, *World Resources 1998–99.*

Increasing family income does not always translate into better welfare for children, since men in many cultures control most financial assets. As the UN Conference in Cairo noted, the best way to improve child survival often is to ensure the rights of mothers. Opportunities for women's education, for instance, as well as land reform, political rights, opportunities to earn an independent income, and improved health status of women often are better indicators of family welfare than is rising gross national product (fig. 4.14).

FAMILY PLANNING AND FERTILITY CONTROL

Family planning allows couples to determine the number and spacing of their children. It doesn't necessarily mean fewer children—people may use family planning to have the maximum number of children possible—but it does imply that the parents will control their reproductive lives and make rational, conscious decisions about how many children they will have and when those children will be born, rather than leaving it to chance. As the desire for smaller families becomes more common, birth control often becomes an essential part of family planning. In this context, **birth control** usually means any method used to reduce births, including celibacy, delayed marriage, contraception, methods that prevent embryo implantation, and induced abortions.

Traditional Fertility Control

Evidence suggests that people in every culture and every historic period have used a variety of techniques to control population size. Studies of hunting and gathering people, such as the !Kung, or San, of the Kalahari Desert in southwest Africa, indicate that our early ancestors had stable population densities, not because they killed each other or starved to death regularly, but because they controlled fertility.

For instance, San women breast-feed children for 3 or 4 years. When calories are limited, lactation depletes body fat stores and suppresses ovulation. Coupled with taboos against intercourse while breast-feeding, this is an effective way of spacing children. (However, breast-feeding among well-nourished women in modern societies doesn't necessarily suppress ovulation or prevent conception.) Other ancient techniques to control population size include celibacy, folk medicines, abortion, and infanticide. We may find some or all of these techniques unpleasant or morally unacceptable, but we shouldn't assume that other people are too ignorant or too primitive to make decisions about fertility.

Current Birth Control Methods

Modern medicine gives us many more options for controlling fertility than were available to our ancestors. The major categories of birth control techniques include: (1) avoidance of sex during fertile periods (for example, celibacy or using changes in body temperature or cervical mucus to judge when ovulation will occur), (2) mechanical barriers that prevent contact between sperm and egg (for example, condoms, spermicides, diaphragms, cervical caps, and vaginal sponges), (3) surgical methods that prevent release of sperm or egg (for example, tubal ligations in females and vasectomies in males), (4) chemicals that prevent maturation or release of sperm or eggs, or that prevent embryo implantation in the uterus (for example, estrogen plus progesterone, or progesterone alone, for females; gossypol for males) (fig. 4.15), (5) physical barriers to implantation (for example, intrauterine devices), and (6) abortion.

Norplant, the trade name for flexible, matchstick-sized, silicon-rubber implants containing a slow-release analog of progesterone, is popular among women who want a reversible, semipermanent birth-control method. The implants are inserted under the skin, where they release hormones for up to five years. Depo-Provera, a progesterone analog, is given by injections four times a year. Both implants and injections have very low failure rates (0.3 percent, compared to about 1 percent for oral contracep-

FIGURE 4.15 *Modern birth control methods can be highly effective when used properly.*

tives) and eliminate the need to keep track of and take daily pills. They also can be used without knowledge of one's partner, who may oppose birth control. Some women experience increased vaginal bleeding or absence of menstrual periods from the injections or implants. Norplant may be linked to ovarian cysts, and the inserts have been difficult to remove in some cases.

Condom use has more than doubled over the past 20 years, from about 3.5 million users in 1980 to 8 million in 2000. While condoms have about a 10 percent failure rate when used alone, their effectiveness is increased when used with spermicidal creams. Condoms also are important protection against sexually transmitted diseases. A condom for women consists of two flexible plastic rings connected by a strong, clear polyurethane sheath. One ring fits over the cervix much like a diaphragm, while the other remains outside the vagina. Each condom costs about two dollars and is designed to be discarded after a single use. The 6-month failure rate is about 12 percent, which means that 12 percent of women using only this device will get pregnant in 6 months. A female condom gives women control over their own reproduction, but some women find it difficult or unpleasant to use. None of these methods is perfect, and none suits every contraceptive need. Many require careful, conscientious use. Which choice is best for you depends on your life situation and your plans for the future.

For nearly 20 years, the French drug RU486 (mifepristone or mifegyne) has been used as a "morning after" drug in Europe. RU486 blocks the effects of progesterone in maintaining the lining of the uterine wall. It is usually administered together with misoprostol (a prostaglandin analog), which causes uterine contraction and expulsion of the fetus. Interestingly, RU486 also appears to have promise in treating breast cancer, brain cancer, diabetes, and hypertension. RU486 has been approved for use in the United States, but restrictions on its use may make it unavailable for many women.

Recently, a combination of two drugs previously on the market for other uses has been shown to be as safe and effective as RU486. Methotrexate and misoprostol are administered a week apart to induce abortion.

New Developments in Birth Control

Researchers are testing more than a hundred new contraceptive methods, and some appear to have great promise. One of these is an adhesive patch that dispenses progesterone through the skin. Each patch, about the size of a half dollar, lasts for about a week. It can be worn on the arm, abdomen, or buttocks. Soft, silicone vaginal rings containing slow-release progesterone analogs have all the advantages of Depo-Provera or Norplant, but should be much cheaper and easier to use since the woman can insert and remove them. Researchers are also testing biodegradable implants for both women and men that would not require removal. Development of simple, inexpensive, do-it-yourself tests for levels of estrogen and progesterone in urine may make the rhythm method easier to follow and more reliable for women who cannot or would rather not use other methods of birth control.

A group of drugs known as gonadotropin releasing-hormone agonists show promise in suppressing egg and sperm develop-

ment. A slow-release anti-gonadotropin patch has been proposed for men. Some antipregnancy vaccines (immunization against chorionic gonadotropin—a hormone required to maintain the uterine lining) and antisperm vaccines that would use the immune system to prevent fertilization or embryonic implantation are being tested, but they are years away from the market. Clinical trials of two oral hormones for men (testosterone undeconate and cyproterone acetate) have shown some promise in suppressing sperm production, but it may be years before they are ready to market or before men are willing to take them.

THE FUTURE OF HUMAN POPULATIONS

How many people will be in the world a century from now? Most demographers believe that world population will stabilize sometime during the twenty-first century. When we reach that equilibrium, the total number of humans is likely to be somewhere around 8 to 10 billion, depending on the success of family-planning programs and the multitude of other factors affecting human populations. The United Nations Population Division projects three population scenarios (fig. 4.16). The optimistic (low) projection shows that world population might stabilize by about 2030 and begin to drop slightly. The medium projection suggests that growth might continue to rise slowly until at least 2050. The most pessimistic (high) projection indicates that population growth will climb at present rates for 50 years or more.

Which of these scenarios will we follow? As you have seen in this chapter, population growth is a complex subject. Stabilizing or reducing human populations will require substantial changes from business as usual. One impediment is that the U.S. Congress

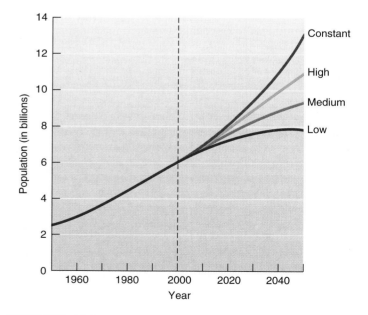

FIGURE 4.16 *Actual and projected world population, 1950 to 2050 under different growth scenarios.*
Source: United Nations Population Division, 2001.

Principles of Environmental Science www.mhhe.com/environmentalscience

refuses to make payments to the United Nations Family Planning Fund because some of the 135 countries that receive UN aid include abortion as part of population control programs. Population issues were conspicuously absent from the agenda of the UN Conference on Environment and Development in Rio de Janeiro (the Earth Summit) in 1992 because of opposition from religious groups and developing countries. At a special UN Conference on Population held in Cairo, Egypt, in 1994, religion again was a complicating factor. This conference did, however, come out strongly in favor of the rights of women and children, and committed all nations to provide family-planning services to the developing world to limit population growth.

An encouraging sign is that worldwide contraceptive use has increased sharply in recent years. About half of the world's married couples used some family-planning techniques in 2000, compared to only 10 percent 30 years earlier, but another 300 million couples say they want but do not have access to family planning. Contraceptive use varies widely by region, with high levels in Latin America and East Asia, but relatively low use in much of Africa (fig. 4.17). Note the differences both in the percentages of couples using birth control in different regions and in the methods used. The IUD and male sterilization are common in Asia, but rarely used in Africa or Latin America. What economic or cultural factors do you see in these patterns? (See the related story "Family Planning in Thailand" at www. mhhe.com/apps.)

The World Health Organization estimates that nearly 1 million conceptions occur daily around the world as a result of some 100 million sex acts. At least half of those conceptions are unplanned or unwanted. Still, birth rates already have begun to fall in East Asia and Latin America. Similar progress is expected in South Asia in a few years. Only Africa will probably continue to grow in the twenty-first century. Coordinated, effective family planning requires governmental action, ranging from policies to assistance.

Successful family-planning programs often require significant societal changes. Among the most important of these are (1) improved social, educational, and economic status for women (birth control and women's rights are often interdependent); (2) improved status for children (fewer children are born if they are not needed as a cheap labor source); (3) acceptance of calculated choice as a valid element in life in general and in fertility in particular (the belief that we have no control over our lives discourages a sense of responsibility); (4) social security and political stability that give people the means and the confidence to plan for the future; and (5) the knowledge, availability, and use of effective and acceptable means of birth control. Concerted efforts to bring about these types of societal changes can be effective. Twenty years of economic development and work by voluntary family-planning groups in Zimbabwe, for example, have lowered total fertility rates from 8.0 to 5.5 children per woman on average. Surveys show that desired family sizes have fallen nearly by half (9.0 to 4.6) and that nearly all women and 80 percent of men in Zimbabwe use contraceptives.

The current world average fertility rate of 2.7 births per woman is less than half what it was 50 years ago. If similar

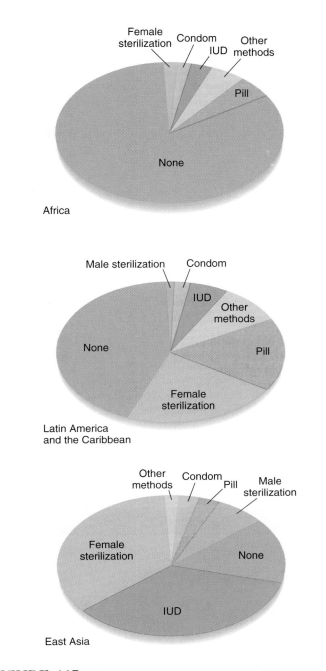

FIGURE 4.17   *Variations in contraceptive use in different regions. Preferences for one form of contraception over another reflect both cultural attitudes and government programs.*
Source: Data from United Nations Family Planning Agency, 1995.

progress could be sustained for the next half century, fertility rates in 2050 would be 1.5 per woman, and total population would be declining rapidly. By the end of the twenty-second century, the total population could be down around 2 billion, and we might come into equilibrium with the carrying capacity of our environment after all. Whether this scenario comes true or not depends on choices that all of us make.

SUMMARY

Human populations have grown at an unprecedented rate over the past three centuries. By 1999, the world population passed 6 billion people. If the current growth rate of 1.3 percent per year persists, the population will double in 54 years. Almost all of that growth will occur in the less-developed countries of Asia, Africa, and Latin America. A serious concern is that the number of humans in the world and our impact on the environment will overload the earth's life-support systems.

Thomas Malthus predicted, some 200 years ago, that populations always outstrip their food supplies unless the negative restraints of starvation, crime, and misery bring numbers into balance with resources. Karl Marx, on the other hand, argued that social justice and equitable distribution of the benefits of technology would provide enough for everyone. In fact, population growth, poverty, injustice, and environmental degradation are all interrelated. We probably can't solve one of these problems without also addressing the others.

The total fertility rate is the number of children born to an average woman in a population. The crude birth rate is the number of births in a year per thousand persons, while the crude death rate is the average number of deaths per thousand persons. The crude birth rate minus the crude death rate gives the rate of natural increase. When this rate reaches a level at which people are just replacing themselves, zero population growth is achieved. For most developed countries, this rate is an average of about 2.1 children per couple.

In the more highly developed countries of the world, population growth has slowed or even reversed in recent years so that without immigration from other areas, populations would be declining. The change from high birth and death rates that accompanies industrialization is called a demographic transition. Many developing countries have already begun this transition. Death rates have fallen, but birth rates remain high. Some demographers believe that, as infant mortality drops and economic development progresses so that people in these countries are more sure of a secure future, the countries will complete the transition to a stable population. Others fear that excessive population growth and limited resources will catch many of the poorer countries in a demographic trap that could prevent them from ever achieving a stable population or a high standard of living.

While larger populations bring many problems, they also may be a valuable resource of energy, intelligence, and enterprise that will make it possible to overcome resource limitation problems. We have many more options now for controlling fertility than were available to our ancestors. Some techniques are safer than those available earlier; many are easier and more pleasant to use. Sometimes, making family-planning programs successful requires significant cultural changes, including improved social, educational, and economic status for women; higher values on individual children; increased responsibility for our own lives; social security and political stability that give people the means and confidence to plan for the future; and knowledge, availability, and use of effective and acceptable means of birth control.

QUESTIONS FOR REVIEW

1. At what point in history did the world population pass its first billion? What factors restricted population before that time, and what factors contributed to growth after that point?

2. How might growing populations be beneficial in solving development problems?

3. Why do some economists consider human resources more important than natural resources in determining a country's future?

4. Where will most population growth occur during the twenty-first century? What conditions contribute to rapid population growth in some countries?

5. Define crude birth rate, total fertility rate, crude death rate, and zero population growth.

6. What is the difference between life expectancy and life span? Why are they different?

7. What is the dependency ratio, and how might it affect the United States in the future?

8. What factors increase or decrease people's desires to have babies?

9. Describe the conditions that lead to a demographic transition.

10. Describe the major choices in modern birth control.

THINKING SCIENTIFICALLY

1. Suppose that you were head of a family-planning agency in India. How would you design a scientific study to determine the effectiveness of different approaches to population stabilization? How would you account for factors such as culture, religion, education, and economics?

2. Why do you suppose that the United Nations gives high, medium, and low projections for future population growth? Why not give a single estimate? What factors would you consider in making these projections?

3. Some demographers claim that the total world population has already begun to slow, while others dispute this claim. How would you recognize a true demographic transition, as opposed to mere random fluctuations in birth and death rates?

4. Why do we usually express crude birth and death rates per thousand people? Why not give the numbers per person or for the entire population?

5. In northern Europe, the demographic transition began in the early 1800s, a century or more before the invention of modern antibiotics and other miracle drugs. What factors do you think contributed to this transition? How would you use historical records to test your hypothesis?

6. In chapter 3, we discussed carrying capacities. What do you think the maximum and optimum carrying capacities for humans are? Why is this a more complex question for humans than it might be for other species? Why is designing experiments in human demography difficult?

KEY TERMS

birth control 91
crude birth rate 83
crude death rates 84
demographic transition 88
demography 80
dependency ratio 86
family planning 91

life expectancy 84
neo-Malthusians 79
pronatalist pressures 86
total fertility rate 83
zero population growth (ZPG)
    83

Welcome to McGraw-Hill's Online Learning Center

Location: http://www.mhhe.com/environmentalscience

Mc Graw Hill

WEB EXERCISE

Exploring Growth Factors in Population Data

How and why populations grow is a key question in environmental science. In this exercise, you examine and graph current world population data to explore which factors most strongly correlate with birth rates. Go to www.mhhe.com/apps.

There you will find an Excel data file named popdata.xls. Double-click on the file name to copy it to your hard disk. If you have Excel on your computer, you should be able to open the data file by double-clicking on it. (Other spreadsheet programs can also read this file, but you must open it from within your program, not by double-clicking.)

1. This file contains population data for the countries of the world, sorted by the United Nations Human Development Index (HDI) rank. First look at the top 20 countries. Where are they? What is the range of income levels (in GNP per capita) of the top 20? What is the range of income for the bottom 20 countries?

2. Now make an X,Y scatter graph of Adult Literacy and Birth Rate. (Detailed instructions for making graphs in Excel are included at the far right side of the spreadsheet page [column N].) How would you describe the relationship between these variables? How would you

explain this relationship? Keep this graph in your spreadsheet while you make three more scatter graphs: (1) GNP per Capita and Birth Rate, (2) Life Expectancy and Birth Rate, and (3) Infant Mortality and Birth Rate. Describe the trends you observe, and explain what they mean.

3. How would you compare the relative amount of scatter in each of your graphs? Why do some curves slope from right to left, while others slope in the opposite direction? If you draw a line through the middle of the dot cluster, some curve smoothly, while others seem to have a break or inflection point. How would you interpret these patterns?

4. Try changing the shape of your graphs. (See instructions on the right side of the spreadsheet to do this.) How does making the graphs taller or wider affect the way your trends look? How could you deliberately manipulate the graph shape to affect other people's interpretation of the data? Is this ethical? Have you ever seen it done?

5. Now make a dot graph of GNP Per Capita and Adult Literacy. Is there a linear relationship between the two variables? Why or why not?

5 Biomes and Biodiversity

The first rule of intelligent tinkering is to save all the pieces.

–Aldo Leopold

Biodiversity is essential for the ecological services on which we all depend.

OBJECTIVES

After studying this chapter, you should be able to

- recognize the characteristics and general distribution of major biomes, and understand the most important factors that determine biome distributions.
- evaluate the degree of disturbance of major biomes, and identify the biomes that are most important for human activities and for biodiversity protection.
- define biodiversity and explain its importance.
- report on the total number and relative distribution of living species on the earth.
- describe how human activities cause biodiversity losses.
- identify regions and ecosystems of high biodiversity.
- evaluate the effectiveness of efforts to protect endangered species.

Welcome to McGraw-Hill's Online Learning Center

Location: http://www.mhhe.com/environmentalscience

LEARNING ONLINE

Visit our webpage at www.mhhe.com/environmentalscience for data sources, further readings, additional case studies, current environmental news, and regional examples within the Online Learning Center to help you understand the material in this chapter. You'll also find active links to information pertaining to this chapter including:

Biomes
Center for Restoration Ecology
Landscape Ecology and Birds
Restoration and Reclamation Review
Biodiversity and WORLDMAP
Ecology and Biodiversity
Biodiversity Profile of India

Using GIS to Protect Biodiversity

As we search for better ways to conserve species and their habitats, a new and increasingly important tool is geographic information systems (GIS). A GIS includes computer programs that map, manage, and analyze information about the environment together with environmental data. When scientists want to overlay, integrate, or model information such as species distribution, habitat and land cover areas, or water resources, computer mapping software and data provide efficient tools for investigation. A GIS can provide opportunities to ask new questions about the environment, or to evaluate and manage resources in ways that we couldn't do before.

One of the most important conservation biology projects using GIS today is the Gap Analysis Program (GAP). The idea behind GAP is that many biologically diverse biomes and habitats are not sufficiently protected. They represent "gaps" in conservation programs. To preserve long-term biodiversity, scientists in this program hope to fill as many of those gaps as possible by identifying and preserving important areas (fig. 5.1).

Finding these gaps requires integrating a great range of information sources. Data from field surveys, land cover information from satellite-captured images, water and soil information, topography, and other factors all help identify areas of ecological significance. A GIS is the most efficient way to incorporate these diverse data sources. Once these and other sources of information are combined into a single, layered digital map, researchers can compare the distribution of critical habitat areas to locations of existing conservation lands. They can identify unprotected lands, and find out who owns them or how they are currently used. Scientists can calculate and compare the extent of protected and unprotected habitat. Using topography and land cover features, they can select priority sites for further field surveys. They can estimate regional populations of threatened species, identify land owners who should be contacted in priority areas, estimate the cost of managing conservation lands, or calculate the density of threatened species per square mile.

The Gap Analysis Program is also an excellent example of collaborative science. Individuals and groups from universities, state governments, conservation organizations, and federal agencies all exchange information and ideas to produce results much greater than they could do individually.

GIS is an important tool in many types of planning: city planners, transportation managers, even farmers use it. Increasingly GIS is a tool that is helping researchers evaluate and manage environmental resources. Biome conservation and biological diversity, the topics of this chapter, are becoming slightly more manageable problems because of improved information availability with GIS and other information technologies, such as remote sensing (see Investigating Our Environment: Remote Sensing of the Earth's Surface, p. 203). See web exercises at the end of this chapter, as well as our website, www.mhhe.com/apps, for more on GIS in conservation.

FIGURE 5.1 *Gap analysis is an important example of using a Geographic Information System (GIS) for conserving biodiversity. Digital data help scientists quantify, analyze, and model environmental problems.*
Source: USGS, Biological Resource Division, Florida Gap Analysis Program.

BIOMES

The biodiversity we see around us occurs partly because the earth has many different combinations of environmental conditions that support the amazing range of organisms around us. We can understand the range of environment types more easily if we classify them into a few broad groups. **Biomes** can be considered life zones, environments with similar climatic, topographic, and soil conditions, and roughly comparable biological communities. Most biomes are identified by the dominant plants of their communities (for example, grassland or deciduous forest). There can be considerable variation in the individual species that make up biological communities at two similar sites. But recognizing biome categories gives us insights about the general kinds of plants and animals that we might expect to find there and how they may be adapted to their particular environment.

Climates and Biomes

Temperature and precipitation are among the most important determinants in biome distribution. If we know the general temperature range and precipitation level, we can predict what kind of biological community is likely to develop on a particular site if that site is free of disturbance for a sufficient time (fig. 5.2). Biome distribution also is influenced by the prevailing landforms of an area. Mountains, in particular, exert major influences on biological communities.

Figure 5.3 shows the distribution of major terrestrial biomes around the world. Many variations exist within each biome type; these could be seen on more local maps. Note, however, that many biomes occupy characteristic ranges of latitude: tundra occurs only in cold regions near the poles, while tropical forests occur only within the tropics—near the equator.

Often, it is the *relationship* between temperature and precipitation that controls biome distribution and conditions. The reason for this is evapotranspiration, or the combination of moisture loss by evaporation and by plant transpiration—water taken up by plant roots and released to the atmosphere as part of plants' respiration and cooling mechanisms. In hot places, available moisture dissipates quickly through both evaporation and transpiration. In cool environments, evapotranspiration proceeds slowly, allowing moisture to accumulate.

This relationship is displayed using climagraphs, or climate diagrams (fig. 5.4). A climagraph shows temperature on the left axis and precipitation on the right for each month of the year. The difference between temperature and precipitation is what makes climagraphs interesting. In figure 5.4, yellow areas show months in which evapotranspiration is greater than available precipitation: if rain falls, it is used or evaporated quickly. Blue areas show

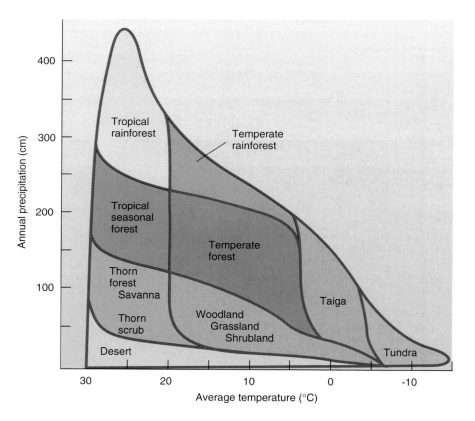

FIGURE 5.2 *Biomes most likely to occur in the absence of human disturbance or other disruptions, according to average annual temperature and precipitation.* Note: *This diagram does not consider soil type, topography, wind speed, or other important environmental factors. Still, it is a useful general guideline for biome location.*

From *Communities and Ecosystems,* 2/e by R. H. Whitaker, © 1975. Reprinted by permission of Prentice Hall, Upper Saddle River, New Jersey.

Principles of Environmental Science   www.mhhe.com/environmentalscience

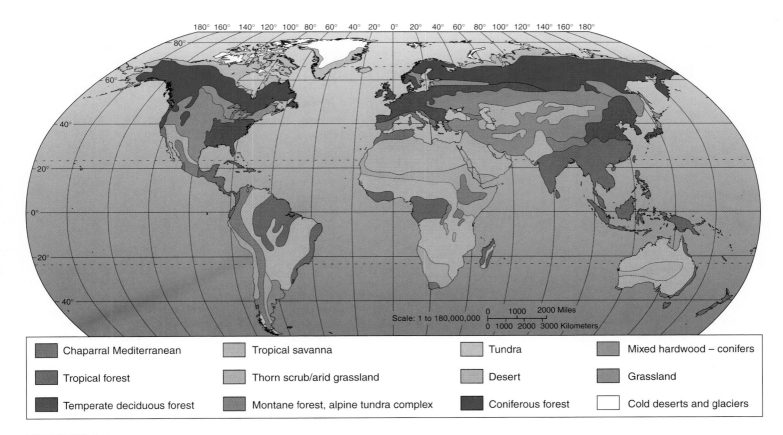

Scale: 1 to 180,000,000

0   1000   2000 Miles
0 1000 2000 3000 Kilometers

Chaparral Mediterranean

Tropical forest

Temperate deciduous forest

Tropical savanna

Thorn scrub/arid grassland

Montane forest, alpine tundra complex

Tundra

Desert

Coniferous forest

Mixed hardwood – conifers

Grassland

Cold deserts and glaciers

FIGURE 5.3   *Major world biomes. Compare this map to figure 5.2 for general climate conditions.*

months with a net excess of precipitation: water accumulates in the soil, and rivers and lakes may become full.

Compare January and February (marked J and F on the horizontal axes) in Taiyuan, China, and Yuma, Arizona. Both places receive only slight moisture, but Taiyuan is cold, while Yuma is about 15°C. Yuma has a net deficit of moisture, colder Taiyuan has a net surplus. You can find similar contrast in April in Tiksi, Russia, and San Diego, USA. Later in the spring, Taiyuan receives more moisture, but rising temperatures increase evapotranspiration. A moisture deficit results for April and May.

Seasonal changes are also evident in a climagraph. Bombay, India, has extreme water deficits in winter, but in summer there is a huge surplus of moisture, which recharges wells and groundwater, fills rivers, and supports plant growth well into the next winter's dry season. Yuma, Arizona, on the other hand, has moisture deficits year-round. There, only specially adapted plants and animals can survive the continuous dry conditions.

Which of the graphs in figure 5.4 most resembles the place where you live? How do seasonal changes affect growing conditions and life-forms in your biome?

As you read the biome descriptions in this chapter and look at the biome photos, think about how water availability, heat, or freezing temperatures might influence the types of organisms that occur in each area.

Deserts

Deserts are dry, with precipitation that is both infrequent and unpredictable from year to year. With little moisture to absorb and store heat, daily and seasonal temperatures can fluctuate widely. Deserts with less than 2.5 cm (1 in) of measurable precipitation support almost no vegetation. Deserts with 2.5 to 5 cm (1 to 2 in) annual precipitation have sparse vegetation (less than 10 percent of the ground is covered), and plants in this harsh climate need a variety of specializations to conserve water and protect tissues from predation. Seasonal leaf production, water-storage tissues, and thick epidermal layers help reduce water loss. Spines and thorns discourage predators while also providing shade (fig. 5.5). Many desert animals avoid the drying sun by feeding at night, and most acquire water from the seeds and vegetation they eat.

Warm, dry, high-pressure atmospheric conditions (see chapter 9) create broad bands of deserts around the world at about 30° north and south latitude. This band includes deserts in the American Southwest, North and South Africa, China, and Australia. The coastal deserts of South America and Africa are among the driest regions in the world. Because vegetation grows so slowly in deserts, recovery after human disturbance or overgrazing can take centuries.

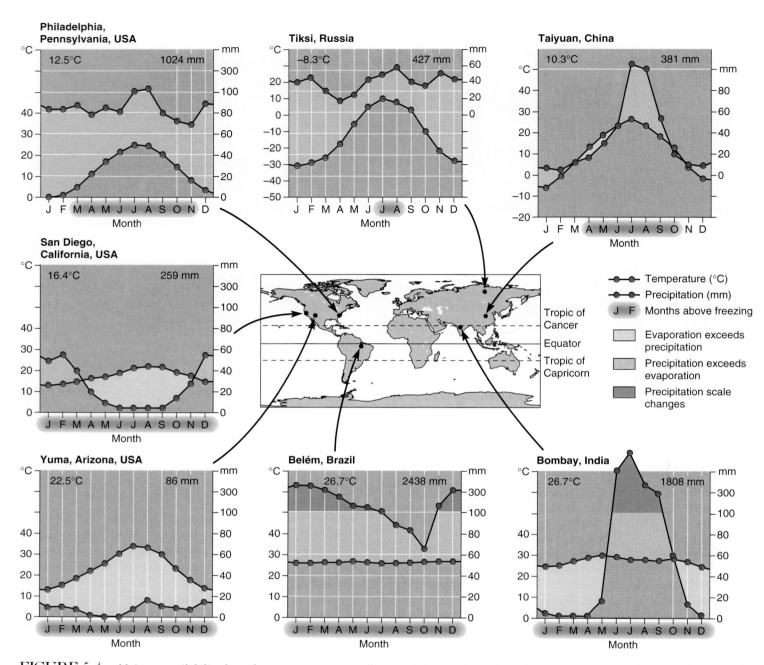

FIGURE 5.4 *Moisture availability depends on temperature as well as precipitation. The horizontal axis on these climate diagrams shows months of the year; vertical axes show temperature* (left side) *and precipitation* (right). *The number of dry months* (shaded yellow) *and wetter months* (blue) *varies with geographic location. Mean annual temperature* (°C) *and precipitation* (mm) *are shown at the top of each graph.*

Grasslands: Prairies and Savannas

The moderately dry continental climates of the Great Plains of central North America, the broad Russian steppes, the African veldt, and the South American pampas support **grasslands,** rich biological communities of grasses, seasonal flowering plants, and open savannas (grasslands with scattered trees) (fig. 5.6). Seasonal cycles of temperature and precipitation contribute to abundant vegetative growth that both protects and enriches the soils of these prairies and plains, making them among the richest farmlands in the world—where there is enough water to support crops. In many parts of the world, including parts of the American Great Plains, grasslands have been artificially created or maintained by native people using fire. In addition to keeping grasslands open and biologically diverse, occasional fires improve grazing (and hunting) and ease of travel.

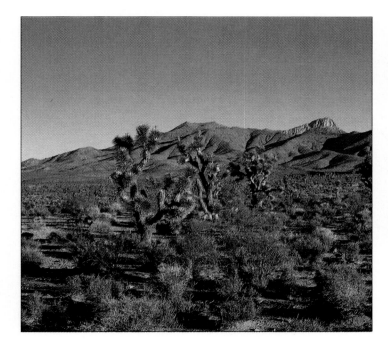

FIGURE 5.5 *Joshua trees in the Joshua Tree National Park are really large yuccas—members of the lily family. Like other plants in this hot, dry, rocky landscape, they are adapted to conserve water and repel enemies.*

FIGURE 5.6 *A combination of dry climate, extreme temperatures, and periodic fires keeps grasslands, like this one in Montana, clear of trees except in protected coulees and streambeds.*

Grasslands have few trees because inadequate rainfall, large daily and seasonal temperature ranges, and frequent grass fires kill woody seedlings. Exceptions are the narrow gallery forests that form corridors along rivers and streams through the grasslands.

Tundra

Climates in high mountain areas or at far northern or southern latitudes often are too harsh for trees. This treeless landscape, called **tundra,** has a very short growing season, cold, harsh winters, and the potential for frost any month of the year. Although water may be abundant on the tundra, for much of the year it is locked up in ice or snow and therefore unavailable to plants. Dominant tundra plants are dwarf shrubs, sedges, grasses, mosses, and lichens (fig. 5.7). Some larger animals are adapted to these conditions, including musk ox and caribou. Most animals migrate south or downhill in the winter. In summer, arctic tundra provides some of the most important bird breeding grounds in the world. (See related story, "Too Many Geese," at www.mhhe.com/apps.)

Tundra has low biological productivity, low diversity, and low resilience. Truck ruts and bulldozer tracks on the tundra landscapes of Alaska will probably take centuries to heal. Arctic tundra winters are long and dark. Only the top several centimeters of the soil thaw out in the summer, and the lower soil is permafrost. This permanently frozen layer prevents snowmelt water from being absorbed into the soil, so the surface soil is waterlogged during the summer. Try to imagine the difficulties that plants encounter in this kind of soil. Most of the year, the soil is

FIGURE 5.7 *Tundra ecosystems have short growing seasons and long, dark winters. Vegetation is mainly short and grows extremely slowly.*

completely frozen, and even during the brief growing season, the permafrost is an impenetrable barrier to deep root growth. In addition, the top layer of soil buckles and heaves in response to cycles of freezing and thawing, toppling plants and disrupting root systems.

Alpine (high mountain) tundra plants endure thin mountain air, strong ultraviolet radiation, hot daytime temperatures and often freezing nights, even in summer. Alpine soil is windswept and often gravelly or rocky.

Conifer Forests

Conifer (cone-bearing) trees dominate several distinctive biomes. The thin, needle-like leaves of conifers have small surface areas and thick, waxy coatings that help trees minimize moisture loss.

FIGURE 5.8 *Mixed conifer and deciduous forests like this one in northern Minnesota are characteristic of the Great Lakes Region. Although other species assemblages are found elsewhere, temperate cool forests around the world share many common characteristics.*

FIGURE 5.9 *In Alaska's Chugach Mountains, a sparse, black spruce forest grades into arctic tundra.*

Consequently, conifers often occur where moisture is limited by sandy soil, low precipitation, or a short growing season. Although not as efficient in carrying out photosynthesis during summer months as the broad, soft leaves of deciduous trees, conifer needles and scales can survive harsh winters or extended droughts and do accomplish some photosynthesis, even under poor conditions. In many areas, frequent fires help keep the forest open and encourage seedling growth.

The **boreal forest,** or northern coniferous forest, stretches in a broad band of mixed coniferous and deciduous trees around the world between about 45° and 60° north latitude, depending on altitude and distance from coastlines or major rivers. Among the dominant conifers are pine, hemlock, spruce, cedar, and fir. Some small deciduous trees are also common, including birches, aspens, and maples, while mosses and lichens form much of the ground cover. In this moist, cool biome, streams and wetlands abound—especially on recently glaciated landscapes. As a result, there are many bogs, fens, and lakes (fig. 5.8). **Taiga,** a Russian word used for the northernmost edge of the boreal forest, is a species-poor, often sparse forest of black spruce and sphagnum moss (peat moss). The taiga of Russia and North America forms a ragged border with the treeless arctic tundra (fig. 5.9).

Warm-climate pine forests occur at lower latitudes. In the United States, southern pine forests grow in a warm, moist climate on sandy soils. The undergrowth includes saw palmetto and various thorny bushes.

Temperate rainforests, such as those in the Pacific Northwest, are also dominated by conifers. Here, mild temperatures and precipitation as high as 250 cm (100 in) per year support luxuriant growth and huge trees, such as the California redwood, the largest tree in the world (fig. 5.10).

Broad-Leaved Deciduous and Evergreen Forests

Forests of broad-leaved trees occur throughout the world where rainfall is plentiful. In temperate regions, the climate supports lush summer plant growth when water is plentiful. In winter, when water is scarce or the ground is frozen, **deciduous** trees shed their leaves to conserve water. Temperate deciduous forests contain rich and varied associations of many tree species, including oak, maple, birch, beech, elm, ash, and other hardwoods. These tall

APPLICATION:	Comparing Biome Climates

Look back at the climate graphs for Yuma, Arizona, a desert region, and Tiksi, Russia, a tundra region (fig. 5.4). How much colder is Tiksi than Yuma in January? In July? Which location has the greater range of temperature through the year? How much do the two locations differ in precipitation during their wettest months?

Compare the temperature and precipitation in these two places to those in the other biomes shown. How wet are the wettest locations? Which biomes have distinct dry seasons? How do rainfall and length of warm seasons explain vegetation conditions in these biomes?

Answers: Tiksi is about 42°C colder in January, 25°C colder in June; Tiksi has the greater range of temperature; there is about 40 mm difference in precipitation in August.

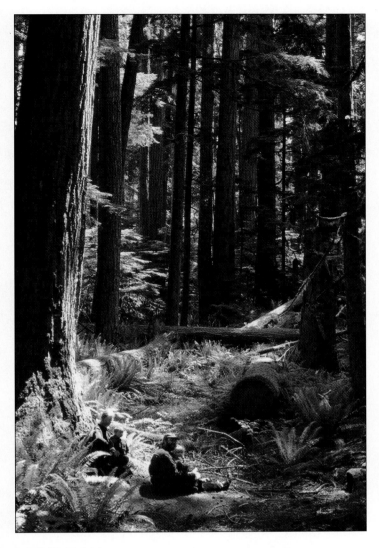

FIGURE 5.10 *Temperate rainforests like this one on Vancouver Island have the largest conifers in North America, as well as high biodiversity.*

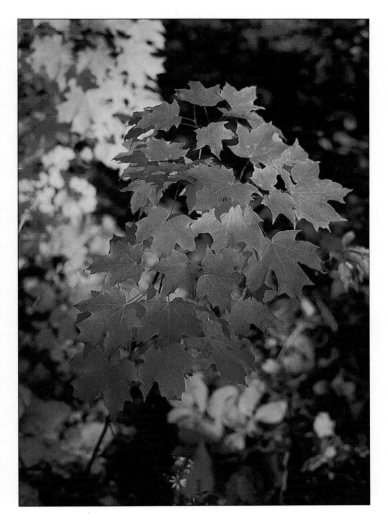

FIGURE 5.11 *The deciduous forests of the eastern United States are stratified assemblages of tall trees, understory shrubs, and small, lovely, ground-covering species.*

trees form a forest canopy over a diverse understory of smaller trees, shrubs, and herbaceous plants, including many annual spring flowers that grow, flower, set seed, and store carbohydrates before they are shaded by the canopy (fig. 5.11). Where the climate is warm year-round, forests are dominated by broadleaf evergreen trees, which keep their leaves year-round. These include the live oaks and cypresses of the southern United States. Broadleaf evergreens also dominate tropical rainforests.

Broad-leaf deciduous forest covered much of Europe until it was cleared a thousand years ago. European settlers cleared similar forests from much of eastern North America, although much is regrowing, with new species assemblages, where agriculture has disappeared. Vast, ancient deciduous forests in eastern Siberia are now being harvested at a rapid rate, perhaps the highest deforestation rate in the world. As the forests disappear, so, too, do Siberian tigers, bears, cranes, and a host of other unique and endangered species.

Mediterranean/Chaparral/Thorn Scrub

Mediterranean climate regions, with hot, dry summers and cool, moist winters, are often dominated by evergreen shrubs with small, leathery, sclerophyllous (hard, waxy) leaves. This community is drought-resistant, often with small or thorny trees as well as shrubs. Fires burn fiercely in this fuel-rich plant assemblage and are a major factor in plant succession. Annual spring flowers often bloom profusely, especially after fires. In California, this landscape is called **chaparral,** Spanish for thicket. Some typical animals include jackrabbits, kangaroo rats, mule deer, chipmunks, lizards, and many bird species. Very similar landscapes are found along the Mediterranean coast and in southwestern Australia, central Chile, and South Africa. Although this biome doesn't cover a very large total area, it contains a high number of unique species and often is considered a biodiversity "hot-spot." It also is highly desired for human habitation, often leading to conflicts with rare and endangered plant and animal species.

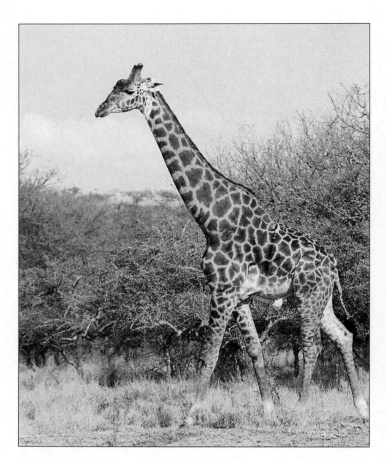

FIGURE 5.12 *A giraffe feeds on acacias in the dry, thorn-scrub landscape of Africa.*

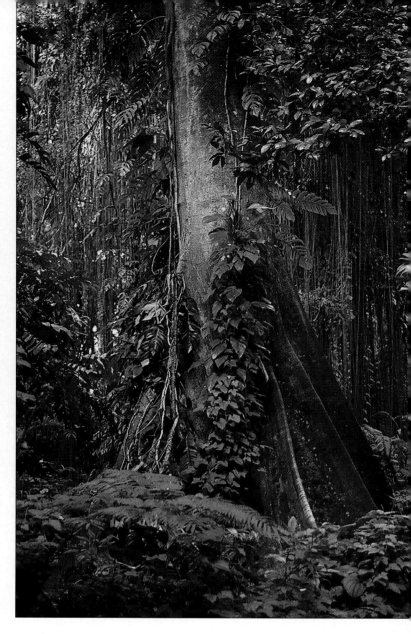

FIGURE 5.13 *Tropical rainforests, like this one in Java, support a luxuriant profusion of species and biomass. Notice the many vines and epiphytes clinging to the buttressed trunk of this giant forest tree.*

Areas that are drier year-round, such as the African Sahel (edge of the Sahara Desert), northern Mexico, or the American intermountain west (or Great Basin) tend to have a more sparse, open scrubland, characterized by sagebrush (*Artemisia* sp.), chamiso (*Adenostoma* sp.), or saltbush (*Atriplex* sp.). In Africa, acacias and other spiny plants dominate this landscape, giving it the name **thorn scrub.** Some typical animals of this biome in America are a wide variety of snakes and lizards, rodents, birds, antelope, and mountain sheep. In Africa, this landscape is home to gazelle, rhinos, and giraffes (fig. 5.12).

Tropical Moist Forests

The humid tropical regions of South and Central America, Africa, Southeast Asia, and some of the Pacific Islands support one of the most complex and biologically rich biome types in the world (fig. 5.13). Although there are several kinds of moist tropical forests, all have ample rainfall and uniform temperatures. Cool **cloud forests** are found high in the mountains where fog and mist keep vegetation wet all the time. **Tropical rainforests** occur where

rainfall is abundant—more than 200 cm (80 in) per year—and temperatures are warm to hot year-round.

The soil of both these tropical moist forest types tends to be old, thin, acidic, and nutrient-poor, yet the number of species present can be mind-boggling. For example, the number of insect species found in the canopy of tropical rainforests has been estimated to be in the millions! It is estimated that one-half to two-thirds of all species of terrestrial plants and insects live in tropical forests. The luxuriant growth in tropical rainforests depends on rapid decomposition and recycling of dead organic material.

Leaves and branches that fall to the forest floor decay and are incorporated almost immediately back into living biomass.

When the forest is removed for logging, agriculture, and mining, the thin soil quickly loses its fertility and erodes in the heavy rain. If cleared areas are extensive, the rainforest community may never recover.

Tropical Seasonal Forests

Many areas in India, Southeast Asia, Australia, West Africa, the West Indies, and South America have tropical regions characterized by distinct wet and dry seasons instead of uniform heavy rainfall throughout the year, although temperatures are hot year-round. These areas have produced communities of tropical seasonal forests: semievergreen or partly deciduous forests tending toward open woodlands and grassy savannas dotted with scattered, drought-resistant tree species.

Tropical dry forests have typically been more attractive than wet forests for human habitation and have suffered greater degradation. Clearing a dry forest with fire is relatively easy during the dry season. Soils of dry forests often have higher nutrient levels and are more agriculturally productive than those of a rainforest. Finally, having fewer insects, parasites, and fungal diseases than a wet forest makes a dry or seasonal forest a healthier place for humans to live. Consequently, these forests are highly endangered in many places. Less than 1 percent of the dry tropical forests of the Pacific coast of Central America or the Atlantic coast of South America, for instance, remain in an undisturbed state.

Freshwater Ecosystems

Freshwater ecosystems cover relatively little total area, but biologically they are very distinctive. They include the standing waters of ponds and lakes as well as the flowing waters of rivers and streams. There also are some unique freshwater ecosystems, including underground rivers and subterranean pools. Freshwater ecosystems are as varied as their individual sites because they are influenced not only by characteristics of local climate, soil, and resident communities, but also by the surrounding terrestrial ecosystems and anything that happens uphill or upstream from them (fig. 5.14). Both freshwater and saltwater ecosystems perform important ecological services such as water purification (see chapter 10).

In lakes, streams, and rivers, availability of oxygen, carbon dioxide, and sunlight for plant growth depend on a number of environmental factors, including (1) substances dissolved in the water, such as oxygen, nitrates, phosphates, potassium compounds, and other by-products of agriculture and industry; (2) suspended matter, such as silt and microscopic algae, that affect water clarity and, therefore, light penetration; (3) depth; (4) temperature; (5) rate of flow; (6) bottom characteristics (muddy, sandy, rocky); (7) internal convection currents; and (8) connection to, or isolation from, other aquatic ecosystems.

Vertical stratification, or layering, results from gradients of light, temperature, nutrients, and oxygen in aquatic systems. Specialized insects, such as water striders, live at the air-water interface.

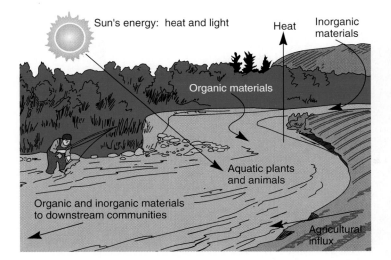

FIGURE 5.14 *The character of freshwater ecosystems is greatly influenced by the immediately surrounding terrestrial ecosystem, and even by ecosystems far upstream or far uphill from a particular site.*

Fish and plankton (microscopic, floating plants, animals, and single-celled organisms such as amoebae) move freely in the water column. Bottom dwellers, such as snails, burrowing worms and insect larvae, and bacteria, make up the benthos, or bottom community. In deep lakes, a sun-warmed upper layer, called the epilimnion, contains much of the lake's photosynthetic activity. The cold, dark lower layer is the hypolimnion (*epi* = on top; *hypo* = below). The **thermocline** is the transition zone between these zones (fig. 5.15).

Wetlands

Wetlands are ecosystems in which the land surface is saturated or covered with standing water at least part of the year. Wetland vegetation is adapted for growth under saturated conditions. There are

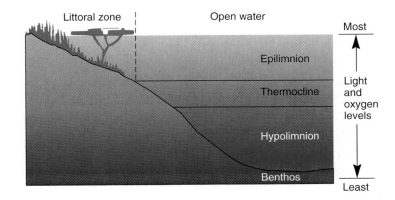

FIGURE 5.15 *The layers of a deep lake are determined mainly by gradients of light, oxygen, and temperature. The epilimnion (upper layer) is affected by surface mixing from wind and thermal convections, while mixing between the hypolimnion (lower layer) and epilimnion is inhibited by a sharp temperature and density boundary known as the thermocline.*

FIGURE 5.16 *Biological communities such as the Okefenokee Swamp in Georgia can be amazingly diverse, complex, and productive. Myriad life-forms coexist in exuberant abundance, interlinked by manifold relationships. Who lives where, and why, are central ecological questions.*

many kinds of wetlands, but we can group them into three major categories: swamps, marshes, and bogs and fens. In general, **swamps** are wetlands with trees (fig. 5.16); **marshes** are wetlands without trees (fig. 5.17); and bogs and fens are waterlogged areas saturated by groundwater or rainwater.

Because the water in marshes and swamps usually is shallow enough to allow full penetration of sunlight, photosynthetic activity is generally high. Biomass production and species diversity are often much greater than in surrounding uplands. Wetlands are especially valued as major breeding, nesting, and migration staging areas for waterfowl and shorebirds. Although wetlands currently occupy less than 5 percent of the 0.9 billion ha (2.2 billion acres) of land in the United States, the Fish and Wildlife Service estimates that one-third of all endangered species spend at least part of their lives in wetlands.

Wetlands also provide important ecological services in storing and purifying water. Wetland storage of flood waters is worth an estimated $3 to $4 billion per year. Wetlands also improve water quality by acting as natural water purification systems, removing silt and absorbing nutrients and toxins. The flow of groundwater through coastal marshes prevents saltwater intrusion

that would otherwise contaminate wells. Coastal wetlands help stabilize shorelines and reduce storm damage.

Worldwide, the most extensive wetland areas are in Canadian and Russian arctic tundra.

Estuaries

Estuaries are bays or semienclosed bodies of brackish (moderately salty) water that form where rivers enter the ocean. Estuaries usually contain rich sediments carried down river, forming shoals and mudflats that nurture a multitude of aquatic life. Estuaries are sheltered from the most drastic ocean action but do experience tidal ebbs and flows. Daily tides may even cause river levels to rise and fall far inland from the river mouth. The combination of physical factors in estuaries makes them very productive and of high species diversity. They are significant "nurseries" for economically important fish, crustaceans (such as crabs and shrimp), and mollusks (such as clams, cockles, and oysters).

Deltas, broad, shallow deposits of riverborne sand and mud, can be part of the larger estuary zone. A steady flow of nutrients makes deltas biologically rich.

Marine Ecosystems

Marine ecosystems have as much variability as those on land. In shallow water or on the surface of the open ocean, abundant sunlight supports photosynthesizing plants, algae, and plankton (tiny, floating organisms). Complex, free-floating ecosystems exist in the open ocean. Enormous algae mats in the Sargasso Sea, for example, support a rich community of fish, turtles, and seabirds. In deep, dark oceans, primary producers are usually absent, so these ecosystems depend on external nutrient sources, such as material sinking from the surface or transported by currents. Along shorelines, biological productivity is increased by mineral nutrients washing off the land surface (see chapter 2). Upwelling currents can also enhance biological productivity: currents along the coast of Peru carry nutrient-rich detritus from the deep ocean floor, supporting abundant fish populations. Often, food webs and communities offshore are intricately connected to those onshore (see fig. 3.22).

FIGURE 5.17 *Coastal wetlands, such as this one at Chincoteague National Wildlife Refuge, are highly productive ecosystems and are vitally important for both terrestrial and aquatic species.*

FIGURE 5.18 *Coral reefs harbor some of the most diverse biological communities on earth, rivaling tropical rainforests in species diversity and surpassing any biome in the number of phylla (broad taxonomic groups) represented. Surprisingly, almost all the organisms you see here are animals.*

Coral reefs are among the world's most productive ecosystems (see fig. 3.21). Reefs accumulate from the skeletons of innumerable tiny, colonial coral animals, which derive energy from photosynthetic algae that live in the reef. Reefs develop in warm, shallow environments, and they support a fabulous diversity of organisms (fig. 5.18). They also serve as nurseries for many fisheries on which humans depend for food.

Reefs are vulnerable to human activities onshore, especially farming and land-clearing that causes erosion or pollution. Silt and mud clouds the water, reducing photosynthetic activity. Destructive fishing practices, aimed at capturing exotic fish for American and European aquarium markets, have also destroyed reefs in many parts of Southeast Asia.

Deep ocean communities are also threatened by fishing. Almost completely unknown before the development of fishing boats with modern sonar devices, these cold, dark environments can contain myriad unfamiliar organisms and marvelously abundant fish. When orange roughy were discovered in the deep ocean canyons near New Zealand and Australia, they were so numerous that 60 tons of the fish could be caught in a 20-minute trawl. But the inhabitants of such cold, dark waters grow slowly. Orange roughy take 20 years to reach reproductive age and 80 to 100 years to reach market size. Further, nets scraping the ocean bottom destroy the complex communities in which the fish live and breed. The fishery collapsed in just a few years. Marine biologists are now proposing international no-fishing zones to protect these delicate, largely unknown, ecosystems.

BIODIVERSITY

The biomes described earlier shelter an astounding variety of living organisms. From the driest desert to the dripping rainforests, from the highest mountain peaks to the deepest ocean trenches, life occurs in a marvelous spectrum of sizes, colors, shapes, life cycles, and interrelationships. The varieties of organisms and complex ecological relationships give the biosphere its unique, productive characteristics. **Biodiversity,** the variety of living things, also makes the world a more beautiful and exciting place to live. Three kinds of biodiversity are essential to preserve ecological systems and functions: (1) *genetic diversity* is a measure of the variety of different versions of the same genes within individual species; (2) *species diversity* describes the number of different kinds of organisms within individual communities or ecosystems; and (3) *ecological diversity* means the richness and complexity of a biological community, including the number of niches, trophic levels, and ecological processes that capture energy, sustain food webs, and recycle materials within this system.

How Many Species Are There?

The 1.4 million species presently known probably represent only a small fraction of the total number that exist (table 5.1). Based on

TABLE 5.1 Approximate Numbers of Known Living Species by Taxonomic Group

Bacteria and cyanobacteria	5,000
Protozoans (single-celled animals)	31,000
Algae (single-celled plants)	27,000
Fungi (molds, mushrooms)	45,000
Multicellular plants	250,000
Sponges	5,000
Jellyfish, corals, and anemones	10,000
Flatworms (tapeworms, flukes)	12,000
Roundworms (nematodes, hookworms)	12,000
Earthworms and leeches	12,000
Clams, snails, slugs, squids, and octopuses	70,000
Insects	750,000
Mites, ticks, spiders, crabs, shrimp, centipedes, and other non-insect arthropods	120,000
Starfish and sea urchins	6,000
Fish and sharks	22,000
Amphibians	4,000
Reptiles	6,000
Birds	9,000
Mammals	4,000
TOTAL	1,400,000

Source: World Resources Institute, 1999.

the rate of new discoveries by research expeditions—especially in the tropics—taxonomists estimate that somewhere between 3 million and 50 million different species may be alive today. In fact, there may be 30 million species of tropical insects alone (fig. 5.19). About 70 percent of all known species are invertebrates (animals without backbones such as insects, sponges, clams, worms, etc.). This group probably makes up the vast majority of organisms yet to be discovered and may constitute 90 percent of all species.

Biodiversity Hot Spots

Most of the world's biodiversity concentrations are near the equator, especially tropical rainforests and coral reefs (fig. 5.20). Of all the world's species, only 10 to 15 percent live in North America and Europe. Many of the organisms in megadiversity countries have never been studied by scientists. The Malaysian Peninsula, for instance, has at least 8,000 species of flowering plants, while Britain, with an area twice as large, has only 1,400 species. There may be more botanists in Britain than there are species of higher plants. South America, on the other hand, has fewer than 100 botanists to study perhaps 200,000 species of plants.

Areas isolated by water, deserts, or mountains can also have high concentrations of unique species and biodiversity. Madagascar, New Zealand, South Africa, and California are all midlatitude areas isolated by barriers that prevent mixing with biological communities from other regions and produce rich, unusual collections of species.

HOW DO WE BENEFIT FROM BIODIVERSITY?

We benefit from other organisms in many ways, some of which we don't appreciate until a particular species or community disappears. Even seemingly obscure and insignificant organisms can play irreplaceable roles in ecological systems or be the source of genes or drugs that someday may be indispensable.

Food

All of our food comes from other organisms. Many wild plant species could make important contributions to human food supplies either as they are or as a source of genetic material to

FIGURE 5.19 *Insects make up more than half of all known species and may comprise more than 90 percent of all species on earth. There may be more than a million species of beetles alone.*

improve domestic crops. Noted tropical ecologist Norman Myers estimates that as many as 80,000 edible wild plant species could be utilized by humans. Villagers in Indonesia, for instance, are thought to use some 4,000 native plant and animal species for food, medicine, and other valuable products. Few of these species have been explored for possible domestication or more widespread cultivation. A 1975 study by the U.S. National Academy of Science found that Indonesia has 250 edible fruits, only 43 of which have been cultivated widely (fig. 5.21).

Drugs and Medicines

Living organisms provide us with many useful drugs and medicines (table 5.2). More than half of all prescriptions contain some natural products. The United Nations Development Programme estimates the value of pharmaceutical products derived from developing world plants, animals, and microbes to be more than $30 billion per year.

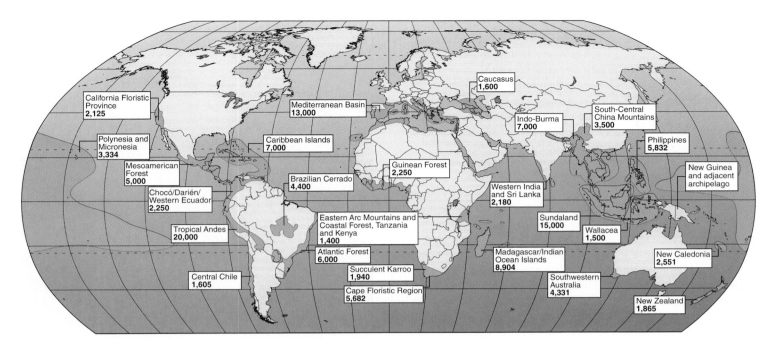

FIGURE 5.20 *Biodiversity "hot spots" identified by Conservation International tend to be in tropical or Mediterranean climates and on islands, coastlines, or mountains where many habitats exist and physical barriers encourage speciation. Numbers represent estimated endemic (locally unique) species in each area.*
Source: Data from Conservation International.

FIGURE 5.21 *Mangosteens from Indonesia have been called the world's best-tasting fruit, but they are practically unknown beyond the tropical countries where they grow naturally. There may be thousands of other traditional crops and world food resources that could be equally valuable but are threatened by extinction.*

TABLE 5.2 Some Natural Medicinal Products

PRODUCT	SOURCE	USE
Penicillin	Fungus	Antibiotic
Bacitracin	Bacterium	Antibiotic
Tetracycline	Bacterium	Antibiotic
Erythromycin	Bacterium	Antibiotic
Digitalis	Foxglove	Heart stimulant
Quinine	Chincona bark	Malaria treatment
Diosgenin	Mexican yam	Birth-control drug
Cortisone	Mexican yam	Anti-inflammation treatment
Cytarabine	Sponge	Leukemia cure
Vinblastine, vincristine	Periwinkle plant	Anticancer drugs
Reserpine	Rauwolfia	Hypertension drug
Bee venom	Bee	Arthritis relief
Allantoin	Blowfly larva	Wound healer
Morphine	Poppy	Analgesic

Consider the success story of vinblastine and vincristine. These anticancer alkaloids are derived from the Madagascar periwinkle (*Catharanthus roseus*). They inhibit the growth of cancer cells and are very effective in treating certain kinds of cancer. Twenty years ago, before these drugs were introduced, childhood leukemias were invariably fatal. Now the remission rate for some childhood leukemias is 99 percent. Hodgkin's disease was 98 percent fatal a few years ago, but is now only 40 percent fatal, thanks to these compounds. The total value of the periwinkle crop is roughly $15 million per year, although Madagascar gets little of those profits.

Pharmaceutical companies are actively prospecting for useful products in many tropical countries. Merck, the world's largest biomedical company, is paying $1 million to the Instituto Nacional de Biodiversidad (INBIO) of Costa Rica for plant, insect, and microbe samples to be screened for medicinal applications. INBIO, a public/private collaboration, trains native people as practical "parataxonimists" to locate and catalog all the native flora and fauna—estimated to be between 500,000 and 1 million species—in Costa Rica (fig. 5.22). Selling data and specimens will finance scientific work and nature protection. This is one of the rare cases where indigenous people are receiving economic benefits from the ecological knowledge they possess and the biodiversity they struggle to protect. Pharmaceutical companies often have been accused of "biopiracy," using traditional knowledge and resources without paying royalties to native peoples.

Ecological Benefits

Human life is inextricably linked to ecological services provided by other organisms. Soil formation, waste disposal, air and water purification, nutrient cycling, solar energy absorption, and food production all depend on biodiversity (see chapters 2 and 7). Total value of these ecological services is at least $33 trillion per year, or more than double total world GNP. In many environments, high diversity may help

FIGURE 5.22   *Costa Rican taxonomists study insect collections as part of an ambitious project to identify and catalog all the species in this small, but highly diverse, tropical country. The knowledge gained may contribute toward valuable commercial products that will provide funds to help preserve biodiversity.*

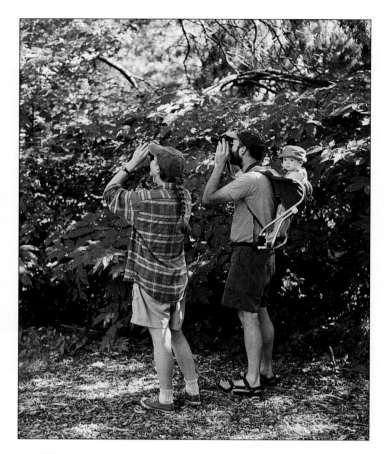

FIGURE 5.23   *Birdwatching and other wildlife observation contribute more than $29 million each year to the U.S. economy.*

biological communities withstand environmental stress better and recover more quickly than those with fewer species.

Because we don't fully understand the complex interrelationships between organisms, we often are surprised and dismayed at the effects of removing seemingly insignificant members of biological communities. For instance, it is estimated that 95 percent of the potential pests and disease-carrying organisms in the world are controlled by natural predators and competitors. Maintaining biodiversity is essential to preserving these ecological services.

Aesthetic and Cultural Benefits

Millions of people enjoy hunting, fishing, camping, hiking, wildlife watching, and other nature-based activities. These activities provide invigorating physical exercise, and contact with nature can be psychologically and emotionally restorative. In many cultures, nature carries spiritual connotations, and a particular species or landscape may be inextricably linked to a sense of identity and meaning. Observing and protecting nature has religious or moral significance for many people. Some religious organizations call for the protection of nature simply because it is God's creation.

Nature appreciation is economically important. The U.S. Fish and Wildlife Service estimates that Americans spend $104 billion every year on wildlife-related recreation (fig. 5.23). This compares to $81 billion spent each year on new automobiles. Forty percent of all adults enjoy wildlife, including 39 million who hunt or fish and 76 million who watch, feed, or photograph wildlife. Many communities are finding that local biodiversity can bring cash to remote areas through ecotourism (see chapter 6).

For many people, the value of wildlife goes beyond the opportunity to photograph or shoot a particular species. They argue that existence value, the value of simply knowing that a species exists, is reason enough to protect and preserve it. Even if we never see a whooping crane or a redwood tree, many of us are happy to know that these species have not been destroyed.

WHAT THREATENS BIODIVERSITY?

Extinction, the elimination of a species, is a normal process of the natural world. Species die out and are replaced by others, often their own descendants, as part of evolutionary change. In undisturbed ecosystems, the rate of extinction appears to be about one species lost every decade. In this century, however, human impacts on populations and ecosystems have accelerated that rate, possibly causing thousands of species, subspecies, and varieties to become extinct every year. Ecologist E. O. Wilson estimates that we are losing 10,000 species or subspecies a year—that makes more than 27 per day! If present trends continue, we may destroy millions of kinds of plants, animals, and microbes in the next few decades. In this section, we will look at some ways we threaten biodiversity.

Natural Causes of Extinction

Studies of the fossil record suggest that more than 99 percent of all species that ever existed are now extinct. Most of those species were gone long before humans came on the scene. Periodically, mass extinctions have wiped out vast numbers of species and even whole families (table 5.3). The best studied of these events occurred at the end of the Cretaceous Period when dinosaurs disappeared, along with at least 50 percent of existing species. An even greater disaster occurred at the end of the Permian Period about 250 million years ago when 90 percent of species and half of all families died out over a period of about 10,000 years—a mere moment in geologic time. Current theories suggest that these catastrophes were caused by climate changes, perhaps triggered when large asteroids struck the earth. Many ecologists worry that global climate change caused by our release of "greenhouse" gases in the atmosphere could have similarly catastrophic effects (see chapter 9).

HUMAN-CAUSED REDUCTIONS IN BIODIVERSITY

The rate at which species are disappearing has increased dramatically over the last 150 years. Between A.D. 1600 and 1850, human activities appear to have eliminated two or three species per decade, about double the natural extinction rate. In the last 150 years, the extinction rate has increased to thousands per decade. If present trends continue, biologist Paul Ehrlich warns, somewhere

	TIME	
HISTORIC PERIOD	(BEFORE PRESENT)	EFFECTS
Ordovician	444 million	25% of all families extinct
Devonian	370 million	19% of all families extinct
Permian	250 million	54% of families, 90% of species extinct
Triassic	210 million	23% of families, 1/2 of species extinct
Cretaceous	65 million	17% of families, 1/2 of species extinct (including dinosaurs but not mammals)
Quaternary	Present	1/3 to 2/3 of all species extinct if present trends continue?

TABLE 5.3 Mass Extinctions

between one-third to two-thirds of all current species could be extinct by the middle of the twenty-first century. Conservation biologists call this the sixth mass extinction, but note that this time it's not asteroids or volcanoes, but human impacts that are responsible.

Habitat Destruction

Human disturbance of natural habitat is the largest single cause of loss of biological biodiversity. Over the past 10,000 years, billions of hectares of forests, woodlands, and grasslands have been converted to commercial forests, croplands, or grazing lands. Humans now use about 10 percent of the world's land surface for crop production, and about twice that amount for pasture and grazing lands. By some estimates, we now preempt about 40 percent of the net primary productivity of the biosphere either by consuming it directly, by interfering with its production or use, or by altering the species composition or physical processes of human-dominated ecosystems. Think of the biome in which you live. How much—if any—of it remains in its original state? Where could you go to experience what the presettlement flora and fauna in your area was like?

Temperate broad-leaved forests are the most completely human-dominated of any major biome (table 5.4). The climate and soils that support such forests are especially congenial for human occupation. Temperate grasslands such as the U.S. and Canadian Great Plains also have been widely converted to agricultural production.

Tropical forest losses are accelerating in many parts of the world. In North America about one percent of original grasslands, a biome that once dominated the middle of the continent, remain intact. Forest losses and preservation efforts are discussed further in chapter 6.

Among dwindling habitat types, wetland and estuary destruction are disproportionately important. These environments occupy relatively little area worldwide, but they support such high biodiversity and biological productivity, and they have been

TABLE 5.4 Human Disturbance, Listed by Extent of Disturbance

BIOME	% HUMAN DOMINATED	TOTAL AREA (10^6 KM^2)
Temperate broadleaved forests	81.9	9.5
Chaparral and thorn scrub	67.8	6.6
Temperate rainforests	46.1	4.2
Tropical dry forests	45.9	19.5
Mixed island systems	41.8	3.2
Temperate grasslands	40.4	12.1
Mixed mountain systems	25.6	12.1
Tropical rainforests	24.9	11.8
Warm deserts/semideserts	12.2	29.2
Temperate conifer forests	11.8	18.8
Cold deserts/semideserts	8.5	10.9
Tropical grasslands	4.7	4.8
Tundra and arctic desert	0.3	20.6

Note: Where undisturbed and human-dominated areas do not add up to 100 percent, the difference represents partially disturbed lands. Different sources give varying numbers for total land area depending on how they consider lakes, inland seas, glaciers, bays, etc. This table estimates an unusually large total land area.

Source: Data from Hannah, Lee, et al., "Human Disturbance and Natural Habitat: A Biome Level Analysis of a Global Data Set," in *Biodiversity and Conservation,* 1995, Vol. 4:128–155.

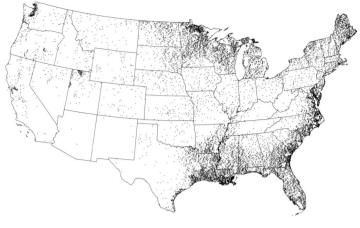

Each blue dot represents 10,000 acres of wetlands

Total: 111,156,000 acres

FIGURE 5.24 *Distribution of wetlands in the conterminous United States, 1997. Less than half of pre-1800 wetlands remain. Drainage for agriculture accounts for more than half of this loss.* Source: Natural Resources Conservation Service, 2001.

drained and disrupted so extensively, that wetland loss is a key factor in biodiversity loss. About half of all original wetlands in the United States have been drained, filled, polluted, or otherwise degraded over the past 250 years. In the Great Plains and in California's central valley, both critical breeding grounds for water birds of the Western Hemisphere, more than 90 percent of wetlands have been drained and converted to farmland. Until the 1940s, U.S. federal land policies encouraged draining and filling of these "worthless swamps," and filling and drainage were largely unregulated into the 1970s. Less than half of the original 90 million ha remain today in the United States (excluding Alaska). Since 1985 it has been illegal to destroy large wetlands without a permit. Permits are often easy to obtain, however, and until recently wetlands smaller than 1.2 ha (3 acres) remained unprotected. The United States continues to lose wetlands at a rate of about 40,000 ha (100,000 acres) per year. A common estimate is that 87 percent of wetland losses are due to conversion to farmland, while 8 percent are due to urban development. One of the fastest rates of wetland loss in North America is in the Gulf Coast area. In Louisiana, which has about 40 percent of the remaining coastal wetlands in the United States, canal dredging for oil and gas production and navigation—coupled with diversion of river sediment that once replenished marshes—are causing losses of about 13,000 ha (50 sq mi) per year (fig. 5.24).

Wildlife and hunting organizations, together with the federal government, have invested extensively in restoring wetland habi-tat across the United States. These efforts have succeeded in reestablishing more than 80,000 ha (200,000 acres) of wetlands since 1995.

Worldwide wetland destruction is also severe. In New Zealand, less than 10 percent of wetlands remain since European settlement. Coastal forests around the world have been logged and drained for farmland. Current destruction is worst in Southeast Asia, where mangrove forests are being cleared for wood and for conversion to shrimp and fish ponds.

Fragmentation

In addition to loss of absolute habitat area, a serious problem is habitat **fragmentation**—the reduction of habitat into smaller and smaller, more scattered patches (fig. 5.25). Fragmentation reduces biodiversity because many species, such as bears and large cats, require large territories to subsist. Other species, such as forest interior birds, reproduce successfully only in deep forest or other habitat far from edges and human settlement (See related story, "Where Have All the Songbirds Gone?" at www.mhhe.com/apps.) Often, invasive species spread quickly into new regions by following fragment edges. New competitors further destabilize remaining native populations.

Fragmentation also divides populations into isolated groups. Small, isolated populations are vulnerable to catastrophic events, such as a single storm or disease outbreak. Very small populations may not have enough breeding adults to be viable even under normal circumstances.

Much of our understanding of fragmentation was elegantly outlined in the theory of **island biogeography,** developed by R. H. MacArthur and E. O. Wilson in the 1960s. Noticing that small

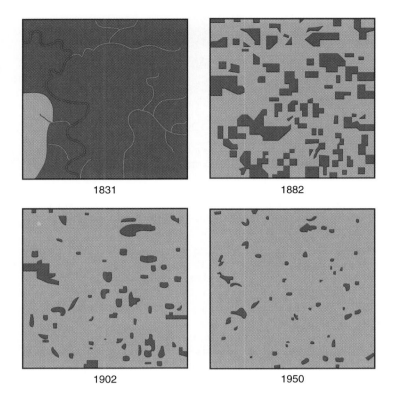

1831   1882

1902   1950

FIGURE 5.25 *Decrease in wooded area of Cadiz Township in southern Wisconsin during European settlement. Green areas represent the amount of land in forest each year.*

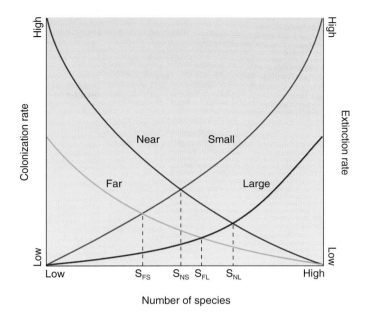

Number of species

FIGURE 5.26 *Predicted species richness on an island resulting from a balance between colonization (immigration) and extinction by natural causes. This island biogeography theory of MacArthur and Wilson (1967) is used to explain why large islands near a mainland (S_{NL}) tend to have more species than small, far islands (S_{FS}).*
Source: Based on MacArthur and Wilson, *The Theory of Island Biogeography*, 1967, Princeton University Press.

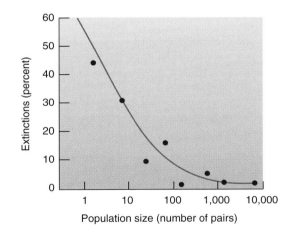

Population size (number of pairs)

FIGURE 5.27 *Extinction rates of bird species on the California Channel Islands as a function of population size over 80 years.*
Source: Data from H. L. Jones and J. Diamond, "Short-term-base Studies of Turnover in Breeding Bird Populations on the California Coast Island," in *Condor,* vol. 78:526–549, 1976.

islands far from a mainland have fewer terrestrial species than larger, nearer islands, MacArthur and Wilson proposed that species diversity is a balance between colonization and extinction rates (fig. 5.26). An island far from a population source naturally has a lower rate of colonization than a nearer island because it is harder for terrestrial organisms to reach. At the same time, a large island can support more individuals of a given species and is, therefore, less likely to suffer extinction due to natural catastrophes, genetic problems, or demographic uncertainty (the chance that all the members of a single generation will be of the same sex).

Island biogeographical effects have been observed in many places. Cuba, for instance, is one hundred times as large and has about ten times as many amphibian species as its Caribbean neighbor, Montserrat. Similarly, in a study of bird species on the California Channel Islands, Jared Diamond observed that on islands with fewer than ten breeding pairs, 39 percent of the populations went extinct over an 80-year period, while only 10 percent of populations numbering between 10 and 100 pairs went extinct in the same time (fig. 5.27). Only one species numbering between 100 and 1,000 pairs went extinct and no species with over 1,000 pairs disappeared over this time.

Many habitat types can be seen as islands: wetlands, forest patches, alpine tundra, and other habitat fragments exist as scattered patches in a larger landscape. Often, size and proximity to other fragments contribute to differences in the species that occupy these "islands."

Hunting and Fishing

Overharvesting is responsible for depletion or extinction of many species. A classic example is the extermination of the American passenger pigeon (*Ectopistes migratorius*). Even though it inhabited only eastern North America, 200 years ago this was the world's most abundant bird with a population of between 3 and 5 billion. This species once accounted for about one-quarter of all

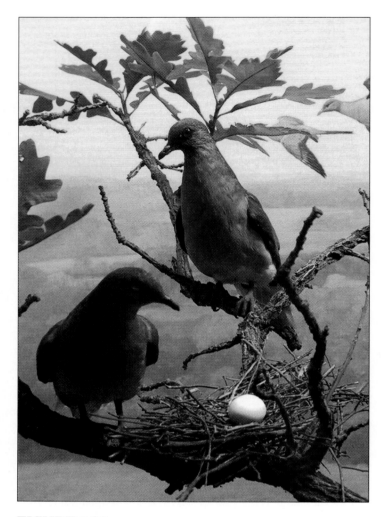

FIGURE 5.28 *A pair of stuffed passenger pigeons* (Ectopistes migratorius). *The last member of this species died in the Cincinnati Zoo in 1914.*

Printed with permission from the Bell Museum of Natural History at the University of Minnesota.

birds in North America. In 1830, John James Audubon saw a single flock of passenger pigeons estimated to be ten miles wide, hundreds of miles long, and thought to contain perhaps a billion birds. In spite of this vast abundance, market hunting and habitat destruction caused the entire population to crash in only about 20 years between 1870 and 1890. The last known wild bird was shot in 1900, and the last existing passenger pigeon, a female named Martha, died in 1914 in the Cincinnati Zoo (fig. 5.28).

Some other well-known overhunting cases include the near extermination of the great whales and the American bison (buffalo). In 1850, some 60 million bison roamed the western plains. In just 40 years that number was reduced to only 400. Many were killed only for their hides or tongues. The U.S. Army is accused of slaughtering bison to deprive native peoples of their principal source of food and shelter, thereby forcing them onto reservations.

We think that about 2.5 million great whales once inhabited the world's oceans. Their blubber was highly prized as a source of

Don't Buy Endangered Species Products

You probably are not shopping for a fur coat from an endangered tiger, but there might be other ways you are supporting unsustainable harvest and trade in wildlife species. To be a sustainable consumer, you need to learn about the source of what you buy. Often plant and animal products are farm-raised, not taken from wild populations. But some commercial products are harvested in unsustainable ways. Here are a few products about which you should inquire before you buy:

Seafood includes many top predators that grow slowly and reproduce only when many years old. Despite efforts to manage many fisheries, the following have been severely, sometimes tragically, depleted:

- Top predators: swordfish, marlin, shark, bluefin tuna, albacore ("white") tuna
- Groundfish and deepwater fish: orange roughy, Atlantic cod, haddock, pollack (source of most fish sticks, artificial crab, generic fish products), yellowtail flounder, monkfish
- Other species, especially shrimp, yellowfin tuna, and wild sea scallops, are often harvested with methods that destroy other species or habitats

Pets and plants are often collected from wild populations, some sustainably and others not:

- aquarium fish (often harvested by stunning with dynamite and squirts of cyanide, which destroy tropical reefs and many fish)
- reptiles: snakes and turtles, especially, are often collected in the wild.
- plants: orchids and cacti are the best-known, but not the only, group collected in the wild.

Herbal products such as wild ginseng, wild echinacea (purple coneflower), should be investigated before purchasing.

Do buy some of these sustainably harvested products:

- shade-grown (or organic) coffee, nuts, and other sustainably harvested forest products
- pets from the Humane Society, which works to protect stray animals
- organic cotton, linen, and other fabrics
- fish products that have relatively little environmental impact or fairly stable populations: farm-raised catfish, tilapia, trout, and salmon, most mackerel, Pacific pollack, dolphinfish (mahimahi), squids, crabs, and crayfish.
- Wild freshwater fish like bass, sunfish, pike, catfish, and carp, which are usually better managed than most ocean fish.

See also What Can You Do, chapter 6 (page 142).

oil. With the introduction of steamships and explosive harpoons in the nineteenth century, one species after another was driven into near extinction. For the past decade, the International Whaling Commission has prohibited taking of great whales. This hunting ban seems to be having positive effects. There are now about

26,000 California gray whales, more than twice as many as 20 years ago and close to the prehunting number. Around 3,400 humpbacks now visit Hawaii each year compared to only one-third that many in the mid-1980s.

Fish stocks have been seriously depleted by overharvesting in many parts of the world, mostly because of a huge increase in fishing fleet size and efficiency in recent years. Worldwide, 13 of 17 principal fishing zones are now reported to be commercially exhausted or in steep decline. At least three-quarters of all commercial oceanic species are overharvested. The Atlantic cod fishery, for example, once seemed inexhaustible. More than 300,000 metric tons per year were harvested in the 1980s, but then the population crashed. In 1992, Canada banned Atlantic cod fishing, putting 20,000 people out of work and closing down a $700-million-per-year industry. Trawlers scouring the seafloor have destroyed egg-laying habitat, so that some biologists doubt the population will ever recover (see What Can You Do? p. 114).

Overharvesting is not only a historic problem. Conservationists are increasingly concerned about "bush meat" markets. Especially in central Africa and Southeast Asia, well-armed poachers are killing primates and other large mammals for meat. While this is not a new phenomenon, it has become more serious as poachers have obtained more powerful weapons from regional wars and as wildlife populations—and habitat availability—have fallen. Defenders of Wildlife estimates that poachers kill a million metric tons of game for the bush meat trade every year, helping to reduce chimpanzee populations by 95 percent and vastly reducing gorilla and orangutan populations.

Commercial Products and Live Specimens

Despite international bans on trade in products from endangered species, smuggling of furs, hides, horns, live specimens, and folk medicines amounts to millions of dollars each year (fig. 5.29). Developing countries in Asia, Africa, and Latin America with the richest biodiversity in the world are the main sources of wild animals and animal products, while Europe, North America, and some of the wealthy Asian countries are the principal importers. Japan, Taiwan, and Hong Kong buy three-quarters of all cat and snake skins, for instance, while European countries buy a similar percentage of live birds. The United States imports 99 percent of all live cacti and 75 percent of all orchids sold each year.

The profits to be made in wildlife smuggling are enormous. Tiger or leopard fur coats can bring $100,000 in Japan or Europe. Elephant tusks can be worth $100 per kilogram. Bird collectors will pay $10,000 for a rare hyacinth macaw from Brazil or up to $12,000 for a pair of golden-shouldered parakeets from Australia. An endangered albino python might bring $20,000 in Germany. The mortality rate in this live animal trade is enormous. It is generally estimated that 50 animals are caught or killed for every live animal that gets to market. Buyers, many of whom say they love animals, keep this trade going. If you collect rare plants, fish, or birds, make sure that what you buy is commercially grown, not

FIGURE 5.29 *Parts from rare and endangered species for sale on the street in China. Use of animal products in traditional medicine and prestige diets is a major threat to many species.*

FIGURE 5.30 *A diver uses cyanide to stun tropical fish being caught for the aquarium trade. Many fish are killed by the method itself, while others die later during shipment. Even worse is the fact that cyanide kills the coral reef itself.*

wild-caught. If you travel abroad, don't buy products made from rare or endangered species. Doing so not only contributes to their decline but also exposes you to confiscation and serious fines when you get home.

The trade in wild species for pets is an enormous business. Worldwide, some 5 million live birds are sold each year for pets, mostly in Europe and North America. Currently, pet traders import (often illegally) into the United States some two million reptiles, one million amphibians and mammals, 500,000 birds, and 128 million tropical fish each year. About 75 percent of all saltwater tropical aquarium fish sold come from coral reefs of the Philippines and Indonesia, where they are commonly caught with dynamite or cyanide (fig. 5.30). These fishing methods are destroying about 50 km^2 of reefs each year, as well as killing many of the fish bound for the market.

Overharvesting also threatens plants. Wild ginseng has been nearly eliminated in many areas because of the Asian demand for its roots, which are used as an aphrodisiac and folk medicine. Cactus "rustlers" steal cacti by the ton from the American southwest and Mexico. With prices ranging as high as $1,000 for rare specimens, it's not surprising that many are now endangered.

Predator and Pest Control

Some animal populations have been greatly reduced, or even deliberately exterminated, because they are regarded as dangerous to humans or livestock, or because they compete with our use of resources. Every year, U.S. government animal control agents trap, poison, or shoot thousands of coyotes, bobcats, prairie dogs, and other species considered threats to people, domestic livestock, or crops. These efforts cost about $20 million in federal and state funds, and kill some 700,000 birds and mammals, about 100,000 of which are coyotes. Wildlife defenders regard this program as cruel, callous, and mostly ineffective. Ranchers and trappers argue that without predator control, western livestock operations would be uneconomical.

Exotic Species Introductions

Exotic organisms—aliens introduced into habitats where they are not native—are one of the greatest threats to biodiversity worldwide. Exotics can be thought of as biological pollution. Freed from the predators, parasites, pathogens, and competition that kept them in check in their native home, formerly mild-mannered species can turn into superaggressive "weedy" invaders in a new habitat. There are now more than 4,500 alien species in North America. A few of those causing most trouble include:

- Kudzu vine (*Pueraria lobata*) has blanketed large areas of the southeastern United States (fig. 5.31). Long cultivated in Japan for edible roots, medicines, and fibrous leaves and stems used for paper production, kudzu was introduced by the U.S. Soil Conservation Service in the 1930s to control erosion. Unfortunately, it succeeded too well. In the ideal conditions of its new home, kudzu can grow 18 to 30 m (60 to 100 ft) in a single season. Smothering everything in its path, it kills trees, pulls down utility lines, and causes millions of dollars in damage every year.

- Leafy spurge (*Euphorbia esula*), a relative of the familiar Christmas poinsettia, has spread across millions of hectares of the northern Great Plains since it was first introduced in the 1870s in the seeds of Mennonite farmers from Russia. A short, bushy, green weed with small yellow flowers, the entire plant—leaves, stems, roots, flowers, and seeds—is poisonous. It reduces the carrying capacity of western range by three-quarters when it invades because cattle can't eat it and it crowds out native grasses and forbs. North Dakota alone loses $7 million per year in forage and beef production due to leafy spurge, and spends another $6 million annually in control programs.

FIGURE 5.31 *Invasive exotic species can badly disrupt ecological systems. Kudzu* (top) *blankets millions of acres; Asian longhorned beetles* (bottom) *destroy mature trees and threaten urban parks as well as forests.*

- Purple loosestrife (*Lythrum salicaria*) grows in any wet soil. Originally cultivated by gardeners for its beautiful purple flower spikes, this tall wetland plant escaped into New England marshes about a century ago. Spreading rapidly across the Great Lakes, it now fills wetlands across much of the northern United States and southern Canada. Because it crowds out indigenous vegetation and has few native predators or symbionts, it tends to reduce biodiversity wherever it takes hold.

- Zebra mussels (*Dreissena polymorpha*) probably made their way from their home in the Caspian Sea to the Great Lakes in ballast water of transatlantic cargo ships, arriving sometime around 1985. Gluing themselves to any solid surface, zebra mussels reach enormous densities—up to 70,000 animals per square meter—covering fish spawning beds, smothering native mollusks, and clogging utility intake pipes. Found in all the Great Lakes, zebra mussels also have invaded the Mississippi River and its tributaries. Public and private costs for zebra mussel removal now amount to some $400 million per year. On the positive side, mussels have improved water clarity in Lake Erie at least fourfold by filtering out algae and particulates.

- Asian long-horned beetles (*Anoplophora galabripennis*) are one of the most recent exotic threats in the United States. Probably transported in imported logs or wooden packing crates from China, these wood-eating insects were first spotted in Amityville, New York, in 1996. The beetle larvae burrow into living tree trunks. Sap flow to leaves may be cut off, and trees may fall over simply because tunneling badly weakens the trunks. These shiny black bugs with distinctive white spots and enormous feelers are especially fond of some of our primary shade trees—maples, oaks, and poplars. If they spread widely they could cause billions of dollars in damage. Chicago suburbs cut and burned 2,000 shade trees after discovering a long-horned beetle infestation in 1999.

The flow of organisms isn't just into North America; we also send exotic species to other places. The Leidy's comb jelly, for example, a jellyfish native to the western Atlantic coast, has devastated the Black Sea, now making up more than 90 percent of all biomass at certain times of the year. Australia has seen many disastrous introductions, including European rabbits and African cane toads, which weigh more than 2 kg and are toxic to the touch. Eliminating these alien species is difficult, if not impossible.

Diseases

Disease organisms, or pathogens, may also be considered predators. To be successful over the long term, a pathogen must establish a balance in which it is vigorous enough to reproduce, but not so lethal that it completely destroys its host. When a disease is introduced into a new environment, however, this balance may be lacking, and an epidemic may sweep through the area.

The American chestnut was once the heart of many Eastern hardwood forests. In the Appalachian Mountains, at least one of every four trees was a chestnut. Often over 45 m (150 ft) tall, 3 m (10 ft) in diameter, fast growing, and able to sprout quickly from a cut stump, it was a forester's dream. Its nutritious nuts were important for birds (like the passenger pigeon), forest mammals, and humans. The wood was straight grained, light, rot-resistant, and used for everything from fence posts to fine furniture and its bark was used to tan leather. In 1904, a shipment of nursery stock from China brought a fungal blight to the United States, and within 40 years, the American chestnut had all but disappeared from its native range. Efforts are now underway to transfer blight-resistant genes into the few remaining American chestnuts that weren't reached by the fungus or to find biological controls for the fungus that causes the disease.

An infection called whirling disease is decimating trout populations in many western states. It is caused by an exotic microorganism, *Myxobolus cerebralis,* which destroys cartilage in young fish causing them to swim erratically. The parasite is thought to have come into the United States in 1956 in a shipment of frozen fish.

Pollution

We have known for a long time that toxic pollutants can have disastrous effects on local populations of organisms. Pesticide-linked declines of fish-eating birds and falcons was well documented in the 1970s. Marine mammals, alligators, fish, and other declining populations suggest complex interrelations between pollution and health (chapter 8). Mysterious, widespread deaths of thousands of seals on both sides of the Atlantic in recent years are thought to be linked to an accumulation of persistent chlorinated hydrocarbons, such as DDT, PCBs, and dioxins, in fat, causing weakened immune systems that make animals vulnerable to infections. Similarly, mortality of Pacific sea lions, beluga whales in the St. Lawrence estuary, and striped dolphins in the Mediterranean are thought to be caused by accumulation of toxic pollutants.

Lead poisoning is another major cause of mortality for many wildlife species. Bottom-feeding waterfowl, such as ducks, swans, and cranes, ingest spent shotgun pellets that fall into lakes and marshes. The U.S. Fish and Wildlife Service estimates that between 2 and 3 million waterfowl die each year from lead poisoning. Lead shot has been banned for most waterfowl hunting, and poisonings should decrease as residual contaminants are buried more deeply in sediments.

Genetic Assimilation

Some rare and endangered species are threatened by **genetic assimilation** because they crossbreed with closely related species that are more numerous or more vigorous. For example, hatchery-raised trout often are introduced into streams and lakes where they genetically dilute indigenous stocks. Similarly, black ducks have declined severely in the eastern United States and Canada in recent years. Hunting pressures and habitat loss are factors in this decline, but so is interbreeding with mallards forced into black duck habitat by destruction of prairie potholes in the West.

PROTECTING BIODIVERSITY

As we gradually become aware of how serious biodiversity losses have become, we are slowly adopting laws and treaties to protect these irreplaceable assets (see chapter 6). Parks, wildlife refuges, and nature preserves have produced encouraging progress in this area, but much remains to be done.

Hunting and Fishing Laws

In 1874, a bill was introduced in the U.S. Congress to protect the American bison, whose numbers were already falling to dangerous levels. This initiative failed, however, because most legislators believed that all wildlife—and nature in general—was so abundant and prolific that it could never be depleted by human activity. Without protection from excessive hunting and habitat disruption, the bison population plunged from around 60 million to only a few hundred animals in about 50 years.

By the 1890s, most states had enacted some hunting and fishing restrictions. The general idea behind these laws was to conserve the resource for future human use rather than to preserve wildlife for its own sake. The wildlife regulations and refuges established since that time have been remarkably successful for many species. At the beginning of the twentieth century, there were an estimated half a million white-tailed deer in the United States; now there are some 14 million—more in some places than the environment can support. Wild turkeys and wood ducks were nearly extinct 50 years ago. By restoring habitat, planting food crops, transplanting breeding stock, building shelters or houses, protecting these birds during breeding season, and other conservation measures, populations of these beautiful and interesting birds have been restored to several million each. Snowy egrets, which plume hunters almost wiped out 80 years ago, are now common again.

The Endangered Species Act

Simply identifying species at risk can initiate protection. Globally, the World Conservation Union maintains a "Red List" of threatened species, and increasingly, countries are developing their own lists. In North America, the U.S. Endangered Species Act (ESA) of 1973 and the Committee on the Status of Endangered Wildlife in Canada (COSEWIC) in 1976 represented powerful new approaches to wildlife protection. While earlier regulations had focused almost exclusively on "game" animals, these programs have sought to identify all endangered species and populations and to save as much biodiversity as possible, regardless of its usefulness to humans. **Endangered species** are those considered in imminent danger of extinction, while **threatened species** are those that are likely to become endangered—at least locally—within the foreseeable future. Bald eagles, gray wolves, brown (or grizzly) bears, and sea otters, for instance, together with a number of native orchids and other rare plants, are considered locally threatened even though they remain abundant in other parts of their former range. **Vulnerable species** are naturally rare or have been locally depleted by human activities to a level that puts them at risk. They often are candidates for future listing. For vertebrates, protected categories include species, subspecies, and local races or ecotypes.

The ESA regulates a wide range of activities involving endangered species, including "taking" (harassing, harming, pursuing, hunting, shooting, trapping, killing, capturing, or collecting) species either accidentally or on purpose, and importing, exporting, selling, or possessing endangered species, parts, or products. Violators of the ESA are subject to fines up to $100,000 and one year imprisonment, as well as loss of vehicles and equipment used in the violation. In 1995 the Supreme Court ruled that critical habitat—habitat essential for a species survival—must also be protected, whether on public or private land. Canada's Species at Risk Act (SARA), introduced in 1999, is designed to impose similar fines and criminal penalties for injury to both species at risk and critical habitat on both public and private land.

In 2002, the United States had 1,260 native species on its endangered and threatened species lists and 213 candidate species waiting to be considered. Canada, which generally has less diversity because of its boreal location, has designated 46 endangered and 50 threatened species. The number of listed species in different taxonomic groups reflects much more about the kinds of organisms that humans consider interesting and desirable than the actual number in each group (table 5.5). Compare tables 5.1 and 5.5, for instance. Although invertebrates make up about three-quarters of all known species, they account for only 9 percent of those deemed worthy of protection. The total number of endangered species worldwide is at least five times the number officially listed in the United States.

Listing of new species in the United States has been very slow, generally taking several years from the first petition to final determination. Limited funding, political pressures, listing moratoria, and changing administrative policies have created long delays. Hundreds of species are classified as "warranted (deserving of protection) but precluded" for lack of funds or local support.

When Congress passed the original ESA, it probably intended to protect only a few charismatic species like birds and big game animals. Sheltering obscure species, such as the Delhi Sands flower-loving fly, the Coachella Valley fringe-toed lizard,

TABLE 5.5 Endangered and Threatened Species, U.S. and Foreign, Listed by USFWS	
Mammals	357
Birds	279
Reptiles	123
Amphibians	31
Fish	153
Invertebrates	203
Plants	746

Source: Data from U.S. Fish and Wildlife Service, 2002.

Mrs. Furbisher's lousewort, or the orange-footed pimple-back mussel, most likely never occurred to many who voted for the bill. This raises some interesting ethical questions about the rights and values of seemingly minor species. Although uncelebrated, these species may be indicators of environmental health. Protecting them usually preserves habitat and a host of unlisted species.

Recovery Plans

Once a species is officially listed as endangered, the U.S. Fish and Wildlife Service must prepare a recovery plan detailing how populations will be rebuilt to sustainable levels. Negotiating the costs, politics, local economic interests, and biological species requirements can take years. Recovery is also expensive. The total cost of endangered species protection and recovery in the United States is about $150 million per year. About half that amount is spent on a dozen charismatic species like the grizzly bear and the Florida panther (fig. 5.32). By contrast, the 135 endangered invertebrates and 520 endangered plants get less than $5 million per year altogether. Our funding priorities are based more on emotion and politics than biology.

Some recovery plans have been gratifyingly successful. The American alligator was listed as endangered in 1967 because hunting (for meat, skins, and sport) and habitat destruction had reduced populations to precarious levels. Protection has been so effective that the species is now plentiful throughout its entire southern range. In 1999 and 2000, the bald eagle and peregrine falcon, both reduced to the edge of survival by DDT poisoning, were declared recovered and removed from the endangered species list.

FIGURE 5.32 *Only 40 Florida panthers remain in the Everglades. They have been crossbred with Texas cougars in an attempt to save the population.*

Opponents of the ESA have repeatedly tried to require that economic costs and benefits be incorporated into endangered species planning. An important test of the ESA occurred in 1978 in Tennessee, where construction of the Tellico Dam threatened a tiny fish called the snail darter. As a result of this case, a federal committee (the so-called "God Squad") was given power to override the ESA for economic reasons. Another important test was the case of the northern spotted owl (see chapter 6), whose protection depends on preserving old-growth forest in the Pacific Northwest. Economists argue that the cost of saving owls must be borne by local companies and communities in Washington and Oregon. Environmentalists argue that logging communities declined before the owl debate because of industry mechanization, and that protecting owls also aids many other organisms and resources.

An even more costly recovery program may be required for Columbia River salmon and steelhead endangered by hydropower dams and water storage reservoirs that block their migration to the sea. Opening floodgates to allow young fish to run downriver and adults to return to spawning grounds would have high economic costs to barge traffic, farmers, and utility customers who have come to depend on abundant water and cheap electricity. On the other hand, commercial and sport fishing for salmon is worth $1 billion per year and employs about 60,000 people directly or indirectly.

Reintroductions

An endangered species can be restored by reintroducing it to its former habitat once the major threats to its survival have been removed. Reintroduction is expensive, so it has been attempted for only a few species. It usually requires raising animals in captivity and intensive management after release. For the peregrine falcon, reintroduction programs succeeded in raising populations to stable levels in about two decades of effort. For the falcon, reintroduced populations rebounded because the principal threat, DDT, had been removed from its food chain. In a major triumph for endangered species activists, the bird was removed from the endangered species list in 2001. Reintroduction efforts have been less dramatic for the California condor, which has yet to produce live wild offspring seven years after its first reintroductions.

For peregrines and condors, there is wide agreement that reintroduction is a worthwhile, if expensive, undertaking. For large predators such as wolves and grizzlies, reintroduction programs have caused persistent and painful controversy (see "Reintroducing Wolves to Yellowstone," p. 138). Ecologists argue that such predators are essential for ecosystem stability. Opponents of reintroduction fear threats to livestock and growing human populations near these species' range.

Minimum Viable Populations

A critical question in all recovery programs is the minimum population size required for long-term viability of rare and endangered species. Small populations can lack genetic diversity, and consequently, adaptability or reproductive viability. How is diversity lost in small populations? (1) A *founder effect* occurs when a few individuals establish a new population. The limited genetic diversity from those original founders may not be enough to sustain the population. (2) A *demographic bottleneck* arises when only a few individuals survive some catastrophe. As the population replenishes itself, a limited genetic diversity similar to that in a founder effect results. (3) *Genetic drift* is a reduction in gene frequency in a population due to unequal reproductive success. For example, some individuals breed more than others, and their genes gradually come to dominate the population. (4) *Inbreeding* is mating of closely related individuals. Random, recessive, deleterious mutations—what we consider genetic diseases such as hemophilia in humans—that are usually hidden in a widely outcrossed population can be expressed when inbreeding occurs. This is why humans have laws and cultural taboos prohibiting mating between siblings or first cousins.

Some species seem not to be harmed by lack of genetic diversity. The northern elephant seal, for example, was reduced by overharvesting, a century ago, to fewer than 100 individuals. Today there are more than 150,000 of these enormous animals along the Pacific coast of Mexico and California. No marine mammal has come closer to extinction and then made such a remarkable recovery. All northern elephant seals today appear to be essentially genetically identical and yet they seem to be surviving just fine. Although interpretations are controversial, in highly selected populations where only the most fit individuals reproduce, or in which there are few deleterious genes, inbreeding may not be such a negative factor.

Private Land and Critical Habitat

Private land is essential in endangered species protection. Eighty percent of the habitat for more than half of all listed species is on nonpublic property (fig. 5.33). The U.S. Supreme Court has ruled that destroying habitat is as harmful to endangered species as directly taking (killing) them. Many people are very resistant, however, to restrictions on how they use their own property to protect what they perceive to be insignificant or worthless organisms. This is especially true when the land has potential for economic development. If property is worth millions of dollars as the site of a housing development or shopping center, most owners don't want to be told they have to leave it undisturbed to protect some rare organism. Landowners may be tempted to "shoot, shovel, and shut up," if they discover endangered species on their property. Many feel they should be compensated for lost value caused by ESA regulations.

To avoid what former Interior Secretary Bruce Babbitt called "environmental train wrecks" like the controversy over northern spotted owls, the U.S. Fish and Wildlife Service has been negotiating agreements called **habitat conservation plans (HCP)** with private landowners. Under these plans, landowners are allowed to harvest resources or build on part of their land as long as the species benefits overall. By improving habitat in some areas, funding conservation research, removing predators and competitors, or other steps that benefit the endangered species,

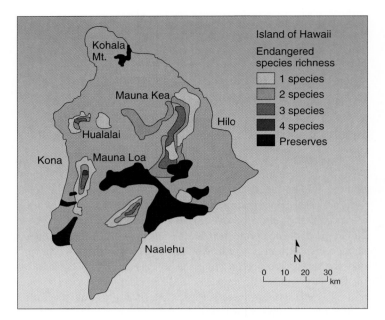

FIGURE 5.33 *Protected lands* (green) *are often different from biologically diverse areas* (red shades), *as shown here on the island of Hawaii.*

developers are allowed to destroy habitat or even "take" endangered organisms.

By early 2002, 379 HCPs had been approved, covering nearly 12 million acres and 200 endangered or threatened species. Details of the ESA can be found at http://endangered.fws.gov. Scientists and environmentalists often are critical of HCPs, claiming these plans often are based more on politics than biology, and that the potential benefits are frequently overstated. Defenders argue that by making the ESA more landowner-friendly, the HCPs will benefit wildlife in the long run.

Reauthorizing the Endangered Species Act

The ESA officially expired in 1992. Since then, Congress has debated many alternative proposals, ranging from outright elimination to substantial strengthening of the act. Perhaps no other environmental issue divides Americans more strongly than the ESA. In the western United States, where traditions of individual liberty and freedom are strong and the federal government is viewed with considerable suspicion and hostility, the ESA seems to many to be a plot to take away private property and trample on individual rights. Many people believe that the law puts the welfare of plants and animals above that of humans. On the other hand, much of the public also supports the ESA and the goal of protecting natural beauty and biodiversity. Scientists widely support its efforts to preserve species and the habitat on which those species—and many others—rely on for life. The political and philosophical differences between the two sides are unlikely to be resolved for a long time. What do you think could be done to bring the two sides closer together?

International Wildlife Treaties

The 1975 Convention on International Trade in Endangered Species (CITES) was a significant step toward worldwide protection of endangered flora and fauna. It regulated trade in living specimens and products derived from listed species, but has not been foolproof. Species are smuggled out of countries where they are threatened or endangered, and documents are falsified to make it appear they have come from areas where the species are still common. Investigations and enforcement are especially difficult in developing countries, where wildlife is disappearing most rapidly. Still, eliminating markets for endangered wildlife is an effective way to stop poaching. Appendix I of CITES lists 700 species threatened with extinction by international trade.

SUMMARY

Major ecosystem types called biomes are characterized by similar climates, soil conditions, and biological communities. Among the major terrestrial biomes are deserts, tundra, grasslands, temperate deciduous forests, temperate coniferous forests, tropical moist forests, and tropical seasonal forests. Aquatic ecosystems include oceans and seas, rivers, lakes, estuaries, marshes, swamps, bogs, fens, and reefs. Moisture and temperature are generally the most critical determinants for terrestrial biomes. Periodic natural disturbances, such as fires, play a major role in maintaining some biomes.

Humans have disturbed, preempted, or damaged much—perhaps half or more—of all terrestrial biomes and now dominate about 40 percent of all net primary productivity on the land. Some of this disturbance and domination is permanent, but we have opportunities to apply ecological knowledge and practical experience to restoring and repairing ecosystems.

Biomes shelter the world's biodiversity. Humans benefit from biodiversity in many ways, but we also threaten biodiversity through land conversion and other activities. Extinctions occur naturally, including ancient mass extinctions and evolutionary replacement. Among the threats from humans are over-harvesting of animals and plants for food and commercial products. Millions of live wild plants and animals are collected for pets, houseplants, and medical research. Among the greatest damage we do to biodiversity are habitat destruction, the introduction of exotic species and diseases, pollution of the environment, and genetic assimilation.

The potential value of the species that may be lost if environmental destruction continues could be enormous. It is also possible that the changes we are causing could disrupt vital ecological services on which we all depend for life.

The first hunting and fishing laws in the United States were introduced more than a century ago to restrict overexploitation and to preserve species for future uses. The Endangered Species Act and CITES represent a new attitude toward wildlife in which we protect organisms just because they are rare and endangered. Now we are expanding our concern from individual species to protecting habitat, threatened landscapes, and entire biogeographical regions. Social, cultural, and economic factors must also be considered if we want to protect biological resources on a long-term, sustainable basis.

QUESTIONS FOR REVIEW

1. Throughout the central portion of North America is a large biome once dominated by grasses. Describe how physical conditions and other factors control this biome.

2. What is taiga, and where is it found? Why might logging in taiga be more disruptive than in southern coniferous forests?

3. Figure 5.24 shows the distribution of wetlands in 48 U.S. states. Which regions have the most wetlands? What are four different kinds of wetlands?

4. Which major biomes have been most heavily disturbed by human activities? Where is the largest area of undisturbed habitat in the world?

5. What is the range of estimates of the total number of species on the earth? Why is the range so great?

6. What group of organisms has the largest number of species?

7. Define extinction. What is the natural rate of extinction in an undisturbed ecosystem?

8. List at least five products derived from wild organisms.

9. What is the current rate of extinction, and how does this compare to past rates?

10. Compare the scope and effects of the Endangered Species Act and CITES.

11. Describe eight ways that humans directly or indirectly cause biological losses.

12. What is gap analysis, and how is it related to ecosystem management and design of nature preserves?

THINKING SCIENTIFICALLY

1. Genetic diversity, or diversity of genetic types, is believed to enhance stability in a population. Most agricultural crops are genetically very uniform. Why might the usual importance of genetic diversity *not* apply to food crops? Why *might* it apply?

2. Explain the graph in figure 5.26. What is predicted by the points at which the colored lines cross?

3. Scientists need to be cautious about their theories and assumptions. What arguments could you make for *and* against the statement that humans are causing extinctions unlike any in the history of the earth?

4. A conservation organization has hired you to lead efforts to reduce the loss of biodiversity in a tropical country. Which of the following problems would you focus on first and why: habitat destruction and fragmentation, hunting and fishing activity, harvesting of wild species for commercial sale, or introduction of exotic organisms?

5. Many ecologists and resource scientists work for government agencies to study resources and resource management. Do these scientists serve the public best if they try to do pure science, or if they try to support the political positions of democratically elected representatives, who, after all, represent the positions of their constituents?

6. You are a forest ecologist living and working in a logging community. An endangered salamander has recently been discovered in your area. What arguments would you make for and against adding the salamander to the official endangered species list?

7. A variety of environments are adapted to disturbance, including fire-adapted grasslands, fire-dependent forest types, and flood plains. How do you explain the difference between disturbance-adapted ecosystems and ecosystems that simply never mature and never reach climax conditions?

KEY TERMS

biodiversity 107
biomes 98
boreal forest 102
chaparral 103
cloud forests 104
conifer 101
coral reefs 107
deciduous 102
deserts 99
endangered species 118
estuaries 106
exotic organisms 116
extinction 111
fragmentation 112
genetic assimilation 117

grasslands 100
habitat conservation plans (HCP) 120
island biogeography 112
marshes 106
swamps 106
taiga 102
temperate rainforests 102
thermocline 105
thorn scrub 104
threatened species 118
tropical rainforests 104
tundra 101
vulnerable species 118
wetlands 105

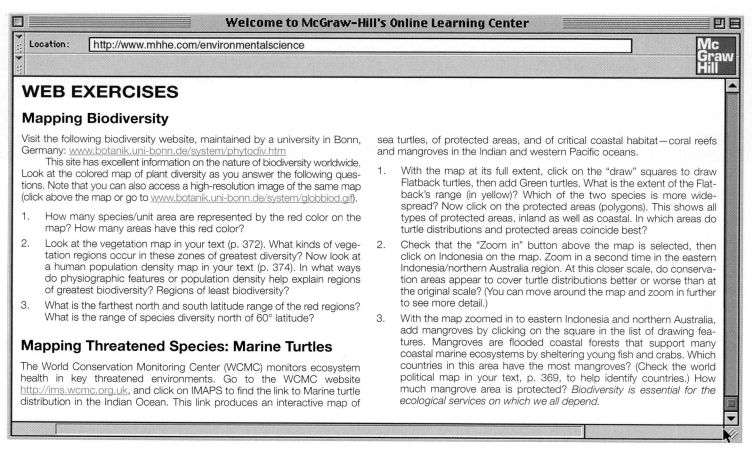

Welcome to McGraw-Hill's Online Learning Center

Location: http://www.mhhe.com/environmentalscience

WEB EXERCISES

Mapping Biodiversity

Visit the following biodiversity website, maintained by a university in Bonn, Germany: www.botanik.uni-bonn.de/system/phytodiv.htm
This site has excellent information on the nature of biodiversity worldwide. Look at the colored map of plant diversity as you answer the following questions. Note that you can also access a high-resolution image of the same map (click above the map or go to www.botanik.uni-bonn.de/system/globbiod.gif).

1. How many species/unit area are represented by the red color on the map? How many areas have this red color?

2. Look at the vegetation map in your text (p. 372). What kinds of vegetation regions occur in these zones of greatest diversity? Now look at a human population density map in your text (p. 374). In what ways do physiographic features or population density help explain regions of greatest biodiversity? Regions of least biodiversity?

3. What is the farthest north and south latitude range of the red regions? What is the range of species diversity north of 60° latitude?

Mapping Threatened Species: Marine Turtles

The World Conservation Monitoring Center (WCMC) monitors ecosystem health in key threatened environments. Go to the WCMC website http://ims.wcmc.org.uk, and click on IMAPS to find the link to Marine turtle distribution in the Indian Ocean. This link produces an interactive map of sea turtles, of protected areas, and of critical coastal habitat—coral reefs and mangroves in the Indian and western Pacific oceans.

1. With the map at its full extent, click on the "draw" squares to draw Flatback turtles, then add Green turtles. What is the extent of the Flatback's range (in yellow)? Which of the two species is more widespread? Now click on the protected areas (polygons). This shows all types of protected areas, inland as well as coastal. In which areas do turtle distributions and protected areas coincide best?

2. Check that the "Zoom in" button above the map is selected, then click on Indonesia on the map. Zoom in a second time in the eastern Indonesia/northern Australia region. At this closer scale, do conservation areas appear to cover turtle distributions better or worse than at the original scale? (You can move around the map and zoom in further to see more detail.)

3. With the map zoomed in to eastern Indonesia and northern Australia, add mangroves by clicking on the square in the list of drawing features. Mangroves are flooded coastal forests that support many coastal marine ecosystems by sheltering young fish and crabs. Which countries in this area have the most mangroves? (Check the world political map in your text, p. 369, to help identify countries.) How much mangrove area is protected? *Biodiversity is essential for the ecological services on which we all depend.*

Forests precede people; deserts succeed them.

—François-René de Chateaubriand

The Indian Peaks Wilderness in Colorado protects an area of surpassing beauty and provides an opportunity for solitude and nonmotorized recreation.

OBJECTIVES

After studying this chapter, you should be able to

- discuss how and why old-growth temperate and tropical forests are being disturbed, as well as ways they might be preserved.
- explain why conservationists criticize large clear-cuts and forest road building.
- recount how overgrazing can lead to desertification in arid lands.
- understand the origins and current problems of national parks in America and other countries.
- evaluate the tension between conservation and economic development, and how the Man and Biosphere program and ecotourism projects address this tension.
- evaluate some of the controversy over wilderness area management, as well as how people in developing countries feel about the concept of wilderness.
- explain the need for, and problems with, wildlife refuges and nature preserves in the United States and elsewhere in the world.

Welcome to McGraw-Hill's Online Learning Center

Location: http://www.mhhe.com/environmentalscience

LEARNING ONLINE

Visit our webpage at www.mhhe.com/environmentalscience for data sources, further readings, additional case studies, current environmental news, and regional examples within the Online Learning Center to help you understand the material in this chapter. You'll also find active links to information pertaining to this chapter including:

The World's Biomes: Forests
Below-Cost Timber Sales on National Forests
Below-Cost Timber Sales: Overview
The Monarch Butterfly Sanctuary Foundation
National Parks Conservation Association
National Wildlife Refuge System
Man and the Biosphere Species Databases
NetCoast—Integrated Coastal Zone Management
Yellowstone Wolf Tracker

Protecting Forests to Preserve Rain

How could cutting down trees in low-elevation forests affect tiny toads and beautiful birds on mountaintops many kilometers away? Recent climatological research in Costa Rica suggests a connection that may help solve a decade-old mystery. It may also suggest new conservation policies for the tropics as well as other parts of the world.

Monteverde, Costa Rica, is home to one of the world's best-known cloud forests. This lush mountaintop, continually bathed in fog and mist, is carpeted with a rich profusion of plants and is home to a host of rare birds, insects, and amphibians that inhabit the cool, moist forest (fig. 6.1). For decades, the Monteverde Cloud Forest Reserve has been an El Dorado for biologists and ecotourists eager to see the beautiful resplendent quetzal (*Pharomachrus mocinno*), Costa Rica's national symbol.

The reserve also is famous for the last known sighting of the glowing golden toad (*Bufo periglenes*), which disappeared abruptly in the late 1980s. Until 1987, the tiny toads congregated by the hundreds in rain pools on the cloud forest floor. In 1988, however, they were scarce, and the next year, only a few scattered individuals could be found. Since 1990, not a single golden toad has been seen anywhere in the world. What happened to them? No one knows for sure. Air pollution, waterborne contaminants, and diseases have been suggested as possible causes for their disappearance, but no conclusive evidence for any of these has been found. The forest itself still seems pristine and intact, having been protected from logging and grazing to preserve the watershed for lower-elevation villages.

Now scientists are realizing that mountaintops, like islands, may not be as isolated and independent as they seem. Satellite images show that the life-giving clouds that sustain Costa Rican mountain flora and fauna are disappearing, and the reason appears to be logging in far-distant lowlands. The clouds are formed as moisture-laden Caribbean winds sweep up the eastern slope of Costa Rica's central Cordillera de Tilarán range. When they reach high elevations, the winds cool and moisture condenses to form the fog and mist on which the biological community depends. Much of the moisture carried by the wind comes from the lowland rainforest over which it blows before reaching the mountains. When those forests are cut down, plant transpiration declines markedly. Furthermore, the pastures and croplands created by deforestation warm the air and dry it out even more.

After a century of logging and fires, only 18 percent of Costa Rica's lowland forest east of the mountains remains untouched. As a result, the air reaching the mountains is drier than normal, and clouds that once covered the mountaintops are sparse or even completely absent some times in the year. The bottoms of the clouds also are at much higher elevations, and areas once bathed continually in mist and fog are now dry for days at a time. If breeding pools disappear, the golden toads can't reproduce and

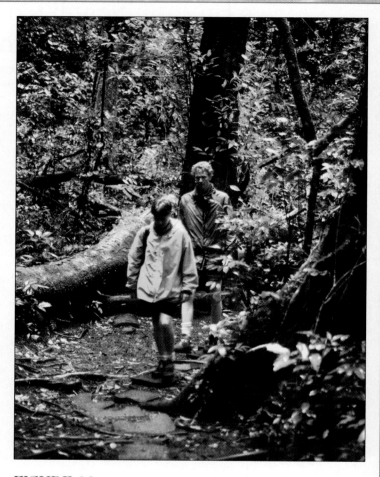

FIGURE 6.1   *The cool, moist cloud forests of Monteverde, Costa Rica, harbor many rare and beautiful species of plants and animals, but this unique habitat may be threatened by climate change caused by logging in far-away lowland forests.*

the species quickly goes extinct. Similarly, if wild avocadoes, the quetzal's favorite food, fail to fruit because of dry conditions, the bird's survival may be threatened. If these trends continue, biologists worry that other rare and beautiful species that depend on the cloud forests may also disappear.

Although plants and animals within the Monteverde reserve are protected from many types of abuse and misuse, we are discovering that park boundaries are not always enough to preserve a functioning ecosystem. Simply fencing a forest may not be sufficient. Increasingly, we are recognizing that successful management of forests, parks, and reserves requires an understanding of the social, economic, and ecological context in which a reserve exists. In this chapter, we'll examine the strategies that have been used to conserve forests, rangelands, and areas of scenic beauty and biological importance.

WORLD FORESTS

As chapter 5 shows, forests and grasslands—especially in temperate regions—are among the most heavily disturbed by human activities of any terrestrial biomes. These ecosystems produce valuable materials, such as lumber, paper pulp, and domestic livestock, that are important in human culture. They also play vital roles in regulating climate, controlling water runoff, providing wildlife habitat, purifying the air, and a host of other ecological services. Furthermore, these terrestrial biomes have scenic, cultural, and historic values that deserve to be protected. In many cases, these competing land uses are incompatible. Much of the attention of environmental groups over the past century has been devoted to protecting forests, prairies, and other landscapes. This chapter will focus on how we now use and abuse these biological communities, as well as some of the ways they are being protected and conserved.

Forest Distribution

Figure 6.2 shows the main forest types around the world, while table 6.1 presents the UN Food and Agriculture Program estimates of total forest area by continent as well as annual rates of forest change. The FAO Forest Resources Assessment defines forest as an area where trees cover 10 percent or more of the land. The 3.9 billion ha (9.6 billion acres) encompassed by this definition include open woodlands, thorn scrub, and savannas, as well as **closed-canopy forests** where tree crowns create a more continuous ground cover. The largest forest area in this summary is in Europe, which includes Russia's vast boreal forest. The world's largest tropical forest area is in South America's Amazon basin. Note that the highest annual rates of forest losses are in Africa (especially tropical West Africa) and in South America. Southeast Asia also has high rates of deforestation, a great tragedy considering the high level of biological diversity in the area.

The total aboveground forest biomass in the world is estimated to be about 422 billion metric tons, one-third of which is in South America. This biomass represents both an important economic resource as well as a huge carbon sink. Clearing and burning of forests releases much of this carbon into the atmosphere and contributes significantly to global climate change (see chapter 9).

Among the forests of greatest ecological importance are the remnants of primeval forests that are home to much of the world's biodiversity, endangered species, and indigenous human cultures. Sometimes called frontier forests, **old-growth forests** are those that cover a large enough area and have been undisturbed by human activities long enough that trees can live out a natural life cycle and ecological processes can occur in relatively normal fashion. That doesn't mean that all trees need be enormous or thousands of years old. In some old-growth forests, most trees may live

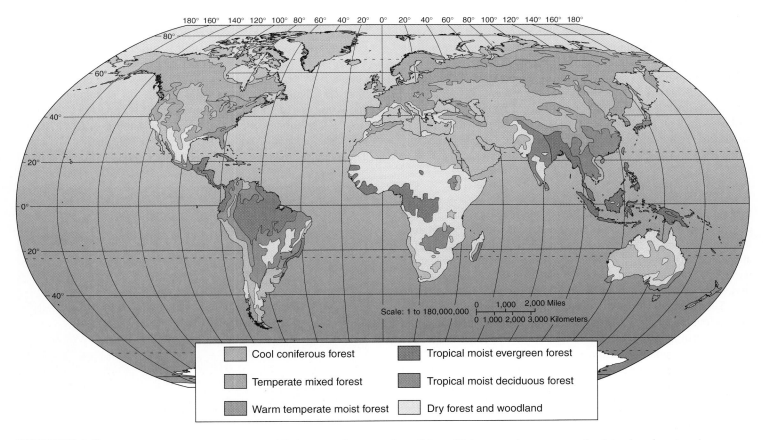

FIGURE 6.2 *Main vegetation zones of the world's forests under natural conditions. Light tan regions are grassland, tundra, desert, or ice.*

CHAPTER 6   Environmental Conservation: Forests, Grasslands, Parks, and Nature Preserves

125

TABLE 6.1 Forest Area by Region, 2000

REGION	LAND AREA (MILLION HA)	FOREST AREA (MILLION HA)	PERCENT OF LAND AREA	PERCENT OF ALL FORESTS	ANNUAL NET CHANGE (MILLION HA)
Africa	2,978	650	22	17	−5.3
Asia	3,085	548	18	14	−0.4
Europe*	2,260	1,039	46	27	+0.9
North and Central America	2,137	549	26	14	−0.6
Oceania**	849	198	23	5	−0.4
South America	1,755	886	51	23	−3.7
World Total	13,064	3,869	30	100	−9.4

*Includes Russia

**Oceania consists of Australia, New Zealand, and many small island nations.

Source: FAO Global Forest Resources Assessment, 2002.

less than a century before being killed by disease or some natural disturbance like a fire. Nor does it mean that humans have never been present. Where human occupation entails relatively little impact, an old-growth forest may have been inhabited by people for a very long time. Even forests that have been logged or converted to cropland often can revert to old-growth characteristics if left alone long enough.

While forests still cover about half the area they once did worldwide, only one-quarter of those forests retain old-growth features. The largest remaining areas of old-growth forest are in Russia, Canada, Brazil, Indonesia, and Papua New Guinea. Together, these five countries account for more than three-quarters of all relatively undisturbed forests in the world. In general, remoteness rather than laws protect those forests. Although official data describe only about one-fifth of Russian old-growth forest as threatened, rapid deforestation—both legal and illegal—especially in the Russian Far East, probably put a much greater area at risk.

Forest Products

Wood plays a part in more activities of the modern economy than does any other commodity. There is hardly any industry that does not use wood or wood products somewhere in its manufacturing and marketing processes. Think about the amount of junk mail, newspapers, photocopies, and other paper products that each of us in developed countries handles, stores, and disposes of in a single day. Total annual world wood consumption is about 3.7 billion metric tons or about 3.7 billion m^3. This is more than steel and plastic consumption combined. International trade in wood and wood products amounts to more than $100 billion each year (fig. 6.3). Developed countries produce less than half of all industrial wood but account for about 80 percent of its consumption. Less-developed countries, mainly in the tropics, produce more than half of all industrial wood but use only 20 percent.

The United States, the former Soviet Union, and Canada are the largest producers of both industrial wood (lumber and panels)

and paper pulp. Much of the industrial logging in North America and Europe occurs in managed forests, where cut trees are grown as a crop. In contrast, tropical hardwoods in Southeast Asia, Africa, and Latin America are being cut at an unsustainable rate, mostly from old-growth forests.

More than half of the people in the world depend on firewood or charcoal as their principal source of heating and cooking fuel (fig. 6.4). Consequently, fuelwood accounts for slightly more than half of all wood harvested worldwide. Unfortunately, burgeoning populations and dwindling forests are causing wood shortages in many less-developed countries. About 1.5 billion people who depend on fuelwood as their primary energy source have less than they need. At present rates of population growth and wood consumption, the annual deficit is expected to increase from 500 million m^3 in 2000 to 2,600 million m^3 in 2025. At that point, the demand would be twice the available fuelwood supply. The average amount of wood used for cooking and heating in 63

FIGURE 6.3   *Total commercial wood production amounts to about $100 billion per year.*

Principles of Environmental Science   www.mhhe.com/environmentalscience

TABLE 6.2	Changes in Forest Area 1990–2000 (Million Hectares per Year)		
REGION	DEFORESTATION	REFORESTATION AND NATURAL REGENERATION	NET CHANGE
Tropics	–14.2	+1.9	–12.3
Non-tropics	–0.4	+3.3	+2.9
World total	–14.6	+5.2	–9.4

Source: FAO Forest Resources Assessment, 2002.

less-developed countries is about 1 m^3 per person per year, roughly equal to the amount of wood that each American consumes each year as paper products alone.

Approximately one-quarter of the world's forests are managed for wood production. Ideally, forest management involves scientific planning for sustainable harvests, with particular attention paid to forest regeneration. In temperate regions, according to the FAO, more land is being replanted or allowed to regenerate naturally than is being permanently deforested (table 6.2). Much of this reforestation, however, is in large plantations of single-species, single-use, intensive cropping called **monoculture forestry.** Although this produces rapid growth and easier harvest-

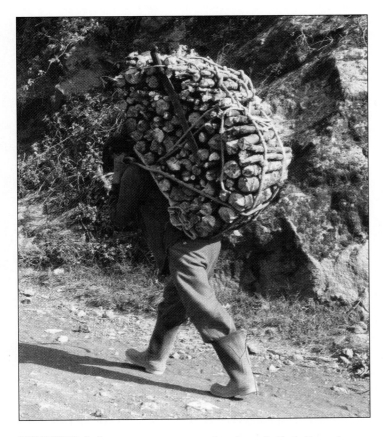

FIGURE 6.4 *Firewood accounts for almost half of all wood harvested worldwide and is the main energy source for nearly half of all humans.*

ing than a more diverse forest, a dense, single-species stand often supports little biodiversity and does poorly in providing the ecological services, such as soil erosion control and clean water production, that may be the greatest value of native forests.

Some of the countries with the most successful reforestation programs are in Asia. China, for instance, cut down most of its forests 1,000 years ago and has suffered centuries of erosion and terrible floods as a consequence. Recently, however, timber cutting in the headwaters of major rivers has been outlawed, and a massive reforestation project has begun. An average 4.5 million ha per year were replanted in the 1990s. Korea and Japan also have had very successful forest restoration programs. After being almost totally denuded during World War II, both countries are now about 70 percent forested.

Tropical Forests

Some of the richest and most diverse terrestrial ecosystems on the earth are tropical forests. Although they now occupy less than 10 percent of the earth's land surface, these forests are thought to contain more than two-thirds of all higher plant biomass and at least half of all the plant, animal, and microbial species in the world.

Diminishing Forests

A century ago, an estimated 12.5 million km^2 of tropical lands were covered with closed-canopy forest. This was an area larger than the entire United States. The Food and Agriculture Organization (FAO) of the United Nations estimates that about around 9.2 million ha or about 0.6 percent of the remaining tropical forest is cleared each year.

There is considerable debate about current rates of deforestation in the tropics. In 1999, satellite data showed more than 31,000 fires in a single month in Brazil. Remote sensing experts calculated that 8 million ha (about 20 million acres) per year were being cut and burned in the Amazon basin alone (fig. 6.5). However, there are different definitions of deforestation. Some scientists insist that it means a complete change from forest to agriculture, urban areas, or desert. Others include any area that has been logged, even if the cut was selective and regrowth will be rapid. Furthermore, savannas, open woodlands, and succession following natural disturbance are hard to distinguish from logged areas. Consequently, estimates for total tropical forest losses range from about 5 million to more than 20 million ha per year. The FAO

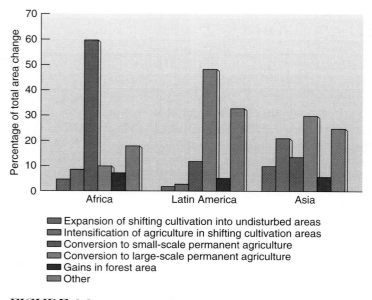

FIGURE 6.6   *Tropical forest losses by different processes.*

FIGURE 6.5   *Cutting and burning of tropical rainforest results in wildlife destruction, habitat loss, soil erosion, rapid water runoff, and waste of forest resources. It also contributes to global climate change.*

estimates of 12.3 million ha deforested per year are generally the most widely accepted. To put that figure in perspective, it means that about one acre—or the area of a football field—is cleared every second, on average, around the clock.

By most accounts, Brazil has the highest total deforestation rate in the world, but it also has by far the largest tropical forests. Indonesia and Malaysia together may be losing as much primary forest each year as Brazil, even though their original amount was far lower. In 1997, forest fires on Borneo and Sumatra exacerbated by severe drought, burned 20,000 km^2 (8,000 mi^2). The fires reportedly were set both to clear land for agriculture and to hide illegal logging.

In Africa, the coastal forests of Senegal, Sierra Leone, Ghana, Madagascar, Cameroon, and Liberia already have been mostly destroyed. Haiti was once 80 percent forested; today, essentially all that forest has been destroyed, and the land lies barren and eroded. India, Burma, Kampuchea (Cambodia), Thailand, and Vietnam all have little old-growth lowland forest left. In Central America, nearly two-thirds of the original moist tropical forest has been destroyed, mostly within the last 30 years and primarily due to conversion of forest to cattle range. (See related story on "Disappearing Butterfly Forests" at www.mhhe.com/apps.)

Causes of Deforestation

A variety of factors contribute to deforestation, and, as figure 6.6 shows, different forces predominate in various parts of the world. Logging for valuable tropical hardwoods, such as teak and mahogany, is generally the first step. Although loggers may take only one or two of the largest trees per hectare, the canopy of tropical forests is usually so strongly linked by vines and interlocking branches that felling one tree can bring down a dozen others. Building roads to remove logs kills more trees, but even more importantly, it allows entry to the forest by farmers, miners, hunters, and others who cause further damage (fig. 6.7).

In Africa, conversion of forest into small-scale agriculture accounts for nearly two-thirds of all tropical forest destruction. In Latin America, poor, landless farmers often start the deforestation, but are bought out—or driven out—after a few years by large-scale farmers or ranchers. The thin, nutrient-poor tropical forest soils frequently are worn out after a few years of cropping, and much of this land becomes unproductive pasture or impoverished scrubland. Shifting cultivation (sometimes called "slash and burn" or milpa farming) is often blamed for forest destruction. This practice can be sustainable where population densities are low and individual plots are allowed to regenerate for a decade or two between cultivation periods. In some Asian countries, however, growing populations and shrinking forests lead to short rotation cycles, and cropping intensification can lead to permanent deforestation.

As the opening story of this chapter shows, forest clearing can change regional rainfall patterns. A computer model created by Pennsylvania State University scientists suggests this phenomenon might create a kind of chain reaction. As forests are cut down, plant transpiration and rainfall decrease. Drought kills more vegetation, and fires become more numerous and extensive. In a worst-case scenario, an area as large as the entire Amazonian forest might be permanently damaged in just a few decades.

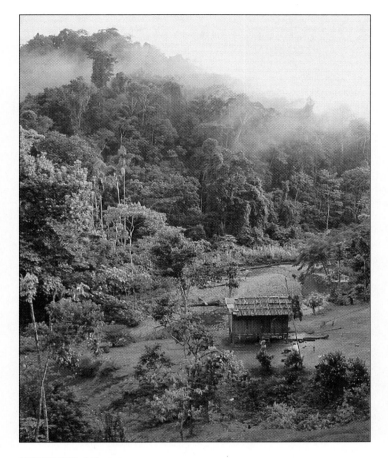

FIGURE 6.7 *Logging roads open up forest access to landless settlers, miners, and hunters, who drive away native species and indigenous people.*

Forest Protection

What can be done to stop this destruction and encourage tropical forest protection? While much of the news is discouraging, there are some hopeful signs for forest conservation in the tropics. Many countries now recognize that forests are valuable resources. Intensive scientific investigations are underway to identify the best remaining natural areas (see Investigating Our Environment, p. 130).

About 12 percent of all world forests are in some form of protected status, but the effectiveness of that protection varies greatly. Costa Rica has one of the best plans for forest guardianship in the world. Attempts are being made there to not only rehabilitate the land (make an area useful to humans), but also to restore the ecosystems to naturally occurring associations. One of the best known of these projects is Dan Janzen's work in Guanacaste National Park. Like many dry, tropical forests, the northwestern part of Costa Rica had been almost totally converted to ranchland. By controlling fires, however, Janzen and his coworkers are bringing back the forest. One of the keys to this success is involving local people in the project. Janzen also permits grazing in the park. The original forest evolved, he reasons, together with ancient grazing animals that are now extinct. Horses and cows can play a valuable role as seed dispersers.

People also are working on the grassroots level to protect and restore forests in other countries. India, for instance, has a long history of nonviolent, passive resistance movements—called *satyagrahas*—to protest unfair government policies. These protests go back to the beginning of Indian culture and often have been associated with forest preservation. Gandhi drew on this tradition in his protests of British colonial rule in the 1930s and 1940s. During the 1970s, commercial loggers began large-scale tree felling in the Garhwal region in the state of Uttar Pradesh in northern India. Landslides and floods resulted from stripping the forest cover from the hills. The firewood on which local people depended was destroyed, and the way of life of the traditional forest culture was threatened. In a remarkable display of courage and determination, the village women wrapped their arms around the trees to protect them, sparking the Chipko Andolan movement (literally, movement to hug trees). They prevented logging on 12,000 km^2 of sensitive watersheds in the Alakanada Basin. Today, the Chipko Andolan movement has grown to more than 4,000 groups working to save India's forests.

Debt-for-Nature Swaps

Those of us in developed countries also can contribute toward saving tropical forests. Financing nature protection is often a problem in developing countries, where the need is greatest. One promising approach is called **debt-for-nature swaps.** Banks, governments, and lending institutions now hold nearly $1 trillion in loans to developing countries. There is little prospect of ever collecting much of this debt, and banks are often willing to sell bonds at a steep discount—perhaps as little as ten cents on the dollar. Conservation organizations buy debt obligations on the secondary market at a discount and then offer to cancel the debt if the debtor country agrees to protect or restore an area of biological importance.

There have been many such swaps. Conservation International, for instance, bought $650,000 of Bolivia's debt for $100,000—an 85 percent discount. In exchange for canceling this debt, Bolivia agreed to protect nearly 1 million ha (2.47 million acres) around the Beni Biosphere Reserve in the Andean foothills. Ecuador and Costa Rica have had a different kind of debt-for-nature swap. They exchanged debt for local currency bonds that fund activities of local private conservation organizations in the country. This has the dual advantage of building and supporting indigenous environmental groups while protecting the land. Critics, however, charge that these swaps compromise national sovereignty and do little to reduce Third World debt or to change the situations that led to environmental destruction in the first place.

Temperate Forests

Tropical countries aren't unique in harvesting forests at an unsustainable rate. Northern countries like the United States and Canada also have allowed controversial forest management practices in many areas. For many years, the official policy of the U.S. Forest Service was "multiple use," which implied that the forests could be used for everything that we might want to do there simultaneously. Some uses are incompatible, however. Bird-watching, for

INVESTIGATING Our Environment

Rapid Forest Assessment

Glancing quickly from side to side as she walks a measured transect through a pristine Peruvian rainforest, a botanist gains a rapid overview of plant diversity as she watches for rare and endangered species. Known as a **relevé** (pronounced relavay, from the French word for survey), this technique provides a speedy assessment of vegetation types and biodiversity. In a nearby clearing, an ornithologist plays prerecorded bird calls to lure birds into view, while in another part of the forest, a zoologist checks live traps baited with peanut butter to see what small mammals have been caught during the night. Later that evening as they gather at their campsite, scientists from a dozen different disciplines write up their field notes, and identify unknown specimens.

All these biological explorers are part of a Rapid Assessment Team assembled by Conservation International, a global environmental organization, to identify and explore the world's most important biodiversity hot spots. Usually given only 2 to 4 weeks in each location, the teams attempt to document as many different species of flora and fauna as possible and to judge both the relative biological wealth of an area and its potential for preservation. Over the past decade, more than a hundred scientists from many different countries have identified 25 areas, ranging from Madagascar, to Brazil's Atlantic Forest, to the mountains of China's Yunnan Province, with unusual concentrations of biological resources.

Representing less than 2 percent of the planet's land area, these biodiversity hot spots account for at least 44 percent of all vascular plant species and 38 percent of all vertebrate animals. Each area also contains high levels of endemic species (those found nowhere else). Unfortunately, they also all face threats from destructive human activities ranging from logging, mining, dam building, grazing, and agriculture. On average, less than 30 percent of the original natural habitat remains in each area, and some are less than 10 percent undisturbed. The crucial need to identify and protect what remains of our biological heritage is the reason for the urgency of the Rapid Assessment Program. We need to identify and save as much as possible before it's too late.

As Conservation International's president, Russell Mittermeir, says, "Given the scientific consensus that we are in the midst of a species extinction crisis, we're saying it's time to put aside the gloom and doom and get down to the business of prioritizing and targeting conservation investments in order to have the greatest impact."

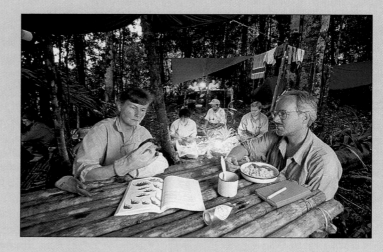

A Rapid Assessment Team studies the day's research results in camp at night.

---

example, isn't very enjoyable in a dirt bike racing course. And protecting species that need unbroken old-growth forest isn't feasible when you cut down the forest.

Old-Growth Forests

The most contentious forestry issues in the United States and Canada in recent years have centered on logging in old-growth forests in the Pacific Northwest. Only a century ago, most of the coastal ranges of Washington, Oregon, northern California, British Columbia, and southeastern Alaska were clothed in a lush forest of huge trees, many a thousand years old or more. These forests have incredibly high levels of biodiversity, and they accumulate more total biomass in standing vegetation per unit area than any other ecosystem on earth (fig. 6.8). Many endemic species, such as the northern spotted owl (fig. 6.9), Vaux's swift, and the marbled murrelet, are so highly adapted to the unique conditions of these ancient forests that they live nowhere else.

Today, less than 10 percent of the old-growth forest in the United States remains intact, and 80 percent of what is left is

FIGURE 6.8 *The temperate rainforests of the Pacific Northwest have the highest biomass per hectare of any landscape in the world.*

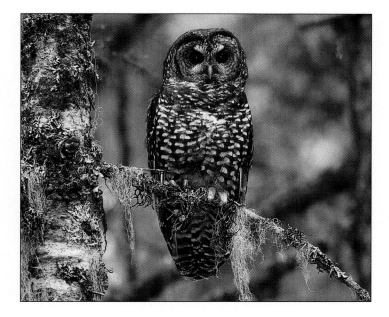

FIGURE 6.9 *Only about 2,000 pairs of northern spotted owls remain in the old-growth forests of the Pacific Northwest. Cutting old-growth forests threatens endangered species, but reduced logging threatens the jobs of many timber workers.*

scheduled to be cut down in the near future. British Columbia has felled at least 60 percent of its richest and most productive ancient forests and is now cutting some 240 million ha (600 million acres) annually, about ten times the rate of old-growth harvest in the United States. At these rates, the only remaining ancient forests in North America in 50 years will be a fringe around the base of the mountains in a few national parks.

In 1989, environmentalists sued the U.S. Forest Service over logging rates in Washington and Oregon, arguing that northern spotted owls are endangered and must be protected under the Endangered Species Act. Federal courts agreed and ordered some 1 million ha (2.5 million acres) of ancient forest set aside to preserve the last 2,000 pairs of owls. This is about half the remaining virgin forest in Washington and Oregon. The timber industry claimed that 40,000 jobs would be lost, although economic studies show that most logging jobs were lost as a result of log exports and mechanization. Still, outrage in the logging communities was loud and clear. Convoys of logging trucks converged on protest sites, while angry crowds burned environmentalists in effigy. Bumper stickers urged, "Save a logger; eat an owl."

A compromise forest management plan was finally agreed upon that allows some continued logging but protects the most valuable forests, including selected old-growth preserves, riparian (streamside) buffer strips, and prime wildlife habitat. This plan may not be enough protection, however, to ensure the survival of endangered salmon and trout populations in northwestern rivers. An even bigger battle may be shaping up as specific fish populations are considered for listing as endangered species. Several salmon and steelhead trout populations were listed as endangered under the Clinton administration. These listings have been chal-

lenged by the timber industry, but the $1 billion-per-year fishing industry in the Pacific Northwest may be at least as powerful a constituency as were those concerned about spotted owls.

Harvest Methods

Most lumber and pulpwood in the United States and Canada currently are harvested by **clear-cutting,** in which every tree in a given area is cut, regardless of size (fig. 6.10). This method is effective for producing even-age stands of sun-loving species such as aspen or pines, but often increases soil erosion, and eliminates habitat for many forest species when carried out on large blocks. It was once thought that good forest management required immediate removal of all dead trees and logging residue. Research has shown, however, that standing snags and coarse woody debris play important ecological roles, including soil protection, habitat for a variety of organisms, and nutrient recycling.

Some alternatives to clear-cutting include **shelterwood harvesting,** where mature trees are removed in a series of two or more cuts, and **strip-cutting,** where all the trees in a narrow corridor are harvested. For many forest types, the least disruptive harvest method is **selective cutting,** in which only a small percentage of the mature trees are taken in each 10- or 20-year rotation. Ponderosa pine, for example, are usually selectively cut to thin stands and improve growth of the remaining trees. A forest managed by selective cutting can retain many of the characteristics of age distribution and ground cover of a mature, old-growth forest. (See related story "Forestry for the Seventh Generation" at www.mhhe.com/apps.)

Should We Ban Logging and Roads in National Forests?

An increasing number of people in the United States are calling for an end to all logging on federal lands. In 2002, more than 220 scientists signed a letter to President Bush arguing that the value of timber produced from public lands is miniscule compared with the environmental costs of harvests. American taxpayers, they

FIGURE 6.10 *This huge clear-cut in Washington's Gifford Pinchot National Forest threatens species dependent on old-growth forest and exposes steep slopes to soil erosion. Restoring something like the original forest will take hundreds of years.*

argued, subsidize logging directly by charging less for timber sales than it costs to administer them, as well as indirectly because logging reduces the economic value of the forest for other uses. Just 4 percent of the nation's timber comes from national forests, they claimed, and this harvest adds about $4 billion to the American economy per year. In contrast, recreation, fish and wildlife, clean water, and other ecological services provided by the forest, by their calculations, are worth at least $224 billion each year. Timber industry officials, on the other hand, dispute these claims, arguing that logging not only provides jobs and supports rural communities, but also keeps forests healthy. What do you think? Could we make up for decreased timber production from public lands by more intensive management of private holdings, and by substitution or recycling of wood products? Are there alternative ways you could suggest to support communities now dependent on timber harvesting? Would prescribed fires be an acceptable method to regenerate forests?

Roads on public lands are another controversy. Over the past 40 years, the Forest Service has expanded its system of logging roads more than tenfold, to a current total of nearly 550,000 km (343,000 mi), or more than ten times the length of the interstate highway system. Government economists regard road building as a benefit because it opens up the country to motorized recreation and industrial uses. Wilderness enthusiasts and wildlife supporters, however, see this as an expensive and disruptive program. In 2001 the Clinton administration announced a plan to protect 23.7 million ha (58.5 million acres) of de facto wilderness from roads. Timber, mining, and oil companies protested this rule. In 2002 a federal judge in Idaho ruled that blocking road construction would do "irreparable harm" to the forest. He also ruled that the 600 public meetings held over a 12-month period to discuss this plan, and the record 1.6 million written comments gathered (90 percent of which supported increased protection) did not represent adequate public input. At the time of this writing, this ruling was being appealed. The Bush administration announced interim plans to give local communities much greater voice in how forests should be managed. Environmentalists fear that local control will result in damaging forest policies.

Fire Management

For more than 70 years, fire fighting has been a high priority for forest managers. To most people, a blackened, smoking, burned forest appears ruined forever (fig. 6.11). Recent studies, however, of fire's ecological role suggest that our attempts to suppress all fires may have been misguided. Many biological communities are fire-adapted and require periodic burning for regeneration. Furthermore, eliminating fire from these forests has allowed woody debris to accumulate, greatly increasing the chances of a really big fire. Forests that once were characterized by 50 to 100 mature, fire-resistant trees per hectare, interspersed with open meadows, now have a thick tangle of up to 2,000 small, spindly, mostly dead, saplings in the same area. The U.S. Forest Service says that 33 million ha (73 million acres), or about 40 percent of all federal forest lands, are at risk of severe fires. To make matters worse, Amer-

FIGURE 6.11   *To many people, a recently burned forest seems ruined forever. Fire, however, is often a natural force for regeneration and renewal. Within a year after the 1988 fire in Yellowstone National Park, millions of lodgepole pine seedlings had germinated to replenish the forest.*

icans increasingly live in remote areas where wildfires are highly likely. Because there haven't been fires in many of these places in living memory, many people assume there is no danger, but, by some estimates, 40 million U.S. residents now live in areas with a high wildfire risk.

Throughout the western United States, 2002 was the second worst wildfire summer in recorded history. Dry weather, high winds, and excess fuel combined to create 68,230 fires that burned about 2.7 million ha (6.7 million acres), or more than double the ten-year average. The total cost for fighting these fires was $1.6 billion.

The dilemma is how to undo years of fire suppression and fuel buildup. Ideally, a series of small, periodic fires should be allowed to clean out debris, but you have to get rid of excess biomass before such a regime can be put in place. Some local residents blame environmentalists for the fires, saying that opposition to logging was responsible for most of the damage. Environmentalists deny this, claiming that locals block prescribed burning and refuse to remove trees and brush from around their houses. The U.S. Forest Service has proposed a massive new program of forest thinning (removing small trees and dead wood) from some 16 million ha (40 million acres) of national forest. Carried out over a 15-year period, this program could cost as much as $12 billion. Critics worry that this is really an excuse for logging big, valuable, fire-resistant trees that will result in growth of fire-prone brush and young trees. If you were in charge of a national park or forest, how would you go about restoring forest health and also protecting human lives and property? (See the related story "The Quincy Library Group" at www.mhhe.com/apps.)

Ecosystem Management

In the 1990s, the U.S. Forest Service began to shift its policies from a timber production focus to **ecosystem management**, which

what can you do?

Lowering Your Forest Impacts

For most urban residents, forests—especially tropical forests—seem far away and disconnected from our everyday life. There are things that each of us can do, however, to protect forests.

- Reuse and recycle paper. Make double-sided copies. Save office paper, and use the back for scratch paper.
- Use email. Store information in digital form, rather than making hard copies of everything.
- If you build, conserve wood. Use wafer board, particle board, laminated beams, or other composites, rather than plywood and timbers made from old-growth trees.
- Buy products made from "good wood" or other certified sustainably harvested wood.
- Don't buy products made from tropical hardwoods such as ebony, mahogany, rosewood, or teak, unless the manufacturer can guarantee that the hardwoods were harvested from agroforestry plantations or sustainable-harvest programs.
- Don't patronize fast-food restaurants that purchase beef from cattle grazing on deforested rainforest land. Don't buy coffee, bananas, pineapples, or other cash crops if their production contributes to forest destruction.
- Do buy Brazil nuts, cashews, mushrooms, rattan furniture, and other nontimber forest products harvested sustainably by local people from intact forests. Remember that tropical rainforest is not the only biome under attack. Contact the Taiga Rescue Network (www. taigarescue.org) for information about boreal forests.
- If you hike or camp in forested areas, practice minimum-impact camping. Stay on existing trails, and don't build more or bigger fires than you absolutely need. Use only downed wood for fires. Don't carve on trees or drive nails into them.
- Write to your congressional representatives, and ask them to support forest protection and environmentally responsible government policies. Contact the U.S. Forest Service, and voice your support for recreation and nontimber forest values.

attempts to integrate sustainable ecological, economic, and social goals in a unified, systems approach. Some of the principles of this new philosophy include:

- Managing across whole landscapes, watersheds, or regions over ecological time scales
- Considering human needs and promoting sustainable economic development and communities
- Maintaining biological diversity and essential ecosystem processes
- Utilizing cooperative institutional arrangements

- Generating meaningful stakeholder and public involvement and facilitating collective decision-making
- Adapting management over time, based on conscious experimentation and routine monitoring

Some critics argue that we don't understand ecosystems well enough to make practical decisions in forest management on this basis. They argue we should simply set aside large blocks of untrammeled nature to allow for chaotic, catastrophic, and unpredictable events. Others see this new approach as a threat to industry and customary ways of doing things. Still, elements of ecosystem management appear in the National Report on Sustainable Forests prepared by the U.S. Forest Service. Based on the Montreal Working Group criteria and indicators for forest health, this report suggests goals for sustainable forest management (table 6.3).

RANGELANDS

After forests, grasslands are among the biomes most heavily used by humans. Prairies, savannas, steppes, open woodlands, and other grasslands occupy about one-quarter of the world's land surface. Much of the U.S. Great Plains and the Prairie Provinces of Canada fall in this category (fig. 6.12). The 3.8 billion ha (12 million mi^2) of pastures and grazing lands in this biome make up about twice the area of all agricultural crops. When you add to this about 4 billion ha of other lands (forest, desert, tundra, marsh, and thorn scrub) used for raising livestock, more than half of all land is used at least occasionally for grazing. More than 3 billion cattle, sheep, goats, camels, buffalo, and other domestic animals on these lands make a valuable contribution to human nutrition. Sustainable pastoralism can increase productivity while maintaining biodiversity in a grassland ecosystem.

Because grasslands, chaparral, and open woodlands are attractive for human occupation, they frequently are converted to cropland, urban areas, or other human-dominated landscapes. Worldwide the rate of grassland disturbance each year is three times that of tropical forest. Although they may appear to be uniform and monotonous to the untrained eye, native prairies

TABLE 6.3	Draft Criteria for Sustainable Forestry

1. Conservation of biological diversity
2. Maintenance of productive capacity of forest ecosystems
3. Maintenance of forest ecosystem health and vitality
4. Maintenance of soil and water resources
5. Maintenance of forest contribution to global carbon cycles
6. Maintenance and enhancement of long-term socioeconomic benefits to meet the needs of legal, institutional, and economic framework for forest conservation and sustainable management

Source: Data from USFS, 2002.

FIGURE 6.12   *This short-grass prairie in northern Montana is too dry for trees but, nevertheless, supports a diverse biological community.*

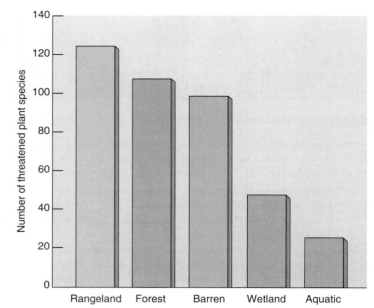

FIGURE 6.13   *Threatened plant species in the United States in 1990 by landscape type. Rangelands have more threatened and endangered species than any other North American biome.*
Source: Data from Curtis A. Flather et al., "Species Endangerment Patterns in the United States," *USDA, Forest Service Technical Report* 241, 1994, Table 3, Fort Collins, CO.

can be highly productive and species-rich. More threatened American plant species occur in rangelands than any other major biome (fig. 6.13).

Range Management

By carefully monitoring the numbers of animals and the condition of the range, ranchers and **pastoralists** (people who live by herding animals) can adjust to variations in rainfall, seasonal plant conditions, and nutritional quality of forage to keep livestock healthy and avoid overusing any particular area. Conscientious management can actually improve the quality of the range.

When grazing lands are abused by overgrazing—especially in arid areas—rain runs off quickly before it can soak into the soil to nourish plants or replenish groundwater. Springs and wells dry up. Seeds can't germinate in the dry, overheated soil. The barren ground reflects more of the sun's heat, changing wind patterns, driving away moisture-laden clouds, and leading to further desiccation. This process of conversion of once-fertile land to desert is called **desertification.**

This process is ancient, but in recent years it has been accelerated by expanding populations and political conditions that force people to overuse fragile lands. According to the International Soil Reference and Information Centre in the Netherlands, nearly three-quarters of all rangelands in the world show signs of either degraded vegetation or soil erosion. The highest percentage of moderate, severe, and extreme land degradation is in Mexico and Central America. Can we reverse this process? In some places, people are reclaiming deserts and repairing the effects of neglect and misuse.

Rangelands in the United States

The health of most public grazing lands in the United States is not good. Political and economic pressures encourage managers to increase grazing allotments beyond the carrying capacity of the range. Lack of enforcement of existing regulations and limited funds for range improvement have resulted in overgrazing, loss of native forage species, and erosion. The Natural Resources Defense Council claims that only 30 percent of public rangelands are in fair condition, and 55 percent are poor or very poor (fig. 6.14).

Overgrazing has allowed populations of unpalatable or inedible species, such as sage, mesquite, cheatgrass, and cactus, to build up on both public and private rangelands. Wildlife conservation groups regard cattle grazing as the most ubiquitous form of ecosystem degradation and the greatest threat to endangered species in the southwestern United States. They call for a ban on cattle and sheep grazing on all public lands, noting that it provides only 2 percent of the total forage consumed by beef cattle and supports only 2 percent of all livestock producers.

Like federal timber management policy, grazing fees charged for use of public lands often are far below market value and represent an enormous hidden subsidy to western ranchers. Holders of grazing permits generally pay the government less than 25 percent the amount of leasing comparable private land. The 31,000 permits on federal range bring in only $11 million in grazing fees but cost $47 million per year for administration and maintenance. The $36 million difference amounts to a massive "cow welfare" system of which few people are aware.

On the other hand, ranchers defend their way of life as an important part of western culture and history. Although few cattle go directly to market from their ranches, they produce almost all the beef calves subsequently shipped to feedlots. And without a

FIGURE 6.14 *On the right side of this fence is lightly grazed pasture, consisting of a dense growth of grasses and small, broad-leaved plants. On the left side is overgrazed rangeland, where the vegetation has become sparse.*

viable ranch economy, they claim, even more of the western landscape would be subdivided into small ranchettes to the detriment of both wildlife and environmental quality. What do you think? How much should we subsidize extractive industries to preserve rural communities and traditional occupations?

New Approaches to Ranching

Where a small number of livestock are free to roam a large area, they generally eat the tender, best-tasting grasses and forbs first (wouldn't you?), leaving the tough, unpalatable species to flourish and gradually dominate the vegetation. In some places, farmers and ranchers find that short-term, intensive grazing helps to maintain forage quality. As South African range specialist, Allan Savory, observed, wild ungulates (hoofed animals) like gnus or zebras in Africa or bison (buffalo) in America often tend to form dense herds that graze briefly but intensively in a particular location before moving on to the next area. Rest alone doesn't necessarily improve pastures and rangelands. Short-duration, **rotational grazing**—confining animals to a small area for a short time (often only a day or two) before shifting them to a new location—simulates the effects of wild herds. Forcing livestock to eat everything equally, to trample the ground thoroughly, and to fertilize heavily with manure before moving on helps to keep weeds in check and encourages growth of more desirable forage species. This approach doesn't work everywhere, however. Many plant communities in the U.S. desert Southwest, for example, apparently evolved in the absence of large, hoofed animals and can't withstand intensive grazing.

Another approach to ranching in some areas is to raise wild species, such as red deer, impala, wildebeest, or oryx. These animals forage more efficiently, resist harsh climates, often are more pest- and disease-resistant, and fend off predators better than usual domestic livestock. Native species also may have different feeding preferences and need for water and shelter than cows, goats, or sheep. The African Sahel, for instance, can only provide enough grass to raise about 20 to 30 kg (44 to 66 lbs) of beef per hectare.

Ranchers can produce three times as much meat with wild native species in that same area because these animals browse on a wider variety of plant materials.

In the United States, ranchers find that elk, American bison, or a variety of African species take less care and supplemental feeding than cattle or sheep and result in a better financial return because their lean meat can bring a better market price than beef or mutton (fig. 6.15). Media mogul Ted Turner has become both the biggest private landholder in the United States and the owner of more American bison than anyone other than the government.

PARKS AND NATURE PRESERVES

Since ancient times, sacred groves have been set aside for religious purposes and hunting preserves or pleasuring grounds for royalty. These lands tended to be open only to elite members of society, but they helped to preserve biodiversity and natural landscapes. Perhaps the first public parks open to ordinary citizens were the grand esplanades and the tree-sheltered agora that served as a gathering place in the planned Greek city. The modern design movement that incorporated Greek and Roman city planning principles in the redevelopment of European cities in the sixteenth and seventeenth centuries brought nature and beauty into the city in the form of parks, boulevards, and public gardens. It has only been in the past 130 years or so that we have begun to think about preserving wild places for the sake of wildlife and scenic beauty.

Parks serve a variety of purposes. They can teach us about our past and provide sanctuaries where nature is allowed to evolve in its own way. They are havens not only for wild plants and animals but also for the human spirit. Canada and the United States have greater total amounts of land dedicated to protected areas than any country except Denmark (which protects vast areas of Greenland's ice and snow) and Australia (which has designated great expanses of outback as aboriginal lands and parks). Although Mexico's parks are newer and less extensive than its wealthy neighbors, they contain far more biological and cultural diversity than do parks in either the United States or Canada.

FIGURE 6.15 *Native species, such as these American bison, are a valuable part of grassland ecosystems. They also may be the best way to harvest biomass for human consumption.*

FIGURE 6.16 *Yellowstone National Park, established in 1872, is regarded as the first national park in the world. Although initial interest focused on the park's natural curiosities and wonders, Yellowstone has come to be appreciated for its beauty and wilderness values.*

FIGURE 6.17 *Golden Gate National Recreation Area, the largest urban park in the world, gets more than 13 million recreational visits per year.*

North American Parks

In 1872, President Ulysses S. Grant signed an act designating about 800,000 ha (almost 2 million acres) of land in the Wyoming, Montana, and Idaho territories as Yellowstone National Park, the first national park in the world. Although the initial interest of both the founders and early visitors to Yellowstone were the curiosity of geysers, hot springs, and other odd features (fig. 6.16), the park was large enough to encompass and preserve real wilderness. As it became apparent that places of wild nature, scenic beauty, and cultural importance were rapidly disappearing with the closing of the North American frontier, the drive to set aside more national parks accelerated. Canada's Banff National Park was established in 1885. In the United States, Crater Lake, Mesa Verde, Grand Canyon, Glacier, and Rocky Mountain National Parks all were created between 1900 and 1915.

The U.S. national park system has grown to more than 280,000 km² (108,000 mi²) in 376 parks, monuments, historic sites, and recreation areas. Each year about 300 million visitors enjoy this system. The most heavily visited units are the urban recreation areas, parkways, and historic sites (fig. 6.17). What most people imagine when they think of a national park, however, are the great wilderness parks of the West. While these parks preserve areas of great natural beauty, they often have far more rocks, ice, and snow than biodiversity or ecological complexity. When the U.S. public domain was being carved up a century ago, the most productive land generally ended up in private hands.

Most of our national parks are intended to preserve an area as we believe it looked before any human habitation. Often, this is an illusion because people have inhabited and affected the land for thousands of years. Furthermore, the idea of maintaining a static equilibrium in nature is based on ideas of climax communities discarded by most ecologists today. What would it mean for park management if we conceive of biological communities as con-

stantly changing mosaics of chance associations of plants and animals (including humans)?

Canada has a total of 1,471 parks and protected areas occupying about 150,000 km². Among this group are national parks, provincial parks, outdoor recreation parks, and historic parks. They range in size from vast wilderness expanses, such as Wood Buffalo National Park in northern Manitoba or Ellesmere National Park Reserve in Nunavut Territory (fig. 6.18), to tiny pockets of cultural or natural history occupying only a few hectares. Kluane National Park in the Yukon, together with the Tatshenshini-Alsek Wilderness in British Columbia, the adjoining Wrangell-St. Elias National Park, and Glacier Bay National Park in Alaska encompass an area of about 10 million ha (24.7 million acres), roughly the size of Belgium or ten times as big as Yellowstone National Park. While many ecological reserves in Canada enforce strictly controlled access, other protected areas encourage intensive recreation, allow hunting, logging, or mining, and permit environmental manipulation for management purposes.

Park Problems

Originally, the great wilderness parks of Canada and the United States were fortresses protected from development or exploitation by legal boundaries and diligent park rangers. Most were buffered from human impacts by their remote location and the wild lands surrounding them. Today, the situation has changed. Many parks have become islands of nature surrounded and threatened by destructive land uses and burgeoning human populations that crowd park boundaries. Forests are clear-cut right up to the edges of some parks, while mine drainage contaminates streams and groundwater. Garish tourist traps clustered at park entrances detract from the beauty and serenity that most visitors seek.

Some popular parks such as Yosemite and Grand Canyon have become so crowded that managers have had to restrict automobile access and institute mass-transit systems to reduce traffic jams and pollution. Sometimes trails are so packed with tourists that they seem more like city streets than wilderness refuges. Visitors demand services such as golf courses, laundries, video arcades, bars, grocery stores, and other facilities that intrude on the solitude and natural beauty that parks were established to protect. Park rangers often spend more time on crime prevention and crowd control than natural history.

In many parks, dune buggies, dirt bikes, and off-road vehicles (ORV) run over fragile landscapes, disturbing vegetation and wildlife and destroying the aesthetic experience of those who come to enjoy nature. The U.S. Park Service has announced plans to exclude snowmobiles and personal watercraft (jet skis and waverunners) from many parks, but these plans have been challenged by manufacturers and owners' groups. Yellowstone National Park is case in point. On a typical winter weekend, more than 1,000 snowmobiles enter the park. Air pollution from their engines often is so bad that rangers at the west Yellowstone entry point are forced to wear respirators. Although the phase-out plan was based on 22 public hearings and 65,000 written comments (a majority of which favored a ban), Secretary of the Interior, Gale Norton, ordered the Park Service to scrap its plans and start a new study. Environmental groups charged that this is just a stalling technique toward eventually overturning the ban altogether.

Air pollution has become a serious threat to other parks as well. Sulfate concentrations in Shenandoah and Great Smoky Mountains National Parks are five times human health standards, and ozone levels in Acadia National Park in Maine exceed primary air quality standards by as much as 50 percent on some summer days. Visitors to the Grand Canyon once could see mountains 160 km (100 mi) away; now the air is so smoggy you can't see from one rim to the other during one-third of the year because of soot and sulfate aerosols from nearby power plants in Utah and Arizona.

Mining and oil interests continue to push for permission to dig and drill in the parks, especially on the 3 million acres of private **inholdings** (private lands) in the parks. Oil or gas wells recently have been drilled in Padre Island National Seashore, Big Cypress National Preserve, and the Upper Missouri River Breaks National Monument.

Wildlife Issues

Wildlife is at the center of many arguments regarding whether the purpose of the parks is to preserve nature or to provide entertainment for visitors (fig. 6.19). In the early days of the parks, "bad" animals (such as wolves and mountain lions) were killed so that populations of "good" animals (such as deer and elk) would be high. Rangers cut trees to improve views, put out salt blocks to lure animals to good viewing points close to roads, and otherwise manipulated nature to provide a more enjoyable experience for the guests.

Critics of this policy claim that favoring some species over others has unbalanced ecosystems and created a sad illusion of a natural system. They claim that excessively large deer and elk populations in Yellowstone and Grand Teton National Parks, for instance, have degraded the range so badly that massive die-offs

FIGURE 6.18 *Hikers explore Ellesmere National Park in Canada's new Nunavut Territory. Although rich in scenery and solitude, this park has very little biodiversity.*

FIGURE 6.19 *Wild animals have always been one of the main attractions in national parks. Many people lose all common sense when interacting with big, dangerous animals. This is not a petting zoo.*

occurred in severe winters. Park rangers tried hiring professional hunters to reduce the herds, but a storm of protest was raised. Sportsmen want to be able to hunt the animals themselves, animal lovers don't want the animals to be killed at all, and wilderness advocates don't like the precedent of hunting in national parks. Proposals to reduce elk and bison populations by reintroducing predators such as wolves and mountain lions have been highly controversial (see Case Study, p. 138).

CASE STUDY

REINTRODUCING WOLVES TO YELLOWSTONE

A little more than a century ago, an estimated 100,000 gray, or timber, wolves *(Canis lupis)* roamed the western United States. As farmers and ranchers moved west, however, wolves were poisoned, shot, and trapped wherever they could be found. The last wolves were eliminated from the northern Rocky Mountains in the early 1900s. Without a predator to restrain their numbers, elk and deer populations expanded rapidly. In Yellowstone National Park, for instance, the elk herd grew to some 25,000 animals, probably four or five times the vegetation's carrying capacity. Vegetation was overgrazed, and populations of smaller animals, such as ground squirrels, declined. For several decades, ecologists urged that wolves be reintroduced to control prey populations. These proposals brought howls of angry protest from local ranchers, who see wolves as sinister killers that threaten children, pets, livestock, and the ranching way of life. It took more than 20 years to get approval for wolf reintroduction.

In 1995, wolves were trapped in western Canada and relocated to Yellowstone. Once in the park, wolves became established surprisingly quickly. Taking advantage of the abundant food supply, they tripled their population in just three years. The effects on the ecosystem were immediate and striking. Biodiversity increased noticeably. Fewer elk, deer, and moose meant more food for squirrels, gophers, voles, and mice. Abundant small prey, in turn, led to increased numbers of eagles, hawks, fox, pine martens, and weasels. Large animal carcasses left by the wolves provided a feast for scavengers, such as bears, ravens, and magpies. Nearly half the coyotes, which had become common in the wolf's absence, were killed by their larger cousins. This helped small mammals that once were coyote prey. Plants such as grasses, forbs, willows, and aspen flourished in the absence of grazing and browsing pressure. Rangers and naturalists were delighted that the ecosystem was back in balance once again, while tourists were thrilled to catch a glimpse of a wolf or to hear them howl.

By the summer of 2000, researchers counted about 125 wolves in nine packs in Yellowstone, just one pack short of the recovery goal of ten packs for three consecutive years. Not everyone was happy, however, with the reintroduction program. Ranchers continue to regard wolves as a threat to their way of life. And some animal rights groups object to a policy that allows any wolf to be shot if it strays from the park and attacks livestock.

Wolf reintroduction raises some important questions about the purposes of

Are wolves beautiful, thrilling symbols of wild nature or ruthless killers? Reintroduction of these top predators into Yellowstone National Park has enthusiastic support from environmental groups but passionate opposition from local ranchers and hunters.

parks. Are they recreational areas or refuges for nature? What responsibilities do parks have to their neighbors (and vice versa)? What role should science play in this dispute? Suppose you were a wildlife biologist charged with designing an acceptable wolf management program in Yellowstone. What strategies would you use, and where would you start?

Parks As Ecosystems

One of the biggest problems with managing parks and nature preserves is that boundaries usually are based on political rather than ecological considerations. Airsheds, watersheds, and animal territories or migration routes often extend far beyond official boundaries and yet profoundly affect communities that we are attempting to preserve. Yellowstone and Grand Teton National Parks in northwestern Wyoming are examples of this dilemma. Although about 1 million ha (2.47 million acres) in total size, these parks probably cannot preserve viable populations of large predators, such as grizzly bears and mountain lions, completely within their borders. Management policies in the surrounding national forests and private lands seriously affect conditions in the park (fig. 6.20). The natural **biogeographical area** (an entire ecosystem and its associated land, water, air, and wildlife resources) must be managed as a unit if we are to preserve all its values.

The International Union for the Conservation of Nature and Natural Resources (IUCN) (also known as the World Conservation Union) divides protected areas into five categories with increasing levels of protection and decreasing human impacts (table 6.4). Many of our parks have tried to meet all these goals

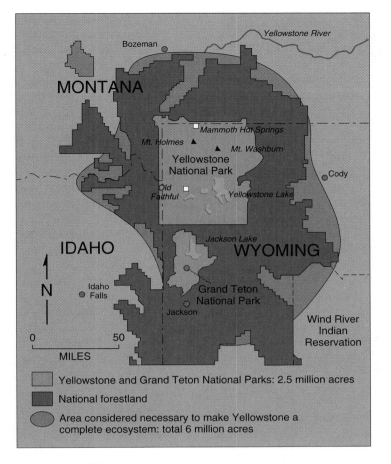

FIGURE 6.20 *This map shows the Yellowstone ecosystem complex or biogeographical region, which extends far beyond the park's boundaries. Park managers and ecologists believe that it is necessary to manage the entire region if the park itself is to remain biologically viable.*

Map labels:
Yellowstone River
Bozeman
MONTANA
Mammoth Hot Springs
Mt. Holmes ▲   ▲ Mt. Washburn
Yellowstone National Park
Cody
Old Faithful   Yellowstone Lake
IDAHO
Jackson Lake
WYOMING
N
Idaho Falls
Grand Teton National Park
Jackson
Wind River Indian Reservation
0   50   MILES

Legend:
☐ Yellowstone and Grand Teton National Parks: 2.5 million acres
■ National forestland
⬤ Area considered necessary to make Yellowstone a complete ecosystem: total 6 million acres

		ALLOWED HUMAN
TABLE 6.4	**IUCN Categories of Protected Areas**	

CATEGORY	ALLOWED HUMAN IMPACT OR INTERVENTION
1. Ecological reserves and wilderness areas	Little or none
2. National parks	Low
3. Natural monuments and archeological sites	Low to medium
4. Habitat and wildlife management areas	Medium
5. Cultural or scenic landscapes, recreation areas	Medium to high

Source: Data from World Conservation Union, 1990.

TABLE 6.5	**World Nature Preserves by Area and Type**

BIOME	NUMBER OF AREAS	PERCENT OF AREA PROTECTED
Tropical dry forests	907	21
Tropical humid forests	355	12
Temperate deciduous forests	682	6
Temperate coniferous forests	114	7
Deserts	215	17
Tundra	31	26
Grasslands	116	3
Mountain regions	318	7
Islands	74	<1
Lakes and wetlands	10	<1

Source: Norman Myers, 1998.

simultaneously but may have to select those of highest value. Most parks limit the number of overnight visitors. The time may come when park permits will need to be reserved years in advance, and visits to certain parks will be limited to once in a lifetime! How would you feel about such a policy? Would you rather visit a pristine, uncrowded park only once or a less perfect place whenever you wanted?

One solution to congestion and overuse of our parks is to create new ones to distribute the load. This would also allow us to protect and enjoy other magnificent areas that deserve preservation in their own right. At the end of his term in office, President Bill Clinton created ten new parks and national monuments protecting nearly 2 million ha (about 5 million acres) in western states. Environmentalists celebrated protection of these areas; local residents, however, generally opposed what they feared would be increased restrictions on their use of the land and its resources. The Bush Administration promised to give local residents and officials unprecedented input in devising management plans for these new parklands and monuments.

Canada's Green Plan, released in 1990, called for a doubling of protected areas to a total of 12 percent of total land surface. Representative samples of every ecoregion should be included in this network, including cultural landscapes and marine ecosystems.

WORLD PARKS AND PRESERVES

The idea of setting aside nature preserves has spread rapidly over the past 50 years as people around the world have become aware of the growing scarcity of wildlife and wild places. Still, in many cultures wilderness and wild lands are regarded as useless wastelands that should be put to some productive use if possible. So far, more than 530 million ha (nearly 4 percent of the earth's land) are designated as parks, wildlife refuges, and nature preserves worldwide (table 6.5).

The largest number of protected areas is in tropical dry forests and savannas (principally in Africa) and temperate deciduous forests (mainly in North America and Europe), but many of these preserves are too small to maintain significant biological populations over the long term. Vast stretches of arctic tundra in a few parks and preserves in Alaska, Canada, Greenland, and Scandinavia make up about 26 percent of all protected areas. Deserts and tropical humid forests also are well represented, but grasslands, aquatic ecosystems, and islands are badly underrepresented.

According to the United Nations Environment Programme, North and Central America have the largest fraction—33 percent of all protected land or nearly 10 percent of their land area—designated for protection of any continent. The former Soviet Union, with 17 percent of the world's land, has only 3 percent of the officially protected area. The rapid destruction of Siberian forests and terrible pollution problems in Russia and its former allies raise serious concerns about these natural areas. The IUCN has identified an additional 3,000 areas, totaling about 3 billion ha, worthy for national park or wildlife refuge status. The most significant of these areas are designated world **biosphere reserves** or world heritage sites. Currently, about 300 of these special refuges have been designated in 75 countries.

Among the individual countries with the most ambitious plans to protect natural resources are Costa Rica, Tanzania, Rwanda, Botswana, Benin, Senegal, the Central African Republic, Zimbabwe, Butan, and Switzerland, each of which has designated 10 percent or more of its land as ecological protectorates. Brazil has even more ambitious plans, calling for some 231,600 km^2, or 18 percent of the country, to be protected in nature preserves. So far, however, many of these areas are parks in name only. Lacking guards, visitor centers, administrative personnel, or even boundary fences, they are open to vandals and thieves to loot as they will.

Marine Preserves

As ocean fish stocks become increasingly depleted globally (see chapter 9), biologists are calling for protected areas where marine organisms are sheltered from destructive harvest methods. Research has shown that "no-take" refuges not only protect the species living within them, but also serve as nurseries for nearby areas. In a study of 100 of these refuges around the world, researchers have found that, on average, the number of organisms inside refuges was twice as high as surrounding areas where harvest was allowed. In addition, the biomass of organisms was three times as great and individual animals were, on average, 30 percent larger inside the refuge compared to outside. The size necessary for a safe haven to protect flora and fauna and replenish the area around it depends on the species involved, but some marine biologists have called on nations to protect at least 20 percent of their near-shore territory as marine refuges.

Coral reefs are among the most threatened marine ecosystems in the world. Recent remote sensing surveys show that living coral covers only about 285,000 km^2 (110,000 mi^2), or an area about the size of Nevada. This is less than half of previous estimates, and 90 percent of all reefs face threats from sea temperature change, destructive fishing methods, coral mining, sediment runoff, and other human disturbance. In many ways, coral reefs are the old-growth rainforests of the ocean. Biologically rich, these sensitive habitats can take a century or more to recover from damage. If current trends continue, some researchers predict that in 50 years there will be no viable coral reefs anywhere in the world.

What can be done to reverse this trend? Among the new national monuments created by President Bill Clinton are three that preserve coral reefs: the U.S. Virgin Islands Coral Reef National Monument, The Pacific Ocean Preserve northwest of Hawaii, and the Tortugas Refuge within the Florida Keys National Marine Sanctuary. Together, these refuges protect 90 percent of all coral in U.S. territory. All are closed to commercial fishing, and the Tortugas reefs are completely protected as a "no-take" zone. Unfortunately, most reefs lack such protection. A survey of marine biological resources identified the ten richest and most threatened "hot spots," which are the Philippines, the Gulf of Guinea and Cape Verde Islands (off the west coast of Africa), Indonesia's Sunda Islands, the Mascarene Islands in the Indian Ocean, South Africa's coast, southern Japan and China, the western Caribbean, and the Red Sea and Gulf of Aden. We urgently need a system of underwater national parks and preserves like those established to protect terrestrial biomes.

Protecting Natural Heritage

Even parks and preserves with protective systems in place are not always safe from exploitation or changes in political priorities. Many problems threaten natural resources and environmental quality in these areas. In Greece, the Pindus National Park is threatened by plans to build a hydroelectric dam in the center of the park. Furthermore, excessive stock grazing and forestry exploitation in the peripheral zone are causing erosion and loss of wildlife habitat. In Colombia, the Paramillo National Park also is threatened by dam building. Oil exploration along the border of the Yasuni National Park in Ecuador pollutes water supplies, while miners and loggers in Peru have invaded portions of Huascarán National Park. In Palau, coral reefs identified as a potential biosphere reserve are damaged by dynamiting, while on some beaches in Indonesia, every egg laid by endangered sea turtles is taken by egg hunters. These are just a few of the many problems around the world. Often, countries with the most important biomes lack funds, trained personnel, and experience to manage some of the areas under their control.

The IUCN has developed a **world conservation strategy** for natural resources that includes the following three objectives: (1) to maintain essential ecological processes and life-support systems (such as soil regeneration and protection, nutrient recycling, and water purification) on which human survival and development depend; (2) to preserve genetic diversity essential for breeding programs to improve cultivated plants and domesticated animals; and (3) to ensure that any utilization of wild species and ecosystems is sustainable.

These goals are further elaborated in the ecological plan of action adopted by the IUCN and shown in table 6.6.

TABLE 6.6 IUCN Ecological Plan of Action

1. Launch a consciousness-raising exercise to bring the issue of biological resources to the attention of policymakers and the public at large.
2. Design national conservation strategies that take explicit account of the values at stake.
3. Expand our network of parks and preserves to establish a comprehensive system of protected areas.
4. Undertake a program of training in the fields relevant to biological diversity to improve the scientific skills and technological grasp of those charged with its management.
5. Work through conventions and treaties to express the interest of the community of nations in the collective heritage of biological diversity.
6. Establish a set of economic incentives to make species conservation a competitive form of land use.

Source: Data from International Union for Conservation and National Resources.

Size and Design of Nature Preserves

What is the optimum size and shape of a nature preserve? For many years, conservation biologists have disputed whether it is better to have a **s**ingle **l**arge **o**r **s**everal **s**mall reserves (the SLOSS debate). Ideally, a reserve should be large enough to support viable populations of endangered species, keep ecosystems intact, and isolate critical core areas from damaging external forces. But as gaps are opened in habitat by human disturbance and areas are fragmented into isolated islands, edge effects may eliminate core characteristics everywhere.

To satisfy the conflicting needs and desires of humans and nature, we may need a spectrum of preserves with decreasing levels of interference and management, ranging from: (1) recreation areas, designed primarily for human entertainment, aesthetics, and enjoyment; (2) historic areas, intended to preserve a landscape as we imagine it looked in some previous time, such as presettlement or pioneer days; (3) conservation reserves, set aside to maintain essential ecological functions, preserve biodiversity, or protect a particular species or group of organisms; (4) pristine research areas, to serve as a baseline of undisturbed nature; and (5) inviolable preserves, for sensitive species and from which all human entrance is strictly prohibited.

For some species with small territories, several small, isolated refuges can support viable populations and provide insurance against a disease or other calamity that might wipe out a single population. But small preserves can't support species such as elephants or tigers that need large amounts of space. Given human needs and pressures, however, big preserves aren't always possible. Establishing **corridors** of natural habitat to allow movement of species from one area to another (fig. 6.21) can help maintain genetic exchange and prevent the high extinction rates often

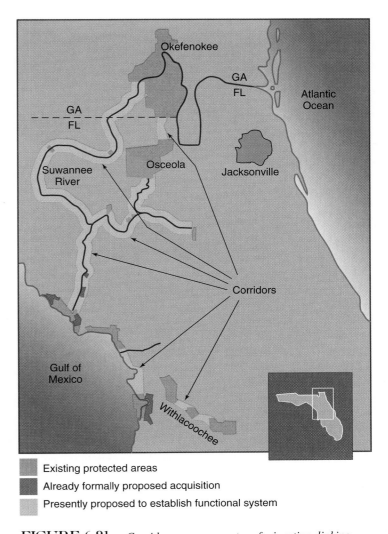

Existing protected areas
Already formally proposed acquisition
Presently proposed to establish functional system

FIGURE 6.21 *Corridors serve as routes of migration, linking isolated populations of plants and animals in scattered nature preserves. Although individual preserves may be too small to sustain viable populations, connecting them through river valleys and coastal corridors can facilitate interbreeding and provide an escape route if local conditions become unfavorable.*

Source: R. F. Noss and L. D. Harris, "Nodes, Networks and MUMs: Preserving Diversity at All Levels," in *Environmental Management,* vol. 10:299–309, 1986.

characteristic of isolated and fragmented areas. Some animals, however, don't use corridors and so large, contiguous areas may be the only way to save them.

An interesting experiment funded by the World Wildlife Fund and the Smithsonian Institution is being carried out in the Brazilian rainforest to determine the effects of shape and size on biological reserves. Some 23 test sites, ranging in size from 1 ha (2.47 acres) to 10,000 ha have been established. Some areas are surrounded by clear-cuts and newly created pastures (fig. 6.22), while others remain connected to the surrounding forest. Selected species are regularly inventoried to monitor their dynamics after disturbance. As was expected, some species disappear very quickly, especially from small areas. Sun-loving species flourish

FIGURE 6.22 *How small can a nature preserve be? In an ambitious research project, scientists in the Brazilian rainforest are carefully tracking wildlife in plots of various sizes, either connected to existing forests or surrounded by clear-cuts. As you might expect, the largest and most highly specialized species are the first to disappear.*

in the newly created forest edges, but deep-forest, shade-loving species move out, particularly when size or shape reduces the distance from the edge to the center below a certain minimum. This demonstrates the importance of surrounding some reserves with buffer zones that maintain the balance of edge and shade species.

Conservation and Economic Development

Many of the most seriously threatened species and ecosystems of the world are in the developing countries, especially in the tropics. This situation concerns us all because these countries are the guardians of biological resources that may be vital to all of us. Unfortunately, where political and economic systems fail to provide people with land, jobs, and food, disenfranchised citizens turn to legally protected lands, plants, and animals for their needs. Immediate human survival always takes precedence over long-term environmental goals. Clearly, the struggle to save species and unique ecosystems cannot be divorced from the broader struggle to achieve a new world order in which the basic needs of all are met.

People in some developing countries are beginning to realize that the biological richness of their environment may be their most valuable resource and that its preservation is vital for sustainable development. **Ecotourism** (tourism that is ecologically and socially sustainable) can be more beneficial to many of these countries over the long term than extractive industries, such as logging and mining (fig. 6.23). What can you do? box (on this page) suggests some ways to ensure that your vacations are responsible ecotourism.

what can you do?

Being a Responsible Ecotourist

1. *Pretrip preparation.* Learn about the history, geography, ecology, and culture of the area you will visit. Understand the do's and don'ts that will keep you from violating local customs and sensibilities.

2. *Environmental impact.* Stay on designated trails, and camp in established sites, if available. Take only photographs and memories and leave only goodwill wherever you go.

3. *Resource impact.* Minimize your use of scarce fuels, food, and water resources. Do you know where your wastes and garbage go?

4. *Cultural impact.* Respect the privacy and dignity of those you meet, and try to understand how you would feel in their place. Don't take photos without asking first. Be considerate of religious and cultural sites and practices. Be as aware of cultural pollution as you are of environmental pollution.

5. *Wildlife impact.* Don't harass wildlife or disturb plant life. Modern cameras make it possible to get good photos from a respectful, safe distance. Don't buy ivory, tortoise shell, animal skins, feathers, or other products taken from endangered species.

6. *Environmental benefit.* Is your trip strictly for pleasure, or will it contribute to protecting the local environment? Can you combine ecotourism with work on cleanup campaigns or delivery of educational materials or equipment to local schools or nature clubs?

7. *Advocacy and education.* Get involved in letter writing, lobbying, or educational campaigns to help protect the lands and cultures you have visited. Give talks at schools or to local clubs after you get home to inform your friends and neighbors about what you have learned.

At the 1982 World Congress on National Parks in Bali, 500 scientists, managers, and politicians discussed the design and location of biological reserves and the ecological, economic, and social factors that impinge on wildlife preservation. They concluded that conservation and rural development are not necessarily incompatible. In many cases, sustainable production of food, fiber, medicines, and water in rural areas depends on ecosystem services derived from adjacent conservation reserves. Tourism associated with wildlife watching and outdoor recreation can be a welcome source of income for underdeveloped countries. If local people share in the benefits of saving wildlife, they probably will cooperate, and the programs will be successful. To reformulate Thoreau's famous dictum, "In broadly shared economic progress is preservation of the wild."

Indigenous Communities and Biosphere Reserves

Areas chosen for nature preservation are often traditional lands of indigenous people who cannot simply be ordered out. Finding ways

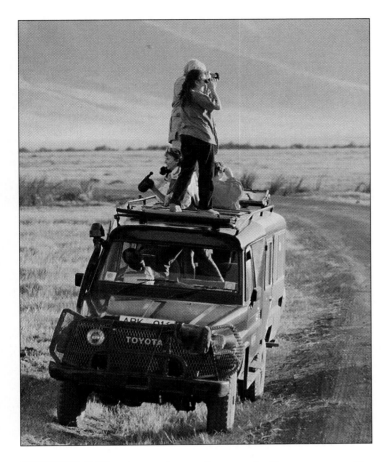

FIGURE 6.23 *Ecotourism can be a sustainable resource use. If local communities share in the revenue, ecotourism gives them an incentive to value and protect biodiversity and natural beauty.*

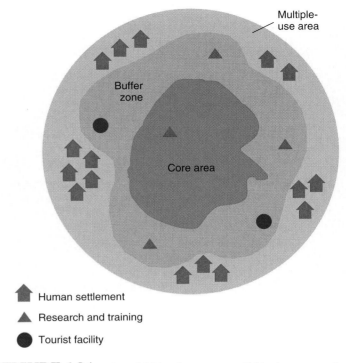

FIGURE 6.24 *A model biosphere reserve. Critical ecosystem is preserved in the core. Research and tourism are allowed in the buffer zone, while sustainable resource harvesting and permanent habitations are located in the multiple-use area around the perimeter.*

to integrate human needs with those of wildlife is essential for local acceptance of conservation goals in many countries. In 1986, UNESCO (United Nations Educational, Scientific, and Cultural Organization) initiated its **Man and Biosphere (MAB) program** that encourages division of protected areas into zones with different purposes. Critical ecosystem functions and endangered wildlife are protected in a central core region, where limited scientific study is the only human access allowed. Ecotourism and research facilities are located in a relatively pristine buffer zone around the core, while sustainable resource harvesting and permanent habitation are allowed in multiple-use peripheral regions (fig. 6.24).

Mexico's 545,000-ha (2,100-mi²) Sian Ka'an Reserve on the Caribbean coast is a good example of an MAB reserve. The core area includes 528,000 ha (1.3 million acres) of coral reef and adjacent bays, marshes, and lowland tropical forest. More than 335 bird species have been observed within the reserve, along with endangered manatees, five types of jungle cats, spider and howler monkeys, and four species of increasingly rare sea turtles. Approximately 25,000 people live in communities in peripheral regions around the reserve, and the resort developments of Cancún are located just to the north. In addition to tourism, the economic base of the area includes lobster fishing, small-scale farming, and coconut cultivation.

The Amigos de Sian Ka'an, a local community organization, played a central role in establishing the reserve and is working to protect the resource base while it improves living standards for local people. New intensive farming techniques and sustainable harvesting of forest products enable people to make a living without destroying the resource base. Better lobster harvesting techniques developed at the reserve have improved the catch without depleting native stocks. Local people now see the preserve as a benefit rather than an imposition from outside. Unfortunately, the government has very limited funds to develop or patrol the reserve. (See a related story "Ecotourism on the Roof of the World" at www. mhhe.com/apps.)

WILDERNESS AREAS

Although indigenous people had lived in the Americas for thousands of years before the first Europeans arrived, introduced diseases killed up to 90 percent of the existing population so that the continent appeared a vast, empty wilderness to early explorers. As historian Frederick Jackson Turner pointed out in a series of articles and speeches at the beginning of the twentieth century, a belief that wilderness was not only a source of wealth but also the origin of strength, self-reliance, wisdom, and character is deeply embedded in our culture. The frontier was seen as a place for continuous generation of democracy, social progress, economic growth, and national energy. A number of authors, including

FIGURE 6.25 *Wilderness provides opportunities for solitude, self-propelled recreation, and undisturbed nature.*

Henry Thoreau, Aldo Leopold, Sigurd Olson, Edward Abbey, and Wallace Stegner, have written about the physical, mental, and social benefits for modern people of rediscovering solitude and challenge in wilderness (fig. 6.25).

Beginning with designated primitive and roadless areas in a few national forests in the 1920s, the United States has established a system of 264 wilderness areas, encompassing nearly 40 million ha (100 million acres). These wilderness areas are found in national parks, national forests, wildlife refuges, and lands administered by the Bureau of Land Management (BLM).

The 1964 Wilderness Act, which is the basis for much of this system, defined **wilderness** as "an area of undeveloped land affected primarily by the forces of nature, where man is a visitor who does not remain; it contains ecological, geological, or other features of scientific or historic value; it possesses outstanding opportunities for solitude or a primitive and unconfined type of recreation; and it is an area large enough so that continued use will not change its unspoiled, natural conditions." Most of the areas meeting these standards are in the western states and Alaska.

Additional wilderness areas continue to be evaluated for protected status. Conservation groups argue that another 80 million ha (200 million acres), half of it in Alaska, should be added to the wilderness system. A prolonged battle has been waged over these de facto wilderness areas, pitting environmental groups who want more wilderness against loggers, miners, ranchers, and others who want less wilderness. The arguments for saving wilderness are that it provides (1) a refuge for endangered wildlife, (2) an opportunity for solitude and primitive recreation, (3) a baseline for ecological research, and (4) an area where we have chosen simply to leave things in their natural state.

Among the most common arguments against more wilderness are that timber, energy resources, and critical minerals contained on these lands are essential for economic development in local communities, while people want access for motorized recreation, such as off-road vehicles, snowmobiles, and motorboats. For many people living in remote areas, jobs, personal freedom, and local control of resources seem more important than abstract values of wilderness. They often see themselves as an embattled minority trying to protect an endangered, traditional way of life against a wealthy elite who want to lock up huge areas for recreation or aesthetic purposes. Wilderness proponents point out that 96 percent of the country already is open for resource exploitation; the remaining 4 percent is mostly land that developers didn't want anyway.

For many people, especially those in developing countries, the idea of pristine wilderness untouched by humans is neither very important nor very interesting. In most places, all land is occupied fully—if sparsely—by indigenous people. To them, the area is home, no matter how empty it may look to outsiders. From this perspective, preserving biological diversity, scenic beauty, and other natural resources may be a good idea, but excluding humans and human features from the land does not necessarily make it more valuable. In fact, saving cultural heritage, working landscapes, and historical evidence of early human occupation can often be among the most important reasons to protect an area.

WILDLIFE REFUGES

In 1901 President Teddy Roosevelt established 51 national **wildlife refuges,** the first in an important but troubled system for wildlife preservation in the United States. There are now 511 wildlife refuges in this system, encompassing nearly 40 million ha of land and water and representing every major biome in North America. They range in size from less than 1 ha (2.47 acres) for the tiny Mille Lacs Refuge in Minnesota to 7.3 million ha (18 million acres) in the Arctic National Wildlife Refuge of Alaska. Altogether, about 1 percent of U.S. surface area is designated as wildlife refuge. Every state has at least one refuge. Ostensibly, wildlife refuges are our most direct efforts to protect wildlife. Many of these sanctuaries are managed, however, primarily to produce game species for hunters.

Over the years, a number of problematic uses have become accepted in wildlife refuges, including oil drilling, cattle grazing, snowmobiling, motorboating, off-road vehicle use, timber harvesting, hay cutting, trapping, and military exercises. A U.S. General Accounting Office report found that 60 percent of all refuges allow activities that are harmful to wildlife. The biggest current battle over U.S. wildlife refuges concerns proposals for oil and gas drilling in the Arctic National Wildlife Refuge on the north slope of Alaska's Brooks Range.

Refuges also face threats from external activities. More than three-quarters of all U.S. refuges have water pollution problems, two-thirds of which are serious enough to affect wildlife. A notorious example is the former Kesterson Wildlife Refuge in

Principles of Environmental Science

California, where selenium-contaminated irrigation water drained from farm fields turned the marsh into a death trap for wildlife, rather than a sanctuary. Eventually, the marsh had to be drained and capped with clean soil to protect wildlife. Subsequent research has shown that at least twenty wildlife refuges in western states have toxic metal pollution caused by agricultural and industrial activities outside the refuge.

International Wildlife Preserves

As we saw earlier in this chapter, most developing countries rarely have separate systems of parks and wildlife refuges. Many nature preserves are set up primarily to protect wildlife, however. An outstanding example of both the promise and the problems in managing parks in the less-developed countries is seen in the Serengeti ecosystem in Kenya and Tanzania. This area of savanna, thorn woodland, and volcanic highland lying between Lake Victoria and the Great Rift Valley in East Africa is home to the highest density of ungulates (hoofed grazing animals) in the world. Over 1.5 million wildebeests (or gnus) graze on the savanna in the wet season, when grass is available, and then migrate through the woodlands into the northern highlands during the dry season. The ecosystem also supports hundreds of thousands of zebras, gazelles, impalas, giraffes, and other beautiful and intriguing animals. The herbivores, in turn, support lions and a variety of predators and scavengers, such as leopards, hyenas, cheetahs, wild dogs, and vultures. This astounding diversity and abundance is surely one of the greatest wonders of the world.

Tanzania's Serengeti National Park was established in 1940 to protect 15,000 km^2 (5,700 mi^2), an area about the size of Connecticut or twice as big as Yellowstone National Park. It is bordered on the east by the much smaller Ngorongoro Conservation Area and Lake Manyara National Park. Kenya's Masa Mara National Reserve borders the Serengeti on the north. Rapidly growing human populations push against the boundaries of the park on all sides. Herds of domestic cattle compete with wild animals for grass and water. Agriculturalists clamor for farmland, especially in the temperate highlands along the Kenya-Tanzania border. So many tourists flock to these parks that the vegetation is ground to dust by hundreds of sight-seeing vans, and wildlife find it impossible to carry out normal lives.

Perhaps the worst problem in Africa is **poaching** by illegal hunters who massacre wildlife for valuable meat, horns, and tusks. Where there once were about 1 million rhinos in Africa, the population had dropped by the mid-1980s to fewer than 10,000 animals. Antipoaching efforts have allowed the population to recover to about 15,000 currently.

FIGURE 6.26   *Elephant populations have been decimated throughout the world. The biggest threat is from poachers, who kill for their valuable ivory.*

Elephants are under a similar assault. Thirty years ago, there were no elephants in the Serengeti, but perhaps 3 million in all of Africa (fig. 6.26). Since then, about 80 percent of the African elephants have been killed—mainly for their ivory—at a rate of 100,000 each year. The 2,000 elephants now in Serengeti National Park have been driven there by hunting pressures elsewhere. Fortunately, the elephants find refuge in the park and add to the pleasure of tourists who come to see the wildlife; however, they are changing the ecosystem. Crowded into inadequate space, the elephants smash down the acacia trees, turning the woodland and mixed savanna into continuous grassland. This is beneficial for some animals, but not for others.

Poachers continue to pursue the elephants and rhinos, even in the park. Armed with high-powered rifles and even machine guns and bazookas from the many African wars in the last decade, the poachers take a terrible toll on the wildlife. Park rangers try to stop the carnage, but they often are outgunned by the poachers. The parks are beginning to resemble war zones, with fierce, lethal firefights rather than peace and tranquility.

Still, as ecotourism grows in importance in national budgets, both local residents and governments are coming to recognize that protecting wildlife and natural resources makes economic sense. Although wildlife has been decimated in many areas, vast numbers remain an inspiring spectacle elsewhere. With careful stewardship, we may be able to protect much of what's left of nature's diversity and abundance.

SUMMARY

Temperate broad-leaved forests and grasslands (biomes that make up most of North America and Europe) are among the most heavily human-dominated habitat type on the earth. The most biologically rich forests are the old-growth forests filled with ancient trees and relatively undisturbed ecological processes. While forests still cover about half the area worldwide that they did before humans became the dominant species on the planet, only about 22 percent of those forests retain old-growth characteristics. The most biologically diverse forests on the earth are the temperate rainforests of North America's Pacific Northwest and the warm rainforests of the tropics. Both are threatened by logging for their valuable wood, but their loss would be a cultural and scientific disaster.

An innovative program for habitat and biodiversity protection involves debt-for-nature swaps in which long-term debt is forgiven if a country or a company agrees to protect natural areas. Many such arrangements have been worked out for developing countries. Conservationists are especially concerned about huge clear-cuts, in which all vegetation is removed over a large area. This method drives out wildlife and exposes soil to erosion. Seventy years of fire suppression in U.S. forests has allowed fuel to build up so that we now worry about huge conflagrations. Prescribed fires to reduce fuel loads would help solve the problem, but the public fears that even small fires could escape and threaten them.

Grasslands are among the most human-dominated of any biome. More than three billion domestic animals raised on pastures and grazing lands make a valuable contribution to human nutrition. Sustainable pastoralism can maintain ecosystem health on many grasslands, but too often overgrazing leads to habitat degradation and desertification. In some areas, rotational grazing and ranching of wild species can be useful approaches.

Parks, wildlife refuges, wilderness areas, and nature preserves occupy a small percentage of the earth's total land area but protect valuable cultural resources and representative samples of the earth's species and ecosystems. Parks are havens for wildlife and places for healthful outdoor recreation. Many are overcrowded, misused, and neglected, however. Pollution and incompatible uses outside parks threaten the values that we seek to protect. Wildlife is at the center of many park controversies. Is the purpose of parks to preserve wild nature or provide entertainment for visitors?

Marine refuges can protect the sensitive, old-growth habitats along coastlines and islands that serve as nurseries for a large percentage of oceanic biodiversity. No-take zones in these refuges not only serve as a sanctuary for threatened species but also replenish nearby areas open to harvesting. Coral reefs are among the most biologically rich but threatened habitats in the world.

The optimum size for nature preserves depends on the terrain and the values they are intended to protect, but in general, the larger the reserve, the more species it can protect. Establishing corridors to link separate areas can be a good way to increase effective space and to allow migration from one area to another. Economic development and nature protection can go hand in hand. Ecotourism may be the most lucrative and long-lasting way to use resources in many developing countries.

Areas chosen for preservation often are lands of indigenous people. Careful planning and zoning can protect nature and also allow sustainable use of resources. Man and Biosphere (MAB) reserves provide for multiple use in some areas but strict conservation in others. Wilderness areas are defined as those where man is a visitor who does not remain. They possess outstanding opportunities for solitude and primitive recreation. Wildlife refuges were intended to be sanctuaries for wildlife, but over the years, many improbable and damaging uses have become established in them.

QUESTIONS FOR REVIEW

1. How could protecting lowland forests help preserve mountaintop fauna?

2. Which continents have the largest forest area, and which have the highest net rate of forest change?

3. Why is there disagreement over the amount of deforestation in the tropics?

4. What is a debt-for-nature swap? How does it work?

5. What is a clear-cut, and why are they criticized?

6. Summarize the state of rangelands in the United States.

7. List some problems and threats from inside and outside national parks and preserves.

8. Why is the reintroduction of wolves into Yellowstone National Park a controversial issue?

9. Describe the IUCN categories of protected areas and the amount of human impact or intervention allowed in each. Can you name some parks or preserves in each category?

10. List the three main points in the IUCN world conservation strategy and the six steps in the action plan to meet these goals.

THINKING SCIENTIFICALLY

1. If you were on a Rapid Assessment Team surveying an area for possible protection in a nature preserve, how would you weigh abundance (many individuals of a few species) against diversity (few individuals of many different species)?

2. Conservationists argue that we could reduce or redirect the demand for wood products; timber companies claim that continued production is essential for jobs and the economy. What evidence would you need to make a decision in this case, and where might you get it?

3. Brazil needs cash to pay increasing foreign debts and to fund needed economic growth. Why shouldn't it harvest its forests and mineral resources to gain the foreign currency it wants and needs? If we want Brazil to save its forests, what can or should we do to encourage conservation?

4. The U.S. government has kept timber prices and grazing leases low to maintain low lumber and meat prices for consumers and to help support rural communities and traditional ways of life. If you were a forest supervisor, how would you weigh those human interests against the ecological values of forests and rangelands?

5. There is considerable uncertainty about the extent of desertification of grazing lands or destruction of tropical rainforests. If you were evaluating these data, what evidence would you want to see, or how would you appraise conflicting evidence?

6. Is "contrived" naturalness a desirable feature in parks and nature preserves? How much human intervention do you think is acceptable in trying to make nature more beautiful, safe, comfortable, or attractive to human visitors? Think of some specific examples that you would or would not accept.

7. If you were superintendent of Yellowstone National Park, how would you determine the park's carrying capacity for elk? If there are too many elk, how would you thin the herd?

8. Why do you suppose that dry tropical forests and tundra are well represented in protected areas, whereas grasslands and wetlands are rarely protected? Consider social, cultural, and economic as well as biogeographical reasons in your answer.

9. Oil and gas companies want to drill in several parks, monuments, and wildlife refuges. Do you think this should be allowed? If so, under what conditions or restrictions?

KEY TERMS

biogeographical area 138
biosphere reserves 140
clear-cutting 131
closed-canopy forest 125
corridors 141
debt-for-nature swaps 129
desertification 134
ecosystem management 132
ecotourism 142
inholdings 137
Man and Biosphere (MAB) program 143
monoculture forestry 127

old-growth forests 125
pastoralists 134
poaching 145
relevé 130
rotational grazing 135
selective cutting 131
shelterwood harvesting 131
strip-cutting 131
wilderness 144
wildlife refuges 144
world conservation strategy 140

Location: http://www.mhhe.com/environmentalscience

Mc Graw Hill

WEB EXERCISES

Managing Columbia River Salmon

Until about a century ago, up to 16 million salmon and steelhead trout migrated every year up the Columbia River system to their breeding grounds in small headwater streams and lakes. Both native people and wildlife in the Pacific Northwest depended on this prodigious bounty. For a few months each year, there were more fish than anyone could eat. The runs were probably never uniform, however. For reasons that we don't fully understand, the number of fish returning from the ocean often varied widely from year to year.

Dam building, logging, urbanization, pollution, overfishing, and a host of other human activities have caused disastrous declines in the salmon fishery over the past century. Total numbers are down nearly 90 percent from historic highs, and only about one-tenth of the fish in the river are now wild stocks; the rest are hatchery-reared.

Visit our website at www.mhhe.com/apps for more information about salmon runs in the Columbia River. Analyze the graph and data presented there, and then answer the following questions:

1. Why do you think salmon catches were so low in the 1870s? Were salmon scarce then? What is the difference between the maximum salmon harvest and the closest low point on the graph?

2. Cycles of maximum and minimum salmon runs are not uniform in periodicity or amplitude, but if you were asked to give an estimate of how frequently big runs occurred, what answer would you give?

3. Cover up data on the graph for years after 1935. If you had been a fisheries manager in 1935, what would you have predicted about the abundance of fish in the next 10 or 20 years, based on previous history?

4. How would you design an experiment to determine which of the factors mentioned in the essay about salmon had the greatest effect on their demise?

5. Do you think that there was any hope, when seines, traps, and set nets were prohibited in 1950, that the salmon would ever return?

7 Food and Agriculture

We abuse the land because we regard it as a commodity belonging to us. When we see land as a community to which we belong, we may begin to use it with love and respect.

—Aldo Leopold

Soil conservation practices can protect air and water quality and preserve wildlife habitat.

OBJECTIVES

After studying this chapter, you should be able to

- describe world food supplies and some causes of chronic hunger in the midst of growing food surpluses.
- explain some major human nutritional requirements, as well as the consequences of deficiencies in those nutrients.
- differentiate between famine and chronic undernutrition, and understand the relation between natural disasters and social or economic forces in triggering food shortages.
- sketch the roles of living organisms, physical forces, and other factors in creating and maintaining fertile soil.
- differentiate between the sources and effects of land degradation, including erosion, nutrient depletion, waterlogging, and salinization.
- analyze some of the promises and perils of genetic engineering.
- explain the need for water, energy, and nutrients for sustained crop production, as well as some limits on our use of these resources.
- recognize the potential for low-input, sustainable, regenerative agriculture.

Golden Rice

Every year more than a million children die and another 350,000 go blind from the effects of vitamin A deficiency. Can modern science help find a solution to this tragedy? Swiss researcher Ingo Potrykus thinks so. For the past decade, Potrykus, of the Swiss Federal Institute of Technology in Zurich, and his colleague Peter Beyer, of the University of Freiburg in Germany, have been working on a crop that could improve the lives of millions of the poorest people in the world. Using genetic engineering techniques, they moved the genes that make daffodils yellow into *Oryza sativa,* the rice species eaten as a staple food by about half the people in the world (fig. 7.1).

The problem is that the poorest people often live on only a bowl or two of rice per day. Rice is low in beta carotene, the precursor for vitamin A. While rice, like all green plants, makes light-absorbing carotenes in its leaves, no known member of the *Oryza* genus stores these compounds in its seeds. Thus ordinary techniques of plant breeding don't offer a way to enrich the crop. Extracting carotene genes from daffodils, and promoters (DNA segments that regulate gene expression) from bacteria, Potrykus and Beyer constructed a plasmid (a naked DNA loop) that was innoculated into the soil bacterium *Agrobacterium tumefaciens.* The transgenic agrobacteria were then incubated with rice embryos in a plant tissue culture medium. As the bacteria infect rice cells, they also transfer the genes that encode instructions for making beta carotene. When the embryos grow into mature plants, they make beta carotene in all their tissues—including the endosperm that fills the golden rice grains. These plants can be crossbred with rice varieties adapted to a variety of local conditions and made available to poor farmers around the world.

In 2000, after nearly 10 years of work and $2.6 million in support from the Swiss government, the European Union, and the Rockefeller Foundation, Potrykus and his colleagues had grain ready to distribute for free to developing countries. To their surprise, however, the researchers found themselves embroiled in heated controversy. While they felt they had made a scientific breakthrough to save millions of lives and incalculable suffering, bioengineering opponents regarded their creation as unholy tinkering with nature and a sinister threat to our diet, traditional farming, and the health of the world environment. Opponents of genetically modified organisms (GMOs) objected that this rice would be a first step toward making farmers dependent on multinational seed producers. They also questioned whether there might be unforeseen ecological consequences of releasing newly created plants around the world. Activists threatened to destroy experimental crops and to mount protests against companies or governments attempting to introduce what they considered to be "Frankenfoods," the unnatural creations of agricultural biotechnology.

FIGURE 7.1 *Rice makes up a major part of the diet for half the people in the world. Unless supplemented with other sources of vitamins, minerals, and protein, however, rice alone doesn't comprise a healthy diet.*

Even some who don't oppose genetic engineering per se argue that technological solutions aren't the answer to poverty and malnutrition. There are better ways, they claim, to provide a nutritious diet to poor children. Furthermore, the families whose children are the most likely to suffer vitamin deficiencies often are landless peasants who can't grow the golden grain for themselves. Will they be able to buy the new varieties, or will they continue to subsist on the cheapest, least nutritious foods that barely keep them alive?

This dilemma illustrates the complexity of food security issues, and it raises several important questions for environmental science. (1) Will there be enough nutritious food for everyone in the world? (2) Will that food be safe to eat? (3) What will be the environmental consequences of raising the food we need? Science plays an important role in answering these questions, but it can't be separated from controversies about human society and access to resources. In this chapter, we'll look at world food supplies, hunger, nutrition, and agricultural systems that contribute to or help solve these difficult questions.

NUTRITION AND FOOD SUPPLIES

All of us need food. Despite dire predictions that runaway population growth will soon lead to global famines (see chapter 4), total world food supplies have more than kept up with increasing human numbers over the past two centuries. But can we maintain this trend as human populations continue to grow? And what will happen to our environment if we have to feed 9 or 10 billion people in another 50 years? While these questions may seem abstract to those of us in wealthy countries, they are life and death issues for a majority of the world's residents.

Chronic Hunger and Food Security

Simply producing large amounts of food doesn't guarantee that everyone will have enough. The UN Food and Agricultural Organization (FAO) estimates that we already grow enough food to provide everyone on earth with well more than the 2,770 calories per day needed (on average) for a healthy, active life. Yet an estimated 815 million people are **chronically undernourished**, meaning they routinely receive less than 90 percent of the minimum caloric intake needed for normal growth and development, and a healthy, productive life. One person in five in the world's developing countries fits this description. These countries account for 95 percent of those who are persistently hungry. According to the FAO, 55 percent of the 12 million child deaths each year are related to malnutrition. The UN World Food Summit in Rome in 2002 called on rich countries to commit $24 billion per year to halve the number of hungry people by 2015. The Summit also recognized the basic human right of physical and economic access to safe, nutritious food.

Figure 7.2 shows countries with the highest risk of food shortages. As you can see, most of sub-Saharan Africa, South and Southeast Asia, and parts of Latin America fall in this category. The greatest food emergencies in 2002 were in southern Africa, where at least 14.4 million people face starvation due to bad weather and political turmoil. In Afghanistan, after two decades of war and the worst drought in history, some 5 million people (many of them war refugees) have little or no access to food. The former Soviet Union and its allies, including North Korea, have had serious crop losses in recent years due to bad weather and political mismanagement. Good data are scarce, but millions of people are thought to be at risk in these countries.

Figure 7.3 shows chronic hunger trends in developing regions over the past 20 years. Notice that Asia—which accounts for about 61 percent of the world population—has by far the highest total number of undernourished people. This region also has made the greatest progress in feeding its population, having reduced the number living in chronic hunger by about 200 million people in spite of having added nearly 1 billion residents over the past two decades. Malnourishment has fallen in Asia from 30 percent of the population in 1980 to only 10 percent in 2000. Sub-Saharan Africa has had the greatest increase in hunger as well as the highest percentage of hungry people. About one-third of this region's residents are now considered chronically undernourished.

Average annual food consumption doesn't mean much to someone who regularly goes to bed hungry. **Food security**, defined as the ability to obtain sufficient food on a day-to-day basis, is much more important to individuals. Poverty and lack of social status are the greatest threats to personal food security. Many of the 1.4 billion people in the world who have to live on

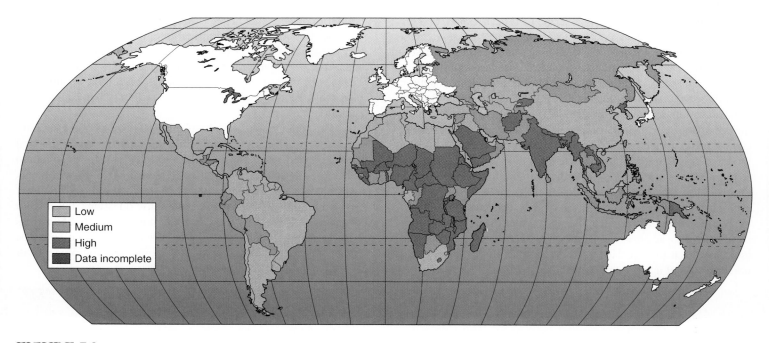

FIGURE 7.2   *Countries with populations at risk for inadequate nutrition. The United States, Canada, Europe, Japan, and Australia have little risk.*
Source: FAO, 2002.

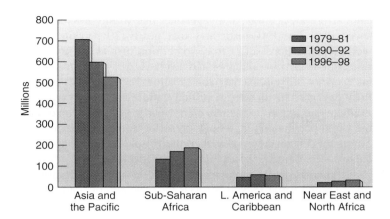

FIGURE 7.3 *Chronically undernourished population in developing regions.*
Source: FAO, 2002.

FIGURE 7.4 *Children wait for their daily ration of porridge at a feeding station in Somalia. When people are driven from their homes by hunger or war, social systems collapse, diseases spread rapidly, and the situation quickly becomes desperate.*

less than $1 per day all too often can't buy the food they need and don't have access to resources to grow it for themselves. Food insecurity occurs at multiple scales. In the poorest countries, hunger may affect nearly everyone. In other places, although the average food availability may be adequate, some individual communities or families may not have enough to eat. And within families, males often get both the largest share as well as the most nutritious food, while women and children—who need food most—all too often get the poorest diet.

Famines and Acute Food Shortages

Famines are characterized by large-scale food shortages, massive starvation, social disruption, and economic chaos. Starving people eat their seed grain and slaughter their breeding stock in a desperate attempt to keep themselves and their families alive. Even if better conditions return, they have often sacrificed their productive capacity and will take a long time to recover. Famines often trigger mass migrations, as starving people travel to refugee camps in search of food and medical care (fig. 7.4). Many die on the way or fall prey to robbers.

What causes these terrible tragedies? Environmental conditions are usually the immediate trigger, but politics and economics are often equally important in preventing people from getting the food they need. Drought, floods, insect outbreaks, and other natural disasters cause crop failures and create food shortages. But the Nobel Prize-winning work of Harvard economist Amartya K. Sen shows that these factors have often been around for a long time, and local people usually have ways to get through hard times if they aren't thwarted by inept or corrupt governments and greedy elites. National politics, commodity hoarding, price gouging, poverty, wars, landlessness, and other social factors often conspire so that poor people can neither grow their own food nor find jobs to earn money to buy the food they need. Professor Sen points out that armed conflict and political oppression almost always are at the root of famine. No democratic country with a relatively free press, he says, has ever had a major famine.

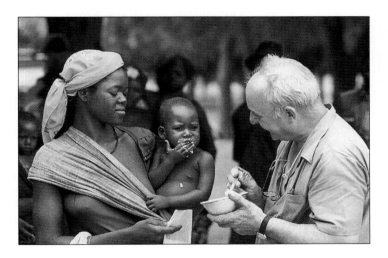

FIGURE 7.5 *We rarely learn from the media about the underlying causes of famine. Often, our attention is focused on images of people from wealthy nations generously helping those who are suffering. This makes us feel good but doesn't get to the root of the problems.*

The aid policies of rich countries often serve more to get rid of surplus commodities and make us feel good about our generosity than to eliminate the root causes of starvation (fig. 7.5). Crowding people into feeding camps generally is the worst thing to do for them. The stress of getting there kills many of them, and the density and lack of sanitation in the camps expose them to epidemic diseases. There are no jobs in refugee camps, so people can't support themselves if they try. Social chaos and family breakdown expose the weakest to robbery and violence. Having left their land and tools behind, people can't replant crops when the weather returns to normal.

Malnutrition and Obesity

In addition to energy (calories), we also need specific nutrients in our diet, such as proteins, vitamins, and certain trace minerals. You might have more than enough calories and still suffer from **malnourishment**, a nutritional imbalance caused by a lack of specific dietary components or an inability to absorb or utilize essential nutrients. Those of us in richer countries often eat too much meat, salt, and fat, and too little fiber, vitamins, trace minerals, and other components lost from highly processed foods. On average, we consume about one-third more calories than we need and get too little exercise. Food provides solace in stressful lives, and we're constantly bombarded with advertising tempting us to consume more. According to the U.S. Surgeon General, 61 percent of all adult Americans are overweight, up from 40 percent only a decade ago, and about one-third of us are seriously overweight or **obese**—generally considered to be a body mass greater than 30 kg/m^2, or roughly 30 pounds above normal for an average person.

Being overweight substantially raises your risk of hypertension, diabetes, heart attacks, stroke, gallbladder disease, osteoarthritis, respiratory problems, and certain cancers. Every year, about 300,000 in the United States die from illnesses related to obesity. This number is close to the 400,000 annual deaths from diseases related to smoking. Paradoxically, food insecurity and poverty can contribute to obesity. In one study, more than half of the women who reported not having enough to eat were overweight compared to one-third of food-secure women. Lack of time for cooking and access to healthy food choices, along with ready availability of fast-food snacks and calorie-laden drinks, lead to dangerous dietary imbalances for many people. This trend isn't limited to richer countries. Obesity is spreading around the world (fig. 7.6). South Pacific Islanders, for example, may have the highest rates of obesity in the world. In the Cook Islands, 52 percent of men are reported to be clinically obese. For the first time in history, there are probably more overweight people (more than 1 billion) than underweight.

In the poorest countries, people often eat the wrong kinds of food, not because they taste good or are trendy, but because they can't afford meat, fruits, and vegetables that would provide a balanced diet. The FAO estimates that nearly 30 percent of the world's populations suffer from some form of malnutrition. This results in devastating illnesses and deaths as well as reduced mental capacity, developmental abnormalities, and stunted growth. Fifty-five percent of the 12 million child deaths each year, for example, are related to malnutrition. Nearly 3 billion people (half the world population) are thought to suffer from vitamin, mineral, or protein deficiencies. Altogether, these problems bring an incalculable loss of human potential and social capital.

Anemia (low hemoglobin levels in the blood, usually caused by dietary iron deficiency) is the most common nutritional problem in the world. According to the FAO, more than 2 billion people (52 percent are pregnant women and 39 percent are children under five) suffer from iron deficiencies. The problem is most severe in India, where more than 80 percent of all pregnant women are anemic. Anemia increases the risk of maternal deaths from hemorrhage in childbirth and affects childhood development. Red meat, eggs, legumes, and green vegetables all are good sources of dietary iron.

Iodine is essential for synthesis of thyroxin, an endocrine hormone that regulates metabolism and brain development, among other things. Chronic iodine deficiency causes goiter (a swollen thyroid gland, fig. 7.7), stunted growth, and reduced mental ability. The FAO estimates that 740 million people—mainly in South and Southeast Asia—suffer from iodine deficiency and that 177 million children have stunted growth and development. Adding a few pennies worth of iodine to our salt has largely eliminated this problem in developed countries.

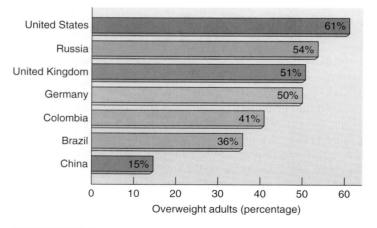

FIGURE 7.6 *While much of the world is hungry, people in wealthier countries are at risk from eating too much.*
Source: WorldWatch Institute, 2001.

FIGURE 7.7 *Goiter, a swelling of the thyroid gland at the base of the neck, is often caused by an iodine deficiency. It is a common problem in many parts of the world, particularly where soil iodine is low and seafood is unavailable.*

Starchy foods like maize (corn), polished rice, and manioc (tapioca), which form the bulk of the diet for many poor people, tend to be low in several essential vitamins as well as minerals. According to the FAO, vitamin A deficiencies affect between 100 and 140 million children at any given time. As mentioned in the "Golden Rice" story at the beginning of this chapter, about a million of those children die and another 350,000 go blind every year from the effects of this vitamin shortage. Folic acid (found in dark green, leafy vegetables) is essential for early fetal development. Folic acid deficiencies have been linked to neurological problems in babies, including microencephaly (an abnormally small head) or even anencephaly (lacking a brain).

Protein also is essential for normal growth and development. The two most widespread human protein deficiency diseases are kwashiorkor and marasmus. **Kwashiorkor** is a West African word meaning "displaced child." (A young child is displaced—and deprived of nutritious breast milk—when a new baby is born.) This condition mainly occurs in young children who eat mainly cheap starchy food and don't get enough good quality protein. Children with kwashiorkor often have reddish-orange hair, puffy, discolored skin, and a bloated belly. **Marasmus** (from the Greek "to waste away") is caused by a diet low in both calories and protein. A child suffering from severe marasmus is generally thin and shriveled, like a tiny, very old starving person (fig. 7.8). Children with both these deficiencies have low resistance to infections and are likely to suffer from stunted growth, mental retardation, and other developmental problems. Altogether, the FAO estimates that the annual losses due to deaths and diseases caused by calorie and nutrient deficiencies are equivalent to 46 million years of productive life (see more in chapter 8 on disability-adjusted life years).

Eating a Balanced Diet

What's the best way to be sure you're getting a healthy diet? Generally, it isn't necessary to take synthetic dietary supplements. Eating a balanced diet with a good variety of foods should give you all the nutrients you need. For years, Americans were advised to eat daily servings of four major food groups: meat, dairy products, grains, and fruits and vegetables. Placing meat and dairy products first was the result of lobbying by those industries. The U.S. Department of Agriculture's recommendation for a balanced diet now emphasizes grains, fruits, and vegetables, with only sparing servings of meat, dairy, fats, and sweets (fig. 7.9). Combined with regular, moderate exercise, this food selection should provide most people with all the nutrition they need.

MAJOR FOOD SOURCES

Of the thousands of edible plants and animals in the world, only about a dozen types of seeds and grains, three root crops, twenty or so common fruits and vegetables, six mammals, two domestic fowl, and a few fish and other forms of marine life make up almost all of the food humans eat. Table 7.1 shows annual production of some important foods in human diets. In this section, we will highlight sources and characteristics of those foods.

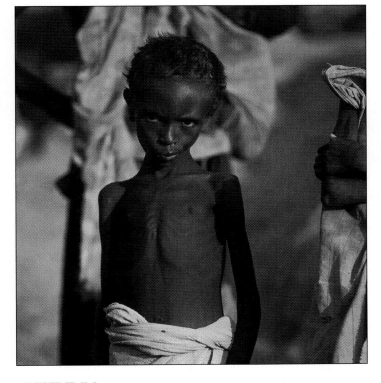

FIGURE 7.8 *Marasmus is caused by combined energy (calorie) and protein deficiencies. Children with marasmus have the wizened look and dry, flaky skin of an old person.*

TABLE 7.1 Some Important Food Resources, 2001/2002 World Production (Million Metric Tons)

Wheat	581
Paddy rice	599[1]
Maize (corn)[2]	597
Oats and barley	170
Soybeans	155
Sorghum and millet	84
Roots and tubers	638
Fruit and vegetables	884
Meat and milk	214
Fish and seafood	113

Notes:

[1] Milled and polished, the rice harvest is 397 million metric tons

[2] Most maize, oats, barley, and soybeans are fed to livestock or used for processed food products.

Source: FAO Sofa, 2002.

Fats and sweets
(use sparingly)

Milk group
(2–3 servings)

Meat group
(2–3 servings)

Vegetable groups
(3–5 servings)

Fruit groups
(2–4 servings)

Bread and cereal group
(6–11 servings)

FIGURE 7.9   *The food pyramid recommended by the U.S. Department of Agriculture. You should eat two to four times as much bread, cereal, rice, and pasta as milk, meat, eggs, or nuts. Fats, oils, and sweets should be eaten sparingly.*
Source: USDA.

Major Crops

The three crops on which humanity depends for the majority of its nutrients and calories are wheat, rice, and maize. Together, about 1,777 million metric tons of these three grains are grown each year. Wheat and rice are especially important, since they are the staple foods for most of the 5 billion people in the developing countries. These two grass species supply around 60 percent of the calories that humans consume directly.

Potatoes, barley, oats, and rye are staples in mountainous regions and high latitudes (northern Europe, North Asia) because they grow well in cool, moist climates. Cassava, sweet potatoes, and other roots and tubers grow well in warm, wet areas and are staples in Amazonia, Africa, Melanesia, and the South Pacific. Sorghum and millet are drought resistant and are staples in the dry regions of Africa.

Fruits and vegetables—including vegetable oils—make a surprisingly large contribution to human diets. They are especially welcome because they typically contain high levels of vitamins, minerals, dietary fiber, and complex carbohydrates (fig. 7.10).

Meat, Milk, and Seafood

Meat and milk are prized by people nearly everywhere, but their distribution is highly inequitable. Although people in the industrialized, more-developed countries of North America, Europe, and

FIGURE 7.10   *Fruits and vegetables are important dietary components because of their vitamin, mineral, and fiber content.*

Japan make up only 20 percent of the world's population, they consume 80 percent of all meat and milk in the world. The 80 percent of the world's people in less-developed countries raise 60 percent of the 3 billion domestic ruminants and 6 billion poultry in the world but consume only 20 percent of all animal products.

Make a list of all the food you ate on a recent day. Insofar as you are able, map where each of those food items came from. For raw materials, you may have to make an educated guess. If you live in North America, for instance, the wheat in baked goods probably came from the upper Great Plains or Prairie Provinces; corn syrup in prepared foods probably came from the corn belt; cane sugar is mostly from Florida; beet sugar is mostly from the Red River Valley of the north. What was the average distance traveled by each food item from its source to your table that day? What do you know about the growing conditions or social circumstances at the source of your food? Are there pesticide controls, for instance, where your food originated? How many calories did you eat that day?

FIGURE 7.11 *Up to 100,000 cattle may be confined in a single feedlot operation. Note the animals in the background standing on a manure mound.*

About 90 percent of the grain grown in North America is used to feed dairy and beef cattle, hogs, poultry, and other animals. It is used especially to fatten beef cattle during the last three months before they are slaughtered. As mentioned in chapter 2, there is a great loss of energy with each step up the food chain. This means we could feed far more people if we ate more grain directly, rather than feeding it to livestock. Every 16 kg of grain and soybeans fed to beef cattle in feedlots produce about 1 kg of edible meat. The other 15 kg are used by the animal for energy or body parts we do not eat, or they are eliminated. If we were to eat the grain directly, we would get 21 times more calories and eight times more protein than we get by eating the meat it produces. Hogs and poultry are about two and four times as efficient, respectively, as cattle in converting feed to edible meat.

Confinement of livestock in "concentrated animal feeding operations" raises a number of social and environmental questions. Up to 10,000 hogs enclosed in giant barns, or 100,000 cattle in a single feedlot complex can cause serious local air and water pollution (fig. 7.11). Animal waste is typically stored in enormous open lagoons, which can leak or rupture, contaminating both surface waters and groundwater supplies. When Hurricane Floyd dumped torrential rains on North Carolina in 1999, an estimated 10 million m^3 (2.5 billion gal) of hog and poultry waste overflowed from these storage lagoons into local rivers. The high density of animals in these facilities and the rich diet they're fed to speed weight gain requires a constant use of antibiotics and growth hormones. More than half of all antibiotics used in the United States are administered to livestock, mostly in confinement facilities. A big part of the rapid rise in antibiotic-resistant pathogens is due to this massive and constant dosing with drugs.

Fish and other seafood contribute about 94 million metric tons of high-quality food to the world's diet. This is an important protein source in many countries, making up about one-half of the animal protein and one-fourth of total dietary protein in Japan, for instance. Unfortunately, overharvesting and habitat destruction threaten most of the world's wild fisheries. Annual catches of ocean fish rose by about 4 percent annually between 1950 and 1988. Since 1989, however, 13 of 17 major marine fisheries have declined dramatically or become commercially inviable. According to the United Nations, 70 percent of the world's edible ocean fish, crustaceans, and mollusks are declining and in urgent need of managed conservation.

Aquaculture (growing fish or shellfish in artificial ponds or net-enclosures in a lake or ocean) was once promoted as the "blue revolution" that would mean as much or more to human nutrition as the miracle grains of the green revolution. Fish can be grown in farm ponds that take very little space but contribute much needed animal protein to the diets of poor people. Salmon, shrimp, oysters, and other high-value crops can be cultivated in enclosed estuaries or tanks built along coastlines, but they also can create serious environmental problems, such as the destruction of biologically important mangrove forests and the contamination of groundwater and coastal regions with animal wastes, pesticides, and excess food. In one of the latest developments in aquaculture, seafood is being raised in closed-loop systems in urban warehouses. This provides the freshest possible product to consumers and eliminates costly shipping. Because the facilities are completely enclosed, growing conditions can be carefully controlled and pollutants eliminated. (See the related story "Are Shrimp Safe to Eat?" at www.mhhe.com/apps.)

SOIL: A RENEWABLE RESOURCE

Growing the food and fiber needed to support human life is a complex enterprise that requires knowledge from many different fields and cooperation from many different groups of people. In this section, we survey some of the principles of soil science and look at some of the inputs necessary for continued agricultural production.

Of all the earth's crustal resources, the one we take most for granted is soil. We are terrestrial animals and depend on soil for life, yet most of us think of it only in negative terms. English is unique in using "soil" as an interchangeable word for earth and excrement. "Dirty" has a moral connotation of corruption and impurity. Perhaps these uses of the word enhance our tendency to abuse soil without scruples; after all, it's only dirt.

The truth is that **soil** is a marvelous substance, a living resource of astonishing beauty, complexity, and frailty. It is a complex mixture of weathered mineral materials from rocks, partially decomposed organic molecules, and a host of living organisms. It can be considered an ecosystem by itself. Soil is an essential component of the biosphere, and it can be used sustainably, or even enhanced, under careful management.

There are at least 20,000 different soil types in the United States and many thousands more worldwide. They vary because of the influences of parent material, time, topography, climate, and organisms on soil formation. There are young soils that, because they have not weathered much, are rich in soluble nutrients. There are old soils, like the red soils of the tropics, from which rainwater has washed away most of the soluble minerals and organic matter, leaving behind clay and rust-colored oxides.

To understand the potential for feeding the world on a sustainable basis, we need to know how soil forms, how it is lost, and what we can do to protect and rebuild good agricultural soil. With careful husbandry, soil can be replenished and renewed indefinitely. Many farming techniques deplete soil nutrients, however, and expose the soil to the erosive forces of wind and moving water. As a result, in many places we are essentially "mining" this resource and using it much faster than it is being replaced.

Building good soil is a slow process. Under the best circumstances, good **topsoil** accumulates at a rate of about 10 tons per ha (2.5 acres) per year—enough soil to make a layer about 1 mm deep when spread over a hectare. Under poor conditions, it can take thousands of years to build that much soil. Perhaps one-third to one-half of the world's current croplands are losing topsoil faster than it is being replaced. In some of the worst spots, erosion carries away about 2.5 cm (1 in.) of topsoil per year. With losses like that, agricultural production has already begun to fall in many areas.

Soil Organisms

Without soil organisms, the earth would be covered with sterile mineral particles far different from the rich, living soil ecosystems on which we depend for most of our food. The activity of the myriad organisms living in the soil helps to create structure, fertility, and tilth (condition suitable for tilling or cultivation).

Soil organisms usually stay close to the surface, but that thin living layer can contain thousands of species and billions of individual organisms per hectare. Algae live on the surface, while bacteria and fungi flourish in the top few centimeters of soil (fig. 7.12). A single gram of soil (about one-half teaspoon) can contain hundreds of millions of these microscopic cells. Worms and nematodes process plant roots and litter. Bacteria and fungi decompose organic detritus and recycle nutrients that plants can use for additional growth. The sweet aroma of freshly turned soil is caused by actinomycetes, bacteria that grow in fungus-like strands and give us the antibiotics streptomycin and tetracycline.

FIGURE 7.12 *Soil ecosystems include numerous consumer organisms, as depicted here: (1) snail, (2) termite, (3) nematodes and nematode-killing constricting fungus, (4) earthworm, (5) wood roach, (6) centipede, (7) carabid (ground) beetle, (8) slug, (9) soil fungus, (10) wireworm (click beetle larva), (11) soil protozoans, (12) earthworm, (13) sow bug, (14) ants, (15) mite, (16) springtail, (17) pseudoscorpion, and (18) cicada nymph.*

Principles of Environmental Science

Soil Profiles

Most soils are stratified into horizontal layers called **soil horizons** that reveal much about the soil's history and usefulness. The thickness, color, texture, and composition of each horizon are used to classify the soil. A cross-sectional view of the horizons in a soil is called a soil profile. Figure 7.13 shows the series of horizons generally seen in a soil profile. Soil scientists give each horizon a letter or descriptive name. Soils usually have one to seven or more horizons with different properties, depending on the soil type and history of a specific area.

The soil surface is often covered with a layer of leaf litter, crop residues, or other fresh or partially decomposed organic material (O horizon). Under this organic layer is the surface horizon, usually an A horizon or topsoil, composed of mineral particles mixed with organic material. The A horizon ranges from a thickness of several meters under virgin prairie to zero in some deserts. The surface horizon contains most of the living organisms and organic material in the soil, and it is in this layer that most plants spread their roots to absorb water and nutrients. The surface horizon often blends into another horizon (the E horizon), which is subject to leaching (removal of soluble nutrients) by water percolating through it. This zone of leaching may have a very different appearance and composition from horizons above or below it.

Beneath the surface horizons, the subsurface horizons, or **subsoil**, usually have a lower organic content and higher concentrations of mineral particles. Under the subsoil is the parent material, or C horizon, made of weathered rock fragments with very little organic material. Weathering of this layer produces new soil particles and allows downward expansion of the horizons above. About 70 percent of all the parent horizon material in the United States was transported to its present site by geologic forces (glaciers, wind, and water) and is not directly related to the bedrock below it.

Soil Types

Soils are classified according to their structure and composition into orders, suborders, great groups, subgroups, families, and series. The richest farming soils are the mollisols (formed under grasslands) and alfisols (formed under moist, deciduous forests). North America is fortunate to have extensive areas of these fertile soils.

WAYS WE USE AND ABUSE SOIL

Only about 11 percent of the earth's land area (14.66 million km^2 out of a total of 132.4 million km^2) is currently in agricultural production. Perhaps four times as much land could potentially be converted to cropland, but much of this land serves as a refuge for cultural or biological diversity or suffers from constraints, such as steep slopes, shallow soils, poor drainage, tillage problems, low nutrient levels, metal toxicity, or excess soluble salts or acidity, that limit the types of crops that can be grown there (fig. 7.14).

FIGURE 7.13 *Soil profile showing possible soil horizons. The actual number, composition, and thickness of these layers vary in different soil types.*

O HORIZON
Surface litter: fallen leaves and partially decomposed organic debris

A HORIZON
Topsoil: organic matter (humus), living organisms, inorganic minerals

E HORIZON
Zone of leaching: dissolved or suspended materials move downward

B HORIZON
Subsoil: accumulation of iron, aluminum, humic compounds, and clay leached down from the A and E horizons

C HORIZON
Weathered parent material: partially broken-down inorganic minerals

Bedrock

Land Resources

In parts of Canada and the United States, temperate climates, abundant water, and high soil fertility produce high crop yields that contribute to high standards of living. Other countries, although rich in land area, lack suitable soil, topography, water, or climate to sustain our levels of productivity.

Arable land has shrunk from 0.38 ha (0.94 acre) per person in 1970 to 0.23 ha (0.56 acre) in 2000. If current population projections are correct, the amount of cropland per person will decline to 0.15 ha (0.37 acre) by 2050. In Asia, cropland will be even more scarce—0.09 ha (0.22 acre) per person—in 50 years. If you live on a typical quarter-acre suburban lot, look at your yard and imagine feeding yourself for a year on what you could produce there.

FIGURE 7.14 *In many areas, soil or climate constraints limit agricultural production. These farmworkers in Chad weed a field that is too dry and sandy for good crop yields.*

In the developed countries, 95 percent of recent agricultural growth in the twentieth century has come from improved crop varieties or increased fertilization, irrigation, and pesticide use, rather than from bringing new land into production. In fact, less land is being cultivated now than 100 years ago in North America, or 600 years ago in Europe. As more effective use of labor, fertilizer, and water and improved seed varieties have increased in the more-developed countries, productivity per unit of land has increased, and marginal land has been retired, mostly to forests and grazing lands. In many developing countries, land continues to be cheaper than other resources, and new land is still being brought under cultivation, mostly at the expense of forests and grazing lands. Still, at least two-thirds of recent production gains have come from new crop varieties and more intense cropping, rather than expansion into new lands.

The largest increases in cropland over the last 30 years occurred in South America and Oceania, where forests and grazing lands are rapidly being converted to farms. Many developing countries are reaching the limit of lands that can be exploited for agriculture without unacceptable social and environmental costs, but others still have considerable potential for opening new agricultural lands. East Asia, for instance, already uses about three-quarters of its potentially arable land. Most of its remaining land has severe restrictions for agricultural use. Further increases in crop production will probably have to come from higher yields per hectare. Latin America, by contrast, uses only about one-fifth of its potential land, and Africa uses only about one-fourth of the land that theoretically could grow crops. However, there would be serious ecological trade-offs in putting much of this land into agricultural production.

While land surveys tell us that much more land in the world could be cultivated, not all of that land necessarily should be farmed. Much of it is more valuable in its natural state. The soils over much of tropical Asia, Africa, and South America are old, weathered, and generally infertile. Most of the nutrients are in the standing plants, not in the soil. In many cases, clearing land for agriculture in the tropics has resulted in tragic losses of biodiversity and the valuable ecological services that it provides. Ultimately, much of this land is turned into useless scrub or semidesert.

On the other hand, there are large areas of rich, subtropical grassland and forest that are well watered, have good soil, and could become productive farmland without unduly reducing the world's biological diversity. Argentina, for instance, has pampas grasslands about twice the size of Texas that closely resemble the American Midwest a century ago in climate and potential for agricultural growth. Some of this land could probably be farmed with relatively little ecological damage if it were done carefully.

Land Degradation

Agriculture both causes and suffers from environmental degradation. The International Soil Reference and Information Centre in the Netherlands estimates that, every year, 3 million ha (7.4 million acres) of cropland are ruined by erosion, 4 million ha are turned into deserts, and 8 million ha are converted to nonagricultural uses, such as homes, highways, shopping centers, factories, reservoirs, etc. Over the past 50 years, some 1.9 billion ha of agricultural land (an area greater than that now in production) have been degraded to some extent. About 300 million ha of this land are strongly degraded (meaning that soil has deep gullies, severe nutrient depletion, or poor crop growth, or that restoration is difficult and expensive). Some 910 million ha—about the size of China—are moderately degraded. Nearly 9 million ha of former croplands are so degraded that they no longer support any crops at all. The causes of this extreme degradation vary: in Ethiopia, it is water erosion; in Somalia, it is wind; and in Uzbekistan, salt and toxic chemicals are responsible.

Definitions of degradation are based on both biological productivity and our expectations about what the land should be like. Often this is a subjective judgment, and it is difficult to distinguish between human-caused deterioration and natural conditions, like drought. We generally consider the land degraded when the soil is impoverished or eroded, water runs off or is contaminated more than normal, vegetation is diminished, biomass production is decreased, or wildlife diversity diminishes. On farmlands, this results in lower crop yields. On ranchlands, it means that fewer livestock can be supported per unit area. On nature reserves, it means lower biological diversity.

The amount and degree of land degradation vary by region and country. About 20 percent of land in Africa and Asia is degraded, but most is in either the light or moderate category. In Central America and Mexico, by contrast, 25 percent of all vegetated land suffers moderate to extreme degradation. Figure 7.15 shows some areas of greatest concern for soil degradation, and figure 7.16 shows the mechanisms for this problem. Water and wind erosion provide the motive force for the vast majority of all soil degradation, worldwide. Chemical degradation includes nutrient depletion, salt accumulation, acidification, and pollution. Physical degradation includes compaction by heavy machinery or trampling

Principles of Environmental Science   www.mhhe.com/environmentalscience

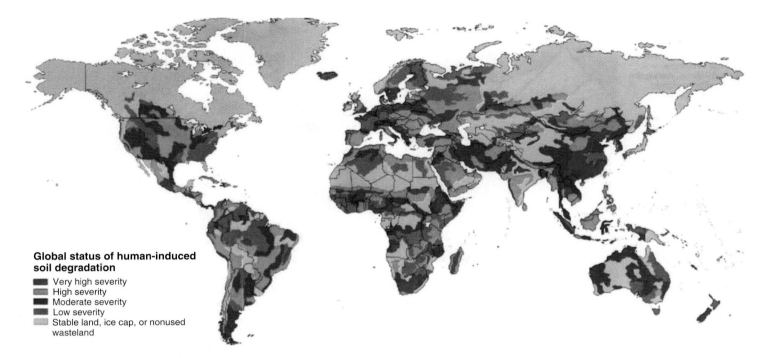

Global status of human-induced soil degradation

- ■ Very high severity
- ■ High severity
- ■ Moderate severity
- ■ Low severity
- ■ Stable land, ice cap, or nonused wasteland

FIGURE 7.15 *Areas of concern for soil degradation. Can you identify conditions where you live? Do you agree with this assessment?*
Source: FAO, 2002.

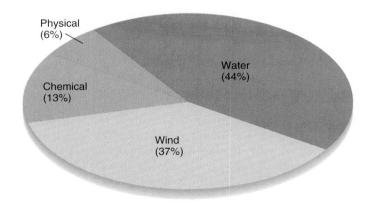

Physical (6%)
Water (44%)
Chemical (13%)
Wind (37%)

FIGURE 7.16 *Causes of global soil degradation.*
Source: Data from FAO, 2002.

by cattle, water accumulation from excess irrigation and poor drainage, and laterization (solidification of iron and aluminum-rich tropical soil when exposed to sun and rain).

Erosion: The Nature of the Problem

Erosion is an important natural process, resulting in the redistribution of the products of geologic weathering, and is part of both soil formation and soil loss. The world's landscapes have been sculpted by erosion. When the results are spectacular enough, we enshrine them in national parks, as we did with the Grand Canyon.

Where erosion has worn down mountains and spread soil over the plains or deposited rich alluvial silt in river bottoms, we gladly farm it. Erosion is a disaster only when it occurs in the wrong place at the wrong time.

In some places, erosion occurs so rapidly that anyone can see it happen. Deep gullies are created where water scours away the soil, leaving fenceposts and trees sitting on tall pedestals as the land erodes away around them. In most places, however, erosion is more subtle. It is a creeping disaster that occurs in small increments. A thin layer of topsoil washes off of fields year after year until, eventually, nothing is left but poor-quality subsoil that requires more and more fertilizer and water to produce any crop at all.

The net effect, worldwide, of this general, widespread, topsoil erosion is a reduction in crop production equivalent to removing about 1 percent of the world's cropland each year. Many farmers are able to compensate for this loss by applying more fertilizer and by bringing new land into cultivation. Continuation of current erosion rates, however, could reduce agricultural production by 25 percent in Central America and Africa and by 20 percent in South America by 2020. The total annual soil loss from croplands is thought to be 25 billion metric tons. About twice that much soil is lost from rangelands, forests, and urban construction sites each year.

In addition to reduced land fertility, erosion results in sediment loading of rivers and lakes, siltation of reservoirs, smothering of wetlands and coral reefs, and clogging of water intakes and waterpower turbines.

Mechanisms of Erosion

Wind and water are the main agents that move soil around (see fig. 7.16). Water flowing across a gently sloping, bare field removes a

FIGURE 7.17 *Severe gullying is cutting deep trenches in this pastureland. Lost topsoil reduces productivity here, while it causes siltation downstream.*

FIGURE 7.18 *Annual row crops leave soil bare and exposed to erosion for most of the year, especially when fields are plowed immediately after harvest, as this one always is.*

thin, uniform layer of soil in what is known as **sheet erosion.** When little rivulets of running water gather together and cut small channels in the soil, the process is called **rill erosion.** If rills enlarge to form bigger channels or ravines that are too large to be removed by normal tillage operations, we call the process **gully erosion** (fig. 7.17). Streambank erosion refers to the washing away of soil from the banks of established streams, creeks, or rivers, often as a result of removing trees and brush along streambanks and of cattle damaging the banks.

Most soil erosion on agricultural land is rill erosion. Large amounts of soil can be transported a little bit at a time without being very noticeable. A farm field can lose 20 metric tons of soil per hectare during winter and spring runoff in rills so small that they are erased by the first spring cultivation. That represents a loss of only a few millimeters of soil over the whole surface of the field, hardly apparent to any but the most discerning eye. But it doesn't take much mathematical skill to see that, if you lose soil twice as fast as it is being replaced, it eventually will run out.

Wind can equal or exceed water in erosive force, especially in a dry climate and on relatively flat land. When plant cover and surface litter are removed from the land by agriculture or grazing, wind lifts loose soil particles and sweeps them away. Windborne dust is sometimes transported from one continent to another. Scientists in Hawaii can tell when spring plowing begins in China because dust from Chinese farmland is carried by winds all the way across the Pacific Ocean.

Some areas of the United States and Canada have very high erosion rates. The U.S. Department of Agriculture reports that 69 million ha (170 million acres) of U.S. farmland and range are eroding at rates that reduce long-term productivity. Five tons per acre (11 metric tons per hectare) is generally considered the maximum tolerable rate of soil loss because that is generally the highest rate at which soil forms under optimum conditions. Some farms lose soil at twice that rate or more.

Intensive farming practices are largely responsible for this situation. Row crops, such as corn and soybeans, leave soil exposed for much of the growing season (fig. 7.18). Deep plowing and heavy herbicide applications create weed-free fields that look neat but are subject to erosion. Because big machines cannot easily follow contours, they often go straight up and down the hills, creating ready-made gullies for water to follow. Farmers sometimes plow through grass-lined watercourses (low areas where water runs off after a rain) and pull out windbreaks and fencerows to accommodate the large machines and to get every last square meter into production. Consequently, wind and water carry away the topsoil.

Erosion Hot Spots

Data on soil conditions and soil erosion often are incomplete, but it is evident that many places have problems as severe as, or perhaps worse than, those found in the United States and Canada. China, for example, has a large area of loess (windblown silt) deposits on the North China Plain that once was covered by forest and grassland. The forests were cut down, and the grasslands were converted to cropland. This plateau is now scarred by gullies 30 to 40 m (100 to 130 ft) deep, and the soil loss is thought to be at least 480 metric tons per ha per year. This would be equivalent to 3 cm (1.2 in.) of topsoil lost per year.

One way to estimate soil loss is to measure the sediment load carried by rivers draining an area. China's Huang (Yellow) River, which flows through the loess plateau, transports 1.6 billion tons of sediment every year. Although its drainage basin is only one-fifth as big as that of the Mississippi River, the Huang carries more than four times as much soil each year. This suggests that the average soil loss per ha in northern China may be twenty times that in the United States.

Haiti is another country with severely degraded soil. Once covered with lush tropical forest, the land has been denuded for firewood and cropland. Erosion has been so bad that some experts now say the country has absolutely no topsoil left, and poor peasant farmers have difficulty raising any crops at all. Economist Lester Brown of Worldwatch Institute warns that the country may never recover from this ecodisaster.

OTHER AGRICULTURAL RESOURCES

Soil is only part of the agricultural resource picture. Agriculture is also dependent upon water, nutrients, favorable climates to grow crops, productive crop varieties, and the mechanical energy to tend and harvest the crops.

Water

All plants need water to grow (fig. 7.19). Agriculture accounts for the largest single share of global water use. Some 73 percent of all fresh water withdrawn from rivers, lakes, and groundwater supplies is used for irrigation (see chapter 10). Irrigation increases yields of most crops by 100 to 400 percent. In 2000, according to the FAO, 42 percent of all arable land in Asia was irrigated, 31 percent in the Near East and North Africa, 14 percent in Latin America, but only 4 percent in sub-Saharan Africa. In Pakistan, 80 percent of food is produced on irrigated land; in China, 70 percent is, while in Ghana, Malawi, and Mozambique, the amount is less than 2 percent. Helping poor countries harvest and use water efficiently may be one of the best things we can do to reduce hunger.

Some countries are water rich and can readily afford to irrigate farmland, while other countries are water poor and must use water very carefully. The efficiency of irrigation water use varies greatly. High evaporative and seepage losses from unlined and uncovered canals in some places can mean that as much as 80 percent of water withdrawn for irrigation never reaches its intended destination. Poor farmers may overirrigate because they lack the technology to meter water and distribute just the amount needed.

Excessive use not only wastes water, it often results in **waterlogging**. Waterlogged soil is saturated with water, and plant roots die from lack of oxygen. **Salinization**, in which mineral salts accumulate in the soil, is often a problem when irrigation water dissolves and mobilizes salts in the soil. As the water evaporates, it leaves behind a salty crust on the soil surface that is lethal to most plants. Flushing with excess water can wash away this salt accumulation, but the result is even more saline water for downstream users.

The FAO estimates that 20 percent of all irrigated land is damaged to some extent by waterlogging or salinity. Water conser-

FIGURE 7.19 *Downward-facing sprinklers on this center-pivot irrigation system deliver water more efficiently than upward-facing sprinklers.*

vation techniques can greatly reduce problems arising from excess water use. Conservation also makes more water available for other uses or for expanded crop production where water is in short supply (see chapter 10).

Fertilizer

In addition to water, sunshine, and carbon dioxide, plants need small amounts of inorganic nutrients for growth. The major elements required by most plants are nitrogen, potassium, phosphorus, calcium, magnesium, and sulfur. Calcium and magnesium often are limited in areas of high rainfall and must be supplied in the form of lime. Lack of nitrogen, potassium, and phosphorus often limits plant growth even more. Adding these elements in fertilizer usually stimulates growth and greatly increases crop yields. A good deal of the doubling in worldwide crop production since 1950 has come from increased inorganic fertilizer use. In 1950, the average amount of fertilizer used was 20 kg per ha. In 2000, this had increased to an average of 90 kg per ha worldwide.

Farmers may overfertilize because they are unaware of the specific nutrient content of their soils or the needs of their crops. While European farmers use more than twice as much fertilizer per hectare as do North American farmers, their yields are not proportionally higher. Phosphates and nitrates from farm fields and cattle feedlots are a major cause of aquatic ecosystem pollution. Nitrate levels in groundwater have risen to dangerous levels in many areas where intensive farming is practiced. Young children are especially sensitive to the presence of nitrates. Using nitrate-contaminated water to mix infant formula can be fatal for newborns.

What are some alternative ways to fertilize crops? Manure and green manure (crops grown specifically to add nutrients to the soil) are important natural sources of soil nutrients. Nitrogen-fixing bacteria living symbiotically in root nodules of legumes are valuable for making nitrogen available as a plant nutrient (see chapter 2). Interplanting or rotating beans or some other leguminous crop with such crops as corn and wheat are traditional ways of increasing nitrogen availability.

There is considerable potential for increasing the world's food supply by increasing fertilizer use in low-production countries if ways can be found to apply fertilizer more effectively and to reduce pollution. Africa, for instance, uses an average of only 19 kg of fertilizer per ha (17 lb per acre), or about one-fourth of the world average. It has been estimated that the developing world could at least triple its crop production by raising fertilizer use to the world average.

Energy

Farming as it is generally practiced in the industrialized countries is highly energy-intensive. Fossil fuels supply almost all of this energy. Between 1920 and 1980, direct energy use on farms rose as gasoline and diesel fuels were consumed by increasing mechanization of agriculture. Indirect energy use, in the form of synthetic fertilizers, pesticides, and other agricultural chemicals, increased even more, especially after World War II. On intensively fertilized farms, indirect energy use may be five times that of direct energy. The energy price shocks of the 1970s encouraged energy conservation that has reduced farm energy use, even though total food production has continued to rise.

After crops leave the farm, additional energy is used in food processing, distribution, storage, and cooking. It has been estimated that the average food item in the American diet travels 2,000 km (1,250 mi) between the farm that grew it and the person who consumes it. The energy required for this complex processing and distribution system may be five times as much as is used directly in farming. Altogether the food system in the United States consumes about 16 percent of the total energy we use. Most of our foods require more energy to produce, process, and get to market than they yield when we eat them.

Clearly, unless we find some new energy sources, our present system is unsustainable. As fossil fuels become more scarce, we may need to adopt farming methods that are self-supporting. Is it possible that we may need to go back to using draft animals that can eat crops grown on the farm? Could we reintroduce natural methods of pest control, fertilization, crop drying, and irrigation? Or can we develop alternative energy sources to run our farming enterprise?

Pest Control

Biological pests reduce crop yields and spoil as much as half of the crops harvested every year in some areas. Modern agriculture largely depends on toxic chemicals to kill or drive away these pests, but there are many serious concerns about the types and amounts of pesticides now in use.

There are many traditional methods for protecting crops from pests, but our war against unwanted organisms entered a new phase in the twentieth century with the invention of synthetic organic chemicals such as DDT (dichlorodiphenyltrichloroethane). These chemicals have been an important part of our increased crop production and also have helped to control many disease-causing organisms. It's estimated that up to half our current crop yields might be lost if we had no pesticides to protect them. Indiscriminate and profligate pesticide use, however, also has caused many problems, such as killing nontarget species, creating new pests of

FIGURE 7.20 *Spraying pesticides by air is quick and cheap, but toxins often drift to nearby fields.*

organisms that previously were not a problem, and causing widespread pesticide resistance among pest species (fig. 7.20). Often highly persistent and mobile in the environment, many pesticides have moved through air, water, and soil, and bioaccumulated or bioconcentrated in food chains, nearly exterminating several top predators, such as peregrine falcons.

Your exposure to pesticides is probably higher than you might suspect. A study by the U.S. Department of Agriculture found that 73 percent of conventionally grown foods (out of 94,000 samples assayed) had residues from at least one pesticide. By contrast, only 23 percent of the organic samples of the same food groups had any residues. It is now clear that humans suffer adverse health effects from persistent organic pollutants (POPs) like pesticides. In fact, most of the "dirty dozen" POPs recently banned globally are pesticides. See chapter 8 for more on health effects of POPs.

In 2002, Swiss researchers published results of a two-decade-long comparison of organic and conventional farming. Crops grown without synthetic fertilizers or pesticides produced, on average, about 20 percent lower yield than similar crops grown by conventional methods, but more than made up the difference with lower operating costs, premium crop prices, less ecological damage, and healthier farm families.

Some alternatives for reducing our dependence on dangerous chemical pesticides include management changes, such as using cover crops and mechanical cultivation, and planting mixed polycultures rather than vast monoculture fields. Consumers may have to learn to accept less than perfect fruits and vegetables if we want to avoid toxic chemicals in our diet. Biological controls, such as insect predators, pathogens, or natural poisons specific for a particular pest, can help reduce chemical use. Genetic breeding and biotechnology can produce pest-resistant crop and livestock strains as well. Integrated pest management (IPM) is a relatively new science that combines all of these alternative methods, together with judicious use of synthetic pesticides under precisely controlled conditions. There are ways that you can minimize pesticides in your diet (see What Can You Do? page 163).

what can you do?

Reducing the Pesticides in Your Food

If you want to reduce the amount of pesticide residues and other toxic chemicals in your diet, follow these simple rules.

- Wash and scrub all fresh fruits and vegetables thoroughly under running water.

- Peel fruits and vegetables when possible. Throw away the outer leaves of leafy vegetables, such as lettuce and cabbage.

- Store food carefully so it doesn't get moldy or pick up contaminants from other foods. Use food as soon as possible to ensure freshness.

- Cook or bake foods that you suspect have been treated with pesticides to break down chemical residues.

- Trim the fat from meat, chicken, and fish. Eat lower on the food chain when possible to reduce bioaccumulated chemicals.

- Don't pick and eat berries or other wild foods that grow on the edges of roadsides where pesticides may have been sprayed.

- Grow your own fruits and vegetables without using pesticides or with minimal use of dangerous chemicals.

- Ask for organically grown food at your local grocery store, or shop at a farmer's market or co-op where you can get such food.

NEW CROPS AND GENETIC ENGINEERING

Although at least 3,000 species of plants have been used for food at one time or another, most of the world's food now comes from only 16 widely grown crops. Many new or unconventional varieties might be valuable human food supplies, however, especially in areas where conventional crops are limited by climate, soil, pests, or other problems. Among the plants now being investigated as potential additions to our crop roster are winged beans, a perennial plant that grows well in hot climates where other legumes will not grow (fig. 7.21). It is totally edible (pods, mature seeds, shoots, flowers, leaves, and tuberous roots), resistant to diseases, and enriches the soil. Another promising crop is triticale, a hybrid between wheat (*Triticum*) and rye (*Secale*) that grows in light, sandy, infertile soil. It is drought resistant, has nutritious seeds, and is being tested for salt tolerance for growth in saline soils or irrigation with seawater. Some traditional crop varieties grown by Native Americans, such as tepary beans, amaranth, and Sonoran panic grass, are being collected by seed conservator Gary Nabhan both as a form of cultural revival for native people and as possible food crops for harsh environments.

The Green Revolution

So far, however, the major improvements in farm production have come from technological advances and modification of a few

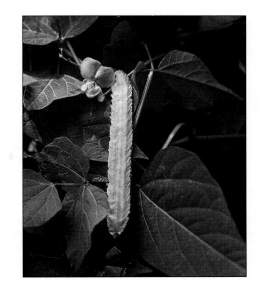

FIGURE 7.21 *Winged beans bear fruit year-round in tropical climates and are resistant to many diseases that prohibit growing other bean species. Whole pods can be eaten when they are green, or dried beans can be stored for later use. Beans are a good protein source in a vegetarian diet.*

well-known species. Yield increases often have been spectacular. A century ago, when all maize (corn) in the United States was open pollinated, average yields were about 25 bushels per acre (bu/acre). In 2000, the average yield from rainfed fields in Iowa was 138 bu/acre, and irrigated maize in Arizona averaged 208 bu/acre. The highest yield ever recorded in field production was 370 bu/acre on an Illinois farm, but theoretical calculations suggest that 500 bu/acre (32 metric tons per hectare) could be possible. Most of this gain was accomplished by use of synthetic fertilizers along with conventional plant breeding: geneticists laboriously hand-pollinating plants, and looking for desired characteristics in the progeny.

Starting about 50 years ago, agricultural research stations began to breed tropical wheat and rice varieties that would provide food for growing populations in developing countries. The first of the "miracle" varieties was a dwarf, high-yielding wheat developed by Norman Borlaug (who received a Nobel Peace Prize for his work) at a research center in Mexico. At about the same time, the International Rice Institute in the Philippines developed dwarf rice strains with three or four times the production of varieties in use at the time. The spread of these new high-yield varieties around the world has been called the **green revolution.** It is one of the main reasons that world food supplies have more than kept pace with the growing human population over the past few decades.

Most green revolution breeds really are "high responders," meaning that they yield more than other varieties if given optimum levels of fertilizer, water, and protection from pests and diseases (fig. 7.22). Under suboptimum conditions, on the other hand, high responders may not produce as well as traditional varieties. Poor farmers who can't afford the expensive seed, fertilizer, and water required to become part of this movement usually are left

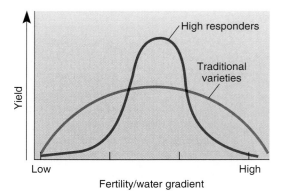

FIGURE 7.22 *Green revolution miracle crops are really high responders, meaning that they have excellent yields under optimum conditions. For poor farmers who can't afford the fertilizer and water needed by high responders, traditional varieties may produce better yields.*

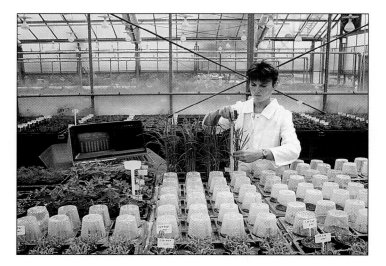

FIGURE 7.23 *A researcher measures growth of genetically engineered rice plants. Superior growth and yield, pest resistance, and other useful traits can be moved from one species to another by molecular biotechnology.*

out of the green revolution. In fact they may be driven out of farming altogether as rising land values and falling commodity prices squeeze them from both sides.

Genetic Engineering

Genetic engineering, splicing a gene from one organism into the chromosome of another, has the potential to greatly increase both the quantity and quality of our food supply. It is now possible to build entirely new genes, and even new organisms, often called "transgenic" organisms or **genetically modified organisms (GMOs),** by taking a bit of DNA from here, a bit from there, and even synthesizing artificial sequences to create desired characteristics in engineered organisms (fig. 7.23).

Proponents predict dramatic benefits from this new technology. Research is now underway to improve yields and create crops that resist drought, frost, or diseases. Other strains are being developed to tolerate salty, waterlogged, or low-nutrient soils, allowing degraded or marginal farmland to become productive. All of these could be important for reducing hunger in developing countries. Plants that produce their own pesticides might reduce the need for toxic chemicals; while engineering for improved protein or vitamin content could make our food more nutritious. Attempts to remove specific toxins or allergens from crops also could make our food safer. Crops such as bananas and potatoes have been altered to contain oral vaccines that can be grown in developing countries where refrigeration and sterile needles are unavailable. Plants have been engineered to make industrial oils and plastics. Animals, too, are being genetically modified to grow faster, gain weight on less food, and produce pharmaceuticals such as insulin in their milk. It may even be possible to create animals with human cell-recognition factors that could serve as organ donors.

Opponents worry that moving genes willy-nilly could create a host of problems, some of which we can't even imagine. GMOs themselves might escape and become pests or they might interbreed wild relatives. In either case, we may create super weeds or

reduce native biodiversity. Constant presence of pesticides in plants could accelerate pesticide resistance in insects or leave toxic residues in soil or our food. Genes for toxicity or allergies could be transferred along with desirable genes, or novel toxins could be created as genes are mixed together. This technology may be available only to the richest countries or the wealthiest corporations, making family farms uncompetitive and driving developing countries even further into poverty.

Both the number of GMO crops and the acreage devoted to growing them is increasing rapidly. Between 1987 and 2002, more than 8,000 field tests for some 750 different crop varieties were carried out in the United States (fig. 7.24). Hawaii is the most popular site for GMO testing, with about 1,500 field releases over the past 14 years. Illinois, Iowa, California, and Puerto Rico each have had more than 1,000 tests during that time. Worldwide, 53 million ha (131

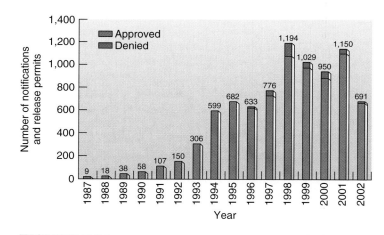

FIGURE 7.24 *Transgenic crop field releases in the United States, 1987–2002 (2002 data only through May).*
Source: Information Systems for Biotechnology, Virginia Tech University.

million acres) were planted with GMO crops in 2002. The United States accounted for 68 percent of that acreage, followed by Argentina with 23 percent. Canada, Australia, Mexico, China, and South Africa together make up 9 percent of all transgenic cropland.

Some transgenic crops have reached mainstream status. About three-fourths of all soybeans, one-third of all maize (corn), and 71 percent of the cotton grown in the United States are now GMOs. You've already probably eaten some genetically modified food. Estimates are that at least 60 percent of all processed food in America contains GMO ingredients. Since the United States, Argentina, and Brazil account for 90 percent of international trade in maize and soybeans (in which GMOs often are mixed with other grains), a large fraction of the world most likely has been exposed to some GMO products.

Pest Resistance and Weed Control

Biotechnologists have created plants with genes for endogenous insecticides. *Bacillus thuringiensis* (Bt), a bacterium, makes toxins lethal to Lepidoptera (butterfly family) and Coleoptera (beetle family). The genes for some of these toxins have been transferred into crops such as maize (to protect against European cutworms), potatoes (to fight potato beetles), and cotton (for protection against boll weevils). This allows farmers to reduce insecticide spraying. Arizona cotton farmers, for example, report reducing their use of chemical insecticides by 75 percent.

Most European nations have not approved Bt-containing varieties. Fear of transgenic products has been a sticking point in agricultural trade between the United States and the European Union.

In 2001, Starlink, a Bt-containing maize variety that had not yet been approved for human consumption, was found mixed into a wide variety of consumer products. By the time food was made from Starlink, no Bt could be detected, only the genes that code for it. There is no evidence that Bt is harmful to humans, but the form in Starlink may possibly be allergenic. In fact, Bt is one of the few insecticides approved for organic farmers. Nevertheless, a public outcry ensued and seed companies, farmers, shippers, and food processors spent more than $1 billion trying to remove it from the food chain. What do you think: was the fear of Starlink justified?

Entomologists worry that Bt plants churn out toxin throughout the growing season, regardless of the level of infestation, creating perfect conditions for selection of Bt resistance in pests. Already 500 species of insect, mite, or tick are resistant to one or more pesticides. Having Bt constantly available can only aggravate this dilemma. One solution is to plant at least a part of every field in non-Bt crops that will act as a refuge for nonresistant pests. The hope is that interbreeding between these "wild-type" bugs and those exposed to Bt will dilute out recessive resistance genes. Deliberately harboring pests and letting them munch freely on crops is something that many farmers find hard to do. In addition, devoting a significant part of their land to nonproductive crops lowers the total yield and counteracts the profitability of engineered seed.

The other major group of transgenic crops is engineered to tolerate high doses of herbicides. The two dominant products in this category are Monsanto's "Roundup Ready" crops—so-called because they can withstand treatment with Monsanto's best-selling herbicide, Roundup (glyphosate)—and AgrEvo's "Liberty Link" crops, which resist that company's Liberty (glufosinate) herbicide. Because crops with these genes can grow in spite of high herbicide doses, farmers can spray fields heavily to exterminate weeds. This allows for conservation tillage and leaving more crop residue on fields to protect topsoil from erosion—both good ideas—but it also can mean using more herbicide in higher doses than might otherwise be done.

Is Genetic Engineering Safe?

Ever since scientists discovered how to move genes from one organism to another, critics have worried about irresponsible use of this technology or unforeseen consequences arising from novel combinations of genetic material. Greenpeace and several other environmental groups have campaigned against transgenic organisms, labeling them "Frankenfoods." Industry groups, on the other hand, describe their critics as Neo-Luddites, who blindly oppose any new technology. How can we decide what to believe in this welter of claims and counterclaims?

At the base of some people's unease with genetic modification is a feeling that it simply isn't right to mess with nature. It doesn't seem proper to recombine genes any way we want. Doing so raises specters of gruesome monsters stitched together from spare parts, like Mary Shelley's Frankenstein. Is this merely a fear of science, or is it a valid ethical issue?

The U.S. Food and Drug Administration declined to require labeling of foods containing GMOs, saying that these new varieties are "substantially equivalent" to related varieties bred via traditional practices. After all, proponents say, we have been moving genes around for centuries through plant and animal breeding. All domesticated organisms should be classified as genetically modified, some people argue. Biotechnology may be a more precise way of creating novel organisms than normal breeding procedures, its supporters suggest. We're moving only a few genes—or even part of one gene—at a time with biotechnology, rather than hundreds of unknown genes through classical techniques.

But what if GMOs escape and cross-breed with native species? In a 10-year study of genetically modified crops, a group from Imperial College, London, concluded that GMOs tested so far do not survive well in the wild, and are no more likely to invade other habitats than their unmodified counterparts.

On the other hand, there is already evidence that some GMO crops can spread their genes to nearby fields. Normal rape seed (canola oil) varieties in Canada were found to contain genes from genetically modified varieties in nearby fields. It isn't clear, however, if the genes involved will have any adverse affect or whether they will persist for many generations. In 2001, researchers reported finding traces of genetically modified corn in wild relatives in Mexico, which supposedly has banned planting of GMOs, especially in regions where species were thought to have originated. This report was later withdrawn due to criticism of the techniques used to detect genetic markers, but it gave an interesting insight into the politics and sociology of science.

The first genetically modified animal designed to be eaten by humans is an Atlantic salmon (*Salmo salar*) containing extra growth hormone genes from an oceanic pout (*Macrozoarces americanus*). The greatest worry from this fish is not that it will introduce extra hormones into our diet—that's already being done by chickens and beef that get extra growth hormone via injections or their diet—but rather the ecological effects if the fish escape from captivity. The transgenic fish grow seven times faster and are more attractive to the opposite sex than a normal salmon. If they escape from captivity, they may outcompete already endangered wild relatives for food, mates, and habitat. Fish farmers say they will grow only sterile females and will keep them in secure net pens. Opponents point out that salmon frequently escape from aquaculture operations, and that if just a few fertile transgenic fish break out it could be catastrophic for wild stocks.

Many scientists argue that we should take a better-safe-than-sorry "precautionary approach" that errs—if at all—on the side of safety. In 2002, the 5,000-member Society of Toxicology issued a position paper concluding that, "Based on available tests, there's no reason to suspect that transgenic plants differ in any substantive way from traditional varieties." On the other hand, a panel of biologists and agricultural scientists convened by the U.S. National Academy of Sciences urged the government to more carefully, and more publicly, review the potential environmental impacts of genetically modified plants and animals before approving them for commercial use.

Finally, there are social and economic implications of GMOs. Will they help feed the world or will they lead to a greater consolidation of corporate power and economic disparity? Might higher yields and fewer losses to pests and diseases allow poor farmers in developing countries to stop using marginal land and avoid cutting down forests to create farmland? Is this simply a technological fix or could it help promote agricultural sustainability? Critics suggest that there are simpler and cheaper ways other than high tech crop varieties to provide vitamin A to children in developing countries or to increase the income of poor, rural families. Adding a cow or a fishpond or training people in water harvesting or regenerative farming techniques (as we'll discuss in the next section) may have a longer-lasting impact than selling them expensive new seeds.

On the other hand, if we hope to reduce malnutrition and feed 8 billion people in 50 years, maybe we need all the tools we can get. Where do you stand in this debate? What additional information would you need to reach a reasoned judgment about the risks and benefits of this new technology?

SUSTAINABLE AGRICULTURE

How, then, shall we feed the world? Can we make agriculture compatible with sustainable ecological and social systems? Having discussed some of the problems that beset modern agriculture, we now consider some suggested ways to overcome problems and make farming and food production just and lasting enterprises. This goal is usually termed **sustainable agriculture,** or **regenerative farming,** both of which aim to produce food and fiber on a sustainable basis and repair the damage caused by destructive practices. Some alternative methods are developed through scien-

tific research; others are discovered in traditional cultures and practices nearly forgotten in our mechanization and industrialization of agriculture.

Soil Conservation

With careful husbandry, soil is a renewable resource that can be replenished and renewed indefinitely. Since agriculture is the area in which soil is most essential and also most often lost through erosion, agriculture offers the greatest potential for soil conservation and rebuilding. Some rice paddies in Southeast Asia, for instance, have been farmed continuously for a thousand years without any apparent loss of fertility. The rice-growing cultures that depend on these fields have developed management practices that return organic material to the paddy and carefully nurture the soil's ability to sustain life.

While American agriculture hasn't reached that level of sustainability, there is evidence that soil conservation programs are having a positive effect. In one Wisconsin study, erosion rates in one small watershed were 90 percent less in 1975–1993 than they were in the 1930s. Among the most important elements in soil conservation are land management, ground cover, climate, soil type, and tillage system.

Managing Topography

Water runs downhill. The faster it runs, the more soil it carries off the fields. Comparisons of erosion rates in Africa have shown that a 5 percent slope in a plowed field has three times the water runoff volume and eight times the soil erosion rate of a comparable field with a 1 percent slope. Water runoff can be reduced by leaving grass strips in waterways and by **contour plowing,** that is, plowing across the hill rather than up and down. Contour plowing is often combined with **strip-farming,** the planting of different kinds of crops in alternating strips along the land contours (fig. 7.25). When

FIGURE 7.25   *Contour plowing and strip-cropping protect the soil from erosion and help maintain fertility, as well as providing a beautiful landscape.*

FIGURE 7.26 *Rice terraces in Indonesia. Some rice paddies have been cultivated for hundreds or even thousands of years without any apparent loss of productivity.*

| | TABLE 7.2 | Soil Cover and Soil Erosion |

CROPPING SYSTEM	AVERAGE ANNUAL SOIL LOSS (TONS/HECTARE)	PERCENT RAINFALL RUNOFF
Bare soil (no crop)	41.0	30
Continuous corn	19.7	29
Continuous wheat	10.1	23
Rotation: corn, wheat, clover	2.7	14
Continuous bluegrass	0.3	12

Source: Based on 14 years of data from Missouri Experiment Station, Columbia, Missouri.

one crop is harvested, the other is still present to protect the soil and keep water from running straight downhill. The ridges created by cultivation make little dams that trap water and allow it to seep into the soil, rather than running off. In areas where rainfall is very heavy, tied ridges are often useful. This method involves a series of ridges running at right angles to each other, so that water runoff is blocked in all directions and is encouraged to soak into the soil.

Terracing involves shaping the land to create level shelves of earth to hold water and soil (fig. 7.26). The edges of the terrace are planted with soil-anchoring plant species. This is an expensive procedure, requiring either much hand labor or expensive machinery, but it makes farming of very steep hillsides possible. The rice terraces in the Chico River Valley in the Philippines rise as much as 300 m (1,000 ft) above the valley floor. They are considered one of the wonders of the world.

Planting **perennial species** (plants that grow for more than two years) is the only suitable use for some lands and some soil types. Establishing forest, grassland, or crops such as tea, coffee, or other crops that do not have to be cultivated every year may be necessary to protect certain unstable soils on sloping sites or watercourses. The $180 billion U.S. Farm Bill passed in 2002 includes $17 billion (over 10 years) to encourage soil, water, and wildlife conservation.

Providing Ground Cover

Annual row crops, such as corn or beans, generally cause the highest erosion rates because they leave soil bare for much of the year (table 7.2). Often, the easiest way to provide cover that protects soil from erosion is to leave crop residues on the land after harvest. They not only cover the surface to break the erosive effects of wind and water, but they also reduce evaporation and soil temperature in hot climates and protect ground organisms that help aerate and rebuild soil. In some experiments, 1 ton of crop residue per acre (0.4 ha) increased water infiltration 99 percent, reduced runoff 99

percent, and reduced erosion 98 percent. Leaving crop residues on the field also can increase disease and pest problems, however, and may require increased use of pesticides and herbicides.

Where crop residues are not adequate to protect the soil or are inappropriate for subsequent crops or farming methods, such **cover crops** as rye, alfalfa, or clover can be planted immediately after harvest to hold and protect the soil. These cover crops can be plowed under at planting time to provide green manure. Another method is to flatten cover crops with a roller and drill seeds through the residue to provide a continuous protective cover during early stages of crop growth.

In some cases, interplanting of two different crops in the same field not only protects the soil but also is more efficient use of the land, providing double harvests. Native Americans and pioneer farmers, for instance, planted beans or pumpkins between the corn rows. The beans provided nitrogen needed by the corn, pumpkins crowded out weeds, and both crops provided foods that nutritionally balance corn. Traditional swidden (slash-and-burn) cultivators in Africa and South America often plant as many as 20 different crops together in small plots. The crops mature at different times so that there is always something to eat, and the soil is never exposed to erosion for very long. Shade-grown coffee and cocoa can play important roles in conserving biodiversity (see Case Study, p. 168).

Mulch is a general term for a protective ground cover that can include manure, wood chips, straw, seaweed, leaves, and other natural products. For some high-value crops, such as tomatoes, pineapples, and cucumbers, it is cost-effective to cover the ground with heavy paper or plastic sheets to protect the soil, save water, and prevent weed growth. Israel uses millions of square meters of plastic mulch to grow crops in the Negev Desert.

Reduced Tillage

Farmers have traditionally used a moldboard plow to till the soil, digging a deep trench and turning the topsoil upside down. In the 1800s, it was shown that tilling a field fully—until it was "clean"—increased crop production. It helped control weeds and pests,

SHADE-GROWN COFFEE AND COCOA

Has it ever occurred to you that your purchases of coffee and chocolate may be contributing to the protection or destruction of tropical forests? Both coffee and cocoa are examples of food products grown exclusively in the Third World but consumed almost entirely in the First World (vanilla and bananas are some other examples). Coffee grows in cool, mountain areas of the tropics, while cocoa is native to the warm, moist lowlands. Both are small trees of the forest understory, adapted to low light levels.

Until a few decades ago, most of the world's coffee and cocoa were grown

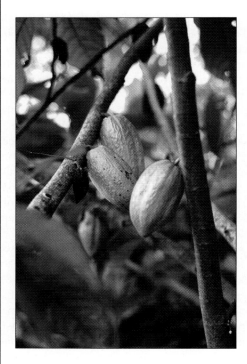

Cocoa pods grow directly on the trunk and large branches of cocoa trees.

under a canopy of large forest trees. Recently, however, new varieties of both crops have been developed that can be grown in full sun. Because more coffee or cocoa trees can be crowded into these fields, and they get more solar energy than in a shaded plantation, yields for sun-grown crops are higher.

There are costs, however, in this new technology. Sun-grown trees die earlier from the stress and diseases common in these fields. Furthermore, ornithologists have found that the number of bird species can be cut in half in full-sun plantations, and the number of individual birds may be reduced by 90 percent. Shade-grown coffee and cocoa generally require fewer pesticides (or sometimes none) because the birds and insects residing in the forest canopy eat many of the pests. Shade-grown plantations also need less chemical fertilizer because many of the plants in these complex forests add nutrients to the soil. In addition, shade-grown crops rarely need to be irrigated because heavy leaf fall protects the soil, while forest cover reduces evaporation.

Currently, about 40 percent of the world's coffee and cocoa plantations have been converted to full-sun varieties and another 25 percent are in process. Traditional techniques for coffee and cocoa production are worth preserving. Thirteen of the world's 25 biodiversity hot spots occur in coffee or cocoa regions. If all the 20 million ha (49 million acres) of coffee and cocoa plantations in these areas are converted to monocultures, an incalculable number of species will be lost.

The Brazilian state of Bahia is a good example of both the ecological importance of these crops and how they might help preserve forest species. At one time, Brazil produced much of the world's cocoa, but in the early 1900s, the crop was introduced into West Africa. Now Côte d'Ivoire alone grows more than 40 percent of the world total, and the value of Brazil's harvest has dropped by 90 percent. Côte d'Ivoire is aided in this competition by a labor system that reportedly includes widespread child slavery. Even adult workers in Côte d'Ivoire get only about $165 per year (if they get paid at all) compared to a minimum wage of $850 per year in Brazil. As African cocoa production ratchets up, Brazilian landowners are converting their plantations to pastures or other crops.

The area of Bahia where cocoa was once king is part of Brazil's Atlantic forest, one of the most threatened forest biomes in the world. Only 8 percent of this forest remains undisturbed. Although cocoa plantations don't represent the full diversity of intact forests, they protect a surprisingly large sample of what once was there. And shade-grown cocoa can provide an economic rationale for preserving that biodiversity. Brazilian cocoa will probably never compete with that from other areas for lowest cost. There is room in the market, however, for specialty products. If consumers were willing to pay a small premium for organic, fair-trade, shade-grown chocolate and coffee it might provide the incentive needed to preserve biodiversity. Wouldn't you like to know that your chocolate or coffee wasn't grown with child slavery, and is helping protect plants and animal species that might otherwise go extinct?

reducing competition; it brought fresh nutrients to the surface, providing a good seedbed; and it improved surface drainage and aerated the soil. This is still true for many crops and many soil types, but it is not always the best way to grow crops. We are finding that less plowing and cultivation often improves water management, preserves soil, saves energy, and increases crop yields.

There are three major **reduced tillage systems.** *Minimum till* involves reducing the number of times a farmer disturbs the

FIGURE 7.27 *In ridge-tilling, a chisel plow is used to create ridges on which crops are planted and shallow troughs filled with crop residue. Less energy is used on plowing and cultivation and weeds are suppressed and moisture is retained by the ground cover, or crop residue, left on the field.*

FIGURE 7.28 *On the Minar family's 230-acre dairy farm near New Prague, Minnesota, cows and calves spend the winter outdoors in the snow, bedding down on hay. Dave Minar is part of a growing counterculture that is seeking to keep farmers on the land and bring prosperity to rural areas.*

soil by plowing, cultivating, etc. This often involves a disc or chisel plow rather than a traditional moldboard plow. A chisel plow is a curved, chisel-like blade that doesn't turn the soil over but creates ridges on which seeds can be planted. It leaves up to 75 percent of plant debris on the surface between the rows, preventing erosion (fig. 7.27). *Conserv-till* farming uses a coulter (a sharp disc like a pizza cutter), which slices through the soil, opening up a furrow or slot just wide enough to insert seeds. This disturbs the soil very little and leaves almost all plant debris on the surface. *No-till* planting is accomplished by drilling seeds into the ground directly through mulch and ground cover. This allows a cover crop to be interseeded with a subsequent crop.

Farmers who use these conservation tillage techniques often must depend on pesticides (insecticides, fungicides, and herbicides) to control insects and weeds. Increased use of toxic agricultural chemicals is a matter of great concern. Massive use of pesticides is not, however, a necessary corollary of soil conservation. It is possible to combat pests and diseases with integrated pest management that combines crop rotation, trap crops, natural repellents, and biological controls.

Low-Input Sustainable Agriculture

Throughout the United States, Canada, and other crop-exporting countries, such as Brazil and New Zealand, huge, highly mechanized farms are becoming increasingly common. Giant grain fields stretch over thousands of acres, while large numbers of cows, hogs, or chickens live their entire lives in enormous, warehouse-size confinement barns. Often owned by multinational agricultural conglomerates, these corporate farms are operated by scientific standards that might make your local hospital seem backward. Milk cows, for instance, are tracked by computer, each one automatically given an individual diet, antibiotics, and hormone injec-

tions according to its pedigree, growth rate, or milk production. Farm managers travel by helicopter to survey vast spreads. One "farm" in California covers 2,000 ha (almost 5,000 acres) and has an annual cash flow of around $50 million.

In contrast to this trend toward industrialized agriculture, other farmers are going back to a size and style that their grandparents might have used a century ago. Finding that they can't—or don't want to—compete in capital-intensive production, these folks are making money and staying in farming by returning to small-scale, low-input farming. Rather than milking 1,000 cows in a highly mechanized operation, some farmers now raise only a couple of dozen cows and let them graze freely in pastures. Animals live outdoors in the winter (fig. 7.28). Milk production follows natural yearly cycles, rather than being controlled by synthetic hormones. The yield is lower, but so are costs.

Where 20,000 chickens living in a single immense shed or 1,000 hogs raised in an industrial warehouse must constantly be dosed with antibiotics to keep them healthy, animals allowed free range tend to have fewer diseases and require less care. Vast amounts of manure generated by industrial-scale factory farming are stored in football-field-size lagoons and spread on fields. Odors from ammonia, hydrogen sulfide (rotten egg odor), and other air pollutants make life miserable for neighbors. In 1999, heavy rains from Hurricane Floyd caused hundreds of North Carolina hog manure lagoons to overflow. Around 10 billion liters (2.5 billion gal) of manure washed down rivers to the sea, creating a large dead zone in Pamlico Sound. In 1999, the U.S. Environmental Protection Agency announced that it would require air pollution permits for factory farms.

In contrast to these gigantic operations, one hog farmer in Iowa, whose 40 sows graze in the pasture along with his 80 dairy cows, spends 30 percent less per animal for feed, 70 percent less for veterinary bills, and half as much for buildings and equipment as neighboring confinement operations.

Although there isn't complete agreement on what *sustainable* actually means, low-input farming generally involves more crop rotation and less pesticides (or none), as well as less inorganic fertilizer, water, heavy machinery, and fossil-fuel energy than the high-intensity, monoculture operation practiced by conventional farms. One of the biggest experiments in low-input farming is currently taking place in Cuba, where loss of aid from the former Soviet Union coupled with a trade embargo imposed by the United States have forced farmers to turn to organic, nonmechanized agriculture.

Organic farming has made great progress in Europe as well. In 2002, tiny Liechtenstein produced more than one-third of its food organically. In Switzerland, 10 percent of all food was organic, and in Austria, it was 8 percent organic. Prices for organic food in many European cities are only slightly higher than for conventional crops, and most consumers regard the choices available and quality of organic products to be better than for chemically-treated crops.

Total revenue from sustainable agriculture often is less than that of conventional operations, but profits may actually be higher. A study in North Dakota, for instance, found that conventional farms averaged $20 more per acre in total yield but that low-input farming had a net income that was $2 per acre higher because total costs were less.

These alternative farming methods are regarded by proponents as environmentally beneficial. Using fewer chemicals and manipulating the soil less reduces erosion and pollution and improves the long-range health of the land. Although this movement is too new to yet have good human health statistics, it seems likely that low-input farming could be good for the farmers, too.

Preserving small-scale, family farms also helps preserve rural culture. As Marty Strange of the Center for Rural Affairs in Nebraska asks, "Which is better for the enrollment in rural schools, the membership of rural churches, and the fellowship of rural communities—two farms milking 1,000 cows each or twenty farms milking 100 cows each?" Family farmers help keep rural towns alive by purchasing machinery at the local implement dealer, gasoline at the neighborhood filling station, and groceries at the mom-and-pop grocery store.

FIGURE 7.29 *Concern about the possible effects of pesticide residues in our diet has led many consumers to request organically grown food that is produced without pesticides. Farmers' markets, such as the one shown here, can be a good place to find locally produced organic food.*

One way to support local agriculture is to shop at a farmers' market (fig. 7.29). The produce is fresh, and profits go directly to the person who grows the crop. A local food co-op or owner-operated grocery store also is likely to buy from local farmers and to feature pesticide-free foods. Many co-ops and buyers' associations sign contracts directly with producers to grow the types of food that consumers want to eat. This benefits both parties. Producers are guaranteed a local market for organic food or specialty items. Consumers can be assured of quality and can even be involved in production.

If you belong to a co-op or grow your own food, you may have to accept fruits and vegetables that are not absolutely perfect. You can trim away a few bad spots or wash off an insect or two. If you overlook a worm or a bug, it may be less harmful to you than the toxic chemicals that are used to make supermarket foods sterile and cosmetically perfect!

SUMMARY

Over the past 30 years, the total amount of food in the world has increased faster than the average rate of population growth, so there is now more food per person than there was in the 1970s, even though the total number of people has doubled.

While there now is enough food to supply everyone in the world with more than the minimum daily food requirements, food is inequitably distributed. The United Nations Food and Agriculture Organization estimates that more than 800 million people are chronically undernourished or malnourished, and at least 15 million people (most of them in Africa) face the immediate threat of starvation. Additional millions survive on a deficient diet but suffer from the resulting stunted growth, mental retardation, and developmental disorders. Around half of the 12 million child deaths each year are related to malnutrition.

Among the essential dietary ingredients for good health are adequate calories, proteins, vitamins, and minerals. Marasmus and kwashiorkor are protein-deficiency diseases. Anemia and goiter are caused by mineral deficiencies that affect millions of people worldwide. Vitamin A deficiency causes the deaths of at least a million children and blinds another 350,000 every year.

The three major crops that provide the main source of calories and nutrients for most of the world's people are rice, wheat, and maize. Some new crops or unrecognized traditional crops hold promise for increasing the nutritional status of the poorer people of the world. Scientific improvement of existing crops and modernization of agriculture (irrigation, fertilizer, and better management) are potential sources of greater agricultural production, but also raise some troubling questions about safety and equity.

Fertile, tillable soil for growing crops is an indispensable resource for our continued existence on earth. Soil is a complex system of inorganic minerals, air, water, dead organic matter, and a myriad of different kinds of living organisms. There are hundreds of thousands of different kinds of soil, each produced by a unique history, climate, topography, bedrock, transported material, and community of living organisms.

It is estimated that 25 billion metric tons of soil are lost from croplands each year because of wind and water erosion. Perhaps twice as much is lost from rangelands, forests, and other areas. This erosion causes pollution and siltation of rivers, reservoirs, estuaries, wetlands, and offshore reefs and banks. The net effect of this loss is worldwide crop reduction equivalent to losing 15 million ha (37 million acres), or 1 percent of the world's cropland, each year.

The United States and Canada have very high rates of soil erosion. Soil losses exceed soil formation on at least 40 percent of U.S. cropland. About half of the topsoil that existed in North America before European settlement has been lost.

It is possible that food production could be expanded considerably, even on existing farmland, given the proper inputs of fertilizer, water, high-yield crops, and technology. This will be essential if human populations continue to grow as they have during the twentieth century. Whether it will be possible to supply agricultural inputs and expand crop production remains to be seen.

Many new and alternative farming methods reduce soil erosion, avoid dangerous chemicals, improve yields, and make agriculture just and sustainable. Returning to low-input, regenerative, "organic" farming may be more sustainable and more healthful than our current practices. Growing your own food or buying locally grown food at co-ops, at farmers' markets, or through a producers' or buyers' association can provide healthy, wholesome food and also support sustainable agriculture.

QUESTIONS FOR REVIEW

1. How many people in the world are chronically undernourished, and how many die each year from starvation and nutritionally related diseases?

2. How many calories does the average person need per day to maintain a healthy, active life?

3. What are proteins, vitamins, and minerals, and why do we need them in our diet?

4. What is the composition of soil? What is humus? Why are soil organisms so important?

5. What are four kinds of erosion? Why are they a problem?

6. What are some possible effects of overirrigation?

7. What do we mean by "green revolution?"

8. What is genetic engineering, or biotechnology, and how might it help or hurt agriculture?

9. What is sustainable agriculture?

10. What can farmers do to increase agricultural production without increasing land use?

THINKING SCIENTIFICALLY

1. Suppose that you were engaged in biotechnology or genetic engineering; what environmental safeguards would you impose on your own research? Are there experiments that would be ethically off-limits for you?

2. Debate the claim that famines are caused more by human actions (or inactions) than by environmental forces. What scientific evidence would you need to have to settle this question? What hypotheses could you test to help resolve the debate?

3. Should farmers be forced to use ecologically sound techniques that serve farmers' best interests in the long run, regardless of short-term consequences? How could we mitigate hardships brought about by such policies?

4. How would you provide assurance to consumers, government officials, and other interested parties that new genetically engineered products are environmentally and socially safe? What experiments would you set up to test for unknown unknowns?

5. Former U.S. President Jimmy Carter said, "Responsible biotechnology is not the enemy; starvation is." Do you agree?

6. Some rice paddies in Southeast Asia have been cultivated continuously for a thousand years or more without losing fertility. How could we in other countries adapt these techniques to our own agricultural systems?

7. Shade-grown coffee and chocolate might help protect endangered tropical forest species. If you were buying these products for your local co-op, what characteristics would you look for, and how much of a premium would you pay?

KEY TERMS

anemia 152
chronically undernourished 150
contour plowing 166
cover crops 167
famines 151
food security 150
genetic engineering 164
genetically modified organisms (GMOs) 164
green revolution 163
gully erosion 160
kwashiorkor 153
malnourishment 152
marasmus 153
mulch 167

obese 152
perennial species 167
reduced tillage systems 168
regenerative farming 166
rill erosion 160
salinization 161
sheet erosion 160
soil 156
soil horizons 157
strip-farming 166
subsoil 157
sustainable agriculture 166
terracing 167
topsoil 156
waterlogging 161

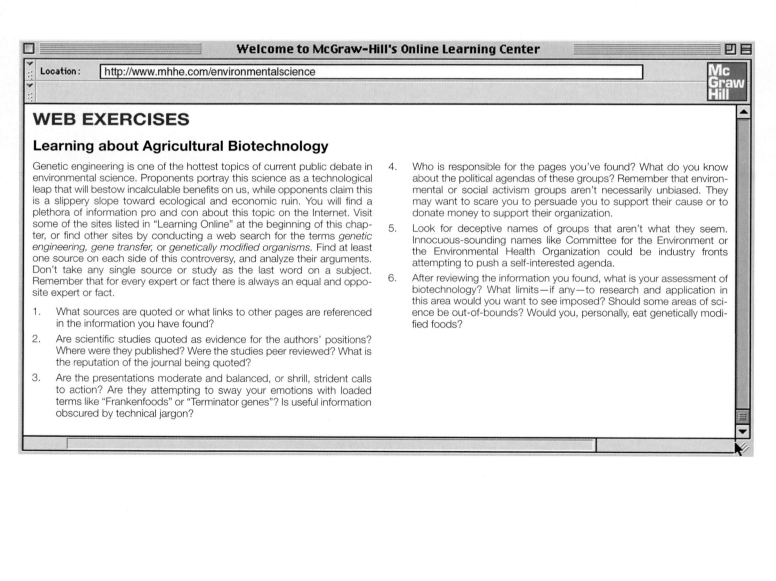

Welcome to McGraw-Hill's Online Learning Center

Location: http://www.mhhe.com/environmentalscience

WEB EXERCISES

Learning about Agricultural Biotechnology

Genetic engineering is one of the hottest topics of current public debate in environmental science. Proponents portray this science as a technological leap that will bestow incalculable benefits on us, while opponents claim this is a slippery slope toward ecological and economic ruin. You will find a plethora of information pro and con about this topic on the Internet. Visit some of the sites listed in "Learning Online" at the beginning of this chapter, or find other sites by conducting a web search for the terms *genetic engineering, gene transfer,* or *genetically modified organisms*. Find at least one source on each side of this controversy, and analyze their arguments. Don't take any single source or study as the last word on a subject. Remember that for every expert or fact there is always an equal and opposite expert or fact.

1. What sources are quoted or what links to other pages are referenced in the information you have found?

2. Are scientific studies quoted as evidence for the authors' positions? Where were they published? Were the studies peer reviewed? What is the reputation of the journal being quoted?

3. Are the presentations moderate and balanced, or shrill, strident calls to action? Are they attempting to sway your emotions with loaded terms like "Frankenfoods" or "Terminator genes"? Is useful information obscured by technical jargon?

4. Who is responsible for the pages you've found? What do you know about the political agendas of these groups? Remember that environmental or social activism groups aren't necessarily unbiased. They may want to scare you to persuade you to support their cause or to donate money to support their organization.

5. Look for deceptive names of groups that aren't what they seem. Innocuous-sounding names like Committee for the Environment or the Environmental Health Organization could be industry fronts attempting to push a self-interested agenda.

6. After reviewing the information you found, what is your assessment of biotechnology? What limits—if any—to research and application in this area would you want to see imposed? Should some areas of science be out-of-bounds? Would you, personally, eat genetically modified foods?

> Do I dare to eat a peach? Do I dare?
> –*T. S. Eliot*

Polish mothers give oxygen treatments to their children who suffer from air pollution–related respiratory disease. In some parts of Eastern Europe and the former USSR, up to 90 percent of all children suffer from environmentally linked diseases.

OBJECTIVES

After studying this chapter, you should be able to

- define health and disease in terms of some major environmental factors that affect humans.
- understand some of the risks of bioterrorism and emergent diseases.
- distinguish between toxic and hazardous chemicals, and between chronic and acute exposures and responses.
- compare the relative toxicity of some natural and synthetic compounds, and report on how such ratings are determined and what they mean.
- evaluate the major environmental risks we face and how risk assessment and risk acceptability are determined.

Welcome to McGraw-Hill's Online Learning Center

Location: http://www.mhhe.com/environmentalscience

LEARNING ONLINE

Visit our webpage at www.mhhe.com/environmentalscience for data sources, further readings, additional case studies, current environmental news, and regional examples within the Online Learning Center to help you understand the material in this chapter. You'll also find active links to information pertaining to this chapter including:

World Health Organization
Pan American Health Organization
Centers for Disease Control and Prevention
National Institutes of Health (NIH)
U.S. National Library of Medicine

Terrorism and Biological Warfare

The tragic events of September 11, 2001, are a wake-up call to how vulnerable we are to terrorist attacks. Antiterrorism experts warn, however, that crashing jet airplanes into large buildings isn't the easiest or most effective way to cause panic and destruction. Biological agents can be a much greater threat. A single gram of the most virulent strains of weaponized smallpox or anthrax, for example, could contain 250 million infectious doses. Under ideal dispersal conditions, about half the people exposed to these germs could become ill and one-third might die.

Using pathogens as weapons isn't a new idea. More than 2,000 years ago, Sythian archers dipped their arrowheads in rotting corpses to increase their deadliness. During the fourteenth century, Tartar soldiers hurled plague victim's corpses over the walls of besieged cities in an effort to start epidemics. In 1763, British General Jeffrey Amherst ordered smallpox-infected blankets distributed to Delaware Indians during the French and Indian War. During World War II, the Japanese dropped paper bags filled with plague-infested fleas on Chinese cities, perhaps killing thousands of people.

At the end of World War II, the United States and the Soviet Union embarked on massive biological and chemical warfare research programs. Both countries spent billions of dollars over the next 50 years developing and producing these weapons. The top-secret Soviet laboratory at Stepnogorsk in Kazakhstan, for example, is once thought to have employed 30,000 workers. Giant fermentation vats—some four stories tall and holding 20,000 liters of culture media—could produce 300 metric tons of supervirulent anthrax spores in a single 220-day production cycle. This is enough anthrax (if optimally distributed) to kill nine or ten times the total world population.

Using molecular biology techniques, these research labs combined genes and created new and more dangerous organisms than ever found in nature (fig. 8.1). Among the most lethal agents known to have been tested are anthrax, plague, botulism, and tularemia, along with viral diseases such as smallpox, and Ebola, and Marburg hemorrhagic fevers. Both the United States and Russia have promised destruction of their biological and chemical stockpiles, but thousands of talented former Soviet scientists are living in poverty. Both knowledge about weapon development and weapons themselves are widely thought to be for sale to the highest bidder. Algeria, India, Iran, Iraq, Israel, Libya, North Korea, and Syria all are known to have active research programs in biological and chemical warfare.

For more than 20 years, experts have warned that America is vulnerable to biological attack. In 2001, these warnings took on a new urgency. Just a week after the assaults of September 11, letters containing anthrax spores were mailed to the *New York Post* and NBC News in New York City, and Senator Tom Daschle in Washington, D.C. In all, 18 people were infected in this assault, 5 people died, and millions of us were suddenly aware of our vulnerability. For the first time, we realize that a small group of fanat-

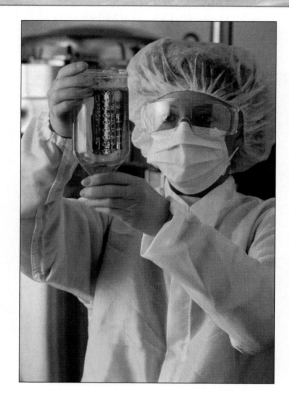

FIGURE 8.1 *Using the tools of molecular biology, terrorists could create new and more potent bioweapons.*

ical terrorists could easily contaminate our air, water, and food with lethal pathogens or chemical toxins.

Mailing individual letters isn't the most likely delivery mechanism for biological weapons. Antiterrorism experts point out, for example, that most American office buildings have unprotected, ground-level air intakes. A few grams of aerosolized germ spores poured into one of these air ducts could spread through a 100-floor skyscraper in less than a half hour, and infect thousands of people without anyone being aware of it. By the time symptoms began to appear a few days later, victims could have spread throughout the city, and if an infectious agent was used, countless more people could be exposed. Medical and emergency personnel would be overwhelmed and chaos would reign. Although there's some hesitation about being very explicit about when, where, and how attacks might occur for fear of giving terrorists useful information, emergency planners believe that the public needs to know of these dangers so we can evaluate risks.

While some people might regard biological warfare beyond the scope of environmental science, it introduces issues of pathogenicity, toxicity, routes of exposure, and the movement, distribution, and persistence of dangerous materials in our environment that are applicable to our everyday lives. In this chapter, we'll look at some principles of environmental health that help in understanding both biowarfare and everyday threats that all of us could face.

TYPES OF ENVIRONMENTAL HEALTH HAZARDS

What is health? The World Health Organization (WHO) defines **health** as a state of complete physical, mental, and social well-being, not merely the absence of disease or infirmity. By that definition, we all are ill to some extent. Likewise, we all can improve our health to live happier, longer, more productive, and more satisfying lives if we pay attention to what we do.

What is a disease? A **disease** is a deleterious change in the body's condition in response to an environmental factor that could be nutritional, chemical, biological, or psychological. Diet and nutrition, infectious agents, toxic chemicals, physical factors, and psychological stress all play roles in **morbidity** (illness) and mortality (death). To understand how these factors affect us, let's look at some of the major categories of environmental health hazards.

Infectious Organisms

For most of human history, the greatest health threats have always been pathogenic (disease-causing) organisms. Although cardiovascular diseases (heart attacks and strokes), cancer, injuries (intentional and unintentional), and other ills of modern life now have become the leading killers almost everywhere in the world, infectious diseases still kill at least 18 million people every year, or about one-third of all disease-related mortality. Most of these deaths are in the poorer countries of the world where better nutrition, improved sanitation, and inexpensive vaccinations could save millions of lives each year. How does that affect us? As we all now know, our world is increasingly interconnected. New, extremely virulent diseases like Ebola, as well as drug-resistant forms of more familiar diseases such as tuberculosis and cholera, can arise in remote areas and then spread rapidly throughout the world.

Before you learn about all the terrible illnesses that afflict people, it's worth pointing out that in many ways most people are healthier than ever before. One sign of this is the fact that people are living longer almost everywhere. As chapter 4 points out, life expectancies have increased around the world from about 40 years to 65 years on average over the past century. The fact that cancers and cardiovascular diseases have become the leading causes of death attests to our now living long enough to suffer these diseases of old age and affluence. Although developing countries lag behind the wealthier, industrialized nations in education and health care, progress in these areas is being made among some of the world's poorest people.

Of the broad categories of communicable diseases (table 8.1), acute respiratory infections such as pneumonia and influenza cause the most deaths each year, followed closely by chronic obstructive lung infections such as tuberculosis. Perinatal conditions including infections and other problems associated with birth take some 2.4 million lives annually, while a group of childhood diseases including pertussis (whooping cough), poliomyelitis, diphtheria, measles, and tetanus kill about 1.4 million children each year. In addition, some 2 billion new cases of diarrhea each year are responsible for about 2 million deaths, mostly in children under 5 years old. Both bacteria and protozoans can cause diarrhea (fig. 8.2).

TABLE 8.1 Leading Causes of Death

DISEASE OR GROUP*	MILLIONS/YEAR
Cardiovascular diseases	16.7
Cancer and tumors	6.9
Injuries and trauma	5.1
Respiratory infections	3.9
Chronic lung disease	3.5
HIV/AIDS	2.9
Maternal and perinatal conditions	2.9
Diarrheal diseases	2.1
Digestive diseases	1.9
Tuberculosis	1.6
Childhood diseases	1.4
Malaria	1.1
Neuropsychiatric disorders	0.9
Diabetes	0.8
Kidney and prostate disease	0.8
Congenital abnormalities	0.7

*See text for further description.
Source: Data from World Health Organization, 2002.

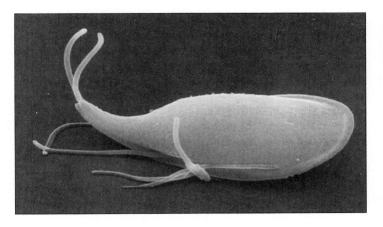

FIGURE 8.2 Giardia, *a parasitic intestinal protozoan, is reported to be the largest single cause of diarrhea in the United States. It is spread from human feces through food and water. Even remote wilderness areas have* Giardia *outbreaks due to careless campers.*

Although virtually unknown 15 years ago, acquired immune deficiency syndrome (AIDS) has now become the fifth greatest cause of contagious deaths. The WHO estimates that 30 million people are now infected with this virus and 2.9 million die each year. Over the next 20 years, an additional 65 million people could die of this terrible disease. Two-thirds of all current AIDS patients are in sub-Saharan Africa. In countries such as Zimbabwe, Botswana, and Zambia, health officials predict that more than two-thirds of current

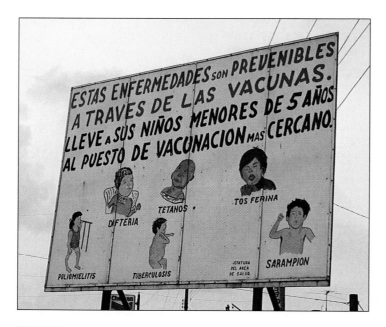

FIGURE 8.3 *About 1.4 million children die each year from easily prevented childhood diseases. This Guatemalan billboard urges that children be vaccinated against polio, diphtheria, TB, tetanus, pertussis (whooping cough), and scarlet fever (l to r).*

15-year-olds will die of AIDS before age 50. A whole generation of young, productive adults has died, and some villages consist only of the very old or very young. No one is left to support families or keep the economy and society functioning. There may be 10 million African children living with AIDS, or orphans whose parents have died from the disease. Heterosexual sex is the main cause of HIV transmission in Africa, and 55 percent of all HIV-positive adults there are women. One sign of hope is that several pharmaceutical companies have offered to sell anti-AIDS drugs in Africa for a fraction of the cost in richer countries. This may make it more feasible to treat AIDS sufferers and to prevent transmission from mothers to newborn children. AIDS is now spreading rapidly in South and Southeast Asia, where a much larger population is at risk. The UN warns that 10 million people could have AIDS in China within a decade, primarily due to unsanitary blood donation programs.

Considerable progress is being made in eliminating many of these diseases (fig. 8.3). Smallpox, for instance, was eradicated in 1977, and polio is now limited to just three countries in South Asia, according to the WHO. A massive inoculation campaign in 1998 and 1999 immunized more than 450 million children against polio. Much more could be done, however. Jeffrey Sachs, of Harvard University, estimates that $30 to $40 per person per year would cover antiretroviral therapy for AIDS, bed nets for malaria, monitored drug therapy for tuberculosis, oral rehydration for life-threatening diarrhea, antibiotics for acute respiratory infection, and midwives in attendance at childbirth. Such a program would save millions of lives every year, and would go a long way toward eliminating the anger and despair that motivate terrorists and instigate wars. The economic stimulus from avoiding disease and enhancing productivity, Dr. Sachs believes, would easily pay for this investment. Cur-

rently, all the rich countries together donate about $6 billion per year for health care in the developing world. This is less than 0.2 percent of the $25 trillion annual GNP of the donor countries. The United States is at the bottom of the list in terms of aid per GNP. Just preventing pathogens from entering our country would be worth a greater investment in health care abroad.

Morbidity and Quality of Life

Death rates don't tell us everything about the burden of disease. Obviously many people who suffer from illness don't die but have a severely diminished quality of life. When people are sick, work isn't done, crops aren't planted or harvested, meals aren't cooked, and children can't study and learn. The billions of people who suffer from various worms, flukes, internal parasites, or infections like the common cold at any given time rarely die from their affliction, but the suffering and debilitation is real, nonetheless.

Disability-Adjusted Life Years or **DALYs**, are a measure created by the World Bank to quantify the disruption to quality of life and economic productivity caused by illness or disability. This is an attempt to evaluate the total social burden of disease, not just how many people die. Clearly, many more years of quality life are lost when a child dies of neonatal tetanus than when a 70-year-old dies of pneumonia. Similarly, a child permanently paralyzed by polio will have many more years of suffering and lost potential than will a senior citizen who has a stroke. According to the World Bank, communicable diseases cause about 40 percent of the 1.2 billion DALYs lost each year (table 8.2).

TABLE 8.2	Some Leading Causes of Disease Burdens
DISEASE OR GROUP	**DALY BURDEN (MILLIONS OF YEARS)***
Injuries and trauma	231
Neuropsychiatric conditions	165
Cardiovascular	155
Malignant neoplasms	90
Respiratory infections	85
Maternal and perinatal diseases	75
Diarrheal diseases	74
Chronic lung disease	71
Childhood diseases	55
Digestive diseases	46
HIV/AIDS	40
Tuberculosis	39
Nutritional deficiencies	35
Congenital abnormalities	34
Musculoskeletal problems	22

*Disability-adjusted life years = Years lost from premature death and disability.
Source: Data from World Bank, 2002.

The heaviest burden of illness is borne by the poorest people who can afford neither a healthy environment nor adequate health care. About 90 percent of all DALY losses occur in the developing world where less than one-tenth of all health care dollars is spent. Women in sub-Saharan Africa, for example, suffer six times the disease burden per 1,000 population as do women in most European countries. The 1.3 billion people worldwide who live on less than $1 per day generally lack access to adequate housing, sanitation, and safe drinking water, all of which increase their exposure to pathogens. Malnutrition exacerbates many diseases.

At any given time, somewhere around 2 billion people suffer from worms, flukes, protozoa, and other internal parasites. While victims rarely die from these infections, their quality of life clearly suffers. Progress is being made, however, in eradicating some dreadful parasitic diseases. The pharmaceutical company, SmithKline Beecham, has pledged to freely distribute the drug albendazole to about 1 billion people in an effort to eliminate the grossly disfiguring disease known as elephantiasis that now afflicts about 120 million people. This effort will cost the company about $500 million over the next two decades. Drancunculiasis or guinea worm, another dreadful tropical disease, is being eliminated through community health education and provision of safe drinking water. (See related story "Fighting the Fiery Serpent" at http://www.mhhe.com/biosci/pae/environmentalscience/cunningham/topics.html.)

Emergent Diseases and Environmental Change

Although many diseases such as measles, pneumonia, and pertussis (whooping cough) have probably inflicted humans for millennia, at least 30 new infectious diseases have appeared in the past two decades, while many well-known diseases have reappeared in more virulent, drug-resistant forms (fig. 8.4). An **emergent disease** is one never known before or one that has been absent for at least 20 years. Ebola fever is a good example of an emergent disease. A kind of viral hemorrhagic fever, Ebola is extremely contagious and often kills up to 90 percent of those who are exposed to it. The disease was unknown until about 20 years ago, but is thought to have been present in monkeys or other primates. Killing and eating chimps, gorillas, and other primates is thought to be the route of infection in humans. AIDS is another disease that appears to have suddenly moved from other primates to humans. How pathogens suddenly move across species barriers to become highly contagious and terribly lethal is one of the most important questions in environmental health.

Some of the most devastating epidemics have occurred when travelers bring new germs to a naive population lacking immunity. An example was the plague, or Black Death, which swept through Europe and Western Asia repeatedly in the fourteenth and fifteenth centuries. During the first—and worst—episode between 1347 and 1355, about half the population of Europe died. In some cities the mortality rate was as high as 80 percent. It's hard to imagine the panic and fear this disease caused. An even worse disaster occurred when Europeans brought smallpox, measles, and other infectious diseases to the Americas. By some calculations, up to 90 percent of the native people perished as diseases swept through their population. One reason European explorers thought the land was an empty wilderness was that these diseases spread out ahead of them, killing everyone in their path.

Probably the largest loss of life from an individual disease in a single year was the great influenza epidemic of 1918. No one knows

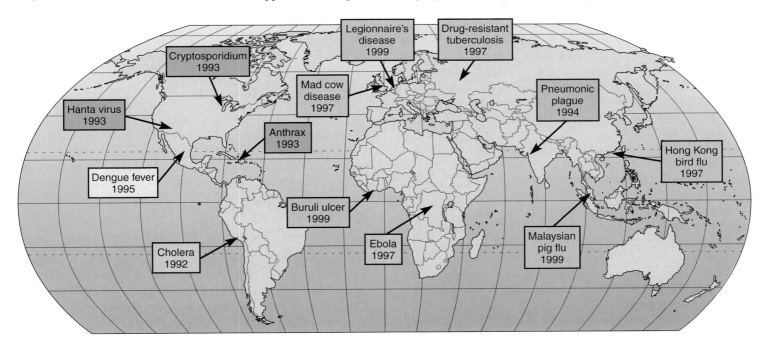

FIGURE 8.4 *Some outbreaks of highly lethal, infectious diseases in the 1990s. Why are supercontagious organisms emerging in so many different places?*
Source: Data from United States Centers for Disease Control and Prevention.

where the disease came from, but war refugees and soldiers returning home quickly spread the virus around the globe. Somewhere between 30 and 40 million people succumbed to this virus in less than 12 months. This was more than twice the total number killed in all the battles of World War I, which was occurring at the time.

A number of factors contribute currently to the appearance and spread of these highly contagious diseases. With 6 billion people now inhabiting the planet, human densities are much higher, enabling germs to spread further and faster than ever before. Expanding populations push into remote areas and encounter new pathogens and parasites. We are causing environmental change on a massive scale: cutting forests, creating unhealthy urban surroundings, and causing global climate change, among other things. Habitat changes and elimination of predators favor disease-carrying organisms such as mice, rats, cockroaches, and mosquitoes. And we are eating other species in a way we may never have done before. A survey published in 2002 found that more than one-fifth of the monkey meat sold in the markets of Cameroon was infected with HIV's ancestor, SIV.

Another important factor in the spread of many diseases is the speed and frequency of modern travel. Millions of people go every day from one place to another by airplane, boat, train, or automobile. Very few places on earth are more than 24 hours by jet plane from any other place. Many highly virulent diseases take several days for symptoms to appear. A person carrying these germs might fly half-way around the world without being detected. Could this be a scenario for a terrorist attack: deliberately infect a group of volunteers with smallpox or ebola and put them on planes bound for a dozen major cities?

Emerging Ecological Diseases

Humans aren't the only ones to suffer from new and devastating diseases. Domestic animals and wildlife also experience sudden and widespread epidemics. In 1998, for example, a distemper virus killed half the seals in Western Europe. It's thought that toxic pollutants and hormone-disrupting environmental chemicals might have made seals and other marine mammals susceptible to infections. In 2002, more dead seals were found in Denmark, raising fears that distemper might be reappearing.

Chronic wasting disease (CWD) is spreading through deer and elk populations in North America. Caused by a strange protein called a prion, CWD is one of a family of irreversible, degenerative neurological diseases known as transmissible spongiform encephalopathies (TSE) that include mad cow disease in cattle, scrapie in sheep, and Creutzfelt-Jacob disease in humans. CWD probably started when elk ranchers fed contaminated animal by-products to their herds. Infected animals were sold to other ranches, and now the disease has spread to wild populations. First recognized in 1967 in Saskatchewan, CWD has been identified in wild deer populations and ranch operations in at least eight American states.

The Canadian government is estimated to have spent $65 million in an attempt to stop the spread of CWD. In 2002, Wisconsin encouraged hunters to kill some 20,000 deer in an area near Madison in an effort to contain the disease. No humans are known

FIGURE 8.5   *Black band disease kills a large brain coral. Pathogenic human bacteria have been found associated with this disease and may play a role in making corals vulnerable.*

to have contracted TSE from deer or elk, but there is a concern that we might see something like the mad cow disaster that inflicted Europe in the 1990s. At least 125 people died, and nearly 5 million European cattle and sheep were slaughtered in an effort to contain that disease.

Starting in the early 1970s, an illness called black-band disease has been attacking corals throughout the Caribbean. A cyanophyte alga (*Phormidium corallyticum*) actually kills the coral polyps. As the black ring of dead polyps spreads through the colony, it leaves behind a bleached coral skeleton (fig. 8.5). Researchers have found pathogenic bacteria from human feces associated with dying corals and think that they may play a role in triggering the algal attack. Coliform bacteria are present on reefs far from any human occupation. These pathogens may be carried by dust storms from as far away as Africa. Coral subjected to environmental stressors such as nutrient imbalance or elevated seawater temperatures may be susceptible to infection.

All of these diseases have in common human-made environmental changes that stress biological communities and upset normal ecological relationships. How many other ways might we be altering the world around us, and what might the consequences be both for ourselves and other species?

Antibiotic and Pesticide Resistance

Malaria, the most deadly of all insect-borne diseases, is an example of the return of a disease that once was thought to be nearly vanquished. Malaria now claims about a million lives every year—90 percent in Africa, and most of them children. With the advent of modern medicines and pesticides, malaria had nearly been wiped out in many places but recently has come roaring back. The protozoan parasite that causes the disease is now resistant to most antibiotics, while the mosquitoes that transmit it have developed resistance to many insecticides. Spraying of DDT in India and Sri Lanka, for instance, reduced malaria from millions of infections per year to only

FIGURE 8.6 *How microbes acquire antibiotic resistance.* (a) *Random mutations make a few cells resistant. When challenged by antibiotics, only those cells survive to give rise to a resistant colony.* (b) *Sexual reproduction (conjugation) or plasmid transfer moves genes from one strain or species to another.*

a few thousand in the 1950s and 1960s. Now South Asia is back to its pre-DDT level of about half a million new cases of malaria every year. Other places that never had malaria cases now have them as a result of climate change and habitat alteration.

Why have vectors such as mosquitoes, and pathogens such as the malaria parasite become resistant to pesticides and antibiotics? Part of the answer is natural selection and the ability of many organisms to evolve rapidly. Another factor is the human tendency to use control measures carelessly. When we discovered that DDT and other insecticides could control mosquito populations, we spread them everywhere. This not only harmed wildlife and beneficial insects, but it created perfect conditions for natural selection. Many pests and pathogens were exposed only minimally to control measures, allowing those with natural resistance to survive and spread their genes through the population (fig. 8.6). After repeated cycles of exposure and selection, many microorganisms and their vectors have become insensitive to almost all our weapons against them.

As chapter 7 discusses, raising huge numbers of cattle, hogs, and poultry in densely packed barns and feedlots is another reason for widespread antibiotic resistance in pathogens. Confined animals are dosed constantly with antibiotics and steroid hormones to keep them disease free and to make them gain weight faster. More than half of all antibiotics used in the United States each year are fed to livestock. A significant amount of these antibiotics and hormones are excreted in urine and feces, which are spread, untreated, on the land or discharged into surface water where they contribute further to the evolution of supervirulent pathogens.

At least half of the 100 million antibiotic doses prescribed for humans every year in the United States are unnecessary or are the wrong ones. Furthermore, many people who start a course of antibiotic treatment fail to carry it out for the time prescribed. For your own health and that of the people around you, if you are taking an antibiotic, follow your doctor's orders. Take all of the prescribed doses and don't stop taking the medicine as soon as you start feeling better.

Toxic Chemicals

Dangerous chemical agents are divided into two broad categories: hazardous and toxic. **Hazardous** means dangerous. This category includes flammables, explosives, irritants, sensitizers, acids, and caustics. Many chemicals that are hazardous in high concentrations are relatively harmless when diluted. **Toxins** are poisonous. This means they react with specific cellular components to kill cells. Because of this specificity, toxins often are harmful even in dilute concentrations. Toxins can be either general poisons that kill many kinds of cells, or they can be extremely specific in their target and mode of action. Ricin, for instance, is a protein found in castor beans and one of the most toxic organic compounds known. Three hundred picograms (trillionths of a gram) injected intravenously is enough to kill an average mouse. A single molecule can kill a cell. This is about 200 times more lethal than any of the dioxins, which often are claimed to be the most toxic substances known.

Table 8.3 shows some of the air toxins of greatest concern to the U.S. Environmental Protection Agency.

Allergens are substances that activate the immune system. Some allergens act directly as **antigens;** that is, they are recognized as foreign by white blood cells and stimulate the production of specific antibodies. Other allergens act indirectly by binding to other materials and changing their structure or chemistry so they become antigenic and cause an immune response.

TABLE 8.3   Air Toxins Considered of Greatest Concern by the U.S. EPA*

Acetaldehyde	Formaldehyde
Acrolein	Hydrazine
Arsenic compounds	Lead compounds
Benzene	Manganese compounds
Beryllium compounds	Mercury compounds
1, 3-butadiene	Methylene chloride
Cadmium compounds	Nickel compounds
Carbon tetrachloride	Perchloroethylene
Chloroform	Polychlorinated biphenyls (PCBs)
Coke oven emissions	Propylene dichloride
1, 3-dichloropropene	Quinoline
Diesel particulate matter	1, 1, 2, 2-tetrachloroethane
Ethylene dibromide	Trichloroethylene
Ethylene dichloride	Vinyl chloride

*Toxins are listed in alphabetical order rather than by most serious threat.
Source: Data from Environmental Protection Agency, 2002.

Formaldehyde is a good example of a widely used chemical that is a powerful sensitizer. It is directly allergenic and can also trigger reactions to other substances. Widely used in plastics, wood products, insulation, glue, and fabrics, formaldehyde concentrations in indoor air can be thousands of times higher than in normal outdoor air. Some people suffer from what is called **sick building syndrome**: headaches, allergies, and chronic fatigue caused by poorly vented indoor air contaminated by molds, carbon monoxide, nitrogen oxides, formaldehyde, and other toxic chemicals released by carpets, insulation, plastics, and building materials. The Environmental Protection Agency estimates that poor indoor air quality may cost the nation $60 billion dollars a year in absenteeism and reduced productivity.

Neurotoxins are a special class of metabolic poisons that specifically attack nerve cells (neurons). The nervous system is so important in regulating body activities that disruption of its activities is especially fast-acting and devastating. Different types of neurotoxins act in different ways. Heavy metals, such as lead and mercury, kill nerve cells and cause permanent neurological damage. Anesthetics (ether, chloroform, halothane, etc.) and chlorinated hydrocarbons (DDT, Dieldrin, Aldrin) disrupt nerve cell membranes necessary for nerve action. Organophosphates (Malathion, Parathion) and carbamates (carbaryl, zeneb, maneb) inhibit acetylcholinesterase, an enzyme that regulates signal transmission between nerve cells and the tissues or organs they innervate (for example, muscle). Most neurotoxins are both acute and extremely toxic. More than 850 compounds are now recognized as neurotoxins.

Mutagens are agents, such as chemicals and radiation, that damage or alter genetic material (DNA) in cells. This can lead to birth defects if the damage occurs during embryonic or fetal growth. Later in life, genetic damage may trigger neoplastic (tumor) growth. When damage occurs in reproductive cells, the results can be passed on to future generations. Cells have repair mechanisms to detect and restore damaged genetic material, but some changes may be hidden, and the repair process itself can be flawed. It is generally accepted that there is no "safe" threshold for exposure to mutagens. Any exposure has some possibility of causing damage.

Teratogens are chemicals or other factors that specifically cause abnormalities during embryonic growth and development. Some compounds that are not otherwise harmful can cause tragic problems in these sensitive stages of life. Perhaps the most prevalent teratogen in the world is alcohol. Drinking during pregnancy can lead to **fetal alcohol syndrome**—a cluster of symptoms including craniofacial abnormalities, developmental delays, behavioral problems, and mental defects that last throughout a child's life. Even one alcoholic drink a day during pregnancy has been associated with decreased birth weight.

Carcinogens are substances that cause **cancer**—invasive, out-of-control cell growth that results in malignant tumors. Cancer rates rose in most industrialized countries during the twentieth century, and cancer is now the second leading cause of death in the United States, killing more than half a million people in 2000. Twenty-three of the 28 compounds listed in table 8.3 are probable or possible human carcinogens. According to the U.S. EPA, more than 200 million people live in areas where the combined upper limit lifetime cancer risk from these carcinogens exceeds 10 in 1 million, or 10 times the risk normally considered acceptable.

The American Cancer Society calculates that one in two males and one in three females in the United States will have some form of cancer in their lifetime. Some authors blame this cancer increase on toxic synthetic chemicals in our environment and diet. Others argue that it is attributable mainly to lifestyle (smoking, sunbathing, drinking alcohol) or simply living longer.

If the number of deaths from cancer is adjusted for age, the only major types that have become more prevalent in recent years are prostate cancer in men and lung cancer in women (fig. 8.7). If we

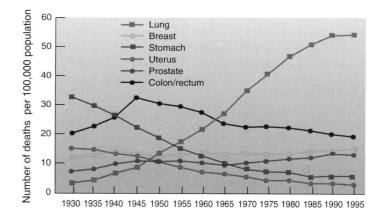

FIGURE 8.7   *Age-adjusted cancer death rates in the United States. Although the total incidence of cancer is up in recent years, better treatment means that more people survive. When adjusted for an aging population, mortality for most major cancers has been stable or falling. One exception is lung cancer, 90 percent of which is attributable to increased smoking, especially among women.*

look at total incidence rather than death rates, however, the picture changes considerably. Recently, the number of new cases of breast, testis, and skin cancer, for example, have increased significantly, but so has the success in treating these diseases. Better diagnosis and treatment, rather than lower incidence, is probably the reason that mortality for cancers such as stomach, uterus, and colon has decreased over the past 30 years.

Endocrine Hormone Disrupters

One of the most recently recognized environmental health threats are **endocrine hormone disrupters,** chemicals that disrupt normal endocrine hormone functions. Hormones are chemicals released into the blood stream by glands in one part of the body to regulate development and function of tissues and organs elsewhere in the body. You undoubtedly have heard about sex hormones and their powerful effects on how we look and behave, but these are only one example of the many regulatory hormones that rule our lives.

We now know that some of the most insidious effects of persistent chemicals such as DDT and PCBs are that they interfere with normal growth, development, and physiology of a variety of animals—presumably including humans—at very low doses. In some cases, picogram concentrations (trillionths of a gram per liter) may be enough to cause developmental abnormalities in sensitive organisms. These chemicals are sometimes called environmental estrogens or androgens, because they often cause sexual dysfunction (reproductive health problems in females or feminization of males, for example). They are just as likely, however, to disrupt thyroxin functions or those of other important regulatory molecules as they are to obstruct sex hormones.

Diet

Diet also has an important effect on health. For instance, there is a strong correlation between cardiovascular disease and the amount of salt and animal fat in one's diet. Fat intake was once thought to be associated with breast cancer, but long-term studies in several countries suggest that the linkage may not be correct. Still, fat intake may be linked to other cancers. Highly processed foods, fat, and smoke-cured, high-nitrate meats also seem to be associated with cancer.

Fruits, vegetables, whole grains, complex carbohydrates, and dietary fiber (plant cell walls), on the other hand, often have beneficial health effects. Certain dietary components, such as pectins; vitamins A, C, and E; substances produced in cruciferous vegetables (cabbage, broccoli, cauliflower, brussels sprouts); and selenium, which we get from plants, seem to have anticancer effects.

Eating too much food is a significant dietary health factor in developed countries and among the well-to-do everywhere. Nearly two-thirds of all Americans are considered overweight. The U.S. Surgeon General estimates that some 300,000 deaths every year are related to obesity. This rivals the number associated with smoking. Cutting back on the number of calories consumed

National Health Recommendations and Diet Goals

1. Balance the food you eat with physical activity to maintain or improve your weight.
2. Choose a diet with plenty of grain products, vegetables, and fruits.
3. Choose a diet low in fat, saturated fat, and cholesterol.
4. Eat a variety of foods.
5. Choose a diet moderate in salt and sodium.
6. Choose a diet moderate in sugars.
7. If you drink alcoholic beverages, do so in moderation.

Source: Data from U.S. Department of Health and Human Services, 1995.

TABLE 8.4	**A Shopper's Guide to Pesticides in Produce**

RANK	TWELVE MOST CONTAMINATED FOODS
1.	Strawberries
2.	Bell peppers (green and red)
3.	Spinach
4.	Cherries (U.S.)
5.	Peaches
6.	Cantaloupe (Mexican)
7.	Celery
8.	Apples
9.	Apricots
10.	Green beans
11.	Grapes (Chilean)
12.	Cucumbers

Source: Environmental Working Group, 2002.

reduces the strain on bones, muscles, and other organs, and has additional beneficial effects, including reduction of cardiovascular disease, diabetes, and—perhaps—cancer.

Table 8.4 shows the 12 fruits and vegetables found by the U.S. Food and Drug Administration to be most commonly contaminated by pesticide residues. More than half the total dietary pesticide risk was accounted for by these crops. The pesticides in these foods are classified as probable human carcinogens, neurotoxins, and endocrine system disrupters (see What Can You Do? above).

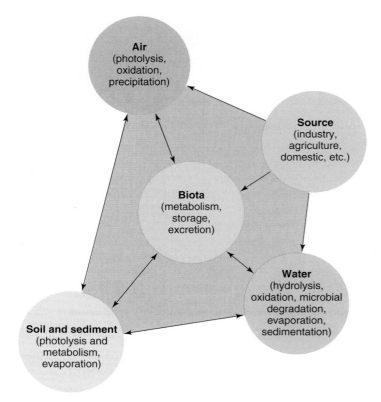

FIGURE 8.8 *Movement and fate of chemicals in the environment. Processes that modify, remove, or sequester compounds are shown in parentheses. Toxins also move directly from a source to soil and sediment.*

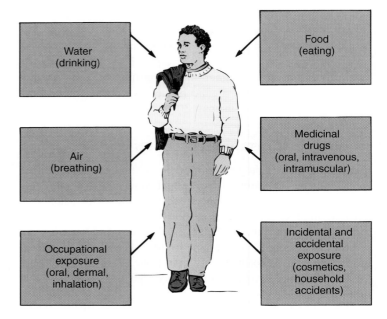

FIGURE 8.9 *Routes of exposure to toxic and hazardous environmental factors.*

MOVEMENT, DISTRIBUTION, AND FATE OF TOXINS

There are many sources of toxic and hazardous chemicals in the environment. Many factors related to each chemical itself, its route or method of exposure, and its persistence in the environment, as well as the characteristics of the target organism, determine the danger of the chemical. We can think of an ecosystem as a set of interacting compartments among which a chemical moves, based on its molecular size, solubility, stability, and reactivity (fig. 8.8). The routes used by chemicals to enter our bodies also play important roles in determining toxicity (fig. 8.9). In this section, we will consider some of these characteristics and how they affect environmental health.

Solubility

Solubility is one of the most important characteristics in determining how, where, and when a toxic material will move through the environment or through the body to its site of action. Chemicals can be divided into two major groups: those that dissolve more readily in water and those that dissolve more readily in oil. Water-soluble compounds move rapidly and widely through the environment because water is ubiquitous. They also tend to have ready access to most cells in the body because aqueous solutions bathe all our cells. Molecules that are oil- or fat-soluble (usually organic molecules)

generally need a carrier to move through the environment and into or within the body. Once inside the body, however, oil-soluble toxins penetrate readily into tissues and cells because the membranes that enclose cells are themselves made of similar oil-soluble chemicals. Once inside cells, oil-soluble materials are likely to accumulate and to be stored in lipid deposits, where they may be protected from metabolic breakdown and persist for many years.

Bioaccumulation and Biomagnification

Cells have mechanisms for **bioaccumulation,** the selective absorption and storage of a great variety of molecules. This allows them to acquire nutrients and essential minerals, but at the same time, can also result in the absorption and buildup of harmful substances. Toxins that are rather dilute in the environment can reach dangerous levels inside cells and tissues through this process.

The effects of toxins also are magnified through food webs. **Biomagnification** occurs when the toxic burden of a large number of organisms at a lower trophic level is accumulated and concentrated by a predator in a higher trophic level. Phytoplankton and bacteria in aquatic ecosystems, for instance, take up heavy metals or toxic organic molecules from water or sediments (fig. 8.10). Their predators—zooplankton and small fish—collect and retain the toxins from many prey organisms, building up higher toxin concentrations. The top carnivores in the food chain—game fish, fish-eating birds, and humans—can accumulate such high toxin levels that they suffer adverse health effects. One of the first-known examples of bioaccumulation and biomagnification involved DDT, which accumulated through food chains so that, by the 1960s, it was shown to be interfering with reproduction of peregrine falcons, bald eagles, brown pelicans, and other predatory birds at the top of their food chains (fig. 8.11).

FIGURE 8.10 *Bioaccumulation and biomagnification. Organisms lower on the food chain take up and store toxins from the environment. They are eaten by larger predators, who are eaten, in turn, by even larger predators. The highest members of the food chain can accumulate very high levels of the toxin.*

Sensitivity and Developmental Stage

Some people are much more sensitive to toxins than others. Age, sex, body weight, nutritional or immunological status, and ability to deactivate toxins or repair damage all help determine how we react to a given dose. Children, especially, are highly susceptible to environmental contaminants. A given dose has a greater effect on a small child than a full-grown adult. Furthermore, in the process of growing and maturing, childhood represents a crucial stage at which irreversible and devastating damage can occur. Most of our testing for toxicity has focused on cancer, but pediatricians suggest that we should also look at subtle but important effects on children's development. Studies have shown that, even at low levels, materials such as lead, mercury, pesticides, and polychlorinated biphenyls can have deleterious effects.

In a study of Mexican farm children, for instance, researchers found significantly reduced cognitive abilities among those exposed to high levels of crop pesticides (fig. 8.12). These children also exhibited abnormal social behavior, including high levels of anger, aggression, autism, and attention deficit disorder. Noting that these conditions have become much more widespread in America in recent years, the U.S. Environmental Protection Agency has begun to focus more attention on pollution and chil-

FIGURE 8.11 *Brown pelicans almost completely disappeared from the eastern United States in the 1960s as a result of excess pesticide use.*

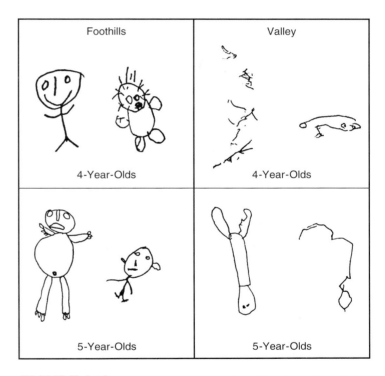

FIGURE 8.12 *Representative examples of drawings of people by 4- and 5-year-old Yaqui children relatively unexposed to pesticides (foothills) and those heavily exposed (valley).*
Source: From E. A. Guillette et al., "An Anthropological Approach to the Evaluation of Preschool Children Exposed to Pesticides in Mexico" in *Environmental Health Perspective,* 106(6):347–353, 1998. U.S. Dept. of Health and Human Services.

dren's health. In 2000, a popular insecticide called Dursban (chlorpyrifos) contained in many household roach and ant killers was banned because of the risks it poses to children. A debate is currently raging over the plasticizers used in making some soft plastics (see the related story "Environmental Estrogens" at www.mhhe.com/apps.)

TABLE 8.5 The "Dirty Dozen" Persistent Organic Pollutants

COMPOUND(S)	USES
Aldrin	Insecticide used on corn, potatoes, cotton, and for termite control
Clordane	Insecticide used on vegetables, small grains, maize, sugarcane, fruits, nuts, and cotton
Dieldrin	Insecticide used on cotton, corn, potatoes, and for termite control
DDT	Insecticide, now used primarily for disease vector control
Endrin	Insecticide used on field crops such as cotton and grains and as a rodenticide
Hexachlorobenzene (HCB)	Fungicide used for seed treatment and as an industrial chemical
Heptachlor	Insecticide used against soil insects, termites, and grasshoppers
Mirex	Insecticide used to combat fire ants, termites, mealybugs, and as a fire retardant
Toxaphene	A mixture of chemicals used as an insecticide on cotton as well as tick and mite control in livestock and fish eradication
Polychlorinated biphyenyls (PCBs)	Industrial chemicals used as insulators in electrical transformers, solvents, paper coatings, and plasticizers
Dioxins	A large family of by-products of chlorinated chemical production and incineration
Furans	A large group of by-products of chlorinated chemical production and incineration

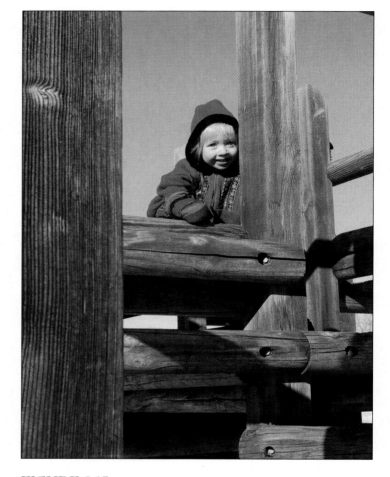

FIGURE 8.13 *CCA (chromium, copper, arsenate) pressure-treated wood, commonly used in playground equipment, may not be safe for children.*

Persistence

Some chemical compounds are very unstable and degrade rapidly under most environmental conditions so that their concentrations decline quickly after release. Most modern herbicides and pesticides, for instance, quickly lose their toxicity. Other substances are more persistent and last for long times. Some of the most useful chemicals, such as chlorofluorocarbons, PVC plastics, chlorinated hydrocarbon pesticides, and asbestos, are valuable because they are resistant to degradation. This stability, however, also causes problems because these materials persist in the environment and have unexpected effects far from the sites of their original use.

In 2000, negotiators from 121 nations agreed to ban or phase out the 12 most notorious persistent organic pollutants, or POPs, for most uses (table 8.5). Interestingly, most of these chemicals are pesticides and most were discussed 40 years ago in Rachael Carson's *Silent Spring*.

Just when we pass regulations to ban one set of POPs, new ones are discovered. Among those of greatest current concern are:

- Polybrominated diphenyl ethers (PBDE), widely used as flame retardant in textiles, foam in upholstery, and plastic in appliances and computers, were first reported accumulating in women's breast milk in Sweden in the 1990s. It was subsequently found in humans and other species everywhere from Canada to Israel. Nearly 150 million metric tons (330 million lbs) of PBDEs are used every year worldwide. The toxicity and environmental persistence of PBDE is much like that of PCBs, to which it is closely related chemically. The dust at ground zero in New York City after September 11 was heavily laden with PBDE. The European Union has already banned this compound.

- Chromated copper arsenate (CCA) is the chemical used in "pressure-treated lumber" to make it insect and rot resistant. All three metals are poisonous, which is why it makes the lumber last in outdoor applications such as decks, sidewalks, and playground equipment. Florida, alone, uses 30,000 tons of CCA per year. Unfortunately, the toxic metals leach out of the wood over time, making it especially bad in products or locations—such as backyards or playgrounds—frequented by children (fig. 8.13). Arsenic, the most toxic of the metals, causes loss of muscle function as well as lung, bladder, and

INVESTIGATING Our Environment

When Are Disease Clusters Significant?

When actor Michael J. Fox revealed that he suffers from Parkinson's disease, he opened some doors that may help us understand the cause of this mysterious illness. Parkinson's occurs when neurons degenerate in the substantia nigra, a part of the midbrain that produces dopamine, a chemical messenger essential for normal muscle control. The classic symptoms of the disease are tremors, muscle rigidity and weakness, and a shuffling gait. These symptoms usually progress slowly, but eventually the escalating paralysis is fatal.

Mr. Fox's illness is unusual in two ways. First, he began showing symptoms at age 30, much earlier than most Parkinson's victims. Moreover, three others who worked with him at a television studio in Vancouver, British Columbia, in the late 1970s also have contracted the disease. Typically, Parkinson's afflicts one in 300 people. To have four victims out of the 125 people on the studio staff makes the incidence in this group more than eight times above average. Furthermore, one of the victims also discovered she had the illness at age 38. Was there something in the air or water—a virus or toxic chemical perhaps—that triggered this disease cluster, or are the similarities just a coincidence?

There are good reasons to suspect environmental involvement in Parkinson's. Other clusters of the disease have been identified in heavily industrialized areas or where heavy use of herbicides and pesticides has occurred. In the 1980s, 65 people in California died after taking a synthetic heroin contaminated with methyl phenyl tetrahydropyridine (MPTP). This drug caused symptoms indistinguishable from Parkinson's (they could be temporarily reversed by administering l-dopa, which is converted to dopamine in the brain), and quickly led to paralysis, coma, and death. MPTP is very similar in its chemical structure and characteristics to paraquat, a widely used herbicide.

Evidence also suggests infectious agents may be involved in Parkinson's. After World War I, people who had contracted sleeping sickness later developed what came to be known as post-encephalitic parkinsonism. Parkinson's is also more prevalent among people such as teachers, doctors, nurses, and others who have relatively high exposure to infections. In one notable study, several members in each of six Canadian families were shown to have developed Parkinson's within a few years of each other even though they were very different ages. In this study, the common thread seemed to be the health food diet eaten by those who contracted the disease.

Suppose you notice what seems like an abnormally large number of people in your family, neighborhood, or workplace have a particular illness. How can you know whether this is a significant connection or simply a statistical fluke? It may take a great deal of work to establish the meaning of a disease cluster, but this is an important way to study how illnesses start and spread through populations. The Centers for Disease Control and Prevention (CDC) in Atlanta searches for patterns in disease distribution to identify epidemics and to zero in on their causes. Rapid analysis of statistical and geographical data is particularly important in recognizing and responding to terrorist attacks as described at the beginning of this chapter.

Perhaps, as in Mr. Fox's case, celebrity status will help educate the public about Parkinson's. He might also illuminate environmental factors that will help us understand the causes of this puzzling disease. Finally, sharing knowledge of his condition with former co-workers calls our attention to the importance of cluster studies in understanding environmental health issues.

---

skin cancer. It also is an endocrine hormone disrupter. Nine countries, including Sweden, Germany, Vietnam and Indonesia, have banned CCA-treated wood. Nonarsenic alternatives exist, but the $4 billion/year U.S. wood-treating industry has resisted change. Nevertheless, two major U.S. manufacturers agreed, in 2002, to begin a phase-out of CCA.

- Perflurooctane sulfonate (PFOS), the major ingredient in a popular stain and spill repellent made by the 3M Company, was found to be extremely widely distributed in the global environment, from remote Pacific Islands to the Arctic. No serious adverse effect of this compound is known, but the fact that it is long lasting and is accumulating in the environment creates worries that it may become a problem. The manufacturer agreed in 2000 to a phase-out.

- Atrazine, one of the most widely used herbicides in the Midwestern corn belt, was shown to cause abnormal development and sexual dysfunction in frogs. Farmers who use Atrazine heavily have high rates of certain lymphomas, but these farmers generally are exposed to other toxic compounds as well. A study of farm families in northwestern Minnesota found considerably higher rates of birth defects than urban families. "The data is associative," said Dr. Vince Gary, who directed the research. No definitive cause and effect can be shown with any particular pesticide, but the sensitivity of other species raises concerns (see Investigating Our Environment, above).

Chemical Interactions

Some materials produce *antagonistic* reactions—that is, they interfere with the effects or stimulate the breakdown of other chemicals. For instance, vitamins E and A can reduce the response to some carcinogens. Other materials are *additive* when they occur together in exposures. Rats exposed to both lead and arsenic show twice the toxicity of only one of these elements. Perhaps the greatest concern is synergistic effects. **Synergism** is an interaction in which one substance multiplies the effects of another. For example, occupational asbestos exposure increases lung cancer rates 20-fold. Smoking increases lung cancer rates by the same amount. Asbestos workers who also smoke, however, have a 400-fold increase in cancer rates. How many other toxic chemicals are we exposed to that are below threshold limits individually but combine to give toxic results?

MECHANISMS FOR MINIMIZING TOXIC EFFECTS

A fundamental concept in toxicology is that every material can be poisonous under some conditions, but most chemicals have some safe level or threshold below which their effects are undetectable or insignificant. Each of us consumes lethal doses of many chemicals over the course of a lifetime. One hundred cups of strong coffee, for instance, contain a lethal dose of caffeine. Similarly, 100 aspirin tablets, or 10 kg (22 lbs) of spinach or rhubarb, or a liter of alcohol would be deadly if consumed all at once. Taken in small doses, however, most toxins can be broken down or excreted before they do much harm. Furthermore, the damage they cause can be repaired. Sometimes, however, mechanisms that protect us from one type of toxin or at one stage in the life cycle become deleterious with another substance or in another stage of development. Let's look at how these processes help to protect us from harmful substances, as well as how they can go awry.

Metabolic Degradation and Excretion

Most organisms have enzymes that process waste products and environmental poisons to reduce their toxicity. In mammals, most of these enzymes are located in the liver, the primary site of detoxification of both natural wastes and introduced poisons. Sometimes, however, these reactions work to our disadvantage. Compounds such as benzepyrene, for example, that are not toxic in their original form are processed by these same liver enzymes into cancer-causing carcinogens. Why would we have a system that makes a chemical more dangerous? Evolution and natural selection are expressed through reproductive success or failure. Defense mechanisms that protect us from toxins and hazards early in life are "selected for" by evolution. Factors or conditions that affect post-reproductive ages (like cancer or premature senility) usually don't affect reproductive success or exert "selective pressure."

We also reduce the effects of waste products and environmental toxins by eliminating them from our body through excretion. Volatile molecules, such as carbon dioxide, hydrogen cyanide, and ketones, are excreted via breathing. Some excess salts and other substances are excreted in sweat. Primarily, however, excretion is a function of the kidneys, which can eliminate significant amounts of soluble materials through urine formation. Toxin accumulation in the urine can damage this vital system, however, and the kidneys and bladder often are subjected to harmful levels of toxic compounds. In the same way, the stomach, intestine, and colon often suffer damage from materials concentrated in the digestive system and may be afflicted by diseases and tumors.

Repair Mechanisms

In the same way that individual cells have enzymes to repair damage to DNA and protein at the molecular level, tissues and organs that are exposed regularly to physical wear-and-tear or to toxic or hazardous materials often have mechanisms for damage repair.

Our skin and the epithelial linings of the gastrointestinal tract, blood vessels, lungs, and urogenital system have high cellular reproduction rates to replace injured cells. With each reproduction cycle, however, there is a chance that some cells will lose normal growth controls and run amok, creating a tumor. Thus, any agent, such as smoking or drinking, that irritates tissues is likely to be carcinogenic. And tissues with high cell-replacement rates are among the most likely to develop cancers.

MEASURING TOXICITY

In 1540, the Swiss scientist Paracelsus said, "The dose makes the poison," by which he meant that almost everything is toxic at some level. This remains the most basic principle of toxicology. Sodium chloride (table salt), for instance, is essential for human life in small doses. If you were forced to eat a kilogram of salt all at once, however, it would make you very sick. A similar amount injected into your bloodstream would be lethal. How a material is delivered—at what rate, through which route of entry, and in what medium—plays a vitally important role in determining toxicity.

This does not mean that all toxins are identical, however. Some are so poisonous that a single drop on your skin can kill you. Others require massive amounts injected directly into the blood to be lethal. Measuring and comparing the toxicity of various materials are difficult because species differ in sensitivity, and individuals within a species respond differently to a given exposure. In this section, we look at methods of toxicity testing and at how results are analyzed and reported.

Animal Testing

The most commonly used and widely accepted toxicity test is to expose a population of laboratory animals to measured doses of a specific substance under controlled conditions. This procedure is expensive, time consuming, and often painful and debilitating to

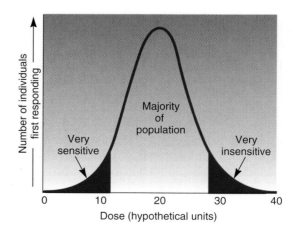

FIGURE 8.14 *Probable variations in sensitivity to a toxin within a population. Some members of a population may be very sensitive to a given toxin, while others are much less sensitive. The majority of the population falls somewhere between the two extremes.*

the animals being tested. It commonly takes hundreds—or even thousands—of animals, several years of hard work, and hundreds of thousands of dollars to thoroughly test the effects of a toxin at very low doses. More humane toxicity tests using computer simulations of model reactions, cell cultures, and other substitutes for whole living animals are being developed. However, conventional large-scale animal testing is the method in which scientists have the most confidence and on which most public policies about pollution and environmental or occupational health hazards are based.

In addition to humanitarian concerns, several other problems in laboratory animal testing trouble both toxicologists and policymakers. One problem is differences in toxin sensitivity among the members of a specific population. Figure 8.14 shows a typical dose/response curve for exposure to a hypothetical toxin. Some individuals are very sensitive to the toxin, while others are insensitive. Most, however, fall in a middle category, forming a bell-shaped curve. The question for regulators and politicians is whether we should set pollution levels that will protect everyone, including the most sensitive people, or only aim to protect the average person. It might cost billions of extra dollars to protect a very small number of individuals at the extreme end of the curve. Is that a good use of resources?

Dose/response curves are not always symmetrical, making it difficult to compare toxicity of unlike chemicals or different species of organisms. A convenient way to describe toxicity of a chemical is to determine the dose to which 50 percent of the test population is sensitive. In the case of a lethal dose (LD), this is called the **LD50** (fig. 8.15).

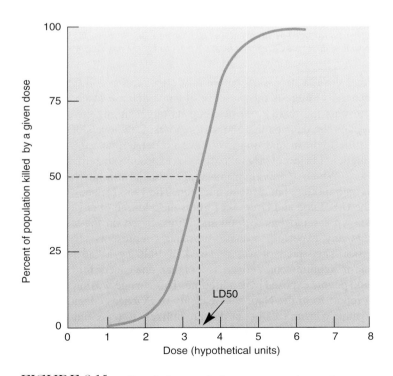

FIGURE 8.15 *Cumulative population response to increasing doses of a toxin. The LD50 is the dose that is lethal to half the population.*

The earliest studies of human toxicology came from experiments in which volunteers (usually students or prisoners) were given measured doses of suspected toxins. Today, it is considered neither ethical nor humane to deliberately expose individuals to danger, even if they volunteer. Toxicology is now done in either retrospective or prospective studies. In a **retrospective study,** you identify a group of people who have been exposed to some suspected risk factor and then compare their health to that of a control group who are as nearly identical as possible to the experimental group, except for exposure to that particular factor. Unfortunately, people often can't remember where they were or what they were doing many years ago. In a **prospective study,** you identify a study group and a control group, and then keep track of everything they do and how it affects their health. Then you watch and wait for years to see if a response appears in the study group but not in the controls. This kind of study is expensive because you may need a very large group to study a rare effect, and it is still difficult to distinguish between many simultaneous variables.

Suppose that you and your classmates have been chosen to be part of a prospective study of the health risks of a particular soft drink.

1. The researchers can't afford to keep records of everything that you do or are exposed to over the next 20 or 30 years. What do you think would be the most important factors and/or effects to monitor?

2. In a study group of a hundred students, how many would have to get sick to convince you that the soft drink was a risk factor? (Remember that there are many potential variables in the health of your study group.)

Unrelated species can react very differently to the same toxin, not only because body sizes vary but also because of differences in physiology and metabolism. Even closely related species can have very dissimilar reactions to a particular toxin. Hamsters, for instance, are nearly 5,000 times less sensitive to some dioxins than are guinea pigs. Of 226 chemicals found to be carcinogenic in either rats or mice, 95 caused cancer in one species but not the other. These variations make it difficult to estimate the risks for humans, since we don't consider it ethical to perform controlled experiments in which we deliberately expose people to toxins.

Toxicity Ratings

It is useful to group materials according to their relative toxicity. A moderate toxin takes about 1 g per kilogram of body weight (about 2 oz for an average human) to make a lethal dose. Very toxic materials take about one-tenth that amount, while extremely toxic substances take one-hundredth as much (only a few drops) to kill most people. Supertoxic chemicals are extremely potent; for some, a few micrograms (millionths of a gram—an amount invisible to the naked eye) make a lethal dose. These materials are not all synthetic. One of the most toxic chemicals known, for instance, is ricin, a protein found in castor bean seeds. It is so toxic that 0.3 billionths of a gram given intravenously will kill a mouse. If aspirin were this toxic for humans, a single tablet, divided evenly, could kill 1 million people.

Many carcinogens, mutagens, and teratogens are dangerous at levels far below their direct toxic effect because abnormal cell growth exerts a kind of biological amplification. A single cell, perhaps altered by a single molecular event, can multiply into millions of tumor cells or an entire organism. Just as there are different levels of direct toxicity, however, there are different degrees of carcinogenicity, mutagenicity, and teratogenicity. Methanesulfonic acid, for instance, is highly carcinogenic, while the sweetener saccharin is a possible carcinogen whose effects may be vanishingly small.

Acute Versus Chronic Doses and Effects

Most of the toxic effects that we have discussed so far have been **acute effects.** That is, they are caused by a single exposure to the toxin and result in an immediate health crisis of some sort. Often, if the individual experiencing an acute reaction survives this immediate crisis, the effects are reversible. **Chronic effects,** on the other hand, are long lasting, perhaps even permanent. A chronic effect can result from a single dose of a very toxic substance, or it can be the result of a continuous or repeated sublethal exposure.

We also describe long-lasting exposures as chronic, although their effects may or may not persist after the toxin is removed. It usually is difficult to assess the specific health risks of chronic exposures because other factors, such as aging or normal diseases, act simultaneously with the factor under study. It often requires very large populations of experimental animals to obtain statistically significant results for low-level chronic exposures. Toxicologists talk about "megarat" experiments in which it might take a million rats to determine the health risks of some supertoxic chemicals at very low doses. Such an experiment would be terribly expensive for even a single chemical, let alone for the thousands of chemicals and factors suspected of being dangerous.

An alternative to enormous studies involving millions of animals is to give massive amounts—usually the maximum tolerable dose—of a toxin being studied to a smaller number of individuals and then to extrapolate what the effects of lower doses might have been. This is a controversial approach because it is not clear that responses to toxins are linear or uniform across a wide range of doses.

Figure 8.16 shows three possible results from low doses of a toxin. Curve *a* shows a baseline level of response in the population, even at zero dose of the toxin. This suggests that some other factor in the environment also causes this response. Curve *b* shows a straight-line relationship from the highest doses to zero exposure. Many carcinogens and mutagens show this kind of response. Any exposure to such agents, no matter how small, carries some risks. Curve *c* shows a threshold for the response where some min-

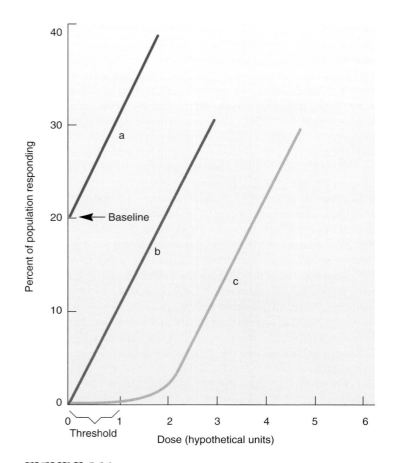

FIGURE 8.16 *Three possible dose-response curves at low doses.* (a) *Some individuals respond, even at zero dose, indicating that some other factor must be involved.* (b) *Response is linear down to the lowest possible dose.* (c) *Threshold must be passed before any response is seen.*

imal dose is necessary before any effect can be observed. This generally suggests the presence of some defense mechanism that prevents the toxin from reaching its target in an active form or repairs the damage that the toxin causes. Low levels of exposure to the toxin in question may have no deleterious effects, and it might not be necessary to try to keep exposures to zero.

Which, if any, environmental health hazards have thresholds is an important but difficult question. The 1958 Delaney Clause to the U.S. Food and Drug Act forbids the addition of any amount of known carcinogens to food and drugs, based on the assumption that any exposure to these substances represents unacceptable risks. This standard was replaced in 1996 by a "no reasonable harm" requirement, defined as less than one cancer for every million people exposed over a lifetime. This change was supported by a report from the National Academy of Sciences concluding that synthetic chemicals in our diet are unlikely to represent an appreciable cancer risk. We will discuss risk analysis in the next section.

Detection Limits

You may have seen or heard dire warnings about toxic materials detected in samples of air, water, or food. A typical headline

Principles of Environmental Science www.mhhe.com/environmentalscience

announced recently that 23 pesticides were found in 16 food samples. What does that mean? The implication seems to be that any amount of dangerous materials is unacceptable and that counting the numbers of compounds detected is a reliable way to establish danger. We have seen, however, that the dose makes the poison. It matters not only what is there, but how much, where it is located, how accessible it is, and who is exposed. At some level, the mere presence of a substance is insignificant.

Toxins and pollutants may seem to be more widespread now than in the past, and this is surely a valid perception for many substances. The daily reports we hear of new materials found in new places, however, are also due, in part, to our more sensitive measuring techniques. Twenty years ago, parts per million were generally the limits of detection for most chemicals. Anything below that amount was often reported as "zero" or "absent," rather than more accurately as "undetected." A decade ago, new machines and techniques were developed to measure parts per billion. Suddenly, chemicals were found where none had been suspected. Now we can detect parts per trillion or even parts per quadrillion in some cases. Increasingly sophisticated measuring capabilities may lead us to believe that toxic materials have become more prevalent. In fact, our environment may be no more dangerous; we are just better at finding trace amounts.

RISK ASSESSMENT AND ACCEPTANCE

Even if we know with some certainty how toxic a specific chemical is in laboratory tests, it still is difficult to determine **risk** (the probability of harm times the probability of exposure) if that chemical is released into the environment. As you already have seen, many factors complicate the movement and fate of chemicals both around us and within our bodies. Furthermore, public perception of relative dangers from environmental hazards can be skewed so that some risks seem much more important than others.

Assessing Risks

A number of factors influence how we perceive relative risks associated with different situations.

- People with social, political, or economic interests—including environmentalists—tend to downplay certain risks and emphasize others that suit their own agendas. We do this individually as well, building up the dangers of things that don't benefit us, while diminishing or ignoring the negative aspects of activities we enjoy or profit from.

- Most people have difficulty understanding and believing probabilities. We feel that there must be patterns and connections in events, even though statistical theory says otherwise. If the coin turned up heads last time, we feel certain that it will turn up tails next time. In the same way, it is difficult to understand the meaning of a 1-in-10,000 risk of being poisoned by a chemical.

- Our personal experiences often are misleading. When we have not personally experienced a bad outcome, we feel it is

more rare and unlikely to occur than it actually may be. Furthermore, the anxieties generated by life's gambles make us want to deny uncertainty and to misjudge many risks.

- We have an exaggerated view of our own abilities to control our fate. We generally consider ourselves above-average drivers, safer than most when using appliances or power tools, and less likely than others to suffer medical problems, such as heart attacks. People often feel they can avoid hazards because they are wiser or luckier than others.

- News media give us a biased perspective on the frequency of certain kinds of health hazards, overreporting some accidents or diseases, while downplaying or underreporting others. Sensational, gory, or especially frightful causes of death, such as murders, plane crashes, fires, or terrible accidents, receive a disproportionate amount of attention in the public media. Heart diseases, cancer, and stroke kill nearly 15 times as many people in the United States as do accidents and 75 times as many people as do homicides, but the emphasis placed by the media on accidents and homicides is nearly inversely proportional to their relative frequency, compared to either cardiovascular disease or cancer. This gives us an inaccurate picture of the real risks to which we are exposed.

- We tend to have an irrational fear or distrust of certain technologies or activities that leads us to overestimate their dangers. Nuclear power, for instance, is viewed as very risky, while coal-burning power plants seem to be familiar and relatively benign; in fact, coal mining, shipping, and combustion cause an estimated 10,000 deaths each year in the United States, compared to none known so far for nuclear power generation. An old, familiar technology seems safer and more acceptable than does a new, unknown one.

Accepting Risks

How much risk is acceptable? How much is it worth to minimize and avoid exposure to certain risks? Most people will tolerate a higher probability of occurrence of an event if the harm caused by that event is low. Conversely, harm of greater severity is acceptable only at low levels of frequency. A 1-in-10,000 chance of being killed might be of more concern to you than a 1-in-100 chance of being injured. For most people, a 1-in-100,000 chance of dying from some event or some factor is a threshold for changing what they do. That is, if the chance of death is less than 1 in 100,000, we are not likely to be worried enough to change our ways. If the risk is greater, we will probably do something about it. The Environmental Protection Agency generally assumes that a risk of 1 in 1 million is acceptable for most environmental hazards. Critics of this policy ask, acceptable to whom?

For activities that we enjoy or find profitable, we are often willing to accept far greater risks than this general threshold. Conversely, for risks that benefit someone else, we demand far higher protection. For instance, your chance of dying in a motor vehicle accident in any given year is about 1 in 5,000, but that doesn't deter many people from riding in automobiles. Your chances of dying from lung cancer if you smoke one pack of cigarettes per day is about 1 in 1,000. By comparison, the risk from drinking water with the EPA limit of trichloroethylene is about 2 in 1 billion. Strangely, many people demand water with zero levels of trichloroethylene, while continuing to smoke cigarettes.

More than 1 million Americans are diagnosed with skin cancer each year. Some of these cancers are lethal, and most are disfiguring, yet only one-third of teenagers routinely use sunscreen. Tanning beds more than double your chances of cancer, especially if you're young, but about 10 percent of all teenagers admit regularly using these devices.

Table 8.6 lists some activities estimated to increase your chances of dying in any given year by 1 in 1 million. These are statistical averages, of course, and there clearly are differences in where one lives or how one rides a bicycle that affect the danger level of these activities. Still, it is interesting how we readily accept some risks, while shunning others.

Our perception of relative risks is strongly affected by whether risks are known or unknown, whether we feel in control of the outcome, and how dreadful the results are. Risks that are unknown or unpredictable and results that are particularly gruesome or disgusting seem far worse than those that are familiar and socially acceptable.

Studies of public risk perception show that most people react more to emotion than to statistics. We go to great lengths to avoid some dangers, while gladly accepting others. Factors that are involuntary, unfamiliar, undetectable to those exposed, catastrophic, or that have delayed effects or are a threat to future generations are especially feared, while those that are voluntary, familiar, detectable, or immediate cause less anxiety. Even though the actual number of deaths from automobile accidents, smoking, or alcohol, for instance, are thousands of times greater than those from pesticides, nuclear energy, or genetic engineering, the latter preoccupy us far more than the former.

TABLE 8.6 Activities Estimated to Increase Your Chances of Dying in Any Given Year by 1 in 1 Million

ACTIVITY	RESULTING DEATH RISK
Smoking 1.4 cigarettes	Cancer, heart disease
Drinking 0.5 liter of wine	Cirrhosis of the liver
Spending 1 hour in a coal mine	Black lung disease
Living 2 days in New York or Boston	Air pollution
Traveling 6 minutes by canoe	Accident
Traveling 10 mi by bicycle	Accident
Traveling 150 mi by car	Accident
Flying 1,000 mi by jet	Accident
Flying 6,000 mi by jet	Cancer caused by cosmic radiation
Living 2 months in Denver	Cancer caused by cosmic radiation
Living 2 months in a stone or brick building	Cancer caused by natural radioactivity
Having one chest X ray	Cancer caused by radiation
Living 2 months with a cigarette smoker	Cancer, heart disease
Eating 40 tablespoons of peanut butter	Cancer from aflatoxin
Living 5 years at the site boundary of a typical nuclear power plant	Cancer caused by radiation from routine leaks
Living 50 years 5 mi from a nuclear power plant	Cancer caused by accidental radiation release
Eating 100 charcoal-broiled steaks	Cancer from benzopyrene

Source: From William Allman, "Staying Alive in the Twentieth Century," *Science 85,* 5(6): 31, October 1985. Reprinted by permission.

ESTABLISHING PUBLIC POLICY

Risk management combines principles of environmental health and toxicology with regulatory decisions based on socioeconomic, technical, and political considerations (fig. 8.17). The biggest problem in making regulatory decisions is that we are usually exposed to many sources of harm, often unknowingly. It is difficult to separate the effects of all these different hazards and to evaluate their risks accurately, especially when the exposures are near the threshold of measurement and response. In spite of often vague and contradictory data, public policymakers must make decisions.

The case of the sweetener saccharin is a good example of the complexities and uncertainties of risk assessment in public health. Studies in the 1970s at the University of Wisconsin and the Canadian Health Protection Branch suggested a link between saccharin and bladder cancer in male rats. Critics of these studies pointed out that humans would have to drink 800 cans of diet soda per day to get a saccharin dose equivalent to that given to the rats. Further-

Risk Assessment   **Risk Management**

FIGURE 8.17   *Risk assessment organizes and analyzes data to determine relative risk. Risk management sets priorities and evaluates relevant factors to make regulatory decisions.*
Source: Data from D. E. Patton, "USEPA's Framework for Ecological Risk Assessment" in *Human Ecological Risk Assessment,* Vol. 1, No. 4.

FIGURE 8.18   *"Do you want to stop reading those ingredients while we're trying to eat?"*
Reprinted with permission of the *Star-Tribune,* Minneapolis-St. Paul.

more, they argued this response may be unique to male rats. In 2000, the U.S. Department of Health concluded a large study that found no association between saccharin consumption and cancer in humans. The U.S. Congress then passed a health bill removing all warnings from saccharin-containing products. Still some groups, like the Center for Science in the Public Interest, consider this sweetener a potential carcinogen and warn us to avoid it if possible.

Experiments testing the toxicity of saccharin in rats merely give a range of probable toxicities in humans. The lower end of this range indicates that only one person in the United States would die from using saccharin every 1,000 years. That is clearly inconsequential. The higher estimate, however, indicates that 3,640 people would die each year from the same exposure. Is that too high a cost for the benefits of having saccharin available to people who must restrict sugar intake? How does the cancer risk compare to the dangers of obesity, cardiovascular disease, and other problems caused by eating too much sugar? What alternatives might there be to saccharin? A popular but more expensive alternative (aspartame, derived from the amino acid aspartic acid), which bears the trade name Nutrasweet, also is controversial because of uncertainties about its safety.

In setting standards for environmental toxins, we need to consider (1) combined effects of exposure to many different sources of damage, (2) different sensitivities of members of the population, and (3) effects of chronic as well as acute exposures. Some people argue that pollution levels should be set at the highest amount that does not cause measurable effects. Others demand that pollution be reduced to zero if possible, or as low as is technologically feasible. It may not be reasonable to demand that we

be protected from every potentially harmful contaminant in our environment, no matter how small the risk. As we have seen, our bodies have mechanisms that enable us to avoid or repair many kinds of damage so that most of us can withstand some minimal level of exposure without harm (fig. 8.18).

On the other hand, each challenge to our cells by toxic substances stresses our bodies. Although each individual stress may not be life threatening, the cumulative effects of all the environmental stresses, both natural and human-caused, to which we are exposed may seriously shorten or restrict our lives. Furthermore, some individuals in any population are more susceptible to those stresses than others. Should we set pollution standards so that no one is adversely affected, even the most sensitive individuals, or should the acceptable level of risk be based on the average member of the population?

Finally, policy decisions about hazardous and toxic materials also need to be based on information about how such materials affect the plants, animals, and other organisms that define and maintain our environment. In some cases, pollution can harm or destroy whole ecosystems, with devastating effects on the life-supporting cycles on which we depend. In other cases, only the most sensitive species are threatened. Table 8.7 shows the Environmental Protection Agency's assessment of relative risks to human welfare. This ranking reflects a concern that our exclusive focus on reducing pollution to protect human health has neglected risks to natural ecological systems. While the case-by-case approach in which we evaluate the health risks of individual chemicals has been beneficial, we have often missed broader ecological problems that may be of greater ultimate importance.

TABLE 8.7   Relative Risks to Human Welfare

RELATIVELY HIGH-RISK PROBLEMS	RELATIVELY MEDIUM-RISK PROBLEMS	RELATIVELY LOW-RISK PROBLEMS
Habitat alteration and destruction	Herbicides/pesticides	Oil spills
Species extinction and loss of biological diversity	Toxics and pollutants in surface waters	Groundwater pollution
Stratospheric ozone depletion	Acid deposition	Radionuclides
Global climate change	Airborne toxics	Thermal pollution

Source: Data from Environmental Protection Agency.

SUMMARY

Health is a state of physical, mental, and social well-being, not merely the absence of disease or infirmity. Nearly every human disease probably has some connection to environmental factors. For most people in the world, the greatest health threat in the environment is still, as it always has been, from pathogenic organisms. Although diseases of affluence and old age now are the leading causes of death worldwide, bacteria, viruses, and other infectious agents still kill millions of people each year, and cause immense suffering and economic losses. Highly lethal emergent diseases, such as Ebola and AIDS, along with new drug-resistant forms of old diseases, are an increasing worry everywhere in the world. Some of these extremely virulent new pathogens could be used as bioweapons, and might represent the greatest risks for terrorist attacks.

Hazardous and toxic materials are serious health threats nearly everywhere. Allergens, mutagens, and carcinogens represent some of the chronic effects of persistent organic pollutants. The most notorious "dirty dozen" POPs have been banned, but new ones, including PBDE, CCA, and PFOs have recently been widely discovered.

The distribution and fate of materials in the environment depend on their physical characteristics and the processes that transport, alter, destroy, or immobilize them. Uptake of toxins into organisms can result in accumulation in tissues and transfer from one organism to another.

Estimating the potential health risk from exposure to specific environmental factors is difficult because information on the precise dose, length and method of exposure, and possible interactions between the chemical in question and other potential toxins to which the population may have been exposed is often lacking. In addition, individuals have different levels of sensitivity and response to a particular toxin and are further affected by general health condition, age, and sex.

Estimates of health risks for large, diverse populations exposed to very low doses of extremely toxic materials are inexact because of biological variation, experimental error, and the necessity of extrapolating from results with small numbers of laboratory animals. In the end, we are left with unanswered questions. Which are the most dangerous environmental factors that we face? How can we evaluate the hazards of all the natural and synthetic chemicals that now exist? What risks are acceptable? We have not yet solved these problems or answered all the questions raised in this chapter, but these issues need to be discussed and considered seriously.

QUESTIONS FOR REVIEW

1. What are some of the most serious infectious diseases in the world? How are they transmitted?

2. What is the difference between toxic and hazardous? Give some examples of materials in each category.

3. Define carcinogenic, mutagenic, teratogenic, neurotoxic, and endocrine hormone disrupters.

4. How do stress, diet, and lifestyle affect environmental health? What diseases are most clearly related to these factors?

5. How do the physical and chemical characteristics of materials affect their movement, persistence, distribution, and fate in the environment?

6. Why is pressure-treated lumber considered dangerous to children?

7. How do organisms reduce or avoid the damaging effects of environmental hazards?

8. Define LD50. Why is it more accurate than simply reporting toxic dose?

9. What is the difference between acute and chronic toxicity?

10. What are the relative risks of smoking, driving a car, and drinking water with the maximum permissible levels of trichloroethylene? Are these relatively equal risks?

THINKING SCIENTIFICALLY

1. What are the greatest risks that you personally, as well as your country collectively, face from bioterrorism? How would you compare the risks from terrorism to the other health risks described in this chapter?

2. Former Surgeon General C. Everett Koop estimates that enforcing current Environmental Protection Agency rules restricting use of solvents such as trichloroethylene will save about one life per year in the United States and will cost about $200 million. By contrast, inoculating all poor children against the most common infectious diseases would cost about $200 per year for every life saved. Why do we insist on the former action but not the latter?

3. Analyze the claim that we are exposed to thousands of times more natural carcinogens in our diet than industrial ones. Is this a good reason to ignore pollution?

4. Some people seem to have a poison paranoia about synthetic chemicals. Why do we tend to assume that natural chemicals are benign and industrial chemicals are evil?

5. Are good health and a clean environment a basic human right, or merely something for which we should strive?

6. What are the underlying premises in the discussion of assessing risk in this chapter? Could conflicting conclusions be drawn from the facts presented here? What is your perception of risk from your environment?

7. Table 8.6 equates activities such as smoking 1.4 cigarettes, having one chest X ray, and riding 10 mi on a bicycle. How was this relationship derived? Do you agree with it? Do some items on this list require further clarification?

8. Who were the stakeholders in the saccharin controversy, and what were their interests or biases? Was the U.S. Congress justified in refusing to ban saccharin? Should diet soft-drink cans have warning labels similar to those on cigarettes?

9. Should pollution levels be set to protect the average person in the population or the most sensitive? Why not have zero exposure to all hazards?

10. What level of risk is acceptable to you? Are there some things for which you would accept more risk than others?

KEY TERMS

acute effects 188
allergens 179
antigens 179
bioaccumulation 182
biomagnification 182
cancer 180
carcinogens 180
chronic effects 188
Disability-Adjusted Life Year
   (DALY) 176
disease 175
emergent disease 177
endocrine hormone
   disrupters 181

fetal alcohol syndrome 180
hazardous 179
health 175
LD50 187
morbidity 175
mutagens 180
neurotoxins 180
prospective study 187
retrospective study 187
risk 189
sick building syndrome 180
synergism 185
toxins 179
teratogens 180

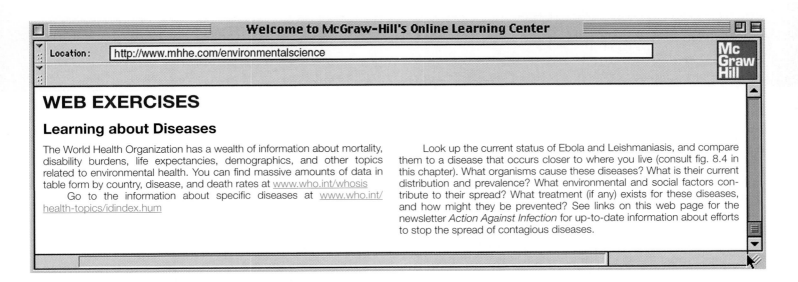

Welcome to McGraw-Hill's Online Learning Center

Location: http://www.mhhe.com/environmentalscience

WEB EXERCISES

Learning about Diseases

The World Health Organization has a wealth of information about mortality, disability burdens, life expectancies, demographics, and other topics related to environmental health. You can find massive amounts of data in table form by country, disease, and death rates at www.who.int/whosis
   Go to the information about specific diseases at www.who.int/health-topics/idindex.hum

   Look up the current status of Ebola and Leishmaniasis, and compare them to a disease that occurs closer to where you live (consult fig. 8.4 in this chapter). What organisms cause these diseases? What is their current distribution and prevalence? What environmental and social factors contribute to their spread? What treatment (if any) exists for these diseases, and how might they be prevented? See links on this web page for the newsletter *Action Against Infection* for up-to-date information about efforts to stop the spread of contagious diseases.

9 Air: Climate and Pollution

If you think education is expensive, try ignorance.
–Derek Bok

A composite satellite image shows three views of Hurricane Andrew as it moved across Florida and along the Gulf Coast.

OBJECTIVES

After studying this chapter, you should be able to

- summarize the structure and composition of the atmosphere.
- explain how jet streams, seasonal winds, and global climate patterns determine local weather.
- understand El Niño cycles and the causes of natural climate change.
- debate the hypothesis that human actions are altering the global climate.
- describe the major categories and sources of air pollution.
- evaluate the dangers of stratospheric ozone depletion.
- understand how air pollution damages human health, vegetation, and building materials.
- judge how air quality around the world has improved or degraded, and suggest what we might do about problem areas.

Welcome to McGraw-Hill's Online Learning Center

Location: http://www.mhhe.com/environmentalscience

LEARNING ONLINE

Visit our webpage at www.mhhe.com/environmentalscience for data sources, further readings, additional case studies, current environmental news, and regional examples within the Online Learning Center to help you understand the material in this chapter. You'll also find active links to information pertaining to this chapter including:

National Oceanic and Atmospheric Administration (NOAA)
Tropical Prediction Center/National Hurricane Center
National Weather Service Page
UM Weather: Weather Sites
Air Pollution
United States Environmental Protection Agency
National Vehicle and Fuel Emissions Laboratory
United States Environmental Protection Agency: Ozone Depletion

What's Happening to Our Weather?

What do skinny polar bears and drowned seal pups, Peruvian cholera epidemics, melting of Mt. Kilimanjaro's famous snows, Chinese drinking-water shortages, unusually severe Bangladeshi floods, coastal erosion in Louisiana, and the disappearance of Edith's Checkerspot butterfly in Southern California have in common? All these phenomena are thought to be signs of human-caused global climate change, which may well be the most critical issue in environmental science today.

The problem is that we are adding greenhouse gases—pollutants that trap in the earth's heat—to the atmosphere at a faster rate than at any time over the past several thousand years. Essentially, we are conducting a giant experiment to see what will happen if we alter atmospheric chemistry. So far, the results don't look very good. All around us, evidence suggests we are modifying our climate on both a local and global scale. The twentieth century was the warmest in the last 1,000 years. The 1990s were the warmest decade and 1998 was the single warmest year of the past millennium.

Polar regions are changing even faster than the rest of the globe. According to Environment Canada, parts of the Arctic coast have warmed as much as 7.5°C (13.5°F) since 1970. Sea ice forms later in the fall and melts earlier in the spring, giving polar bears a shorter seal-hunting season. Hudson's Bay polar bears now weigh as much as 100 kg (220 lbs) less than in the 1960s. In 2002, early melting of ice floes in Canada's Gulf of St. Lawrence appeared to have drowned nearly all of the 200,000 to 300,000 harp seal pups normally born there.

Glaciers are disappearing on every continent. Mt. Kilimanjaro has lost 85 percent of its ice cap since 1915. By 2015, all permanent ice on the mountaintop is expected to be gone. Alpine glaciers feed rivers, such as the Indus, Ganges, Yangtze, Yellow, and Mekong, that supply drinking water and irrigation to more than a billion people in South and East Asia, where water is already becoming a source of conflict. Ocean warming is causing severe storms and heavy monsoon rains that result in flooding in Bangladesh as well as erosion in Louisiana, where rising sea levels have inundated low-lying costal marshes. And Edith's Checkerspot is only one of many species of mammals, birds, amphibians, fish, insects, and plants that are reported to have moved their territory or migration patterns, or to have disappeared altogether as a result of changing climate.

Higher temperatures apparently are allowing disease-causing bacteria, viruses, and fungi to move into new areas where they may harm species as diverse as lions, snails, butterflies, and humans. Climate changes are thought to have contributed to an epidemic of avian malaria that wiped out thousands of birds in Hawaii, the spread of distemper in African lions, and the bleaching of coral reefs around the globe. Unusually warm water off the coast of South America is thought to be responsible for reappearance of cholera in humans during the 1990s after nearly a century of absence.

What do all these changes mean? Taken individually, it's hard to say; they may just be random events in a notoriously variable system. Viewed altogether, however, it seems increasingly evident that we are changing our climate with results we don't yet fully comprehend. Learning something about our atmosphere and how it produces our weather and climate is essential if we are to understand how changing climate might affect us, and what we might do to counter those effects. In this chapter, we'll look at how greenhouse gases and other air pollutants affect human and natural systems, and we'll examine some of the international politics of this crucial topic.

THE ATMOSPHERE AND CLIMATE

We live at the bottom of a virtual ocean of air that extends upward about 500 km (300 mi). In the lowest 10 to 12 km, a layer known as the troposphere, the air moves ceaselessly, flowing and swirling, and continually redistributing heat and moisture from one part of the globe to another. The composition and behavior of the troposphere and other layers control our **weather** (daily temperature and moisture conditions in a place) and our **climate** (long-term weather patterns).

The earth's earliest atmosphere probably consisted mainly of hydrogen and helium. Over billions of years, most of that hydrogen and helium diffused into space. Volcanic emissions added carbon, nitrogen, oxygen, sulfur, and other elements to the atmosphere. Virtually all of the molecular oxygen (O_2) we breathe was probably produced by photosynthesis in blue-green bacteria, algae, and green plants.

Clean, dry air is mostly nitrogen and oxygen (table 9.1). Water vapor concentrations vary from near zero to 4 percent, depending on air temperature and available moisture. Minute particles and liquid droplets—collectively called **aerosols**—also are suspended in the air. Atmospheric aerosols play important roles in the earth's energy budget and in producing rain.

The atmosphere has four distinct zones of contrasting temperature, due to differences in absorption of solar energy (fig. 9.1). The layer of air immediately adjacent to the earth's surface is called the **troposphere** (*tropein* means to turn or change, in Greek). Within the troposphere, air circulates in great vertical and horizontal **convection currents,** constantly redistributing heat and moisture around the globe. The troposphere ranges in depth from about 18 km (11 mi) over the equator to about 8 km (5 mi) over the poles, where air is cold and dense. Because gravity holds most air molecules close to the earth's surface, the troposphere is much more dense than the other layers: it contains about 75 percent of the total

TABLE 9.1 Present Composition of the Lower Atmosphere*

GAS	SYMBOL OR FORMULA	PERCENT BY VOLUME
Nitrogen	N_2	78.08
Oxygen	O_2	20.94
Argon	Ar	0.934
Carbon dioxide	CO_2	0.035
Neon	Ne	0.00182
Helium	He	0.00052
Methane	CH_4	0.00015
Krypton	Kr	0.00011
Hydrogen	H_2	0.00005
Nitrous oxide	N_2O	0.00005
Xenon	Xe	0.000009

*Average composition of dry, clean air.

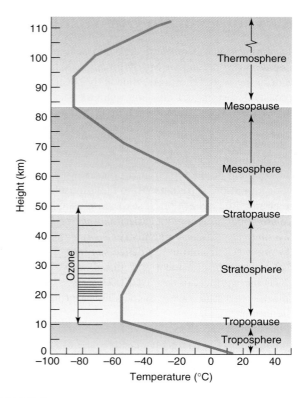

FIGURE 9.1 *Temperatures change drastically in the four layers of the atmosphere. Bars in the ozone graph represent relative concentrations of stratospheric ozone with altitude.*
Source: Courtesy of Dr. William Culver, St. Petersburg Junior College.

mass of the atmosphere. Air temperature drops rapidly with increasing altitude in this layer, reaching about –60°C (–76°F) at the top of the troposphere. A sudden reversal of this temperature gradient creates a sharp boundary called the tropopause, which limits mixing between the troposphere and upper zones.

The **stratosphere** extends from the tropopause up to about 50 km (31 mi). It is vastly more dilute than the troposphere, but it has similar composition—except that it has almost no water vapor and nearly 1,000 times more **ozone** (O_3). This ozone absorbs some wavelengths of ultraviolet solar radiation, known as UV-B (290–330 nm, see fig. 2.9). This absorbed energy makes the atmosphere warmer toward the top of the stratosphere. Since UV radiation damages living tissues, this UV absorption in the stratosphere also protects life on the surface. Recently discovered depletion of stratospheric ozone, especially over Antarctica, is allowing increased amounts of UV radiation to reach the earth's surface. If observed trends continue, this radiation could cause higher rates of skin cancer, genetic mutations, crop failures, and disruption of important biological communities, as you will see later in this chapter.

Unlike the troposphere, the stratosphere is relatively calm. There is so little mixing in the stratosphere that volcanic ash or human-caused contaminants can remain in suspension there for many years.

Above the stratosphere, the temperature diminishes again, creating the mesosphere, or middle layer. The thermosphere (heated layer) begins at about 50 km. This is a region of highly ionized (electrically charged) gases, heated by a steady flow of high-energy solar and cosmic radiation. In the lower part of the thermosphere, intense pulses of high-energy radiation cause electrically charged particles (ions) to glow. This phenomenon is what we know as the *aurora borealis* and *aurora australis,* or northern and southern lights.

No sharp boundary marks the end of the atmosphere. Pressure and density decrease with distance from the earth until they become indistinguishable from the near vacuum of interstellar space.

Energy and the "Greenhouse Effect"

The sun supplies the earth with an enormous amount of energy, but that energy is not evenly distributed over the globe. Incoming solar radiation (insolation) is much stronger near the equator than at high latitudes. Of the solar energy that reaches the outer atmosphere, about one-quarter is reflected by clouds and atmospheric gases, and another quarter is absorbed by carbon dioxide, water vapor, ozone, methane, and a few other gases (fig. 9.2). This energy absorption warms the atmosphere slightly. About half of incoming solar radiation (insolation) reaches the earth's surface. Most of this energy is in the form of light or infrared (heat) energy (see fig. 2.10). Some of this energy is reflected by bright surfaces, such as snow, ice, and sand. The rest is absorbed by the earth's surface and by water. Surfaces that *reflect* energy have a high **albedo** (reflectivity). Most of these surfaces appear bright to us because they reflect light as well as other forms of radiative energy. Surfaces that *absorb* energy have a low albedo and generally appear dark. Black soil, asphalt pavement, and dark green vegetation, for example, have low albedos (table 9.2).

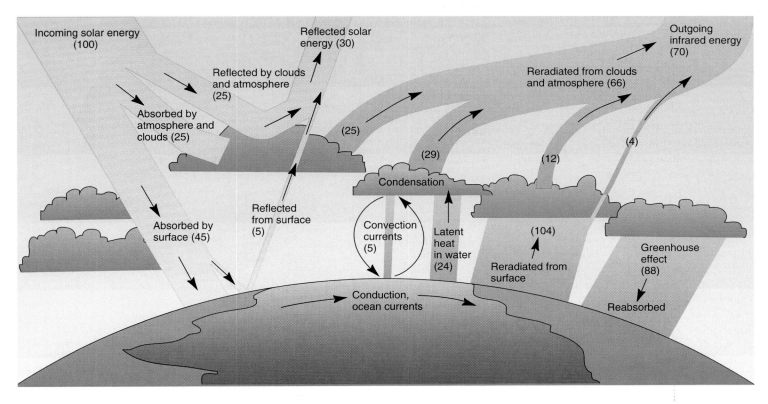

FIGURE 9.2 *Energy balance between incoming and outgoing radiation. The atmosphere absorbs or reflects about half of the solar energy reaching the earth. Most of the energy reemitted from the earth's surface is long-wave, infrared energy. Most of this infrared energy is absorbed by aerosols and gases in the atmosphere and is reradiated toward the planet, keeping the surface much warmer than it would otherwise be. This is known as the greenhouse effect. The numbers shown are arbitrary units. Note that for 100 units of incoming solar energy, 100 units are reradiated to space, but more than 100 units are radiated from the earth's surface because of the greenhouse effect.*

TABLE 9.2	Albedo (Reflectivity) of Surfaces
SURFACE	**ALBEDO (%)**
Fresh snow	80–85
Dense clouds	70–90
Water (low sun)	50–80
Sand	20–30
Water (sun overhead)	5
Forest	5–10
Black soil	3
Earth/atmosphere average	30

APPLICATION: **How Much Heat Is Released in a 1-in. Rainstorm in Your Neighborhood?**

Most Americans measure rainfall in inches, but centimeters are easier to calculate. A 1-in. rainfall is about 2.54 cm of rain. Suppose your neighborhood is a 1 km × 1 km square. If rain releases 580 cal/cm³, how much heat is released by this rainstorm?

Answer: $2.54 \text{ cm of rain} \times \frac{100,000 \text{ cm}}{\text{km}} \times \frac{100,000 \text{ cm}}{\text{km}} \times \frac{580 \text{ cal}}{\text{cm}^3 \text{ of rain}} = 17.2 \times 10^{12} \text{ cal.}$

Absorbed energy heats the absorbing surface (such as an asphalt parking lot in summer), evaporates water, or provides the energy for photosynthesis in plants. Following the second law of thermodynamics, absorbed energy is gradually reemitted as lower-quality heat energy. A brick building, for example, absorbs energy in the form of light and reemits that energy in the form of heat.

The change in energy quality is very important because the atmosphere selectively absorbs longer wavelengths. Most solar energy comes in the form of intense, high-energy light or near-infrared wavelengths (fig. 2.10). This short-wavelength energy passes relatively easily through the atmosphere to reach the earth's surface. Energy re-released from the earth's warmed surface ("terrestrial energy") is lower-intensity, longer-wavelength energy in the far-infrared part of the spectrum. Atmospheric gases, especially carbon dioxide and water vapor, absorb much of this long-wavelength energy, re-releasing it in the lower atmosphere and letting it leak out to space only slowly. This terrestrial energy provides most of the heat in the lower atmosphere. If the atmosphere were as transparent to infrared radiation as it is to visible light, the earth's average surface temperature would be about –18°C (0°F)—33°C (59°F) colder than it is now.

This phenomenon is called the **"greenhouse effect"** because the atmosphere, loosely comparable to the glass of a greenhouse, transmits sunlight while trapping heat inside. The greenhouse effect is a natural atmospheric process that is necessary for life as we know it. However, too much greenhouse effect, caused by burning of fossil fuels and deforestation, may cause harmful environmental change.

Convection and Atmospheric Pressure

Much of the incoming solar energy is used to evaporate water. Every gram of evaporating water absorbs 580 calories of energy as it transforms from liquid to gas. Globally, water vapor contains a huge amount of stored energy, known as **latent heat**. When water vapor condenses, returning from a gas to a liquid form, the 580 calories of heat energy are released. Imagine the sun shining on the Gulf of Mexico in the winter. Warm sunshine and plenty of water allow continuous evaporation that converts an immense amount of solar (light) energy into latent heat stored in evaporated water. Now imagine a wind blowing the humid air north from the Gulf toward Canada. The air cools as it moves north (especially if it encounters cold air moving south). Cooling causes the water vapor to condense. Rain (or snow) falls as a consequence. Note that it is not only water that has moved from the Gulf to the Midwest: 580 calories of heat have also moved with every gram of moisture. The heat and water have moved from a place with strong incoming solar energy to a place with much less solar energy and much less water. The redistribution of heat and water around the globe are essential to life on earth.

Uneven heating, with warm air close to the equator and colder air at high latitudes, also produces pressure differences that cause wind, rain, storms, and everything else we know as weather. As the sun warms the earth's surface, the air nearest the surface warms and expands, becoming less dense than the air above it. The warm air must then rise above the denser air. Vertical convection currents result, which circulate air from warm latitudes to cool latitudes and vice versa. These convection currents can be as small and as localized as a narrow column of hot air rising over a sun-heated rock, or they can cover huge regions of the earth. At the largest scale, the convection cells are described by a simplified model known as Hadley cells, which redistribute heat globally (fig. 9.3).

Where air rises in convection currents, air pressure at the surface is low. Where air is sinking, or subsiding, air pressure is high. On a weather map these high and low pressure centers, or rising and sinking currents of air, move across continents. In most of North America, they generally move from west to east. Rising air tends to cool with altitude, releasing latent heat that causes further rising. Very warm and humid air can rise very vigorously, especially if it is rising over a mass of very cold air. Storms associated with low pressure and rising air are known as cyclonic storms. These include some of the most violent storms we know: hurricanes, tornadoes, and intense rain and hail are forms of cyclonic storms (fig. 9.4).

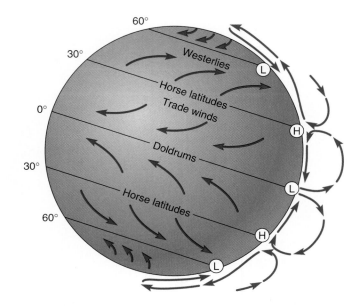

FIGURE 9.3 *General circulation patterns redistribute heat and moisture around the globe. The approximate locations of vertical convection currents, generally referred to as Hadley cells, are noted on the right side. Low pressure belts (L) occur at latitudes where air rises. High pressure belts (H) occur where air sinks. Dominant winds, such as trade winds and westerlies, also occur in latitudinal bands.*

Pressure differences are an important cause of wind. There is always someplace with sinking (high pressure) air and someplace with low pressure (rising) air. Air moves from high-pressure centers toward low-pressure areas, and we call this movement wind.

Why Does It Rain?

To understand why it rains, remember two things: water condenses as air cools, and air cools as it rises. Any time air is rising, clouds, rain, or snow might form. Cooling occurs because of changes in pressure with altitude: air cools as it rises (as pressure decreases); air warms as it sinks (as pressure increases). Air rises in convection currents where solar heating is intense, such as over the equator. Moving masses of air also rise over each other and cool. Air also rises when it encounters mountains. If the air is moist (if it has recently come from over an ocean or an evaporating forest region, for example), condensation and rainfall are likely as the air is lifted (fig. 9.5). Regions with intense solar heating, frequent colliding air masses, or mountains tend to receive a great deal of precipitation.

Where air is sinking, on the other hand, it tends to warm because of increasing pressure. As it warms, available moisture evaporates. Rainfall occurs relatively rarely in areas of high pressure. High pressure and clear, dry conditions occur where convection currents are sinking. High pressure also occurs where air sinks after flowing over mountains. Figure 9.3 shows sinking, dry air at about 30° north and south latitudes. If you look at a world map, you will see a band of deserts at approximately these latitudes.

Principles of Environmental Science

FIGURE 9.4 *Tornadoes are local cyclonic storms caused by rapid mixing of cold, dry air and warm, wet air. Wind speeds in the swirling funnel can reach 320 km/hr (200 mph).*

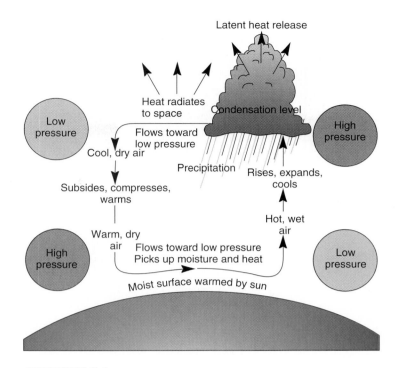

FIGURE 9.5 *Convection currents and latent energy cause atmospheric circulation and redistribution of heat and water around the globe.*

Another ingredient is usually necessary to initiate condensation of water vapor: condensation nuclei. Tiny particles of smoke, dust, sea salts, spores, and volcanic ash all act as condensation nuclei. These particles form a surface on which water molecules can begin to coalesce. Without them even supercooled vapor can remain in gaseous form. Even apparently clear air can contain large numbers of these particles, which are generally too small to be seen by the naked eye.

The Coriolis Effect and Jet Streams

Large-scale winds tend not to move in a straight line across the earth's surface. In the Northern Hemisphere, they generally bend clockwise (right), and in the Southern Hemisphere, they bend counterclockwise (left). This curving pattern results from the fact that the earth rotates in an eastward direction as the winds move above the surface. The apparent curvature of the winds is known as the **Coriolis effect.** On a global scale, this effect produces steady, reliable wind patterns, such as the trade winds and the midlatitude Westerlies (see fig. 9.3). Ocean currents similarly curve clockwise in the Northern Hemisphere and counterclockwise in the south (see appendix 4, p. 376). On a regional scale, the Coriolis effect produces cyclonic winds, or wind movements controlled by the earth's spin. Cyclonic winds spiral clockwise out of an area of high pressure in the Northern Hemisphere and counterclockwise into a low-pressure zone. If

you look at a weather map in the newspaper, can you find this counterclockwise spiral pattern?

Why does this curving or spiraling motion occur? Imagine you were looking down on the North Pole of the rotating earth. Now imagine that the earth was a merry-go-round in a playground, with the North Pole at its center and the equator around the edge. As it spins counter-clockwise (eastward), the spinning edge moves very fast (a full rotation, 39,800 km, every 24 hours for the real earth, or more than 1,600 km/hour!). Near the center, though, there is very little eastward velocity. If you threw a ball from the edge toward the center, it would be traveling faster (edge speed) than the middle. It would appear, to someone standing on the merry-go-round, to curve toward the right. If you threw the ball from the center toward the edge, it would start out with no eastward velocity, but the surface below it would spin eastward, making the ball end up, to a person on the merry-go-round, west of its starting point. Winds move above the earth's surface much as the ball does. If you were looking down at the South Pole, you would see the earth spinning clockwise, and winds—or thrown balls—would appear to bend left. Incidentally, this effect does not apply to drains in your house. Their movement is far too small to be affected by the spinning of the earth.

At the top of the troposphere are **jet streams,** hurricane-force winds that circle the earth. These powerful winds follow an undulating path approximately where the vertical convection currents known as the Hadley and Ferrell cells meet. The approximate path of one jet stream over the Northern Hemisphere is shown in fig. 9.6. Although we can't perceive jet streams on the ground, they are

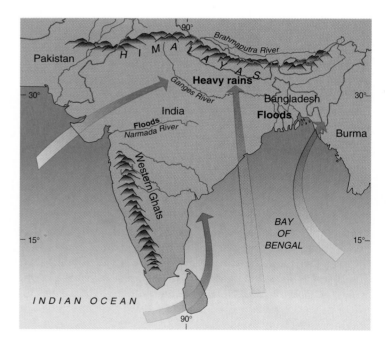

FIGURE 9.6   *The polar-front jet stream circles the pole, isolating arctic air from lower-latitude air. Also called the arctic circumpolar vortex, this large, circulating mass of cold air sends "fingers," or lobes, across North America and Eurasia, spreading storms in their path. If the vortex becomes stalled, weather patterns stabilize, causing droughts in some areas and excess rain elsewhere.*

important to us because they greatly affect weather patterns. Sometimes jet streams dip down near the top of the world's highest mountains, exposing mountain climbers to violent, brutally cold winds.

Ocean Currents

Warm and cold ocean currents strongly influence climate conditions on land. Surface ocean currents result from wind pushing on the ocean surface. As surface water moves, deep water wells up to replace it, creating deeper ocean currents. Differences in water density—depending on the temperature and saltiness of the water—also drive ocean circulation. Huge cycling currents called gyres carry water north and south, redistributing heat from low latitudes to high latitudes (see appendix 4, p. 376, global climate map). The Alaska current, flowing from Alaska southward to California, keeps San Francisco cool and foggy during the summer. The Gulf Stream, one of the best known currents, carries warm Caribbean water north past Canada's maritime provinces to northern Europe. This current is immense, some 800 times the volume of the Amazon, the world's largest river. The heat transported from the Gulf keeps Europe much warmer than it should be for its latitude. As the warm Gulf Stream passes Scandinavia and swirls around Iceland, the water cools and evaporates, becomes dense and salty, and plunges downward, creating a strong, deep, southward current.

Ocean circulation patterns were long thought to be unchanging, but now oceanographers believe that currents can shift abruptly. For example, melting ice at the North Pole could create a flood of cold, fresh water, disrupting the Gulf Stream and causing rapid, drastic cooling in Europe. This shift may have happened in geologic history, and many climatologists fear it could occur again if global warming melts arctic ice again.

Seasonal Winds and Monsoons

While figure 9.3 shows regular global wind patterns, large parts of the world, especially the tropics, receive seasonal winds and rainy seasons that are essential for sustaining both ecosystems and human life. Sometimes these seasonal rains are extreme, causing disastrous flooding. (See the related story "Floods in Mozambique" at www.mhhe.com/apps.)

Sometimes the rains fail, causing crop failures and famine. The most regular seasonal winds and rains are known as **monsoons**. In India and Bangladesh, monsoon rains come when seasonal winds blow hot, humid air from the Indian Ocean (fig. 9.7). Strong convection currents lift this air, causing heavy rain across the subcontinent. When the rising air reaches the Himalayas, it rises even further, creating some of the heaviest rainfall in the world. During the 5-month rainy season of 1970, a weather station in the foothills of the Himalayas recorded 25 m (82 ft) of rain!

Tropical and subtropical regions around the world have seasonal rainy and dry seasons (see the discussion of tropical biomes,

FIGURE 9.7   *Summer monsoon air flows over the Indian subcontinent. Warming air rises over the plains of central India in the summer, creating a low-pressure cell that draws in warm, wet, oceanic air. As this moist air rises over the Western Ghats or the Himalayas, it cools, and heavy rains result. These monsoon rains flood the great rivers, bringing water for agriculture, but also causing much suffering.*

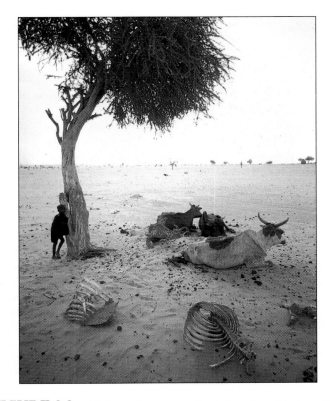

FIGURE 9.8 *Failure of monsoon rains brings drought, starvation, and death to both livestock and people in the Sahel desert margin of Africa. Although drought is a fact of life in Africa, many governments fail to plan for it, and human suffering is much worse than it needs to be.*

chapter 5). The main reason for this variable climate is that the region of most intense solar heating and evaporation shifts through the year. Remember that the earth's axis of rotation is at an angle. Sometimes the sun hits the surface just south of the equator, and sometimes it hits just north of the equator. Wherever the sun shines most directly, evaporation and convection currents—and rainfall and thunderstorms—are very strong. As the earth orbits the sun, different regions fall below the sun, and different regions receive seasonal rain and winds. These seasonal rains support seasonal tropical forests, and they fill some of the world's greatest rivers, including the Ganges and the Amazon. As the year shifts from summer to winter, solar heating weakens, the rainy season ends, and little rain may fall for months.

During the 1970s and 1980s, rains failed repeatedly in parts of the Sahel in northern Africa. Although rains have often been irregular in this region, increasing populations and civil disorder made these droughts result in widespread starvation and death (fig. 9.8).

CLIMATE CHANGE

Climates have always changed, often in cycles and on many different time scales (fig. 9.9). A sudden cooling 65 million years ago is thought to have ended the age of dinosaurs, along with 75 percent of the species existing at the time. There may have been a

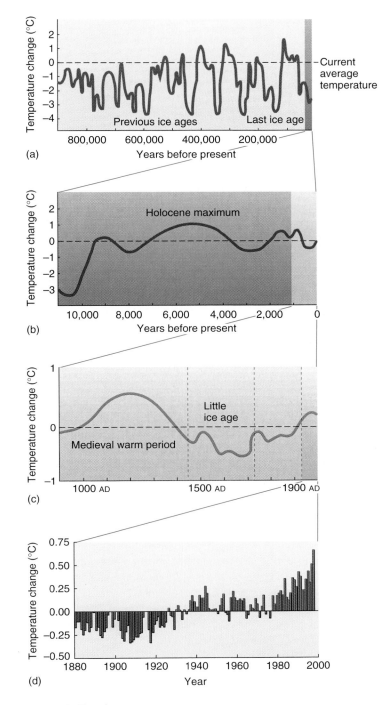

FIGURE 9.9 *Global temperature variations over four time periods: (a) the past 900,000 years, (b) the past 10,000 years, (c) the past 1,000 years, and (d) the most recent 120 years. The horizontal dashed lines at 0 on the vertical axis represent the mean temperature for the twentieth century. Note that both horizontal and vertical scales change in each figure.*

Source: (d) Data from National Climate Data Center/NESDIS/NOAA, 1998.

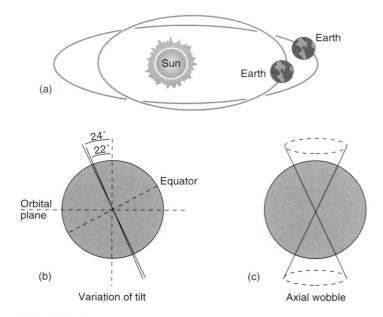

FIGURE 9.10 *Milankovitch cycles that may affect long-term climate conditions: (a) changes in the occupancy of the earth's orbit, (b) shifting tilt of the axis, and (c) wobble of the earth.*

dozen such mass extinctions (see table 5.3). On a shorter time scale, several ice ages, each lasting hundreds of thousands of years, have come and gone in the past 2 million years. Even shorter climate shifts occur, such as the "little ice age," which began in the 1300s and caused crop failures throughout Europe. On a scale of decades there seem to be approximately 20-year drought cycles in North America.

What causes these climate shifts at these different time scales? One explanation is periodic changes in sunlight intensity resulting from **Milankovitch cycles.** These cycles, named for the Serbian scientist who first described them in the 1920s, include (1) the shape of the earth's orbit around the sun elongates and shortens in a 100,000-year cycle, (2) the axis of rotation changes its angle of tilt in a 40,000-year cycle, and (3) over a 26,000-year period, the axis wobbles like an out-of-balance spinning top (fig. 9.10). Worldwide expansion of glaciers every 100,000 years or so could result from the orbital shifts. Decadal droughts, meanwhile, may be explained by 11-year sunspot cycles, shifts in stormy activity on the sun's surface.

Climate does not necessarily change gradually. Atmospheric gas bubbles from ice cores drilled in the Greenland ice cap suggest that during the last major interglacial period 135,000 to 115,000 years ago, temperatures flipped suddenly over a period of years or decades, rather than centuries. Meteor impacts and sudden shifts in ocean currents, which redistribute heat around the globe, could have also caused rapid climate changes. Some climatologists suspect that rapid releases of methane trapped beneath the ocean floor, a "big burp," may have produced a sudden warming that caused the mass extinction 183 million years ago. This theory raises questions about the stability of methane deposits currently stored beneath the seafloor (see chapter 2).

Two important shorter-term climate changes, El Niño and anthropogenic global warming, are discussed in the next section.

El Niño/Southern Oscillation

El Niño, La Niña, and the *Southern Oscillation* are all terms referring to a major ocean-current/climate connection that affects weather throughout the Pacific—and possibly throughout the world. The core of this system is a huge pool of warm surface water in the Pacific Ocean that sloshes slowly back and forth between Indonesia and South America like water in a giant bathtub. Most years, this pool is held in the western Pacific by steady equatorial trade winds that push ocean surface currents westward (fig. 9.11). From Southeast Asia to Australia, this concentration of warm, equatorial water provides latent heat (water vapor) that drives strong upward convection (low pressure) in the atmosphere. Heavy rain results, supporting dense tropical forests. On the other side of the Pacific, the westward-moving surface waters are replaced by cold water welling up along the South American coast. Cold, nutrient-rich waters support dense schools of anchovies and other fish. While the trade winds blow westward on the ocean's surface, returning winds high in the troposphere flow back from Indonesia to Chile and to Mexico and southern California. There the returning air sinks, creating dry, desert conditions.

Every three to five years, for reasons that we don't fully understand, the Indonesian low-pressure system collapses, and the mass of warm surface water surges back east across the Pacific. One theory is that the high cirrus clouds block enough sunshine to cool the ocean surface in Asia. Convection would then weaken, and

FIGURE 9.11 *The El Niño/La Niña/Southern Oscillation cycle. Every three to five years, surface trade winds that normally push warm water westward toward Indonesia weaken and allow this pool of water to flow eastward toward South America.*

INVESTIGATING Our Environment

Satellite Images

How do we measure widespread phenomena like El Niños, climate change, and weather? Some information can be gathered from ground-based observatories, but these facilities only tell us about a scattering of isolated points. To detect patterns on a global scale, a more distant view is useful, and that is why so much environmental science now involves remote sensing.

Remote sensing is a general term for data gathered from satellites, or sometimes by sensors on airplanes. Just as a camera is a sensor that captures light energy reflected from objects, sensors on satellites capture energy wavelengths reflected from the earth's surface, from clouds, or from the atmosphere. Some satellites detect infrared wavelengths, which indicate warm areas in the oceans,

on ice caps, or on land. Others might detect selected wavelengths of visible light, such as green or blue, or microwaves, which provide images of cloud cover and precipitation (see fig. 2.10). Our current understanding of the Antarctic ozone hole (see fig. 9.21) comes from satellites that detect near ultraviolet wavelengths, which are absorbed by ozone in the atmosphere.

In May 2002, NASA launched *Aqua,* the newest of its Earth Observing System (EOS) satellites. This one is designed to monitor climate change by detecting wavelengths reflected from water vapor—the principal heat-trapping gas—as well as water and ice on the earth's surface. Soil moisture, vegetation moisture, surface water, atmospheric ice crystals, glaciers, and clouds can all be monitored. All these aspects of moisture on the earth help provide information

on climate conditions and climate variation over time. Understanding moisture and temperature variations are also key to predicting ecosystem changes as climate conditions fluctuate.

There are currently more than 7,000 satellites orbiting the earth, but only a minority of these are designed for gathering environmental information. Many reflect radio waves or microwaves for TV and telephone signals; others are spy satellites collecting highly detailed images of strategic areas. The closest satellites, including *Aqua,* are 200–600 miles above the earth's surface. TV satellites are more than 22,000 miles up! When you look up at the sky at night, you might see some of the closest satellites, gathering data to give us the "big picture" of environmental change as they trace a path across the sky.

---

trade winds—and ocean currents—would reverse, flowing eastward instead of westward. Another theory is that eastward-flowing deep currents called baroclinic waves periodically interfere with coastal upwelling, warming the sea surface off South America and eliminating the temperature gradient across the Pacific.

Fishermen in Peru were the first to notice irregular cycles of rising ocean temperatures because the fish disappeared when the water warmed. They named this event El Niño (Spanish for the Christ child) because they often occur around Christmas time. The counterpart to El Niño, when the eastern tropical Pacific cools, has come to be called La Niña (little girl). Together, these cycles are called the El Niño Southern Oscillation (ENSO).

ENSO cycles have far-reaching effects. During an El Niño year, the northern jet stream—which is normally over Canada—splits and is drawn south over the United States. This pulls moist air from the Pacific and Gulf of Mexico inland, bringing intense storms and heavy rains from California across the midwestern states. The intervening La Niña years bring hot, dry weather to these same areas. Oregon, Washington, and British Columbia, on the other hand, tend to have warm, sunny weather in El Niño years rather than their usual rain. Droughts in Australia and Indonesia during El Niño episodes cause disastrous crop failures and forest fires, including one in Borneo in 1983 that burned 3.3 million ha (8 million acres).

Some climatologists believe that ENSO events are becoming stronger or more frequent because of global climate change. There are signs that warm ocean-surface temperatures are spreading, which could contribute to El Niño strength or frequency. On the other hand, increased cloud cover over warmer oceans could raise global albedo, and strong convection currents generated by these storms could pump heat into the stratosphere. This might have an overall cooling effect and act as a safety valve for global warming.

Human-Caused Global Climate Change

As the opening vignette in this chapter shows, many scientists regard anthropogenic (human-caused) global climate change to be the most important environmental issue of our times. The possibility that humans might alter world climate is not a new idea. In 1895, Svante Arrhenius, who subsequently received a Nobel Prize for his work in chemistry, predicted that CO_2 released by coal burning could cause global warming. Most of Arrhenius' contemporaries dismissed his calculations as impossible, but his work seems remarkably prescient now.

The first evidence that human activities are increasing atmospheric CO_2 came from an observatory on top of the Mauna Loa volcano in Hawaii. The observatory was established in 1957 as part of an International Geophysical Year, and was intended to provide data on air chemistry in a remote, pristine environment. Surprisingly, measurements showed CO_2 levels increasing about 0.5 percent per year, rising from 315 ppm in 1958 to 372 ppm in 2002 (fig. 9.12). This increase isn't a straight line, however. Because a majority of the world's land and vegetation are in the Northern Hemisphere, northern seasons dominate the signal. Every May, CO_2 levels drop slightly as plant growth on northern continents use CO_2 in photosynthesis. During the northern winter, levels rise again as respiration releases CO_2.

In 1988 the United Nations Environment Program and World Meteorological Organization formed the **Intergovernmental Panel on Climate Change (IPCC).** This panel brings together scientists from many nations and a wide variety of fields to assess the current state of knowledge about climate change. The IPCC's First Assessment Report played an important role in adoption of the UN Framework Convention on Climate Change at the Rio

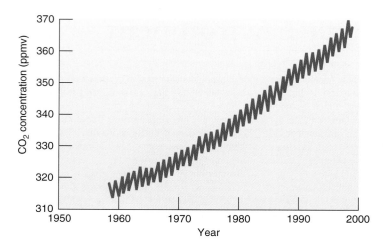

FIGURE 9.12 *Carbon dioxide concentrations at Mauna Loa Observatory in Hawaii have increased dramatically since 1958. Annual fluctuations reflect differences in photosynthesis and respiration between winter and summer.*

Source: Data from Dave Keeling and Tim Whorf, Scripps Institute of Oceanography.

Earth Summit in 1992. In February 2001, the IPCC released its third climate report, which said "with a high degree of confidence" that "recent changes in the world's climate have had discernible impacts on physical and biological systems." For this report, more than 700 scientists representing 100 countries reviewed results from some 3,000 scientific studies, which showed changes in 420 different physical and biological systems.

Noting that the earth's surface temperature has risen by about 0.6°C (1.1°F) over the past century, with most of that warming occurring in the past two decades, the IPCC concluded that human activities must be at least partially responsible. Separating the impact of human activity from natural climate variation is extremely difficult. Nonetheless, the IPCC claims, the observed global warming is unlikely to be the result of natural variability alone. "We have altered the chemical composition of the atmosphere through the buildup of greenhouse gases—primarily carbon dioxide, methane, and nitrous oxide," the report declared. "The heat-trapping property of these gases is undisputed although uncertainties exist about exactly how earth's climate responds to them."

Climate Skeptics

Not everyone agrees with the IPCC conclusions or its method of reaching them. Dr. Frederick Seitz, president emeritus of Rockefeller University and former president of the U.S. National Academy of Sciences, for example, charged that bureaucrats and politicians altered the report in ways that didn't represent the consensus agreed upon by scientists. Others claim we don't know enough yet to make predictions about what may be happening to our global climate. While much of the dissenting views come from industry lobbying groups, some prominent scientists also hold that dire predictions about climate change may be overblown (see What Do You Think? p. 205). John Cristy of the University of

Alabama, for example, says that the upper trophosphere isn't warming as fast as the earth's surface, perhaps showing that our current ideas about global climate are incorrect.

Richard Lindzen of MIT contends that current computer models don't adequately account for cloud effects. Higher evaporation rates, he believes, could produce more clouds that reflect sunlight and balance the greenhouse effect. And Vincent Gray, a climate consultant from New Zealand, contends that CO_2 concentrations measured at Mauna Loa have increased at a remarkably constant rate over the past 42 years, despite the fact that worldwide fossil fuel consumption has increased almost 50 percent during that time. Something is missing from our models, he claims.

Sources of Greenhouse Gases

Since preindustrial times atmospheric concentrations of CO_2, CH_4 and N_2O have climbed by over 31 percent, 151 percent, and 17 percent, respectively. Carbon dioxide is by far the most important cause of anthropogenic climate change (table 9.3). Burning fossil fuels, making cement, burning forests and grasslands, and other human activities release nearly 30 billion tons of CO_2 every year, on average, containing some 8 billion tons of carbon. About 3 billion tons of this excess carbon is taken up by terrestrial ecosystems, and around 2 billion tons are absorbed by the oceans, leaving an annual atmospheric increase of some 3 billion tons per year. If current trends continue, CO_2 concentrations could reach about 500 ppm (approaching twice the preindustrial level of 280 ppm) by the end of the twenty-first century.

Although rarer than CO_2, methane absorbs 20 to 30 times as much infrared and is accumulating in the atmosphere about twice as fast as CO_2. Methane is released by ruminant animals, wet-rice paddies, coal mines, landfills, and pipeline leaks. Chloroflurocarbons (CFCs) also are powerful infrared absorbers. CFC releases have declined since many of their uses were banned, but the CFCs already in the atmosphere will persist for many years. Nitrous oxide is produced by burning organic material and by soil denitrification. As table 9.3 shows, CFCs and N_2O together are thought to account for only about 17 percent of human-caused global warming.

Aerosols have a tendency to counteract the effects of greenhouse gases, on local and sometimes global scales. Sulfate aerosols and soot produced by some of the same activities that

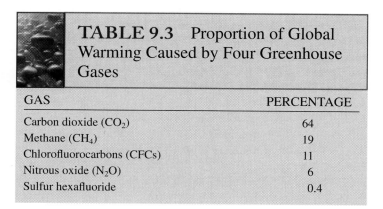

TABLE 9.3 Proportion of Global Warming Caused by Four Greenhouse Gases

GAS	PERCENTAGE
Carbon dioxide (CO_2)	64
Methane (CH_4)	19
Chlorofluorocarbons (CFCs)	11
Nitrous oxide (N_2O)	6
Sulfur hexafluoride	0.4

Science and Uncertainty in Climate Change

Why is there so much disagreement about whether human activities are causing global climate change? A big part of the problem is that weather and climate are complex and highly variable. Making predictions about what we might expect over the long term is difficult even in normal conditions. Many scientists agree, for example, that mean global temperatures will probably warm $1.5°-6°C$ ($2.7°-11°F$) over the next century. They can't say, however, whether particular regions will be warmer or cooler; and for many regions they are even unable to state whether a wetter or a drier climate is more likely. Virtually all published estimates of how the climate could change are the results of computer models of the atmosphere known as "general circulation models." These complicated mathematical models are able to simulate many features of the climate, but they are still not accurate enough to provide reliable forecasts of specific climatic conditions. In fact, different models often yield contradictory results. Given the unreliability of these models, researchers trying to understand the future impacts of climate change generally analyze different scenarios from several different climate models. The hope is that, by using a wide variety of different climate models, one's analysis can include the entire range of scientific uncertainty. Climatologists warn us that projections of climate change in specific areas are not forecasts but are reasonable examples of what might happen.

The U.S. EPA lists several reasons for caution in interpreting climate models:

1. The cooling effects of sulfate aerosols (from volcanoes and burning fossil fuels) could make a big difference in local conditions. Introducing aerosol effects into some models, for instance, changes precipitation predictions from a decline of 15 percent to an increase of 18 percent.

2. Most models focus on a doubling of CO_2, but we may very well be forced to cope with three or four times the current CO_2 concentrations over the longer term.

3. Nature may produce major surprises. If ocean currents are disrupted, for example, some parts of the world may actually need to deal with a substantially cooler climate rather than warmer.

4. Climate may not change in a "linear" fashion. This assumed gradualness underlies the assumption that society will have time to adjust. Some evidence suggests that climate has changed in a rather abrupt and chaotic fashion in the past.

5. Climate model simulations of today's regional climates are often inaccurate. This could be due simply to a lack of computational power to allow for local variations, or it could indicate a serious flaw in the models.

In the face of scientific uncertainty, interpretations of the costs and benefits of actions will vary. If you think major climate change is highly probable, then inaction is perilous. If you think that changes, if any, are likely to be small and gradual, then changing current policies may be unacceptable. For example, some economists (using models themselves) project that strict emission controls would cost the U.S. economy 10,000 jobs and $30 billion per year. It would be cheaper, they argue, to adapt to climate change than to try to prevent it. Ecological economists, meanwhile, argue that the costs of action need not be so high. They point out that we could cut emissions by 30 percent simply by improving efficiency and using newer technology. Further, leading the development of efficient technology will create jobs. These calculations also involve uncertainty, of course; economic models differ just as widely in their predictions as climate models do.

It's important to think about who bears the costs and who enjoys the benefits of any particular policy. Mandating energy conservation would be good for those selling renewable energy, but painful for those whose livelihoods depend on fossil fuels. And assuming that it will be cheaper to adapt to climate changes in the future rather than change our habits now, means shifting the costs to future generations. Is that fair?

If you were in Congress and had to vote on whether or not to mandate greenhouse gas reductions, how would you use scenarios from climate models? What additional information would you need to make a decision in this issue? How would you weigh the risks of inaction compared to the immediate pain of making changes in our society and technology?

---

release CO_2 can shade urban air, for example, and reduce air temperatures by as much or more than greenhouse gases increase them. The 1991 eruption of Mt. Pinatubo in the Philippines ejected enough ash and sulfate particles to cool the global climate about $1°C$, but this cooling lasted for only about a year.

The United States, with less than 5 percent of the world's population, produces 28 percent of all anthropogenic CO_2. China, with 1.3 billion people, is second in total CO_2 emissions, but fourteenth in per capita production (fig. 9.13). Japan and Western Europe, which by most measures have standards of living at least equal to the United States, produce only half as much CO_2 per person.

Current Evidence of Climate Change

Increasingly, evidence from around the world suggests that global climate is already changing as a result of human actions. Precipitation has increased by about 1 percent over the world's continents in the last century. High latitudes are tending to see more rainfall, while precipitation has actually declined in some tropical areas (fig. 9.14). Warmer temperatures apparently are causing more water evaporation. We expect this to produce more severe storms. Hurricanes, typhoons, tornadoes, severe rainstorms, and flooding seem to be more frequent and more damaging, although part of the increased damage results from more people living in dangerous

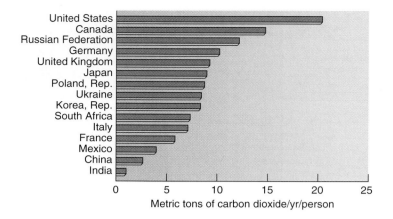

FIGURE 9.13 *Annual per capita CO$_2$ emissions for 15 countries with highest industrial output. Although China is fourteenth in per capita emissions, it is second to the United States in total emissions due to its large population.*

Source: Data from *World Resources 1998–99*, World Resources Institute.

locations along coasts and in floodplains (see the opening photograph in this chapter).

Satellite images and surface temperature data show that growing seasons are now as much as three weeks longer in a band across northern Eurasia than they were 30 years ago (see fig 1.15). Arctic sea ice is 40 percent thinner now, and the edge of the Arctic sea ice pack averages about 500 km (312 mi) further north than it was a century ago. Arctic wildlife, including polar bears, walruses, beluga whales, caribou, and musk oxen, are declining or changing their migration and feeding patterns as a result of changing climatic conditions. In the south, the West Antarctic Ice Sheet has

shrunk 25 percent over the past 25 years, including several giant icebergs—the largest about the size of Delaware—that have calved off in recent years. In 2002, a huge section of the Larsen B ice shelf along the coast of the Antarctic Peninsula suddenly collapsed and broke into innumerable small bergs. Although the ice shelves were known to be breaking up, the disintegration of an area larger than Rhode Island in just a few weeks was unprecedented. Here too, wildlife is suffering from loss of habitat and diminishing food supplies. Adele penguins, for instance, have declined precipitously across the Antarctic as the ice on which they rest and lay their eggs is disappearing.

Alpine glaciers everywhere are retreating rapidly (fig. 9.15). In 1972, Venezuela had six glaciers; now it has only two. All the glaciers in Montana's Glacier National Park will be gone by 2070, if current trends continue. Over the next 50 years, at least half of all alpine glaciers in the world could disappear. Without this resource, agriculture, industry, power generation, and drinking-water supplies will suffer. About 1.7 billion people now live in areas where water supplies are tight. This number could increase to 5.4 billion by 2025. Sea level has risen worldwide approximately 15–20 cm (6–8 in) in the past century. About one-quarter of this increase is due to melting alpine glaciers; roughly half is caused by thermal expansion of ocean water.

Many wild plant and animal species are being forced out of their current ranges as the climate warms. In Europe and North America, for example, 57 butterfly species have either died out at the southern end of their range, or extended their northern limits, or both. Given enough time and a route for migration, these species might adapt to new conditions, but we now are forcing them to move at least ten times the rate many achieved at the end of the last ice age (fig. 9.16). The disappearance of amphibians such as the beautiful golden toads from the cloud forests of Costa

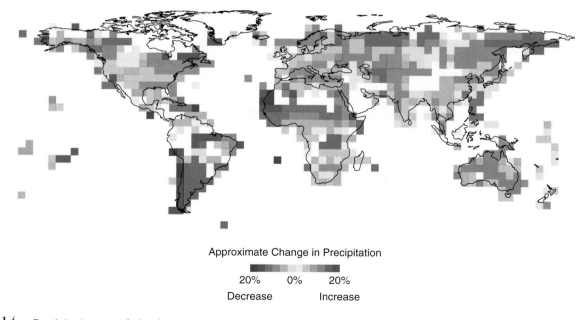

FIGURE 9.14 *Precipitation records for the twentieth century show more precipitation at high latitudes and less in equatorial zones. Green indicates more rain; brown indicates less.*

Principles of Environmental Science

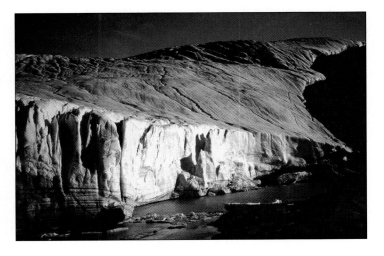

FIGURE 9.15 *Alpine glaciers are retreating rapidly nearly everywhere in the world. For the billions of people whose drinking water comes from glacial-fed rivers, this could be a serious threat.*

FIGURE 9.16 *Change in suitable range for sugar maple trees, according to the climate-change model.*
Source: Data from Margaret Davis and Catherine Zabriskie in *Global Warming and Biological Diversity,* ed. by Peters and Lovejoy, 1991, Yale University Press.

Rica or western toads from Oregon's Cascade Range are thought to have been caused at least in part by changing weather patterns.

Around the world, coral reefs are bleaching because of higher water temperatures. Rivaling tropical rainforests in their biological diversity and net productivity, coral reefs already are threatened by human activities such as dynamite fishing and limestone excavation. If sea temperatures continue to rise, most coral reefs will be wiped out in the next 50 years. Similarly, the Cape Floral Kingdom of South Africa, one of the most unique terrestrial ecosystems in the world, could disappear due to changing weather patterns.

Winners and Losers

Local climate changes could well have severe effects on human societies, agriculture, and natural ecosystems. In many cases, both organisms and human infrastructure will not be able to move or adapt quickly enough to accommodate the rates at which climate seems to be changing. Tropical areas will probably not get much warmer than they are now, but middle and high latitudes could experience unaccustomed heat.

Does anyone win in these scenarios? Residents of northern Canada, Siberia, and Alaska probably would enjoy warmer temperatures, longer growing seasons, and longer ice-free shipping seasons from their ports. Stronger monsoons bringing more moisture to parts of Central Africa and South Asia could increase crop yields. But soils in some areas may not be suitable for crops no matter how agreeable the climate. Carbon dioxide is a fertilizer for plants, and higher CO_2 levels bring faster growth with less water for many plant species. Sherwood Idso of the U.S. Department of Agriculture says that increased CO_2 concentrations could trigger lush plant growth that would make crops abundant.

Enough water is stored in glacial ice caps in Greenland and Antarctica to raise sea levels around 100 meters (300 feet) if they all melt. About one-third of the world's population now live in areas that would be flooded if that happens. Even the 75 cm (30 in) sea-level rise expected by 2050 will flood much of south Florida, Bangladesh, Pakistan, and many other low-lying coastal areas. Most of the world's largest urban areas are on coastlines. Wealthy cities such as New York or London can probably afford to build dikes to keep out rising seas, but poorer cities such as Jakarta, Calcutta, or Manila might simply be abandoned as residents flee to higher ground. Several small island countries such as the Maldives, the Bahamas, Kiribati, and the Marshall Islands could become uninhabitable if sea levels rise a meter or more. In 2002, the South Pacific nation of Tuvalu announced that it was abandoning its island homeland. All 11,000 residents will move to New Zealand, perhaps the first of many climate change refugees.

Insurance companies worry that the $2 trillion in insured property along U.S. coastlines could be at risk from a combination of high seas and catastrophic storms. Some 87,000 homes in the United States within 150 m (500 ft) of a shoreline are in danger of coastal erosion or flooding in the next 50 years. Accountants warn that loss of land and structures to flooding and coastal erosion together with damage to fishing stocks, agriculture, and water supplies could raise worldwide insurance claims from about $40 billion in 2000 (already ten times more than they had been 50 years earlier) to more than $300 billion in 2050. Some of this increase in insurance claims is that more people are living in dangerous places, but extra-severe storms only exacerbate this problem.

Infectious diseases are likely to increase as the insects, rodents, and ticks that carry them spread to new areas. Already we have seen diseases such as malaria, dengue fever, and West Nile virus appear in parts of North America where they had never before been reported. Coupled with the movement of hundreds of millions of environmental refugees and greater crowding as people are forced out of areas made uninhabitable by rising seas and changing climates, the spread of epidemic diseases will also increase.

Permafrost is already beginning to melt in the Arctic, causing roads to buckle and houses to sink into the mud. An ominous possibility from this melting might be the release of vast stores of methane hydrate now locked in frozen ground and in sediments in the ocean floor. Together with the increased oxidation of high-latitude peat lands potentially caused by warmer and dryer conditions, release of these carbon stores could add as much CO_2 to the atmosphere as all the fossil fuels ever burned. We could trigger a disastrous positive feedback loop in which the effects of warming cause even more warming.

Another potentially catastrophic outcome of global climate change is that greater terrestrial runoff from increased rainfall might suddenly change ocean circulation patterns that now moderate northern climates (see appendix 4, p. 376). On the other hand, increased ocean evaporation might intensify snowfall at high latitudes so that arctic glaciers and snow pack would increase rather than decrease. Ironically, the increased albedo (reflectivity) of colder, snow-covered surfaces might then trigger a new ice age. As Wallace Broecker of the Lamont-Doherty Earth Observatory said, "Climate is an angry beast and we are poking it with sticks."

International Climate Negotiations

One of the centerpieces of the 1992 United Nations "Earth Summit" meeting in Rio de Janeiro was the Framework Convention on Climate Change, which set an objective of stabilizing greenhouse gas emissions to reduce the threats of global warming. At a follow-up conference in Kyoto, Japan in 1997, 160 nations agreed to roll back CO_2, methane, and nitrous oxide emissions about 5 percent below 1990 levels by 2012. Three other greenhouse gases, hydrofluorocarbons, perfluorocarbons, and sulfur hexafluoride, would also be reduced, although from what level was not decided. Known as the **Kyoto Protocol,** this treaty sets different limits for individual nations depending on their output before 1990. Poorer nations like China and India were exempted from emission limits to allow development to increase their standard of living. Wealthy countries created the problem, the poorer nations argue; and the wealthy should deal with it.

In 2002 all the nations of the European Union plus Japan ratified the Kyoto agreement. If Russia also ratifies, as it is expected to do, the 55-percent threshold will be reached and the protocol will become binding. The United States, however, has refused so far to comply with the agreement it helped broker. In spite of his campaign promises, President Bush has remained adamant that he will not honor U.S. promises at Kyoto. He contends that reducing emissions would be too costly for the U.S. economy. "We're going to put the interests of our own country first and foremost," he said after German Chancellor Gerhard Schröeder made a visit to plead for cooperation. This go-it-alone attitude provoked a stunned and angry reaction among other countries, some of which began to talk about trade sanctions as a way of forcing U.S. cooperation. Rather than reduce emissions, however, the United States has continued to increase them. If current energy consumption patterns persist, government scientists calculate, the United States will be 43 percent above 1990 emission levels by 2020 rather than 7 percent below them as was promised.

In 2002 the U.S. EPA released a report essentially agreeing with the IPCC that greenhouse gases produced by human activities are causing global warming. It noted that climate changes and rising sea levels are especially threatening to sensitive landscapes such as mountain meadows and coastal barrier islands. President Bush dismissed the report as merely a product of government bureaucracy, and insisted that many questions remain about the science of climate change. He said he intends to base his policies on voluntary reductions and tax credits. At the same time, he rolled back requirements that would have forced old power plants to meet more stringent air quality standards if they are remodeled.

Controlling Greenhouse Emissions

In spite of U.S. intransigence, progress already is being made in many places toward reducing greenhouse emissions. The United Kingdom, for example, had already rolled CO_2 emissions back to 1990 levels by 2000, and vowed to reduce them 20 percent more within a decade. Britain already has started to substitute natural gas for coal, promote energy efficiency in homes and industry, and raise its already high gasoline tax. Plans are to "decarbonize" British society and to decouple GNP growth from CO_2 emissions. A revenue-neutral carbon levy is expected to lower CO_2 releases by 5 million tons per year by 2010, and at least 10 percent of all new energy in the U.K. will be from renewable sources, the government promised.

Germany, also, has reduced its CO_2 emissions at least 10 percent by switching from coal to gas and by encouraging energy efficiency throughout society. Atmospheric scientist Steve Schneider calls this a "no regrets" policy; even if we don't need to stabilize our climate, many of these steps save money, conserve resources, and have other environmental benefits. Nuclear power also is being promoted as an alternative to fossil fuels. It's true that nuclear reactions don't produce greenhouse gases, but security worries and unresolved problems of how to store wastes safely make this option unacceptable to many people.

Renewable energy sources offer a better solution to climate problems, many people believe. Chapter 11 discusses options for conserving energy and switching to renewable sources such as solar, wind, geothermal, biomass, and fuel cells. Denmark, the world's leader in wind power, now gets 20 percent of its electricity from windmills. Plans are to generate half of the nation's electricity from offshore wind farms by 2030. Even China reduced its CO_2 emissions 17 percent between 1997 and 1999 through greater efficiency in coal burning and industrial energy use. The U.S. energy plan, meanwhile, is to burn more coal, drill for oil in the Arctic National Wildlife Refuge, and build more nuclear plants.

In addition to reducing its output, there are options for capturing and storing CO_2. Planting trees can be effective if they're allowed to mature to old-growth status or are made into products like window frames and doors that will last for many years before they're burned or recycled. Farmland also can serve as a carbon sink if farmers change their crop mixture and practice minimum-till cultivation that keeps carbon in the soil. In 1988 the late John H. Martin suggested that phytoplankton growth in the ocean may be limited by iron deficiency. This theory was tested in 2000,

Principles of Environmental Science

when researchers spread about 3.5 tons of dissolved iron over 75 km^2 (about 30 mi^2) of the South Pacific. Monitored by satellite, a ten-fold rise in chlorophyll concentration eventually spread over about 1,700 km^2 (660 mi^2) and was calculated to have pulled several thousand tons of CO_2 out of the air. Still, some ecologists warn about such large-scale tinkering. We don't know whether the carbon fixed by phytoplankton growth will be stored in sediments or simply eaten by predators and returned immediately to the atmosphere. It's possible that dying algae might create an anoxic zone that would devastate oceanic food webs.

Another way to store CO_2 is to inject it into underground strata or deep ocean waters. Since 1996, Norway's Statoil has been pumping more than 1 million metric tons of CO_2 into an aquifer 1,000 m (3,000 ft) below the seafloor at a North Sea gas well. It is economical to do so because otherwise the company would have to pay a $50 per ton carbon tax on its emissions. Around the world, deep, briny aquifers could store a century worth of CO_2 output at current fossil fuel consumption rates. It might also be possible to pump liquid CO_2 into deep ocean trenches, where it would form lakes contained by enormous water pressures. There are worries, however, about what this might do to deep ocean fauna and what might happen if earthquakes or landslides caused a sudden release of this CO_2. In Canada, an Alberta power plant is injecting CO_2 into a coal seam too deep to be mined. The CO_2 releases natural gas, which is burned to produce electricity and more CO_2. Proponents of **carbon management,** as these various projects are called, argue that it may be cheaper to clean up fossil fuel effluents than to switch to renewable energy sources.

Most attention is focused on CO_2 because it lasts in the atmosphere, on average, for about 120 years. Methane and other greenhouse gases are much more powerful infrared absorbers, but remain in the air for a much shorter time. NASA's Dr. James Hansen made a controversial suggestion, however, that the best short-term attack on warming might come by focusing not on carbon dioxide but on methane. Reducing gas pipeline leaks would conserve this valuable resource as well as help the environment. Methane from landfills, oil wells, and coal mines that once would have been simply vented into the air is now being collected and used to generate electricity. Rice paddies are a major methane source. Changing flooding schedules and fertilization techniques can reduce anoxia (oxygen depletion) that produces marsh gas. Finally, ruminant animals (like cows, camels, and buffalo) create large amounts of digestive system gas. Modified diets can reduce flatulence significantly.

Soot, while not mentioned in the Kyoto protocol, might also be important in global warming. Dark, airborne particles absorb both ultraviolet and visible light, converting them to heat energy. According to some calculations, reducing soot from diesel engines, coal-fired generators, forest fires, and wood stoves might reduce net global warming by 40 percent within three to five years. Curbing soot emissions would also have beneficial health effects.

Some individual cities and states have announced their own plans to combat global warming. Among the first of these were Toronto, Copenhagen, and Helsinki, which pledged to reduce CO_2 emissions 20 percent from 1990 levels by 2010. Portland, Oregon,

was the first U.S. city to implement a CO_2-reduction strategy. States with plans to limit greenhouse gas emissions include California, New York, and New Hampshire. Some corporations are following suit. British Petroleum has set a goal of cutting CO_2 releases from all its facilities by 10 percent before 2010. In 2000, BP, Alcan, DuPont, and other companies joined with the Environmental Defense Fund to launch a partnership for climate action, pledging to meet or exceed Kyoto requirements. Each of us can make a contribution in this effort. Chapter 11 discusses ways we can cut our individual energy consumption, the principal cause of greenhouse gas emissions.

CLIMATE AND AIR POLLUTION

As the global warming debate shows, human activities can influence climate processes. In this section, we will discuss ways that human activities affect air quality and environmental processes on a more local scale.

According to the Environmental Protection Agency (EPA), Americans release some 147 million metric tons of air pollution (not counting carbon dioxide or wind-blown soil) each year. Worldwide emissions of these pollutants are around 2 billion metric tons per year. Even remote, pristine wilderness areas are now affected. Over the past 20 years, however, air quality has improved in most cities in Western Europe, North America, and Japan. Many young people might be surprised to learn that, a generation ago, most American cities were much dirtier than they are today. The EPA estimates that, since 1990, when regulation of the most hazardous materials began, air toxics emissions have been reduced more than 1 million tons per year. This is almost ten times the reductions achieved in the previous 20 years. Since the 1970s, the levels of major pollutants monitored by the EPA have decreased in the United States, despite population growth of more than 30 percent (table 9.4). Pollution reductions have resulted mainly from greater efficiency and pollution control technologies in factories, power plants, and automobiles. Our success in controlling some of the

TABLE 9.4 Reductions in U.S. Pollution Emissions, 1970–1997

POLLUTANT	PERCENT CHANGE IN ANNUAL EMISSIONS
Carbon monoxide	−32
Lead[1]	−98
Nitrogen oxides[2]	+11
Ozone	−37
Particulate matter	−75
Sulfur dioxide	−35

[1] Reduction mainly due to elimination of leaded gasoline.

[2] Increase mainly from more vehicle emissions.

Source: U.S. Environmental Protection Agency.

most serious air pollutants gives us hope for similar progress in other environmental problems.

While developed countries have been making progress, however, air quality in the developing world has been getting much worse. Especially in the burgeoning megacities of rapidly industrializing countries (see chapter 14), air pollution often exceeds World Health Organization standards all the time. In Lahore, Pakistan, and Xi'an, China, for instance, airborne dust, smoke, and dirt often are ten times higher than levels considered safe for human health (fig. 9.17).

Studies of air pollutants over southern Asia reveal a 3-km (2-mile)-thick toxic cloud of ash, acids, aerosols, dust, and photochemical reactants regularly covers the entire Indian subcontinent for much of the year. Nobel laureate Paul Crutzen estimates that up to 2 million people in India alone die each year from atmospheric pollution. Produced by forest fires, the burning of agricultural wastes, and dramatic increases in the use of fossil fuels, the Asian smog layer cuts the amount of solar energy reaching the earth's surface beneath it by up to 15 percent. Meteorologists suggest that the cloud—80 percent of which is human-made—could disrupt monsoon weather patterns and cut rainfall over northern Pakistan, Afghanistan, western China, and central Asia by up to 40 percent. Shifting monsoon flows may also have contributed to catastrophic floods in Nepal, Bangladesh, and eastern India that killed at least 1,000 people in 2002, and left more than 25 million homeless.

When this "Asian Brown Cloud" drifts out over the Indian Ocean at the end of the monsoon season, it cools sea temperatures and may be changing El Niño/Southern Oscillation patterns in the Pacific Ocean as well. As UN Environment Programme executive director, Klaus Töpfer, said, "There are global implications because a pollution parcel like this, which stretches three km high, can travel half way round the globe in a week."

Major Kinds of Pollutants

In the United States, principal pollutants are identified and regulated by the Clean Air Act of 1970 (table 9.5). Primary pollutants are released in a harmful form (see fig. 9.18). Secondary pollutants, by contrast, become hazardous after reactions in the air. **Photochemical oxidants** (compounds formed with solar energy) and atmospheric acids are probably the most important secondary pollutants. **Fugitive emissions** are those that do not go through a smokestack. By far the most massive example of this category is dust from soil erosion, strip mining, rock crushing, and building construction (and destruction). Leaking valves and pipe joints contribute as much as 90 percent of the hydrocarbons and volatile organic chemicals emitted from oil refineries and chemical plants.

Conventional or **"criteria" pollutants** are a group of seven major pollutants (sulfur dioxide, carbon monoxide, particulates, volatile organic compounds, nitrogen oxides, ozone, and lead) that contribute the largest volume of air-quality degradation and also are considered the most serious threat of all air pollutants to human health and welfare. Since 1970 the Clean Air Act has mandated that the EPA set allowable limits for concentrations of these pollutants in **ambient air** (the air around us) especially in cities.

FIGURE 9.17 *While air quality is improving in many industrialized countries, newly developing countries have growing pollution problems. Xi'an, China, often has particulate levels above 300 μg/m³.*

FIGURE 9.18 *Smokestacks of this coal-fired power plant emit a wide variety of toxic chemicals.*

TABLE 9.5 Principal Pollutants (including but not limited to criteria pollutants)

POLLUTANT	PRINCIPAL FORMS	MAIN ANTHROPOGENIC SOURCES	NATURAL SOURCES	U.S. EMISSIONS PER YEAR (MILLION METRIC TONS)	ANTHROPOGENIC PORTION
SULFUR					
Sulfur dioxide	SO_2 (also H_2SO_4, sulfuric acid)	Electric utilities (coal), industry (oil, gas, coal), metal ore smelting	Sea spray, volcanoes, biogenic H_2S	114	69%
CARBON					
Carbon monoxide	CO	Transportation (incomplete fuel burning)	Plant metabolism	2510	37%
Carbon dioxide	CO_2	Fossil fuel combustion	Cellular respiration	394	6%
Particulates	Dust, aerosols	Construction, unpaved roads, agriculture	Dust, volcanic ash, sea spray	945	38%
VOLATILE ORGANIC COMPOUNDS (VOCS)					
Methane	CH_4	Transportation, power plants, chemical plants, petroleum refineries, landfills, rice paddies, digestion in livestock	Plants, decay in bogs, termite mounds, ruminate animals	405	69%
Nonmethane hydrocarbons	Terpenes, isoprenes, benzene	Burning gas, oil, coal	Plants	952	10%
NITROGEN					
Nitrogen oxides	NO, NO_2, N_2O, NO_3 (collectively, NO_X)	Transportation, power plants, industry, other fuel uses, fertilizer	Lightning, soil microbes	230	60%
SECONDARY POLLUTANTS					
Photochemical oxidants	O_3, N_2O, PAN (peroxyacetyl nitrate)	Reactions among VOCs and other pollutants initiated by sunlight	Lightning, forest fires	<1	Most
METALS					
Lead	Elemental lead	Leaded gasoline, metal ore processing	Soil, rocks	4	Most
Mercury	Elemental mercury	Coal power plants, garbage incinerators	Soil, rocks	<1	Most
Arsenic	Elemental arsenic	Accelerated rock weathering, industry, ore smelting	Soil, rocks	<1	High
HALONS					
Chlorine	HCl, methyl chloride (CH_3Cl), PCBs, CFCs	Industrial chemicals, plastics, coal burning, biomass burning	Sea spray, biomass burning	2	35%
Bromine	Methyl bromide (CH_3Br)	Fumigants, coal and biomass burning	Sea spray	>0.27	30–60%

The EPA also monitors **unconventional pollutants,** compounds that are produced in less volume than conventional pollutants but that are especially toxic or hazardous. Among these are asbestos, benzene, beryllium, mercury, polychlorinated biphenyls (PCBs), and vinyl chloride. Most of these materials have no natural source in the environment (to any great extent) and are, therefore, only anthropogenic in origin.

Aesthetic degradation is another important form of pollution. Noise, odors, and light pollution may not be life threatening, but they reduce the quality of our lives. In most urban areas, it is difficult or impossible to see stars in the sky at night because of dust in the air and stray light from buildings, outdoor advertising, and streetlights.

Sources and Problems of Major Pollutants

Most conventional pollutants are produced primarily from burning fossil fuels, especially in coal-powered electric plants and in cars and trucks, as well as in processing natural gas and oil. Others, especially sulfur and metals, are by-products of mining and manufacturing processes. Of the 188 air toxics listed in the Clean Air Act, about two-thirds are volatile organic compounds (VOCs), and most of the rest are metal compounds. In this section we will discuss the characteristics and origin of the major pollutants.

Sulfur dioxide (SO_2) is a colorless, corrosive gas that damages both plants and animals. Once in the atmosphere, it can be further oxidized to sulfur trioxide (SO_3), which reacts with water vapor or dissolves in water droplets to form sulfuric acid (H_2SO_4), a major component of acid rain. Sulfur dioxide and sulfate ions are probably second only to smoking as causes of air-pollution–related health damage. Sulfate particles and droplets also reduce visibility in the United States by as much as 80 percent.

Nitrogen oxides (NO_x) are highly reactive gases formed when combustion initiates reactions between atmospheric nitrogen and oxygen. The initial product, nitric oxide (NO), oxidizes further in the atmosphere to nitrogen dioxide (NO_2), a reddish brown gas that gives photochemical smog its distinctive color. Because these gases convert readily from one form to the other, the general term NO_x is used to describe these gases. Nitrogen oxides combine with water to form nitric acid (HNO_3), which is also a major component of acid precipitation. Excess nitrogen in water is causing eutrophication of inland waters and coastal seas. It may also encourage growth of weedy species that crowd out native plants.

Carbon monoxide (CO) is less common than the principal form of atmospheric carbon, carbon dioxide (CO_2) but more dangerous. CO is a colorless, odorless, but highly toxic gas produced mainly by incomplete combustion of fuel (coal, oil, charcoal, wood, or gas). CO inhibits respiration in animals by binding irreversibly to hemoglobin. In the United States, two-thirds of the CO emissions are created by internal combustion engines in transportation. Land-clearing fires and cooking fires also are major sources. About 90 percent of the CO in the air is consumed in photochemical reactions that produce ozone.

Particulate material includes dust, ash, soot, lint, smoke, pollen, spores, algal cells, and many other suspended materials.

TABLE 9.6 Photochemical Oxidant Production

STEPS	PHOTOCHEMICAL PRODUCTS
1. NO + VOC →	NO_2 (nitrogen dioxide)
2. NO_2 + UV sunlight →	NO + O (nitric oxide + atomic oxygen)
3. O + O_2 →	O_3 (ozone)
4. NO_2 + VOC →	PAN (peroxyacetyl nitrate)

Aerosols, or extremely minute particles or liquid droplets suspended in the air, are included in this class. Particulates often are the most apparent form of air pollution, since they reduce visibility and leave dirty deposits on windows, painted surfaces, and textiles. Breathable particles smaller than 2.5 μm are among the most dangerous of this group because they can damage lung tissues. Asbestos fibers and cigarette smoke are among the most dangerous respirable particles in urban and indoor air because they are carcinogenic.

Volatile organic compounds (VOCs) are organic gases. Plants, bogs, and termites are the largest sources of VOCs, especially isoprenes (C_5H_8), terpenes ($C_{10}H_{15}$), and methane (CH_4). These volatile hydrocarbons are generally oxidized to CO and CO_2 in the atmosphere.

More dangerous synthetic organic chemicals, such as benzene, toluene, formaldehyde, vinyl chloride, phenols, chloroform, and trichloroethylene, are released by human activities. Principal sources are incompletely burned fuels from vehicles, power plants, chemical plants, and petroleum refineries. These chemicals play an important role in the formation of photochemical oxidants.

Photochemical oxidants are products of secondary atmospheric reactions driven by solar energy (table 9.6). One of the most important of these reactions involves formation of singlet (atomic) oxygen by splitting nitrogen dioxide (NO_2). This atomic oxygen then reacts with another molecule of O_2 to make ozone (O_3). Although ozone is important in the stratosphere, in ambient air, it is highly reactive and damages vegetation, animal tissues, and building materials. Ozone's acrid, biting odor is a distinctive characteristic of photochemical smog.

Lead and Other Toxic Elements

Toxic metals and halogens are chemical elements that are toxic when concentrated and released in the environment. Principal metals of concern are lead, mercury, arsenic, nickel, beryllium, cadmium, thallium, uranium, cesium, and plutonium. Halogens (fluorine, chlorine, bromine, and iodine) are highly reactive toxic elements. Most of these materials are mined and used in manufacturing. Metals commonly occur as trace elements in fuels, especially coal.

Lead and mercury are widespread neurotoxins that damage the nervous system. By some estimates, 20 percent of all inner-city children suffer some degree of developmental retardation from high environmental lead levels. Long-range transport of lead

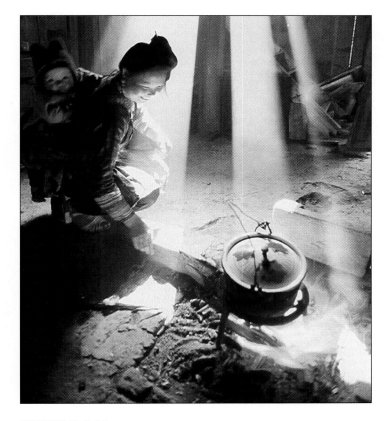

FIGURE 9.19 *Some 2.5 billion people, mainly women and children, spend hours each day in poorly ventilated kitchens and living spaces where carbon monoxide, particulates, and cancer-causing hydrocarbons often reach dangerous levels.*

and mercury through the air is causing bioaccumulation in remote aquatic ecosystems, such as arctic lakes and seas. Chlorine is a toxic halogen widely used in bleach, plastics, and other products. Methyl bromide (a powerful fungicide used in agriculture) and **chlorofluorocarbons** (propellants and refrigerants) are also implicated in ozone depletion.

Indoor Air Pollution

We have spent a considerable amount of effort and money to control the major outdoor air pollutants, but we have only recently become aware of the dangers of indoor air pollutants. The EPA has found that indoor concentrations of toxic air pollutants are often higher than outdoors. Furthermore, people generally spend more time inside than out and therefore are exposed to higher doses of these pollutants. In some cases, indoor air in homes has chemical concentrations that would be illegal outside or in the workplace. Under some circumstances, compounds such as chloroform, benzene, carbon tetrachloride, formaldehyde, and styrene can be seventy times higher in indoor air than in outdoor air. Molds, pathogens, and other biohazards also represent serious indoor pollutants.

Cigarette smoke is without doubt the most important air contaminant in developed countries, in terms of human health. The U.S. surgeon general has estimated that 430,000 people die

each year in the United States from emphysema, heart attacks, strokes, lung cancer, or other diseases caused by smoking. These diseases are responsible for 20 percent of all mortality in the United States, or four times as much as infectious agents. Total costs for early deaths and smoking-related illnesses are estimated to be $100 billion per year. Eliminating smoking probably would save more lives than any other pollution-control measure.

In the less-developed countries of Africa, Asia, and Latin America, where such organic fuels as firewood, charcoal, dried dung, and agricultural wastes make up the majority of household energy, smoky, poorly ventilated heating and cooking fires represent the greatest source of indoor air pollution (fig. 9.19). The World Health Organization (WHO) estimates that 2.5 billion people—nearly half the world's population—are adversely affected by pollution from this source. In particular women and small children spend long hours each day around open fires or unventilated stoves in enclosed spaces.

INTERACTIONS BETWEEN CLIMATE PROCESSES AND AIR POLLUTION

Physical processes in the atmosphere transport, concentrate, and disperse air pollutants. To comprehend the global effects of air pollution, it is necessary to understand how climate processes interact with pollutants. Global warming (discussed previously), in which pollutants are altering the earth's energy budget, is the best-known case of interaction between anthropogenic pollutants and the atmosphere. In this section we will survey other important climate-pollution interactions.

Long-Range Transport

Global circulation patterns carry pollutants far from their source of origin. The importance of long-range transport for ecosystems and for climate has only recently been appreciated. Florida researchers recently showed that as much as half of the fine reddish dust visible in Miami's air originates in the Sahara Desert. Dust, sulfur dioxide, and ozone from China is regularly found on the west coast of North America. DDT and other toxins often travel with this windborne dust. Recently atmospheric scientists were surprised to find a dense haze of soot and sulfate aerosols—industrial particulates—over the Indian Ocean. Covering an area larger than the continental United States and reaching from the Bay of Bengal to the Arabian Sea, this huge cloud of pollutants is produced in South Asia and blown south by the winter monsoon. In summer, it is blown back north to fall as dirty acid rain.

Long-range transport can concentrate pollutants at high latitudes (near the poles). Since the 1950s, pilots flying in the high Arctic have reported dense layers of reddish-brown haze clouding the arctic atmosphere. Aerosols of sulfates, soot, dust, and toxic heavy metals, such as vanadium, manganese, and lead, travel to the pole from the industrialized parts of Europe, Russia, and the United States (fig. 9.20). These contaminants, trapped by winds that circle the pole, concentrate at high latitudes, eventually fall out in snow and ice, and then enter the food chain. The Inuit people of Broughton Island in Canada's high arctic have higher levels of polychlorinated biphenyls (PCBs) in their blood than any other known population, except victims of industrial accidents.

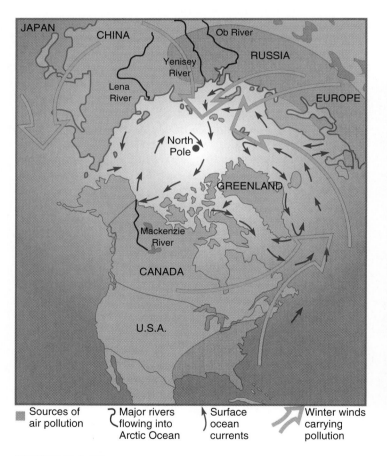

FIGURE 9.20 *Circumpolar winds transport air pollution from heavily industrialized regions in Europe, Russia, and America to the Arctic, where high levels of smog accumulate. The average transit time from Russia to Canada is only about three days.*

Stratospheric Ozone Depletion

Long-range pollution transport and the chemical reactions of atmospheric gases and pollution produce the phenomenon known as the ozone hole (fig. 9.21). The ozone "hole," really a thinning of ozone concentrations in the stratosphere, was discovered in 1985 but has probably been developing since at least the 1960s. Chlorine-based aerosols, especially chlorofluorocarbons (CFCs) and other halon gases, are the principal agents of ozone depletion. Nontoxic, nonflammable, chemically inert, and cheaply produced, CFCs were extremely useful as industrial gases and in refrigerators, air conditioners, styrofoam inflation, and aerosol spray cans for many years. From the 1930s until the 1980s, CFCs were used all over the world and widely dispersed through the atmosphere.

Although ozone is a pollutant in the ambient air, ozone in the stratosphere is important because it absorbs much of the ultraviolet radiation entering the atmosphere. UV radiation harms plant and animal tissues, including the eyes and the skin. A 1 percent loss of ozone could result in about a million extra human skin cancers per year worldwide if no protective measures are taken. Excessive UV exposure could reduce agricultural production and disrupt ecosystems. Scientists worry, for example, that high UV levels in Antarctica could reduce populations of plankton, the tiny floating organisms that form the base of a food chain that includes fish, seals, penguins, and whales in Antarctic seas.

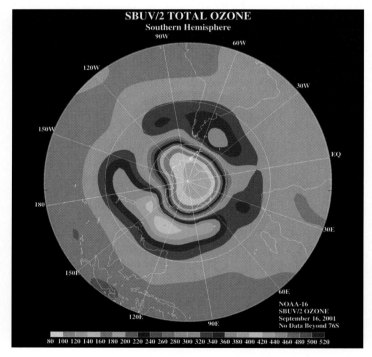

FIGURE 9.21 *This satellite image from 16 September 2001 shows ozone depletion over the South Pole. Outside the polar vortex, green and yellow areas show elevated ozone levels. Over Antarctica (outlined in white), however, stratospheric ozone levels are reduced by 80 percent or more from normal levels. The dark circle over the South Pole is an artifact of the measuring system.*

TABLE 9.7 Stratospheric Ozone Destruction by Chlorine Atoms and UV Radiation

STEP	PRODUCTS
1. $CFCl_3$ (chlorofluorocarbon) + UV energy	$CFCl_2 + Cl$
2. $Cl + O_3$	$ClO + O_2$
3. O_2 + UV energy	$2O$
4. $ClO + 2O$	$O_2 + Cl$
5. Return to step 2	

Average area of ozone hole during Antarctic spring

FIGURE 9.22 *Growth of the ozone "hole," the region of depleted stratospheric ozone, over Antarctica. Vertical lines represent minimum and maximum area during the Antarctic spring.*
Source: Goddard Space Flight Center, NASA, 2001.

Antarctica's exceptionally cold winter temperatures (–85° to –90°C) help break down ozone. During the long, dark, winter months, strong winds known as the circumpolar vortex isolate Antarctic air and allow stratospheric temperatures to drop low enough to create ice crystals at high altitudes—something that rarely happens elsewhere in the world. Ozone and chlorine-containing molecules are absorbed on the surfaces of these ice particles. When the sun returns in the spring, it provides energy to liberate chlorine ions, which readily bond with ozone, breaking it down to molecular oxygen (table 9.7). It is only during the Antarctic spring (September through December) that conditions are ideal for rapid ozone destruction. During that season temperatures are still cold enough for high-altitude ice crystals, but the sun gradually becomes strong enough to drive photochemical reactions. As the Antarctic summer arrives, temperatures moderate somewhat, the circumpolar vortex breaks down, and air from warmer latitudes mixes with Antarctic air, replenishing ozone concentrations in the ozone hole. Slight decreases worldwide result from this mixing, however. Ozone re-forms naturally, but not nearly as fast as it is destroyed. Since the chlorine atoms are not themselves consumed in reactions with ozone, they continue to destroy ozone for years, until they finally precipitate or are washed out of the air. Almost every year since it was discovered, the Antarctic ozone hole has grown. In 1998, the region of ozone depletion covered 27.3 million km² (10.5 million mi², or about the size of North America, fig. 9.22).

Ominously, this phenomenon is now spreading to other parts of the world as well. About 10 percent of all stratospheric ozone worldwide was destroyed during the spring of 1998, and levels over the Arctic averaged 40 percent below normal. Ozone depletion has been observed over the North Pole as well, although it is not as concentrated as that in the South.

Signs of Progress

The discovery of stratospheric ozone losses brought about a remarkably quick international response. In 1987, an international meeting in Montreal, Canada, produced the Montreal Protocol, the first of several major international agreements on phasing out most use of CFCs by 2000. As evidence accumulated, showing that losses were larger and more widespread than previously thought, the deadline for the elimination of all CFCs (halons, carbon tetrachloride, and methyl chloroform) was moved up to 1996, and a $500-million fund was established to assist poorer countries in switching to non-CFC technologies. Fortunately, alternatives to CFCs for most uses already exist. The first substitutes will be hydrochlorofluorocarbons (HCFCs), which release much less chlorine per molecule. Eventually, we hope to develop halogen-free molecules that work just as well and are no more expensive than CFCs.

There is some evidence that the CFC ban is already having an effect. CFC production in industrialized countries has fallen nearly 80 percent since 1989 (fig. 9.23), and CFCs are now being removed from the atmosphere more rapidly than they are being added. In 50 years or so, stratospheric ozone levels are expected to be back to normal. Unfortunately, there may be a downside to stopping ozone destruction. Ozone is a potent greenhouse gas. Some models suggest that lower ozone levels have been offsetting the effects of increased CO_2. When ozone is restored, global warming may be accelerated. As so often is the case, when we disturb one environmental factor, we affect others as well.

Urban Climates

Temperature inversions can concentrate dangerous levels of pollutants within cities. Normally, air temperatures decrease with elevation above the earth's surface. An inversion reverses these conditions, with cool, dense air lying below a warmer, lighter layer. In these very stable conditions, convection currents cannot develop, and there is no mechanism for dispersing pollutants. The most stable inversion conditions are usually created by rapid nighttime cooling in a valley or basin where air movement is restricted. Los Angeles is a classic example of an area with conditions that

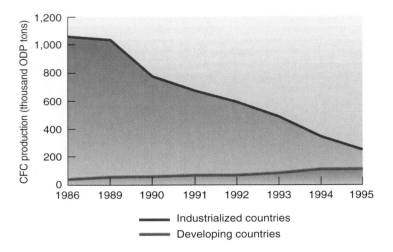

FIGURE 9.23
Production of chlorofluorocarbons in industrialized countries has fallen sharply since the Helsinki convention was passed in 1989.

Source: Data from Sebastian Oberthur, *Production and Consumption of Ozone-Depleting Substances, 1986–1995* (Deutsche Gesellschaft für Technische Zusammenarbeit, Bonn Germany, 1997, p. 30).

create temperature inversions and photochemical smog. Mountains surround the city on three sides, reducing wind movement, the climate is sunny, and heavy traffic creates high pollution levels. Skies are generally clear at night, allowing rapid radiant heat loss, and the ground cools quickly. Surface air layers are cooled by conduction, while upper layers remain relatively warm. Density differences retard vertical mixing. During the night, cool, humid, onshore breezes slide in under the contaminated air, squeezing it up against the cap of warmer air above and concentrating the pollutants accumulated during the day.

Morning sunlight initiates photochemical oxidation in the concentrated aerosols and gaseous chemicals in the inversion layer. A toxic brew of hazardous chemicals, especially ozone and nitrogen dioxide, quickly develops. On summer days, ozone concentrations in the Los Angeles basin can reach 0.34 ppm or more by late afternoon, and the pollution index can be 300, the stage considered a health hazard.

Heat islands and *dust domes* occur in cities even without inversion conditions. With their low albedo, concrete and brick surfaces in cities absorb large amounts of solar energy. A lack of vegetation or water results in very slight evaporation (latent heat production); instead, available solar energy is turned into heat. As a result, temperatures in cities are frequently 3° to 5°C (5° to 9°F) warmer than in the surrounding countryside, a condition known as an urban "heat island." Tall buildings create convective updrafts that sweep pollutants into the air. Stable air masses created by this heat island over the city concentrate pollutants in a "dust dome."

EFFECTS OF AIR POLLUTION

Air pollution is equally serious for ecosystem health and for human health. In this section we will review the most important effects of air pollution.

Human Health

Consequences of breathing dirty air include increased probability of heart attacks, respiratory diseases, and lung cancer. This can mean as much as a 5- to 10-year decrease in life expectancy if you live in the worst parts of Los Angeles or Baltimore, for example. Of course the intensity and duration of your exposure, as well as your age and general health, are extremely important: you are much more likely to be at risk if you are very young, very old, or already suffering from some respiratory or cardiovascular disease. Bronchitis and emphysema are common chronic conditions resulting from air pollution. The U.S. Office of Technology Assessment estimates that 250,000 people suffer from pollution-related bronchitis and emphysema in the United States, and some 50,000 excess deaths each year are attributable to complications of these diseases, which are probably second only to heart attack as a cause of death.

Conditions are often much worse in developing countries. The United Nations estimates that at least 1.3 billion people around the world live in areas where the air is dangerously polluted. In the "black triangle" region of Poland, Hungary, the Czech Republic, and Slovakia, for example, respiratory ailments, cardiovascular diseases, lung cancer, infant mortality, and miscarriages are as much as 50 percent higher than in cleaner parts of those countries. And in China, city dwellers are four to six times more likely than country folk to die of lung cancer. The World Health Organization estimates that 4 million people die each year from diseases exacerbated by air pollution.

How does air pollution cause these health effects? Because they are strong oxidizing agents, sulfates, SO_2, NO_x, and O_3 irritate and damage delicate tissues in the eyes and lungs. Fine, suspended, particulate materials penetrate deep into the lungs, causing irritation, scarring, and even tumor growth. Heart stress results from impaired lung functions. Carbon monoxide binds to hemoglobin, reducing oxygen flow to the brain. Headaches, dizziness, and heart stress result. Lead also binds to hemoglobin, damaging critical neurons in the brain and resulting in mental and physical impairment and developmental retardation.

Plant Pathology

In the early days of industrialization, fumes from furnaces, smelters, refineries, and chemical plants often destroyed vegetation and created desolate, barren landscapes around mining and manufacturing centers. The copper-nickel smelter at Sudbury, Ontario, is a spectacular example. Starting in 1886, open-bed roasting was used to purify sulfide ores of nickel and copper. The resulting sulfur dioxide and sulfuric acid destroyed almost all plant life within about 30 km (18.6 mi) of the smelter. Rains washed away the exposed soil, leaving a barren moonscape of blackened bedrock (fig. 9.24). Recently, emission controls have been introduced, and the environment is beginning to recover.

There are two probable ways that air pollutants damage plants. They can damage sensitive cell membranes, much as irritants do in human lungs. Toxic levels of oxidants produce discoloration

FIGURE 9.24 *Sulfur dioxide emissions and acid precipitation from the International Nickel Company copper smelter* (background) *killed all vegetation over a large area near Sudbury, Ontario. Even the pink granite bedrock was burned black. The installation of scrubbers has dramatically reduced sulfur emissions. The ecosystem farther away from the smelter is slowly beginning to recover.*

(destruction of chlorophyll) and then necrotic (dead) spots. Often, however, the symptoms are vague and difficult to separate from diseases or insect damage. Pollutants can also act as hormones, disrupting plant metabolism, growth, and development. Ethylene, a volatile organic compound released in automobile exhaust and from petroleum refineries and chemical plants, often injures plants around roads and factories.

Certain combinations of environmental factors have synergistic effects in which the injury caused by exposure to two factors together is more than the sum of exposure to each factor individually. For instance, white pine seedlings exposed to subthreshold concentrations of ozone and sulfur dioxide individually do not suffer any visible injury. If the same concentrations of pollutants are given together, however, visible damage occurs.

Pollutant levels too low to produce visible symptoms of damage may still have important effects. Field studies show that yields in some crops, such as soybeans, may be reduced as much as 50 percent by currently existing levels of oxidants in ambient air. Some plant pathologists suggest that ozone and photochemical oxidants are responsible for as much as 90 percent of agricultural, ornamental, and forest losses from air pollution. The total costs of this damage may be as much as $10 billion per year in North America alone.

Visibility Reduction

We have only recently realized that pollution affects rural areas as well as cities. Even supposedly pristine places like our national parks are suffering from air pollution. Grand Canyon National Park, where maximum visibility used to be 300 km (185 mi), is now so smoggy on some winter days that visitors can't see the

opposite rim only 20 km (12.5 mi) across the canyon. Mining operations, smelters, and power plants (some of which were moved to the desert to improve air quality in cities like Los Angeles) are the main culprits. Huge regions are affected by pollution. A gigantic "haze blob" as much as 3,000 km (about 2,000 mi) across covers much of the eastern United States in the summer, cutting visibility as much as 80 percent. People become accustomed to these conditions and don't realize that the air once was clear. Studies indicate, however, that if all human-made sources of air pollution were shut down, the air would clear up in a few days, and there would be about 150 km (90 mi) visibility nearly everywhere, rather than the 15 km to which we have become accustomed.

Acid Deposition

Acid precipitation, the deposition of wet, acidic solutions or dry, acidic particles from the air, became widely recognized as a pollution problem only in the last 20 years. But the concept has been recognized since the 1850s. We describe acidity in terms of pH, with substances below pH 7 being acidic (see chapter 2). Normal, unpolluted rain generally has a pH of about 5.6 due to carbonic acid created when rainwater reacts with CO_2 in the air. Downwind of industrial areas, rainfall acidity can reach levels below pH 4.3, more than ten times as acidic as normal rain. Acid fog, snow, mist, and dew can deposit damaging acids on plants, in water systems, and on buildings. Furthermore, fallout of dry sulfate and nitrate particles can account for as much as half of the acidic deposition in some areas.

A vigorous program of pollution control has been undertaken by Canada, the United States, and several European countries since the widespread recognition of acid rain. SO_2 and NO_x emissions from power plants have decreased dramatically over the past three decades over much of Europe and eastern North America as a result of pollution control measures. However, rain falling in these areas remains acidic (fig. 9.25). Apparently, alkaline dust that would once have neutralized acids in air has been depleted by years of acid rain and is now no longer effective.

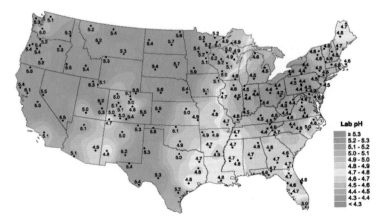

FIGURE 9.25 *Acid precipitation over the United States, 2000.*
Source: National Acid Deposition Program 2001.

FIGURE 9.26   *A Fraser fir forest on Mount Mitchell, North Carolina, killed by acid rain, insect pests, and other stressors.*

FIGURE 9.27   *Atmospheric acids, especially sulfuric and nitric acids, have almost completely eaten away the face of this medieval statue. Each year, the total losses from air pollution damage to buildings and materials amount to billions of dollars.*

Aquatic Effects of Acid Deposition

Aquatic ecosystems in Scandinavia were among the first discovered to be damaged by acid precipitation. Prevailing winds from Germany, Poland, and other parts of Europe deliver acids generated by industrial and automobile emissions—principally H_2SO_4 and HNO_3. The thin, acidic soils and nutrient-poor lakes and streams in the mountains of southern Norway and Sweden have been severely affected by this acid deposition. Most noticeable is the reduction of trout, salmon, and other game fish, whose eggs and fry die below pH 5. Aquatic plants, insects, and invertebrates also suffer. Some 18,000 lakes in Sweden are now so acidic that they will no longer support game fish or other sensitive aquatic organisms. Large parts of Europe and eastern North America are similarly affected by acid precipitation.

Forest Damage

Since the early 1980s observers have documented forest declines apparently related to acid damage in both Europe and North America. A detailed 1980 ecosystem inventory on Camel's Hump Mountain in Vermont found that seedling production, tree density, and viability of spruce-fir forests at high elevations had declined about 50 percent in 15 years. On Mount Mitchell in North Carolina, nearly all the trees above 2,000 m (6,000 ft) are losing needles, and about half of them are dead (fig. 9.26). Damage has been reported throughout Europe, from the Netherlands to Switzerland, as well as in China and the states of the former Soviet Union. In 1985 West German foresters estimated that about half the total forest area in West Germany (more than 4 million ha) was declining. The loss to the forest industry is estimated to be about 1 billion DM (deutsche marks) per year.

High-elevation forests are most severely affected. Mountain tops often have thin, acidic soils under normal conditions, with little ability to buffer, or neutralize, acid precipitation. Acidic fog and mist frequently rest on mountain tops, lengthening plants'

exposure to acidity. Mountain tops also catch more rain and snow than lowlands, so they are more exposed to acidic precipitation.

The most visible mechanisms of forest decline include direct damage to plant tissues and seedlings. Nutrient availability in forest soils is also depleted, however. Acids can also dissolve and mobilize toxic metals such as aluminum, and weakened trees become susceptible to diseases and insect pests.

Buildings and Monuments

In cities throughout the world, air pollution is destroying some of the oldest and most glorious buildings and works of art. Smoke and soot coat buildings, paintings, and textiles. Acids dissolve limestone and marble, destroying features and structures of historic buildings (fig. 9.27). The Parthenon in Athens, the Taj Mahal in Agra, the Coliseum in Rome, medieval cathedrals in Europe, and the Washington Monument in Washington, D.C., are slowly dissolving and flaking away because of acidic fumes in the air. On a more mundane level, air pollution and acid precipitation corrode steel in reinforced concrete, weakening buildings, roads, and bridges. Limestone, marble, and some kinds of sandstone flake and crumble. The Council on Environmental Quality estimates that U.S. economic losses from architectural damage caused by air pollution amount to about $4.8 billion in direct costs and $5.2 billion in property-value losses each year.

AIR POLLUTION CONTROL

"Dilution is the solution to pollution" was one of the early approaches to air pollution control. Tall smokestacks were built to send emissions far from the source, where they became unidentifiable and largely untraceable. But dispersed and diluted pollutants are now the source of some of our most serious pollution problems. We are finding that there is no "away" to which we can throw our waste products.

Reducing Production

The most effective strategy for controlling pollution is to minimize polluting activities. Since most air pollution in the developed world is associated with transportation and energy production, the most effective strategy would be conservation: reducing electricity consumption, insulating homes and offices, and developing better public transportation could all greatly reduce air pollution in the United States, Canada, and Europe. Alternative energy sources, such as wind and solar power, produce energy with little or no pollution, and these and other technologies are becoming economically competitive (see chapter 12). In addition to conservation, pollution can be controlled by technological innovation.

Particulate removal involves filtering air emissions. Filters trap particulates in a mesh of cotton cloth, spun glass fibers, or asbestos-cellulose. Industrial air filters are generally giant bags 10 to 15 m long and 2 to 3 m wide. Effluent gas is blown through the bag, much like the bag on a vacuum cleaner. Every few days or weeks, the bags are opened to remove the dust cake. Electrostatic precipitators are the most common particulate controls in power plants. Ash particles pick up an electrostatic surface charge as they pass between large electrodes in the effluent stream. Charged particles then collect on an oppositely charged collecting plate. These precipitators consume a large amount of electricity, but maintenance is relatively simple, and collection efficiency can be as high as 99 percent. The ash collected by both of these techniques is a solid waste (often hazardous due to the heavy metals and other trace components of coal or other ash source) and must be buried in landfills or other solid-waste disposal sites.

Sulfur removal is important because sulfur oxides are among the most damaging of all air pollutants in terms of human health and ecosystem viability. Switching from soft coal with a high sulfur content to low-sulfur coal is the surest way to reduce sulfur emissions. High-sulfur coal is frequently politically or economically expedient, however. In the United States, Appalachia, a region of chronic economic depression, produces most high-sulfur coal. In China, much domestic coal is rich in sulfur. Switching to cleaner oil or gas would eliminate metal effluents as well as sulfur. Cleaning fuels is an alternative to switching. Coal can be crushed, washed, and gasified to remove sulfur and metals before combustion. This improves heat content and firing properties, but may replace air pollution with solid-waste and water pollution problems; furthermore, these steps are expensive.

Sulfur can also be removed to yield a usable product instead of simply a waste disposal problem. Elemental sulfur, sulfuric acid, or ammonium sulfate can all be produced using catalytic converters to oxidize or reduce sulfur. Markets have to be reasonably close and fly ash contamination must be reduced as much as possible for this procedure to be economically feasible.

Nitrogen oxides (NO_x) can be reduced in both internal combustion engines and industrial boilers by as much as 50 percent by carefully controlling the flow of air and fuel. Staged burners, for example, control burning temperatures and oxygen flow to prevent formation of NO_x. The catalytic converter on your car uses platinum-palladium and rhodium catalysts to remove up to 90 percent of NO_x, hydrocarbons, and carbon monoxide at the same time (fig. 9.28).

Hydrocarbon controls mainly involve complete combustion or controlling evaporation. Hydrocarbons and volatile organic compounds are produced by incomplete combustion of fuels or by solvent evaporation from chemical factories, paints, dry cleaning, plastic manufacturing, printing, and other industrial processes. Closed systems that prevent escape of fugitive gases can reduce many of these emissions. In automobiles, for instance, positive

FIGURE 9.28 *Elements of a modern automobile emission-control system. A closed-loop, electronically controlled carburetor or fuel injector carefully meters fuel-air ratios to optimize combustion. Exhaust oxygen sensors measure completeness of fuel burning. Positive crankcase ventilation captures oil "blow-by" and unburned exhaust gases and recycles them to the cylinder.*

crankcase ventilation (PCV) systems collect oil that escapes from around the pistons and unburned fuel and channels them back to the engine for combustion. Controls on fugitive losses from industrial valves, pipes, and storage tanks can have a significant impact on air quality. Afterburners are often the best method for destroying volatile organic chemicals in industrial exhaust stacks.

CLEAN AIR LEGISLATION

Throughout history, countless ordinances have prohibited emission of objectionable smoke, odors, and noise. Air pollution traditionally has been treated as a local problem, however. The Clean Air Act of 1963 was the first national legislation in the United States aimed at air pollution control. The act provided federal grants to states to combat pollution but was careful to preserve states' rights to set and enforce air quality regulations. It soon became obvious that some pollution problems cannot be solved on a local basis.

In 1970, an extensive set of amendments essentially rewrote the Clean Air Act. These amendments identified the "criteria" pollutants discussed earlier in this chapter, and established primary and secondary standards for ambient air quality. Primary standards (table 9.8) are intended to protect human health, while secondary standards are set to protect materials, crops, climate, visibility, and personal comfort.

Since 1970 the Clean Air Act has been modified, updated, and amended. The most significant amendments were in the 1990 update. Amendments have involved acrimonious debate, with bills sometimes languishing in Congress from one session to the next because of disputes over burdens of responsibility and cost and definitions of risk. Some of the principal problems that have been addressed in amendments include the following:

- Acid rain: reductions in sulfur dioxide and nitrogen oxides
- Urban smog: lower vehicle emissions, alternative fuels in the smoggiest cities
- Toxic air pollutants: controls on 188 airborne toxins from about 250 types of sources
- Ozone protection: phasing out CFCs, carbon tetrachloride, and other compounds (Smuggling from developing countries such as Mexico, where CFCs are still cheap and legal, has become a serious problem.)
- Marketing pollution rights: allowing corporations to avoid pollution reduction by buying, selling, and "banking" pollution rights from other factories
- Fugitive emissions of volatile organic compounds (VOCs): detailed standards for manufacturing, use, and storage methods
- Ambient ozone, soot, and dust: reduced-sulfur gasoline, cleaner smokestacks and cars
- NO_x emissions: improved standards for sport-utility vehicles, personal watercraft, and other polluting vehicles

A 2002 report concluded that simply by enforcing existing clean air legislation, the United States could save at least another 6,000 lives per year and prevent 140,000 asthma attacks.

TABLE 9.8 National Ambient Air Quality Standards (NAAQS)

POLLUTANT	PRIMARY (HEALTH-BASED) AVERAGING TIME	STANDARD CONCENTRATION
Particulates	Annual geometric mean[a]	$50 \ \mu g/m^3$
	24 hours	$150 \ \mu g/m^3$
SO_2	Annual arithmetic mean[b]	$80 \ \mu g/m^3$ (0.03 ppm)
	24 hours	$365 \ \mu g/m^3$ (0.14 ppm)
CO	8 hours	$10 \ mg/m^3$ (9 ppm)
	1 hour	$40 \ mg/m^3$ (35 ppm)
NO_2	Annual arithmetic mean	$100 \ \mu g/m^3$ (0.05 ppm)
O_3	Daily maximum 1-hour average	$235 \ \mu g/m^3$ (0.12 ppm)
Lead	Maximum quarterly average	$1.5 \ \mu g/m^3$

[a]The geometric mean is obtained by taking the nth root of the product of n numbers. This tends to reduce the impact of a few very large numbers in a set.

[b]An arithmetic mean is the average determined by dividing the sum of a group of data points by the number of points.

California has gone further than the federal government in making specific plans for air pollution control. The South Coast Air Quality Management District has adopted 160 rules to clean the air in the Los Angeles Basin. If these measures are successful, smog-causing emissions could be reduced by 70 percent. The number of days when the air is considered hazardous to breathe would decrease from 150 per year to none.

Reaching these goals will require substantial lifestyle changes for most Californians. Aerosol hair sprays, deodorants, charcoal lighter fluid, gasoline-powered lawnmowers, and drive-through burger stands could be banned. More than 3,000 consumer products, including automotive polishes, spot removers, herbicides, lubricants, and floor-wax strippers, would need to meet new pollution limits. Paints and cleaning solutions would have to contain fewer volatile solvents.

CURRENT CONDITIONS AND FUTURE PROSPECTS

Although the United states has not yet achieved the Clean Air Act goals in many parts of the country, air quality has improved dramatically in the last decade in terms of the major large-volume pollutants. For 23 of the largest U.S. cities, the number of days each year in which air quality reached the hazardous level is down 93 percent from a decade ago. Of 97 metropolitan areas that failed to meet clean air standards in the 1980s, 41 were in compliance in

Comparison of 1970 and 1994 emissions
(24% decrease for all pollutants)

FIGURE 9.29 *Air pollution trends in the United States, 1970–1998. Although population and economic activity increased during this period, emissions of all "criteria" air pollutants except for nitrogen oxides and particulate matter decreased significantly.*
Source: Data from Environmental Protection Agency, 2002.

1991–1992. For many cities, this was the first time they met air quality goals in 20 years. Still, the EPA estimates that some 86 million Americans breathe unhealthy air at least part of the time.

The EPA estimates that between 1970 and 1994, lead fell 98 percent, SO_2 declined 35 percent, and CO shrank 32 percent (fig. 9.29). Filters, scrubbers, and precipitators on power plants and other large stationary sources are responsible for most of the particulate and SO_2 reductions. Catalytic converters on automobiles are responsible for most of the CO and O_3 reductions.

The only conventional "criteria" pollutants that have not dropped significantly are particulates and NO_x. Because automobiles are the main source of NO_x, cities, such as Los Angeles, where pollution comes largely from traffic, still have serious air quality problems. Particulate matter (mostly dust and soot) is produced by agriculture, fuel combustion, metal smelting, concrete manufacturing, and other activities. Industrial cities, such as Baltimore, MD, and Baton Rouge, LA, also have continuing problems. In 1999, for the first time in many years, Houston passed Los Angeles as having the worst air quality in the country. Eighty-five other urban areas are still considered nonattainment regions. In spite of these local failures, however, 80 percent of the United States now meets the National Ambient Air Quality Standards. This improvement in air quality is perhaps the greatest environmental success story in our history.

Air Pollution in Developing Countries

The outlook is not so encouraging in other parts of the world. The major metropolitan areas of many developing countries are growing at explosive rates to incredible sizes (see chapter 14), and environmental quality is abysmal in many of them. Mexico City remains notorious for bad air. Pollution levels exceed WHO health standards 350 days per year, and more than half of all city children have lead levels in their blood high enough to lower intelligence and retard development. Mexico City's 131,000 industries and 2.5 million vehicles spew out more than 5,500 tons of air pollutants daily. Santiago, Chile, averages 299 days per year on which suspended particulates exceed WHO standards of 90 mg/m³.

While there are few statistics on China's pollution situation, it is known that many of China's 400,000 factories have no air pollution controls. Experts estimate that home coal burners and factories emit 10 million tons of soot and 15 million tons of sulfur dioxide annually and that emissions have increased rapidly over the past 20 years. Sheyang, an industrial city in northern China, is thought to have the world's worst continuing particulate problem, with peak winter concentrations over 700 mg/m³ (nine times U.S. maximum standards). Airborne particulates in Sheyang exceed WHO standards on 347 days per year. Beijing, Xi'an, and Guangzhou are nearly as bad (see fig. 9.17). The high incidence of cancer in Shanghai is thought to be linked to air pollution.

As political walls came down across Eastern Europe and the Soviet Union at the end of the 1980s, horrifying environmental conditions in these centrally planned economies were revealed. Inept industrial managers, a rigid bureaucracy, and lack of democracy have created ecological disasters. Where governments own, operate, and regulate industry, there are few checks and balances or incentives to clean up pollution. Much of the Eastern bloc depends heavily on soft, brown coal for its energy, and pollution controls are absent or highly inadequate.

Southern Poland, the northern Czech Republic, and Slovakia are covered most of the time by a permanent cloud of smog from factories and power plants. Acid rain is eating away historic buildings and damaging already inadequate infrastructures. Home gardening has been banned in Katowice, Poland, because vegetables raised there have unsafe levels of lead and cadmium. For miles around the infamous Romanian "black town" of Copsa Mica, the countryside is so stained by soot that it looks as if someone poured black ink over everything. Birth defects afflict 10 percent of infants in northern Bohemia. Workers in factories there get extra hazard pay—burial money, they call it. Life expectancy in these industrial towns is as much as 10 years less than the national average.

Signs of Hope

Not all is pessimistic, however. There have been some spectacular successes in air pollution control. Sweden and West Germany (countries affected by forest losses due to acid precipitation) cut their sulfur emissions by two-thirds between 1970 and 1985. Austria and Switzerland have gone even further, regulating even motorcycle emissions. The Global Environmental Monitoring System (GEMS) reports declines in particulate levels in 26 of 37 cities worldwide. Sulfur dioxide and sulfate particles, which cause acid rain and respiratory disease, have declined in 20 of these cities.

Ten years ago, Cubatao, Brazil, was described as the "Valley of Death," one of the most dangerously polluted places in the world. A steel plant, a huge oil refinery, and fertilizer and chemical factories churned out thousands of tons of air pollutants every year. Trees died on the surrounding hills. Birth defects and respiratory diseases were alarmingly high. Since then, however, the citizens of Cubatao have made remarkable progress in cleaning up their environment. The end of military rule and restoration of democracy allowed residents to publicize their complaints. The environment became an important political issue. The state of São Paulo invested about $100 million and the private sector spent twice as much to clean up most pollution sources in the valley. Particulate pollution was reduced 75 percent, ammonia emissions were reduced 97 percent, hydrocarbons that cause ozone and smog were cut 86 percent, and sulfur dioxide production fell 84 percent. Fish are returning to the rivers, and forests are regrowing on the mountains. Progress is possible! We hope that similar success stories will be obtainable elsewhere.

SUMMARY

The nature of atmospheric gases make our air relatively transparent to visible light that warms the earth's surface and supports photosynthesis. The same gases trap outgoing energy, keeping the earth warm enough for life as we know it. Excessive "greenhouse gases," though, appear to be causing global warming, which will have complex effects.

When air is warmed by the earth's surface, it expands and rises, creating convection currents. These vertical updrafts carry water vapor aloft and initiate circulation patterns that redistribute energy and water from areas of surplus to areas of deficit. Pressure gradients created by this circulation drive great air masses around the globe and generate winds that determine both immediate weather and long-term climate.

The earth's climate changes over time. The El Niño/Southern Oscillation (ENSO) is an important example. ENSO cycles involve a complex interaction between oceans and atmosphere that has far-reaching climatic, ecological, and social effects. Anthropogenic global warming is another important climate change. International agreements, including the Kyoto Protocol, have sought to limit CO_2 production, but so far, economic concerns have prevented adoption in America, the world's largest CO_2 producer. Numerous changes, especially at high latitudes, suggest that global warming already has begun, but doubters claim these may all be merely natural climate anomalies.

Air pollution interacts with climate in important ways: climate patterns concentrate or distribute pollution around the globe, and air pollution can alter climate conditions. Among the most important of these processes are long-range transport of pollutants and photochemical reactions in trapped inversion layers over urban areas. The Clean Air Act regulates major types and sources of air pollution in the United States, and this law has dramatically improved ambient air quality in the past 30 years.

Encouraging improvements have been made in ambient outdoor air quality over most of the United States in the last few decades. We have made considerable progress in designing and installing pollution-control equipment to reduce the major conventional pollutants. Much remains to be done, especially in developing countries and in Eastern Europe, but air pollution control is, perhaps, our greatest success in environmental protection and an encouraging example of what can be accomplished in this field.

QUESTIONS FOR REVIEW

1. What is the greenhouse effect?

2. What is the ENSO cycle?

3. What are some theories explaining major climatic changes?

4. Define primary and secondary air pollutants.

5. What are the seven "criteria" pollutants in the original Clean Air Act? Why were they chosen?

6. What is an atmospheric inversion, and how does it trap air pollutants?

7. What is acid deposition? What causes it?

8. Which of the conventional or "criteria" pollutants has decreased most in the recent past, and which has decreased least?

THINKING SCIENTIFICALLY

1. El Niño is a natural climate process, but some scientists suspect it is changing because of global warming. What sort of evidence would you look for to decide if El Niño has gotten stronger recently?

2. If you look at the map of long-range transport (fig. 9.20), which regions are most likely to produce pollutants that will collect in the Arctic?

3. Global warming appears to be warming night temperatures more than day temperatures, and it is warming winter weather more than summer. Given what you know about global warming, how can you explain this?

4. Explain the process of photochemical oxidation. How does it work? What are important examples?

5. One of the problems with the Kyoto Protocol and with the Clean Air Act is that economists and scientists define problems differently and have contrasting priorities. How would an economist and an ecologist explain disputes over the Kyoto Protocol differently?

6. Many analysts think economists and scientists can reach common terms for defining and solving issues like the Clean Air Act renewal. How might their conflicting definitions be reshaped to make the discussion more successful?

KEY TERMS

acid precipitation 217
aerosols 195
albedo 196
ambient air 210
carbon management 209
carbon monoxide 212
chlorofluorocarbons 213
climate 195
convection currents 195
conventional or "criteria" pollutants 210
Coriolis effect 199
El Niño 202
fugitive emissions 210
greenhouse effect 198
Intergovernmental Panel on Climate Change (IPCC) 203

jet streams 199
Kyoto Protocol 208
latent heat 198
Milankovitch cycles 202
monsoons 200
nitrogen oxides 212
ozone 196
particulate material 212
photochemical oxidants 210
stratosphere 196
sulfur dioxide 212
troposphere 195
unconventional pollutants 212
volatile organic compounds 212
weather 195

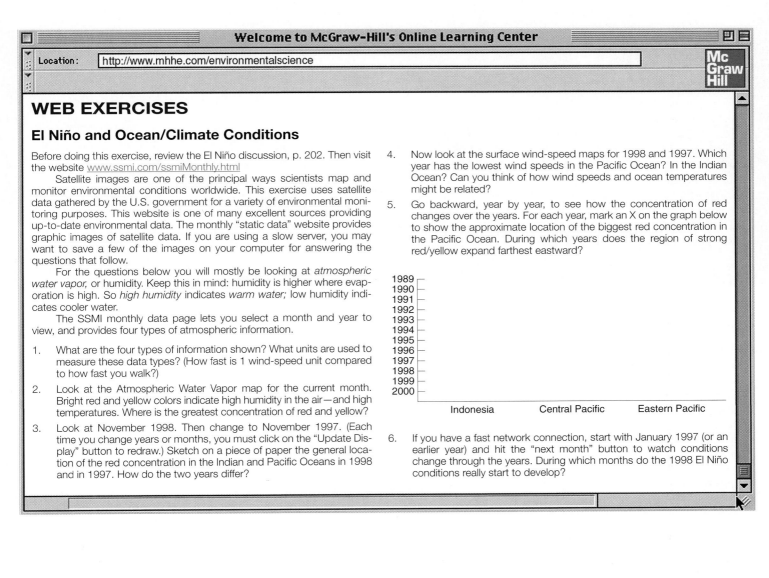

Welcome to McGraw-Hill's Online Learning Center

Location: http://www.mhhe.com/environmentalscience

WEB EXERCISES

El Niño and Ocean/Climate Conditions

Before doing this exercise, review the El Niño discussion, p. 202. Then visit the website www.ssmi.com/ssmiMonthly.html

Satellite images are one of the principal ways scientists map and monitor environmental conditions worldwide. This exercise uses satellite data gathered by the U.S. government for a variety of environmental monitoring purposes. This website is one of many excellent sources providing up-to-date environmental data. The monthly "static data" website provides graphic images of satellite data. If you are using a slow server, you may want to save a few of the images on your computer for answering the questions that follow.

For the questions below you will mostly be looking at *atmospheric water vapor,* or humidity. Keep this in mind: humidity is higher where evaporation is high. So *high humidity* indicates *warm water;* low humidity indicates cooler water.

The SSMI monthly data page lets you select a month and year to view, and provides four types of atmospheric information.

1. What are the four types of information shown? What units are used to measure these data types? (How fast is 1 wind-speed unit compared to how fast you walk?)

2. Look at the Atmospheric Water Vapor map for the current month. Bright red and yellow colors indicate high humidity in the air—and high temperatures. Where is the greatest concentration of red and yellow?

3. Look at November 1998. Then change to November 1997. (Each time you change years or months, you must click on the "Update Display" button to redraw.) Sketch on a piece of paper the general location of the red concentration in the Indian and Pacific Oceans in 1998 and in 1997. How do the two years differ?

4. Now look at the surface wind-speed maps for 1998 and 1997. Which year has the lowest wind speeds in the Pacific Ocean? In the Indian Ocean? Can you think of how wind speeds and ocean temperatures might be related?

5. Go backward, year by year, to see how the concentration of red changes over the years. For each year, mark an X on the graph below to show the approximate location of the biggest red concentration in the Pacific Ocean. During which years does the region of strong red/yellow expand farthest eastward?

```
1989 ┌─
1990 ├─
1991 ├─
1992 ├─
1993 ├─
1994 ├─
1995 ├─
1996 ├─
1997 ├─
1998 ├─
1999 ├─
2000 ├─
     └─────────────────────────────────────
      Indonesia    Central Pacific    Eastern Pacific
```

6. If you have a fast network connection, start with January 1997 (or an earlier year) and hit the "next month" button to watch conditions change through the years. During which months do the 1998 El Niño conditions really start to develop?

10 Water: Resources and Pollution

If there is magic on this planet, it is in water.
–Loren Eisley

By 2025, experts predict that two-thirds of the world's population will live in countries suffering from serious water shortages.

OBJECTIVES

After studying this chapter, you should be able to

- describe the important sources of water and the major ways we use it.
- appreciate the causes and consequences of water shortages around the world and what they mean in people's lives in water-poor countries.
- debate the merits of proposals to increase water supplies and manage demand.
- apply some water conservation methods in your own life.
- define water pollution and describe the sources and effects of some major types.
- appreciate why access to sewage treatment and clean water are important to people in developing countries.
- explain ways to control water pollution, including technological and legal solutions.

The Three Gorges Dam

When finished in 2009, the Three Gorges Dam on the Chang Jiang (Yangtze River) will be the largest dam in the world. Spanning 2.0 km (1.2 mi) and standing 185 m (607 ft) above normal river level, the dam will create a reservoir more than 644 km (400 mi) long. The dam is designed to control floods, ease navigation, and generate 18,200 megawatts of electricity to support industrialization and modernization in China's heartland.

The dam is flooding the scenic Three Gorges, one of China's most picturesque and historically significant areas. It already has displaced more than a million people, and will flood 1,300 cities and villages and more than 8,000 archeological and cultural sites. Water stored in the reservoir will make possible a long-discussed plan to build aqueducts to carry water from southern China to the dry plains around Beijing (fig. 10.1). Planners expect the dam to reduce annual floods that for centuries have caused misery and death for the 300 million people who live in the Yangtze River Valley.

Environmentalists criticize the dam because it will reduce fish stocks, eliminate important agricultural lands, and disrupt habitats and migration patterns of critically endangered species such as the Yangtze River dolphin and the Chinese sturgeon. Currently, about a trillion liters (roughly 260 billion gallons) of untreated sewage is dumped into the Yangtze and flushed out to sea every year. Critics claim the reservoir will become a stagnant cesspool, dangerous to both aquatic life and to the millions of people who depend on the river for their drinking water.

Tremendous volumes of sediment will accumulate as the silt-laden Yangtze River slows in the reservoir. To reduce sediment buildup, dam operators plan to let spring floods flow through silt doors at the bottom of the dam. They hope this flow will scour out the bottom of the reservoir. Critics doubt this will work.

Geologists worry about catastrophic dam failure because the dam is built over an active seismic fault. Engineers are confident the dam can withstand the maximum expected earthquake, but China has a poor record of dam safety. More than 3,200 Chinese dams have failed since 1949. Probably the worst series of dam failures in world history occurred in Henan Province in 1975, when heavy monsoon rains caused 62 large, modern dams to fall like a line of dominoes. Some 230,000 people died in the massive flooding that followed. Even if the dam is able to withstand earthquakes, giant waves spawned by upstream landslides could easily

FIGURE 10.1 *Flowing more than 4,800 km (3,000 mi) from the high Tibetian Plateau, across the southern Gobi Desert, to the Yellow Sea, the Huang He, or Yellow River, is the major water source for much of the arid North China Plain. Currently, agricultural, industrial, and domestic water withdrawals drain the river dry for several months each year.*

cause a calamitous dam failure. In 1986, a landslide just a few miles upstream from the dam site dumped 15,000 m^3 of rock and soil into the river, creating an 80-m-high (260-ft) wave. If a similar wave hits the dam when the reservoir is full, some engineers predict a flood "of biblical proportions" that could kill millions of people downstream. The Chinese government has banned timber cutting and farming on steep upstream hillsides in an effort to control both landslides and sediment loads in the river.

Critics of this project claim that a series of smaller dams on tributary streams might have been much cheaper and less disruptive than the current project. Original estimates were that the Three Gorges Dam would cost $11 billion. By 2002, the costs for construction, relocation, and landscape stabilization had risen to $75 billion, and the project is not yet finished (see chapter 12).

This complex project illustrates the importance of water resources for modern societies. The United Nations warns that water supplies are likely to become one of the most pressing resource issues of the twenty-first century. By 2025, two-thirds of all humans could be living in countries where water supplies are inadequate. Our attempts to reengineer those shrinking resources are increasingly in conflict with natural systems. In this chapter, we'll look at where our fresh water comes from, what we do with it, and how we might protect its quality and extend its usefulness.

WATER RESOURCES

Water is a marvelous substance—flowing, swirling, seeping, constantly moving from sea to land and back again. It shapes the earth's surface and moderates our climate. Water is essential for life. It is the medium in which all living processes occur (see chapter 2). Water dissolves nutrients and distributes them to cells, regulates body temperature, supports structures, and removes waste products. About 60 percent of your body is water. You could sur-

vive for weeks without food, but only a few days without water. An American family of four consumes more than 1,000 m^3 (264,000 gal) of water per year. Families in other countries subsist on a tiny fraction of that amount.

Where Does Our Water Come From?

The water we use cycles endlessly through the environment. The total amount of water on our planet is immense—more than 1,404 million km^3 (370 billion billion gal) (table 10.1). This water evaporates from moist surfaces, falls as rain or snow, passes through living organisms, and returns to the ocean in a process known as the hydrologic cycle (see fig. 2.18). Every year, about 500,000 km^3, or a layer 1.4 m thick, evaporates from the oceans. More than 90 percent of that moisture falls back on the ocean. The 47,000 km^3 carried onshore joins some 72,000 km^3 that evaporate from lakes, rivers, soil, and plants to become our annual, renewable freshwater supply. Plants play a major role in the hydrologic cycle, absorbing groundwater and pumping it into the atmosphere by transpiration (transport plus evaporation). In tropical forests, as much as 75 percent of annual precipitation is returned to the atmosphere by plants.

Solar energy drives the hydrologic cycle by evaporating surface water, which becomes rain and snow. Because water and sunlight are unevenly distributed around the globe, water resources are very uneven. At Iquique in the Chilean desert, for instance, no rain has fallen in recorded history. At the other end of the scale, 22 m (72 ft) of rain was recorded in a single year at Cherrapunji in India. Figure 10.2 shows broad patterns of precipitation around the world. Most of the world's rainiest regions are tropical, where heavy rainy seasons occur, or in coastal mountain regions. Most of the driest areas are in the high-pressure bands of deserts (see chapter 9). Deserts occur on every continent just outside the tropics (the Sahara, the Namib, the Gobi, the Sonoran,

TABLE 10.1 Units of Water Measurement

One cubic kilometer (km^3) equals 1 billion cubic meters (m^3), 1 trillion liters, or 264 billion gal.

One acre-foot is the amount of water required to cover an acre of ground 1 ft deep. This is equivalent to 325,851 gal, or 1.2 million liters, or 1,234 m^3, approximately the amount consumed annually by a family of four in the United States.

One cubic foot per second of river flow equals 28.3 liters per second, or 449 gal per minute.

and many others). Rainfall is also slight at very high latitudes, another high-pressure region.

Mountains also influence moisture distribution. The windward sides of mountain ranges, including the Pacific Northwest and the flanks of the Himalayas, are typically wet and have large rivers; on the leeward sides of mountains, in areas known as the rain shadow, dry conditions dominate, and water can be very scarce. The windward side of Mount Waialeale on the island of Kauai, for example, is one of the wettest places on earth, with an annual rainfall near 12 m (460 in.). The leeward side, only a few kilometers away, has an average yearly rainfall of only 46 cm (18 in.).

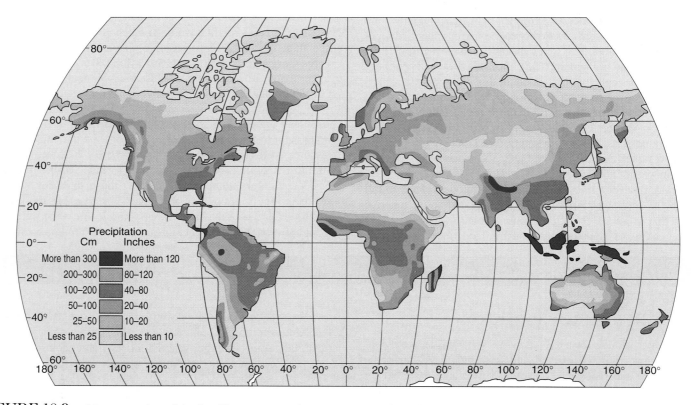

FIGURE 10.2   *Mean annual precipitation. Note wet areas that support tropical rainforests occur along the equator, while the major world deserts occur in zones of dry, descending air between 20 and 40 degrees North and South.*

TABLE 10.2 Earth's Water Compartments

COMPARTMENT	VOLUME (1,000 KM³)	PERCENT OF TOTAL WATER	AVERAGE RESIDENCE TIME
Total	1,386,000	100	2,800 years
Oceans	1,338,000	96.5	3,000 to 30,000 years*
Ice and snow	24,364	1.76	1 to 100,000 years*
Saline groundwater	12,870	0.93	Days to thousands of years*
Fresh groundwater	10,530	0.76	Days to thousands of years*
Fresh lakes	91	0.007	1 to 500 years*
Saline lakes	85	0.006	1 to 1,000 years*
Soil moisture	16.5	0.001	2 weeks to 1 year*
Atmosphere	12.9	0.001	1 week
Marshes, wetlands	11.5	0.001	Months to years
Rivers, streams	2.12	0.0002	1 week to 1 month
Living organisms	1.12	0.0001	1 week

*Depends on depth and other factors
Source: UNEP, 2002.

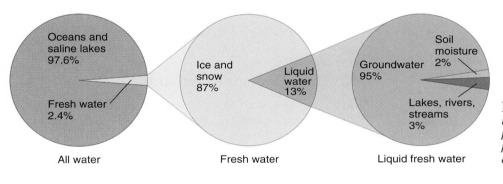

FIGURE 10.3 *The easily accessible water in lakes, rivers, and streams represents only 3 percent of all liquid fresh water, which is 13 percent of all fresh water, which is 2.4 percent of all water on the earth.*

MAJOR WATER COMPARTMENTS

The distribution of water often is described in terms of interacting compartments in which water resides, sometimes briefly and sometimes for eons (table 10.2). The length of time water typically stays in a compartment is its **residence time.** On average, a water molecule stays in the ocean for about 3,000 years, for example, before it evaporates and starts through the hydrologic cycle again. Nearly all the world's water is in the oceans (fig. 10.3). Oceans play a crucial role in moderating the earth's temperature, and over 90 percent of the world's living biomass is contained in the oceans. What we mainly need, though, is fresh water. Of the 2.4 percent that is fresh, most is locked up in glaciers or in groundwater. Amazingly, only about 0.1 percent of the world's water is in a form accessible to us and to other organisms that rely on fresh water (fig. 10.4).

Groundwater

Groundwater is one of our most important freshwater resources. Originating as precipitation that percolates into layers of soil and

FIGURE 10.4 *Water is essential for life, yet only about 0.1 percent of the world's total supply is accessible fresh water.*

FIGURE 10.5 *Precipitation that does not evaporate or run off over the surface percolates through the soil in a process called infiltration. The upper layers of soil hold droplets of moisture between air-filled spaces. Lower layers, where all spaces are filled with water, make up the zone of saturation, or groundwater.*

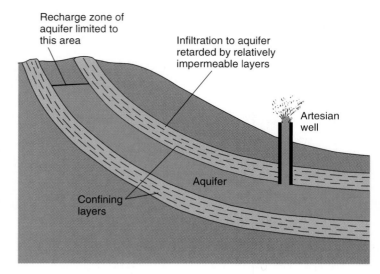

FIGURE 10.6 *An aquifer is a porous, water-bearing layer of sand, gravel, or rock. This aquifer is confined between layers of rock or clay and bent by geological forces, creating hydrostatic pressure. A break in the overlying layer creates an artesian well or spring.*

rock, groundwater makes up the largest compartment of liquid, fresh water (fig. 10.5). The groundwater within 1 km of the surface is more than 100 times the volume of all the freshwater lakes, rivers, and reservoirs combined.

Plants get moisture from a relatively shallow layer of soil containing both air and water, known as the *zone of aeration.* Depending on rainfall amount, soil type, and surface topography, the zone of aeration may be a few centimeters or many meters deep. Lower soil layers, where all soil pores are filled with water, make up the *zone of saturation,* the source of water in most wells; the top of this zone is the **water table.**

Geologic layers that contain water are known as **aquifers.** Aquifers may consist of porous layers of sand or gravel, or of cracked or porous rock. Below an aquifer, relatively impermeable layers of rock or clay keep water from seeping out at the bottom. Instead water seeps more or less horizontally through the porous layer. Depending on geology, it can take anywhere from a few hours to several years for water to move a few hundred meters through an aquifer. If impermeable layers lie above an aquifer, pressure can develop within the water-bearing layer. A well or conduit puncturing the aquifer flows freely at the surface and is called an *artesian well* or spring.

Areas where surface water filters into an aquifer are **recharge zones** (fig. 10.6). Most aquifers recharge extremely slowly, and road and house construction or water use at the surface can further slow recharge rates. Contaminants can also enter aquifers through recharge zones. Urban or agricultural runoff in recharge zones is often a serious problem. About 2 billion people—approximately one-third of the world's population—depend on groundwater for drinking and other uses. Every year

700 km^3 are withdrawn by humans, mostly from shallow, easily polluted aquifers.

Rivers, Lakes, and Wetlands

Fresh, flowing surface water is one of our most precious resources. Rivers contain a minute amount of water at any one time. Most rivers would begin to dry up in weeks or days if they were not constantly replenished by precipitation, snowmelt, or groundwater seepage.

The volume of water carried by a river is its **discharge,** or the amount of water that passes a fixed point in a given amount of time. This is usually expressed as liters or cubic feet of water per second. The 16 largest rivers in the world carry nearly half of all surface runoff on the earth, and a large fraction of that occurs in a single river, the Amazon, which carries ten times as much water as the Mississippi (table 10.3).

Lakes contain nearly 100 times as much water as all rivers and streams combined, but much of this water is in a few of the world's largest lakes. Lake Baikal in Siberia, the Great Lakes of North America, the Great Rift Lakes of Africa, and a few other lakes contain vast amounts of water, not all of it fresh. Worldwide, lakes are almost as important as rivers in terms of water supplies, food, transportation, and settlement.

Wetlands—bogs, swamps, wet meadows, and marshes—play a vital and often unappreciated role in the hydrological cycle. Their lush plant growth stabilizes soil and holds back surface runoff, allowing time for infiltration into aquifers and producing even, year-long stream flow. In the United States, about 20 percent of the 1 billion ha of land area was once wetland. When wetlands are disturbed, their natural water-absorbing capacity is reduced, and surface waters run off quickly, resulting in floods and erosion during the rainy season and low stream flow the rest of the year.

TABLE 10.3 World's Ten Largest Rivers		
RIVER	LOCATION	ANNUAL DISCHARGE (M³/SECOND)*
Amazon	Brazil, Peru	175,000
Orinoco	Venezuela, Colombia	45,300
Congo	Congo	39,200
Yangtze	Tibet, China	28,000
Brahmaputra	South Asia	19,000
Mississippi	United States	18,400
Mekong	Southeast Asia	18,300
Paraná	Paraguay, Argentina	18,000
Yenisey	Russia	17,200
Lena	Russia	16,000

*1 m³ = 264 gal.
Source: Data from World Resource Institute.

The Atmosphere

The atmosphere contains only 0.001 percent of the total water supply, but it is the most important mechanism for redistributing water around the world. An individual water molecule resides in the atmosphere for about 10 days, on average. Some water evaporates and falls within hours. Water can also travel halfway around the world before it falls, replenishing streams and aquifers on land.

WATER AVAILABILITY AND USE

Clean, fresh water is essential for nearly every human endeavor. Perhaps more than any other environmental factor, the availability of water determines the location and activities of humans on the earth (fig. 10.7). **Renewable water supplies** are resources that are replenished regularly—mainly surface water and shallow groundwater. Renewable water is most plentiful in the tropics, where rainfall is heavy, followed by midlatitudes, where rainfall is regular.

Water-Rich and Water-Poor Countries

Water availability is usually measured in terms of renewable water per capita, so population density, as well as total water volumes, dictate renewable supplies for human use. The highest per capita water supplies generally occur in countries with moist climates and low population densities. Iceland, for example, has about 160 million gal per person per year. In contrast, Kuwait, where temperatures are extremely high and rain almost never falls, has less than 3,000 gal per person per year from renewable natural sources. Almost all of Kuwait's water comes from imports and desalinized seawater. In countries such as Libya and Israel, where water is one of the most crucial environmental resources, renewable supplies

FIGURE 10.7 *Water has always been the key to survival. Who has access to this precious resource and who doesn't has long been a source of tension and conflict.*

APPLICATION:	Mapping the Water-Rich and Water-Poor Countries

The top ten water-rich countries, in terms of water availability per capita, and the ten most water-poor countries are listed below. Locate these countries on the political map (page 369). Describe the patterns: Where are the water-rich countries concentrated? (Hint: Does latitude matter?) Where are the water-poor countries most concentrated?

Water-rich countries: Iceland, Surinam, Guyana, Papua New Guinea, Gabon, Solomon Islands, Canada, Norway, Panama, Brazil

Water-poor countries: Kuwait, Egypt, United Arab Emirates, Malta, Jordan, Saudi Arabia, Singapore, Moldavia, Israel, Oman

Answers: Water-rich countries (per capita) are either in the far north where populations are low, or in the tropics. Water-poor countries are in the desert belt at about 15° to 25° latitude or are densely populated island nations (Malta, Singapore).

do not meet basic needs. These countries manage by "mining" groundwater, depleting sources that are probably unrenewable on a human time scale.

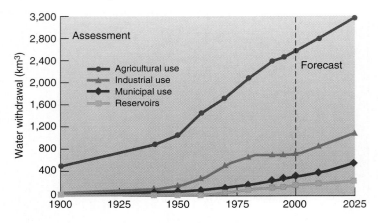

FIGURE 10.8 *Water withdrawals have increased more than seven-fold over the past century, and are expected to continue to rise in the future.*
Source: UNESCO, 2001.

Interannual variability in rainfall is an important issue in water availability. In some areas, such as the African Sahel region, abundant rainfall occurs some years but not others. Usually natural ecosystems can survive these changes, but human societies, and ecosystems greatly altered by grazing, farming, or urban development, can be badly destabilized by rainfall fluctuations. Some of the world's earliest civilizations, such as the Sumerians and Babylonians of Mesopotamia, were based on communal efforts to divert floods during wet seasons or wet years, and to store water in dry years. Many climatologists now worry that the greenhouse effect (see chapter 9) will bring about more serious or frequent droughts in dry parts of the world.

Water Use

In contrast to energy resources, which usually are consumed when used, water can be used over and over if it is not too badly conta-

FIGURE 10.9 *Rolling sprinklers allow farmers to irrigate crops on uneven terrain. In some areas, irrigation consumes 90 percent of all available water.*

FIGURE 10.10 *Drip irrigation delivers measured amounts of water exactly where the plants need and can use it. This technique can save up to 90 percent of irrigation water usage and reduces salt buildup. It also is an efficient way to deliver soluble nutrients.*

minated. Water **withdrawal** is the total amount of water taken from a water body. Much of this water could be returned to circulation in a reusable form. Water **consumption,** on the other hand, is loss of water due to evaporation, absorption, or contamination.

The natural cleansing and renewing functions of the hydrologic cycle replace the water we need if natural systems are not overloaded or damaged. Water is a renewable resource, but renewal takes time. The rate at which many of us now use water may make it necessary to conscientiously protect, conserve, and replenish our water supply.

Quantities of Water Used

Water use has been increasing about twice as fast as population growth over the past century. Water withdrawals are expected to continue to grow as more land is irrigated to feed an expanding population (fig. 10.8). Conflicts increase as different countries, economic sectors, and other stakeholders compete for the same, limited water supply. Water wars may well be the major source of hostilities in the twenty-first century. (See the related story "Water Conflicts in Israel" at www.mhhe.com/apps.)

Worldwide, agriculture claims about 70 percent of total water withdrawal, ranging from 93 percent of all water used in India to only 4 percent in Kuwait, which cannot afford to spend its limited water on crops. In many developing countries and in parts of the United States, the most common type of irrigation is to simply flood the whole field or run water in rows between crops. As much as half the water can be lost through evaporation or seepage from unlined irrigation canals bringing water to fields. Sprinklers are more efficient in distributing water, but they are more costly and energy intensive (fig. 10.9). Water-efficient drip irrigation can save significant amounts of water but currently is used on only about 1 percent of the world's croplands (fig. 10.10).

Industry uses about one-fourth of water withdrawals worldwide. Some European countries use 70 percent of water for industry; less industrialized countries use as little as 5 percent. Cooling water for power plants is by far the largest single industrial use of water, typically accounting for 50 to 75 percent of industrial withdrawal.

Domestic, or household, water use accounts for only about 6 percent of world water use. This includes water for drinking, cooking, and washing. The amount of water used per household varies enormously, however, depending on a country's wealth.

FRESHWATER SHORTAGES

Clean drinking water and basic sanitation are necessary to prevent communicable diseases and to maintain a healthy life. For many of the world's poorest people, one of the greatest environmental threats to health remains the continued use of polluted water. In 2002 the United Nations estimated that at least 1.1 billion people lacked access to safe drinking water and 2.4 billion didn't have adequate sanitation. These deficiencies result in hundreds of millions of cases of water-related illness and more than 5 million deaths every year. As populations grow, more people move into cities, and agriculture and industry compete for increasingly scarce water supplies, water shortages are expected to become even worse. By 2025, two-thirds of the world's people will be living in water-stressed countries—defined by the United Nations as consumption of more than 10 percent of renewable freshwater resources. One of the highest priorities announced at the UN World Summit in Johannesburg in 2002, is to reduce by one-half the proportion of people without reliable access to clean water and improved sanitation.

A Precious Resource

The World Health Organization considers 1,000 m³ (264,000 gal) per person per year to be a necessary amount of water for human health and development. Some 45 countries, most of them in Africa or the Middle East, cannot meet the minimum essential water needs of all their citizens. In some countries, the problem is access to *clean* water. In Mali, for example, 88 percent of the population lacks clean water; in Ethiopia, it is 94 percent. Rural people often have less access to clean water than do city dwellers. Causes of water shortages include natural deficits, overconsumption by agriculture or industry, and inadequate funds for purifying and delivering good water.

More than two-thirds of the world's households have to fetch water from outside the home (fig. 10.11). This is heavy work, done mainly by women and children, and sometimes taking two hours a day. Improved public systems bring many benefits to these poor families.

Availability does not always mean affordability. A typical poor family in Lima, Peru, for instance, uses one-sixth as much water as a middle-class American family but pays three times as much for it. If they followed government recommendations to boil all water to prevent cholera, up to one-third of the poor family's income could be used just in acquiring and purifying water.

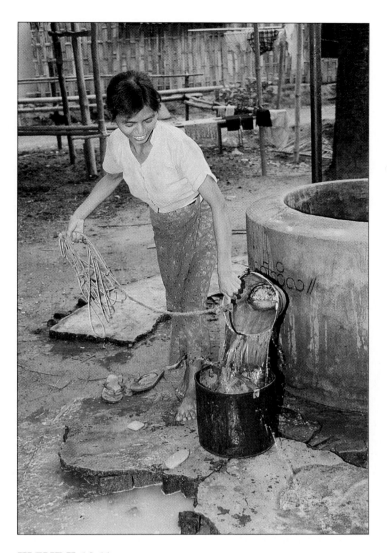

FIGURE 10.11 *Drawing water from a village well is a daily chore in many developing countries.*

Investments in rural development have brought significant improvements over the past decade. Since 1990, nearly 800 million people—about 13 percent of the world's population—have gained access to clean water. The percentage of rural families with safe drinking water has risen from less than 10 percent to nearly 75 percent.

Depleting Groundwater

Groundwater provides nearly 40 percent of the fresh water for agricultural and domestic use in the United States. Nearly half of all Americans and about 95 percent of the rural population depend on groundwater for drinking and other domestic purposes. Overuse of these supplies dries up wells, natural springs, and even groundwater-fed wetlands, rivers, and lakes. Pollution of aquifers through dumping of contaminants on recharge zones, leaks through abandoned wells, or deliberate injection of toxic wastes can make this valuable resource unfit for use.

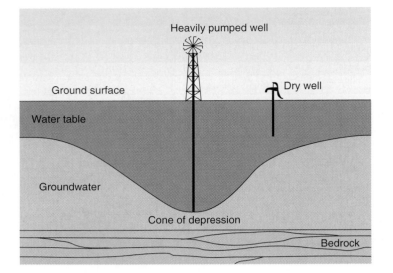

FIGURE 10.12   *A cone of depression forms in the water table under a heavily pumped well. This may dry up nearby shallow wells or make pumping so expensive that it becomes impractical.*

In many areas of the United States, groundwater is being withdrawn from aquifers faster than natural recharge can replace it. On a local level, this causes a cone of depression in the water table (fig. 10.12). On a broader scale, heavy pumping can deplete a whole aquifer. The Ogallala Aquifer underlies eight arid states from Texas to North Dakota. This porous bed of sand, gravel, and sandstone once held more water than all the freshwater lakes, streams, and rivers on the earth. Excessive pumping for irrigation has removed so much water that wells have dried up in many places, and farms, ranches, even whole towns are being abandoned. Recharging many such aquifers will take thousands of years. Using "fossil" water like this is essentially water mining. For all practical purposes, these aquifers are nonrenewable resources.

Water withdrawal also allows aquifers to collapse. Subsidence, or sinking of the ground surface, follows. The San Joaquin Valley in California has sunk more than 10 m in the last 50 years because of excessive groundwater pumping. Where aquifers become compressed, recharge becomes impossible.

Another consequence of aquifer depletion is saltwater intrusion. Along coastlines and in areas where saltwater deposits are left from ancient oceans, overuse of freshwater reservoirs often allows saltwater to intrude into aquifers used for domestic and agricultural purposes.

Can We Increase Water Supplies?

On a human time scale, the amount of water on the earth is fixed. Many efforts have been made to redistribute water resources, however. Towing icebergs from Antarctica has been proposed, and creating rain in dry regions has been accomplished, with mixed success, by cloud seeding—distributing condensation nuclei in humid air to help form rain drops. Desalination is locally important: in the arid Middle East, where energy and money are available

but water is scarce, desalination is sometimes the principal source of water. Most efforts, however, have involved dams, canals, water diversions, and desalination. There are, however, several ways to increase local supplies, which we will discuss next.

Dams, Reservoirs, and Canals

Dams and canals are a fundamental basis of civilization; they can also be a source of environmental disaster and injustice. Some of the great civilizations (Sumeria, Egypt, China, and the Incan culture of South America) were organized around the large-scale redistribution of water from rivers to irrigated farm fields. More than half the world's 227 largest rivers have been blocked by dams or diversion structures with adverse effects on freshwater ecosystems (fig. 10.13). Of the 50,000 large dams in the world, 90 percent were built in the twentieth century, and half of those are in China. Economically speaking, at least one-third of those dams should never have been built.

Dams and Justice

While many people benefit from the water and hydroelectricity provided by dams and diversion projects, other stakeholders, including wildlife and ecosystems, suffer. Towns and farms have been starved by the huge dams and diversions. Fishing enthusiasts, whitewater boaters, and others mourn the loss of rivers

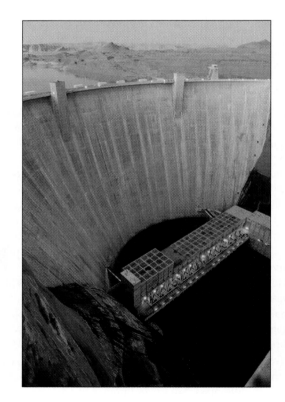

FIGURE 10.13   *The Glen Canyon Dam on the Colorado River drowned a spectacular canyon and changed sediment deposition patterns in the Grand Canyon downstream.*

CASE STUDY

WATER WARS ON THE KLAMATH

Until a century ago, the Klamath River was the third most productive salmon fishery south of the Canadian border. The 35-km (22-mi)-long Upper Klamath Lake, the largest lake in the Pacific Northwest, once contained great schools of C'wam and Qapdo (Lost River and shortnose suckers). Millions of hectares of marshes, lakes, and dry steppe in the Klamath Basin teemed with waterfowl and wildlife. At least 80 percent of the birds following the Pacific Flyway stopped to feed in the area. Native American tribes, including the Klamath and Paiute in the upper basin, and the Yurok and Hupa downstream, depended on the abundant fish and wildlife for their survival.

In 1905 the newly formed United States Bureau of Reclamation was directed to "reclaim the sunbaked prairies and worthless swamps" in the upper Klamath Basin. Spending $50 million, the bureau built 7 dams, 45 pumping stations, and more than 1,600 km (1,000 mi) of canals and ditches. The project drained three-quarters of the wetlands in the upper basin, and provided irrigation water to 90,000 ha (220,000 acres) of cropland. Promises of cheap land and subsidized water lured some 1,400 farmers to the valley to grow potatoes, alfalfa, sugar beets, mint, onions, and cattle. In the 1990s, irrigators in the Klamath Basin used almost 1 million acre-feet (325 billion gallons or 1.2 trillion liters) of surface water per year.

These water diversions left salmon stranded in dried-up streams, while oxygen-depleted lake water was contaminated with agricultural runoff and clogged with algae. Downstream, Native American tribes and commercial fishermen, who once had brought in about 500,000 kg (roughly a million pounds) of salmon per year, saw their catches decline by as much as 90 percent. More than 7,000 jobs were lost when the fisheries collapsed. In 1997, the C'wam and Qapdo were declared endangered, and Coho salmon were listed as threatened. A coalition of commercial and sports fishermen, environmentalists, and native people sued the government for damaging fish and wildlife resources. A federal judge agreed, and ordered the Bureau of Reclamation to reduce irrigation flow and to maintain minimum water levels in rivers and lakes.

A severe drought in 2001 precipitated a crisis. For the first time in its history, the bureau closed the gates on its irrigation canals and cut off water to the farmers. Outraged at being denied the water they had come to believe was their inalienable right, angry farmers broke open headgate locks and let water flow to their drying fields. News media flocked to the site to film the confrontation. Conservative politicians condemned government actions and pledged to uphold property rights and to eliminate the Endangered Species Act. Progressives supported fishing and tribal rights. And environmentalists called for more water for the Lower Klamath Wildlife Refuge, the first in the United States established to protect migratory waterfowl, and home to the largest population of wintering bald eagles in the lower 48 states.

Sympathizing with the farmers, Interior Secretary Gale Norton went to Upper Klamath Lake and personally opened the gates to give them more water. Favoring a market-based approach, she suggested a "water bank," in which farmers could voluntarily sell surplus water to Indian tribes or the Fish and Wildlife Service. The tribes and others objected that, while this approach might work when water was abundant, there probably wouldn't be any surplus in dry years.

Conservationists argue that two-thirds of Klamath County's irrigated acreage is watered with inefficient flood irrigation in which about half the water is lost to evaporation. Farmers could save water by installing efficient drip or sprinkler systems, or they could abandon water-demanding crops such as alfalfa and potatoes and convert to dryland agriculture. Farmers demanded compensation for the "taking" of water they considered promised to them in perpetuity.

This case illustrates some of the complexity and importance of water resources. With growing demands on dwindling supplies, it becomes increasingly difficult to satisfy all the needs we have for this precious commodity. If you were Secretary of the Interior, how would you weigh the competing claims for water in the Klamath Basin?

The Klamath River rises in the Cascade Mountains and flows about 400 km (250 mi) across Oregon and Washington to the Pacific Ocean.

drowned in reservoirs or dried up by diversion projects. These projects also have been criticized for using public funds to increase the value of privately held farmland and for encouraging agricultural development and urban growth in arid lands, where other uses might be more appropriate (see Case Study, p. 233).

Worldwide, large dams often flood towns and farmlands, raising international outcry. In India, the Sardar Sarovar Dam on the sacred Narmada River has been the focus of decades of protest. Many of the one million villagers and tribal people being displaced by this project have engaged in mass resistance and civil disobedience, while police have tried to remove them forcibly. In Nepal, construction of the 240-m (850-ft)-high Tehri Dam on the Bhagirathi River has stirred fears that a strong earthquake in this active seismic region might cause the dam to collapse and cause a catastrophic flood downstream. This dam is only one of 17 high dams that Nepal and India plan for the Himalaya Mountains.

Canada's James Bay project built by Hydro-Quebec has diverted three major rivers flowing west into Hudson Bay and flooded more than 10,000 km² (4,000 mi²) of forest and tundra to generate 26,000 megawatts of electrical power. In 1984, 10,000 caribou drowned while trying to follow ancient migration routes across the newly flooded land. The loss of traditional hunting and fishing sites has been culturally devastating for native Cree people. In addition, mercury leeched out of rocks in newly submerged lands has entered the food chain, and many residents show signs of mercury poisoning.

Environmental Costs of Dams

Dams ensure a year-round water supply, but they also waste tremendous amounts of water from evaporation and through seepage into porous rock beds. Some dams built in the western United States lose more water than they make available. Evaporative losses from Lake Mead and Lake Powell on the Colorado River are about 1 km³ (264 billion gal) per year. The salts left behind by evaporation and agricultural runoff nearly double the salinity of the river. Mexico had to sue the United States to force construction of a $350 million desalination plant at Yuma, Arizona, to make the water partially usable again.

Dams also collect silt, decreasing the effectiveness of reservoirs and starving streambeds and sandbars downstream. As the turbulent Colorado River slows in the reservoirs created by Glen Canyon and Boulder Dams, it drops its load of suspended material. More than 10 million metric tons of silt per year collect behind these dams. Imagine a line of 20,000 dump trucks backed up to Lake Mead and Lake Powell every day, dumping dirt into the water. Within as little as 100 years, these reservoirs could be full of silt and useless for either water storage or hydroelectric generation (fig. 10.14).

In Egypt, the Aswân High Dam was built to irrigate thousands of hectares of farmland, but the dam loses much of the Nile's river to evaporation. Without the annual floods that carried rich silt to farmlands for thousands of years, many farming areas are becoming infertile, and the famous Nile Delta—and its rich fisheries—are disappearing.

Dams and river channelization also drown or destroy free-flowing rivers. One of the first and most divisive battles over this

FIGURE 10.14 *This dam is now useless because its reservoir has filled with silt and sediment.*

loss was in the Hetch Hetchy Valley in Yosemite National Park. In the early 1900s, San Francisco wanted to dam the Tuolumne River to produce hydroelectric power and provide water for the city water system. This project was supported by many prominent San Francisco citizens because it represented an opportunity for both clean water and municipal power. John Muir, founder of the Sierra Club and protector of Yosemite Park, said: "These temple destroyers, devotees of ravaging commercialism, seem to have perfect contempt for Nature, and, instead of lifting their eyes to the God of the mountains, lift them to the Almighty Dollar. Dam Hetch Hetchy! As well dam for water-tanks the people's cathedrals and churches, for no holier temple has ever been consecrated by the heart of man." After a prolonged and bitter fight, the developers won, and the dam was built.

Price Mechanisms and Water Policy

Throughout most of U.S. history, water policies have generally worked against conservation. In the well-watered eastern United States, water policy was based on riparian use rights: those who lived along a river bank had the right to use as much water as they liked as long as they didn't interfere with its quality or availability to neighbors downstream. In the drier western regions, where water is often a limiting resource, water law is based primarily on the Spanish system of prior appropriation rights, or "first in time are first in right." But the appropriated water had to be put to "beneficial" use by being consumed. This creates a policy of "use it or lose it." Water left in a stream, even if essential for recreation, aesthetic enjoyment, or to sustain ecological communities, is not being appropriated or put to "beneficial" (that is, economic) use. Under this system, water rights can be bought and sold, but water owners frequently are reluctant to conserve water for fear of losing their rights.

In most federal "reclamation" projects, customers have been charged only for the immediate costs of water delivery. The costs of building dams and distribution systems were subsidized, and the potential value of competing uses was routinely ignored.

Farmers in California's Central Valley, for instance, for many years paid only about one-tenth of what it cost the government to supply water to them. This didn't encourage conservation. Subsidies created by underpriced water amounted to as much as $500,000 per farm per year in some areas.

WATER MANAGEMENT AND CONSERVATION

Watershed management and conservation are often more economical and environmentally sound ways to prevent flood damage and store water for future use than building huge dams and reservoirs. A **watershed,** or catchment, is all the land drained by a stream or river. It has long been recognized that retaining vegetation and ground cover in a watershed helps to hold back rainwater and lessens downstream floods. In 1998, Chinese officials acknowledged that unregulated timber cutting upstream on the Yangtze contributed to massive floods that killed 30,000 people. Similarly, after disastrous floods in the upper Mississippi valley in 1993, it was suggested that, rather than allowing residential, commercial, or industrial development on floodplains, these areas should be reserved for water storage, aquifer recharge, wildlife habitat, and agriculture.

Sound farming and forestry practices can reduce runoff. Retaining crop residue on fields reduces flooding, and minimizing plowing and forest cutting on steep slopes protects watersheds. Wetlands conservation preserves natural water storage capacity and aquifer recharge zones. A river fed by marshes and wet meadows tends to run consistently clear and steady, rather than in violent floods.

A series of small dams on tributary streams can hold back water before it becomes a great flood. Ponds formed by these dams provide useful wildlife habitat and stock-watering facilities. They also catch soil where it could be returned to the fields. Small dams can be built with simple equipment and local labor, eliminating the need for massive construction projects and huge dams.

In 1998, U.S. Forest Service chief, Mike Dombeck, announced a major shift in his agency's priorities. "Water," he said, "is the most valuable and least appreciated resource the national forests provide. More than 60 million people in 33 states obtain their drinking water from national forest lands. Protecting watersheds is far more economically important than logging or mining, and will be given the highest priority in forest planning."

You Can Make a Difference: Domestic Conservation

We could probably save as much as half of the water we now use for domestic purposes without great sacrifice or serious changes in our lifestyles. Simple steps, such as taking shorter showers, fixing leaks, and washing cars, dishes, and clothes as efficiently as possible, can go a long way toward forestalling the water shortages that many authorities predict. Isn't it better to adapt to more conservative uses now when we have a choice than to be forced to do it by scarcity in the future?

what can you do?

Saving Water and Preventing Pollution

Each of us can conserve much of the water we use and avoid water pollution in many simple ways.

- Don't flush every time you use the toilet. Take shorter showers, and shower instead of taking baths.
- Don't let the faucet run while brushing your teeth or washing dishes. Draw a basin of water for washing and another for rinsing dishes. Don't run the dishwasher when half full.
- Use water-conserving appliances: low-flow showers, low-flush toilets, and aerated faucets.
- Fix leaking faucets, tubs, and toilets. A leaky toilet can waste 50 gal per day. To check your toilet, add a few drops of dark food coloring to the tank and wait 15 minutes. If the tank is leaking, the water in the bowl will change color.
- Put a brick or full water bottle in your toilet tank to reduce the volume of water in each flush.
- Dispose of used motor oil, household hazardous waste, batteries, etc., responsibly. Don't dump anything down a storm sewer that you wouldn't want to drink.
- Avoid using toxic or hazardous chemicals for simple cleaning or plumbing jobs. A plunger or plumber's snake will often unclog a drain just as well as caustic acids or lye. Hot water and soap can accomplish most cleaning tasks.
- If you have a lawn, or know someone who does, use water, fertilizer, and pesticides sparingly. Plant native, low-maintenance plants that have low water needs.
- Use recycled (gray) water for lawns, house plants, and car washing.

Conserving appliances, such as low-volume shower heads and efficient dishwashers, can reduce water consumption greatly (see What Can You Do?, above). If you live in an arid part of the country, you might consider whether you really need a lush, green lawn that requires constant watering, feeding, and care. Planting native ground cover in a "natural lawn" or developing a rock garden or landscape in harmony with the surrounding ecosystem can be both ecologically sound and aesthetically pleasing (fig. 10.15). Cultivated lawns, golf courses, and parks in the United States use more water, fertilizer, and pesticides per hectare than any other kind of land.

Toilets are our greatest water user (table 10.4). Usually each flush uses several gallons of water to dispose of a few ounces of waste. Each person in the United States uses about 50,000 liters (13,000 gal) of drinking-quality water annually to flush toilets. Low-flush toilets and composting toilets can drastically reduce this water use. Low-flow shower heads can reduce our second-largest household water use.

FIGURE 10.15 *By using native plants in a natural setting, residents of Phoenix save water and fit into the surrounding landscape.*

TABLE 10.4 Typical Household Water Use in the United States

WATER USE	PERCENT	GALLONS/ PERSON/YEAR
Toilet flushing	38	13,000
Bathing	31	10,600
Laundry and dishes	20	6,800
Drinking and cooking	6	2,000
Teeth brushing, etc.	5	1,700
Total	100	34,200

These steps are so important that a number of cities (including Los Angeles, Orlando, Austin, and Phoenix) ordered that water-saving toilets, showers, and faucets be installed in all new buildings. The motivation was twofold: to relieve overburdened sewer systems and to conserve water.

Significant amounts of water also can be reclaimed and recycled (see the discussion of sewage treatment following). California uses more than 555 million m³ (450,000 acre-feet) of recycled water annually. That's equivalent to about two-thirds of the water consumed by Los Angeles every year.

Signs of Progress

Growing recognition that water is a precious and finite resource has changed policies and encouraged conservation across the United States. Despite a growing population, the United States is now saving some 144 million liters (38 million gal) per day—a tenth the volume of Lake Erie—compared to per capita consumption rates of 20 years ago. With 37 million more people in the United States now than in 1980, we get by with 10 percent less water. New requirements for water-efficient fixtures and low-flush toilets in many cities help to conserve water on the home front. More efficient irrigation methods on farms also are a major reason for the downward trend. New sprinkler systems have small spray heads just a foot or so above the plant tops and apply water much more directly. Even better is drip irrigation, which applies water directly to plant roots (see fig. 10.10). California and Florida farmers currently water about 500,000 ha (1.3 million acres) with this technique.

Charging a higher proportion of real costs to users of public water projects has helped encourage conservation, and so have water marketing policies that allow prospective users to bid on water rights. Both the United States and Australia have had effective water pricing and allocation policies that encourage the most socially beneficial uses and discourage wasteful water uses.

It will be important, as water markets develop, to be sure that environmental, recreational, and wildlife values are not sacrificed to the lure of high-bidding industrial and domestic uses.

WATER POLLUTION

Any physical, biological, or chemical change in water quality that adversely affects living organisms or makes water unsuitable for desired uses can be considered pollution. There are natural sources of water contamination, such as poison springs, oil seeps, and sedimentation from erosion, but here we will focus primarily on human-caused changes that affect water quality or usability.

Point and Nonpoint Source Pollution

Pollution control standards and regulations usually distinguish between point and nonpoint pollution sources. Factories, power plants, sewage treatment plants, underground coal mines, and oil wells are classified as **point sources** because they discharge pollution from specific locations, such as drain pipes, ditches, or sewer outfalls (fig. 10.16). These sources are discrete and identifiable, so they are relatively easy to monitor and regulate. It is generally possible to divert effluent from the waste streams of these sources and treat it before it enters the environment.

In contrast, **nonpoint sources** of water pollution are scattered or diffuse, having no specific location where they discharge into a particular body of water. They are much harder to monitor and regulate than point sources because their sources are hard to identify. Nonpoint sources include runoff from farm fields and feedlots, golf courses, lawns and gardens, construction sites, logging areas, roads, streets, and parking lots (fig. 10.17). Whereas point sources may be fairly uniform and predictable throughout the year, nonpoint sources are often highly episodic. The first heavy rainfall after a dry period may flush high concentrations of gasoline, lead, oil, and rubber residues off city streets, for instance, while subsequent runoff may be much cleaner.

Perhaps the ultimate in diffuse, nonpoint pollution is atmospheric deposition of contaminants carried by air currents and precipitated into watersheds or directly onto surface waters as rain,

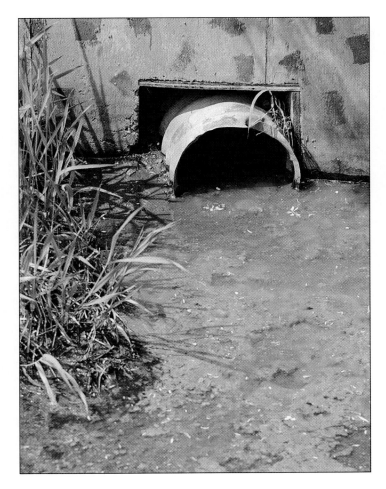

FIGURE 10.16 *Point-source pollution is easier to identify and control than nonpoint-source pollution.*

FIGURE 10.17 *This bucolic scene looks peaceful and idyllic, but allowing cows to trample stream banks is a major cause of bank erosion and water pollution. Nonpoint sources such as this have become the leading unresolved cause of stream and lake pollution in the United States.*

snow, or dry particles. The Great Lakes, for example, have been found to be accumulating industrial chemicals, such as PCBs (polychlorinated biphenyls) and dioxins, as well as agricultural toxins, such as the insecticide toxaphene, that cannot be accounted for by local sources alone. The nearest sources for many of these chemicals are sometimes thousands of kilometers away.

Biological Pollution

Although the types, sources, and effects of water pollutants are often interrelated, it is convenient to divide them into major categories for discussion (table 10.5). Here, we look at some of the important sources and effects of different pollutants.

Infectious Agents

The most serious water pollutants in terms of human health worldwide are pathogenic organisms (see chapter 8). Among the most important waterborne diseases are typhoid, cholera, bacterial and amoebic dysentery, enteritis, polio, infectious hepatitis, and schistosomiasis. Malaria, yellow fever, and filariasis are transmitted by insects that have aquatic larvae. Altogether, at least 25 million deaths each year are blamed on these water-related diseases. Nearly two-thirds of the mortalities of children under 5 years old are associated with waterborne diseases.

The main source of these pathogens is untreated or improperly treated human wastes. Animal wastes from feedlots or fields near waterways and food processing factories with inadequate waste treatment facilities also are sources of disease-causing organisms.

In developed countries, sewage treatment plants and other pollution-control techniques have reduced or eliminated most of the worst sources of pathogens in inland surface waters. Furthermore, drinking water is generally disinfected by chlorination, so epidemics of waterborne diseases are rare in these countries. The United Nations estimates that 90 percent of the people in developed countries have adequate (safe) sewage disposal, and 95 percent have clean drinking water.

The situation is quite different in less-developed countries, where billions of people lack adequate sanitation and access to clean drinking water. Conditions are especially bad in remote, rural areas, where sewage treatment is usually primitive or nonexistent, and purified water is either unavailable or too expensive to obtain. The World Health Organization estimates that 80 percent of all sickness and disease in less-developed countries can be attributed to waterborne infectious agents and inadequate sanitation.

Fecal Coliform Bacteria and Oxygen Demand

Detecting specific pathogens in water is difficult, time consuming, and costly, so water quality is usually described in terms of

TABLE 10.5 Major Categories of Water Pollutants

CATEGORY	EXAMPLES	SOURCES
A. CAUSES HEALTH PROBLEMS		
1. Infectious agents	Bacteria, viruses, parasites	Human and animal excreta
2. Organic chemicals	Pesticides, plastics, detergents, oil, and gasoline	Industrial, household, and farm use
3. Inorganic chemicals	Acids, caustics, salts, metals	Industrial effluents, household cleaners, surface runoff
4. Radioactive materials	Uranium, thorium, cesium, iodine, radon	Mining and processing of ores, power plants, weapons production, natural sources
B. CAUSES ECOSYSTEM DISRUPTION		
1. Sediment	Soil, silt	Land erosion
2. Plant nutrients	Nitrates, phosphates, ammonium	Agricultural and urban fertilizers, sewage, manure
3. Oxygen-demanding wastes	Animal manure and plant residues	Sewage, agricultural runoff, paper mills, food processing
4. Thermal	Heat	Power plants, industrial cooling

concentrations of **coliform bacteria**—any of the many types that live in the colon or intestines of humans and other animals. The most common of these is *Escherichia coli* (or *E. coli*), which lives symbiotically in many animals, but other bacteria, such as *Shigella, Salmonella,* or *Listeria,* can cause fatal diseases. If any coliform bacteria are present in a water sample, infectious pathogens are usually assumed to be present also. Therefore, the Environmental Protection Agency (EPA) considers water with any coliform bacteria at all to be unsafe for drinking.

The amount of oxygen dissolved in water is a good indicator of water quality and of the kinds of life it will support. An oxygen content above 6 parts per million (ppm) will support game fish and other desirable forms of aquatic life. At oxygen levels below 2 ppm, water will support mainly worms, bacteria, fungi, and other detritus feeders and decomposers. Oxygen is added to water by diffusion from the air, especially when turbulence and mixing rates are high, and by photosynthesis of green plants, algae, and cyanobacteria. Therefore, turbulent, rapidly flowing water is constantly aerated, so it often recovers quickly from oxygen-depleting

processes. Oxygen is removed from water by respiration and chemical processes that consume oxygen.

Adding organic materials, such as sewage or paper pulp, stimulates activity and oxygen consumption by decomposers. Consequently, **biochemical oxygen demand (BOD),** or the amount of dissolved oxygen consumed by aquatic microorganisms, is a standard measure of water contamination. Alternatively, chemical oxygen demand (COD) is a measure of all organic matter in water. In addition, **dissolved oxygen (DO) content** can be measured directly, with high DO levels indicating good-quality water.

Downstream from a point source, such as a municipal sewage plant discharge, a characteristic decline and restoration of water quality can be detected either by measuring DO content or by observing the types of flora and fauna that live in successive sections of the river. The oxygen decline downstream is called the **oxygen sag** (fig. 10.18). Upstream from the pollution source, oxygen levels support normal populations of clean-water organisms. Immediately below the source of pollution, oxygen levels begin to fall as decomposers metabolize waste materials. Trash fish, such as carp, bullheads, and gar, are able to survive in this oxygen-poor environment, where they eat both decomposer organisms and the waste itself.

Farther downstream, the water may become so oxygen depleted that only the most resistant microorganisms and invertebrates can survive. Eventually, most of the nutrients are used up, decomposer populations are smaller, and the water becomes oxygenated once again. Depending on the volumes and flow rates of the effluent plume and the river receiving it, normal communities may not appear for several miles downstream.

Plant Nutrients and Cultural Eutrophication

Water clarity (transparency) is affected by sediments, chemicals, and the abundance of plankton organisms, and is a useful measure of water quality and water pollution. Rivers and lakes that have clear water and low biological productivity are said to be **oligotrophic** (*oligo* = little + *trophic* = nutrition). By contrast, **eutrophic** (*eu* + *trophic* = well nourished) waters are rich in organisms and organic materials. Eutrophication, an increase in nutrient levels and biological productivity, often accompanies successional changes (see chapter 5) in lakes. Tributary streams bring in sediments and nutrients that stimulate plant growth. Over time, ponds or lakes often fill in, becoming marshes or even terrestrial biomes. The rate of eutrophication depends on water chemistry and depth, volume of inflow, mineral content of the surrounding watershed, and the biota of the lake itself.

Human activities can greatly accelerate eutrophication, an effect called **cultural eutrophication.** Cultural eutrophication is mainly caused by increased nutrient input into a water body. Increased productivity in an aquatic system sometimes can be beneficial. Fish and other desirable species may grow faster, providing a welcome food source. Often, however, eutrophication produces "blooms" of algae or thick growths of aquatic plants stimulated by elevated phosphorus or nitrogen levels (fig. 10.19). Bacterial populations then increase, fed by larger amounts of organic matter. The water often becomes cloudy or turbid, and has

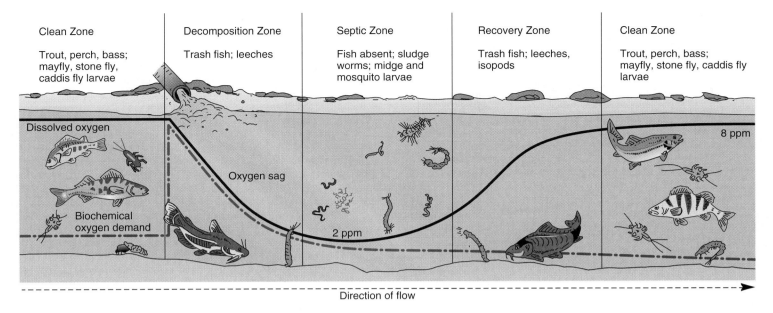

Clean Zone	Decomposition Zone	Septic Zone	Recovery Zone	Clean Zone
Trout, perch, bass; mayfly, stone fly, caddis fly larvae	Trash fish; leeches	Fish absent; sludge worms; midge and mosquito larvae	Trash fish; leeches, isopods	Trout, perch, bass; mayfly, stone fly, caddis fly larvae

Dissolved oxygen

Oxygen sag

Biochemical oxygen demand

8 ppm

2 ppm

Direction of flow

FIGURE 10.18 *Oxygen sag downstream of an organic source. A great deal of time and distance may be required for the stream and its inhabitants to recover.*

unpleasant tastes and odors. Cultural eutrophication can accelerate the "aging" of a water body enormously over natural rates. Lakes and reservoirs that normally might exist for hundreds or thousands of years can be filled in a matter of decades.

Eutrophication also occurs in marine ecosystems, especially in nearshore waters and partially enclosed bays or estuaries. Partially enclosed seas, such as the Black, Baltic, and Mediterranean Seas, tend to be in especially critical condition. During the tourist season, the coastal population of the Mediterranean, for example, swells to 200 million people. Eighty-five percent of the effluents from large cities go untreated into the sea. Beach pollution, fish kills, and contaminated shellfish result. Extensive "dead zones" often form where rivers dump nutrients into estuaries and shallow seas. The largest in the world occurs during summer months in the Gulf of Mexico at the mouth of the Mississippi River. This hypoxic (oxygen-depleted) zone can cover 18,000 km^2 (7,000 mi^2, or about the area of New Jersey).

Toxic Tides

Excessive nutrients support blooms of deadly aquatic microorganisms in polluted nearshore waters. Red tides—and other colors, depending on the species involved—have become increasingly common where nutrients and wastes wash down rivers.

One of the most feared of these organisms is *Pfiesteria piscicida,* an extraordinarily poisonous dinoflagellate that only recently has been recognized as a killer of fish and shellfish in polluted rivers and estuaries. In North Carolina's Pamlico Sound, *Pfiesteria* kills hundreds of thousands to millions of fish annually. Dinoflagellates are peculiar organisms with complex life cycles and many different shapes. *Pfiesteria* can change into at least two dozen distinct forms and sizes, depending on water temperature, turbulence, and food supply. The right conditions can cause a pop-

FIGURE 10.19 *Eutrophic lake. Nutrients from agriculture and domestic sources have stimulated growth of algae and aquatic plants. This reduces water quality, alters species composition, and lowers the lake's recreational and aesthetic values.*

ulation explosion. If fish blunder into this profuse swarm, *Pfiesteria* quickly turn into a toxic, swimming form that attacks with soluble poisons. These toxins produce skin lesions and paralyze fish, so they can't escape. The predatory *Pfiesteria* feed on both the flesh and the oozing sores.

Humans are harmed if they eat contaminated seafood or even if they breathe airborne *Pfiesteria* cells or secretions. Symptoms of *Pfiesteria* poisoning include headaches, blurred vision, aching joints, difficulty breathing, memory loss, and long-term damage to the brain, liver, and other organs. (See the related story "A Flood of Pigs" at www.mhhe.com/apps.)

INVESTIGATING Our Environment

How Do We Test Water?

How do we know how clean our water is? A wide variety of tests give us information about water quality.

To test for coliform bacteria, a measured water sample is passed through a filter and the filter is then placed in a dish containing a nutrient medium that supports bacterial growth. After 24 hours of incubation, any bacteria present will have produced small colonies. If any colonies are found in drinking-water samples, the U.S. Environmental Protection Agency considers the water unsafe and requiring disinfection. At 200 colonies per 100 ml, water may be swimmable, but some cities and states allow higher levels. If the limit is exceeded, the contaminated pool, river, or lake usually is closed to swimming.

To test for BOD, oxygen levels in a water sample are measured, then measured again after five days to see how much oxygen has been consumed. COD tests use a strong oxidizing agent (dichromate ion in 50 percent sulfuric acid) to completely break down all organic matter in a water sample. This method is much faster than the BOD test, but normally gives much higher results because it oxidizes compounds not ordinarily metabolized by bacteria. To measure DO levels directly, an oxygen electrode is used. The DO content of water depends on factors other than pollution (for example, temperature and aeration), but it is usually more directly related to whether aquatic organisms survive than is BOD.

Sediment is determined by weighing the amount of solids present in a known volume of sample. The water sample is evaporated in an oven or vacuum evaporator, and the resulting residue is weighed. Turbidity is generally determined by using a colorimeter to measure water transparency. Water clarity also can be measured using a Secchi disk, a round metal plate with a distinct black-and-white pattern painted on its surface. The disk is lowered into the water and the depth at which it can no longer be seen is determined.

Inorganic compounds such as nitrates and phosphates usually can be measured by simple, inexpensive chemical analyses. Generally, these procedures involve adding a reactant that changes color in the presence of the compound to be assayed. The color can be measured either by comparison to a color chart or with a spectrophotometer. Dissolved oxygen (DO) is most easily measured with an oxygen meter and probe. The probe is filled with a salt solution and has a selectively permeable membrane that allows DO to pass from the water into the salt solution. As DO diffuses into the salt solution, it changes the electric potential of the salt solution, which is measured by a meter that converts the signal to milligrams per liter.

Conductivity is a measure of the ability of water to pass an electrical current. It is affected by the presence of inorganic ions but generally not by organic molecules. Conductivity is measured by applying voltage across two electrodes in a probe immersed in the water sample. The drop in voltage caused by the resistance of the water is used to calculate the conductivity per centimeter.

Alkalinity is a measure of the capacity of water to neutralize acids. Alkaline compounds in the water, such as bicarbonates, carbonates, and hydroxides remove H^+ ions and lower the acidity of the water (which means increased pH). Alkalinity usually is measured by titration with sulfuric acid until the sample reaches a certain pH (which is measured either colorimetrically or with a pH meter). The amount of acid used corresponds to the total alkalinity of the sample.

Biological monitoring uses indicator organisms to assess environmental conditions. Biological observation is more representative because it reveals cumulative effects as opposed to chemical measurement, which is representative only of the actual time of sampling. Benthic macro invertebrates, such as aquatic insects, insect larvae, and crustaceans that live in the bottom portion of a waterway for part of their life cycle, often are used as bioindicators of water quality because they are ubiquitous, relatively sedentary, and long-lived. However, like toxicity testing, biomonitoring doesn't tell you why animals are present or absent.

For more accurate measurements, sophisticated and expensive equipment exists that can detect extremely small amounts of contaminants in water. Atomic absorption spectroscopy and plasma atomic emission spectrometry, for example, use very sensitive measurements of absorption or emission of specific wavelengths of light to quantify inorganic elements in water. They can measure 30 or 40 elements at levels of parts per million in a matter of minutes.

Mass spectrometry is a powerful analytical technique that is used to identify compounds and to elucidate the structure and chemical properties of molecules. Detection of compounds can be accomplished with very minute quantities (as little as 10^{-12}g). Molecules are converted into ions, sorted according to their mass-to-charge (m/z) ratios, and then collected by a detector that converts the ion flux to a proportional electrical current. The data system converts this information into a mass spectrum. Usually gas/liquid chromatographs or other devices are connected to mass spectrometers to separate components of complex mixtures prior to mass analysis. These machines are most often used to analyze organic compounds, such as pesticides or PCBs in water.

Inorganic Pollutants

Some toxic inorganic chemicals are naturally released into water from rocks by weathering processes (see chapter 11). Humans accelerate the transfer rates in these cycles thousands of times above natural background levels through the mining, processing, using, and discarding of minerals.

Among the chemicals of greatest concern are heavy metals, such as mercury, lead, tin, and cadmium. Supertoxic elements, such as selenium and arsenic, also have reached hazardous levels in some waters. Other inorganic materials, such as acids, salts, nitrates, and chlorine, that are nontoxic at low concentrations may become concentrated enough to lower water quality and adversely affect biological communities.

Metals

Many metals, such as mercury, lead, cadmium, and nickel, are highly toxic in minute concentrations. Because metals are highly persistent, they accumulate in food chains and have a cumulative effect in humans.

A mercury-poisoning disaster appears to be in process now in Brazil, Ecuador, and Bolivia. Since the 1980s, thousands of

Principles of Environmental Science

garimperios, or prospectors, have invaded the jungles along the Amazon River and its tributaries to pan for gold. They use mercury to trap the gold and separate it from sediments, and then boil off the mercury with a blowtorch. Miners and their families suffer nerve damage from breathing the toxic fumes. Estimates are that 130 tons of mercury per year are deposited in the Amazon, which will be impossible to clean up.

Lead poisoning has been known since Roman times to be dangerous to human health. Lead pipes are a serious source of drinking water pollution, especially in older homes or in areas where water is acidic and, therefore, leaches more lead from pipes. Even lead solder in pipe joints and metal containers can be hazardous.

Mine drainage and leaching of mining wastes are serious sources of metal pollution in water. A survey of water quality in eastern Tennessee found that 43 percent of all surface streams and lakes and more than half of all groundwater used for drinking supplies was contaminated by acids and metals from mine drainage. In some cases, metal levels were 200 times higher than what is considered safe for drinking water.

Nonmetallic Salts

Desert soils often contain high concentrations of soluble salts, including toxic selenium and arsenic. You have probably heard of poison springs and seeps in the desert, where percolating groundwater brings these compounds to the surface. Irrigation and drainage of desert soils mobilize these materials on a larger scale and can result in serious pollution problems, as in Kesterson Marsh in California, where selenium poisoning killed thousands of migratory birds in the 1980s.

Salts such as sodium chloride (table salt) that are nontoxic at low concentrations also can be mobilized by irrigation and concentrated by evaporation, reaching levels that are toxic for plants and animals. Salinity levels in the Colorado River and surrounding farm fields have become so high in recent years that millions of hectares of valuable croplands have had to be abandoned. In northern states, millions of tons of sodium chloride and calcium chloride are used to melt road ice in the winter. Leaching of road salts into surface waters has a devastating effect on some aquatic ecosystems.

Acids and Bases

Acids are released as by-products of industrial processes, such as leather tanning, metal smelting and plating, petroleum distillation, and organic chemical synthesis. Coal mining is an especially important source of acid water pollution. Sulfur compounds in coal react with oxygen and water to make sulfuric acid. Thousands of kilometers of streams in the United States have been acidified by acid mine drainage, some so severely that they are essentially lifeless.

Acid precipitation (see chapter 9) also acidifies surface-water systems. In addition to damaging living organisms directly, these acids leach aluminum and other elements from soil and rock, further destabilizing ecosystems.

Organic Chemicals

Thousands of different natural and synthetic organic chemicals are used in the chemical industry to make pesticides, plastics, pharmaceuticals, pigments, and other products that we use in everyday life. Many of these chemicals are highly toxic (see chapter 8). Exposure to very low concentrations (perhaps even parts per quadrillion in the case of dioxins) can cause birth defects, genetic disorders, and cancer. Some can persist in the environment because they are resistant to degradation and toxic to organisms that ingest them.

The two principal sources of toxic organic chemicals in water are (1) improper disposal of industrial and household wastes, and (2) pesticide runoff from farm fields, forests, roadsides, golf courses, and private lawns. The EPA estimates that about 500,000 metric tons of pesticides are used in the United States each year. Much of this material washes into the nearest waterway, where it passes through ecosystems and may accumulate in high levels in nontarget organisms. The bioaccumulation of DDT in aquatic ecosystems was one of the first of these pathways to be understood. Dioxins and other chlorinated hydrocarbons (hydrocarbon molecules that contain chlorine atoms) have been shown to accumulate to dangerous levels in the fat of salmon, fish-eating birds, and humans and to cause health problems similar to those resulting from toxic metal compounds.

Hundreds of millions of tons of hazardous organic wastes are thought to be stored in dumps, landfills, lagoons, and underground tanks in the United States (see chapter 13). Many, perhaps most, of these sites have leaked toxic chemicals into surface waters or groundwater or both. The EPA estimates that about 26,000 hazardous waste sites will require cleanup because they pose an imminent threat to public health, mostly through water pollution.

Sediment and Thermal Pollution

Sediment is a natural and necessary part of river systems. Sediment fertilizes floodplains and creates fertile deltas. But human activities, chiefly farming and urbanization, greatly accelerate erosion and increase sediment loads in rivers. Silt and sediment are considered the largest source of water pollution in the United States, being responsible for 40 percent of the impaired river miles in EPA water quality surveys. Cropland erosion contributes about 25 billion metric tons of soil, sediment, and suspended solids to world surface waters each year. Forest disturbance, road building, urban construction sites, and other sources add at least 50 billion additional tons.

This sediment fills lakes and reservoirs, obstructs shipping channels, clogs hydroelectric turbines, and makes purification of drinking water more costly. Sediments smother gravel beds in which insects take refuge and fish lay their eggs. Sunlight is blocked so that plants cannot carry out photosynthesis, and oxygen levels decline. Murky, cloudy water also is less attractive for swimming, boating, fishing, and other recreational

FIGURE 10.20 *Sediment and industrial waste flow from this drainage canal into Lake Erie.*

uses (fig. 10.20). Sediment washed into the ocean clogs estuaries and coral reefs.

Thermal pollution, usually effluent from cooling systems of power plants or other industries, alters water temperature. Raising or lowering water temperatures from normal levels can adversely affect water quality and aquatic life. Water temperatures are usually much more stable than air temperatures, so aquatic organisms tend to be poorly adapted to rapid temperature changes. Lowering the temperature of tropical oceans by even one degree can be lethal to some corals and other reef species. Raising water temperatures can have similar devastating effects on sensitive organisms. Oxygen solubility in water decreases as temperatures increase, so species requiring high oxygen levels are adversely affected by warming water.

Humans also cause thermal pollution by altering vegetation cover and runoff patterns. Reducing water flow, clearing streamside trees, or adding sediment all make water warmer and alter the ecosystems in a lake or stream.

Warm water plumes from power plants often attract fish and birds, which find food and refuge there, especially in cold weather. This artificial environment can be a fatal trap, however. Florida's manatees, an endangered mammal, are attracted to the abundant food supply and warm water in power plant thermal plumes. Often they are enticed into spending the winter much farther north than they normally would. On several occasions, a midwinter power plant breakdown has exposed a dozen or more of these rare animals to a sudden, deadly, thermal shock.

WATER QUALITY TODAY

Surface-water pollution is often both highly visible and one of the most common threats to environmental quality. In more developed countries, reducing water pollution has been a high priority over the past few decades. Billions of dollars have been spent on control programs, and considerable progress has been made. Still, much remains to be done.

Progress in the United States and Canada

Like most developed countries, the United States and Canada have made encouraging progress in protecting and restoring water quality in rivers and lakes over the past 50 years. In 1948, only about one-third of Americans were served by municipal sewage systems, and most of those systems discharged sewage with little or no treatment. Today, sewage plants at least partially treat almost all household wastes in these countries.

In the United States, the 1972 Clean Water Act established a National Pollution Discharge Elimination System (NPDES), which requires that any industry, municipality, or other entity dumping wastes in surface waters have a permit. The permit requires disclosure of what is being dumped and gives regulators valuable data and evidence for litigation, and it can be easily revoked if violations occur. As a consequence, only about 10 percent of our water pollution now comes from industrial or municipal point sources.

Since the Clean Water Act was passed, the United States has spent more than $180 billion in public funds and perhaps ten times as much in private investments on water pollution control. Most of that effort has been aimed at point sources, especially to build or upgrade thousands of municipal sewage treatment plants. As a result, nearly everyone in urban areas is now served by municipal sewage systems, and no major city discharges raw sewage into a river or lake, except as overflow during heavy rainstorms.

This campaign has led to significant improvements in surface-water quality in many places. Fish and aquatic insects have returned to waters that formerly were depleted of life-giving oxygen. Swimming and other water-contact sports are again permitted in rivers, lakes, and at ocean beaches that once were closed by health officials.

The Clean Water Act goal of making all U.S. surface waters "fishable and swimmable" has not been fully met, but in 1999, the EPA reported that 91.4 percent of all monitored river miles and 87.5 percent of all assessed lake acres are suitable for their designated uses. (However, in some waters, the designated use is only boating; water doesn't have to be very clean to put a boat on it.)

Canada's 1970 Water Act has produced comparable results. Seventy percent of all Canadians in towns with over 1,000 population are now served by some form of municipal sewage treatment. In Ontario, the vast majority of those systems include tertiary treatment. After 10 years of controls, phosphorus levels in the Bay of Quinte in the northeast corner of Lake Ontario have dropped nearly by half, and algal blooms that once turned waters green are less frequent and less intense than they once were. Elimination of mercury discharges from a pulp and paper mill on the Wabigoon-English River system in western Ontario has resulted in a dramatic decrease in mercury contamination that poisoned local native people 20 years ago.

Watershed Approaches

In 1998, the EPA instituted a new regulatory approach to water quality assurance. Rather than issue standards on a river by river

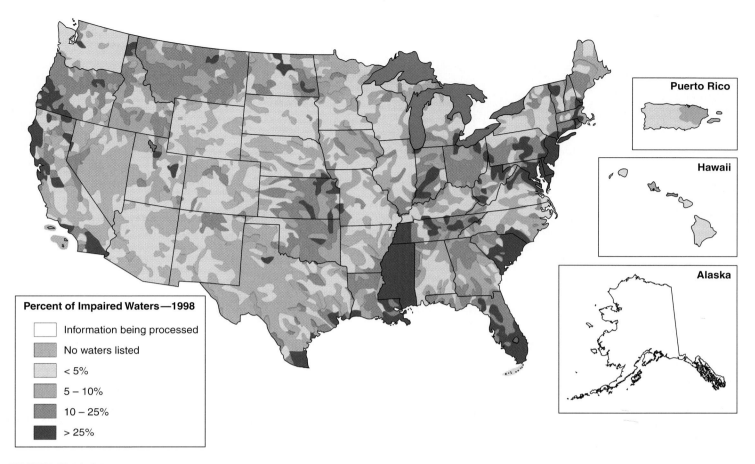

Percent of Impaired Waters—1998

- Information being processed
- No waters listed
- < 5%
- 5 – 10%
- 10 – 25%
- > 25%

FIGURE 10.21 *Percent of impaired U.S. rivers by watershed in 1998.*
Source: Data from Environmental Protection Agency, August 1999.

approach or factory by factory permit discharge, the focus is being changed to watershed-level monitoring and protection. Some 4,000 watersheds are monitored for water quality (fig. 10.21). (You can find information about your watershed at www.mhhe.com/apps.)

The intent of this program is to give the public more and better information about the health of their watersheds. In addition, states will have greater flexibility as they identify impaired water bodies and set priorities, and new tools will be used to achieve goals. States are required to identify waters not meeting water quality goals and to develop **total maximum daily loads (TMDL)** for each pollutant and each listed water body. A TMDL is the amount of a particular pollutant that a water body can receive from both point and nonpoint sources. It considers seasonal variation and includes a margin of safety.

By 1999, all 56 states and territories had submitted TMDL lists, and the EPA had approved most of them. Of the 3.5 million mi (5.6 million km) of rivers monitored, 65 percent were considered to be in good condition, while 35 percent were classified as impaired, meaning they failed one or more clean water goal (fig. 10.22). Of 40 million lake acres (99 million ha), 45 percent (in about 20,000 lakes) failed to meet their goal. To give states more flexibility in planning, the EPA has proposed new rules that

include allowances for reasonably foreseeable increases in pollutant loadings to encourage "smart growth." In the future, TMDLs also will include load allocations from all nonpoint sources, including air deposition and natural background levels.

Remaining Problems in the United States and Canada

The greatest impediments to achieving national goals in water quality in both the United States and Canada are sediment, nutrients, and pathogens, especially from nonpoint discharges of pollutants. These sources are harder to identify and to reduce or treat than are specific point sources. About three-fourths of the water pollution in the United States comes from soil erosion, fallout of air pollutants, and surface runoff from urban areas, farm fields, and feedlots. In the United States, runoff carries away as much as 25 percent of the 46,800,000 metric tons (52 million tons) of fertilizer spread on farmland each year (fig. 10.23).

Cattle in feedlots produce some 129.6 million metric tons (144 million tons) of manure each year, and the runoff from these sites is rich in viruses, bacteria, nitrates, phosphates, and other contaminants. A single cow produces about 30 kg (66 lb) of

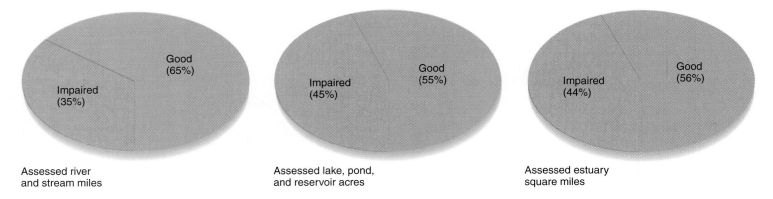

Assessed river
and stream miles

Assessed lake, pond,
and reservoir acres

Assessed estuary
square miles

FIGURE 10.22 *Of the 23 percent of river miles assessed, nearly two-thirds are considered in good condition, while 35 percent are impaired, meaning they don't support one or more of intended uses.*
Source: EPA.

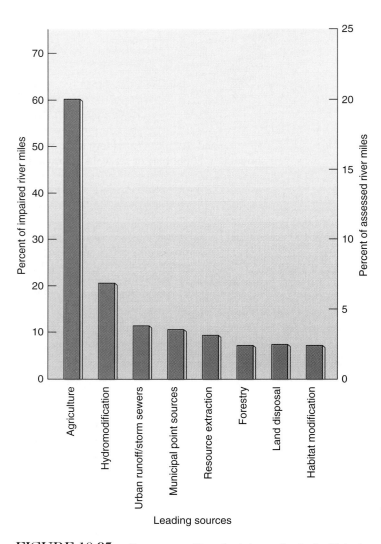

FIGURE 10.23 *Percentage of impaired river miles in the United States by source of damage. The total adds up to more than 100 percent because one river can be affected by many sources.*
Source: The Quality of Our Nation's Waters. U.S. EPA, 1998.

manure per day, or about as much as that produced by ten people. Some feedlots have 100,000 animals, with no facilities for holding or treating runoff. Imagine drawing your drinking water downstream from such a facility. Pets also can be a problem. It is estimated that the wastes from about a half million dogs in New York City are disposed of primarily through storm sewers and therefore do not go through sewage treatment. Inorganic nutrients—nitrates and phosphates—are also a growing problem. They decreased from point sources but have increased about fourfold since 1972 from nonpoint sources.

Problems and Progress in Other Countries

Japan, Australia, and most of Western Europe also have improved surface-water quality in recent years. Sewage treatment in the wealthier countries of Europe generally equals or surpasses that in the United States. Sweden, for instance, serves 98 percent of its population with at least secondary sewage treatment (compared with 70 percent in the United States), and the other 2 percent have primary treatment. Poorer countries have much less to spend on sanitation. Spain serves only 18 percent of its population with even primary sewage treatment; in Ireland, it is only 11 percent, and in Greece, less than 1 percent. Most of the sewage, both domestic and industrial, is dumped directly into the ocean.

The fall of the Iron Curtain in 1989 revealed appalling environmental conditions in much of the former Soviet Union and Eastern Europe. The countries closest geographically and socially to Western Europe, the Czech Republic, Hungary, East Germany, and Poland have made massive investments and encouraging progress toward cleaning up environmental problems. Parts of Russia, the Balkans, and Central Asia, however, remain some of the most polluted places on the earth. In Russia, for example, only about half the tap water is fit to drink. In cities like St. Petersburg, even boiling and filtering isn't enough to make municipal water safe.

The Danube River, draining a landscape with a long history of unregulated mining and heavy industry, carries chromium, copper, mercury, lead, zinc, and oil to the Black Sea. Bratislava, the

Principles of Environmental Science

www.mhhe.com/environmentalscience

capital of Slovakia, alone dumps 73 million m^3 (about 2 billion gal) of industrial and municipal wastes into the river each year. Bombing in Serbia during the Kosovo War in 1999 further exacerbated the situation, releasing oil, industrial chemicals, agricultural fertilizers, and pesticides into the Danube.

In the less-developed countries of South America, Africa, and Asia, sewage treatment is usually either totally lacking or woefully inadequate. In urban areas, 95 percent of all sewage is discharged untreated into rivers, lakes, or the ocean.

Surface waters suffer from a combination of factors, especially lack of money for pollution control, burgeoning populations, rapid urbanization, and the shift of heavy industries (especially the dirtier ones) from developed countries, where pollution laws are strict, to less-developed countries, where regulations may be nonexistent (fig. 10.24). In India, two-thirds of surface waters are considered dangerous to human health. The Yamuna River in New Delhi has 7,500 coliform bacteria per 100 ml (37 times the level considered safe for swimming in the United States) *before* entering the city. The coliform count increases to an incredible 24 million cells per 100 ml as the river leaves the city!

In the Philippines, domestic sewage makes up 60 to 70 percent of the total volume of Manila's Pasig River. Thousands of people use the river not only for bathing and washing clothes, but also as their source of drinking and cooking water. China treats only 2 percent of its sewage. Of 78 monitored rivers in China, 54 are reported to be seriously polluted.

There are also some encouraging pollution control stories. In 1997, Minamata Bay in Japan, long synonymous with mercury poisoning, was declared officially clean again. The Rhine River, which winds through five countries before emptying into the North Sea, has long been a major commercial artery into the heart of Europe. Nearly 20 million people get their drinking water from the river or its tributaries. By the 1970s, the Rhine was so polluted that dozens of fish species disappeared, and swimming was discouraged along most of its length. Today, new land-use practices, environmental accords, and better management of urban and industrial waste have raised oxygen concentrations from less than 2 mg/liter to nearly 10 mg/liter, or about 90 percent of saturation, in long stretches of the river. Many species of invertebrates and fish, including salmon, have returned to the river.

Groundwater Pollution

About half the people in the United States, including 95 percent of those in rural areas, depend on underground aquifers for their drinking water. Industrial, agricultural, and domestic contaminants threaten this vital resource.

The EPA estimates that every day some 4.5 trillion liters (1.2 trillion gal) of contaminated water seep into the ground in the United States from septic tanks, cesspools, municipal and industrial landfills, waste disposal sites, surface impoundments, agricultural fields, forests, and wells (fig. 10.25). The most toxic contaminants come from waste disposal sites. The greatest volume and area of contamination come from agricultural chemicals and wastes. Because deep underground aquifers often have residence

FIGURE 10.24   *Ditches in this Haitian slum serve as open sewers into which all manner of refuse and waste are dumped. The health risks of living under these conditions are severe.*

times of thousands of years, many contaminants are extremely stable once underground.

In farm country, especially in the Midwest's corn belt, fertilizers and pesticides commonly contaminate aquifers and wells. Herbicides such as atrazine and alachlor are widely used on corn and soybeans, and show up in about half of all wells in Iowa, for example. Nitrates from fertilizers often exceed safety standards in rural drinking water. These high nitrate levels are dangerous to infants (nitrates combine with hemoglobin in the blood and result in "blue-baby" syndrome).

Leaky underground storage tanks were once common at filling stations and factories. Most have now been removed and replaced by more modern tanks, but a great deal of soil in American cities remains contaminated by previous careless storage and disposal of petroleum products. Considering that a single gallon (3.8 liters) of gasoline can make a million gallons of water undrinkable, soil contamination remains a serious problem.

Every year, epidemiologists estimate that around 1.5 million Americans fall ill from infections caused by fecal contamination. In 1993, for instance, a pathogen called cryptosporidium got into the Milwaukee public water system, making 400,000 people sick and killing at least 100 people. The total costs of these diseases amount to billions of dollars per year. Preventive measures, such as protecting water sources and aquifer recharge zones and updating treatment and distribution systems, would cost far less.

Ocean Pollution

Although we don't use ocean waters directly, ocean pollution is serious and one of the fastest growing water pollution problems. Coastal bays, estuaries, shoals, and reefs are often overwhelmed by pollution. Dead zones and poisonous algal blooms are increasingly widespread. Toxic chemicals, heavy metals, oil, sediment, and plastic refuse affect some of the most attractive and productive

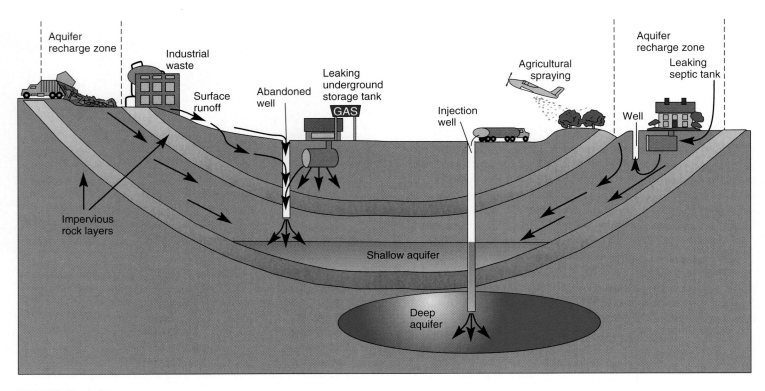

FIGURE 10.25 *Sources of groundwater pollution. Septic systems, landfills, and industrial activities on aquifer recharge zones leach contaminants into aquifers. Wells provide a direct route for injection of pollutants into aquifers.*

ocean regions. The potential losses caused by this pollution amount to billions of dollars each year. In terms of quality of life, the costs are incalculable.

Discarded plastic flotsam and jetsam are becoming an ubiquitous mark of human impact on the oceans. Since plastic is light-weight and nonbiodegradable, it is carried thousands of miles on ocean currents and lasts for years. Even the most remote beaches of distant islands are likely to have bits of polystyrene foam containers or polyethylene packing material that were discarded half a world away. It has been estimated that some 6 million metric tons of plastic bottles, packaging material, and other litter are tossed from ships every year into the ocean, where they ensnare and choke seabirds, mammals, and even fish (fig. 10.26).

Oil pollution affects beaches and open seas around the world (fig. 10.27). Oceanographers estimate that somewhere between 3 million and 6 million metric tons of oil are discharged into the world's oceans each year from oil tankers, fuel leaks, intentional discharges of fuel oil, and coastal industries. About half of this amount is due to maritime transport. Of this portion, most is not from dramatic, headline-making accidents like the 1989 *Exxon Valdez* spill in Alaska, but from routine, open-sea bilge pumping and tank cleaning. These activities are illegal but very common.

The transport of huge quantities of oil creates opportunities for major oil spills through a combination of human and natural hazards. Military conflict in the Middle East destabilizes shipping routes. More important, drilling and transport in stormy seas cause spills. Plans to drill for oil along the seismically active Cal-

FIGURE 10.26 *A deadly necklace. Marine biologists estimate that cast-off nets, plastic beverage yokes, and other packing residue kill hundreds of thousands of birds, mammals, and fish each year.*

ifornia and Alaska coasts have been controversial because of the damage that spills could cause to these biologically rich coastal ecosystems.

Fortunately, awareness of ocean pollution is growing. Oil spill cleanup technologies and response teams are improving,

Principles of Environmental Science   www.mhhe.com/environmentalscience

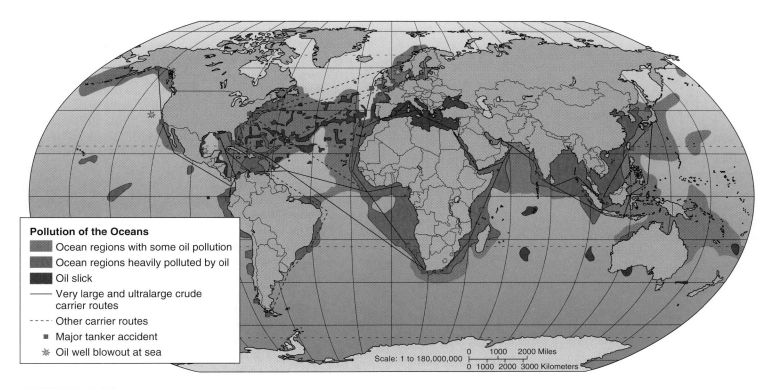

FIGURE 10.27   *Location of oil pollution in the ocean. Major shipping routes of large oil tankers are very likely to be contaminated by oil.*

although most oil is eventually decomposed by natural bacteria. Efforts are growing to control waste plastic. Sixteen states now require that six-pack yokes be made of biodegradable or photodegradable plastic, limiting their longevity as potential killers. International concern about ocean ship waste is increasing, and some shipping companies have been prosecuted and fined for dumping fuel oil. Beach pollution—mainly plastic debris, but also sewage waste, oil, and chemical contaminants—is becoming more common, but it is also more frequently reported in the mainstream media. Volunteer efforts are helping to reduce beach pollution locally: in one day, volunteers in Texas gathered more than 300 tons of plastic refuse from Gulf Coast beaches.

POLLUTION CONTROL

The cheapest and most effective way to reduce pollution is to avoid producing it or releasing it in the first place. Eliminating lead from gasoline has resulted in a widespread and significant decrease in the amount of lead in U.S. surface waters. Studies have shown that as much as 90 percent less road deicing salt can be used in many areas without significantly affecting the safety of winter roads. Careful handling of oil and petroleum products can greatly reduce the amount of water pollution caused by these materials. Although we still have problems with persistent chlorinated hydrocarbons spread widely in the environment, the banning of DDT and PCBs in the 1970s has resulted in significant reductions in levels in wildlife.

Industry can reduce pollution by recycling or reclaiming materials that otherwise might be discarded in the waste stream. These approaches usually have economic as well as environmental benefits. Companies can extract valuable metals and chemicals and sell them, instead of releasing them as toxic contaminants into the water system. Both markets and reclamation technologies are improving as awareness of these opportunities grows. In addition, modifying land use is an important component of reducing pollution.

Nonpoint Sources and Land Management

Farmers have long contributed a huge share of water pollution, especially in the developed world. Increasingly, though, farmers are finding ways to save money and water quality at the same time. Soil conservation practices on farmlands (see chapter 7) maintain soil fertility, as well as protect water quality. Precise application of fertilizer, irrigation water, and pesticides saves money and reduces water contamination. Preserving wetlands that act as natural processing facilities for removing sediment and contaminants helps protect surface and groundwaters.

In urban areas, reducing waste that enters storm sewers is essential. It is getting easier for city residents to recycle waste oil and to properly dispose of paint and other household chemicals that they once dumped into storm sewers or the garbage. Urbanites can also minimize use of fertilizers and pesticides. Regular street sweeping greatly reduces nutrient loads (from decomposing leaves and debris) in rivers and lakes. Runoff can also be diverted away from streams and lakes. Many cities are separating storm sewers and municipal sewage lines to avoid overflow during storms.

A good example of watershed management is seen in Chesapeake Bay, America's largest estuary. Once fabled for its abundant oysters, crabs, shad, striped bass, and other valuable fisheries, the bay had deteriorated seriously by the early 1970s. Citizens' groups, local communities, state legislatures, and the federal government together established an innovative pollution control program that made the bay the first estuary in America targeted for protection and restoration.

Among the principal objectives of this plan is reducing nutrient loading through land-use regulations in the bay's six watershed states to control agricultural and urban runoff. Pollution prevention measures, such as banning phosphate detergents, also are important, as are upgrading wastewater treatment plants and improving compliance with discharge and filling permits. Efforts are underway to replant thousands of hectares of sea grasses and to restore wetlands that filter out pollutants. Since the 1980s, annual phosphorous discharges into the bay have dropped 40 percent. Nitrogen levels, however, have remained constant or have even risen in some tributaries. Although progress has been made, the goals of reducing both nitrogen and phosphate levels by 40 percent and restoring viable fish and shellfish populations are still decades away. Still, as EPA Administrator Carol Browner says, it demonstrates the "power of cooperation" in environmental protection. (See a related story "Watershed Protection in the Catskills" at www.mhhe.com/apps.)

Sewage Treatment

As we have already seen, human and animal wastes usually create the most serious health-related water pollution problems. More than 500 types of disease-causing (pathogenic) bacteria, viruses, and parasites can travel from human or animal excrement through water.

Natural Processes

In the poorer countries of the world, most rural people simply go out into the fields and forests to relieve themselves as they have always done. Where population densities are low, natural processes eliminate wastes quickly, making this an effective method of sanitation. The high population densities of cities, however, make this practice unworkable. Even major cities of many less-developed countries are often littered with human waste that has been left for rains to wash away or for pigs, dogs, flies, beetles, or other scavengers to consume. This is a major cause of disease, as well as being extremely unpleasant. Studies have shown that a significant portion of the airborne dust in Mexico City is actually dried, pulverized human feces.

Where intensive agriculture is practiced—especially in wet rice paddy farming in Asia—it has long been customary to collect "night soil" (human and animal waste) to be spread on the fields as fertilizer. This waste is a valuable source of plant nutrients, but it is also a source of disease-causing pathogens in the food supply.

Until about 50 years ago, most rural American families and quite a few residents of towns and small cities depended on a pit toilet or "outhouse" for waste disposal. Untreated wastes tended to seep into the ground, however, and pathogens sometimes contaminated drinking water. The development of septic tanks and properly constructed drain fields represented a considerable improvement in public health. Septic systems allow solids to settle in a tank, where bacteria decompose them; liquids percolate through soil, where soil bacteria presumably purify them. Where population densities are not too high, this can be an effective method of waste disposal. With urban sprawl, however, groundwater pollution often becomes a problem.

Municipal Sewage Treatment

Over the past 100 years, sanitary engineers have developed ingenious and effective municipal wastewater treatment systems to protect human health, ecosystem stability, and water quality. This topic is an important part of pollution control, and is a principal responsibility of every municipal government.

Primary treatment physically separates large solids from the waste stream with screens and settling tanks (fig. 10.28a). Settling tanks allow grit and some dissolved (suspended) organic solids to fall out as sludge. Water drained from the top of settling tanks still carries up to 75 percent of the organic matter, including many pathogens. These are treated by **secondary treatment,** in which aerobic bacteria break down dissolved organic compounds. In secondary treatment effluent is aerated, often with sprayers or in an aeration tank, in which air is pumped through the microorganism-rich slurry (fig. 10.28b). Fluids can also be stored in a sewage lagoon, where sunlight, algae, and air process waste more cheaply but more slowly. Effluent from secondary treatment processes is usually disinfected with chlorine, UV light, or ozone to kill harmful bacteria before it is released to a nearby waterway.

Tertiary treatment removes dissolved metals and nutrients, especially nitrates and phosphates, from the secondary effluent. Although wastewater is usually free of pathogens and organic

(a) Primary

Bar screen

Grit chamber

Primary sedimentation tank

(b) Secondary

Aeration tank

Final settling tank

Sludge return

Air pump

Sludge line

Sludge is incinerated, composted, or used as soil amendment

Cl

Tertiary filter

Disinfection tank

(c) Tertiary

FIGURE 10.28 *Activated sludge wastewater treatment. (a) Primary treatment removes only solids and suspended sediment. (b) Secondary treatment, through aeration of activated sludge, followed by sludge removal and chlorination of effluent, kills pathogens and removes most organic material. (c) During tertiary treatment, passage through a trickling bed evaporator and/or a tertiary filter further removes inorganic nutrients, oxidizes any remaining organics, and reduces effluent volume.*

material after secondary treatment, it still contains high levels of these inorganic nutrients. If discharged into surface waters, these nutrients stimulate algal blooms and eutrophication. Allowing effluent to flow through a wetland or lagoon can remove nitrates and phosphates. Alternatively, chemicals often are used to bind and precipitate nutrients (fig. 10.28c).

Sewage sludge would be valuable fertilizer if it were not contaminated by metals, toxic chemicals, and pathogenic organisms. The toxic content of most sewer sludge necessitates disposal by burial in a landfill or incineration. Sludge disposal is a major cost in most municipal sewer budgets.

In many American cities, sanitary sewers are connected to storm sewers, which carry runoff from streets and parking lots. Storm sewers are routed to the treatment plant rather than discharged into surface waters because runoff from streets, yards, and

industrial sites generally contains a variety of refuse, fertilizers, pesticides, oils, rubber, tars, lead (from gasoline), and other undesirable chemicals. Unfortunately heavy storms often overload the system, especially where the system is old and already overtaxed. As a result large volumes of raw sewage and toxic surface runoff are dumped directly into receiving waters. To prevent this overflow, cities are spending hundreds of millions of dollars to separate storm and sanitary sewers.

Low-Cost Waste Treatment

A number of alternative treatment systems have been developed. One of the most attractive is using natural or artificial wetlands to process wastes. Arcata, California, for instance, needed an expensive sewer plant upgrade. Instead, the city transformed a 65-ha (160-acre) garbage dump into a series of ponds and marshes that

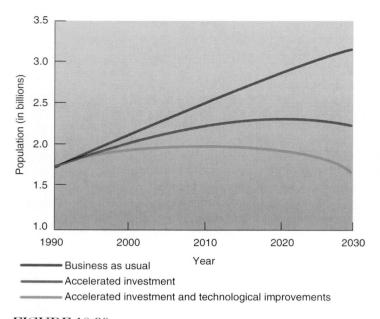

FIGURE 10.29 *World population without adequate sanitation—three scenarios in the year 2030. If business as usual continues, more than 3 billion people will lack safe sanitation. Accelerated investment in sanitation services could lower this number. Higher investment, coupled with technological development, could keep the number of people without adequate sanitation from growing, even though the total population increases.*

Source: World Bank estimates based on research paper by Dennis Anderson and William Cavendish, "Efficiency and Substitution in Pollution Abatement: Simulation Studies in Three Sectors."

serve as a simple, low-cost, waste treatment facility. Arcata saved millions of dollars and improved the environment simultaneously. The marsh is a haven for wildlife and has become a prized recreation area for the city. Eventually, the purified water flows into Humboldt Bay, where marine life flourishes.

Similar wetland waste treatment systems are now operating in many developing countries. Effluent from these operations can be used to irrigate crops or raise fish for human consumption if care is taken to first destroy pathogens. Usually 20 to 30 days of exposure to sun, air, and aquatic plants is enough to make the water safe. These systems make an important contribution to human food supplies. A 2,500-ha (6,000-acre) waste-fed aquaculture facility in Calcutta, for example, supplies about 7,000 metric tons of fish annually to local markets.

The World Bank estimates that some 3 billion people will be without sanitation services by the middle of the twenty-first century under a business-as-usual scenario (fig. 10.29). With investments in innovative programs, however, sanitation could be provided to about half those people, and a great deal of misery and suffering could be avoided.

Remediation

Just as there are many sources of water contamination, there are many ways to clean it up. New developments in environmental engineering are providing promising solutions to many water pollution problems. Containment methods keep dirty water from spreading. Chemicals can be added to toxic wastewater to precipitate, immobilize, or solidify contaminants. Many pollutants can be destroyed or detoxified by chemical reactions that oxidize, reduce, neutralize, hydrolyze, precipitate, or otherwise change their chemical composition. Where chemical techniques are ineffective, physical methods may work. Solvents and other volatile organic compounds, for instance, can be stripped from solution by aeration and then burned in an incinerator.

Often, living organisms can clean contaminated water effectively and inexpensively. We call this **bioremediation.** Restored wetlands, for instance, along stream banks or lake margins can effectively filter out sediment and remove pollutants. They generally cost far less than mechanical water-treatment facilities and provide wildlife habitat as well.

Lowly duckweed (*Lemna* sp.), the green scum you often see covering the surface of eutrophic ponds, grows fast and can remove large amounts of organic nutrients from water. Large duckweed lagoons are being used as inexpensive, low-tech, sewage treatment plants in developing countries. The duckweed can also be harvested and used as feed, fuel, or fertilizer. Up to 35 percent of its dry mass is protein—about twice as much as alfalfa, a popular animal feed.

Ocean Arks International in Falmouth, Massachusetts, has been developing holistic systems for water purification that are combinations of different plants and animals, including algae, rooted aquatic plants, clams, snails, and fish, each chosen to provide a particular service in a contained environment. Technically, the water that has flowed through such a system is drinkable, although few people feel comfortable doing so. More often, the final effluent is used to flush toilets or for irrigation. Called ecological engineering, this novel approach can save resources and money, and it can serve as a valuable educational tool. (See a related story "International Accord to Clean Up the Rhine River" at www.mhhe.com/apps.)

WATER LEGISLATION

Water pollution control has been among the most broadly popular and effective of all environmental legislation in the United States. It has not been without controversy, however. Table 10.6 describes some of the most important water legislation in the United States.

The Clean Water Act

Passage of the U.S. Clean Water Act of 1972 was a bold, bipartisan step that made clean water a national priority. Along with the Endangered Species Act and the Clean Air Act, this is one of the most significant and effective pieces of environmental legislation ever passed by the U.S. Congress. It also is an immense and complex law, with more than 500 sections regulating everything from urban runoff, industrial discharges, and municipal sewage treatment to land-use practices and wetland drainage.

The ambitious goal of the Clean Water Act was to return all U.S. surface waters to "fishable and swimmable" conditions. For

TABLE 10.6 Some Important U.S. Water Quality Legislation

1. *Federal Water Pollution Control Act* (1972). Establishes uniform nationwide controls for each category of major polluting industries.

2. *Marine Protection Research and Sanctuaries Act* (1972). Regulates ocean dumping and established sanctuaries for protection of endangered marine species.

3. *Ports and Waterways Safety Act* (1972). Regulates oil transport and the operation of oil handling facilities.

4. *Safe Drinking Water Act* (1974). Requires minimum safety standards for every community water supply. Among the contaminants regulated are bacteria, nitrates, arsenic, barium, cadmium, chromium, fluoride, lead, mercury, silver, and pesticides; radioactivity and turbidity also are regulated. This act also contains provisions to protect groundwater aquifers.

5. *Resource Conservation and Recovery Act (RCRA)* (1976). Regulates the storage, shipping, processing, and disposal of hazardous wastes and sets limits on the sewering of toxic chemicals.

6. *Toxic Substances Control Act (TOSCA)* (1976). Categorizes toxic and hazardous substances, establishes a research program, and regulates the use and disposal of poisonous chemicals.

7. *Comprehensive Environmental Response, Compensation, and Liability Act (CERCLA)* (1980) and *Superfund Amendments and Reauthorization Act (SARA)* (1984). Provide for sealing, excavation, or remediation of toxic and hazardous waste dumps.

8. *Clean Water Act* (1985) (amending the 1972 Water Pollution Control Act). Sets as a national goal the attainment of "fishable and swimmable" quality for all surface waters in the United States.

9. *London Dumping Convention* (1990). Calls for an end to all ocean dumping of industrial wastes, tank-washing effluents, and plastic trash. The United States is a signatory to this international convention.

point sources, the act requires discharge permits and use of the best practicable control technology (BPT). For toxic substances, the act sets national goals of best available, economically achievable technology (BAT) and zero discharge goals for 126 priority toxic pollutants. As discussed earlier, these regulations have had a positive effect on water quality. While not yet swimmable or fishable everywhere, surface-water quality in the United States has significantly improved on average over the past quarter century. Perhaps the most important result of the act has been investment of $54 billion in federal funds and more than $128 billion in state and local funds for municipal sewage treatment facilities.

Opponents of federal regulation have tried repeatedly to weaken or eliminate the Clean Water Act. They regard restriction of their "right" to dump toxic chemicals and wastes into wetlands and waterways to be an undue loss of freedom. They resent being forced to clean up municipal water supplies, and call for cost/benefit analysis that places greater weight on economic interests in all environmental planning.

Supporters of the Clean Water Act would like to see a shift away from "end-of-the-pipe" focus on effluent removal and more attention to changing industrial processes so toxic substances aren't produced in the first place. Environmentalists also would like to see stricter enforcement of existing regulations, mandatory minimum penalties for violations, more effective community right-to-know provisions, and increased powers for citizen lawsuits against polluters.

SUMMARY

The hydrologic cycle constantly purifies and redistributes fresh water, providing an endlessly renewable resource. More than 97 percent of all water in the world is salty ocean water. Lakes, rivers, and other surface freshwater bodies make up less than 0.01 percent of all the water in the world, but they provide more than half of all water for human use and for habitat and nourishment for aquatic ecosystems that play a vital role in the chain of life.

In the United States, about one-tenth of the water we withdraw from our resources is for direct personal use. Our two largest water uses are agricultural irrigation and industrial cooling. Only about half the water we withdraw is consumed or degraded so that it is unsuitable for other purposes; much could be reused or recycled. Water conservation and recycling would have both economic and environmental benefits.

Water shortages in many parts of the world result from rising demand, unequal distribution, and increased contamination. Water storage and transfer projects are a response to flooding and water shortages. Giant dams and diversion projects can have environmental and social costs far above the benefits they provide. Among the problems they pose are evaporation and infiltration losses, siltation of reservoirs, and loss of recreation and wildlife habitat. Many conservationists prefer watershed management and small dams as means of flood control and water storage.

Any physical, biological, or chemical change in water quality that adversely affects living organisms or makes water unsuitable for desired uses can be considered pollution. World-wide, the most serious water pollutants for human health are pathogenic organisms from human and animal wastes. We have traditionally taken advantage of the capacity of ecosystems to destroy these organisms, but as population density has grown, these systems have become overloaded and ineffective.

In industrialized nations, toxic chemical wastes have become an increasing problem. Agricultural and industrial chemicals have been released or spilled into surface waters and are seeping into groundwater supplies. The extent of this problem is probably not yet fully appreciated.

Pollution levels in the ocean are increasing. Major causes of ocean pollution are oil spills from tanker bilge pumping or acci-dents, and oil well blowouts. Surface runoff and sewage outfalls discharge fertilizers, pesticides, organic nutrients, and toxic chemicals that have a variety of deleterious effects on marine ecosystems.

Silt and sediment make up the greatest quantity of water pollutants. Biomass production by aquatic organisms, land erosion, and refuse discharge all contribute to this problem. Salts and metals from highway and farm runoff and industrial activities also damage water quality. In some areas, drainage from mines and tailings piles delivers sediment and toxic materials to rivers and lakes.

Appropriate land-use practices and careful disposal of industrial, domestic, and agricultural wastes are essential for control of water pollution. Natural processes and living organisms have a high capacity to remove or destroy water pollutants, but these systems become overloaded and ineffective when pollution levels are too high. Reducing pollution sources is often the best solution to our pollution problems.

QUESTIONS FOR REVIEW

1. Describe the path a molecule of water might follow through the hydrologic cycle from the ocean to land and back again.
2. Define aquifer. How does water get into an aquifer?
3. What fraction of the world's water is fresh?
4. What is drip irrigation and what are its benefits?
5. Describe some problems associated with dam building and water diversion projects.
6. Define water pollution.
7. Describe eight major sources of water pollution in the United States. What pollution problems are associated with each source?
8. What is eutrophication? What causes it?
9. What is an impaired river mile? What percentage of U.S. river miles are impaired?
10. Describe primary, secondary, and tertiary processes for sewage treatment. What is the quality of the effluent from each of these processes?
11. Describe remediation techniques and how they work.

THINKING SCIENTIFICALLY

1. What changes might occur in the hydrologic cycle if our climate were to warm or cool significantly?
2. Why does it take so long for deep ocean waters to circulate through the hydrologic cycle? What happens to substances that contaminate deep ocean water or deep aquifers in the ground?
3. How precise is the estimate that 2 billion people lack access to clean water? If, in 2025, two-thirds of the world's population live in countries with severe water shortages, what might be the number of people who lack access to clean water? Why is this number ambiguous?
4. Do you think that water pollution is worse now than it was in the past? What considerations go into a judgment like this? How do your personal experiences influence your opinion?
5. What additional information would you need to make a judgment about whether conditions are getting better or worse? How would you weigh different sources, types, and effects of water pollution?
6. Under what conditions might sediment in water or cultural eutrophication be beneficial? How should we balance positive and negative effects?
7. Suppose that part of the silt in a river is natural and part is human-caused. How might you evaluate what proportion of the silt is caused by humans?

Principles of Environmental Science

KEY TERMS

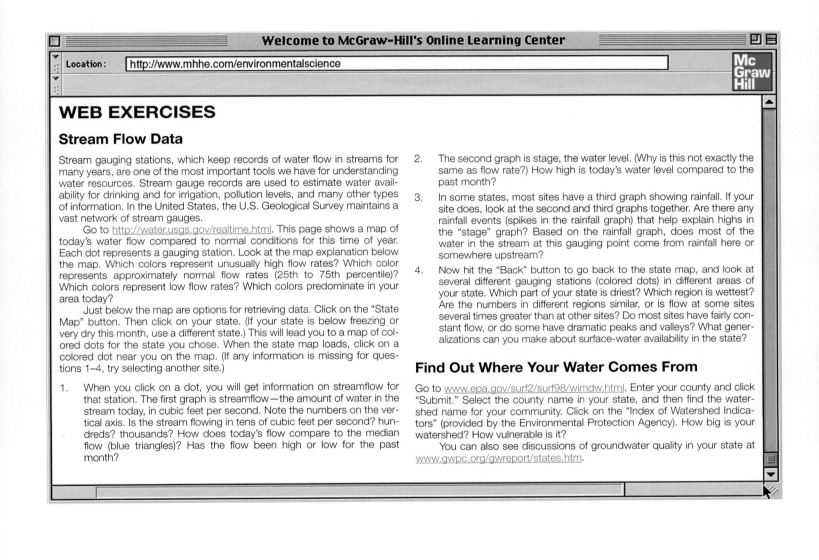

WEB EXERCISES

Stream Flow Data

Stream gauging stations, which keep records of water flow in streams for many years, are one of the most important tools we have for understanding water resources. Stream gauge records are used to estimate water availability for drinking and for irrigation, pollution levels, and many other types of information. In the United States, the U.S. Geological Survey maintains a vast network of stream gauges.

Go to http://water.usgs.gov/realtime.html. This page shows a map of today's water flow compared to normal conditions for this time of year. Each dot represents a gauging station. Look at the map explanation below the map. Which colors represent unusually high flow rates? Which color represents approximately normal flow rates (25th to 75th percentile)? Which colors represent low flow rates? Which colors predominate in your area today?

Just below the map are options for retrieving data. Click on the "State Map" button. Then click on your state. (If your state is below freezing or very dry this month, use a different state.) This will lead you to a map of colored dots for the state you chose. When the state map loads, click on a colored dot near you on the map. (If any information is missing for questions 1–4, try selecting another site.)

1. When you click on a dot, you will get information on streamflow for that station. The first graph is streamflow—the amount of water in the stream today, in cubic feet per second. Note the numbers on the vertical axis. Is the stream flowing in tens of cubic feet per second? hundreds? thousands? How does today's flow compare to the median flow (blue triangles)? Has the flow been high or low for the past month?

2. The second graph is stage, the water level. (Why is this not exactly the same as flow rate?) How high is today's water level compared to the past month?

3. In some states, most sites have a third graph showing rainfall. If your site does, look at the second and third graphs together. Are there any rainfall events (spikes in the rainfall graph) that help explain highs in the "stage" graph? Based on the rainfall graph, does most of the water in the stream at this gauging point come from rainfall here or somewhere upstream?

4. Now hit the "Back" button to go back to the state map, and look at several different gauging stations (colored dots) in different areas of your state. Which part of your state is driest? Which region is wettest? Are the numbers in different regions similar, or is flow at some sites several times greater than at other sites? Do most sites have fairly constant flow, or do some have dramatic peaks and valleys? What generalizations can you make about surface-water availability in the state?

Find Out Where Your Water Comes From

Go to www.epa.gov/surf2/surf98/wimdw.html. Enter your county and click "Submit." Select the county name in your state, and then find the watershed name for your community. Click on the "Index of Watershed Indicators" (provided by the Environmental Protection Agency). How big is your watershed? How vulnerable is it?

You can also see discussions of groundwater quality in your state at www.gwpc.org/gwreport/states.htm.

Environmental Geology and Earth Resources

Learn geology or die!

—Louis Agassiz

Yucca Mountain, Nevada (long ridge in center of photo) has been chosen as the United States' first permanent high-level nuclear waste storage site.

OBJECTIVES

After studying this chapter, you should be able to

- understand some basic geologic principles, including how plate-tectonic movements affect conditions for life on the earth.
- explain how the three major rock types form and how the rock cycle works.
- summarize economic mineralogy and strategic minerals.
- discuss the environmental effects of mining and mineral processing.
- recognize the geologic hazards of earthquakes, volcanoes, floods, and erosion.

Welcome to McGraw-Hill's Online Learning Center

Location: http://www.mhhe.com/environmentalscience

LEARNING ONLINE

Visit our webpage at www.mhhe.com/environmentalscience for data sources, further readings, additional case studies, current environmental news, and regional examples within the Online Learning Center to help you understand the material in this chapter. You'll also find active links to information pertaining to this chapter including:

USGS
National Geophysical Data Center (NGDC)
MTU Volcanoes Page—World Reference Map
USGS Cascades Volcano Observatory
USGS, Volcano Hazards Program

Radioactive Waste Disposal at Yucca Mountain

In July 2002 the U.S. Senate voted to designate Yucca Mountain, Nevada, as the permanent resting place for 77,000 metric tons of high-level nuclear waste, most of it spent fuel from the nation's 107 nuclear power plants (fig. 11.1). Nuclear power proponents celebrated: after twenty years of research and lobbying, the site was finally approved. With a repository on the horizon, plans for new nuclear plants could proceed, and existing plants could look forward to clearing out 50,000 tons of spent nuclear fuel currently in temporary, sometimes inadequate, storage at reactor sites. The state of Nevada, meanwhile, immediately filed a lawsuit to stop the plan. Nevada Governor Kenny Guinn charged that the Department of Energy (DOE) had lowered its scientific standards for evaluating the geological integrity of the site. Opponents charged that heavy-handed eastern politics, not sound geology, was behind the decision to put the entire nation's nuclear waste in Nevada.

The stakes at Yucca Mountain are high. The DOE and the nuclear power industry have invested more than $4 billion in research, testing, and promotion of the site. The lack of permanent storage has been an obstacle to construction of nuclear power plants since the 1980s. Further, cleanup is beginning at old military installations, such as Hanford, Washington; and Rocky Flats, Colorado. These are some of the nation's worst toxic waste sites, and permanent storage is needed for their radioactive materials.

Geology is key to siting a waste repository of any kind. For high-level radioactive wastes, the DOE needed to find a site where a labyrinth of deep tunnels could keep extremely dangerous materials isolated and secure for 10,000 years, more than twice the length of recorded human history. (The waste will remain highly radioactive for more than 500,000 years, but the DOE considers itself unlikely to be responsible for the site for that long.) This time span requires a geologically stable area—no active faults like those in California, no volcanoes like those in Washington and Oregon. Bedrock needs to be relatively impermeable, not riddled with underground channels and sinkholes as in Florida or parts of Texas. There must be no groundwater, which would soak storage casks and mobilize radioactive materials, possibly allowing tainted water to reach the surface. This rules out the central Great Plains and most of the eastern United States, where plentiful rainfall keeps water tables high and aquifers full.

Yucca Mountain fits these requirements better than most places in the United States. But there is conflicting evidence, and conflicting interpretations of evidence, about whether even Yucca Mountain is good enough. DOE geologists insist that the only aquifers in the area are 300 m below the site; other geologists say there is evidence in the rocks that aquifers have risen in the past, suggesting that they might rise again to the level of the repository tunnels. Critics of the site point out that the area has more than 30 known faults. Geologists disagree on how long these faults have been dormant, and on how long it will be before they shift again.

FIGURE 11.1 *Yucca Mountain, Nevada, is the designated storage site for high-level radioactive waste.*

In addition, there are seven dormant volcanoes in the area. While these are likely to remain dormant for 10,000 years, their activity in several hundred thousand years is hard to predict. Opponents of the site also worry about the potential dangers of shipping waste, potentially 28,000 truckloads and 10,000 rail cars over the site's 30-year lifespan. Defenders point out that high-level waste is currently stored at more than 130 temporary sites, and those sites are clearly unsafe for the long term.

Unless you live along an earthquake fault or near a volcano, you might think of the earth as solid, stable, and unchanging. The extraordinary difficulty of finding a nuclear waste repository reminds us that, in fact, the earth is constantly shifting and changing—although usually on a time scale much slower than we can perceive. The Yucca Mountain controversy also highlights the importance of understanding geological processes and forces if we are to solve environmental, policy, and energy problems. In this chapter we'll look at how rocks are formed, where mineral resources are found, how the earth moves, and how these forces affect us all.

A DYNAMIC PLANET

Although we think of the ground under our feet as solid and stable, the earth is a dynamic and constantly changing structure. Titanic forces stir inside the earth, causing continents to split, move apart, and then crash into each other in slow but inexorable collisions. In this section, we will look at the structure of our planet and some of the forces that shape it.

A Layered Sphere

The **core,** or interior of the earth, is composed of a dense, intensely hot mass of metal—mostly iron—thousands of kilometers in diameter (fig. 11.2). Solid in the center but more fluid in the outer core, this immense mass generates the magnetic field that envelops the earth.

Surrounding the molten outer core is a hot, pliable layer of rock called the **mantle.** The mantle is much less dense than the core because it contains a high concentration of lighter elements, such as oxygen, silicon, and magnesium.

The outermost layer of the earth is the cool, lightweight, brittle rock **crust.** The crust below oceans is relatively thin (8–15 km), dense, and young (less than 200 million years old) because of constant recycling. Crust under continents is relatively thick (25–75 km), light, and as old as 3.8 billion years because additional material is continually being added. It also is predominantly granitic while oceanic crust is mainly dense basaltic rock. Table 11.1 compares the composition of the whole earth (dominated by the dense core) and the crust.

Tectonic Processes and Shifting Continents

Hot enough to flow, the upper layer of the mantle has huge convection currents that break the overlying crust into a mosaic of huge blocks called **tectonic plates** (fig. 11.3). These plates slide slowly across the earth's surface like immense icebergs, in some places breaking up into smaller pieces, in other places crashing ponderously into each other to create new, larger landmasses. Ocean basins form where continents crack and pull apart. The Atlantic Ocean, for example, is growing slowly as Europe and Africa move away from the Americas. **Magma** (molten rock) forced up through the cracks forms new oceanic crust that piles up underwater in **midocean ridges.** Creating the largest mountain range in the world, these ridges wind around the earth for 74,000 km (46,000 mi) (see fig. 11.3). Although concealed from our view, this jagged range boasts higher peaks, deeper canyons, and sheerer cliffs than any continental mountains. Slowly spreading from these fracture zones, ocean plates push against continental plates.

Earthquakes are caused by the grinding and jerking as plates slide past each other. Mountain ranges like those on the west coast of North America and in Japan are pushed up at the margins of colliding continental plates. The Himalayas are still rising as the Indian subcontinent grinds slowly into Asia. Southern California is slowly sailing north toward Alaska. In about 30 million years, Los Angeles will pass San Francisco, if either still exists by then.

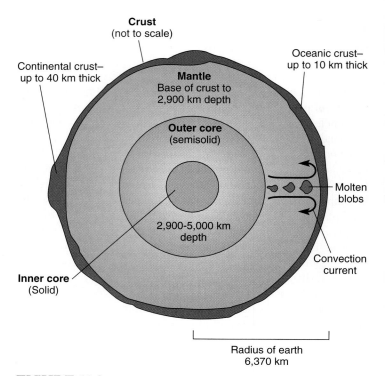

FIGURE 11.2 *The layered earth. The intensely hot, liquid or semisolid outer core consists primarily of molten metal. Around it is a solid but pliable mantle, hot enough to bend and flow like warm taffy. Floating on top of the mantle is a thin crust of rock that breaks into large, slowly moving tectonic plates. The crust appears ten times thicker in this drawing than it is in reality.*

TABLE 11.1 Eight Most Common Chemical Elements (Percent) in Whole Earth and Crust

WHOLE EARTH		CRUST	
Iron	33.3	Oxygen	45.2
Oxygen	29.8	Silicon	27.2
Silicon	15.6	Aluminum	8.2
Magnesium	13.9	Iron	5.8
Nickel	2.0	Calcium	5.1
Calcium	1.8	Magnesium	2.8
Aluminum	1.5	Sodium	2.3
Sodium	0.2	Potassium	1.7

When an oceanic plate collides with a continental landmass, the continental plate usually rides up over the seafloor, while the oceanic plate is subducted, or pushed down into the mantle, where it melts and rises back to the surface as magma (fig. 11.4). Deep ocean trenches mark these subduction zones, and volcanoes form where the magma erupts through vents and fissures in the overlying crust. Trenches and volcanic mountains ring the Pacific Ocean

FIGURE 11.3 *Map of tectonic plates. Plate boundaries are dynamic zones, characterized by earthquakes and volcanism and the formation of great rifts and mountain ranges. Arrows indicate direction of subduction where one plate is diving beneath another. These zones are sites of deep trenches in the ocean floor and high levels of seismic and volcanic activity.*
Sources: U.S. Department of the Interior, U.S. Geological Survey.

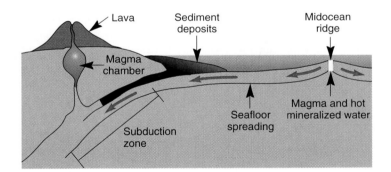

FIGURE 11.4 *Plate tectonic movements drive the rock cycle. Old seafloor and sediment deposits melt in subduction zones. Magma rises to erupt through volcanoes or to recrystallize at depth into igneous rocks. Weathering breaks down surface rocks, and erosion deposits residue in sedimentary formations. Tectonic movements create pressure and heat, which cause metamorphism of both sedimentary and igneous rocks.*

rim from Indonesia to Japan to Alaska and down the west coast of the Americas, forming a so-called "ring of fire" where oceanic plates are being subducted under the continental plates. This ring is the source of more earthquakes and volcanic activity than any other region on the earth.

Over millions of years, continents can drift long distances. Antarctica and Australia once were connected to Africa, for

instance, somewhere near the equator and supported luxuriant forests. Geologists suggest that several times in the earth's history most or all of the continents have gathered to form supercontinents, which have ruptured and reformed over hundreds of millions of years (fig. 11.5). The redistribution of continents has profound effects on the earth's climate and may help explain the periodic mass extinctions of organisms marking the divisions between many major geologic periods.

MINERALS AND ROCKS

A **mineral** is a naturally occurring, inorganic solid with a definite chemical composition and a specific internal crystal structure. A mineral is solid; therefore ice is a mineral (with a distinct composition and crystal structure), but liquid water is not. Similarly molten lava is not crystalline, although it generally hardens to create distinct minerals. Metals (such as iron, copper, aluminum, or gold) come from mineral ores, but once purified, metals are noncrystalline and thus are not minerals. Depending on the conditions in which they were formed, mineral crystals can be microscopically small, like asbestos fibers, or huge, like the tree-size selenite crystals recently discovered in a Chihuahua, Mexico mine.

A **rock** is a solid, cohesive aggregate of one or more minerals. Within the rock, individual mineral crystals (or grains) are

FIGURE 11.5 *Pangaea, the ancient supercontinent of 200 million years ago, combined all the world's continents in a single landmass.*

FIGURE 11.6 *The rock cycle consists of processes of creation, destruction, and metamorphosis. Each of the three rock types can be converted to either of the other types.*

mixed together and held firmly in a solid mass. The grains may be large or small, depending on how the rock was formed, but each grain retains its own unique mineral qualities. Each rock type has a characteristic mixture of minerals, grain sizes, and ways in which the grains are mixed and held together. Granite, for example, is a mixture of quartz, feldspar, and mica crystals. Rocks with a granite-like mineral content but much finer crystals are called rhyolite; chemically similar rocks with large crystals are called pegmatite.

Rock Types and How They Are Formed

Although rocks appear hard and permanent, they are part of a relentless cycle of formation and destruction. They are crushed, folded, melted, and recrystallized by dynamic processes related to those that shape the large-scale features of the earth's crust. We call this cycle of creation, destruction, and metamorphosis the **rock cycle** (fig. 11.6). Understanding something of how this cycle works helps explain the origin and characteristics of different types of rocks.

There are three major rock classifications: igneous, metamorphic, and sedimentary. **Igneous rocks** (from *igni,* the Latin word for fire) are solidified from hot, molten magma or lava. Most rock in the earth's crust is igneous. Magma extruded to the surface from volcanic vents cools quickly to make finely crystalline rocks such as basalt, rhyolite, or andesite. Magma that cools slowly in subsurface chambers or is intruded between overlying layers makes coarsely crystalline rocks, such as gabbro (rich in iron and

silica) or granite (rich in aluminum and silica), depending on the chemical composition of the magma.

Metamorphic rocks form from the melting, contorting, and recrystallizing of other rocks. Deep in the ground, tectonic forces squeeze, fold, heat, and recrystallize solid rock. Under these conditions, chemical reactions can alter both the composition and structure of the component minerals. Metamorphic rocks are classified by their chemical composition and by the degree of recrystallization: some minerals form only under extreme pressure and heat (diamonds or jade, for example); others form under more moderate conditions (graphite or talc). Some common metamorphic rocks are marble (from limestone), quartzite (from sandstone), and slate (from mudstone and shale). Metamorphic rocks often have swirling patterns left by the twisting and folding that created them.

Sedimentary rocks are formed when loose grains of other rocks are consolidated by time and pressure. Sandstone, for example, is solidified from layers of sand, and mudstone consists of extremely hardened mud and clay. Tuff is formed from volcanic ash, and conglomerates are aggregates of sand and gravel. Some sedimentary rocks develop from crystals that precipitate out of extremely salty water. Rock salt, made of the mineral halite, is ground up to produce ordinary table salt (sodium chloride). Salt deposits often form when a body of saltwater dries up, leaving salt crystals behind. Limestone is a rock composed of cemented remains of marine organisms. You can often see the shapes of shells and corals in a piece of limestone. Sedimentary formations often have distinctive layers that show different conditions when they were laid down. Relatively soft sedimentary rocks such as sandstone can be shaped by erosion into striking features (fig. 11.7).

Weathering and Sedimentation

Most crystalline rocks are extremely hard and durable, but exposure to air, water, changing temperatures, and reactive chemical agents slowly breaks them down in a process called **weathering.** Mechanical weathering is the physical breakup of rocks into smaller particles without a change in chemical composition of the constituent minerals. You have probably seen mountain valleys scraped by glaciers, or river and shoreline pebbles that are rounded from being rubbed against one another as they are tumbled by waves and currents.

Chemical weathering is the selective removal or alteration of specific minerals in rocks. This alteration leads to weakening and disintegration of rock. Among the more important chemical weathering processes are oxidation (combination of oxygen with an element to form an oxide or hydroxide mineral) and hydrolysis (hydrogen atoms from water molecules combine with other chemicals to form acids). The products of these reactions are more susceptible to both mechanical weathering and to dissolving in water. For instance, when carbonic acid (formed when rainwater absorbs CO_2) percolates through porous limestone layers in the ground, it dissolves the rock and creates caves.

Particles of rock loosened by wind, water, ice, and other weathering forces are carried downhill, downwind, or downstream until they come to rest again in a new location. The deposition of these materials is called **sedimentation.** Water, wind, and glaciers deposit particles of sand, clay, or silt far from their source. Much of the American Midwest, for instance, is covered with hundreds of meters of sedimentary material left by glaciers (till, or rock debris deposited by glacial ice), wind (loess, or fine dust deposits), river deposits of sand and gravel, and ocean deposits of sand, silt, clay, and limestone.

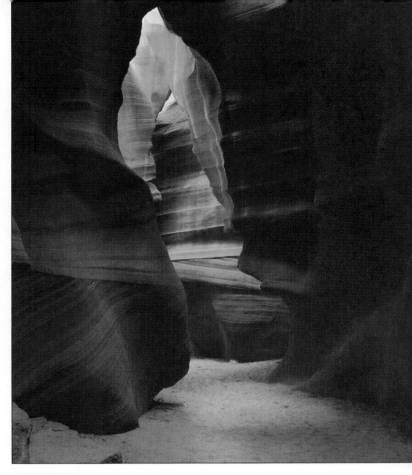

FIGURE 11.7   *The sandstone of Antelope Canyon, Arizona, is sculpted by water.*

ECONOMIC GEOLOGY AND MINERALOGY

Economic mineralogy is the study of minerals that are valuable for manufacturing and trade. Most economic minerals are metal ores, minerals with unusually high concentrations of metals. Lead, for example, often comes from the mineral galena (PbS), and copper from sulfide ores such as bornite (Cu_5FeS_4). Nonmetallic geological resources include graphite, feldspar, quartz crystals, diamonds, and other crystals that are valued for their usefulness or beauty. Metals have been so important in human affairs that major epochs of human history are commonly known by the dominant materials and the technology to use them (Stone Age, Bronze Age, Iron Age, etc.). The mining, processing, and distribution of these materials have broad implications for both our culture and our environment. Most economically valuable crustal resources exist everywhere in small amounts; the important thing is to find them concentrated in economically recoverable levels.

Public policy in the United States has encouraged mining on public lands as a way of boosting the economy and utilizing natural resources. Today these policies are controversial, and there are efforts to recover public revenue from these publicly owned resources. (See the related essay "Should We Revise the 1872 Mining Law?" at www.mhhe.com/apps.)

Metals

Metals are malleable substances that are useful and valuable because they are strong, relatively light, and can be reshaped for many purposes. The availability of metals and the methods to extract and use them have determined technological developments, as well as economic and political power for individuals and nations.

The metals consumed in greatest quantity by world industry include iron (740 million metric tons annually), aluminum (40 million metric tons), manganese (22.4 million metric tons), copper and chromium (8 million metric tons each), and nickel (0.7 million metric tons). Most of these metals are consumed in the United States, Japan, and Europe, in that order. They are produced primarily in South America, South Africa, and Russia. It is easy to see how these facts contribute to a worldwide mineral trade network that has become crucially important to the economic and social stability of all nations involved (fig. 11.8). Table 11.2 shows the primary uses of these metals.

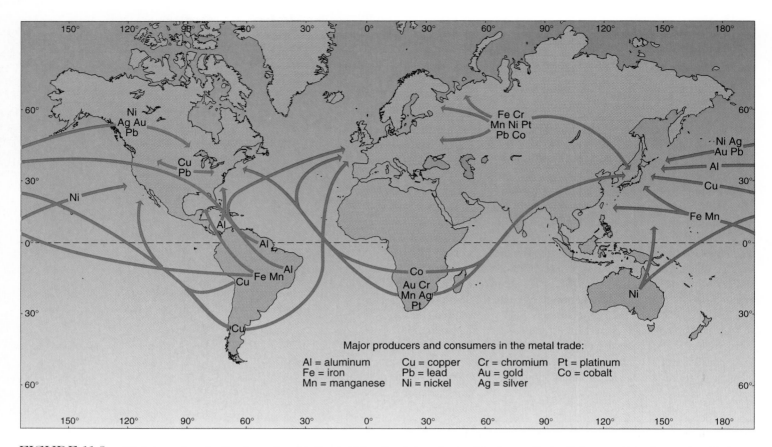

FIGURE 11.8 *Global metal trade. Metals produced in South Africa, South America, and the former Soviet Union are shipped to markets in the United States, Europe, and Japan, creating a global economic network on which both consumers and producers depend.*

TABLE 11.2 Primary Uses of Some Major Metals Consumed in the United States

METAL	USE
Aluminum	Packaging foods and beverages (38%), transportation, electronics
Chromium	High-strength steel alloys
Copper	Building construction, electric and electronic industries
Iron	Heavy machinery, steel production
Lead	Leaded gasoline, car batteries, paints, ammunition
Manganese	High-strength, heat-resistant steel alloys
Nickel	Chemical industry, steel alloys
Platinum group	Automobile catalytic converters, electronics, medical uses
Gold	Medical, aerospace, electronic uses; accumulation as monetary standard
Silver	Photography, electronics, jewelry

Nonmetal Mineral Resources

Nonmetal minerals are a broad class that covers resources from silicate minerals (gemstones, mica, talc, and asbestos) to sand, gravel, salts, limestone, and soils. Sand and gravel production for road and building construction comprise by far the greatest volume and dollar value of all nonmetal mineral resources and a far greater volume than all metal ores. Sand and gravel are used mainly in brick and concrete construction, paving, as loose road filler, and for sandblasting. High-purity silica sand is our source of glass. These materials usually are retrieved from surface pit mines and quarries, where they have been deposited by glaciers, winds, or ancient oceans.

Limestone, like sand and gravel, is mined and quarried for concrete and crushed for road rock. It also is cut for building stone, pulverized for use as an agricultural soil additive that neutralizes acidic soil, and roasted in lime kilns and cement plants to make plaster (hydrated lime) and cement.

Evaporites (materials deposited by evaporation of chemical solutions) are mined for halite, gypsum, and potash. These are often found at or above 97-percent purity. Halite, or rock salt, is used for water softening and melting ice on winter roads in some northern areas. Refined, it is a source of table salt. Gypsum (calcium sulfate) now makes our plaster wallboard, but it has been used to cover walls ever since the Egyptians plastered their frescoed tombs along the

Nile River some 5,000 years ago. Potash is an evaporite composed of a variety of potassium chlorides and potassium sulfates. These highly soluble potassium salts have long been used as a soil fertilizer.

Sulfur deposits are mined mainly for sulfuric acid production. In the United States, sulfuric acid use amounts to more than 200 lbs per person per year, mostly because of its use in industry, car batteries, and some medicinal products.

Fuels: Oil, Coal, and Uranium

Modern society functions largely on energy produced from geologic deposits of oil, coal, and natural gas. Nuclear energy, which runs on uranium, makes up a large portion of our electricity resources. Energy production from these sources is discussed in chapter 12. Oil, coal, and gas are organic, created over millions of years as extreme heat and pressure transform the remains of ancient organisms. They are not minerals because they have no crystalline structure, but they can be considered part of economic mineralogy because they are such important geologic resources. In addition to providing energy, oil is the source material for plastics, and natural gas is used to make agricultural fertilizers.

The search for deposits of these organic deposits is one of the most important parts of economic mineralogy. Debates over exploitation and ownership of these resources play important roles in national and international politics. The Persian Gulf War of 1990 was fought over control of vast underground oil deposits, as has been Russia's war in Chechnya. Debates over exploiting potential oil reserves in the Alaskan National Wildlife Refuge (ANWR) have raged in the United States for decades. (See the related essays "Exploiting Oil in ANWR" and "Oil and the War in Chechnya?" at www.mhhe.com/apps.) Recently, Canada's oil shale and tar sands have become another source of oil (see opening story, chapter 12).

Strategic Metals and Minerals

World industry depends on about 80 minerals and metals, some of which exist in plentiful supplies. Three-fourths of these resources are abundant enough to meet all of our anticipated needs or have readily available substitutes. At least 18 metals, however, including tin, platinum, gold, silver, and lead, are in short supply.

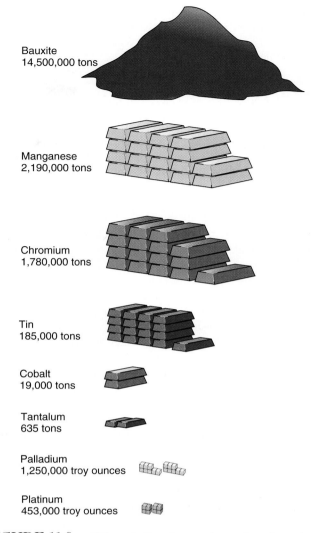

Bauxite
14,500,000 tons

Manganese
2,190,000 tons

Chromium
1,780,000 tons

Tin
185,000 tons

Cobalt
19,000 tons

Tantalum
635 tons

Palladium
1,250,000 troy ounces

Platinum
453,000 troy ounces

FIGURE 11.9 *U.S. stockpiles of strategic metals and metal ores.*

Of these 80 metals and minerals, between one-half and one-third are considered "strategic" resources. **Strategic metals and minerals** are those that a country uses but cannot produce itself. As the term *strategic* suggests, these are materials that a government considers capable of crippling the country's economy or military strength if unstable global economics or politics were to cut off supplies. For this reason, wealthy industrial nations stockpile strategic resources, especially metals, in times when prices are low and a supply is available. Figure 11.9 shows some of the major stockpiles of strategic metals and ores in the United States.

For less wealthy mineral- and metal-producing nations, there is another side to strategic minerals. Many less-developed countries depend on steady mineral exports for most of their foreign exchange. Zambia, for instance, relies on cobalt production for 50 percent of its national income. If a steady international market is not maintained, such producer nations could be devastated. From the small-nation producer's point of view, metal or mineral exports, like concentration on any single product or industry, are an unstable economic foundation. Often no option exists, however, if a producer is to participate in the world economy. The environmental consequences of mining may not be a high priority under these circumstances.

ENVIRONMENTAL EFFECTS OF RESOURCE EXTRACTION

Each of us depends daily on geologic resources mined from sites around the world. We use scores of metals and minerals, many of which we've never even heard of, in our lights, computers, watches, fertilizers, and cars. Mining and purifying all these resources can have severe environmental and social consequences. The most obvious effect of mining is often the disturbance or removal of the land surface. Farther-reaching effects, though, include air and water pollution. The EPA lists more than 100 toxic air pollutants, from acetone to xylene, released from U.S. mines every year. Nearly 80,000 metric tons of particulate matter (dust) and 11,000 tons of sulfur dioxide are released from nonmetal mining alone. Chemical- and sediment-runoff pollution is a major problem in many local watersheds.

   Resource extraction can affect water quality for several reasons. Gold and other metals are often found in sulfide ores. Sulfide minerals often occur in hydrothermal (hot water) deposits, where sulfur and other elements are mobilized, then deposited as they are heated by deep magma chambers. When sulfur-bearing minerals are exposed to air and water, they produce sulfuric acid, which is highly mobile and strongly acidic. In addition, metal elements often occur in very low concentrations—10 to 20 parts per billion may be economically extractable for gold, platinum, and other metals. Consequently, vast quantities of ore must be crushed and washed to extract metals. Cyanide, mercury, and other toxic substances are used to chemically separate metals from the minerals that contain them, and these substances can easily contaminate lakes and streams. Further, a great deal of water is used in washing crushed ore with cyanide and other solutions. In arid Nevada, the USGS estimates that mining consumes about 230,000 m^3 (60 million gal) per day. After use in ore processing, much of this water contains sulfuric acid, arsenic, heavy metals, and other contaminants. Mine runoff leaking into lakes and streams can damage or destroy aquatic ecosystems (fig. 11.10).

Mining

There are many techniques for extracting geologic materials. The most common methods are open-pit mining, strip-mining, and underground mining. An ancient method of accumulating gold, diamonds, and coal is placer mining, in which pure nuggets are washed from stream sediments. Since the California gold rush of 1849, placer miners have used water cannons to blast away hillsides. This method, which chokes stream ecosystems with sediment, is still used in Alaska, Canada, and many other regions. Another ancient, and much more dangerous, method is underground mining. Ancient Roman, European, and Chinese miners tunneled deep into tin, lead, copper, coal, and other mineral seams.

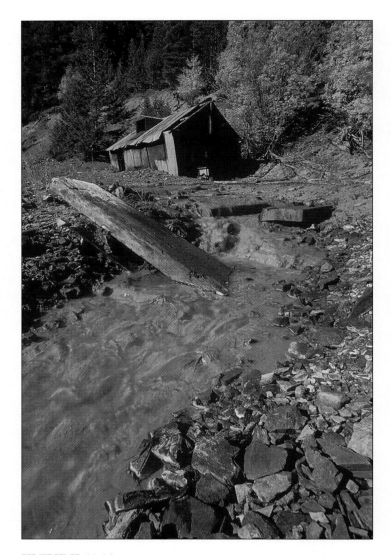

FIGURE 11.10   *Thousands of abandoned mines on public lands poison streams and groundwater with acid, metal-laced drainage. This old mine in Montana drains into the Blackfoot River, the trout stream featured in Norman Maclean's book* A River Runs Through It.

Mine tunnels occasionally collapse, and natural gas in coal mines can explode. Water seeping into mine shafts also dissolves toxic minerals. Contaminated water seeps into groundwater; it is also pumped to the surface, where it enters streams and lakes.

   In underground coal mines, another major environmental risk is fires. Hundreds of coal mines smolder in the United States, China, Russia, India, South Africa, and Europe. The inaccessibility and size of these fires make many impossible to extinguish or control. One mine fire in Centralia, Pennsylvania, has been burning since 1962; control efforts have cost at least $40 million, but the fire continues to expand. China, which depends on coal for much of its heating and electricity, has hundreds of smoldering mine fires; one has been burning for 400 years. According to a recent study from the International Institute for Aerospace Survey in the Netherlands, these fires consume up to 200 million tons of coal every year and emit as much carbon dioxide as all the cars in the United States. Toxic fumes, explosive methane, and other hazardous emissions are also released from these fires.

Open-pit mines are used to extract massive beds of metal ores and other minerals. The size of modern open pits can be hard to comprehend. The Bingham Canyon mine, near Salt Lake City, Utah, is 800 m (2,640 ft) deep and nearly 4 km (2.5 mi) wide at the top. More than 5 billion tons of copper ore and waste material have been removed from the hole since 1906. A chief environmental challenge of open-pit mining is that groundwater accumulates in the pit. In metal mines, a toxic soup results. No one yet knows how to detoxify these lakes, which endanger wildlife and nearby watersheds. (See the related essay "Death in a Mine Pit" at www.mhhe.com/app.)

Half the coal used in the United States comes from strip mines. Since coal is often found in expansive, horizontal beds, the entire land surface can be stripped away to cheaply and quickly expose the coal. The overburden, or surface material, is placed back into the mine, but usually in long ridges called spoil banks. Spoil banks are very susceptible to erosion and chemical weathering. Since the spoil banks have no topsoil (the complex organic mixture that supports vegetation—see chapter 7), revegetation occurs very slowly.

The 1977 federal Surface Mining Control and Reclamation Act (SMCRA) requires better restoration of strip-mined lands, especially where mines replaced prime farmland. Since then, the record of strip-mine reclamation has improved substantially. Complete mine restoration is expensive, often more than $10,000 per hectare. Restoration is also difficult because the developing soil is usually acidic and compacted by the heavy machinery used to reshape the land surface.

Bitter controversy has grown recently over mountaintop removal, a coal mining method mainly practiced in Appalachia. Long, sinuous ridge-tops are removed by giant, 20-story-tall shovels to expose horizontal beds of coal (fig. 11.11). Up to 215 m (700 ft) of ridge-top is pulverized and dumped into adjacent river valleys. The debris can be laden with selenium, arsenic, coal, and other toxic substances. At least 900 km (560 mi) of streams have been buried in West Virginia alone. Environmental lawyers have sued to stop the destruction of streams, arguing that it violates the Clean Water Act (chapter 10). In response, the Bush administration issued a "clarification" of the Clean Water Act, rewriting the law to allow stream filling by any sort of mining or industrial debris, rather than forbidding any stream destruction. Environmentalists charge that the new wording subverts the Clean Water Act, giving a green light to all sorts of stream destruction. West Virginians are deeply divided about mountaintop removal because they live in affected stream valleys but also depend on a coal economy.

The Mineral Policy Center in Washington, D. C., estimates that 19,000 km (12,000 mi) of rivers and streams in the United States are contaminated by mine drainage. The EPA estimates that cleaning up impaired streams, along with 550,000 abandoned mines in the U.S., may cost $70 billion. Worldwide, mine closing and rehabilitation costs are estimated in the trillions of dollars. Because of the volatile prices of metals and coal, many mining companies have gone bankrupt before restoring mine sites, leaving the public responsible for cleanup. In 2002, more than 500

FIGURE 11.11 *Mountaintop removal mining is a relatively new, and deeply controversial, method of extracting Appalachian coal.*

leading mine executives and their critics convened at the Global Mining Initiative, a meeting in Toronto aimed at improving the sustainability of mining. Executives acknowledged that in the future they will increasingly be held liable for environmental damages, and they said they were seeking ways to improve the industry's social and environmental record. Jay Hair, secretary general of the International Council on Mining and Metals, stated that "environmental protection and social responsibility are important," and that mining companies were interested in participating in sustainable development. Mine executives also recognized that, increasingly, big cleanup bills will cut into company values and stock prices. Finding creative ways to keep mines cleaner from the start will make good economic sense, even if it's not easy to do.

Processing

Metals are extracted from ores by heating or with chemical solvents. Both processes release large quantities of toxic materials that can be even more environmentally hazardous than mining. **Smelting**—roasting ore to release metals—is a major source of air pollution. One of the most notorious examples of ecological devastation from smelting is a wasteland near Ducktown, Tennessee. In the mid-1800s, mining companies began excavating the rich copper deposits in the area. To extract copper from the ore, they built huge, open-air, wood fires, using timber from the surrounding forest. Dense clouds of sulfur dioxide released from sulfide ores poisoned the vegetation and acidified the soil over a 50-mi² (13,000-ha) area. Rains washed the soil off the denuded land, creating a barren moonscape.

Sulfur emissions from Ducktown smelters were reduced in 1907 after Georgia sued Tennessee over air pollution. In the 1930s the Tennessee Valley Authority (TVA) began treating the soil and replanting trees to cut down on erosion. Recently, upwards of $250,000 per year has been spent on this effort. While the trees and other plants are still spindly and feeble, more than two-thirds of the area is considered "adequately" covered with vegetation. Similarly, smelting of copper-nickel ore in Sudbury, Ontario, a century ago

Gold mine
Leaching pad
Influent leaching solution
Solution makeup tank
Solution is recirculated to makeup tank
Front-end loader to spread ore
Processing plant
Gold
Solution sprays
Solution trickling to bottom of heap is piped to the plant
Dam
Drain
Ore trucks
Pregnant effluent solution
Pump

FIGURE 11.12 *In a heap-leach operation, huge piles of low-grade ore are heaped on an impervious pad and sprayed continuously with a cyanide solution. As the leaching solution trickles through the crushed ore, it extracts gold and other precious metals. The "pregnant" effluent solution is then pumped to a processing plant where metals are extracted and purified. This technique is highly profitable but carries large environmental risks.*

Source: George Laycock, *Audubon Magazine,* vol. 91(7), July 1989.

caused widespread ecological destruction that is slowly being repaired following pollution-control measures (see fig. 9.24).

Chemical extraction is used to dissolve or mobilize pulverized ore, but it uses and pollutes a great deal of water. A widely used method is **heap-leach extraction,** which involves piling crushed ore in huge heaps and spraying it with a dilute alkaline-cyanide solution (fig. 11.12). The solution percolates through the pile and dissolves gold. The gold-containing solution is then pumped to a processing plant that removes the gold by electrolysis. A thick clay pad and plastic liner beneath the ore heap is supposed to keep the poisonous cyanide solution from contaminating surface or groundwater, but leaks are common.

Once all the gold is recovered, mine operators may simply walk away from the operation, leaving vast amounts of toxic effluent in open ponds behind earthen dams. A case in point is the Summitville Mine near Alamosa, Colorado. After extracting $98 million in gold, the absentee owners declared bankruptcy in 1992, abandoning millions of tons of mine waste and huge leaking ponds of cyanide. The Environmental Protection Agency may spend more than $100 million trying to clean up the mess and keep the cyanide pool from spilling into the Alamosa River.

CONSERVING GEOLOGIC RESOURCES

Conservation offers great potential for extending our supplies of economic minerals and reducing the effects of mining and pro-

cessing. The advantages of conservation are significant: less waste to dispose of, less land lost to mining, and less consumption of money, energy, and water resources.

Recycling

Some waste products already are being exploited, especially for scarce or valuable metals. Aluminum, for instance, must be extracted from bauxite by electrolysis, an expensive, energy-intensive process. Recycling waste aluminum, such as beverage cans, on the other hand, consumes one-twentieth of the energy of extracting new aluminum. Today, nearly two-thirds of all aluminum beverage cans in the United States are recycled, up from only 15 percent 20 years ago. The high value of aluminum scrap ($650 a ton versus $60 for steel, $200 for plastic, $50 for glass, and $30 for paperboard) gives consumers plenty of incentive to deliver their cans for collection. Recycling is so rapid and effective that half of all the aluminum cans now on a grocer's shelf will be made into another can within two months. Table 11.3 shows the energy cost of extracting other materials.

TABLE 11.3 Energy Requirements in Producing Various Materials from Ore and Raw Source Materials

PRODUCT	ENERGY REQUIREMENT (MJ/kg)*	
	NEW	FROM SCRAP
Glass	25	25
Steel	50	26
Plastics	162	n.a.[†]
Aluminum	250	8
Titanium	400	n.a.[†]
Copper	60	7
Paper	24	15

*Megajoules per kilogram

†Not available

Source: Data from E. T. Hayes, *Implications of Materials Processing,* 1997.

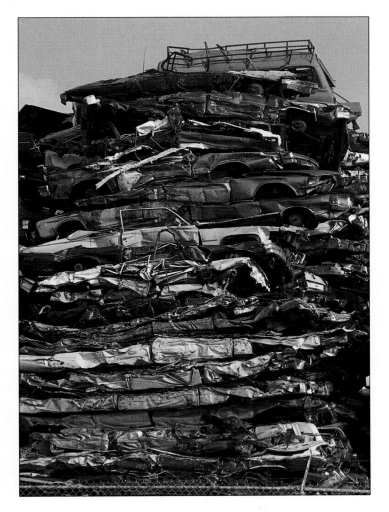

FIGURE 11.13 *The richest metal source we have—our mountains of scrapped cars—offer a rich, inexpensive, and ecologically beneficial resource that can be "mined" for a number of metals.*

Platinum, the catalyst in automobile catalytic exhaust converters, is valuable enough to be regularly retrieved and recycled from used cars (fig. 11.13). Other commonly recycled metals are gold, silver, copper, lead, iron, and steel. The latter four are readily available in a pure and massive form, including copper pipes, lead batteries, and steel and iron auto parts. Gold and silver are valuable enough to warrant recovery, even through more difficult means.

Steel and Iron Recycling: Minimills

While total U.S. steel production has fallen in recent decades—largely because of inexpensive supplies from new and efficient Japanese steel mills—a new type of mill subsisting entirely on a readily available supply of scrap/waste steel and iron is a growing industry. Minimills, which remelt and reshape scrap iron and steel, are smaller and cheaper to operate than traditional integrated mills that perform every process from preparing raw ore to finishing iron and steel products. Minimills produce steel at between $225 and $480 per metric ton, while steel from integrated mills costs $1,425

to $2,250 per metric ton on average. The energy cost is also lower in minimills: 5.3 million BTU/ton of steel compared to 16.08 million BTU/ton in integrated mill furnaces. Minimills now produce about half of all U.S. steel. Recycling is slowly increasing as raw materials become more scarce and wastes become more plentiful.

Substituting New Materials for Old

Mineral and metal consumption can be reduced by new materials or new technologies developed to replace traditional uses. This is a long-standing tradition; for example, bronze replaced stone technology, and iron replaced bronze. More recently, the introduction of plastic pipe has decreased our consumption of copper, lead, and steel pipes. In the same way, the development of fiber-optic technology and satellite communication reduces the need for copper telephone wires.

Iron and steel have been the backbone of heavy industry, but we are now moving toward other materials. One of our primary uses for iron and steel has been machinery and vehicle parts. In automobile production, steel is being replaced by polymers (long-chain organic molecules similar to plastics), aluminum, ceramics, and new high-technology alloys. All of these reduce vehicle weight and cost, while increasing fuel efficiency. Some of the newer alloys that combine steel with titanium, vanadium, or other metals wear much better than traditional steel. Ceramic engine parts provide heat insulation around pistons, bearings, and cylinders, keeping the rest of the engine cool and operating efficiently. Plastics and glass fiber-reinforced polymers are used in body parts and some engine components.

Electronics and communications (telephone) technology, once major consumers of copper and aluminum, now use ultra-high-purity glass cables to transmit pulses of light, instead of metal wires carrying electron pulses. Once again, this technology has been developed for its greater efficiency and lower cost, but it also affects consumption of our most basic metals.

GEOLOGIC HAZARDS

Earthquakes, volcanoes, floods, and landslides are normal earth processes, events that have made our earth what it is today. However, when they affect human populations, their consequences can be among the worst and most feared disasters that befall us.

Earthquakes

Earthquakes are sudden movements in the earth's crust that occur along faults (planes of weakness), where one rock mass slides past another one. When movement along faults occurs gradually and relatively smoothly, it is called creep or seismic slip and may be undetectable to the casual observer. When friction prevents rocks from slipping easily, stress builds up until it is finally released with a sudden jerk. The point on a fault at which the first movement occurs during an earthquake is called the epicenter.

Earthquakes have always seemed mysterious, sudden, and violent, coming without warning and leaving ruined cities and

FIGURE 11.14 *An elevated freeway buckled and collapsed as a result of the 1995 earthquake in Kobe, Japan.*

One of the most notorious effects of earthquakes is the tsunami. These giant sea swells (sometimes improperly called tidal waves) can move at 1,000 km/h (600 mph), or faster, away from the center of an earthquake. When these swells approach the shore, they often reach 15 m or more, and some have been as high as 65 m (nearly 200 ft). A 1960 tsunami coming from a Chilean earthquake still caused 7-m breakers when it reached Hawaii 15 hours later. Tsunamis also can be caused by underwater volcanic explosions or massive seafloor slumping. The eruption of the Indonesian volcano Krakatoa in 1883 created a tsunami 40 m (130 ft) high that killed 30,000 people on nearby islands.

Volcanoes

Volcanoes and undersea magma vents are the sources of most of the earth's crust. Over hundreds of millions of years, gaseous emissions from these sources formed the earth's earliest oceans and atmosphere. Many of the world's fertile soils are weathered volcanic materials. Volcanoes have also been an ever-present threat to human populations (fig. 11.15). One of the most famous historic volcanic eruptions was that of Mount Vesuvius in southern Italy, which buried the cities of Herculaneum and Pompeii in A.D. 79. The mountain had been showing signs of activity before it erupted, but many citizens chose to stay and take a chance on survival. On August 24, the mountain buried the two towns in ash. Thousands were killed by the dense, hot, toxic gases that accompanied the ash flowing down from the volcano's mouth. It continues to erupt from time to time.

Nuees ardentes (French for "glowing clouds") are deadly, denser-than-air mixtures of hot gases and ash like those that inundated Pompeii and Herculaneum. Temperatures in these clouds may exceed 1,000°C, and they move at more than 100 km/h (60 mph). *Nuees ardentes* destroyed the town of St. Pierre on the

dislocated landscapes in their wake. Cities such as Kobe, Japan, or Mexico City, parts of which are built on soft landfill or poorly consolidated soil, usually suffer the greatest damage from earthquakes (fig. 11.14). Water-saturated soil can liquify when shaken. Buildings sometimes sink out of sight or fall down like a row of dominoes under these conditions.

Earthquakes frequently occur along the edges of tectonic plates, especially where one plate is being subducted, or pushed down, beneath another. Earthquakes also occur in the centers of continents, however. In fact, one of the largest earthquakes ever recorded in North America was one of an estimated magnitude 8.8 that struck the area around New Madrid, Missouri, in 1812. Fortunately, few people lived there at the time, and the damage was minimal.

Modern contractors in earthquake zones attempt to prevent damage and casualties by constructing buildings that can withstand tremors. The primary methods used are heavily reinforced structures, strategically placed weak spots in the building that can absorb vibration from the rest of the building, and pads or floats beneath the building on which it can shift harmlessly with ground motion.

FIGURE 11.15 *Pyroclastic flows spill down the slopes of Mayon volcano in the Philippines in this September 23, 1984, image. Because more than 73,000 people evacuated danger zones, this eruption caused no casualties.*

Principles of Environmental Science

FIGURE 11.16 *Volcanic activity at Mount Rainier has produced at least 12 large mudflows in the past 6,000 years (brown areas). Future flows could threaten large populations now in the area.*
Source: Data from T. W. Sisson, USGS Open File Report 95-642.

Caribbean island of Martinique on May 8, 1902. Mount Pelee released a cloud of *nuees ardentes* that rolled down through the town, killing somewhere between 25,000 and 40,000 people within a few minutes. All of the town's residents died except for a single prisoner being held in the town dungeon.

Mudslides are also disasters sometimes associated with volcanoes. The 1985 eruption of Nevado del Ruíz, 130 km (85 mi) northwest of Bogotá, Colombia, caused mudslides that buried most of the town of Armero and devastated the town of Chinchina. An estimated 25,000 people were killed. Heavy mudslides also accompanied the eruption of Mount St. Helens in Washington in 1980. Sediments mixed with melted snow destroyed roads, bridges, and property, but because of sufficient advance warning, there were few casualties. Geologists worry that similar mudflows from an eruption at Mount Rainier would threaten much larger populations (fig. 11.16).

Volcanic eruptions often release large volumes of ash and dust into the air. Mount St. Helens expelled 3 km^3 of dust and ash, causing ash fall across much of North America. This was only a minor eruption. An eruption in a bigger class of volcanoes was that of Tambora in Indonesia in 1815, which expelled 175 km^3 of dust and ash, more than 58 times that of Mount St. Helens. These dust clouds circled the globe and reduced sunlight and air temperatures enough so that 1815 was known as the year without a summer.

It is not just a volcano's dust that blocks sunlight. Sulfur emissions from volcanic eruptions combine with rain and atmospheric moisture to produce sulfuric acid (H_2SO_4). Droplets of H_2SO_4 interfere with solar radiation and can significantly cool the world climate. In 1991, Mount Pinatubo in the Philippines emitted

20 million tons of sulfur dioxide aerosols that remained in the stratosphere for two years. This thin haze cooled the entire earth by 1°C for about 2 years. It also caused a 10 to 15 percent reduction in stratospheric ozone, allowing increased ultraviolet light to reach the earth's surface.

Floods

In most moderately humid climates, stream channels adjust to accommodate average maximum stream flows. Much of the year, the water level may be well below the stream bank height, but heavy rains or sudden snow melt can deliver more water than the stream can carry. Excess water that overflows stream banks and covers adjacent land is considered a **flood.** The severity of floods can be described by the depth of water above the normal stream banks or by how frequently a similar event normally occurs—on average—for a given area. Note that these are statistical averages over long time periods. A "10-year flood" would be expected to occur once in every ten years; a "100-year flood" would be expected to occur once every century. But two 100-year floods can occur in successive years or even in the same year.

The biggest economic loss from floods is usually not the buildings and property they carry away, but rather the contamination they cause. Virtually everything flood waters touch in a house—carpets, furniture, drapes, electronics, even drywall and insulation—must be removed and discarded. Although floods usually don't take as many lives as some other natural disasters, they are among the most costly because they occur relatively frequently and in populated river corridors.

Many human activities increase both the severity and frequency of floods. Paving roads and parking lots reduces water infiltration into the soil and speeds the rate of runoff into streams and lakes. Clearing forests for agriculture and filling cities with buildings similarly increase both the volume and rate of water discharge after a storm.

Under normal conditions, floods are mitigated by **floodplains**—low land that is periodically inundated during normal floods. However, floodplains are usually fertile, flat, and easily farmed. They are convenient for building and close to the river. In much of the developed world, floodplains are widely farmed, developed with cities and houses, and cleared of vegetation. Floodplains have lost much of their ability to absorb flood waters.

Even more than development, though, flood control structures have separated floodplains from rivers. Levees and flood walls are built to contain water within riverbanks, and river channels are dredged and deepened to allow water to recede faster. Every flood-control structure simply transfers the problem downstream, however. The water has to go somewhere. If it doesn't soak into the ground upstream, it will simply exacerbate floods somewhere downstream—leading to more levee development, and then to more flooding further downstream, and so on (fig. 11.17).

Flood Control

More than $25 billion of river-control systems have been built on the Mississippi and its tributaries. These systems have protected

FIGURE 11.17 *The Mississippi River inundated downtown Davenport, Iowa, during the summer floods of 1993. Although sympathetic with the heartbreak and economic losses caused by this flooding, many people argue that floodplains such as this never should have been settled in the first place.*

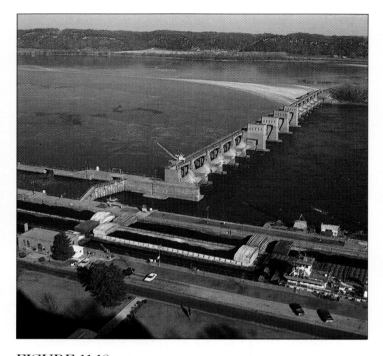

FIGURE 11.18 *Locks, dams, levees, and other flood-control and navigational structures such as this complex on the Mississippi River at Dubuque, Iowa, have been a godsend for commercial shippers but have also been disastrous for riparian ecosystems, backwater sloughs, marshes, and the wildlife that depends on them.*

many communities over the past century (fig. 11.18). In the major floods of 1993, however, this elaborate system helped turn a large flood into a major disaster. Deprived of the ability to spill out over floodplains, the river is pushed downstream to create faster currents and deeper floods until eventually a levee gives way somewhere. Hydrologists calculate that the floods of 1993 were about 3 m (10 ft) higher than they would have been, given the same rainfall in 1900 before the flood-control structures were in place.

Under current rules, the government is obligated to finance most levees and flood-control structures. Many people think that it would be much better to spend this money to restore wetlands, replace ground cover on water courses, build check dams on small streams, move buildings off the floodplain, and undertake other nonstructural ways of reducing flood danger. According to this view, floodplains should be used for wildlife habitat, parks, recreation areas, and other uses not susceptible to flood damage.

The National Flood Insurance Program administered by the Federal Emergency Management Agency (FEMA) was intended to aid people who cannot buy insurance at reasonable rates, but its effects have been to encourage building on the floodplains by making people feel that whatever happens, the government will take care of them. Many people would like to relocate homes and businesses out of harm's way after the recent floods or to improve them so they will be less susceptible to flooding, but owners of damaged property can collect only if they rebuild in the same place and in the same way as before. This serves to perpetuate problems rather than solve them.

Erosion

Gravity constantly pulls downward on every material everywhere on earth. Hillsides, beaches, even relatively flat farm fields can lose material to erosion. Often, water helps to mobilize loose material, and catastrophic landslides, beach erosion, and gully development can occur in a storm. Where houses are built on erodible surfaces, enormous loss of property can result.

Landslides, also called mass wasting or mass movement, occur when masses of material move downslope. Slow and subtle landslips are common, but rockslides, mudslides, and slumping can be swift and dangerous. In the United States alone, landslides and related mass wasting cause over $1 billion in property damage every year. When unconsolidated sediments on a hillside are saturated by a storm or exposed by logging, road building, or house construction, slopes are especially susceptible to sudden landslides.

Often people are unaware of the risks they face by locating on or under unstable hillsides. Sometimes they simply ignore clear and obvious danger. In Southern California, where land prices are high, people often build expensive houses on steep hills and in narrow canyons. Most of the time, this dry environment appears quite stable, but the chaparral vegetation burns frequently and fiercely. When fires expose the soil in late summer, heavy winter rains cause mudslides and debris flows that destroy whole neighborhoods (fig. 11.19).

Gullying is the development of deep trenches on relatively flat ground. Especially on farm fields, which have a great deal of loose soil unprotected by plant roots, rainwater running across the surface can dig deep gullies. Sometimes land becomes useless for

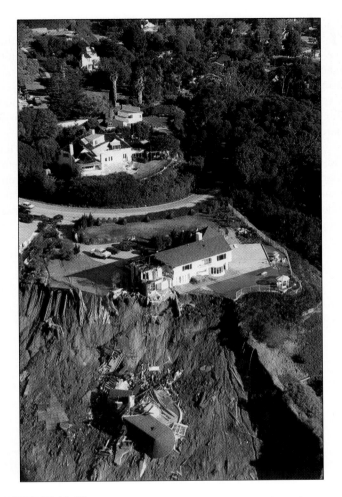

FIGURE 11.19 *Parts of an expensive house slide down the hillside in Pacific Palisades, California. Building at the edge of steep slopes made of unconsolidated sediment in an earthquake-prone region is a risky venture.*

FIGURE 11.20 *Winter storms have eroded the beach and undermined the foundations of homes on this barrier island. Breaking through protective dunes to build such houses damages sensitive plant communities and exposes the whole island to storm sand erosion. Coastal zone management attempts to limit development on fragile sites.*

farming because gullying is so severe and because erosion has removed the fertile topsoil. Agricultural soil erosion has been described as an invisible crisis. Erosion has reduced the fertility of millions of acres of prime farmland in the United States alone.

Beach erosion occurs on all sandy shorelines because the motion of the waves is constantly redistributing sand and other sediments. One of the world's longest and most spectacular sand beaches runs down the Atlantic coast of North America from New England to Florida and around the Gulf of Mexico. Much of this beach lies on some 350 long, thin **barrier islands** that stand between the mainland and the open sea. Behind these barrier islands lie shallow bays or brackish lagoons fringed by marshes or swamps.

Early inhabitants recognized that the shore was a hazardous place to live and settled on the bay side of barrier islands or as far upstream on coastal rivers as was practical. Modern residents, however, place a high value on living where they have an ocean view and ready access to the beach. And they assume that modern

technology makes them immune to natural forces. The most valuable and prestigious property is closest to the shore. Over the past 50 years, more than one million acres of estuaries and coastal marshes have been filled to make way for housing or recreational developments.

Construction directly on beaches and barrier islands can cause irreparable damage to the whole ecosystem. Normally fragile vegetative cover holds the shifting sand in place. Damaging this vegetation with construction, road-building, and breaching dunes with roads, can destabilize barrier islands. Storms then wash away beaches or even whole islands. A single severe storm in 1962 caused $300 million in property damage along the East Coast of the United States and left hundreds of beach homes tottering into the sea (fig. 11.20). FEMA estimates that 25 percent of all coastal homes in the U.S. will have the ground washed out from under them by 2060.

Cities and individual property owners often spend millions of dollars to protect beaches from erosion. Sand is dredged from the ocean floor or hauled in by the truckload, only to wash away again in the next storm. Building artificial barriers, such as groins or jetties, can trap migrating sand and build beaches in one area, but they often starve downstream beaches and make erosion there even worse (fig. 11.21).

As is the case for inland floodplains, government policies often encourage people to build where they probably shouldn't. Subsidies for road building and bridges, support for water and sewer projects, tax exemptions for second homes, flood insurance, and disaster relief are all good for the real estate and construction businesses but invite people to build in risky places. Flood insurance typically costs $300 per year for $80,000 of coverage. In 1998, FEMA paid out $40 billion in

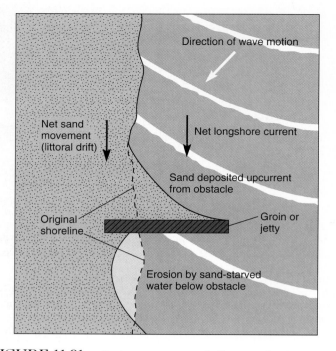

FIGURE 11.21 *Groins are structures built perpendicular to a beach to slow erosion and trap sand. They build up beaches toward prevailing wind and wave direction but starve beaches and increase erosion downstream. Dashed line indicates original shoreline.*

claims, 80 percent of which were flood related. Settlement usually requires that structures be rebuilt exactly where and as they were before. There is no restriction on how many claims can be made, and policies are rarely canceled, no matter what the risk. Some beach houses have been rebuilt—at public expense—three times in a decade. The General Accounting Office found that 2 percent of federal flood policies were responsible for 30 percent of the claims.

The Coastal Barrier Resources Act of 1982 prohibited federal support, including flood insurance, for development on sensitive islands and beaches. In 1992, however, the U.S. Supreme Court ruled that ordinances forbidding floodplain development amount to an unconstitutional "taking" or confiscation of private property.

SUMMARY

The earth is a complex, dynamic system. Although it seems stable and permanent to us, the crust is in constant motion. Huge blocks called tectonic plates slide over the surface of the ductile mantle. They crash into each other in ponderous slow motion, crumpling their margins into mountain ranges and causing earthquakes. Often, one plate slides under another, carrying rock layers down into the mantle, where they melt and flow back toward the surface to form new rocks.

Rocks are classified according to composition, structure, and origin. The three basic types of rock are igneous, metamorphic, and sedimentary. These rock types can be transformed from one to another by way of the rock cycle, a continuous process of weathering, transport, burial in sediments, metamorphism, melting, and recrystallization.

During the cooling and crystallization process that forms rock from magma, minerals and metals can become concentrated enough to become economically important reserves if they are close enough to the surface to be reached by mining. Having a reliable supply of strategically important minerals and metals is vital in industrialized societies.

A few places in the world are especially rich in mineral deposits. Southern Africa and Russia contain some of the world's richest supplies of strategic minerals. Less-developed countries, most of which are in the tropics or the Southern Hemisphere, are often the largest producers of iron, aluminum, copper, gold, and other metals on which the industrialized world depends. The major consumers of these resources are the industrialized countries.

Both mining and processing of metals and mineral resources can have negative environmental effects. Mine drainage has polluted thousands of kilometers of streams and rivers. Fumes from smelters kill forests and spread pollution over large areas. Surface mining results in removal of natural ecosystems, soil disruption, creation of trenches or open pits, and tailings accumulations. It is now required that strip-mined areas be recontoured, but revegetation is often difficult and limited in species composition. Smelting and chemical extraction processes also create pollution problems.

Worldwide, only a small percentage of metals are recycled, although it is not a difficult process technically. Recycling saves energy and reduces environmental damage caused by mining and smelting. It reduces waste production and makes our metal supplies last much longer. Substitution of materials usually occurs when mineral supplies become so scarce that prices are driven up. Many of the strategic metals that we now stockpile may become obsolete when newer, more useful substitutes are found.

Earthquakes, volcanoes, floods, and erosion are natural geological processes. When they affect human lives and property, however, they become some of the worst "natural disasters" known. Some of these natural disasters, especially floods and erosion, are made more serious by human activities and land-use decisions.

QUESTIONS FOR REVIEW

1. Describe the layered structure of the earth.
2. Figure 11.3 shows subduction and rifting zones. Describe what happens at these tectonic plate margins.
3. Why are there so many volcanoes and earthquakes along the "ring of fire" that rims the Pacific Ocean?
4. Define mineral and rock.
5. Describe the rock cycle, and name the three main rock types that it produces.
6. Give some examples of nonmetal mineral resources, and describe how they are used.
7. Give some examples of strategic metals. Where are the largest supplies of these resources located?
8. What are some environmental hazards associated with mineral extraction?
9. Describe some ways we can reduce our consumption of new mineral resources.
10. Describe some of the leading geologic hazards and their effects.

THINKING SCIENTIFICALLY

1. Understanding and solving the environmental problems of mining are basically geologic problems, but geologists need information from a variety of environmental and scientific fields. What are some of the other sciences (or disciplines) that could contribute to solving mine contamination problems?
2. Geologists are responsible for identifying and mapping mineral resources. But mineral resources are buried below the soil and covered with vegetation. How do you suppose geologists in the field find clues about the distribution of rock types?
3. If you had an igneous rock with very fine crystals and one with very large crystals, which would you expect to have formed deep in the ground, and why?
4. Heat and pressure tend to help concentrate metal ores. Explain why such ores might often occur in mountains such as the Andes in South America.
5. The idea of tectonic plates shifting across the earth's surface is central to explanations of geologic processes. Why is this idea still called the "theory" of plate tectonic movement?
6. Geologic evidence from fossils and sediments provided important evidence for past climate change. What sorts of evidence in the rocks and landscape around you suggest that the place where you lived once looked much different than it does today?

KEY TERMS

barrier islands 269
core 256
crust 256
earthquakes 265
flood 267
floodplains 267
heap-leach extraction 264
igneous rocks 258
landslides 268
magma 256
mantle 256
metamorphic rocks 258

midocean ridges 256
mineral 257
rock 257
rock cycle 258
sedimentary rocks 258
sedimentation 259
smelting 263
strategic metals and
    minerals 261
tectonic plates 256
volcanoes 266
weathering 259

WEB EXERCISES

Exploring Recent Earthquakes

Go to the World Data Center for Seismology at http://neic.usgs.gov/. Click on "Current earthquake information" then choose "Current earthquake maps" (the second item in the list) to see the location of recent earthquakes.

1. Where was the largest earthquake in the world in the past month? Where was the most recent one? Click on the map location of the most recent earthquake, and record the following information: When did it occur? Where was it exactly? How deep was it? What was its magnitude?

2. One of the largest earthquakes in recent years was one of magnitude 8.4 off the coast of South America in 2001. Where and when did this occur? How much damage did it cause? Why was the damage less than it otherwise might have been? Hint: Look first for "Large/Significant Earthquakes" by year at http://neic.usgs.gov/current_seismicity.html, then click on 2001 to find the largest. A brief summary is presented under "Large/Noteworthy Earthquakes in 2001." A more detailed analysis is given under "Significant Earthquakes of the World in 2001."

For another view of recent earthquakes, go to www.crustal.ucsb.edu/ics/understanding/ and click on the rotating globe. If you have a broadband connection, increase the speed of rotation to get a smoother motion. Does information given on this page agree with your answer to question 1 above?

Understanding Volcanoes

USGS Hawaii Volcano Observatory is one of the most important places for studying and understanding volcanoes. Visit the observatory's website at http://hvo.wr.usgs.gov/

1. Which of the volcanoes discussed at this website is the largest in the world? Which is the most active?

2. Why are earthquakes discussed on this volcano website?

3. What are the major hazards to people from these volcanoes?

For a good collection of volcano images, go to the USGS Cascades Volcano Observatory at http://vulcan.wr.usgs.gov/Photo/framework.html

Evaluating Erosion on Farmland

The Natural Resources Conservation Service (NRCS), part of the U.S. Department of Agriculture, is the agency that monitors agricultural resources and conditions. Among its data-gathering efforts, the NRCS produces a Natural Resources Inventory (NRI), with maps of farmland conditions across the mainland United States. Visit the NRCS/NRI website at www.nrcs.usda.gov/technical/land/erosion.html

Find the list of erosion maps. Look at several of the maps. First identify the meaning of the colors. Then identify the major concentration of high and low erosion rates.

1. Where is wind erosion worst? Where is water erosion worst?

2. Where are the most acres of highly erodible cropland? What kind of physical features might occur there?

The Three Gorges Dam on China's Yangtze River will be the largest dam in the world. Four people on the lower right give a sense of scale.

OBJECTIVES

After studying this chapter, you should be able to

- summarize our current energy sources, and explain briefly how our energy compares with that of other people in the world.
- analyze the resources and reserves of fossil fuels in the world.
- evaluate the costs and benefits of using coal, oil, and natural gas.
- understand how nuclear reactors work, why they are dangerous, and how they might be made safer.
- appreciate the opportunities for energy conservation available to us.
- understand how active and passive systems capture solar energy and how photovoltaic collectors generate electricity.
- comprehend the promise and problems of using biomass as an energy source.
- explain how hydropower, wind, and other energy from the earth's forces can contribute to our energy supply.

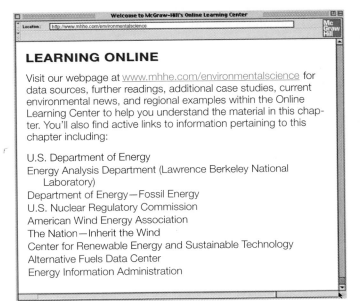

LEARNING ONLINE

Visit our webpage at www.mhhe.com/environmentalscience for data sources, further readings, additional case studies, current environmental news, and regional examples within the Online Learning Center to help you understand the material in this chapter. You'll also find active links to information pertaining to this chapter including:

U.S. Department of Energy
Energy Analysis Department (Lawrence Berkeley National Laboratory)
Department of Energy—Fossil Energy
U.S. Nuclear Regulatory Commission
American Wind Energy Association
The Nation—Inherit the Wind
Center for Renewable Energy and Sustainable Technology
Alternative Fuels Data Center
Energy Information Administration

Oil Drilling in the Arctic

A narrow strip of coastal plain in Alaska's Arctic National Wildlife Refuge (ANWR) presents one of nature's grandest spectacles as well as one of the most contentious environmental battles in recent years (fig. 12.1). For a few months during the brief arctic summer, the tundra teems with wildlife. It is the calving ground of the 130,000 caribou of the Porcupine herd; a nesting area for millions of snow geese, tundra swans, shorebirds and other migratory waterfowl; a denning area for polar bears, arctic foxes, and arctic wolves; and a year-round home to about 350 shaggy musk ox.

The coastal plain may also be the site of the last big, onshore, liquid petroleum field in North America. Although only limited exploratory drilling has been permitted in the refuge, geologists estimate that it may contain as much as 16 billion barrels of oil and several trillion cubic feet of natural gas. Can we extract these valuable fossil fuels without driving away the wildlife and polluting the pristine landscape? The oil industry believes it can access resources without doing lasting environmental harm; biologists and environmentalists doubt this is so.

Critics argue that oil drilling would do irreparable harm in this fragile arctic environment. They claim that excavating gravel for drill pads and pumping millions of gallons of water for ice roads would destroy wetlands on which wildlife depend for summer habitat. The noise and odors of up to 700 workers and their vehicles, construction equipment, drill rigs, and giant power-generating turbines could drive away wildlife. Pointing to the problems of other arctic oil-drilling operations where crews dumped garbage, sewage, and toxic waste into surface pits, environmentalists predict disaster if drilling is allowed in the refuge. Pipeline and drilling spills at Prudhoe Bay have contaminated the tundra and seeped into waterways. Scars from bulldozer tracks made 50 years ago can still be seen clearly today.

Oil company engineers, on the other hand, claim that old, careless ways are no longer permitted in their operations. Wastes now are collected and either burned or injected into deep wells. Heavy equipment is hauled to the sites during the winter when most wildlife is either absent or hibernating. Modern directional drilling will allow up to 50 wells to be placed on a single pad, greatly reducing land impacts.

The amount of economically recoverable oil in ANWR depends not only on geology but also on market prices and shipping costs. At current wholesale prices of $25 per barrel, the U.S. Geological Survey estimates that only 7 billion barrels could be extracted profitably from ANWR. If prices drop below $10 per barrel, as they did in the early 1990s, the economic resource might be only a few hundred million barrels. Under the most optimistic

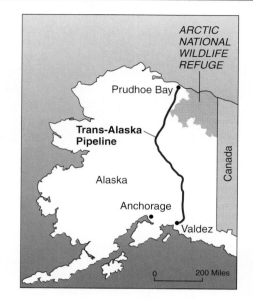

FIGURE 12.1 *The Trans-Alaska Pipeline could transport oil from the adjacent Arctic National Wildlife Refuge if drilling is allowed there.*

scenarios, it will take at least a decade to begin to get this oil to market, and the peak production rate would probably be about 1 million barrels of oil per day in 2030. This oil would meet less than 4 percent of the U.S. daily oil consumption. Improving the average fuel efficiency of American cars and light trucks by just one mile per gallon would save more oil than is ever likely to be recovered from ANWR, and it would do so far faster and cheaper than extracting and transporting arctic oil.

In 2001 the Republican-controlled House of Representatives passed an Energy Security Act that authorized ANWR oil exploration. Citing rising gasoline prices and our growing dependence on foreign oil, supporters of this bill claimed that drilling in ANWR is essential for maintaining a strong economy. The Senate, on the other hand, passed its own Energy Bill in 2002 prohibiting drilling in ANWR and encouraging renewable energy sources. At the time of this writing, contradictions between these bills remained to be resolved.

Both sides in this debate urge adoption of an energy policy, although they disagree on what it should contain. Can we have all the energy we want and still have a tolerable environment? This struggle illustrates the importance of energy supplies in modern life, and some of the environmental costs of obtaining fuels. In this chapter, we will look at the sources and uses of fossil fuels and nuclear power as well as alternative, renewable energy sources.

ENERGY SOURCES AND USES

Being able to utilize external energy to do useful work is one of the most unique characteristics of humans. Access to a source of high-quality energy is essential for continuation of the current way of life. But what do these terms mean? **Work** is the application of force through a distance. **Energy** is the capacity to do work. **Power** is the rate of flow of energy, or the rate at which work is done. The energy we use to move our muscles, to think, and to carry out metabolic functions comes from stored chemical energy (or potential energy) in our food. Food energy is generally measured in calories (cal). One cal is the amount of energy to heat 1 g of water 1°C. A kilocalorie (or food Calorie) is 1,000 cal. In physics, the basic metric unit of force is a newton, which is the force necessary to accelerate 1 kilogram 1 meter per second. A **joule (J)** is the amount of work done when a force of 1 newton is exerted over 1 meter, or 1 amp per second flows through 1 ohm. One J equals 0.238 cal. Some other common energy units are presented in table 12.1.

Current Energy Sources

Fossil fuels (petroleum, natural gas, and coal) now provide about 79 percent of all commercial energy in the world (fig. 12.2). Biomass fuels, such as wood, peat, charcoal, and manure, contribute about 9.5 percent of commercial energy. Other renewable sources—solar, wind, geothermal, and hydroelectricity—make up about 4.5 percent of our commercial power (actually, hydro accounts for most of that). Although not included in these data, private systems (individual houses, businesses, industries) may capture as much renewable energy as that reported in commercial transactions. Similarly, this summary probably underrepresents the importance of biomass because these fuels often are gathered by the people using them, or they are sold in the informal economy and not reported internationally. In many poorer countries, such as Haiti, Bhutan, and Malawi, biomass supplies more than 90 percent of the energy used for heating and cooking. This is an important energy source for poor people but can be a serious cause of forest destruction (see chapter 6).

TABLE 12.1 Some Energy Units

1 joule (J) = the force exerted by a current of 1 amp per second
flowing through a resistance of 1 ohm

1 watt (W) = 1 joule (J) per second

1 kilowatt-hour (kwh) = 1 thousand (10^3) watts exerted for 1 hour

1 megawatt (MW) = 1 million (10^6) watts

1 gigawatt (GW) = 1 billion (10^9) watts

1 petajoule (PJ) = 1 quadrillion (10^{15}) joules

1 British thermal unit (Btu) = energy to heat 1 lb of water 1°F

1 PJ = 947 billion Btu, or 278 million kwh

1 standard barrel (bbl) of oil = 42 gal (160 liters) or 5.8 million Btu

1 metric ton of standard coal = 27.8 million Btu or 4.8 bbl oil

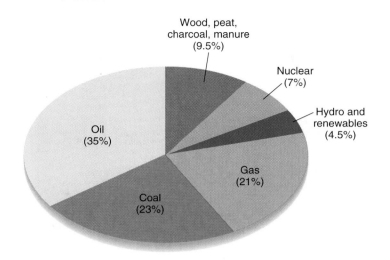

FIGURE 12.2 *Worldwide commercial energy production.*
Source: Data from World Energy Council, 2002.

Nuclear power is roughly equal to hydroelectricity worldwide (about 5 percent of all commercial energy), but it makes up about 20 percent of all electric power in more-developed countries, such as Canada and the United States. We have enough nuclear fuel to produce power for a long time, and it has the benefit of not contributing to global warming (see chapter 9), but as we discuss later in this chapter, safety concerns make this option unacceptable to most people.

Per Capita Consumption

Perhaps the most important facts about fossil fuel consumption are that we in the twenty richest countries consume nearly 80 percent of the natural gas, 65 percent of the oil, and 50 percent of the coal produced each year. Although we make up less than one-fifth of the world's population, we use more than one-half of the commercial energy supply. The United States and Canada, for instance, constitute only 5 percent of the world's population but consume about one-quarter of the available energy. To be fair, however, some of that energy goes to grow crops or manufacture goods that are later shipped to developing countries.

How much energy do you use every year? Most of us don't think about it much, but maintaining the lifestyle we enjoy requires an enormous energy input. On average, each person in the United States and Canada uses more than 300 gigajoules (GJ) (equivalent to about 60 barrels of oil) per year. By contrast, in some of the poorest countries of the world, such as Ethiopia, Nepal, and Bhutan, each person generally consumes less than 1 GJ per year. This means that each of us consumes, on average, almost as much energy in a single day as a person in one of these countries consumes in a year.

Clearly, energy consumption is linked to the comfort and convenience of our lives (fig. 12.3). Those of us in the richer countries enjoy many amenities not available to most people in the world. The link is not absolute, however. Several European countries, including Sweden, Denmark, and Switzerland, have higher

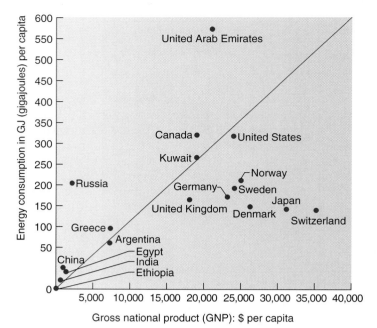

FIGURE 12.3 *Per capita energy use and GNP. In general, higher energy use correlates with a higher standard of living. Denmark and Switzerland, however, use about half as much energy as the United States does and have a higher standard of living by most measures.*
Source: Data from World Resources Institute, 1996–97.

standards of living than does the United States by almost any measure but use about half as much energy. These countries have had effective energy conservation programs for many years. Japan also has a far lower energy consumption rate than might be expected for its industrial base and income level. Because Japan has few energy resources of its own, it has developed very efficient energy conservation measures.

How We Use Energy

The largest share (36.5 percent) of the energy used in the United States is consumed by industry (fig. 12.4). Mining, milling, smelting, and forging of primary metals consume about one-quarter of the industrial energy share. The chemical industry is the second largest industrial user of fossil fuels, but only half of its use is for energy generation. The remainder is raw material for plastics, fertilizers, solvents, lubricants, and hundreds of thousands of organic chemicals in commercial use. The manufacture of cement, glass, bricks, tile, paper, and processed foods also consumes large amounts of energy. Residential and commercial buildings use roughly 34 percent of the primary energy consumed in the United States, mostly for space heating, air conditioning, lighting, and water heating. Small motors and electronic equipment take an increasing share of residential and commercial energy. In many office buildings, computers, copy machines, lights, and human

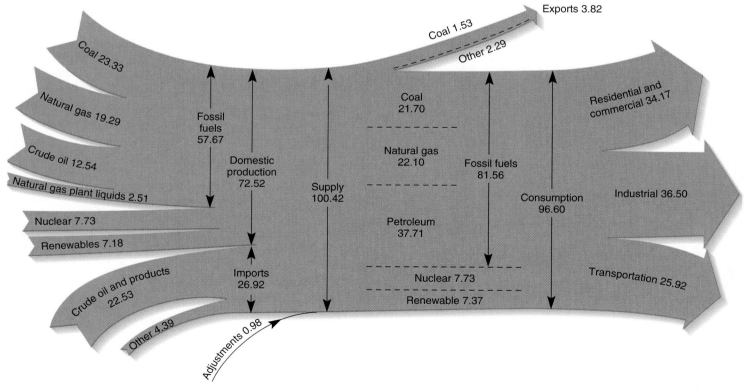

FIGURE 12.4 *U.S. energy flow, 1999. Inputs are shown on the left; summaries of domestic production and imports, as well as proportions of different primary sources, can be found in the middle; and end uses are shown on the right. Quantities are in quadrillion BTUs (quads). Note that U.S. fuel percentages are slightly different from world production in figure 12.2.*
Source: Data from U.S. Energy Flow, 2000. U.S. Department of Energy.

Principles of Environmental Science www.mhhe.com/environmentalscience

Most college students either already own or are likely to buy an automobile and a computer sometime soon. How do these items compare in energy usage? Suppose that you were debating between a high-mileage car such as the Honda Insight or a sport-utility vehicle like a Ford Excursion. How do the energy requirements of these two purchases measure up? To put it another way, how long could you run a computer on the energy you would save by buying an Insight rather than an Excursion? (See Case Study, p. 284, for more information about the Insight.)

Here are some numbers you need to know. The Insight gets about 75 mpg, while the Excursion gets about 12 mpg. A typical American drives about 15,000 mi per year. A gallon of regular, unleaded gasoline contains about 115,000 Btu on average. Most computers use about 100 watts of electricity. One kilowatt-hour (kwh) = 3,413 Btu.

1. How much energy does the computer use if it is left on continuously? (You really should turn it off at night or when it isn't in use, but we'll simplify the calculations.)

   100 watt/h × 24 h/day × 365 days/yr = _____ kwh/yr

2. How much gasoline would you save in an Insight compared to an Excursion?

   a. Excursion:
      15,000 miles/yr ÷ 12 mpg = _____ gal/yr

   b. Insight:
      15,000 miles/yr ÷ 75 mpg = _____ gal/yr

   c. Gasoline savings $(a - b)$ = _____ gal/yr

   d. Energy savings:
      (gal × 115,000 Btu) = _____ Btu/yr

   e. Converting Btu to kwh:
      (Btu × 0.00029 Btu/kwh) = _____ kwh/yr saved

3. How long would the energy saved run your computer?

   kwh/yr saved by Insight ÷ kwh/yr consumed by computer = _____

Answers: 1. The computer consumes 876 kWh/yr if run continuously. *2.* The Excursion consumes 1,250 gal/yr, the Insight uses 200 gal/yr. Savings are 1050 gal/yr = 120 million Btu/yr or 35,000 kWh/yr. *3.* 40 years.

bodies liberate enough waste heat that a supplementary source is not required, even in the winter.

Transportation consumes about 26 percent of all energy used in the United States each year. About 98 percent of that energy comes from petroleum products refined into liquid fuels, and the remaining 2 percent is provided by natural gas and electricity. Almost three-quarters of all transport energy is used by motor vehicles. Nearly 3 trillion passenger miles and 600 billion ton miles of freight are carried annually by motor vehicles in the United States. About 75 percent of all freight traffic in the United States is carried by trains, barges, ships, and pipelines, but because they are very efficient, they use only 12 percent of all transportation fuel.

Finally, analysis of how energy is used has to take into account waste and loss of potential energy. About half of all the energy in primary fuels is lost during conversion to more useful forms, while being shipped to the site of end use, or during use. Electricity, for instance, is generally promoted as a clean, efficient source of energy because, when it is used to run a resistance heater or an electrical appliance, almost 100 percent of its energy is converted to useful work and no pollution is given off. What happens before then, however? We often forget that huge amounts of pollution are released during mining and burning of the coal that fires power plants. Furthermore, nearly two-thirds of the energy in the coal that generated that electricity was lost in thermal conversion in the power plant. About 10 percent more is lost during transmission and stepping down to household voltages. Similarly, about 75 percent of the original energy in crude oil is lost during distillation into liquid fuels, transportation of that fuel to market, storage, marketing, and combustion in vehicles.

Natural gas is our most efficient fuel. Only 10 percent of its energy content is lost in shipping and processing, since it moves by pipelines and usually needs very little refining. Ordinary gas-burning furnaces are about 75 percent efficient, and high-economy furnaces can be as much as 95 percent efficient. Because natural gas has more hydrogen per carbon atom than oil or coal, it produces about half as much carbon dioxide—and therefore contributes half as much to global warming—per unit of energy.

FOSSIL FUELS

Fossil fuels are organic chemicals created by living organisms millions of years ago and buried in sediments, where high pressures and temperatures concentrated and transformed them into energy-rich compounds. Coal is solid, petroleum (crude oil) is liquid, oil shales and tar sands are semisolid tars trapped in rock, and natural gas (methane and other gaseous hydrocarbons) is, as its name suggests, a gas. Most of these fuels were laid down during the Carboniferous period (286 million to 360 million years ago), when the earth's climate was warmer and wetter than it is now, but some are thought to date as far back as Precambrian times, perhaps a billion years ago. Because fossil fuels take so long to form, they are essentially nonrenewable resources.

Coal

World coal deposits are vast, ten times greater than conventional oil and gas resources combined. Coal seams can be 100 m thick and can extend across tens of thousands of square kilometers that were vast, swampy forests in prehistoric times. The total resource is estimated to be 10 trillion metric tons. If all this coal could be extracted, and if coal consumption continued at present levels, this would amount to several thousand years' supply. At present rates of consumption, these **proven-in-place reserves**—those explored

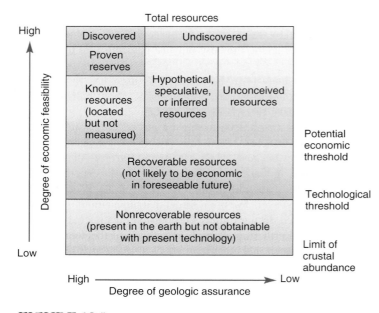

FIGURE 12.5 *Categories of natural resources according to economic and technologic feasibility as well as geologic assurance.*

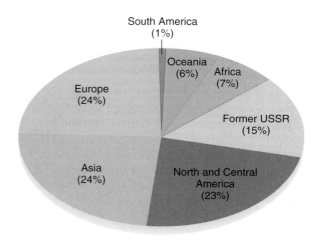

FIGURE 12.6 *Proven-in-place coal reserves, by region as percent of a total 1.96 trillion metric tons.*
Source: Data from World Resources Institute, 1998–99.

and mapped but not necessarily economic at today's prices—will last about 200 years. Proven reserves are generally a small fraction of a total resource (fig. 12.5).

Where are these coal deposits located? They are not evenly distributed throughout the world (fig. 12.6). North America has one-fourth of all proven reserves located in large deposits across the central and mountain regions. China and the former Soviet Union have nearly as much. Eastern and Western Europe have fairly large coal supplies in spite of their relatively small size. Africa and Latin America have very little coal despite their large size. Antarctica is thought to have large coal deposits, but they would be difficult, expensive, and ecologically damaging to mine.

It would seem that the abundance of coal deposits is a favorable situation. But do we really want to use all of the coal? Coal mining is a dirty, dangerous activity. Underground mines are notorious for cave-ins, explosions, and lung diseases, such as black-lung suffered by miners. Surface mines (called strip mines, where large machines strip off overlying sediment to expose coal seams) are cheaper and generally safer for workers than tunneling, but leave huge holes where coal has been removed and vast piles of discarded rock and soil (fig. 12.7). An especially damaging technique employed in Appalachia is called mountaintop removal. Typically, the whole top of a mountain ridge is scraped off to access buried coal. The waste rock pushed down into the nearest valley buries forests, streams, houses, and farms (see chapter 11 for further discussion of this issue). Mine reclamation is now mandated in the United States, but efforts often are only partially successful.

Coal burning releases large amounts of air pollution. Every year the 900 million tons of coal burned in the United States (83 percent for electric power generation) releases 18 million metric tons of sodium dioxide (SO_2), 5 million metric tons of nitrogen oxides (NO_x), 4 million metric tons of airborne particulates, 600,000 metric tons of hydrocarbons and carbon monoxide, and close to a trillion metric tons of carbon dioxide (CO_2). This is about three-quarters of the SO_2, one-third of the NO_x, and about half of the industrial CO_2 released in the United States each year. As described in chapter 9, sulfur and nitrogen oxides combine with water in the air to form sulfuric and nitric acid, making coal burning the largest single source of acid rain in many areas. These air pollutants have many deleterious effects, including human health costs, injury to domestic and wild plants and animals, and damage to buildings and property. Total losses from air pollution are estimated to be between $5 billion and $10 billion per year in the United States alone. By some accounts, at least 5,000 excess human deaths per year can be attributed to coal production and burning.

FIGURE 12.7 *Some giant mining machines stand as tall as a 20-story building and can scoop up thousands of cubic meters of rock per hour. Note the pick-up trucks in the machine's shadow.*

Principles of Environmental Science www.mhhe.com/environmentalscience

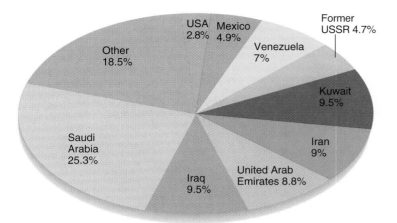

FIGURE 12.8 *World proven recoverable liquid oil reserves by country, 1997. If the Caspian Sea region really has 200 billion bbl of recoverable oil, it could increase world supplies by about 25 percent, or 10 years at current rates of use.*

Source: Data from World Resources Institute, 1998–99.

Oil

The total amount of oil in the world is estimated to be about 4 trillion bbl (600 billion metric tons), half of which is thought to be ultimately recoverable. Some 465 billion bbl of oil already have been consumed. In 2000, the proven reserves were roughly 650 billion bbl, enough to last only 30 years at the current consumption rate of 22 billion bbl per year. It is estimated that another 800 billion bbl either remain to be discovered or are not recoverable at current prices with present technology. As oil resources become depleted and prices rise, it probably will become economical to find and bring this oil to the market unless alternative energy sources are developed. This estimate of the resource does not take into account the very large potential from unconventional semiliquid hydrocarbon resources, such as shale oil and tar sands, which might double the total reserve if they can be mined with acceptable social, economic, and environmental impacts.

By far the largest supply of proven-in-place oil is in Saudi Arabia, which has 250 billion bbl, about one-fourth of the total proven world reserve (fig. 12.8). Altogether, the countries of the Middle East control nearly two-thirds of all known oil reserves—a fact that explains a great deal about American foreign policy in recent years. Among the places considered most likely to contain undiscovered oil, in addition to the Middle East, are the northeast Greenland Shelf, the western Siberian and Caspian areas, and the Niger and Congo delta areas of Africa. Sediments under and around the Caspian Sea have been estimated to hold 200 billion bbl of oil. If this is true, this region would be second only to Saudi Arabia in size of oil reserves.

Altogether, the United States has already used about 40 percent of its original recoverable petroleum resource. Of the 120 billion bbl thought to remain, about 58 billion bbl are proven-in-place. If we stopped importing oil and depended exclusively on indigenous supplies, our proven reserves would last only ten years

at current rates of consumption. The regions of North America with the greatest potential for substantial new oil discoveries are portions of the continental shelf along the California coast, the Beaufort Sea, and the Grand Banks, all of which are prime wildlife areas. Proposals to do exploratory drilling in these areas have been strongly opposed by many people concerned about the potential for long-term environmental damage from drilling activities and oil spills like the *Exxon Valdez* in Prince William Sound. Other people argue that having an inexpensive, plentiful energy supply, even if only for a few decades, is worth the social and environmental costs.

Like coal, petroleum often contains high sulfur levels. Because it is highly corrosive, the sulfur in oil is usually stripped out before the fuel is shipped to market and thus doesn't end up in urban air nearly as often as that from coal. Oil is used primarily for transportation (the gasoline in your car or truck) and heating. Because about one-quarter of all energy used in the United States fuels transportation, and oil provides more than 90 percent of that transportation energy, oil combustion is responsible for between one-quarter and two-thirds of the volatile organic compounds, nitrogen oxides, and carbon monoxide emitted each year.

Oil drilling is far less destructive to the earth's surface than coal mining, but the huge amounts of oil shipped around the world and used as lubricants and fuels result in very serious soil and water pollution. As chapter 10 points out, oceanographers estimate that somewhere between 3 million and 6 million metric tons of oil are discharged into the world's rivers and oceans every year. About half of this comes from oil tanker accidents and dumping of ballast water. The rest is mostly deliberate dumping of used motor oil. Many people believe that a storm sewer is a good place to discard waste oil and don't realize that the sewer probably leads directly into a lake, river, or seashore.

Oil Shales and Tar Sands

Estimates of our recoverable oil supplies usually don't account for the very large potential from unconventional resources. The World Energy Council estimates that oil shales, tar sands, and other unconventional deposits contain ten times as much oil as liquid petroleum reserves. **Tar sands** are composed of sand and shale particles coated with bitumen, a viscous mixture of long-chain hydrocarbons. Shallow tar sands are excavated and mixed with hot water and steam to extract the bitumen, then fractionated to make useful products. For deeper deposits, superheated steam is injected to melt the bitumen, which can then be pumped to the surface like liquid crude. Once the oil has been retrieved, it still must be cleaned and refined to be useful.

Canada and Venezuela have the world's largest and most accessible tar sand resources. Canadian deposits in northern Alberta are estimated to be equivalent to 300 billion bbl of oil or nearly as much as Saudi Arabian oil reserves. Current Alberta production is about 1 billion bbl per day, or more than the output of Alaska's North Slope. By 2010, Alberta is expected to increase its flow to 2 billion bbl per day, or twice the maximum projected output of the Arctic National Wildlife Refuge. Furthermore, because Alberta tar sand beds are twenty times larger and closer to the surface than

FIGURE 12.9   *Natural gas is difficult to store and ship. In remote locations like this offshore platform, it often is flared off.*

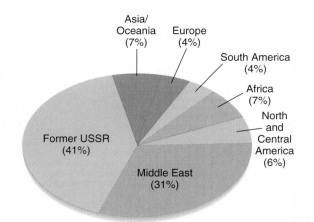

FIGURE 12.10   *Percent of proven-in-place natural gas reserves by region.*
Source: Data from World Resources Institute, 2000–01.

ANWR oil, the Canadian resource will last longer and may be cheaper to extract. Canada is already the largest supplier of oil to the United States, having surpassed Saudi Arabia in 2000.

There are severe environmental costs, however in producing this oil. A typical plant producing 125,000 bbl of oil per day creates about 15 million m^3 of toxic sludge, releases 5,000 tons of greenhouse gases, and consumes or contaminates billions of liters of water each year. Surface mining could destroy millions of hectares of boreal forest. Native Cree, Chipewyan, and Metis people worry about effects on traditional ways of life if forests are destroyed and wildlife and water are contaminated. Many Canadians dislike becoming an energy colony for the United States, and environmentalists argue that investing billions of dollars to extract this resource simply makes us more dependent on fossil fuels.

Oil shales are fine-grained sedimentary rock rich in solid organic material called kerogen. Like tar sands, the kerogen can be heated, liquefied, and pumped out like liquid crude oil. Oil shale beds up to 600 m (1,800 ft) thick underlie much of Colorado, Utah, and Wyoming. If these deposits could be extracted at a reasonable price and with acceptable environmental impacts, they might yield the equivalent of several trillion barrels of oil. Mining and extraction of oil shale uses vast amounts of water (a scarce resource in the arid western United States and creates enormous quantities of waste. The rock matrix expands when heated, resulting in twice or three times the volume of waste as was dug out of the ground. Billions of dollars were spent in the 1980s on pilot projects to produce synthetic oil. When oil prices dropped, these schemes were abandoned. Should we explore these unconventional resources further?

Natural Gas

Natural gas is the world's third largest commercial fuel, making up 23 percent of global energy consumption. Because natural gas produces only half as much CO_2 as an equivalent amount of coal, substitution could help reduce global warming (see chapter 9). Natural gas is difficult to ship across oceans, however, or to store in large quantities. North America is fortunate to have an abundant, easily available supply of gas and a pipeline network to deliver it to market. Many countries cannot afford such a pipeline network, and much of the natural gas produced in conjunction with oil pumping in remote areas is simply burned (flared off), an unfortunate waste of a valuable resource (fig. 12.9).

The republics of the former Soviet Union have 41 percent of known natural gas reserves (mostly in Siberia and the Central Asian republics) and account for 36.5 percent of all production. Both Eastern and Western Europe buy substantial quantities of gas from these wells. Figure 12.10 shows the distribution of proven natural gas reserves in the major producing countries of the world.

The total ultimately recoverable natural gas resources in the world are estimated to be 10,000 trillion ft^3, corresponding to about 80 percent as much energy as the recoverable reserves of crude oil. The proven world reserves of natural gas are 3,200 trillion ft^3 (73,000 million metric tons). Because gas consumption rates are only about half of those for oil, current gas reserves represent roughly a 60-year supply at present usage rates. Proven reserves in the United States are about 200 trillion ft^3, or 6 percent of the world total. This is a ten-year supply at current rates of consumption. Known reserves are more than twice as large.

Large amounts of methane are released from coal deposits. The Rocky Mountain front in Colorado, Wyoming, and Montana could have 10 percent of the total world methane supply. Proposals for 51,000 coal-bed methane wells in Wyoming's Power River Basin, for example, have raised concerns about water pollution and land damage. Other states could face similar problems.

NUCLEAR POWER

In 1953 President Dwight Eisenhower presented his "Atoms for Peace" speech to the United Nations. He announced that the United States would build nuclear-powered electrical generators to provide

FIGURE 12.11 *Two nuclear reactors (domes) at the San Onofre Nuclear Generating Station sit between the beach and Interstate 5, the major route between Los Angeles and San Diego.*

clean, abundant energy. He predicted that nuclear energy would fill the deficit caused by predicted shortages of oil and natural gas. It would provide power "too cheap to meter" for continued industrial expansion of both the developed and the developing world. It would be a supreme example of "beating swords into plowshares." Technology and engineering would tame the evil genie of atomic energy and use its enormous power to do useful work.

However, rapidly increasing construction costs, declining demand for electric power, and safety fears have made nuclear energy much less attractive than promoters expected. Of the 140 reactors on order in 1975, 100 were subsequently canceled. Nuclear power now produces less than 8 percent of the U.S. energy supply (2 percent more than the world average). All of it is used to generate electricity. The nuclear power industry has been campaigning for greater acceptance, arguing that reactors don't release greenhouse gases that cause global warming. Worries about susceptiblity to terrorist attacks make many people fearful of this power source (fig. 12.11).

How Do Nuclear Reactors Work?

The most commonly used fuel in nuclear power plants is U^{235}, a naturally occurring radioactive isotope of uranium. Ordinarily, U^{235} makes up only about 0.7 percent of uranium ore—too little to sustain a chain reaction in most reactors. It must be purified and concentrated by mechanical or chemical procedures. When the U^{235} concentration reaches about 3 percent, the uranium is formed into cylindrical pellets slightly thicker than a pencil and about 1.5 cm long. Although small, these pellets pack an amazing amount of energy. Each 8.5-g pellet is equivalent to a ton of coal or 4 bbl of crude oil.

The pellets are stacked in hollow metal rods approximately 4 m long. About 100 of these rods are bundled together to make a

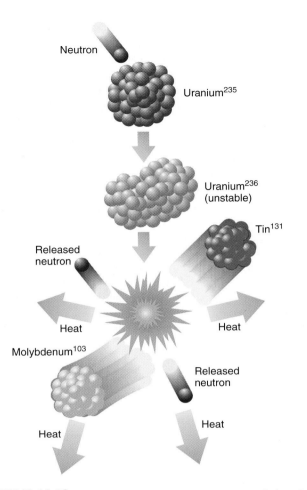

FIGURE 12.12 *The process of nuclear fission is carried out in the core of a nuclear reactor. In the sequence shown here, the unstable isotope, uranium-235, absorbs a neutron and splits to form tin-131 and molybdenum-103. Two or three neutrons are released per fission event and continue the chain reaction. The total mass of the reaction product is slightly less than the starting material. The residual mass is converted to energy (mostly heat).*

fuel assembly. Thousands of fuel assemblies containing about 100 tons of uranium are bundled in a heavy steel vessel called the reactor core. Radioactive uranium atoms are unstable—that is, when struck by a high-energy subatomic particle called a neutron, they undergo **nuclear fission** (splitting), releasing energy and more neutrons. When uranium is packed tightly in the reactor core, the neutrons released by one atom will trigger the fission of another uranium atom and the release of still more neutrons (fig. 12.12). Thus, a self-sustaining **chain reaction** is set in motion, and vast amounts of energy are released.

The chain reaction is moderated (slowed) in a power plant by a neutron-absorbing cooling solution that circulates between the fuel rods. In addition, **control rods** of neutron-absorbing material, such as cadmium or boron, are inserted into spaces between fuel assemblies to shut down the fission reaction or are withdrawn to allow it to proceed. Water or some other coolant is circulated between the fuel rods to remove excess heat. The greatest danger in one of these complex machines is a cooling system failure. If the pumps fail or pipes break during operation, the nuclear fuel quickly overheats and a "meltdown" can result that releases deadly radioactive material. Although nuclear power plants cannot explode like a nuclear bomb, the radioactive releases from a worst-case disaster like the meltdown of the Chernobyl reactor in the Soviet Ukraine in 1986 are just as devastating as a bomb.

Nuclear Reactor Design

Seventy percent of the world's nuclear plants are pressurized water reactors (PWR). Water circulates through the core, absorbing heat as it cools the fuel rods (fig. 12.13). This primary cooling water is heated to 317°C (600°F) and reaches a pressure of 2,235 psi. It then is pumped to a steam generator, where it heats a secondary water-cooling loop. Steam from the secondary loop drives a high-speed turbine generator that produces electricity. Both the reactor vessel and the steam generator are contained in a thick-walled, concrete-and-steel containment building that prevents radiation from escaping and is designed to withstand high pressures and temperatures in case of accidents. Engineers operate the plant from a complex, sophisticated control room containing many gauges and meters that indicate how the plant is running.

Overlapping layers of safety mechanisms are designed to prevent accidents, but these fail-safe controls make reactors very expensive and very complex. A typical nuclear power plant has 40,000 valves, compared to only 4,000 in a fossil fuel–fired plant of similar size. In some cases the controls are so complex that they confuse operators and cause accidents, rather than prevent them. Under normal operating conditions, a PWR releases very little radioactivity and is probably less dangerous for nearby residents than a coal-fired power plant.

In Britain, France, and the former Soviet Union, a common reactor design uses graphite, both as a moderator and as the structural material for the reactor core. In the British MAGNOX design (named after the magnesium alloy used for its fuel rods), gaseous carbon dioxide is blown through the core to cool the fuel assemblies and carry heat to the steam generators. In the Soviet design, called RBMK (the Russian initials for a graphite-moderated, water-cooled reactor), low-pressure cooling water circulates through the core in thousands of small metal tubes.

These designs were originally thought to be very safe because graphite has a high capacity for both capturing neutrons and dissipating heat. Designers claimed that these reactors could not possibly run out of control; unfortunately, they were proven wrong. The small cooling tubes are quickly blocked by steam if the cooling system fails, and the graphite core burns when exposed to air. Burning graphite in the Chernobyl nuclear plant made the fire much more difficult to control than it might have been in another reactor design.

Nuclear Wastes

One of the most difficult problems associated with nuclear power is the disposal of wastes produced during mining, fuel production, and reactor operation. How these wastes are managed may ultimately be the overriding obstacle to nuclear power.

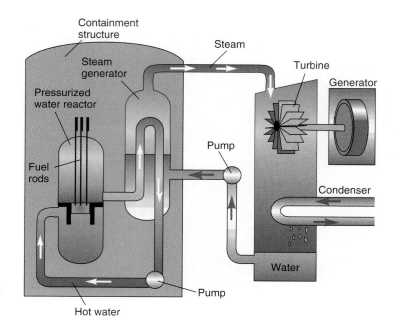

FIGURE 12.13 *Pressurized water nuclear reactor. Water is superheated and pressurized as it flows through the reactor core. Heat is transferred to nonpressurized water in the steam generator. The steam drives the turbogenerator to produce electricity.*

Enormous piles of mine wastes and abandoned mill tailings in uranium-producing countries represent another serious waste disposal problem. Production of 1,000 tons of uranium fuel typically generates 100,000 tons of tailings and 3.5 million liters of liquid waste. There now are approximately 200 million tons of radioactive waste in piles around mines and processing plants in the United States. This material is carried by the wind or washes into streams, contaminating areas far from its original source. Canada has even more radioactive mine waste on the surface than does the United States.

In addition to the leftovers from fuel production, there are about 100,000 tons of low-level waste (contaminated tools, clothing, building materials, etc.) and about 15,000 tons of high-level (very radioactive) wastes in the United States. The high-level wastes consist mainly of spent fuel rods from commercial nuclear power plants and assorted wastes from nuclear weapons production. For the past 20 years, spent fuel assemblies from commercial reactors have been stored in deep, water-filled pools at the power plants. These pools were originally intended only as temporary storage until the wastes were shipped to reprocessing centers or permanent disposal sites.

With internal waste storage pools now full but neither reprocessing nor permanent storage available, a number of utility companies are beginning to store nuclear waste in large, metal dry casks placed outside power plants (fig. 12.14). These projects are meeting with fierce opposition from local residents, who fear the casks will leak. Most nuclear power plants are built near rivers, lakes, or seacoasts. Extremely toxic radioactive materials could spread quickly over large areas if leaks occur. A hydrogen gas explosion and fire in 1997 in a dry storage cask at Wisconsin's

Protective cover

Lid

Neutron shield

Metallic seals

Steel body

Trunnion

Inserted fuel rod

Neutron shield

Outer shell

Cavity with fuel basket

FIGURE 12.14 *Dry cask storage for nuclear waste. Each cask is 17 ft tall, 8.5 ft in diameter, has steel walls 10 in. thick, and weighs 122 tons when fully loaded with 40 fuel assemblies. The casks are expected to last 40 years and cost $700,000 each. Critics worry that the casks will leak, but proponents point out that they can be monitored easily and replaced if necessary.*

Point Beach nuclear plant intensified opponents' suspicions about this form of waste storage.

In 1987, the U.S. Department of Energy announced plans to build the first high-level waste repository on a barren desert ridge near Yucca Mountain, Nevada. After 15 years of research and $4 billion in exploratory drilling, it still isn't certain that the site is safe (see chapter 11). Nevertheless, Congress approved the project in 2002, and President Bush signed the bill. Intensely radioactive wastes are to be buried deep in the ground, where it is hoped that they will remain unexposed to groundwater and earthquakes for the thousands of years required for the radioactive materials to decay to a safe level. Although the area is very dry now, we can't

be sure that it will always remain that way. Total costs now are expected to be at least $35 billion. Although the facility was supposed to open in 1998, the earliest possible date is now 2010.

Some nuclear experts believe that **monitored, retrievable storage** would be a much better way to handle wastes. This method involves holding wastes in underground mines or secure surface facilities where they can be watched. If canisters begin to leak, they could be removed for repacking. Safeguarding the wastes would be expensive, and the sites might be susceptible to wars or terrorist attacks. We might need a perpetual priesthood of nuclear guardians to ensure that the wastes are never released into the environment.

ENERGY CONSERVATION

One of the best ways to avoid energy shortages and to relieve environmental and health effects of our current energy technologies is simply to use less. Conservation offers many benefits both to society and to the environment.

Using Energy More Efficiently

Much of the energy we consume is wasted. This statement is not a simple admonishment to turn off lights and turn down furnace thermostats in winter; it is a technological challenge. Our ways of using energy are so inefficient that most potential energy in fuel is lost as waste heat, becoming a form of environmental pollution. More efficient and less energy-intensive industry, transportation, and domestic practices could save large amounts of energy. In response to federal regulations and high gasoline prices, automobile gas-mileage averages in the United States more than doubled from 13 mpg in 1975 to 28.8 mpg in 1988. Unfortunately, the oil glut and falling fuel prices of the 1990s discouraged further conservation. By 2000 the average slipped to only 24 mpg.

Much more could be done. High-efficiency automobiles are already available. Both Honda and Toyota have low-emission, hybrid gas-electric vehicles that get up to 30.3 km/liter (72 mpg) on the highway (see Case Study, p. 284). Amory B. Lovins of the Rocky Mountain Institute in Colorado estimates that raising the average fuel efficiency of the U.S. car and light-truck fleet by 1 mpg would cut oil consumption about 295,000 bbl per day. In one year, this would equal the total amount the U.S. Department of the Interior hopes to extract from the Arctic National Wildlife Refuge in Alaska.

Many improvements in domestic energy efficiency have occurred in the past decade. Today's average new home uses one-half the fuel required in a house built in 1974, but much more can be done. Household energy losses can be reduced by one-half to three-fourths by using better insulation, installing double- or triple-glazed windows, purchasing thermally efficient curtains or window coverings, and sealing cracks and loose joints. Reducing air infiltration is usually the cheapest, quickest, and most effective way of saving energy because it is the largest source of losses in a typical house. It doesn't take much skill or investment to caulk around doors, windows, foundation joints, electrical outlets, and other sources of air leakage. Mechanical ventilation (including air-to-air heat exchange) is needed to prevent moisture buildup in tightly

It's not surprising that California was the first state to get tough on automobile emissions. Auto exhaust counts for 90 percent of the state's carbon monoxide, 77 percent of its nitrogen oxides, and 55 percent of its smog-producing hydrocarbons. In 1990 the California Air Resources Board shocked the automobile industry by ordering it to start producing emission-free vehicles or face stiff penalties. At the time this order was issued, only battery-powered electric vehicles were available. Although several manufacturers built all-electric autos, the batteries were heavy, expensive, and required more frequent recharging than customers would accept. Even though 90 percent of all daily commutes are less than 80 km (50 mi), most people want the capability to take a long road trip of several hundred kilometers without needing to stop for fuel or recharging.

An alternative that appears to have much more customer appeal is the hybrid gas-electric vehicle. The first of these hybrid cars to be marketed in the United States is the two-seat Honda Insight. Its 3-cylinder, 1.0-liter gas engine is the main power source. A 7-hp electric motor helps during acceleration and hill climbing. When the small battery pack begins to run down, it is recharged by the gas engine, so that the vehicle never needs to be plugged in. More electricity is captured during "regenerative" braking, further increasing efficiency. With a streamlined, lightweight, plastic and aluminum body, the Insight gets 72 mpg (30.3 km/liter) in highway driving and has low-enough emissions to satisfy California requirements. Quick acceleration and nimble handling make the Insight fun to drive. Cost is about $20,000.

Both Toyota and Honda have also introduced 5-passenger hybrids. The Toyota is called a Prius, and the Honda is a Civic hybrid. During most city driving, they depend only on a quiet, emission-free, battery-powered, electric motor. The 1.5-liter gas engine kicks in to help accelerate or when the batteries need recharging. Getting about 50 mpg (22 km/liter) in city driving, both models are among the most efficient cars on the road and can travel more than 625 mi (1,000 km) without refueling. Some drivers are unnerved by the noiseless electric motor of hybrids. Sitting at a stoplight, it makes no sound at all. You might think it was dead, but when the light changes, you glide off silently and smoothly. Both the hybrid Civic and Prius sell for about $20,000 and qualify for a $2,000 Federal Clean-Fuel Vehicle tax deduction. The Sierra Club estimates that in 100,000 mi (160,000 km), a Prius will generate 27 tons of CO_2, a Ford Taurus will generate 64 tons, while the Ford Excursion SUV will produce 134 tons. In 1999 the Sierra Club awarded both the Insight and the Prius an "excellence in engineering" award, the first time this organization has ever endorsed commercial products.

Most American automobile makers are skeptical of hybrid designs and are putting all their research into fuel cells, which may be years from commercial use but have the promise of being truly zero-emission. What do you think? Would you buy a vehicle with a hybrid engine system? Would you take a chance on this new technology to get a quiet, clean, efficient, environmentally friendly means of transportation? Do you think that automobile makers are wise to wait for fuel cells, or should they be producing hybrid vehicles as well?

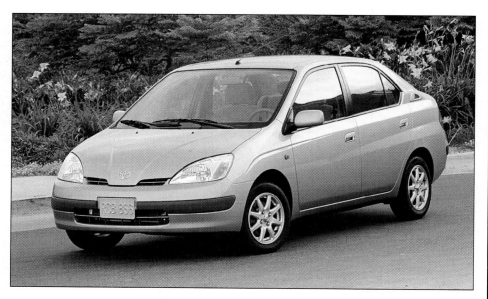

The hybrid gas-electric Toyota Prius seats five adults and gets 52 mpg (22 km/l) in city driving, while meeting super-low emission standards.

TABLE 12.2 Energy-Saving Potential

	MODEL AVERAGE	BEST ON MARKET	BEST PROTOTYPE	SAVINGS (PERCENT)*
Automobile (mi/gal)	24	78	107	71
Home (1,000 J/day)	190	68	11	94
Refrigerator (kwh/day)	4	2	1	75
Gas furnace (million J/day)	210	140	110	48
Air conditioner (kwh/day)	10	5	3	70

*Savings = average model ÷ best prototype – 1.0.

Source: U.S. Department of Energy.

FIGURE 12.15 *This new house, built with the most efficient insulation, windows, and appliances, uses a geothermal pump and is expected to cost only $400 per year for heating, cooling, and hot water, even in Minnesota's cold winters and hot summers.*

sealed homes. Many people are attracted to superefficient houses because of their superior air quality as well as energy savings.

Geothermal heat pumps use water circulated through pipes buried in the ground to aid in heating buildings in the winter and cooling in the summer. Because the ground below a couple of meters maintains a constant temperature year-round, a heat pump can cut energy bills by half or more compared to other energy sources. Buying energy-efficient appliances can also cut your energy consumption substantially (table 12.2). Superinsulated, high-efficiency houses require 90 percent less energy for heating, cooling, water heating, and other daily activities than the average American home (fig. 12.15).

Orienting homes so that living spaces have passive solar gain in the winter and are shaded by trees or roof overhang in the summer also helps conserve energy. In addition to building more energy-efficient homes, there are many personal actions we can take to conserve energy (see What Can You Do?).

Industrial energy savings are another important part of our national energy budget. More efficient electric motors and pumps,

what can you do?

Some Things You Can Do to Save Energy

1. Drive less: make fewer trips, use telecommunications and mail instead of going places in person.
2. Use public transportation, walk, or ride a bicycle.
3. Use stairs instead of elevators.
4. Join a car pool or drive a smaller, more efficient car; reduce speeds.
5. Insulate your house or add more insulation to the existing amount.
6. Turn thermostats down in the winter and up in the summer.
7. Weather-strip and caulk around windows and doors.
8. Add storm windows or plastic sheets over windows.
9. Create a windbreak on the north side of your house; plant deciduous trees or vines on the south side.
10. During the winter, close windows and drapes at night; during summer days, close windows and drapes if using air conditioning.
11. Turn off lights, television sets, and computers when not in use.
12. Stop faucet leaks, especially hot water.
13. Take shorter, cooler showers; install water-saving faucets and showerheads.
14. Recycle glass, metals, and paper; compost organic wastes.
15. Eat locally grown food in season.
16. Buy locally made, long-lasting materials.

new sensors and control devices, advanced heat-recovery systems, and material recycling have reduced industrial energy requirements significantly. In the early 1980s U.S. businesses saved $160 billion per year through conservation. When oil prices collapsed, however, we returned to wasteful ways.

SOLAR ENERGY

The sun is a giant nuclear furnace in space, constantly bathing our planet with a free energy supply. Solar heat drives winds and the hydrologic cycle. All biomass, as well as fossil fuels and our food (both of which are derived from biomass), results from conversion of light energy (photons) into chemical bond energy by photosynthetic bacteria, algae, and plants.

The average amount of solar energy arriving at the top of the atmosphere is 1,330 watts m². About half of this energy is absorbed or reflected by the atmosphere (more at high latitudes than at the equator), but the amount reaching the earth's surface is some 10,000 times all the commercial energy used each year. However, this tremendous infusion of energy comes in a form that, until this century, has been too diffuse and low in intensity to be used except for environmental heating and photosynthesis. Figure 12.16 shows solar energy levels over the United States for typical summer and winter days.

Passive Solar Heat

Our simplest and oldest use of solar energy is **passive heat absorption,** using natural materials or absorptive structures with no moving parts to simply gather and hold heat. For thousands of years, people have built thick-walled stone and adobe dwellings that slowly collect heat during the day and gradually release it at night. After cooling at night, these massive building materials maintain a comfortable daytime temperature within the house, even as they absorb external warmth.

A modern adaptation of this principle is a glass-walled "sunspace" or greenhouse on the south side of a building. Massive energy-storing materials, such as brick walls, stone floors, or barrels of heat-absorbing water, incorporated into buildings collect heat that is released slowly at night. An interior, heat-absorbing wall called a Trombe wall is an effective passive heat collector. Some Trombe walls are built of glass blocks enclosing a water-filled space or water-filled circulation tubes so heat from solar rays can be absorbed and stored, while light passes through to inside rooms.

Active Solar Heat

Active solar systems generally pump a heat-absorbing, fluid medium (air, water, or an antifreeze solution) through a relatively small collector, rather than passively collecting heat in a stationary medium like masonry. Active collectors can be located adjacent to or on top of buildings, rather than being built into the structure (fig. 12.17).

A flat, black surface sealed with a double layer of glass makes a good solar collector. A fan circulates air over the hot surface and into the house through ductwork of the type used in standard forced-air heating. Alternatively, water can be pumped through the collector to pick up heat for space heating or to provide hot water. Water heating consumes 15 percent of the U.S.

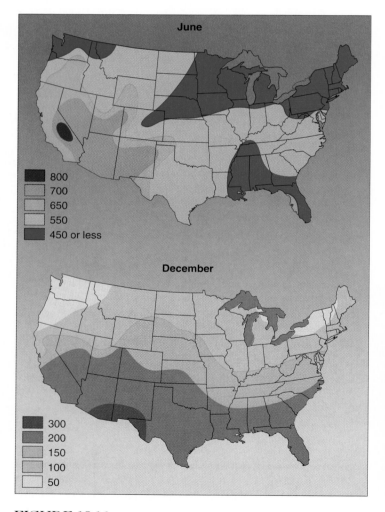

FIGURE 12.16 *Average daily solar radiation in the United States in June and December. One langley, the unit for solar radiation, equals 1 cal/cm² of earth surface (3.69 Btu/ft²).*
Source: Data from National Weather Bureau, U.S. Department of Commerce.

FIGURE 12.17 *A roof-top solar water heater* (top dark panel) *sits above eight photovoltaic electric-generating panels* (left) *on an ecotourist villa at Maho Bay in the U.S. Virgin Islands.*

domestic energy budget, so savings in this area alone can be significant. A simple, flat panel with about 5 m² of surface can reach 95°C (200°F) and can provide enough hot water for an average family of four almost anywhere in the United States. In California, 650,000 homes now heat water with solar collectors. In Greece, Italy, Israel, and other countries where fuels are very expensive, up to 70 percent of domestic hot water comes from solar collectors.

Sunshine doesn't reach us all the time, of course. How can solar energy be stored for times when it is needed? A number of options are available. In a climate where sunless days are rare and seasonal variations are small, a small, insulated water tank is a good solar energy storage system. In areas where clouds block the sun for days at a time or where energy must be stored for winter use, a large, insulated bin containing a heat-storing mass, such as stone, water, or clay, provides solar energy storage (fig. 12.18). During the summer months, a fan blows the heated air from the collector into the storage medium. In the winter a similar fan at the opposite end of the bin blows the warm air into the house. During the summer the storage mass is cooler than the outside air, and it helps cool the house by absorbing heat. During the winter it is warmer and acts as a heat source by radiating stored heat. In many areas six or seven months' worth of thermal energy can be stored in 10,000 gal of water or 40 tons of gravel, about the amount of water in a small swimming pool or the gravel in two average-sized dump trucks.

High-Temperature Solar Energy

Parabolic mirrors are curved reflecting surfaces that collect light and focus it into a concentrated point. There are two ways to use mirrors to collect solar energy to generate high temperatures. One technique uses long, curved mirrors focused on a central tube containing a heat-absorbing fluid (fig. 12.19). Fluid flowing through the tubes reaches much higher temperatures than possible in a basic flat-panel collector. Another high-temperature system uses thousands of smaller mirrors arranged in concentric rings around a tall, central tower. The mirrors, driven by electric motors, track the sun and focus its light on a heat absorber at the top of the "power tower," where molten salt is heated to temperatures as high as 500°C (1,000°F), which then drives a steam-turbine electric generator.

Under optimum conditions, a 50-ha (130-acre) mirror array should be able to generate 100 MW of clean, renewable power. The only power tower in the United States is Southern California Edison's Solar II plant in the Mojave Desert east of Los Angeles. Its 2,000 mirrors focused on a 100-m (300-ft) tall tower generate 10 MW, or enough electricity for 5,000 homes at an operating cost far below that of nuclear power or oil. We haven't had enough experience with these facilities to know how reliable the mirrors, motors, heat absorbers, and other equipment will be over the long run.

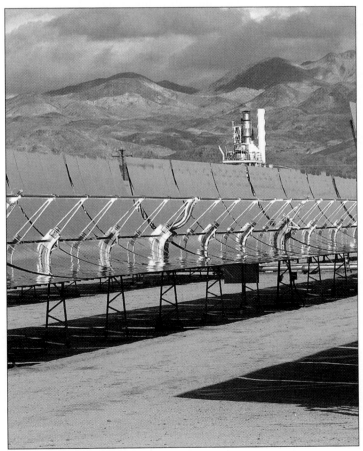

FIGURE 12.19 *Parabolic mirrors focus sunlight on steam-generating tubes at this power plant in the California desert.*

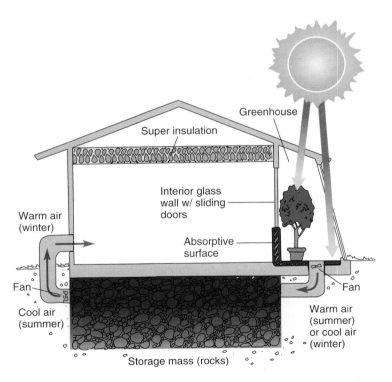

FIGURE 12.18 *Underground massive heat storage unit. Heated air collected behind double- or triple-glazed windows is pumped down into a storage medium of rock, water, clay, or similar material, where it can be stored for a number of months.*

Photovoltaic Energy

The **photovoltaic cell** offers an exciting potential for capturing solar energy in a way that will provide clean, versatile, renewable energy. This simple device has no moving parts, negligible maintenance costs, produces no pollution, and has a lifetime equal to that of a conventional fossil fuel or nuclear power plant.

Photovoltaic cells capture solar energy and convert it directly to electrical current by separating electrons from their parent atoms and accelerating them across a one-way electrostatic barrier formed by the junction between two different types of semiconductor material (fig. 12.20). The photovoltaic effect, which is the basis of these devices, was first observed in 1839 by French physicist Alexandre-Edmond Becquerel, who also discovered radioactivity. His discovery didn't lead to any useful applications until 1954, when researchers at Bell Laboratories in New Jersey learned how to carefully introduce impurities into single crystals of silicon.

These handcrafted, single-crystal cells were much too expensive for any practical use until the advent of the U.S. space program. In 1958, when Vanguard I went into orbit, its radio was powered by six palm-sized photovoltaic cells that cost $2,000 per peak watt of output, more than 2,000 times as much as conventional energy at the time. Since then, prices have fallen dramatically. In 1970 they cost $100 per watt; in 2002 they were about $5 per watt. This makes solar energy cost-competitive with other sources in some areas.

By 2020 photovoltaic cells could be less than $1 per watt of generating capacity. With 15 percent efficiency and a 30-year life, they should be able to produce electricity for around 6¢ per kilowatt hour. At that time, coal-fired steam power will probably cost about half again as much, and nuclear power will likely cost twice as much as photovoltaic energy.

During the last 25 years, the efficiency of energy captured by photovoltaic cells has increased from less than 1 percent of incident light to more than 10 percent under field conditions and over 75 percent in the laboratory. Promising experiments are underway using exotic metal alloys, such as gallium arsenide, and semiconducting polymers of polyvinyl alcohol, which are more efficient in energy conversion than silicon crystals. Every year college students from all over North America race solar-powered cars across the country to test solar technology and publicize its potential. Perhaps you and your classmates would like to build a solar car and enter the race.

One of the most promising developments in photovoltaic cell technology in recent years is the invention of amorphous silicon collectors. First described in 1968 by Stanford Ovshinky, a self-taught inventor from Detroit, these noncrystalline silicon semiconductors can be made into lightweight, paper-thin sheets that require much less material than conventional photovoltaic cells. They also are vastly cheaper to manufacture and can be made in a variety of shapes and sizes, permitting ingenious applications. Roof tiles with photovoltaic collectors layered on their surface already are available. Even flexible films can be coated with amorphous silicon collectors. Silicon collectors already are providing power to places where conventional power is unavailable, such as lighthouses, mountaintop microwave

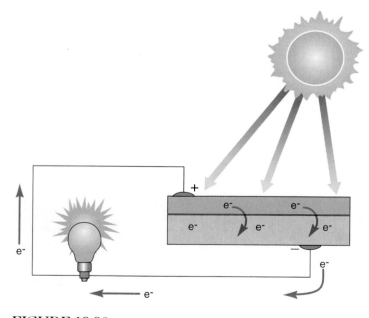

FIGURE 12.20 *The operation of a photovoltaic cell. Boron impurities incorporated into the upper silicon crystal layers cause electrons (e-) to be released when solar radiation hits the cell. The released electrons move into the lower layer of the cell, thus creating a shortage of electrons, or a positive charge, in the upper layer and an oversupply of electrons, or negative charge, in the lower layer. The difference in charge creates an electric current in a wire connecting the two layers.*

APPLICATION: **Payback Times**

It currently costs about $20,000 for enough photovoltaic cells, batteries, and inverters to run all the appliances, lights, and other electrical services of a modern house. If you live in a place with reliable, sustained winds of at least 15 mph (on average) year-round, you could buy a wind generator for about half as much. By contrast, purchasing electricity from a public utility would cost about $600 to $1,200 per year for comparable consumption levels in most parts of the United States or Canada.

1. What would be the payback time for installing solar or wind power under these conditions?

2. What payback would you consider acceptable?

3. How would you factor in our responsibilities to future generations or the value of clean air and clear water in your decision?

Answers: 1. Solar = 33.3 or 16.6 years; wind = 16.6 or 8.3 years. *2.* and *3.* Depends on your values and beliefs.

repeater stations, villages on remote islands, and ranches in the Australian outback.

You probably already use amorphous silicon photovoltaic cells. They are being built into solar-powered calculators, watches, toys, photosensitive switches, and a variety of other consumer products. Japanese electronic companies presently lead in this field, having foreseen the opportunity for developing a market for photovoltaic cells. This market is already more than $100 million per year. Japanese companies now have home-roof arrays capable of providing all the electricity needed for a typical home at prices in some areas competitive with power purchased from a utility.

Think about how solar power could affect your future energy independence. Imagine the benefits of being able to build a house anywhere and having a cheap, reliable, clean, quiet source of energy with no moving parts to wear out, no fuel to purchase, and little equipment to maintain. You could have all the energy you need without commercial utility wires or monthly energy bills. Coupled with modern telecommunications and information technology, an independent energy source would make it possible to live in the countryside and yet have many of the employment and entertainment opportunities and modern conveniences available in a metropolitan area.

Transporting and Storing Electrical Energy

Electrical energy is difficult and expensive to store or to transport over long distances. If we come to depend on solar, wind, or other renewable energy sources, we may need many more high-voltage power lines to transport electricity to distant markets (fig. 12.21). Many property owners resist having power lines cross their land because they are unsightly and may create hazardous electromagnetic fields. The Bush administration has proposed giving eminent domain to utilities to expedite transmission-line siting. An alternative would be to use electricity at remote sites to electrolyze water, then ship hydrogen through buried pipelines. Hydrogen is the ideal fuel for fuel cells, discussed later in this chapter.

Storage is a problem for photovoltaic generation as well as other sources of electric power. Traditional lead-acid batteries are heavy and have low energy densities; that is, they can store only moderate amounts of energy per unit mass or volume. Acid from batteries is hazardous, and lead from smelters or battery manufacturing is a serious health hazard for workers who handle these materials. A typical lead-acid battery array sufficient to store several days of electricity for an average home would cost about $5,000 and weigh 3 or 4 tons. All the components for an electric car are readily available, but as mentioned earlier in the Case Study (p. 284), battery technology limits how far they can go between charges.

Another strategy is to store energy in a form that can be turned back into electricity when needed. Pumped-hydro storage involves pumping water to an elevated reservoir at times when excess electricity is available. The water is released to flow back down through turbine generators when extra energy is needed. Using a similar principle, pressurized air can be pumped into such reservoirs as natural caves, depleted oil and gas fields, abandoned mines, or special tanks. An Alabama power company currently uses off-peak electricity to

FIGURE 12.21 *As we come to depend more on wind, solar, and hydropower energy, we need more transmission lines to transport electricity to urban centers.*

pump air at night into a deep salt mine. By day, the air flows back to the surface through turbines, driving a generator that produces electricity. Cool night air is heated to 1,600°F by compression plus geothermal energy, increasing pressure and energy yield.

Flywheels also are the subject of current experimentation for energy storage. Massive, high-speed flywheels, spinning in a nearly friction-free environment, store large amounts of mechanical energy in a small area. This energy is easily convertible to electrical energy. It is difficult, however, to find materials strong enough to hold together when spinning at high speed. Flywheels have a disconcerting tendency to fail explosively and unexpectedly, sending shrapnel in all directions.

Promoting Renewable Energy

Utility restructuring currently being planned in the United States could include policies to encourage conservation and alternative energy sources. Among the proposed policies are: (1) "distributional surcharges" in which a small per kilowatt-hour charge is levied on all utility customers to help finance renewable energy research and development, (2) "renewables portfolio" standards to require power suppliers to obtain a minimum percentage of their energy from sustainable sources, and (3) **"green pricing"** that allows utilities to profit from conservation programs and charge premium prices for energy from renewable sources.

Some individual states already are pursuing these policies. For example, Iowa has a Revolving Loan Fund supported by a surcharge on investor-owned gas and electric utilities. This fund provides low-interest loans for renewable energy and conservation. Several states have initiated green pricing programs as a way to encourage a transition to sustainable energy. One of the first is in Colorado, where 1,000 customers have agreed to pay $2.50 per month above their regular electric rates to help finance a 10 MW wind farm being built on the Colorado-Wyoming border. Buying a 100 kW "block" of wind power provides the same environmental benefits as planting a half acre of trees or not driving an automobile 4,000 km (2,500 mi) per year. Not all "green pricing" plans are as straightforward as this, however. Some utilities collect the premium rates for facilities that already exist or for renewable energy bought from other utilities at much lower prices.

FUEL CELLS

Rather than store and transport energy, another alternative would be to generate it locally, on demand. **Fuel cells** are devices that use ongoing electrochemical reactions to produce an electrical current. They are very similar to batteries except that, rather than recharging them with an electrical current, you add more fuel for the chemical reaction.

All fuel cells consist of a positive electrode (the cathode) and a negative electrode (the anode) separated by an electrolyte, a material that allows the passage of charged atoms, called ions, but is impermeable to electrons (fig. 12.22). In the most common systems, hydrogen or a hydrogen-containing fuel is passed over the anode, while oxygen is passed over the cathode. At the anode, a reactive catalyst, such as platinum, strips an electron from each hydrogen atom, creating a positively charged hydrogen ion (a proton). The hydrogen ion can migrate through the electrolyte to the cathode, but the electron is excluded. Electrons flow through an external circuit, and the electrical current generated by their passage can be used to do useful work. At the cathode, the electrons and protons are reunited and combined with oxygen to make water.

Fuel cells provide direct-current electricity as long as they are supplied with hydrogen and oxygen. For most uses, oxygen is provided by ambient air. Hydrogen can be supplied as a pure gas, but storing hydrogen gas is difficult and dangerous because of its explosive nature. Liquid hydrogen takes far less space than the gas but must be kept below −250°C (−400°F), not a trivial task for most mobile applications. The alternative is a device called a **reformer** or converter that strips hydrogen from fuels such as natural gas, methanol, ammonia, gasoline, ethanol, or even vegetable oil. Many of these fuels can be derived from sustainable biomass

FIGURE 12.22 *Fuel cell operation. Electrons are removed from hydrogen atoms at the anode to produce hydrogen ions (protons) that migrate through a semipermeable electrolyte medium to the cathode, where they reunite with electrons from an external circuit and oxygen atoms to make water. Electrons flowing through the circuit connecting the electrodes create useful electrical current.*

crops. Even methane effluents from landfills and wastewater treatment plants can be used as a fuel source. Where a fuel cell can be hooked permanently to a gas line, hydrogen can be provided by solar, wind, or geothermal facilities that use electricity to hydrolyze water.

A fuel cell run on pure oxygen and hydrogen produces no waste products except drinkable water and radiant heat. When a reformer is coupled to the fuel cell, some pollutants are released (most commonly carbon dioxide), but the levels are typically far less than conventional fossil fuel combustion in a power plant or automobile engine. Although the theoretical efficiency of electrical generation of a fuel cell can be as high as 70 percent, the actual yield is closer to 40 or 45 percent. This is not much better than a very good fossil fuel power plant or a gas turbine electrical generator. On the other hand, the quiet, clean operation and variable size of fuel cells make them useful in buildings where waste heat can be captured for water heating or space heating. A new 45-story office building at 4 Times Square, for example, has two 200-kW fuel cells on its fourth floor that provide both electricity and heat. This same building has photovoltaic panels on its façade, natural lighting, fresh-air intakes to reduce air-conditioning, and a number of other energy conservation features.

Fuel cells would be ideal zero-emission power sources for vehicles. Fuel cell buses are being tested in Canada, and Honda announced that a fuel cell automobile would be available in California in 2002.

The current from a fuel cell is proportional to the size (area) of the electrodes, while the voltage is limited to about 1.23 volts per cell. A number of cells can be stacked together until the desired power level is reached. A fuel cell stack that provides almost all of the electricity needed by a typical home (along with hot water and

FIGURE 12.23 *Firewood is an important resource in this Siberian village near Lake Baikal.*

FIGURE 12.24 *Harvesting marsh reeds (*Phragmites *sp.) in Sweden as a source of biomass fuel. In some places, biomass from wood chips, animal manure, food-processing wastes, peat, marsh plants, shrubs, and other kinds of organic material make a valuable contribution to energy supplies. Care must be taken, however, to avoid environmental damage in sensitive areas.*

space heating) would be about the size of a refrigerator. A 200-kW unit fills a medium-sized room and provides enough energy for 20 houses or a small factory. Tiny fuel cells running on methanol might soon be used in cell phones, pagers, toys, computers, video-cameras, and other appliances now run by batteries. Rather than buy new batteries or spend hours recharging spent ones, you might just add an eyedropper of methanol every few weeks to keep your gadgets operating.

BIOMASS

Wood fires have been a primary source of heating and cooking for thousands of years (fig. 12.23). As recently as 1850, wood supplied 90 percent of the fuel used in the United States. Wood now provides less than 1 percent of the energy in the United States, but in many of the poorer countries of the world, wood and other biomass fuels still provide up to 95 percent of all energy used. The 1,500 million m³ of fuelwood collected in the world each year is about half of all wood harvested.

In northern industrialized countries, wood burning has increased since 1975 in an effort to avoid rising oil, coal, and gas prices. Most northern areas have adequate wood supplies to meet demands at current levels, but problems associated with wood burning may limit further expansion of this use. Inefficient and incomplete burning of wood in fireplaces and stoves produces smoke laden with fine ash and soot and hazardous amounts of carbon monoxide and hydrocarbons. In valleys where inversion layers trap air pollutants, the effluent from wood fires can be a major source of air quality degradation and health risk. Polycyclic aromatic compounds produced by burning are especially worrisome because they are carcinogenic (cancer-causing). Some communities have passed ordinances limiting the use of woodstoves.

Wood chips, sawdust, wood residue, and other plant materials are being used in some places in the United States and Europe as fuel in industrial boilers (fig. 12.24). In Vermont, for instance, where fossil fuels are expensive and 76 percent of the land is cov-

ered by forest, 250,000 cords of unmarketable cull wood are burned annually to fuel a 50-MW power plant in Burlington. Pollution-control equipment is easier to install and maintain in a central power plant than in individual home units. Wood burning also contributes less to acid precipitation than does coal. Because wood has little sulfur, it produces few sulfur gases. And because it burns at lower temperatures than coal, it also produces fewer nitrogen oxides. Burning wood as a renewable crop doesn't produce any net increase in atmospheric carbon dioxide (and, therefore, doesn't add to global warming) because all the carbon released by burning biomass was taken up from the atmosphere when the biomass was grown.

Fuelwood Crisis in Less-Developed Countries

Two billion people—about 40 percent of the total world population—depend on firewood and charcoal as their primary energy source. Of these people, three-quarters (1.5 billion) do not have an adequate, affordable supply. Many people in the less-developed countries face a daily struggle to find enough fuel to warm their homes and cook their food. The problem is intensifying because rapidly growing populations in many developing countries create increasing demands for firewood and charcoal from a diminishing supply.

As firewood becomes increasingly scarce, women and children, who do most of the domestic labor in many cultures, spend more and more hours searching for fuel. In some places, it now takes eight hours or more just to walk to the nearest fuelwood supply and even longer to walk back with a load of sticks and branches that will only last a few days.

Currently, about half of all wood harvested each year worldwide is used as fuel. Eighty-five percent of that fuel is harvested in developing countries, whereas three-quarters of all industrial roundwood (lumber, poles, beams, and building materials) is harvested and consumed in developed countries. The poorest countries, such as Ethiopia, Bhutan, Burundi, and Bangladesh, depend on biomass for 90 percent of their energy. Often, the harvest is sustainable, consisting of deadwood, branches, trimmings, and shrubs. In Pakistan, for example, some 4.4 million tons of twigs and branches and 7.7 million tons of shrubs and crop residue are consumed as fuel each year with destruction of very few living trees.

In other countries, however, desperate people often chop down anything that will burn. In Haiti, for instance, more than 90 percent of the once-forested land has been almost completely denuded, and people cut down even valuable fruit trees to make charcoal they can sell in the marketplace. It is estimated that the 1,700 million tons of fuelwood now harvested each year globally is at least 500 million tons less than is needed. By 2025 the worldwide demand for fuelwood is expected to be about twice current harvest rates, while supplies will not have expanded much beyond current levels. Some places will be much worse than this average. In some African countries, such as Mauritania, Rwanda, and the Sudan, firewood demand already is ten times the sustainable yield. Reforestation projects, agroforestry, community woodlots, and inexpensive, efficient, locally produced woodstoves could help alleviate expected fuelwood shortages in many places.

Dung and Methane As Fuels

Where wood and other fuels are in short supply, people often dry and burn animal manure. This may seem like a logical use of waste biomass, but it can intensify food shortages in poorer countries. Not putting this manure back on the land as fertilizer reduces crop production and food supplies. In India, for example, where fuelwood supplies have been chronically short for many years, a limited manure supply must fertilize crops and provide household fuel. Cows in India produce more than 800 million tons of dung per year, more than half of which is dried and burned in cooking fires (fig. 12.25). If that dung were applied to fields as fertilizer, it could boost crop production of edible grains by 20 million tons per year, enough to feed about 40 million people.

When cow dung is burned in open fires, more than 90 percent of the potential heat and most of the nutrients are lost. Compare that to the efficiency of using dung to produce methane gas, an excellent fuel. In the 1950s, simple, cheap methane digesters were designed for villages and homes, but they were not widely used. In China, 6 million households use biogas for cooking and lighting. Two large municipal facilities in Nanyang will soon provide fuel for more than 20,000 families. Perhaps other countries will follow China's lead.

Methane gas, the main component of natural gas, is produced by anaerobic decomposition (digestion by anaerobic bacteria) of any moist organic material. Many people are familiar with the fact that swamp gas is explosive. Swamps are simply large methane digesters, basins of wet plant and animal wastes sealed from the air by a layer of water. Under these conditions, organic materials are decomposed by anaerobic (oxygen-free) rather than aerobic (oxygen-using) bacteria, producing flammable gases instead of carbon dioxide. This same process can be reproduced

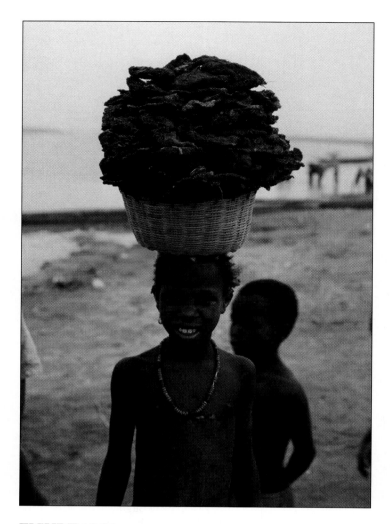

FIGURE 12.25 *In many countries, animal dung is gathered by hand, dried, and used for fuel. This girl in Mali is carrying home fuel for cooking. Using dung as fuel deprives fields of nutrients and reduces crop production.*

artificially by placing organic wastes in a container and providing warmth and water (fig. 12.26). Bacteria are ubiquitous enough to start the culture spontaneously.

Burning methane produced from manure provides more heat than burning the dung itself, and the sludge left over from bacterial digestion is a rich fertilizer, containing bacterial biomass as well as most of the nutrients originally in the dung. Whether the manure is of livestock or human origin, airtight digestion also eliminates some health hazards associated with direct use of dung, such as exposure to fecal pathogens and parasites.

How feasible is methane—from manure or from municipal sewage—as a fuel resource in developed countries? Methane is a clean fuel that burns efficiently. It is produced in a low-technology, low-capital process from any kind of organic waste material: livestock manure, kitchen and garden scraps, and even municipal garbage and sewage. In fact, municipal landfills are active sites of methane production, contributing as much as 20 percent of the annual output of methane to the atmosphere. This is a waste of a

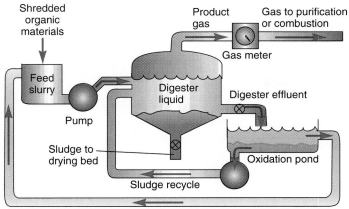

FIGURE 12.26 *Continuous unit for converting organic material to methane by anaerobic fermentation. One kilogram of dry organic matter will produce 1–1.5 m³ of methane, or 2,500–3,600 million cal per metric ton.*

Source: *Solar Energy as a National Energy Resource,* NSF/NASA Solar Energy Panel, National Science Foundation, December 1972.

valuable resource and a threat to the environment because methane absorbs infrared radiation and contributes to the greenhouse effect (see chapter 9). Some municipalities are drilling gas wells into landfills and garbage dumps. Cattle feedlots and chicken farms in the United States are a tremendous potential fuel source. Collectible crop residues and feedlot wastes each year contain 4.8 billion GJ (4.6 quadrillion Btu) of energy, more than all the nation's farmers use. Municipal sewage treatment plants routinely use anaerobic digestion as a part of their treatment process, and many facilities collect the methane they produce and use it to generate heat or electricity for their operations. Although this technology is well developed, its utilization could be much more widespread.

Alcohol from Biomass

Ethanol (grain alcohol) and methanol (wood alcohol) are produced by anaerobic digestion of plant materials with high sugar content, mainly grain and sugarcane. Ethanol can be burned directly in automobile engines adapted to use this fuel, or it can be mixed with gasoline (up to about 10 percent) to be used in any normal automobile engine. A mixture of gasoline and ethanol is often called **gasohol.** Ethanol in gasohol raises octane ratings and is a good substitute for lead antiknock agents, the major cause of lead pollution. It also helps reduce carbon monoxide emissions in automobile exhaust. Gas stations in U.S. cities that exceed EPA air quality standards have been ordered to sell so-called oxygenated fuels containing ethanol or methanol.

Ethanol production could be a solution to grain surpluses and bring a higher price for grain crops than the food market offers. Given our current methods of farming, however, it takes more energy in the form of petroleum-based fertilizer, pesticides, and vehicle fuel than is produced in ethanol. Although the 2002 Farm Bill granted $3 billion in support of ethanol production, this is more a subsidy for grain farmers than a step toward energy independence.

Methanol might have many advantages as a motor vehicle fuel. Since it burns at a lower temperature than gasoline or diesel, the bulky, heavy radiator might be eliminated, making automobiles sleeker and more efficient. Combined with flywheel-energy storage, intermittent burning engines, turbochargers, and other high-tech devices, we might be able to have personal transportation that is both energy efficient and less polluting. Both methanol and ethanol make good fuels for fuel cells. Methanol can be produced from woody crops that require less petroleum-based inputs than grain-based ethanol.

ENERGY FROM THE EARTH'S FORCES

The winds, waves, tides, ocean thermal gradients, and geothermal areas are renewable energy sources. Although available only in selected locations, these sources could make valuable contributions to our total energy supply.

Hydropower

Falling water has been used as an energy source since ancient times. The invention of water turbines in the nineteenth century greatly increased the efficiency of hydropower dams. By 1925, falling water generated 40 percent of the world's electric power. Since then, hydroelectric production capacity has grown fifteen-fold, but fossil fuel use has risen so rapidly that water power is now only one-quarter of total electrical generation. Still, many countries produce most of their electricity from falling water. Norway, for instance, depends on hydropower for 99 percent of its electricity; Brazil, New Zealand, and Switzerland all produce at least three-quarters of their electricity with water power. Canada is the world's leading producer of hydroelectricity, running 400 power stations with a combined capacity exceeding 60,000 MW. First Nations people protest that their rivers are being diverted and lands flooded to generate electricity, most of which is sold to the United States.

The total world potential for hydropower is estimated to be about 3 million MW. If all of this capacity were put to use, the available water supply could provide between 8 and 10 terrawatt hours (1,012 watt-hours) of electrical energy. Currently, we use only about 10 percent of the potential hydropower supply. The energy derived from this source in 1994 was equivalent to about 500 million tons of oil, or 8 percent of the total world commercial energy consumption.

Much of the hydropower development in recent years has been in enormous dams. There is a certain efficiency of scale in giant dams, and they bring pride and prestige to the countries that build them, but they can have unwanted social and environmental effects that spark protests in many countries. China's Three Gorges Dam on the Yangtze River will span 2.0 km (1.2 mi) and will be 185 m (600 ft) tall when completed in 2009 (see opening photograph for this chapter). The reservoir it creates will be 644 km (400 mi) long, and will displace more than 1 million people (see chapter 10).

In tropical climates, large reservoirs often suffer enormous water losses. Lake Nasser, formed by the Aswan High Dam in

Egypt, loses 15 billion m³ each year to evaporation and seepage. Unlined canals lose another 1.5 billion m³. Together, these losses represent one-half of the Nile River flow, or enough water to irrigate 2 million ha of land. The silt trapped by the Aswan High Dam formerly fertilized farmland during seasonal flooding and provided nutrients that supported a rich fishery in the delta region. Farmers now must buy expensive chemical fertilizers, and the fish catch has dropped almost to zero. Schistosomiasis, spread by snails that flourish in the reservoir, is an increasingly serious problem.

Wind Energy

The air surrounding the earth has been called a 20-billion-km³ storage battery for solar energy. The World Meteorological Organization has estimated that 20 million MW of wind power could be commercially tapped worldwide, not including contributions from windmill clusters at sea. This is about 50 times the total present world nuclear generating capacity. Wind power has advantages and disadvantages, as do other nontraditional technologies. Like solar power and hydropower, wind power taps a natural physical force. Like solar power (its ultimate source), wind power is an abundant, nonpolluting resource, and causes minimal environmental disruption. Also like solar power, however, it requires expensive storage during peak production times to offset nonwindy periods.

As the world's conventional fuel prices rise, interest in wind energy is growing rapidly. In the 1980s the United States was a world leader in wind technology, and California hosted 90 percent of all existing windpower generators. Some 17,000 windmills marched across windy mountain ridges at Altamont, Tehachapi, and San Gorgino Passes (fig. 12.27). Poor management, technical flaws, and overdependence on subsidies, however, led to bankruptcy of major corporations, including Kenetech, once the largest turbine producer in the United States. Now Danish, German, and Japanese wind machines are capturing the rapidly growing world market. Wind technology is now Denmark's second-largest export, employing 20,000 people and bringing in about $1 billion (U.S.) per year.

The 30,000 MW of installed wind power currently in operation worldwide demonstrates the economy of wind turbines. Theoretically up to 60 percent efficient, windmills typically produce about 35-percent efficiency under field conditions. Where conditions are favorable, wind power is now cheaper than any other new energy source, with electric prices typically around 5¢ per kilowatt-hour in places with steady winds averaging at least 24 km/h (15 mph). Much of the Great Plains meets this requirement.

The World Energy Council predicts that wind could account for 200,000 MW of electricity by 2020, depending on how seriously politicians take global warming and how many uneconomical nuclear reactors go offline. One thousand MW meets the energy needs of about 50,000 typical U.S. households or is equivalent to about 6 million bbl of oil. Shell Oil suggests that half of all the world's energy could be wind and solar generated by the middle of the twenty-first century.

The standard modern wind turbine uses only two or three propeller blades. More blades on a windmill provide more torque in low-speed winds, so that the traditional midwestern windmill,

FIGURE 12.27 *Wind farms are becoming competitive with other power sources in many areas.*

with 20 or 30 blades, was most appropriate for small-scale use in less reliable wind fields. Fewer blades operate better in high-speed winds, providing more energy for less material cost at wind speeds of 25 to 40 km/h (15 to 25 mph). A two-blade propeller can extract most of the available energy from a large vertical area and has less material to weaken and break in a storm. Three-bladed propellers often are preferred because they are easier to balance and spin more smoothly.

Wind farms are large concentrations of wind generators producing commercial electricity. Construction of wind farms is not limited to mountain ridges. The Great Plains offer tremendous wind-power potential. North Dakota, Texas, Kansas, South Dakota, and Montana all have over 1,000 MW of installed capacity. Seacoasts also provide excellent wind farm sites. Offshore wind farms are being installed by Denmark, the United Kingdom, and the Netherlands, which expect to produce up to 20 percent of their electricity with wind. Large (up to 8 MW each) generators are to be anchored a short distance out to sea, where competition for real estate development is low and winds are high year-round. Germany leads the world in wind power, with 10,000 MW currently installed, and plans to build 22,000 MW by 2010.

Do wind farms have any negative impacts? They generally occupy places with wind and weather too severe for residential or other development. Most wind farms are too far from residential areas to be heard or seen. But they do interrupt the view in remote isolated places and destroy the sense of isolation and natural beauty. Bird kills were a problem for some California wind farms. Careful placement outside of migration corridors and the addition of warning devices can reduce mortality greatly.

Wouldn't wind power take up a huge land area if we were to depend on it for a major part of our energy supply? As table 12.3 shows, the actual space taken up by towers, roads, and other structures on a wind farm is only about one-third as much as would be

Principles of Environmental Science www.mhhe.com/environmentalscience

TABLE 12.3 Jobs and Land Required for Alternative Energy Sources		
TECHNOLOGY	LAND USE (M² PER GIGAWATT-HOUR FOR 30 YEARS)	JOBS (PER TERAWATT-HOUR PER YEAR)
Coal	3,642	116
Photovoltaic	3,237	175
Solar thermal	3,561	248
Wind	1,335	542

Source: Data from Lester R. Brown et al. *Saving the Planet*, 1991, W. W. Norton & Co., Inc.

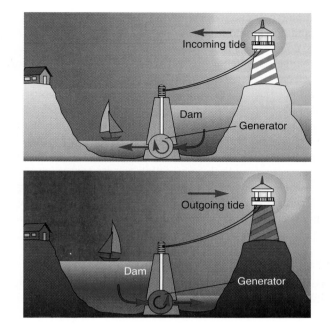

FIGURE 12.28 *Tidal power station. Both incoming and outgoing tides are held back by a dam. The difference in water levels generates electricity in both directions as water runs through reversible turbogenerators.*

consumed by a coal-fired power plant or solar thermal energy system to generate the same amount of energy over a 30-year period. Furthermore, the land under windmills is more easily used for grazing or farming than is a strip-mined coal field or land under solar panels.

When a home owner or community invests independently in wind generation, the same question arises as with solar energy: What should be done about energy storage when electricity production exceeds use? Besides the storage methods mentioned earlier, many private electricity producers believe the best use for excess electricity is to sell it to the public utility grid. In states that allow net energy pricing, you sell excess power from wind or solar systems back to a local utility. Ideally, the utility pays you something close to their average wholesale price. The 1978 Public Utilities Regulatory Policies Act required utilities to buy power generated by small hydro, wind, cogeneration (simultaneous production of useful heat and electricity), and other privately owned technologies at a fair price. Not all utilities yet comply, but some—notably in California, Oregon, Maine, and Vermont—are purchasing significant amounts of private energy.

Geothermal, Tidal, and Wave Energy

The earth's internal temperature can provide a useful source of energy in some places. High-pressure, high-temperature steam fields exist below the earth's surface. Around the edges of continental plates or where the earth's crust overlays magma (molten rock) pools close to the surface, this energy is expressed in the form of hot springs, geysers, and fumaroles. Yellowstone National Park is the largest geothermal region in the United States. Iceland, Japan, and New Zealand also have high concentrations of geothermal springs and vents. Depending on the shape, heat content, and access to groundwater, these sources produce wet steam, dry steam, or hot water.

While few places have geothermal steam, the earth's warmth can help reduce energy costs nearly everywhere. Pumping water through deeply buried pipes can extract enough heat so that

a heat pump will operate more efficiently. Similarly, the relatively uniform temperature of the ground can be used to augment air-conditioning in the summer.

Ocean tides and waves contain enormous amounts of energy that can be harnessed to do useful work. A tidal station works like a hydropower dam, with its turbines spinning as the tide flows through them (fig. 12.28). A high-tide/low-tide differential of several meters is required to spin the turbines. Unfortunately, variable tidal periods often cause problems in integrating this energy source into the electric utility grid. Nevertheless, demand has kept some plants running for many decades.

The first North American tidal generator, producing 20 MW, was completed in 1984 at Annapolis Royal, Nova Scotia. A much larger project has been proposed to dam the Bay of Fundy and produce 5,000 MW of power on the bay's 17-m tides. The total flow at each tide through the Bay of Fundy theoretically could generate energy equivalent to the output of 250 large nuclear power plants. The environmental consequences of such a gargantuan project, however, may prevent its ever being built. The main worries are saltwater flooding of freshwater aquifers when seawater levels rise behind the dam and the flooding and destruction of rich shoals and salt flats, breeding grounds for aquatic species and a vital food source for millions of migrating shorebirds. There also would be heavy siltation, as well as scouring of the seafloor, as water shoots through the dam.

Ocean wave energy can easily be seen and felt on any seashore. The energy that waves expend as millions of tons of water are picked up and hurled against the land, over and over, day after day, can far exceed the combined energy budget for both insolation (solar energy) and wind power in localized areas. Captured and turned into useful forms, that energy could make a substantial contribution to meeting local energy needs.

Numerous attempts have been made to use wave energy to drive electrical generators. Generally, these take the form of a floating bar that moves up and down as the wave passes. When coupled to a dynamo, this mechanical energy can be converted to electricity in the same way that a waterwheel or steam turbine works. England, with a long coastline facing the stormy North Sea, plans to build an extensive system of wave-energy platforms. Unfortunately for developers of this energy source, the stormy coasts where waves are strongest are usually far from major population centers that need the power. In addition, the storms that bring this energy destroy the equipment intended to exploit it.

WHAT'S OUR ENERGY FUTURE?

None of the renewable energy sources discussed in this chapter are likely to completely replace fossil fuels and nuclear power in the near future. They could, however, make a substantial collective contribution toward providing us with the conveniences we crave in a sustainable, environmentally friendly manner. They could also make us energy independent and balance our international payment deficit.

The World Energy Council projects that renewables could provide about 40 percent of world cumulative energy consumption under an idealized "ecological" scenario assuming that political leaders take global warming seriously and pass taxes to encourage conservation and protect the environment (fig. 12.29). This scenario also envisions measures to shift wealth from the north to south, and to enhance economic equity. By the end of the twenty-first century, renewable sources could provide all our energy needs if we take the necessary steps to make this happen.

As noted in this chapter's introduction, rising fuel prices and increasing dependence on imported oil have prompted demands for a U.S. energy policy. Environmentalists point to the dangers of air pollution, global climate change, and other environmental problems associated with burning of fossil fuels. Businesses stress the importance of a reliable energy supply for economic growth. While both call for a new policy, they disagree on what it should contain. Conservatives tend to favor increasing production and easing regulations on power plant operation and transmission-line siting, rather than limiting demand. Progressives, on the other hand, prefer conservation measures such as forcing automakers to increase average fuel efficiency of cars and light trucks and providing heating bill assistance for low-income households.

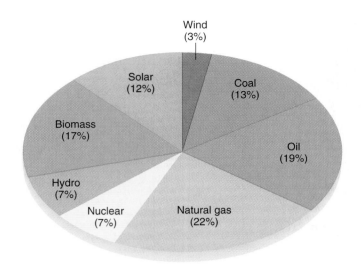

FIGURE 12.29   *Idealized "ecological" scenario for cumulative world energy consumption, 2000–2100.*
Source: Data from World Energy Council, 2002.

In 2001, Vice President Dick Cheney convened a task force to forge a national energy policy. Made up almost entirely of oil, gas, coal, and nuclear power industry representatives, the task force called for $27 billion in subsidies for traditional energy, but only $6 billion for conservation or renewables. Speaking for the task force, Mr. Cheney said that oil, coal, and natural gas would remain the United States' primary energy resources for "years down the road." He strongly supported the use of nuclear power, and claimed that the United States needs to build one large electric-generating plant each week for the next 20 years. He dismissed as 1970s-era thinking the notion that "we could simply conserve or ration our way out" of an energy crisis. "Conservation," he said, "may be a sign of personal virtue, but it is not a sufficient basis for a sound, comprehensive energy policy."

In light of what you've read in this chapter about current and potential sources, what sort of energy policy would you suggest if you were invited to advise government officials? How much conservation is it realistic or fair to ask people to adopt? What's a reasonable percentage of our energy that might come from renewable sources in 20 or 50 years? If you wanted to move toward the "ecological scenario" described above by the World Energy Council, how would you go about implementing that plan? What government actions do you think would be most effective in accomplishing these goals?

SUMMARY

Energy is the capacity to do work. Power is the rate of doing work. Today nearly 80 percent of all commercial energy is generated by fossil fuels, with about 35 percent coming from petroleum. Coal produces 23 percent, and natural gas (methane) accounts for 21 percent of our commercial supply. Petroleum and natural gas were not used in large quantities until the beginning of the twentieth century, but supplies are already running low. Coal supplies will last several more centuries at present rates of usage, but it appears that the fossil fuel age will have been a rather short episode in the total history of humans. Nuclear power provides about 7 percent of commercial energy worldwide.

The environmental damage caused by mining, shipping, processing, and using fossil fuels may necessitate cutting back on our use of these energy sources. Coal is an especially dirty and dangerous fuel, at least as we currently obtain and use it. Coal combustion is a major source of acid precipitation, which is suspected of being a significant cause of environmental damage in many areas. We now recognize that carbon dioxide buildup in the atmosphere has the potential to trap heat and raise the earth's temperature to catastrophic levels.

The greatest worry about nuclear power is the danger that accidents or terrorist attacks could release hazardous radioactive materials into the environment. Several accidents, most notably the "meltdown" at the Chernobyl plant in the Soviet Ukraine in 1986, have convinced many people that this technology is too risky to pursue. Other major worries about nuclear power include where to put the waste products of the nuclear fuel cycle and how to ensure that the wastes will remain safely contained for the thousands of years required for "decay" of the radioisotopes to nonhazardous levels. Yucca Mountain, Nevada, has been chosen for a high-level waste repository, but many experts believe that burying these toxic residues in nonretrievable storage is a mistake.

Several sustainable energy sources could reduce or eliminate our dependence on fossil fuels and nuclear energy. Some of these sources have been used for centuries but have been neglected since fossil fuels came into widespread use. Passive solar heat, fuelwood, windmills, and waterwheels, for instance, once supplied a major part of the external energy for human activities. With increased concern about the dangers and costs associated with conventional commercial energy, these ancient energy sources are being reexamined as part of a more sustainable future for humankind.

Exciting new technologies have been invented to use renewable energy sources. Active solar air and water heating, for instance, require less material and function more quickly than passive solar collection. Parabolic mirrors collect solar energy to produce temperatures high enough to be used as process heat in manufacturing. Wind is now the cheapest form of new energy in many places. It has potential to supply one-third or more of our energy requirements.

Fuel cells use catalysts and semipermeable electrolytes to extract energy from fuels such as hydrogen or methanol at high efficiency and with very low emissions. Ocean thermal electric conversion, tidal and wave power stations, and geothermal steam sources can produce useful amounts of energy in some localities. One of the most promising technologies is direct electricity generation by photovoltaic cells. Since solar energy is available everywhere, photovoltaic collectors could provide clean, inexpensive, nonpolluting, renewable energy, independent of central power grids or fuel supply systems.

Biomass also may have some modern applications. In addition to direct combustion, biomass can be converted into methane or ethanol, which are clean-burning, easily storable, and transportable fuels. These alternative uses of biomass also allow nutrients to be returned to the soil and help reduce our reliance on expensive, energy-consuming artificial fertilizers.

QUESTIONS FOR REVIEW

1. What is energy? What is power?

2. What are the major sources of commercial energy worldwide and in the United States?

3. How does our energy use compare with that of people in other countries?

4. How much coal, oil, and natural gas are in proven reserves worldwide? Where are those reserves located?

5. What are the most important health and environmental consequences of our use of fossil fuels?

6. Describe how nuclear reactors work and why they are dangerous.

7. Describe methods proposed for storing and disposing of nuclear wastes.

8. What is the difference between active and passive solar energy?

9. How do photovoltaic cells work?

10. What is a fuel cell, and how does it work?

11. Describe some problems with wood burning in both industrialized nations and developing nations.

12. What are some advantages and disadvantages of large hydroelectric dams?

THINKING SCIENTIFICALLY

1. If you were considering an alternative energy source for your region and climate, what factors would you need to take into account? Where might you get this information, or how would you measure it yourself?

2. If you were the energy czar of your state or country, where would you invest your budget? Why?

3. What are the advantages and disadvantages of being disconnected from central utility power?

4. Consumers are reluctant to invest in new, sustainable-energy technology because they don't know how it will hold up in the long run. How would you assess the long-term reliability and efficiency of these alternative energy sources?

5. We have discussed a number of different energy sources and energy technologies in this chapter. Each has advantages and disadvantages. If you were an energy policy analyst, how would you compare such different problems as the risk of a nuclear accident versus air pollution effects from burning coal?

6. If your local utility company were going to build a new power plant in your community, what kind would you prefer? Why?

7. The nuclear industry is placing ads in popular magazines and newspapers, claiming that nuclear power is environmentally friendly since it doesn't contribute to the greenhouse effect. How do you respond to that claim?

8. Our energy policy effectively treats some strip-mined lands as national sacrifice areas, since we know that they will never be restored to their original state when mining is finished. How do we decide who wins and who loses in this transaction?

9. Storing nuclear wastes in dry casks outside nuclear power plants is highly controversial. Opponents claim the casks will inevitably leak. Proponents claim the casks can be designed to be safe. What evidence would you consider adequate or necessary to choose between these two positions?

10. Although we have used vast amounts of energy resources in the process of industrialization and development, some would say that it was a necessary investment to get to a point at which we can use energy more efficiently and sustainably. Do you agree? Might we have followed a different path?

KEY TERMS

active solar systems 286
chain reaction 281
control rods 281
energy 275
fossil fuels 275
fuel assembly 281
fuel cells 290
gasohol 293
"green pricing" 290
joule (J) 275
monitored, retrievable
  storage 283

nuclear fission 281
passive heat
  absorption 286
power 275
photovoltaic cell 288
proven-in-place
  reserves 277
reformer 290
tar sands 279
wind farms 294
work 275

Welcome to McGraw-Hill's Online Learning Center

Location: http://www.mhhe.com/environmentalscience

WEB EXERCISES

Fossil Fuel Supplies

Go to http://greenwood.cr.usgs.gov/energy/WorldEnergy/DDS-60/. Click on the world maps link, and look at the "World Petroleum Assessment 2000 Regions with Geologic Provinces" (the first map listed).

1. Where are the major petroleum reserves?

2. Why are most of them along coastlines?

3. What might be the environmental risks of oil drilling and shipping in these regions?

4. Where is there no oil? Why might this be? Find the "Executive Summary" on this page.

5. How many barrels of oil and gas are reported to be in world reserves?

6. Where are the greatest volumes of undiscovered oil expected to be found?

7. Why might we not have explored these places already?

You can also find energy information at http://energy.cr.usgs.gov/energy/WorldEnergy/WEnergy.html

Renewable Energy

Go to www.eren.doe.gov, the renewable energy program of the Department of Energy. Click on the following links: "Solar," then "Million Solar Roofs," then "About Solar Technologies," and finally, "Photovoltaics." Hint: you should be at www.eren.doe.gov/millionroofs/whatispv.html

Either follow the links in the left frame (Learning about PV, Turning Sunlight into Electricity, Why Use PV) or the right frame: various materials, produce electricity, logical power sources, etc.

Take the quiz in "Learning About PV." This is a challenging quiz. If you can get all the answers right, you deserve an A+ in solar power.

Solid and Hazardous Waste

We have no knowledge, so we have stuff; but stuff without knowledge is never enough.

—Greg Brown

A crane unloads a garbage barge at Fresh Kills on Staten Island, the world's largest landfill before it closed in 2001.

OBJECTIVES

After studying this chapter, you should be able to

- identify the major components of the waste stream, and describe how wastes have been—and are being—disposed of in North America and around the world.
- explain the differences between dumps, sanitary landfills, and modern, secure landfills.
- summarize the benefits, problems, and potential of recycling and reusing wastes.
- analyze some alternatives for reducing the waste we generate.
- understand what hazardous and toxic wastes are and how we dispose of them.
- evaluate the options for hazardous-waste management.
- outline some ways we can destroy or permanently store hazardous wastes.

Garbology: The Science of Trash

What can the stuff we throw away every day tell us about our society—and ourselves? That's an interesting question addressed by Professor William Rathje, his colleagues, and students at the University of Arizona. Trained in classical archeology, Rathje decided to use his scientific tools to study contemporary society, rather than ancient ones. Most archeological digs, he reasoned, no matter how dignified or scholarly they seem, are essentially organized trash picking. The broken pottery shards or stone tools carefully dug out of old midden heaps or fire pits by archeologists are really nothing more than the discards of somebody else's daily life. Why not, he reasoned, look at modern garbage in the same way?

Starting in the early 1970s, Rathje and his coworkers systematically collected garbage bags from homes in Tucson and Phoenix, Arizona. Trash was painstakingly sorted, classified, weighed, and recorded. To protect individual privacy, records were kept only by census tracks, rather than by specific households.

After several years of intercepting garbage headed for the landfill, Rathje decided to turn to excavations to look at older refuse. Digging trenches and boring holes as much as 90 ft deep in more than a dozen municipal dumps, Rathje's team has recovered trash dating back to the 1940s. In addition to sorting and classifying larger objects by type, brand, size, material, and condition, the team also washes and filters small items through decreasing mesh-size screens to analyze the composition of "fines" created in tightly packed landfills.

This analysis of what we throw away reveals some interesting things about how modern Americans live. Packaging now makes up about half of all the garbage we discard. Paper bags, metal cans, glass bottles, wrapping paper, and a wide variety of plastic products fall in this category. Some materials, on the other hand, are less prevalent than you might imagine. While some people believe that disposable diapers account for up to 30 percent of all landfill volume, Rathje and his coworkers found that diapers make up only about 1 percent of contemporary trash.

Often, objective research can reveal interesting differences between what people say they do and their actual behavior. In one study in Phoenix, four out of five households claimed that they recycled conscientiously. Researchers found, however, that the garbage of nine out of ten of those houses contained aluminum cans, glass bottles, newsprint, and other easily recyclable materials. Extrapolating to the whole city, Rathje estimated that $2 million in valuable materials were being buried in landfills every year. Newsprint makes up the largest single item in our trash, accounting for 16 percent of what's discarded in a typical city. Together with cardboard, office paper, and other related products, paper accounts for about 38 percent of our waste stream.

One of the biggest surprises from Rathje's work is how well most materials are preserved in landfills. We imagine that landfills are giant compost piles, where everything organic decomposes and disappears in a few months. Quite to the contrary, landfill digs have found newspapers dating from the 1950s that are just as legible today as when they were printed. In one case, pork chops buried for 40 years still looked edible. Because the trash is packed tightly in the landfill and protected from air, water, and sunlight, it is preserved almost perfectly. We feel good about using biodegradable materials as a way of protecting our environment, but when they go into a landfill, they will probably last for centuries.

So what does garbology show us about modern life? We throw away a tremendous amount of stuff. Much of what we discard could be used again or recycled in other useful products if we would only take the time to sort it and send it to a recycling center, rather than just dump it in the trash. But we too often take the easy course, rather than the more environmentally responsible one. In this chapter we'll look further at what we throw away, where it goes, and what some of our options are for reducing the material we discard.

WASTE

Waste is everyone's business. We all produce unwanted by-products and residues in nearly everything we do. According to the Environmental Protection Agency (EPA), the United States produces 11 billion tons of solid waste each year. Nearly half of that amount consists of agricultural waste, such as crop residues and animal manure, which are generally recycled into the soil on the farms where they are produced. They represent a valuable resource as ground cover to reduce erosion and as fertilizer to nourish new crops, but they also constitute the single largest source of nonpoint air and water pollution in the country. About one-third of all solid wastes are mine tailings, overburden from strip mines, smelter slag, and other residues produced by mining and primary metal processing. Much of this material is stored in or near its source of production and isn't mixed with other kinds of wastes. Improper disposal practices, however, can result in serious and widespread pollution.

Industrial waste—other than mining and mineral production—amounts to some 400 million metric tons per year in the United States. Most of this material is recycled, converted to other forms, destroyed, or disposed of in private landfills or deep injection wells. About 60 million metric tons of industrial waste fall in a special category of hazardous and toxic waste, which we will discuss later in this chapter.

Municipal waste—a combination of household and commercial refuse—amounts to about 209 million metric tons per year in the United States. That's just over 2 kg (4.6 lb) per person per day—twice as much per capita as Europe or Japan, and five to ten times as much as most developing countries (fig. 13.1).

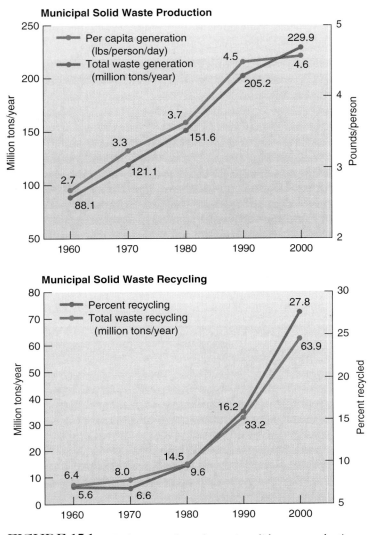

Municipal Solid Waste Production

Municipal Solid Waste Recycling

FIGURE 13.1 *Bad news and good news in solid waste production. Per capita waste has risen steadily to more than 2 kg per person per day. Recycling rates are also rising, however.*
Source: Data from Environmental Protection Agency, 2001.

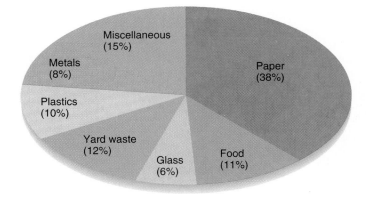

FIGURE 13.2 *Composition of domestic waste in the United States by percent of total weight. Plastics make up only 10 percent by weight but up to 20 percent of total volume.*
Source: Data from Environmental Protection Agency, 2001.

The Waste Stream

Does it surprise you to learn that you generate that much garbage? Think for a moment about how much we discard every year. There are organic materials, such as yard and garden wastes, food wastes, and sewage sludge from treatment plants; junked cars; worn-out furniture; and consumer products of all types. Newspapers, magazines, advertisements, and office refuse make paper one of our major wastes (fig. 13.2). In spite of recent progress in recycling, many of the 200 billion metal, glass, and plastic food and beverage containers used every year in the United States end up in the trash. Wood, concrete, bricks, and glass come from construction and demolition sites, dust and rubble from landscaping and road building. All of this varied and voluminous waste has to arrive at a final resting place somewhere.

The **waste stream** is a term that describes the steady flow of varied wastes that we all produce, from domestic garbage and yard wastes to industrial, commercial, and construction refuse. Many of the materials in our waste stream would be valuable resources if they were not mixed with other garbage. Unfortunately, our collecting and dumping processes mix and crush everything together, making separation an expensive and sometimes impossible task. In a dump or incinerator, much of the value of recyclable materials is lost.

Another problem with refuse mixing is that hazardous materials in the waste stream get dispersed through thousands of tons of miscellaneous garbage. This mixing makes the disposal or burning of what might have been rather innocuous stuff a difficult, expensive, and risky business. Spray-paint cans, pesticides, batteries (zinc, lead, or mercury), cleaning solvents, smoke detectors containing radioactive material, and plastics that produce dioxins and PCBs (polychlorinated biphenyls) when burned are mixed willy-nilly with paper, table scraps, and other nontoxic materials. The best thing to do with household toxic and hazardous materials is to separate them for safe disposal or recycling, as we will see later in this chapter.

WASTE-DISPOSAL METHODS

Where do our wastes go now? In this section, we will examine some historic methods of waste disposal, as well as some future options. Notice that our presentation begins with the least desirable—but most commonly used—measures and proceeds to discuss some preferable options. Keep in mind as you read this that modern waste management reverses this order and stresses the "three Rs" of reduction, reuse, and recycling before destruction or, finally, secure storage of wastes.

Open Dumps

Often, the way people dispose of waste is to simply drop it someplace. Open, unregulated dumps are still the predominant method

FIGURE 13.3 *Scavengers sort through the trash at "Smoky Mountain," one of the huge metropolitan dumps in Manila, Philippines. Some 20,000 people live and work on these enormous garbage dumps. The health effects are tragic.*

of waste disposal in most developing countries. The giant Third World megacities have enormous garbage problems. Mexico City, the largest city in the world, generates some 10,000 tons of trash each day. Until recently, most of this torrent of waste was left in giant piles, exposed to the wind and rain, as well as rats, flies, and other vermin. Manila, in the Philippines, has at least ten huge open dumps. The most notorious is called "Smoky Mountain" because of its constant smoldering fires (fig. 13.3). Thousands of people live and work on this 30-m-high heap of refuse. They spend their days sorting through the garbage for edible or recyclable materials. Health conditions are abysmal, but these people have nowhere else to go. The government would like to close these dumps, but how will the residents be housed and fed? Where else will the city put its garbage?

Most developed countries forbid open dumping, at least in metropolitan areas, but illegal dumping is still a problem. You have undoubtedly seen trash accumulating along roadsides and in vacant, weedy lots in the poorer sections of cities. Is this just a question of aesthetics? Consider the problem of waste oil and solvents. An estimated 200 million liters of waste motor oil are poured into the sewers or allowed to soak into the ground every year in the United States. This is about five times as much as was spilled by the *Exxon Valdez* in Alaska in 1989! No one knows the volume of solvents and other chemicals disposed of by similar methods.

Increasingly, these toxic chemicals are showing up in the groundwater supplies on which nearly half the people in America depend for drinking (see chapter 10). An alarmingly small amount of oil or other solvents can pollute large quantities of drinking or irrigation water. One liter of gasoline, for instance, can make a million liters of water undrinkable. The problem of illegal dumping is likely to become worse as acceptable sites for waste disposal become more scarce and costs for legal dumping escalate. We clearly need better enforcement of antilittering laws, as well as a change in our attitudes and behavior.

Ocean Dumping

The oceans are vast, but not so large that we can continue to treat them as carelessly as has been our habit. Every year some 25,000 metric tons (55 million lbs) of packaging, including half a million bottles, cans, and plastic containers, are dumped at sea. Beaches, even in remote regions, are littered with the nondegradable flotsam and jetsam of industrial society. About 150,000 tons (330 million lbs) of fishing gear—including more than 1,000 km (660 mi) of nets—are lost or discarded at sea each year. Environmental groups estimate that 50,000 northern fur seals are entangled in this refuse and drown or starve to death every year in the North Pacific alone.

Until recently, many cities in the United States dumped municipal refuse, industrial waste, sewage, and sewage sludge in the ocean. Federal legislation now prohibits this dumping. New York City, the last to stop offshore sewage sludge disposal, finally ended this practice in 1992. Still, 60 million to 80 million m3 of dredge spoil—much of it highly contaminated—are disposed of at sea. Some people claim that the deep abyssal ocean plain is the most remote, stable, and innocuous place to dump our wastes. Others argue that we know too little about the values of these remote places or the rare species that live there to smother them with sludge and debris.

Landfills

Over the past 50 years, most American and European cities have recognized the health and environmental hazards of open dumps. Increasingly, cities have turned to **sanitary landfills,** where solid waste disposal is regulated and controlled. To decrease smells and litter and to discourage insect and rodent populations, landfill operators are required to compact the refuse and cover it every day with a layer of dirt (fig. 13.4). This method helps control pollution, but the dirt fill also takes up as much as 20 percent of landfill

FIGURE 13.4 *In a sanitary landfill, trash and garbage are crushed and covered each day to prevent accumulation of vermin and spread of disease. A waterproof lining is now required to prevent leaching of chemicals into underground aquifers.*

APPLICATION: | **How Much Waste Do You Create?**

Collect all the dry trash (excluding food waste) that you discard in a typical day or week. Sort it into major categories: packaging material, junk mail, etc., as well as material type: paper, cardboard, glass, metal, plastic.

1. What is the total weight and volume of your trash?
2. Which categories and materials make up the largest amount?
3. How much of what you discard could be reused or recycled?
4. Are there ways that you could reduce your trash generation?

space. Since 1994, all operating landfills in the United States have been required to control such hazardous substances as oil, chemical compounds, toxic metals, and contaminated rainwater that seep through piles of waste. An impermeable clay and/or plastic lining underlies and encloses the storage area. Drainage systems are installed in and around the liner to catch drainage and to help monitor chemicals that may be leaking. Modern municipal solid-waste landfills now have many of the safeguards of hazardous waste repositories described later in this chapter.

More careful attention is now paid to the siting of new landfills. Sites located on highly permeable or faulted rock formations are passed over in favor of sites with less leaky geologic foundations. Landfills are being built away from rivers, lakes, floodplains, and aquifer recharge zones, rather than near them, as was often done in the past. More care is being given to a landfill's long-term effects so that costly cleanups and rehabilitation can be avoided.

Historically, landfills have been a convenient and relatively inexpensive waste-disposal option in most places, but this situation is changing rapidly. Rising land prices and shipping costs, as well as increasingly demanding landfill construction and maintenance requirements, are making this a more expensive disposal method. The cost of disposing of a ton of solid waste in Philadelphia went from $20 in 1980 to more than $100 in 1990. Union County, New York, experienced an even steeper price rise. In 1987, it paid $70 to get rid of a ton of waste; a year later, that same ton cost $420, or about $10 for a typical garbage bag. The United States now spends about $10 billion per year to dispose of trash. A decade from now, it may cost us $100 billion per year to dispose of our trash and garbage.

Suitable places for waste disposal are becoming scarce in many areas. Other uses compete for open space. Citizens have become more concerned and vocal about health hazards, as well as aesthetics. It is difficult to find a neighborhood or community willing to accept a new landfill. Since 1984, when stricter financial and environmental protection requirements for landfills took effect, more than 1,200 of the 1,500 existing landfills in the United States have closed. Many major cities are running out of local landfill space. They export their trash, at enormous expense, to neighboring communities and even other states. More than half the solid waste from New Jersey goes out of state, some of it up to 800 km (500 mi) away.

Exporting Waste

Although most industrialized nations in the world have agreed to stop shipping hazardous and toxic waste to less developed countries, the practice still continues. In 1999, for example, 3,000 tons of incinerator waste from a plastics factory in Taiwan were unloaded from a ship in the middle of the night and dumped in a field near the small coastal Cambodian village of Bet Trang. The village residents thought they had been blessed with a windfall. They emptied out chunks of crumbling residue so they could use the white, plastic shipping bags as bedding and roofing material. They rinsed out bags to use for rice storage, and they ripped them open with their teeth to get string to use as clotheslines and lashing for their oxcarts. Children played happily on the big pile of dusty, white material.

In the following weeks, unfortunately, the villagers discovered that, rather than a treasure, they had a calamity. The first sign of trouble was when one of the dock workers who unloaded the waste died and five others were hospitalized with symptoms of nerve damage and respiratory distress. Villagers also began to complain of a variety of illnesses. The village was evacuated, and about 1,000 residents of the nearby city of Sihanoukville fled in panic. Subsequent analysis found high levels of mercury and other toxic metals in the residue. The Formosa Plastics Corporation, which shipped the waste, admitted paying a $3 million bribe to Cambodian officials to permit its dumping. They said they couldn't dispose of the waste in Taiwan because of a threat of public protest. Following an international uproar, the plastics company agreed to

INVESTIGATING Our Environment

Environmental Justice

When a new landfill, petrochemical factory, incinerator, or other unwanted industrial facility is proposed for a minority neighborhood, charges of environmental racism often are raised by those who oppose this siting. Everyday experiences tell us that minority neighborhoods are much more likely to have high pollution levels and facilities that you wouldn't want to live near than are middle- or upper-class white neighborhoods. But does this prove that land-use decisions are racist or just that minorities are less politically powerful than middle- or upper-class residents? Could it be that land prices are simply cheaper and public resistance to locating a polluting facility in a place that's already polluted is less than putting it in a cleaner environment? Or does this distinction matter? Perhaps showing that a disproportionate number of minorities live in dirtier places is evidence enough of racism. How would you decide?

One of the first systematic studies showing this inequitable distribution of environmental hazards based on race in the United States was conducted by Robert D. Bullard in 1978. Asked for help by a predominantly black community in Houston that was slated for a waste incinerator, Bullard discovered that all five of the city's existing landfills and six of eight incinerators were located in African-American neighborhoods. In a book entitled *Dumping on Dixie*, Bullard showed that this pattern of risk exposure in minority communities is common throughout the United States. Among his findings:

- Three of the five largest commercial hazardous waste landfills, accounting for about 40 percent of all hazardous-waste disposal in the United States, are located in predominantly African-American or Hispanic communities.

- Sixty percent of African Americans and Latinos and nearly half of all Asians, Pacific Islanders, and Native Americans live in communities with uncontrolled toxic-waste sites.

- The average percentage of the population made up by minorities in communities without a hazardous waste facility is 12 percent. By contrast, communities with one hazardous waste facility have, on average, twice as high (24 percent) a minority population, while those with two or more such facilities average three times as high a minority population (38 percent) as those without one.

But does this prove that race, not class or income, is the strongest determinant of who is exposed to environmental hazards? What additional information might you look for to make this distinction? One of the lines of evidence Dr. Bullard raises is the fact that the discrepancy between the pollution exposure of middle-class blacks and middle class whites is even greater than the difference between poorer whites and blacks. While upper-class whites can "vote with their feet" and move out of polluted and dangerous neighborhoods, Bullard argues, minorities are restricted by color barriers and prejudice to less-desirable locations.

Some additional evidence uncovered by this research is variation in the way toxic waste sites are cleaned up and how polluters are punished in different neighborhoods. White communities see faster responses and get better results once toxic wastes are discovered than do minority communities. For instance, the EPA takes 20 percent longer to place a hazardous waste site in a minority community on the Superfund National Priority List than it does for one in a white community. Penalties assessed against polluters of white communities average six times higher than those against polluters of minority communities. Cleanup is more thorough in white communities as well. Most toxic wastes in white communities are treated, that is, removed or destroyed. By contrast, waste sites in minority neighborhoods are generally only "contained" by putting a cap over them, leaving contaminants in place to potentially resurface or leak into groundwater at a later date.

How would you evaluate these findings? Do they convince you that racism is at work, or do you think that other explanations might be equally likely? Which of these arguments do you find most persuasive, or what other evidence would you need to make a reasoned judgment about whether or not environmental racism is a factor in determining who gets exposed to pollution and who enjoys a cleaner, more pleasant environment?

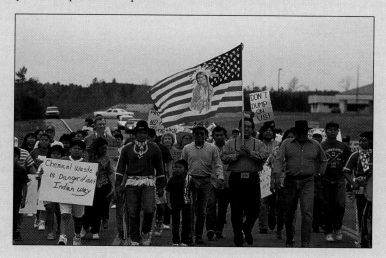

Native Americans march in protest of toxic waste dumping on tribal lands.

go back and pick up the waste. But the villagers who handled the toxic wastes face an uncertain future. Is it safe to reinhabit their homes? Is it wise to have children? Will they suffer long-term health effects from exposure to this material? (See the related story "The Voyage of the Khian Sea" at www.mhhe.com/apps.)

As we will discuss later in this chapter, "garbage imperialism" also operates within richer countries as well. Poor neighborhoods and minority populations are much more likely than richer ones to be the recipients of dumps, waste incinerators, and other locally unwanted land uses (LULUs). In recent years, attention has turned, in the United States, to Indian reservations, which are exempt from some state and federal regulations concerning waste disposal. Virtually every tribe in America has been approached with schemes to store wastes on their reservation (see Investigating Our Environment, p. 304).

Another method of disposing of toxic wastes is to "recycle" them as asphalt or concrete filler for building highways. This is considered a beneficial use, but what happens to the toxins as the roadway is slowly worn away by traffic? Similar waste products are "land farmed," or sold as fertilizer and soil amendments. There are no safety standards for fertilizer composition because it's not intended for human consumption, but much of it is used on crops that humans will eat, or will be fed to livestock that are part of our food chain. For example, Florida has about a billion cubic meters of phosphogypsum, a waste product of phosphate mining that producers want to market as a soil amendment. While it's true that phosphate is an essential plant nutrient, this particular product is also radioactive. In another case in Oregon, metal-rich dust and ash from steel mills is classified as hazardous waste when it leaves the mill. After being mixed with other minerals, however, it becomes fertilizer that will be spread on farm fields. Manufacturers are required to report the "active" ingredient—things like nitrogen, phosphorus, and phosphate—content of their product, but much can go unreported as "inert" matter.

Incineration and Resource Recovery

Landfilling is still the disposal method for the majority of municipal waste in the United States. Faced with growing piles of garbage and a lack of available landfills at any price, however, public officials are investigating other disposal methods. The method to which they frequently turn is burning. Another term commonly used for this technology is **energy recovery,** or waste-to-energy, because the heat derived from incinerated refuse is a useful resource. Burning garbage can produce steam used directly for heating buildings or generating electricity. Internationally, well over 1,000 waste-to-energy plants in Brazil, Japan, and Western Europe generate much-needed energy while also reducing the amount that needs to be landfilled. In the United States more than 110 waste incinerators burn 45,000 metric tons of garbage daily. Some of these are simple incinerators; others produce steam and/or electricity.

Types of Incinerators

Municipal incinerators are specially designed burning plants capable of burning thousands of tons of waste per day. In some plants, refuse is sorted as it comes in to remove unburnable or recyclable materials before combustion. This is called **refuse-derived fuel** because the enriched burnable fraction has a higher energy content than the raw trash. Another approach, called **mass burn,** is to dump everything smaller than sofas and refrigerators into a giant furnace and burn as much as possible (fig. 13.5). This technique avoids the expensive and unpleasant job of sorting through the garbage for nonburnable materials, but it often causes greater problems with air pollution and corrosion of burner grates and chimneys.

In either case, residual ash and unburnable residues representing 10 to 20 percent of the original volume are usually taken to a landfill for disposal. Because the volume of burned garbage is reduced by 80 to 90 percent, disposal is a smaller task. However, the residual ash usually contains a variety of toxic components that make it an environmental hazard if not disposed of properly. Ironically, one worry about incinerators is whether enough garbage will be available to feed them. Some communities in which recycling has been really successful have had to buy garbage from neighbors to meet contractual obligations to waste-to-energy facilities. In other places, fears that this might happen have discouraged recycling efforts.

FIGURE 13.5 *A municipal mass-burn garbage incinerator. Steam produced in the boiler can be used to generate electricity or to heat nearby buildings.*

Incinerator Cost and Safety

The cost-effectiveness of garbage incinerators is the subject of heated debates. Initial construction costs are high—usually between $100 million and $300 million for a typical municipal facility. Tipping fees at an incinerator (the fee charged to haulers for each ton of garbage dumped) are often much higher than those at a landfill. As landfill space near metropolitan areas becomes more scarce and more expensive, however, landfill rates are certain to rise. It may pay in the long run to incinerate refuse so that the lifetime of existing landfills will be extended.

Environmental safety of incinerators is another point of concern. The EPA has found alarmingly high levels of dioxins, furans, lead, and cadmium in incinerator ash. These toxic materials were more concentrated in the fly ash (lighter, airborne particles capable of penetrating deep into the lungs) than in heavy bottom ash. Dioxin levels can be as high as 780 ppb (parts per billion). One part per billion of TCDD, the most toxic dioxin, is considered a health concern. All of the incinerators studied exceeded cadmium standards, and 80 percent exceeded lead standards. Proponents of incineration argue that, if they are run properly and equipped with appropriate pollution-control devices, incinerators are safe to the general public. Opponents counter that neither public officials nor pollution-control equipment can be trusted to keep the air clean. They argue that recycling and source reduction efforts are better ways to deal with waste problems.

The EPA, which generally supports incineration, acknowledges the health threat of incinerator emissions but holds that the danger is very slight. The EPA estimates that dioxin emissions from a typical municipal incinerator may cause one death per million people in 70 years of operation. Critics of incineration claim that a more accurate estimate is 250 deaths per million in 70 years.

One way to reduce these dangerous emissions is to remove batteries containing heavy metals and plastics containing chlorine before wastes are burned. Bremen, West Germany, is one of several European cities now trying to control dioxin emissions by keeping all plastics out of incinerator waste. Bremen is requiring households to separate plastics from other garbage. This is expected to eliminate nearly all dioxins and other combustion by-products and prevent the expense of installing costly pollution-control equipment that otherwise would be necessary to keep the burners operating. Minneapolis has initiated a recycling program for the small "button" batteries used in hearing aids, watches, and calculators in an attempt to lower mercury emissions from its incinerator.

SHRINKING THE WASTE STREAM

Having less waste to discard is obviously better than struggling with disposal methods, all of which have disadvantages and drawbacks. In this section we will explore some of our options for recycling, reuse, and reduction of the wastes we produce.

Recycling

The term *recycling* has two meanings in common usage. Sometimes, we say we are recycling when we really are reusing some-

FIGURE 13.6   *Trucks with multiple compartments pick up residential recyclables at curbside, greatly reducing the amount of waste that needs to be buried or burned. For many materials, however, collection costs are too high and markets are lacking for recycling to be profitable.*

thing, such as refillable beverage containers. In terms of solid waste management, however, **recycling** is the reprocessing of discarded materials into new, useful products (fig. 13.6). Some recycling processes reuse materials for the same purposes; for instance, old aluminum cans and glass bottles are usually melted and recast into new cans and bottles. Other recycling processes turn old materials into entirely new products. Old tires, for instance, are shredded and turned into rubberized road surfacing. Newspapers become cellulose insulation, kitchen wastes become a valuable soil amendment, and steel cans become new automobiles and construction materials.

There have been some dramatic successes in recycling in recent years. New Jersey, for instance, now claims a 60-percent recycling rate, something thought unattainable a decade ago. The high value of aluminum scrap ($700 per ton in 1999) has spurred a large percentage of aluminum recycling nearly everywhere. About two-thirds of all aluminum cans are now recycled; up from only 15 percent 20 years ago. This recycling is so rapid that half of all the aluminum cans now on grocery shelves will be made into another can within two months.

Recycling Challenges

Despite encouraging gains in recycling rates, major challenges exist. Despite the value of aluminum, Americans still throw away nearly 350,000 metric tons of aluminum beverage containers each year. That is enough to make 3,800 Boeing 747 airplanes. Plastics recyclers have developed innovative methods and products, but the low price of virgin plastic, made from oil, is usually less than the cost of transporting and storing used plastics. Consequently, less than 5 percent of the United States' 24 million tons of plastic waste is recycled each year.

Wild fluctuations in commodity prices make it still harder to develop a market for recycled materials. Newsprint, for example, cost $160 a ton in 1995; by 1999 it dropped to just $42 per ton.

Principles of Environmental Science

Contamination is a major obstacle in plastic recycling. Most of the 24 billion plastic soft drink bottles sold every year in the United States are made of PET (polyethylene terphthalate), which can be remanufactured into carpet, fleece clothing, plastic strapping, and nonfood packaging. However, even a trace of vinyl—a single PVC (polyvinyl chloride) bottle in a truckload, for example—can make PET useless. Although most bottles are now marked with a recycling number, it's hard for consumers to remember which is which. Because single-use beverage containers are so costly to recycle, they have been outlawed in Denmark and Finland. A looming worry is the prospect of single-use, plastic beer bottles. Already being test-marketed, these bottles are made of PET but are amber colored to block sunlight and have a special chemical coating to keep out oxygen, which would ruin the beer. The special color, interior coating, and vinyl cap lining will make these bottles incompatible with regular PET, and it will probably cost more to remove them from the waste stream than the reclaimed plastic is worth. Plastic recycling already is down 50 percent from a decade ago because so many soft drink bottles are sold and consumed on the go, and never make it into recycling bins. Throw-away beer bottles could badly set back recycling programs.

Benefits of Recycling

Recycling is usually a better alternative to either dumping or burning wastes. It saves money, energy, raw materials, and land space, while also reducing pollution. Recycling also encourages individual awareness and responsibility for the refuse produced (fig. 13.7). Curbside pickup of recyclables costs around $35 per ton, as opposed to the $80 paid to dispose of them at an average metropolitan landfill. Many recycling programs cover their own expenses with materials sales and may even bring revenue to the community.

Recycling drastically reduces pressure on landfills and incinerators. Philadelphia is investing in neighborhood collection centers that will recycle 600 tons a day, enough to eliminate the need for a previously planned, high-priced incinerator. New York City, down to one available landfill but still producing 27,000 tons of garbage a day, set a target of 50-percent waste reduction to be accomplished by recycling office paper and household and commercial waste. However, New York mayor Michael Bloomberg has discontinued most recycling, arguing that the program is too expensive. This move enraged recycling advocates, who insist recycling can pay for itself simply in avoided disposal costs.

Japan probably has the most successful recycling program in the world. Half of all household and commercial wastes in Japan are recycled, while the rest are about equally incinerated or landfilled. By comparison, the United States landfills more than 60 percent of all solid waste. Japan has also instituted a new law requiring that restaurants and food producers and distributors recycle at least 20 percent of food waste by 2006.

Recycling lowers our demands for raw resources (fig. 13.8). In the United States we cut down 2 million trees every day to produce newsprint and paper products, a heavy drain on our forests. Recycling the print run of a single Sunday issue of the *New York Times* would spare 75,000 trees. Every piece of plastic we make reduces the reserves supply of petroleum and makes us more dependent on foreign oil. Recycling 1 ton of aluminum saves 4 tons of bauxite (aluminum ore) and 700 kg (1,540 lbs) of petroleum coke and pitch, as well as keeping 35 kg (77 lbs) of aluminum fluoride out of the air.

Recycling also reduces energy consumption and air pollution. Plastic bottle recycling could save 50 to 60 percent of the energy needed to make new ones. Making new steel from old scrap offers up to 75 percent energy savings. Producing aluminum from scrap instead of bauxite ore cuts energy use by 95 percent, yet we still throw away more than a million tons of aluminum every year. If aluminum recovery were doubled worldwide, more than a million tons of air pollutants would be eliminated every year.

Reducing litter is an important benefit of recycling. Ever since disposable paper, glass, metal, foam, and plastic packaging

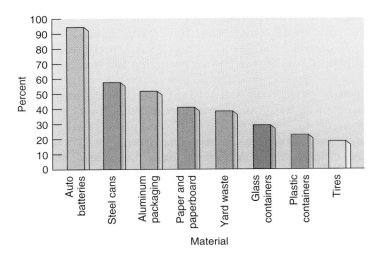

FIGURE 13.7 *Source separation in the kitchen—the first step in a strong recycling program.*

FIGURE 13.8 *Recycling rates of selected materials in the United States.*
Source: Data from Environmental Protection Agency, 2000.

began to accompany nearly everything we buy, these discarded wrappings have collected on our roadsides and in our lakes, rivers, and oceans. Litter is a costly as well as unsightly problem. We pay an estimated 32¢ for each piece of litter picked up by crews along state highways, which adds up to $500 million every year. "Bottle bills" requiring deposits on bottles and cans have reduced littering in many states.

Creating Incentives for Recycling

In many communities citizens have done such a good job of collecting recyclables that a glut has developed. In some cities mountains of some waste materials accumulate in warehouses because there are no markets for them. Too often wastes that we carefully separate for recycling end up being mixed together and dumped in a landfill or incinerator.

Our present public policies often tend to favor extraction of new raw materials. Energy, water, and raw materials are often sold to industries below their real cost to create jobs and stimulate the economy. For instance, in 1999, a pound of recycled clear PET, the material in most soft-drink bottles, sold for about 40¢. By contrast, a pound of off-grade, virgin PET cost 25¢. Setting the prices of natural resources at their real cost would tend to encourage efficiency and recycling. State, local, and national statutes requiring government agencies to purchase a minimum amount of recycled material have helped create a market for used materials. Each of us can play a role in creating markets as well. If we buy things made from recycled materials—or ask for them if they aren't available—we will help recycling programs to succeed.

Composting

Pressed for landfill space, many cities have banned yard waste from municipal garbage. Rather than bury this valuable organic material, they are turning it into a useful product through **composting**: biological degradation or breakdown of organic matter under aerobic (oxygen-rich) conditions. The organic compost resulting from this process makes a nutrient-rich soil amendment that aids water retention, slows soil erosion, and improves crop yields. A home compost pile is an easy and inexpensive way to dispose of organic waste in an interesting and environmentally friendly way. All you need to do is to pile up lawn clippings, vegetable waste, fallen leaves, wood chips, or other organic matter in an out-of-the way place, keep it moist, and turn it over every week or so (fig. 13.9). Within a few months, naturally-occurring microorganisms will decompose the organic material into a rich, pleasant-smelling compost that you can use as a soil amendment.

Some cities and counties have developed successful composting programs. Residents pay a small amount to drop off yard waste, and they can buy back rich, inexpensive fertilizer.

Energy from Waste

Every year, we throw away the energy equivalent of 80 million barrels of oil in organic waste in the United States. In developing countries up to 85 percent of the waste stream is food, textiles,

FIGURE 13.9 *Composting is a good way to convert yard waste, vegetable scraps, and other organic materials into useful garden mulch. Mix everything together, keep it moist and well aerated, and in a few weeks, you will have a rich, odor-free mulch. This three-bin system allows you to have batches in different stages.*

vegetable matter, and other biodegradable materials. Worldwide, at least one-fifth of municipal waste is organic kitchen and garden refuse. In a landfill, much of this matter is decomposed by microorganisms generating billions of cubic meters of methane ("natural gas"), which contributes to global warming if allowed to escape into the atmosphere (see chapter 9). Many cities are drilling methane wells in their landfills to capture this valuable resource. Fuel cells are a good way to use this methane (see chapter 12).

This valuable organic material can be burned in an incinerator rather than being buried in landfills, but there are worries about air pollution from incineration. Organic wastes also can be decomposed in large, oxygen-free digesters to produce methane under more controlled conditions than in a landfill and with less air pollution than mass garbage burning.

Anaerobic digestion also can be done on a small scale. Millions of household methane generators provide fuel for cooking and lighting for homes in China and India (see chapter 12). In the United States some farmers produce all the fuel they need to run their farms—both for heating and to run trucks and tractors—by generating methane from animal manure.

Demanufacturing

Demanufacturing is the disassembly and recycling of obsolete consumer products, such as television sets, computers, refrigerators, washing machines, and air conditioners. There are about 300 million televisions and personal computers in use in the United States. Televisions often are discarded after only about five years, and computers, play-stations, and other electronics become obsolete even faster. Stoves, refrigerators, and other "white goods" have a much longer lifetime—typically about 12 years—but the EPA estimates that Americans dispose of 54 million of these household appliances every year. Many of these consumer products contain both valuable materials and toxins that must be kept

Principles of Environmental Science

out of the environment. Older refrigerators and air conditioners, for example, have chloroflurocarbons (CFCs) that destroy stratospheric ozone and cannot be released into the air. Because new production has been banned, recycled CFCs are worth about $100 per pound. For both reasons, it pays to recycle them.

Computers and other electronic equipment contain both toxic metals (mercury, lead, gallium, germanium, nickel, palladium, beryllium, selenium, arsenic) and valuable ores, such as gold, silver, copper, and steel. The United States now discards 50 million computers per year, or one per person in the course of three and a half years. It's estimated that 90 percent of the cadmium, lead, and mercury contamination in our solid waste stream comes from consumer electronics.

Unfortunately, the cheapest way to demanufacture used electronics is often to ship them to developing countries. The majority of electronic products collected in the United States goes to China, where poor villages, including young children, break them apart to retrieve copper, silver, gold, lead, and other metals. Effects on workers handling toxic materials may be severe, especially for growing children. Soil, groundwater, and surface water have been found with up to 200 times the World Health Organization's standards of electronics-associated contaminants. An estimated 100,000 workers demanufacture electronics in southeastern China alone (fig. 13.10).

In response to such problems, the European Union now requires that recycling fees be built into the purchase price of many electronics. This adds about $20 to the purchase price, less than one percent for many computers. Increasingly, local governments are requiring electronics recycling. Massachusetts is the first state to require that both households and businesses recycle used computer parts. Recycling companies in the United States are beginning to develop comprehensive electronics recycling systems, and these offer growing alternatives to shipping our hazardous waste overseas.

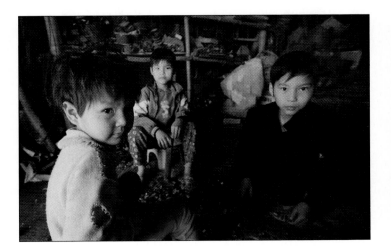

FIGURE 13.10 *Asian villagers, many of them children, are exposed to toxic elements as they process at least half of U.S. recycled computers and electronics.*

Reuse

Even better than recycling or composting is cleaning and reusing materials in their present form, thus saving the cost and energy of remaking them into something else. We do this already with some specialized items. Auto parts are regularly sold from junkyards, especially for older car models. In some areas stained-glass windows, brass fittings, fine woodwork, and bricks salvaged from old houses bring high prices. Some communities sort and reuse a variety of materials received in their dumps (fig. 13.11).

In many cities glass and plastic bottles are routinely returned to beverage producers for washing and refilling. The reusable, refillable bottle is the most efficient beverage container we have. It is better for the environment than remelting and more profitable for local communities. A reusable glass container

FIGURE 13.11 *Reusing discarded products is a creative and efficient way to reduce wastes. This recycling center in Berkeley, California, is a valuable source of used building supplies and a money saver for the whole community.*

makes an average of 15 round-trips between factory and customer before it becomes so scratched and chipped that it has to be recycled. Reusable containers also favor local bottling companies and help preserve regional differences.

Since the advent of cheap, lightweight, disposable food and beverage containers, many small, local breweries, canneries, and bottling companies have been forced out of business by huge national conglomerates. These big companies can afford to ship food and beverages great distances, as long as it is a one-way trip. If they had to collect their containers and reuse them, canning and bottling factories serving large regions would be uneconomical. Consequently, the national companies favor recycling rather than refilling because they prefer fewer, larger plants and don't want to be responsible for collecting and reusing containers. In some circumstances, life-cycle assessment shows that washing and decontaminating containers takes as much energy and produces as much air and water pollution as manufacturing new ones.

In less-affluent nations reuse of all sorts of manufactured goods is an established tradition. Where most manufactured products are expensive and labor is cheap, it pays to salvage, clean, and repair products. Cairo, Manila, Mexico City, and many other cities have large populations of poor people who make a living by scavenging. Entire ethnic populations may survive on scavenging, sorting, and reprocessing scraps from city dumps.

Producing Less Waste

What is even better than reusing materials? Generating less waste in the first place. The "What Can You Do?" box on this page describes some contributions you can make to reduce the volume of our waste stream. Industry also can play an important role in source reduction. In Minnesota, the 3M Company has saved over $500 million since 1975 by changing manufacturing processes, finding uses for waste products, and listening to employees' suggestions. What is waste to one division is a treasure to another.

Excess packaging of food and consumer products is one of our greatest sources of unnecessary waste. Paper, plastic, glass, and metal packaging material make up 50 percent of our domestic trash by volume. Much of that packaging is primarily for marketing and has little to do with product protection (fig. 13.12). Manufacturers and retailers might be persuaded to reduce these wasteful practices if consumers ask for products without excess packaging. Canada's National Packaging Protocol (NPP) recommends that packaging minimize depletion of virgin resources and production of toxins in manufacturing. The preferred hierarchy is (1) no packaging, (2) minimal packaging, (3) reusable packaging, and (4) recyclable packaging. This plan sets an ambitious target of 50 percent reduction in excess packaging.

Where disposable packaging is necessary, we still can reduce the volume of waste in our landfills by using materials that are compostable or degradable. **Photodegradable plastics** break down when exposed to ultraviolet radiation. **Biodegradable plastics** incorporate such materials as cornstarch that microorganisms can decompose. Several states have introduced legislation requiring biodegradable or photodegradable six-pack beverage yokes,

what can you do?

Reducing Waste

1. Buy foods that come with less packaging; shop at farmers' markets or co-ops, using your own containers.
2. Take your own washable, refillable beverage container to meetings or convenience stores.
3. When you have a choice at the grocery store between plastic, glass, or metal containers for the same food, buy the reusable or easier-to-recycle glass or metal.
4. Separate your cans, bottles, papers, and plastics for recycling.
5. Wash and reuse bottles, aluminum foil, plastic bags, etc., for your personal use.
6. Compost yard and garden wastes, leaves, and grass clippings.
7. Help your school develop responsible systems for disposing of electronics and other waste.
8. Write to your senators and representatives, and urge them to vote for container deposits, recycling, and safe incinerators or landfills.

Source: Minnesota Pollution Control Agency

FIGURE 13.12 *How much more do we need? Where will we put what we already have?*
Reprinted with special permission of King Features Syndicate.

fast-food packaging, and disposable diapers. These degradable plastics often don't decompose completely; they only break down to small particles that remain in the environment. In doing so, they can release toxic chemicals. And in modern, lined landfills, they don't decompose at all. Furthermore, they make recycling less feasible and may lead people to believe that littering is okay.

Some environmental groups are beginning to think that we have put too much emphasis on recycling. Many people think that, if they recycle aluminum cans and newspapers, they are doing everything they can for the environment. While recycling is an important part of waste management, we have to remember that it is actually the third "R" in the waste hierarchy. The two preferred methods—reduction and reuse—get lost in our enthusiasm for recycling.

HAZARDOUS AND TOXIC WASTES

The most dangerous aspect of the waste stream we have described is that it often contains highly toxic and hazardous materials that are injurious to both human health and environmental quality (fig. 13.13). We now produce and use a vast array of flammable, explosive, caustic, acidic, and highly toxic chemical substances for industrial, agricultural, and domestic purposes. According to the EPA, U.S. industries generate about 265 million metric tons of officially classified hazardous wastes each year, slightly more than 1 ton for each person in the country. In addition, considerably more toxic and hazardous waste material is generated by industries or processes not regulated by the EPA. Shockingly, at least 40 million metric tons (22 billion lbs) of toxic and hazardous wastes are released into the air, water, and land in the United States each year. The biggest sources of these toxins are the chemical and petroleum industries (fig. 13.14).

What Is Hazardous Waste?

Legally, a **hazardous waste** is any discarded material, liquid or solid, that contains substances known to be (1) fatal to humans or laboratory animals in low doses; (2) toxic, carcinogenic, mutagenic, or teratogenic to humans or other life-forms; (3) ignitable with a flash point less than 60°C; (4) corrosive; or (5) explosive or highly reactive (undergoes violent chemical reactions either by itself or when mixed with other materials). Notice that this definition includes both toxic and hazardous materials as defined in chapter 8. Certain compounds are exempt from regulation as hazardous waste if they are accumulated in less than 1 kg (2.2 lbs) of commercial chemicals or 100 kg of contaminated soil, water, or debris. Even larger amounts (up to 1,000 kg) are exempt when stored at an approved waste treatment facility for the purpose of being beneficially used, recycled, reclaimed, detoxified, or destroyed.

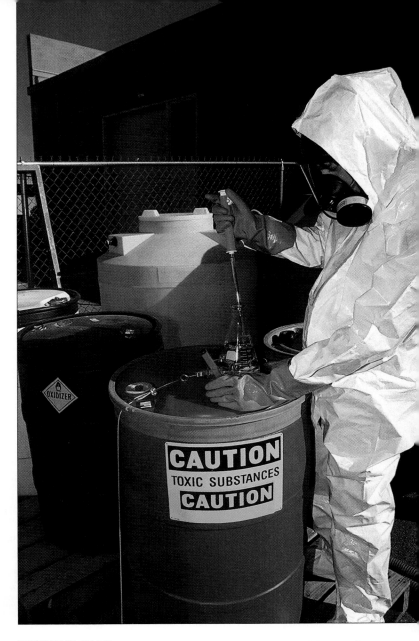

FIGURE 13.13 *Careful handling, sampling, and record keeping for all toxic and hazardous wastes are now required by federal and state laws.*

APPLICATION:	A Personal Hazardous Waste Inventory

Make a survey of your house or apartment to see how many toxic materials you can identify. If you live in a dorm, you may need to survey your parents' house. Read the cautionary labels. Which of the products are considered hazardous? When you use them, do you usually follow all the safety procedures recommended? Where is the nearest hazardous waste collection site for disposal of unwanted household products? If you don't know, call your city administrator or mayor's office to find out.

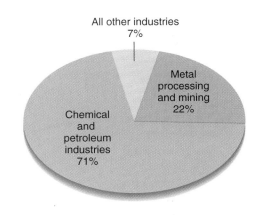

FIGURE 13.14 *Hazardous waste producers in the United States.*

Hazardous-Waste Disposal

Most hazardous waste is recycled, converted to nonhazardous forms, stored, or otherwise disposed of on-site by the generators—chemical companies, petroleum refiners, and other large industrial facilities—so that it doesn't become a public problem. Still, the hazardous waste that does enter the waste stream or the environment represents a serious environmental problem. And orphan wastes left behind by abandoned industries remain a serious threat to both environmental quality and human health. For years little attention was paid to this material. Wastes stored on private property, buried, or allowed to soak into the ground were considered of little concern to the public. An estimated 5 billion metric tons of highly poisonous chemicals were improperly disposed of in the United States between 1950 and 1975 before regulatory controls became more stringent.

Federal Legislation

Two important federal laws regulate hazardous waste management and disposal in the United States. The Resource Conservation and Recovery Act (RCRA, pronounced "rickra") of 1976 is a comprehensive program that requires rigorous testing and management of toxic and hazardous substances. A complex set of rules require generators, shippers, users, and disposers of these materials to keep meticulous account of everything they handle and what happens to it from generation (cradle) to ultimate disposal (grave) (fig. 13.15).

The Comprehensive Environmental Response, Compensation, and Liability Act (CERCLA or Superfund Act), passed in 1980 and modified in 1984 by the Superfund Amendments and Reauthorization Act (SARA), is aimed at rapid containment, cleanup, or remediation of abandoned toxic waste sites. This statute authorizes the EPA to undertake emergency actions when a threat exists that toxic material will leak into the environment. The EPA is empowered to bring suit for the recovery of its costs from potentially responsible parties, such as site owners, operators, waste generators, or transporters.

SARA also established (under title III) community right to know and state emergency response plans that give citizens access to information about what is present in their communities. One of the most useful tools in this respect is the **Toxic Release Inventory,** which requires 20,000 manufacturing facilities to report annually on releases of more than 300 toxic materials. You can find specific information in the inventory about what is in your neighborhood.

The government does not have to prove that anyone violated a law or what role they played in a Superfund site. Rather, liability

Treatment facility

Tracking of hazardous waste through manifest system

Secure landfill

Storage

Transporter

Hazardous waste generator

EPA or state agency office

FIGURE 13.15 *Toxic and hazardous wastes must be tracked from "cradle to grave" by detailed shipping manifests.*

under CERCLA is "strict, joint, and several," meaning that anyone associated with a site can be held responsible for the entire cost of cleaning it up, no matter how much of the mess they made. In some cases property owners have been assessed millions of dollars for removal of wastes left there years earlier by previous owners. This strict liability has been a headache for the real estate and insurance businesses.

CERCLA was amended in 1995 to make some of its provisions less onerous. In cases where treatment is unavailable or too costly and it is likely that a less-costly remedy will become available within a reasonable time, interim containment is now allowed. The EPA also now has the discretion to set site-specific cleanup levels, rather than adhere to rigid national standards.

Superfund Sites

The EPA estimates that there are at least 36,000 seriously contaminated sites in the United States (fig. 13.16). The General Accounting Office (GAO) places the number much higher—perhaps more than 400,000 when all are identified. By 2000 some 1,551 sites had been placed on the National Priority List (NPL) for cleanup with financing from the federal Superfund program. The **Superfund** is a revolving pool designed to (1) provide an immediate response to emergency situations that pose imminent hazards, and (2) clean up or remediate abandoned or inactive sites. Originally a $1.6 billion pool, the Superfund has since been enlarged tenfold. It is financed by fees assessed against hazardous or toxic waste producers and costs charged for remedial actions. By using the Superfund to pay for immediate actions and then collecting later from responsible parties, we can take care of problems without waiting for litigation to run its course.

Total costs for hazardous waste cleanup in the United States are estimated to be between $370 billion and $1.7 trillion, depending on the remedial actions chosen and the degree to which sites are required to be cleaned. For years, disputes over responsibility for cleanup slowed remediation efforts. In the 1990s the federal administration committed itself to progress, and the number of completed NPL sites jumped from 155 to 809, cleaned or mostly clean (fig. 13.17). Between 230 and 500 sites may be added to the list in the coming decade. Continued funding for the Superfund is an issue of hot debate. In 1995 the industry tax supporting the Superfund expired, and Congress has not renewed it. The Bush administration has argued that the public, not industry, should pay for cleanup. Citizen advocates insist companies should clean up their own messes.

What qualifies a site for the NPL? These sites are considered to be especially hazardous to human health and environmental quality because they are known to be leaking or have a potential for leaking supertoxic, carcinogenic, teratogenic, or mutagenic materials (see chapter 8). The ten substances of greatest concern or most commonly detected at Superfund sites are lead, trichloroethylene, toluene, benzene, PCBs, chloroform, phenol, arsenic, cadmium, and chromium. These and other hazardous or toxic materials are known to have contaminated groundwater at 75 percent of the sites now on the NPL. In addition, 56 percent of these sites have contaminated surface waters, and airborne materials are found at 20 percent of the sites.

Where are these thousands of hazardous waste sites, and how did they get contaminated? Old industrial facilities, such as smelters, mills, petroleum refineries, and chemical manufacturing plants, are highly likely to have been sources of toxic wastes. Regions of the country with high concentrations of aging factories, such as the "rust belt" around the Great Lakes or the Gulf Coast petrochemical centers, have large numbers of Superfund sites. Mining districts also are prime sources of toxic and hazardous waste. Within cities, factories and places such as railroad yards, bus repair barns, and filling stations, where solvents, gasoline, oil, and other petrochemicals were spilled or dumped on the ground, often are highly contaminated.

Some of the most infamous toxic waste sites were old dumps where many different materials were mixed together

FIGURE 13.16 *About 1,551 seriously contaminated industrial sites have been placed on the National Priority List to be cleaned up with Superfund money.*

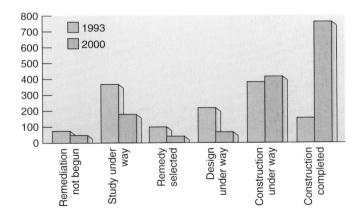

FIGURE 13.17 *Progress on Superfund National Priority List (NPL) sites. After years of little progress, the number of completed sites jumped from 155 in 1993 to 809 in 2002. Over 90 percent of the 1,500 NPL sites are under construction or completed.*
Source: Data from Environmental Protection Agency, 2001.

indiscriminately. For instance, Love Canal in Niagara Falls, New York, was an open dump that both the city and nearby chemical factories used as a disposal site. More than 20,000 tons of toxic chemical waste were buried under what later became a housing development. Another infamous example occurred in Hardeman County, Tennessee, where about a quarter of a million barrels of chemical waste were buried in shallow pits that subsequently leaked toxins into the groundwater.

How Clean Is Clean?

Among the biggest problems in cleaning up hazardous waste sites are questions of liability and the degree of purity required. In many cities, these problems have created large areas of contaminated properties known as **brownfields** that have been abandoned or are not being used to their potential because of real or suspected pollution. Up to one-third of all commercial and industrial sites in the urban core of many big cities fall in this category. In heavy industrial corridors the percentage typically is higher.

For years no one was interested in redeveloping brownfields because of liability risks. Who would buy a property, knowing that they might be forced to spend years in litigation and negotiations and be forced to pay millions of dollars for pollution they didn't create? Even if a site has been cleaned to current standards, there is a worry that additional pollution might be found in the future or that more stringent standards might be applied.

In many cases, property owners complain that unreasonably high levels of purity are demanded in remediation programs. Consider the case of Columbia, Mississippi. For many years a 35-ha (81-acre) site in Columbia was used for turpentine and pine tar manufacturing. Soil tests showed concentrations of phenols and other toxic organic compounds exceeding federal safety standards. The site was added to the Superfund NPL, and remediation was ordered. Some experts recommended that the best solution was to simply cover the surface with clean soil and enclose the property with a fence to keep people out. The total costs would have been about $1 million. Instead, the EPA ordered Reichhold Chemical, the last known property owner, to excavate more than 12,500 tons of soil and haul it to a commercial hazardous-waste dump in Louisiana at a cost of some $4 million. The intention is to make the site safe enough to be used for any purpose, including housing—even though no one has proposed building anything there. According to the EPA, the dirt must be clean enough for children to play in—even eat every day for 70 years— without risk.

Similarly, in places where contaminants have seeped into groundwater, the EPA generally demands that cleanup be carried to drinking water standards. Many critics believe that these pristine standards are unreasonable. Former Congressman Jim Florio, a principal author of the original Superfund Act says, "It doesn't make any sense to clean up a rail yard in downtown Newark so it can be used as a drinking water reservoir." Depending on where the site is, what else is around it, and what its intended uses are, much less stringent standards may be perfectly acceptable.

Brownfield redevelopment is increasingly seen as opportunity for rebuilding cities, creating jobs, increasing the tax base, and preventing needless destruction of open space at urban mar-

gins. In 2002 the EPA established a new brownfields revitalization fund designed to encourage restoration of more sites, and more kinds of sites. In some communities former brownfields are being turned into "eco-industrial parks" that feature environmentally friendly businesses and bring in much-needed jobs to inner-city neighborhoods (see chapter 14).

Options for Hazardous-Waste Management

What shall we do with toxic and hazardous wastes? In our homes, we can reduce waste generation and choose less-toxic materials. Buy only what you need for the job at hand. Use up the last little bit, or share leftovers with a friend or neighbor. Many common materials that you probably already have make excellent alternatives to commercial products.

Produce Less Waste

As with other wastes, the safest and least expensive way to avoid hazardous waste problems is to avoid creating the wastes in the first place. Manufacturing processes can be modified to reduce or eliminate waste production. In Minnesota, the 3M Company reformulated products and redesigned manufacturing processes to eliminate more than 140,000 metric tons of solid and hazardous wastes, 4 billion liters (1 billion gal) of wastewater, and 80,000 metric tons of air pollution each year. They frequently found that these new processes not only spared the environment but also saved money by using less energy and fewer raw materials.

Recycling and reusing materials also eliminates hazardous wastes and pollution. Many waste products of one process or industry are valuable commodities in another. Already, about 10 percent of the wastes that would otherwise enter the waste stream in the United States are sent to surplus material exchanges, where they are sold as raw materials for use by other industries. This figure could probably be raised substantially with better waste management. In Europe at least one-third of all industrial wastes are exchanged through clearinghouses, where beneficial uses are found. This represents a double savings: the generator doesn't have to pay for disposal, and the recipient pays little, if anything, for raw materials.

Convert to Less-Hazardous Substances

Several processes are available to make hazardous materials less toxic. *Physical treatments* tie up or isolate substances. Charcoal or resin filters absorb toxins. Distillation separates hazardous components from aqueous solutions. Precipitation and immobilization in ceramics, glass, or cement isolate toxins from the environment so that they become essentially nonhazardous. One of the few ways to dispose of metals and radioactive substances is to fuse them in silica at high temperatures to make a stable, impermeable glass that is suitable for long- term storage. Plants, bacteria, and fungi can also concentrate or detoxify contaminants (see Investigating Our Environment, p. 315).

Incineration is applicable to mixtures of wastes. A permanent solution to many problems, it is quick and relatively easy, but not necessarily cheap—nor always clean—unless done correctly.

INVESTIGATING Our Environment

Bioremediation

Cleaning up the thousands of hazardous waste sites at factories, farms, and gas stations is an expensive project. In the United States alone, waste cleanup is projected to cost at least $700 billion. Usually hazardous waste remediation (cleanup) involves digging up soil and incinerating it, potentially releasing toxins into the air, or trucking it to a secure landfill. Contaminated groundwater is frequently pumped out of the ground; hopefully, contaminants are retrieved at the same time.

A promising alternative to these methods involves **bioremediation,** or biological waste treatment. Microscopic bacteria and fungi can absorb, accumulate, and detoxify a remarkable variety of toxic compounds. They can also accumulate heavy metals, and some have been developed that can metabolize (break down) PCBs. Aquatic plants such as water hyacinths and cattails can also be used to purify contaminated effluent.

Recently, an increasing variety of plants have been used in phytoremediation (cleanup using plants). Some types of mustard can extract lead, arsenic, zinc, and other metals from contaminated soil. Radioactive strontium and cesium have been extracted from soil near the Chernobyl nuclear power plant using common sunflowers. Poplar trees can absorb and break down toxic organic chemicals. Natural bacteria in groundwater, when provided with plenty of oxygen, can neutralize contaminants in aquifers. Experiments have shown that pumping air *into* groundwater can be a more effective cleanup method than pumping water *out.*

How do plants, bacteria, and fungi do all this? Many of the biophysical details are poorly understood, but in general, plant roots are designed to efficiently extract nutrients, water, and trace minerals from soil and groundwater. The mechanisms involved may aid extraction of metallic and organic contaminants. Some plants also use toxic elements as a defense against herbivores: locoweed, for example, selectively absorbs elements such as selenium, concentrating toxic levels in its leaves. Absorption can be extremely effective. Bracken fern growing in Florida has been found to contain arsenic at concentrations more than 200 times higher than in the soil in which it was growing.

Genetically modified plants are also being developed to process toxins. Poplars have been developed to process toxins, using a gene borrowed from bacteria that transforms a toxic compound of mercury into a safer form. In another experiment, a gene for producing mammalian liver enzymes, which specialize in breaking down toxic organic compounds, was inserted into tobacco plants. The plants succeeded in producing the liver enzymes and breaking down toxins absorbed through their roots.

These remediation methods are not without risks. Insects could consume leaves containing concentrated substances, allowing contaminants to enter the food chain. Some absorbed contaminants are volatilized, or emitted in gaseous form, through pores in plant leaves. Once contaminants are absorbed into plants, the plants themselves are usually toxic and must be landfilled. But the cost of phytoremediation can be less than half the cost of landfilling or treating toxic soil, and the volume of plant material requiring secure storage is a fraction of the volume of the contaminated dirt.

Cleaning up hazardous and toxic waste sites will be a big business for the foreseeable future, in North America and around the world. Innovations such as bioremediation offer promising prospects for business development, as well as for environmental health and saving taxpayer money.

Fast-growing poplar trees have been developed that can absorb and break down toxic compounds.

Wastes must be heated to over 1,000°C (2,000°F) for a sufficient period of time to complete destruction. The ash resulting from thorough incineration is reduced in volume up to 90 percent and often is safer to store in a landfill or other disposal site than the original wastes. Nevertheless, incineration remains highly controversial (fig. 13.18).

Chemical processing can transform materials so they become nontoxic. Included in this category are neutralization, removal of metals or halogens (chlorine, bromine, etc.), and oxidation. The Sunohio Corporation of Canton, Ohio, for instance, has developed a process called PCBx in which chlorine in such molecules as PCBs is replaced with other ions that render the compounds less toxic. A portable unit can be moved to the location of the hazardous wastes, eliminating the need for shipping them.

Store Permanently

Inevitably, there will be some materials that we can't destroy, make into something else, or otherwise cause to vanish. We will have to store them out of harm's way. There are differing opinions about how best to do this.

Retrievable Storage. Dumping wastes in the ocean or burying them in the ground generally means that we have lost control of them. If we learn later that our disposal technique was a mistake, it is difficult, if not impossible, to go back and recover the wastes. For many supertoxic materials, the best way to store them may be in **permanent retrievable storage.** This means placing waste storage containers in a secure building, salt mine, or bedrock cavern, where they can be inspected periodically and retrieved, if necessary, for

FIGURE 13.18 *Actor Martin Sheen joins local activists in a protest in East Liverpool, Ohio, site of the largest hazardous waste incinerator in the United States. About 1,000 people marched to the plant to pray, sing, and express their opposition. Involving celebrities draws attention to your cause. A peaceful, well-planned rally builds support and acceptance in the broader community.*

FIGURE 13.19 *A secure landfill for toxic waste. A thick plastic liner and two or more layers of impervious compacted clay enclose the landfill. A gravel bed between the clay layers collects any leachate, which can then be pumped out and treated. Well samples are tested for escaping contaminants.*

repacking or for transfer if a better means of disposal is developed. This technique is more expensive than burial in a landfill because the storage area must be guarded and monitored continuously to prevent leakage, vandalism, or other dispersal of toxic materials. Remedial measures are much cheaper with this technique, however, and it may be the best system in the long run.

Secure Landfills. One of the most popular solutions for hazardous-waste disposal has been landfilling. Although, as we saw earlier in this chapter, many such landfills have been environmental disasters, newer techniques make it possible to create safe, modern **secure landfills** that are acceptable for disposing of many hazardous wastes. The first line of defense in a secure landfill is a thick bottom cushion of compacted clay that surrounds the pit like a bathtub (fig. 13.19). Moist clay is flexible and resists cracking if the ground shifts. It is impermeable to groundwater and will safely contain wastes. A layer of gravel is spread over the clay liner, and perforated drainpipes are laid in a grid to collect any seepage that escapes from the stored material. A thick polyethylene liner, protected from punctures by soft padding materials, covers the gravel bed. A layer of soil or absorbent sand cushions the inner liner, and the wastes are packed in drums, which then are placed into the pit, separated into small units by thick berms of soil or packing material.

When the landfill has reached its maximum capacity, a cover much like the bottom sandwich of clay, plastic, and soil—in that order—caps the site. Vegetation stabilizes the surface and improves its appearance. Sump pumps collect any liquids that filter through the landfill, either from rainwater or leaking drums. This leachate is treated and purified before being released. Monitoring wells check groundwater around the site to ensure that no toxins have escaped.

Most landfills are buried below ground level to be less conspicuous; however, in areas where the groundwater table is close to the surface, it is safer to build aboveground storage. The same protective construction techniques are used as in a buried pit. An advantage to such a facility is that leakage is easier to monitor because the bottom is at ground level.

Transportation of hazardous wastes to disposal sites is of concern because of the risk of accidents. Emergency-preparedness officials conclude that the greatest risk in most urban areas is not nuclear war or natural disaster but crashes involving trucks or trains carrying hazardous chemicals through densely packed urban corridors. Another worry is who will bear financial responsibility for abandoned waste sites. Hazardous wastes remain toxic long after the businesses that created them are gone. As is the case with nuclear wastes (see chapter 12), we may need new institutions for perpetual care of these wastes.

SUMMARY

We produce enormous volumes of solid waste in industrialized societies, and there is an increasing problem of how to dispose of those wastes in an environmentally safe manner. In this chapter we have looked at the character of our solid and hazardous wastes. We have surveyed the ways we dispose of our wastes and the environmental problems associated with waste disposal.

Solid wastes are domestic, commercial, industrial, agricultural, and mining wastes that are primarily nontoxic. About 60 percent of our domestic and industrial wastes are deposited in landfills; most of the rest is incinerated or recycled. Old landfills were often messy and leaky, but new ones are required to have impermeable clay or plastic linings, drainage, and careful siting. Incineration can destroy organic compounds, but whether incinerators can or will be operated satisfactorily is a matter of debate. Recycling is growing nationwide, encouraged by the economic and environmental benefits it brings. City leaders tend to doubt the viability of recycling programs, but successful programs have been sustained in other countries and in some American cities.

Near major urban centers, land suitable for waste disposal is becoming increasingly scarce and expensive. Annual costs for solid waste disposal currently total nearly $10 billion in the United States and are expected to climb to $100 billion per year in a few decades. Some cities now ship their refuse to other states or even other countries, but worries about toxic and hazardous material in the waste are leading to increasing resistance to shipping or storing it.

Hazardous and toxic wastes, when released into the environment, cause such health problems as birth defects, neurological disorders, reduced resistance to infection, and cancer. Environmental losses include contamination of water supplies, poisoning of the soil, and destruction of habitat. The major categories of hazardous wastes are ignitable, corrosive, reactive, explosive, and toxic. Some materials of the greatest concern are heavy metals, solvents, and synthetic organic chemicals, such as halogenated hydrocarbons, organophosphates, and phenoxy herbicides.

Disposal practices for solid and hazardous wastes have often been unsatisfactory. Thousands of abandoned, often unknown waste disposal sites still leak toxic materials into the environment. Some alternative techniques for treating or disposing of hazardous wastes include not making the material in the first place, incineration, secure landfill, and physical, chemical, or biological treatment to detoxify or immobilize wastes. People are often unwilling to have transfer facilities, storage sites, disposal operations, or transportation of hazardous or toxic materials in or through their cities. Questions of safety and liability remain unanswered in solid and hazardous waste disposal.

QUESTIONS FOR REVIEW

1. What are solid wastes and hazardous wastes? What is the difference between them?

2. Describe the difference between an open dump, a sanitary landfill, and a modern, secure, hazardous-waste disposal site.

3. Why are landfill sites becoming limited around most major urban centers in the United States? What steps are being taken to solve this problem?

4. Describe some concerns about waste incineration.

5. List some benefits and drawbacks of recycling wastes. What are the major types of materials recycled from municipal waste, and how are they used?

6. What is composting, and how does it fit into solid-waste disposal?

7. Describe some ways that we can reduce the waste stream to avoid or reduce disposal problems.

8. What materials are most recycled in the United States?

9. What societal problems are associated with waste disposal? Why do people object to waste handling in their neighborhoods?

10. What are brownfields, and why do cities want to redevelop them?

THINKING SCIENTIFICALLY

1. If you were planning an archeological dig of a modern garbage dump, would you record weight or volume of the objects uncovered? Why would this matter?

2. A toxic waste disposal site has been proposed for the Pine Ridge Indian Reservation in South Dakota. Many tribal members oppose this plan, but some favor it because of the jobs and income it will bring to an area with 70 percent unemployment. If local people choose immediate survival over long-term health, should we object or intervene?

3. Should industry officials be held responsible for dumping chemicals that were legal when they did it but are now known to be extremely dangerous? At what point can we argue that they should have known about the hazards involved?

4. Suppose that your brother or sister has decided to buy a house next to a toxic waste dump because it costs $20,000 less than a comparable house elsewhere. What do you say to him or her?

5. Is there a fundamental difference between incinerating municipal, medical, or toxic industrial waste? Would you oppose an incinerator for one type of waste in your neighborhood but not others? Why or why not?

6. The Netherlands incinerates much of its toxic waste at sea by a shipborne incinerator. Would you support this as a way to dispose of our wastes as well? What are the critical considerations for or against this approach?

7. Some scientists argue that permanent retrievable storage of toxic and hazardous wastes is preferable to burial. How can we be sure that material that will be dangerous for thousands of years will remain secure? If you were designing such a repository, how would you address this question?

KEY TERMS

biodegradable plastics 310
bioremediation 315
brownfields 314
composting 308
demanufacturing 308
energy recovery 305
hazardous waste 311
mass burn 305
permanent retrievable
   storage 315

photodegradable
   plastics 310
recycling 306
refuse-derived fuel 305
sanitary landfills 302
secure landfills 316
Superfund 313
Toxic Release
   Inventory 312
waste stream 301

Welcome to McGraw-Hill's Online Learning Center

Location: http://www.mhhe.com/environmentalscience

Mc Graw Hill

WEB EXERCISES

Finding Superfund Sites in Your Neighborhood

Go to the Superfund website maintained by the EPA at www.epa.gov/superfund/sites/npl/npl.htm

1. Click on your state. How many National Priority List (NPL) sites occur near where you live? Where are they? Which ones did you already know about?

2. Click on a site near you to see a fact sheet. What types of pollution does this site have? From what sources? How old is the problem? What is being done, or what has been done to clean up this site?

Recycling Challenge

The Internet Consumer Recycling Guide, www.obviously.com/recycle/, provides a detailed list of recyclable materials and where to send them. Before you look at this website, take five minutes and write down at least

25 different types of items in the room around you. These items can be small (paper clips, hair clips, books, food packaging, etc.) or large (carpets, furniture, computers, light fixtures, plumbing, etc.). Once you have made the list, mark all those that are recyclable and note how you would recycle them.

   Now look at the Recycling Guide web page, above. Look at all three "guide" pages—The World's Shortest Comprehensive Recycling Guide, the Guide to Recycling Common Materials, and Guide to Hard-to-Recycle Materials. How many additional items on your list can be recycled, according to these guides? How many could be recycled in principle but would be hard to recycle in your community? Why? What recycled products could be made by recycling the materials on your list?

   How many of the items on your list might be available with recycled content? How many of them could you have avoided acquiring in the first place?

Sustainability and Human Developmemt

Gentlemen, there is no more money. We shall have to start using our brains.
–Ernest Rutherford

Cities like Hong Kong are powerful engines for cultural, economic, and political change. How can they be made more sustainable?

OBJECTIVES

After studying this chapter, you should be able to

- explain the difference between neoclassical and ecological economics, and how each discipline views ecological processes and natural resources.
- distinguish between different types and categories of resources.
- discuss internal and external costs, market approaches to pollution control, and cost-benefit analysis.
- analyze the role of business and some possible strategies for achieving future sustainability.
- recognize the push-and-pull factors that lead to urban growth.
- appreciate how cities fail to be sustainable and how they might become more sustainable.
- understand the causes and consequences of sprawl, and how smart growth could solve some urban problems.
- see the connection between sustainable economic development, social justice, and the solution of urban problems.

Curitiba: An Environmental Showcase

A few decades ago Curitiba, Brazil, was like many cities in developing countries: growing rapidly, with increasing air pollution and traffic congestion, and deteriorating and inadequate roads, water systems, housing, and waste disposal systems. Looking at the gradual decay and corruption of other Brazilian cities, Jaime Lerner, a landscape architect and former student activist, ran for Mayor in 1969. For the next 20 years Lerner motivated and organized a campaign of civic action that turned around the city's decline. Curitiba, with a population of about 2 million, has become an outstanding example of sustainable urban development. Environmentally and socially healthy by any country's standards, Curitiba has a worldwide reputation for its innovative urban planning and environmental protection policies.

The heart of Curitiba's environmental plan is education for both children and adults. Signs posted along roadways proclaim "50 kg of paper equals one tree" and "Recycle; it pays." School children study ecology along with Portuguese and math. With the assistance of children who encourage their parents, the city has successfully instituted a complex recycling plan that requires careful separation of different kinds of materials. The city calculates that 1,200 trees per day are saved by paper recycled in this program. "Imagine if the whole of Brazil did this," Lerner exclaims. "We could save 26 million trees per year!" More than 70 percent of the residents now recycle. Food and bus coupons are given out in exchange for recyclables, providing an alternative to welfare, while also protecting the environment.

Another area in which Curitiba is setting an example is transportation. Faced with a population that tripled in two decades, bringing increasing levels of traffic congestion and air pollution, Curitiba had a transportation dilemma. The choices were either to bulldoze freeways through the historic heart of the city or to institute mass transit. The city chose mass transit. The system has special buses, elevated bus stops, and a network of smaller feeder routes that work quickly and efficiently. Now, more than three-quarters of the city's population leave their cars at home every day. The system is so successful that ridership has increased from 25,000 passengers a day 20 years ago to more than 1.5 million per day now, or 70 percent of all trips in the city. The result is not only less congestion and pollution, but major energy savings.

Other measures to clear the air and reduce congestion include a limit on building height and construction of an industrial park outside city boundaries. Maximum use is made of all materials and buildings. Water and energy conservation are practiced widely. Even litter—an ubiquitous component of most Brazilian cities—is absent in Curitiba. The city has more green space per capita than most American cities. To further beautify their city, civic volunteers have planted 1.5 million trees—more than any other place in Brazil.

Although many residents initially were skeptical of this environmental plan, now a remarkable 99 percent of the city's inhabitants would not want to live anywhere else. The World Bank uses Curitiba as an example of what can be done through civic leadership and public participation to clean up the urban environment. Even in a developing country without a great deal of wealth and resources, Curitiba has demonstrated that a remarkable level of sustainability can be achieved. Some people claim that Curitiba, with its cool climate and high percentage of European immigrants, may be a special case among Brazilian cities. Lerner claims that Curitiba has no special features except concern, creativity, and communal efforts to care for its environment.

Curitiba's example illustrates some of the ingredients of sustainable development. In this chapter we will look at sustainability—efforts to reduce our impacts on the environment and improve human well-being at the same time. And we will discuss additional opportunities for, and obstacles to, sustainable resource use.

SUSTAINABILITY AND RESOURCES

Sustainability has become a central theme of environmental science and of human development and resource use. Although the idea of sustainability has many facets, the central idea is that we should use **resources** (anything that is useful for creating wealth or improving our lives) in ways that do not diminish them. Resources and natural amenities, including wildlife, natural beauty, and open space, should be preserved so that future generations can have lifestyles at least as healthy and happy as ours—or perhaps better.

Can sustainability be achieved, and if so, how? This is one of the core questions in environmental science because human impacts will continue to damage our environment unless we find intelligent alternatives and solutions. Key problems of unsustainability have been woven through much of this book. In this chapter we will look at a few of the root problems of unsustainability and some promising steps toward sustainable use of our environment and sustainable development of human communities.

Sustainable Development

By now it is clear that security and living standards for the world's poorest people are inextricably linked to environmental protection. One of the most important questions in environmental science is how we can continue improvements in human welfare within the limits of the earth's natural resources. A possible solution to this dilemma is **sustainable development,** a term popularized in a 1987 report of the World Commission on Environment and Development called *Our Common Future*. In the words of this report, sustainable development means "meeting the needs of the present without compromising the ability of future generations to meet their own needs." Development means improving people's lives. Sustainable development means extending progress, without exhausting resources, beyond the foreseeable future.

Many people argue that endlessly continuing development, continually improving the lives of more and more people, requires constantly increased resource exploitation. Consequently, the very notion of sustainable development should be impossible. Many of our resources are nonrenewable, and the capacity of the biosphere to absorb our wastes is limited. As we will discuss in the following section, the possibility of sustainable development depends partly on how you define resources, your theories about resource use, and your views on the possibility of extending the use of the resources we have.

To be truly enduring, the benefits of sustainable development must be available to all humans, rather than to just the members of a privileged group. Many of the world's people live in poverty and in unhealthy conditions. By 2050, demographers predict, at least two-thirds of the world's population will live in urban areas. These cities have an extraordinary impact on their surrounding environments. But they also are powerful engines for cultural, economic, and political change. It is probably in cities that the struggle to achieve sustainability will be won or lost. Because cities are so important in our struggle for environmental, social, and economic sustainabilty, they are the focus of the second half of this chapter.

Does sustainable development mean raising all the world's people to the same level of consumption that North Americans and Europeans have? No. It does mean trying to raise the level of health, security, political stability, and quality of life around the world, however. Many places, such as Curitiba, have demonstrated that healthy, satisfying lives are possible without the extravagant resource use we currently have in the wealthiest nations.

Can Development Be Sustainable?

While economic growth makes possible a more comfortable lifestyle, it doesn't automatically result in a cleaner environment. People will purchase clean water and sanitation if they can afford to do so. For low-income people, however, more money tends to result in higher air pollution because they can afford to burn more fuel for transportation and heating. Given enough money, people will be able to afford both convenience and clean air. Some environmental problems, such as waste generation and carbon dioxide emissions, continue to rise sharply with increasing wealth because we are slow to notice and curb their effects. If we are able to sustain economic growth, we will need to develop personal restraint or social institutions to deal with these problems.

Sometimes, development has caused environmental, economic, and social disasters. Large-scale hydropower projects, like the Three Gorges Dam in China (see chapter 10) that are intended to generate valuable electrical power, also displace local residents, destroy wildlife, and accumulate water pollutants. Similarly, introduction of "miracle" crop varieties in Asia and huge grazing projects in Africa financed by international lending agencies have crowded out wildlife, diminished the diversity of traditional crops, and destroyed markets for small-scale farmers.

Other development projects, however, work more closely with both nature and local social systems. Socially conscious businesses and environmental, nongovernmental organizations sponsor ventures that allow people in developing countries to grow or make high-value products—often using traditional techniques and designs—that can be sold on world markets for good prices. "Fair trade" coffee importers, for instance, provide a market for sustainably grown coffee, and for small coffee growers, at a fair price. This is an example of economic development that raises incomes in rural Central America and also rewards farmers for treating their environment with care.

Resources in Classical Economics

To understand the problems and promise of sustainability, you need to understand the different kinds of resources we use. The way we treat resources depends largely on how we view and define them. *Classical economics,* developed in the 1700s by philosophers such as Adam Smith (1723–1790) and Thomas Malthus (1766–1834), assumes that resources are finite—that resources such as iron, gold, water, and land exist in fixed amounts. According to this view, as populations grow, scarcity of these resources reduces quality of life, increases competition, and ultimately causes populations to fall again. In a free market, where fully informed buyers and sellers make free, independent decisions to buy and sell, the price of a pound of butter depends on the supply of available butter (it is cheap when plenty is available) and the demand for butter (buyers pay more when they must compete to get the resource, fig. 14.1).

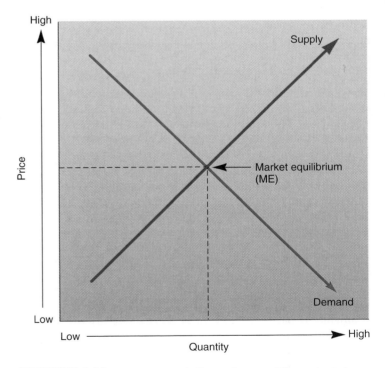

FIGURE 14.1 *Classic supply/demand curves. When price is low, supply is low and demand is high. As prices rise, supply increases but demand falls. Market equilibrium is the price at which supply and demand are equal.*

This price mechanism has been modified by the idea of *marginal costs*. Buyers and sellers evaluate the added return for buying (or making and selling) slightly more of a product, or the marginal cost of slightly greater production or purchase. If the return is greater than the marginal cost, then a sale is made.

The nineteenth-century economist John Stuart Mill assumed that most resources are finite, but he developed the idea of a **steady-state economy.** Rather than boom-and-bust cycles of population and resource use, as envisioned by Malthus, Mill proposed that economies could achieve an equilibrium of resource use and production. Intellectual and moral development would continue, he argued, once this stable, secure state was achieved.

Neoclassical Economics

Neoclassical economics, developed in the nineteenth century, expanded the idea of resources to include labor, knowledge, and capital. Labor and knowledge are resources because they are necessary to create goods and services; they are not finite because every new person can add more labor and energy to an economy. **Capital** is any form of wealth that contributes to the production of more wealth. Money can be invested to produce more money. Mineral resources can be developed and manufactured into goods that return more money. Economists distinguish several kinds of capital:

1. Natural capital: goods and services provided by nature
2. Human capital: knowledge, experience, human enterprise
3. Manufactured (built) capital: tools, buildings, roads, technology

To this list some social theorists would add social capital, the shared values, trust, cooperation, and organization that can develop in a group of people but cannot exist in one individual alone.

Because the point of capital is the production of more capital (that is, wealth), neoclassical economics emphasizes the idea of growth. Growth results from the flow of resources, goods, and services (fig. 14.2). Continued growth is always necessary for continued prosperity, according to this view. Natural resources contribute to production and growth, but they are not critical supplies that limit growth. They are not limiting because resources are considered to be interchangeable and substitutable. As one resource becomes scarce, neoclassical economics predicts that a substitute will be found.

Since production of wealth is central to neoclassical economics, an important measure of growth and wealth is consumption. If a society consumes more oil, more minerals, and more food, it is presumably becoming wealthier. This idea has extended to the idea of *throughput,* the amount of resources a society uses and discards. More throughput is a measure of greater consumption and greater wealth, according to this view. Throughput is commonly measured in terms of **gross national product (GNP),** the sum of all products bought and sold in an economy.

Natural resource economics extends the neoclassical viewpoint to treat natural resources as important waste sinks (absorbers), as well as sources of raw materials. Natural capital (resources) is considered more abundant, and therefore cheaper, than built or human-made capital.

FIGURE 14.2 *The neoclassical model of the economy focuses on the flow of goods, services, and factors of production (land, labor, capital) between business and individual workers/consumers. The social and environmental consequences of these relationships are irrelevant in this view.*

Ecological Economics

Ecological economics, developed in recent decades, applies ecological ideas of system functions and recycling to the definition of resources. Ecological economics also recognizes efficiency in nature, and it acknowledges the importance of ecosystem functions for the continuation of human economies and cultures. In nature, one species' waste is another's food, so that nothing is wasted. We need an economy that recycles materials and uses energy efficiently, much as a biological community does. Ecological economics also treats the natural environment as part of our economy, so that natural capital becomes a key consideration in economic calculations. Ecological functions, such as absorbing and purifying wastewater, processing air pollution, providing clean water, carrying out photosynthesis, and creating soil, are known as **ecological services** (table 14.1). These services are free: we don't pay for them directly (although we often pay indirectly when we suffer from their absence). Therefore, they are often excluded from conventional economic accounting, a situation that ecological economists attempt to rectify (fig. 14.3).

Many ecological economists also promote the idea of a steady-state economy. As with John Stuart Mill's original conception of steady states, these economists argue that economic health can be maintained without constantly growing consumption and throughput. Instead, efficiency and recycling of resources can allow steady prosperity where there is little or no population growth. Low birth rates and death rates (like *K*-adapted species, see chapter 3), political and social stability, and reliance on renewable energy would characterize such a steady-state economy. Like Mill, these economists argue that human and social capital—knowledge, happiness, art, life expectancies, and cooperation—can continue to grow even without constant expansion of resource use.

Both ecological economics and neoclassical economics distinguish between renewable and nonrenewable resources. **Nonrenewable resources** exist in finite amounts: minerals, fossil fuels,

TABLE 14.1 Important Ecological Services

We depend on our environment to continually provide

1. A regulated global energy balance and climate; chemical composition of the atmosphere and oceans; water catchment and groundwater recharge; production, and recycling of organic and inorganic materials; maintenance of biological diversity.

2. Space and suitable substrates for human habitation, crop cultivation, energy conversion, recreation, and nature protection.

3. Oxygen, fresh water, food, medicine, fuel, fodder, fertilizer, building materials, and industrial inputs.

4. Aesthetic, spiritual, historic, cultural, artistic, scientific, and educational opportunities and information.

Source: R. S. de Groot, *Investing in Natural Capital,* 1994.

FIGURE 14.4 *Biological resources are renewable in that they replace themselves by reproduction, but if overused or misused, populations die. If a species is lost, it cannot be re-created. It is permanently lost as a component of its ecosystem and as a resource to humans.*

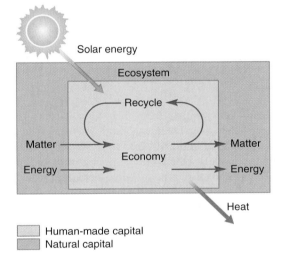

FIGURE 14.3 *An ecological economics view considers natural capital and recycling integral to the economy. Human-made capital is created using limited supplies of natural capital.*
Source: Data from Herman Daly in A. M. Jansson et al., *Investing in Natural Capital,* ISEE.

and also groundwater that recharges extremely slowly are all fixed, at least on a human timescale. **Renewable resources** are naturally replenished and recycled at a fairly steady rate. Fresh water, living organisms, air, and food resources are all renewable (fig. 14.4).

These categories are important, but they are not as deterministic as you might think. Nonrenewable resources such as iron and gold can be extended through more efficient use: cars now use less steel than they once did, and gold is mixed in alloys to extend its use. Substitution also reduces demand for these resources: car parts once made of iron are now made of plastic and ceramics; copper wire, once stockpiled to provide phone lines, is now being replaced with cheap, lightweight fiber-optic cables made from silica (sand). Recycling also extends supplies of nonrenewable

resources. Aluminum, platinum, gold, silver, and many other valuable metals are routinely recycled now, further reducing the demand for extracting new sources. The only limit to recycling is usually the relative costs of extracting new resources compared to collecting used materials. Recoverable sources of nonrenewable resources are also expanded by technological improvements. New methods make it possible to mine very dilute metal ores, for example. Gold ore of extremely low concentrations is now economically recoverable—that is, you can make money on it—even though the price of gold has fallen because of greater efficiency, more discoveries, and resource substitution. Scarcity of resources, seen by classical economists as the trigger for conflict and suffering, can actually provide the impetus for much of the innovation that leads to substitution, recycling, and efficiency.

On the other hand, renewable resources can become exhausted if they are managed badly. This is especially apparent in biological resources, such as the passenger pigeon, American bison, and Atlantic cod. All these species once existed in extraordinary numbers, but within a few years, each was brought to the brink of extinction (or eliminated entirely) by overharvesting.

Scarcity and Limits to Growth

Are we about to run out of essential natural resources? It stands to reason that, if we consume a fixed supply of nonrenewable resources at a constant rate, we'll eventually use up all the economically recoverable reserves. There are many warnings in the environmental literature that our extravagant depletion of nonrenewable resources sooner or later will result in catastrophe, misery, and social decay. Models for exploitation rates of nonrenewable resources—called Hubbert curves, after Stanley Hubbert who developed them—often closely match historic experience for natural resource depletion (fig. 14.5).

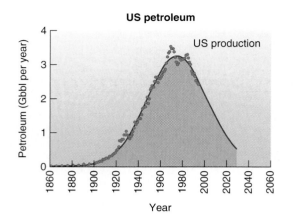

FIGURE 14.5 *U.S. petroleum production. Dots indicate actual production. The bell-shaped curve is a theoretical Hubbert curve for a nonrenewable resource. The shaded area under the curve, representing 220 Gbbl (Gbbl = Gigabarrels or billions of standard 42-gal barrels), is an estimate of the total economically recoverable resource.*

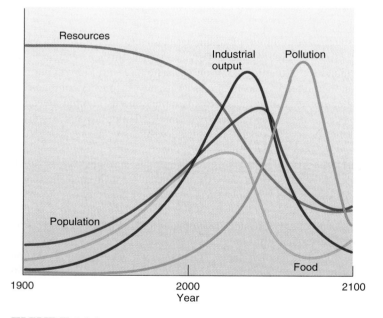

FIGURE 14.6 *A run of one of the world models in* Limits to Growth. *This model assumes business-as-usual for as long as possible until Malthusian limits cause industrial society to crash. Notice that pollution continues to increase well after industrial output, food supplies, and population have all plummeted.*

Many economists, however, contend that human ingenuity and enterprise often allow us to respond to scarcity in ways that postpone or alleviate dire effects of resource use. The question of whether this view is right has to do with the important theme of **limits to growth.**

In the early 1970s an influential study of resource limitations was funded by the Club of Rome, an organization of wealthy business owners and influential politicians. The study was carried

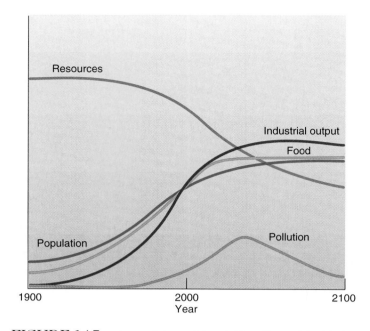

FIGURE 14.7 *A run of the world model from* Beyond the Limits. *This model assumes that population and consumption are curbed, new technologies are introduced, and sustainable environmental policies are embraced immediately, rather than after resources are exhausted.*

out by a team of scientists from the Massachusetts Institute of Technology, headed by Donnela Meadows. The results of this study were published in the 1972 book *Limits to Growth.* A complex computer model of the world economy was used to examine various scenarios of different resource depletion rates, growing population, pollution, and industrial output. Given the Malthusian assumptions built into this model, catastrophic social and environmental collapse seemed inescapable.

Figure 14.6 shows one example of the world model. Food supplies and industrial output rise as population grows and resources are consumed. Once past the carrying capacity of the environment, however, a crash occurs as population, food production, and industrial output all decline precipitously. Pollution continues to grow as society decays and people die, but eventually, it also falls. Notice the similarity between this set of curves and the "boom-and-bust" population cycles described in chapter 3.

Many economists criticized this model because it underestimated technological development and factors that might mitigate the effects of scarcity. In 1992 the Meadows group published updated computer models in *Beyond the Limits* that include technological progress, pollution abatement, population stabilization, and new public policies that work for a sustainable future. If we adopt these changes sooner rather than later, the models show an outcome like that in figure 14.7, in which all factors stabilize sometime in the twenty-first century at an improved standard of living for everyone. Of course, none of these computer models shows what will happen, only what some possible outcomes might be, depending on the choices we make.

Communal Property and the Tragedy of the Commons

One of the difficulties of economics and resource management is that there are many resources we all share and nobody owns. Clean air, fish in the ocean, clean water, wildlife, and open space are all natural amenities that we exploit but that nobody clearly controls.

In 1968 biologist Garret Hardin wrote **"The Tragedy of the Commons,"** an article describing how commonly held resources are degraded and destroyed by self-interest. Using the metaphor of the "commons," or community pastures in colonial New England villages, Hardin theorized that it behooves each villager to put more cows on the pasture. Each cow brings more wealth to the individual farmer, but the costs of overgrazing are shared by the entire community. The individual farmer, then, suffers only part of the cost, but gets to keep all the profits from the extra cows s/he put on the pasture. Consequently, the commons becomes overgrazed, exhausted, and depleted. This dilemma is also known as the "free rider problem." The best solution, Hardin argued, was to give coercive power to the government or to privatize resources so that a single owner controls resource use.

This metaphor has been applied to many resources, especially to human population growth. It benefits every poor villager to produce a few more children, but collectively, these children consume all the resources available, making us all poorer in the end. The same argument has been applied to many resource overuse problems, such as depletion of ocean fisheries, pollution, African famines, and urban crime.

Recent critics have pointed out that what Hardin was really describing was not a commons, or collectively owned and managed resource, but an **open access system,** in which there are no rules to manage resource use. In fact, many common resources have been managed successfully for centuries by cooperative agreements among users. Native American management of wild rice beds, Swiss village-owned mountain forests and pastures, Maine lobster fisheries, and communal irrigation systems in Spain, Bali, Laos, and many other countries have all remained viable under communal management.

Each of these "commons," or **communal resource management systems,** shares a number of features: (1) community members have lived on the land or used the resource for a long time and anticipate that their children and grandchildren will as well, thus giving them a strong interest in sustaining the resource and maintaining bonds with their neighbors; (2) the resource has clearly defined boundaries; (3) the community group size is known and enforced; (4) the resource is relatively scarce and highly variable so that the community is forced to be interdependent; (5) management strategies appropriate for local conditions have evolved over time and are collectively enforced; that is, those affected by the rules have a say in them; (6) the resource and its use are actively monitored, discouraging anyone from cheating or taking too much; (7) conflict resolution mechanisms reduce discord; and (8) incentives encourage compliance with rules, while sanctions for noncompliance keep community members in line.

Rather than being the only workable solution to problems in common pool resources, privatization and increasing external controls often prove to be disastrous. In places where small villages have owned and operated local jointly held forests and fishing grounds for generations, nationalization and commodification of resources generally have led to rapid destruction of both society and ecosystems. Where communal systems once enforced restraint over harvesting, privatization encouraged narrow self-interest and allowed outsiders to take advantage of the weakest members of the community.

COST-BENEFIT ANALYSIS AND NATURAL RESOURCE ACCOUNTING

Decision making about sustainable resource use often entails **cost-benefit analysis (CBA),** the process of accounting and comparing the costs of a project and its benefits. Ideally, this process assigns values to social and environmental effects of a given undertaking, as well as the value of the resources consumed or produced. However, the results of CBA often depend on how resources are accounted for and measured in the first place. CBA is one of the main conceptual frameworks of resource economics, and it is used by decision makers around the world as a way of justifying the building of dams, roads, and airports, as well as in considering what to do about biodiversity loss, air pollution, and global climate change. CBA is a useful way of rational decision making about these projects. It is also widely disputed because it tends to discount the value of natural resources, ecological services, and human communities, and it is used to justify projects that jeopardize all these resources.

In CBA the monetary value of all benefits of a project are counted up and compared to the monetary costs of the project. Usually, the direct expenses of a project are easy to ascertain: how

---

APPLICATION: | **Costs and Benefits**

Figure 14.8 shows an example of the relative costs and benefits of reducing air pollution in Poland.

1. For the first 30 percent of reduction (on the X axis), benefits rise rapidly. What are some of the economic benefits that might be represented by this rising curve?

2. What might be some of the nearly free pollution reduction steps represented by the "costs" curve in the first 30 percent reduction?

3. If you were planning a budget, you would want to maximize benefits and minimize costs. For what percentage of pollution reduction would you aim? At what point would you redirect your budget to other priorities?

Answers: 1. Benefits include improved health, less damage to buildings, crops, and materials, and improved quality of life. *2.* Conservation, improved planning and efficiency in transportation, and energy production; replacing old, inefficient industrial plants and vehicles. *3.* At any reduction up to 70 percent, benefits outweigh costs.

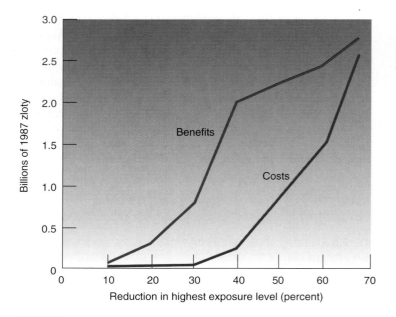

FIGURE 14.8 *Total estimated benefits and costs of reducing particulate air pollution in Tarnobrzeg, Poland. Removing the highest 40 to 50 percent of particulates costs relatively little compared to the large benefits that would be gained. Removing 70 percent of particulates may cost more than the benefits are worth.*

Source: J. Cofala, T. Lis, and H. Balandynowicz, *Cost-Benefit Analysis of Regional Air Pollution Control: Case Study for Tarnobrzeg,* 1991, Polish Academy of Sciences, Warsaw.

TABLE 14.2 Estimated Annual Value of Ecological Services

ECOSYSTEM SERVICES	VALUE (TRILLION $ U.S.)
Soil formation	17.1
Recreation	3.0
Nutrient cycling	2.3
Water regulation and supply	2.3
Climate regulation (temperature and precipitation)	1.8
Habitat	1.4
Flood and storm protection	1.1
Food and raw materials production	0.8
Genetic resources	0.8
Atmospheric gas balance	0.7
Pollination	0.4
All other services	1.6
Total value of ecosystem services	33.3

Source: Adapted from R. Costanza et al., "The Value of the World's Ecosystem Services and Natural Capital," in *Nature,* vol. 387, 1997.

much will you have to pay for land, materials, and labor? The monetary worth of lost opportunities—to swim or fish in a river, or to see birds in a forest—on the other hand, are much harder to appraise, as are inherent values of the existence of wild species or wild rivers. What is a bug or a bird worth, for instance, or the opportunity for solitude or inspiration? Eventually, the decision maker compares all the costs and benefits to see whether the project is justified or whether some alternative action might bring greater benefit at less cost.

Critics of CBA point out its absence of standards, inadequate attention to alternatives, and the placing of monetary values on intangible and diffuse or distant costs and benefits. Who judges how costs and benefits will be estimated? How can we compare things as different as the economic gain from cheap power with loss of biodiversity or the beauty of a free-flowing river? Critics claim that placing monetary values on everything could lead to a belief that only money and profits count and that any behavior is acceptable as long as you can pay for it. Sometimes, speculative or even hypothetical results are given specific numerical values in CBA and then treated as if they are hard facts.

Figure 14.8 shows an example of a cost-benefit analysis for reducing particulate air pollution (soot) in Poland. As you can see, removing the highest 40 percent of particulates is highly cost effective. Approaching 70 percent particulate removal, however, the costs may exceed benefits. Data such as these can be useful to decision makers. On the other hand, this same study showed that controlling sulfur emissions had high costs and negligible benefits. Might this conclusion result from the ways "benefits" were evaluated?

Accounting for Nonmonetary Resources

Values such as wildlife, nonhuman ecological systems, and ecological services can be incorporated with natural resource accounting. In theory this accounting contributes to sustainable resource use because it can put a value on long-term or intangible goods that are necessary but often disregarded in economic decision making. One important part of natural resource accounting is assigning a value to ecological services (table 14.2). Another is using alternative measures of wealth and development. As mentioned earlier, GNP is a widely used measure of wealth that is based on rates of consumption and throughput. GNP doesn't account for natural resource depletion or ecosystem damage, however. The World Resources Institute, for example, estimates that soil erosion in Indonesia reduces the value of crop production about 40 percent per year. If natural capital were taken into account, Indonesian GNP would be reduced by at least 20 percent annually. Similarly, Costa Rica experienced impressive increases in timber, beef, and banana production between 1970 and 1990. But decreased natural capital during this period, represented by soil erosion, forest destruction, biodiversity losses, and accelerated water runoff, add up to at least $4 billion, or about 25 percent of annual GNP.

A number of alternative measures have been developed, though, that put priorities on more sustainable priorities/factors in development and growth. Herman Daly and John Cobb have proposed an Index of Sustainable Economic Welfare (ISW) that takes into account real per capita income, distributional equity, natural resource depletion, environmental damage, and the value of unpaid labor. The United Nations Development Program (UNDP) uses a benchmark called the Human Development Index (HDI) to

Principles of Environmental Science www.mhhe.com/environmentalscience

track social progress. HDI incorporates life expectancy, educational attainment, and standard of living as critical measures of development. Gender issues are accounted for in the Gender Development Index (GDI), which is simply HDI adjusted or discounted for inequality of achievement between men and women. These indexes are not as widely used as GNP, but they provide important alternatives that help assess sustainable development, as well as economic development.

Discount Rates

One of the problems with accounting for resources is that economists (and people in general) typically assign lower values to future resources. The reduced value of a commodity in the future is a **discount rate.** In general, discount rates are equivalent to interest rates. When you buy something on credit, for instance, you are agreeing to pay extra (the cost plus interest) because the value of having the item today is greater than the discounted value of that money months from now.

The choice of discount rates to apply to future benefits becomes increasingly problematic with intangible resources or long time frames. How much will a barrel of oil or a 4,000-year-old redwood tree be worth a couple of centuries from now? Maybe there will be substitutes for oil by then, so the barrel of oil won't be worth much. On the other hand, if we don't find substitutes, that oil or tree may be priceless someday. This forecasting is complicated by the fact that we are making decisions not only for ourselves, but also for future generations.

Which discount rate should be used for future benefits and costs is often a crucial question in large public projects, such as dams and airports. Proponents may choose to use low discount rates that make a venture seem attractive, while opponents prefer higher rates that show the investment to be questionable. What discount rate do you use in comparing future environmental costs to immediate financial returns? To calculate for sustainable resource use, it may be necessary to assume no discount rate—that future resources are as important as current resources. Understanding these nuances is extremely important for environmental activists fighting environmentally destructive schemes.

Internal and External Costs

One of the factors that can make resource-exploiting enterprises look good in cost-benefit analysis is externalizing costs. **Externalizing costs** is the act of disregarding or discounting resources or goods that contribute to producing something, but for which the producer does not actually pay. Usually, external costs are diffuse and difficult to quantify. Generally, they belong to society at large, not to the individual user. When a farmer harvests a crop in the fall, for example, the value of seeds, fertilizer, and the sale of the crop are all tabulated; the value of soil lost to erosion, water quality lost to nonpoint-source pollution, and depleted fish populations are not accounted for. These are most often costs shared by the whole society, rather than borne by the resource user. They are external to the accounting system, and they are generally ignored in cost-benefit analysis—or when the farmer evaluates whether

the year was profitable. Larger enterprises, such as dam building, logging, and road building, generally externalize the cost of ecological services lost along the way.

One way to use the market system to optimize resource use is to make sure that those who reap the benefits of resource use also bear all the external costs. This is referred to as **internalizing costs.** Calculating the value of ecological services or diffuse pollution is not easy, but it is an important step in sustainable resource accounting.

TRADE AND DEVELOPMENT

A sustainable society requires some degree of equitable resource distribution: if most wealth is held by just a few people, the misery and poverty of the majority eventually lead to social instability and instability of resource supplies. Accordingly, wealthy industrial nations have worked harder in recent decades to assist in developing the economies in poorer countries.

International Trade

Expanding trade relations has been promoted as a way to distribute wealth, stimulate economies around the world, and at the same time satisfy the desires of consumers in wealthy countries. According to the economic theory of *comparative advantage,* each place has some sort of goods or services it can sell cheaper, or better, than others can. International trade allows us to take advantage of all the best or cheapest products from around the world. If Egypt can produce cotton cheaper than Texas, then we should buy our cotton from Egypt. If Malaysia can manufacture athletic shoes with labor that costs a few cents an hour, then we

---

APPLICATION:	**Natural Resource Accounting and Discount Rates**

If you want to plan for sustainable resource use, you need to account for the long-term value of resources.

1. How many years would you consider in "long-term" planning? _____

2. To put a price on long-term values, should you use a high or low discount rate? (For example, does a 5 percent or 10 percent discount rate serve long-term sustainability better?) _____

3. If you were considering clearing a woodlot (worth $1,000 today), approximately what would the lot's value be 10 years from now at a 10 percent discount rate? _____

4. What costs are externalized in this accounting? _____

Answers: 1. Depends on your priorities; for yourself, perhaps 10 to 50 years, for a community, decades to centuries. *2.* Low (5% is better than 10%). *3.* $349. *4.* The value of maintained supplies of wood and other resources; biodiversity and wildlife habitat; recreation; temperature and water regulation, and other services.

should buy our shoes from Malaysia. Building factories in Malaysia also stimulates the Malaysian economy, even though workers earn only a small fraction of what a worker in a wealthy country would earn.

A problem with international trade is that it externalizes costs on a grand scale. Tropical hardwood products can be sold extremely cheaply in the United States—as lumber, plywood, shipping pallets, and so on. The environmental costs of producing those hardwood products occur far from the consumer who buys a piece of cheap Brazilian plywood. Making matters worse, the environmental costs of the plywood are usually exported to places where there are few legal controls on pollution and resource extraction. A factory in the United States, for example, is legally bound to minimize its production of air and water pollution. Pollution control can be expensive, and internalizing this cost makes a factory less profitable. A similar factory in Mexico might have far less responsibility for pollution control, making it cheaper, at least in the short term, to produce goods there than in the United States. The World Trade Organization (WTO) protests in Seattle in 1999 were largely about such exporting and externalizing of environmental and social costs of production. (See the related web story "The Battle in Seattle" at www.mhhe.com/apps.)

Another criticism of international trade is that the international banking systems that finance it are set up by and for the wealthy countries. The WTO and the General Agreement on Tariffs and Trade (GATT), for example, regulate 90 percent of all international trade. The WTO and GATT are both made up of relationships and agreements between corporations in a few very wealthy countries. Representatives of less-powerful countries often charge that these agreements trap poorer regions into the role of suppliers of natural resources—timber, mineral ores, fruit, and also cheap labor. These countries are forced to mine their natural capital for only small returns in wealth.

International Development and the World Bank

The World Bank has more influence on the financing and policies of developing countries than any other institution. Of some $25 billion loaned each year for Third World projects by multinational development banks, about two-thirds comes from the World Bank. This institution was founded in 1945 to provide aid to war-torn Europe and Japan. In the 1950s its emphasis shifted to development aid for Third World countries. This aid was justified on humanitarian grounds, but it also conveniently provided markets and political support for growing American and European multinational corporations.

The World Bank is jointly owned by 150 countries, but one-third of its support comes from the United States. Its president has always been an American. The bulk of its $66.8 billion capital comes from private investors. The bank makes loans for development projects, and the investors hope to make a profit on the investment, usually from interest paid on the loans.

Many World Bank projects have been environmentally destructive and highly controversial. In Botswana, for example,

$18 million was provided to increase beef production for export by 20 percent, despite already severe overgrazing on fragile grasslands. The project failed, as had two previous beef production projects in the same area. In Ethiopia, rich floodplains in the Awash River Valley were flooded to provide electric power and irrigation water for cash export crops. More than 150,000 subsistence farmers were displaced, and food production was seriously reduced.

Recently, the World Bank has begun to attempt some environmental and social review of its loans. In part this change comes from demands from the U.S. Congress that projects be environmentally and socially benign. Whether this will improve the World Bank's track record remains to be seen.

Microlending

The World Bank deals in huge loans for massive projects. These are impressive to investors and to the countries that borrow the money. They are also a huge economic gamble, and on average, the economic return on large loans has been quite low.

Recently, smaller, local lending programs have begun to develop. These are aimed at small-scale, widespread development, and their results have been very promising. The first of these was Bangladesh's Grammeen (village) Bank network. These banks make small loans, averaging just $67, to help poor people buy a sewing machine, a bicycle, a loom, a cow, or some other commodity that will help them start, or improve, a home business (fig. 14.9).

FIGURE 14.9 *A small loan to buy a used sewing machine can provide a sustainable livelihood for a family in the developing world.*

Ninety percent of the customers are women, usually with no collateral or steady income. Still, loan repayment rates are 98 percent—compared to only 30 percent at a conventional bank! This program enhances dignity, respect, and cooperation in a village community, and also teaches individual responsibility and enterprise.

Comparable programs have now sprung up around the world. In the United States, more than a hundred different organizations have begun providing microloans and small grants for training. The Women's Self-Employment Project in Chicago, for instance, teaches job skills to single mothers in housing projects. Similarly, "tribal circle" banks on Native American reservations successfully finance microscale economic development projects.

GREEN BUSINESS AND GREEN DESIGN

Business people and consumers are increasingly aware of the unsustainability of producing the goods we use every day. Recently, a number of business innovators have tried to develop green businesses, which produce environmentally and socially sound products. Environmentally conscious or "green" companies, such as the Body Shop, Patagonia, Aveda, Malden Mills, Johnson and Johnson, and others, have shown that operating according to the principles of sustainable development and environmental protection can be good for public relations, employee morale, and sales (table 14.3).

Green business works because consumers are becoming aware of the ecological consequences of their purchases. Increasing interest in environmental and social sustainability has caused an explosive growth of green products. The National Green Pages published by Co-Op America currently lists more than 2,000 green companies. You can find eco-travel agencies, telephone companies that donate profits to environmental groups, entrepreneurs selling organic foods, shade-grown coffee, straw-bale houses, paint thinner made from orange peels, sandals made from recycled auto tires, and a plethora of hemp products, including burgers, ale, clothing, shoes, rugs, and shampoo. Although these eco-entrepre-

neurs represent a tiny sliver of the $7 trillion per year U.S. economy, they often are pioneers in developing new technologies and offering innovative services. Markets also grow over time: organic food marketing has grown from a few funky, local co-ops to a $7 billion market segment. Most supermarket chains now carry some organic food choices. Similarly, natural-care health and beauty products reached $2.8 billion in sales in 1999 out of a $33 billion industry. By supporting these products, you can ensure that they will continue to be available and, perhaps, even help expand their penetration into the market.

Corporations committed to eco-efficiency and clean production include such big names as Monsanto, 3M, DuPont, and Duracell. Applying the famous three Rs—reduce, reuse, recycle— these firms have saved money and gotten welcome publicity. Savings can be substantial. Pollution prevention programs at 3M, for example, have saved $857 million over the past 25 years. In a major public relations achievement, DuPont has cut its emissions of airborne cancer-causing chemicals almost 75 percent since 1987. Small operations can benefit as well. Stanley Selengut, owner of three eco-tourist resorts in the U.S. Virgin Islands, attributes $5 million worth of business to free press coverage about the resorts' green building features and sustainable operating practices.

Design for the Environment

Architects are starting to get on board the green bandwagon, too. Acknowledging that heating, cooling, lighting, and operating buildings is one of our biggest uses of energy and resources, architects like William McDonough are designing "green office" projects. Among McDonough's projects are the Environmental Defense Fund headquarters in New York City, the Environmental Studies Center at Oberlin College in Ohio, the European headquarters for Nike in Hilversum, the Netherlands, and the Gap corporate offices in San Bruno, California (fig. 14.10). Each uses a combination of energy efficient designs and technologies, including natural lighting and efficient water systems. The Gap office building, for example, is intended to promote employee well-being and productivity, as well as efficiency. It has high ceilings,

TABLE 14.3 Goals for an Eco-Efficient Economy

- Introduce no hazardous materials into the air, water, or soil.
- Measure prosperity by how much natural capital we can accrue in productive ways.
- Measure productivity by how many people are gainfully and meaningfully employed.
- Measure progress by how many buildings have no smokestacks or dangerous effluents.
- Make the thousands of complex governmental rules that now regulate toxic or hazardous materials unnecessary.
- Produce nothing that will require constant vigilance from future generations.
- Celebrate the abundance of biological and cultural diversity.
- Live on renewable solar income rather than fossil fuels.

FIGURE 14.10   *The award-winning Gap, Inc. corporate offices in San Bruno, California, demonstrate some of the best features of environmental design. A roof covered with native grasses provides insulation and reduces runoff. Natural lighting, an open design, and careful relation to its surroundings make this a pleasant place to work.*

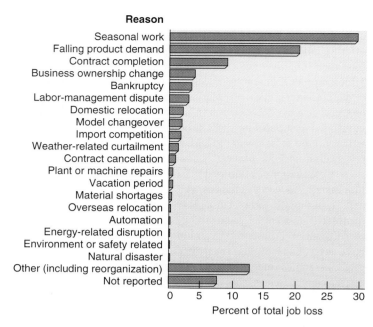

Reason

Seasonal work
Falling product demand
Contract completion
Business ownership change
Bankruptcy
Labor-management dispute
Domestic relocation
Model changeover
Import competition
Weather-related curtailment
Contract cancellation
Plant or machine repairs
Vacation period
Material shortages
Overseas relocation
Automation
Energy-related disruption
Environment or safety related
Natural disaster
Other (including reorganization)
Not reported

0   5   10   15   20   25   30

Percent of total job loss

FIGURE 14.11 *Although opponents of environmental regulation often claim that protecting the environment costs jobs, studies by economist E. S. Goodstein show that only 0.1 percent of all large-scale layoffs in the United States were the result of environmental laws.*
Source: Data from E. S. Goodstein, Economic Policy Institute, Washington, D.C.

abundant skylights, windows that open, a full-service fitness center (including pool), and a landscaped atrium for each office bay that brings the outside in. The roof is covered with native grasses. Warm interior tones and natural wood surfaces (all wood used in the building was harvested by certified sustainable methods) give a friendly feeling. Paints, adhesives, and floor coverings are low toxicity, and the building is one-third more energy efficient than strict California laws require. The pleasant environment helps to improve employee effectiveness and retention. Gap, Inc. estimates that the increased energy and operational efficiency will have a 4- to 8-year payback.

Jobs and the Environment

For years business leaders and politicians have portrayed environmental protection and jobs as mutually exclusive. They claim that pollution control, protection of natural areas and endangered species, and limits on use of nonrenewable resources will strangle the economy and throw people out of work. Ecological economists dispute this claim, however. Their studies show that only 0.1 percent of all large-scale layoffs in the United States in recent years were due to government regulations (fig. 14.11). Environmental protection, they argue, is not only necessary for a healthy economic system, it actually creates jobs and stimulates business.

Japan, already a leader in efficiency and environmental technology, has recognized the multibillion-dollar economic potential of "green" business. The Japanese government is investing $4 billion per year on research and development, and now is selling about $12 billion worth of equipment and services per year world-

what can you do?

Personally Responsible Consumerism

Each of us can do many things to lower our ecological impacts and support green businesses through responsible consumerism and ecological economics.

- Practice living simply. Ask yourself if you really need more material goods to make your life happy and fulfilled.
- Rent, borrow, or barter when you can. Can you reduce the amount of stuff you consume by renting, instead of buying, machines and equipment you actually use only rarely?
- Recycle or reuse building materials: doors, windows, cabinets, appliances. Shop at salvage yards, thrift stores, yard sales, or other sources of used clothes, dishes, appliances, etc.
- Consult the National Green Pages from Co-Op America for a list of eco-friendly businesses. Write to companies from which you buy goods or services, and ask them what they are doing about environmental protection and human rights.
- Buy green products. Look for efficient, high-quality materials that will last and that are produced in the most environmentally friendly manner possible. Subscribe to clean energy programs if they are available in your area. Contact your local utility and ask that it provide this option if it doesn't now.
- Buy locally grown or locally made products made under humane conditions by workers who receive a fair wage.
- Think about the total life-cycle costs of the things you buy, especially big purchases like cars. Try to account for the environmental impacts, energy use, and disposal costs, as well as initial purchase price.
- Stop junk mail. Demand that your name be removed from mass-mailing lists.
- Invest in socially and environmentally responsible mutual funds or green businesses when you have money for investment.

wide. It is already marketing advanced waste incinerators, pollution-control equipment, alternative energy sources, and water-treatment systems. The superefficient "hybrid" gas-electric cars only recently introduced to the U.S. market have been available in Japan for several years. Unfortunately, the United States has been resisting international pollution-control conventions, rather than recognizing the potential for economic growth and environmental protection in the field of "green" business.

URBAN DEVELOPMENT AND SUSTAINABLE CITIES

Understanding sustainable resource use and sustainable development requires paying some attention to cities. Cities are where most of our consumed resources end up. Nearly half the people in

TABLE 14.4 The World's Largest Metropolitan Regions (populations in millions)

1900		1995		2015	
London, England	6.6	Tokyo, Japan	26.9	Tokyo, Japan	29.0
New York, USA	4.2	Mexico City, Mexico	16.6	Mumbai, India†	26.2
Paris, France	3.3	São Paulo, Brazil	16.5	Mexico City, Mexico	25.1
Berlin, Germany	2.4	New York, USA	16.3	Lagos, Nigeria	24.6
Chicago, USA	1.7	Mumbai, India†	15.1	São Paulo, Brazil	20.3
Vienna, Austria	1.6	Shanghai, China	13.6	Shanghai, China	18.0
Tokyo, Japan	1.5	Los Angeles, USA	12.4	New York, USA	17.6
St. Petersburg, Russia	1.4	Calcutta, India	11.9	Calcutta, India	17.3
Philadelphia, USA	1.4	Seoul, S. Korea	11.6	Beijing, China	16.0
Manchester, England	1.3	Beijing, China	11.3	Los Angeles, USA	14.2
Birmingham, England	1.2	Osaka, Japan	10.6	Buenos Aires, Argentina	13.9
Moscow, Russia	1.1	Lagos, Nigeria	10.3	Rio de Janeiro, Brazil	11.9
Peking, China*	1.1	Rio de Janeiro, Brazil	10.2	Osaka, Japan	10.6

*Now spelled Beijing.

†Formerly known as Bombay.

Source: Data from T. Chandler, *Three Thousand Years of Urban Growth,* 1974, Academic Press; and United Nations Populations Division, 1998.

the world now live in urban areas, and in a century, 80 to 90 percent of us are expected to live in urban agglomerations. Cities have extraordinary, often disastrous impacts on the environment. They can also have beneficial environmental effects: resource use is more efficient where people can share goods and services; less energy is spent in transportation and production when people don't have to travel too far. Some of the greatest promise for innovative improvements also comes from cities, where people are close enough together to share ideas and resources, to teach each other new methods of doing things, and to recognize the need for cooperative, collaborative environmental problem solving. In this section we will explore the ways that cities fail to be sustainable and look at some of the ways they might contribute to sustainability.

Urban Growth

Until recently, the vast majority of humanity has always lived in rural areas, where farming, fishing, hunting, timber harvesting, animal herding, mining, or other natural resource-based occupations were the main form of economic activity. Cities—where people live in high enough densities to have specialized professions—have always been the center of most political organization, economic power, and learning, but relatively few people actually lived in cities.

Since the beginning of the Industrial Revolution some 300 years ago, however, cities have grown rapidly in both size and power. In every developing country, the transition from an agrarian society to an industrial one has been accompanied by **urbanization,** an increasing concentration of the population in cities and a transformation of land use and society to a metropolitan pattern of organization.

How big are cities? Some have only a few thousand people, but more and more cities are growing to extraordinary size (table

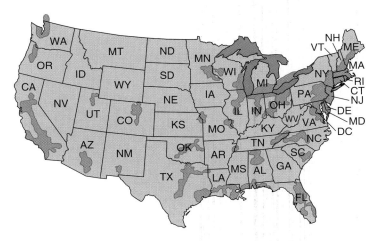

FIGURE 14.12 *Urban core agglomerations (lavender areas) are forming megalopolises in many areas. While open space remains in these areas, the flow of information, capital, labor, goods, and services links each into an interacting system.*
Source: Data from U.S. Census Bureau, 1998.

14.4). Many of the world's large cities have expanded until they merged with their neighbors to form giant urban complexes, also known as **megacities** or megalopolises. In the United States, urban areas between Boston and Washington, D.C., have merged into a nearly continuous megacity (sometimes called Bos-Wash), containing about 35 million people (fig. 14.12). The Tokyo-Yokohama-Osaka-Kobe corridor (referred to as Toko-hama) contains nearly 50 million people.

The United States underwent a dramatic rural-to-urban shift in the nineteenth and early twentieth centuries. In 1850, more than 90 percent of Americans lived in rural areas. A century later, only

TABLE 14.5 Urban Share of Total Population (percent)			
REGION	1975	1995	2025*
Africa	25	34	54
Asia	25	35	55
Europe	67	74	83
Russian Federation	66	76	86
North and Central America	57	68	79
South America	64	78	88
Oceania	72	70	75
World	38	45	61

*Estimated.

Source: Data from *World Resources, 1996–97.*

about one-third of the U.S. population was rural, and less than 5 percent were farmers. Now, many developing countries are experiencing a similar demographic movement. In 1850 only about 2 percent of the world's population lived in cities. By 2002, half the people in the world were urban. Only Africa and South Asia remain predominantly rural, but people there are swarming into cities in ever-increasing numbers. About three-fourths of the people in Europe, North America, and Latin America already live in cities (table 14.5). Some urbanologists predict that by 2100 the whole world will be urbanized to the levels now seen in developed countries.

Cities in Developing Countries

Recent urban growth has been particularly dramatic in large cities of the developing world. In 1900, 13 cities had populations over 1 million; all except Tokyo and Peking were in Europe or North America (see table 14.4). By 1995 there were 235 metropolitan areas of more than 1 million people, but only three of the largest cities were in developed countries. Cities like Mumbai, Jakarta, Manila, Cairo, and Mexico City receive thousands of new immigrants *every day.* Many of these large cities are expected to grow by 50 percent in just the next 25 years (fig. 14.13).

Ninety percent of the population growth over the next 25 years is expected to occur in the less-developed countries of the world (fig. 14.14). Most of that growth will be in the already overcrowded cities of the least-affluent countries, such as India, China, Mexico, and Brazil. The combined population of these cities is expected to jump from its present 6 billion to more than 8 billion by the year 2025. Meanwhile, rural populations in these countries are expected to remain constant or even decline somewhat as rural people migrate into the cities.

Can cities function with 20 or 25 million people? Can they supply the public services necessary to sustain a civilized life? Adding 750,000 new people annually to Mexico City amounts to building a new city the size of Baltimore or San Francisco every

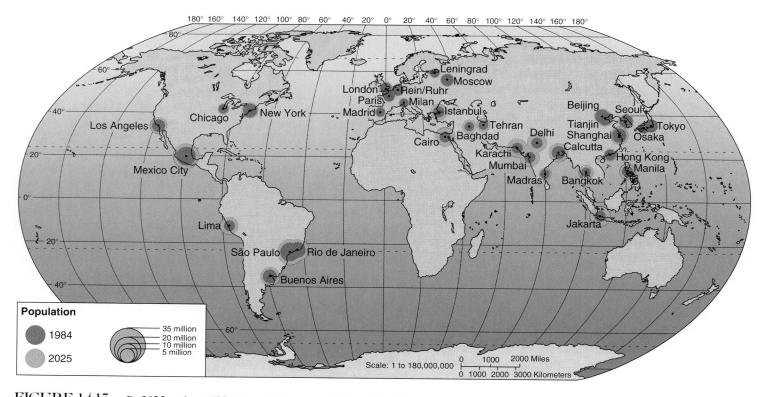

FIGURE 14.13 *By 2025, at least 400 cities will have populations of 1 million or more, and 93 supercities will have populations above 5 million. Three-fourths of the world's largest cities will be in developing countries that already have trouble housing, feeding, and employing their people.*

Principles of Environmental Science

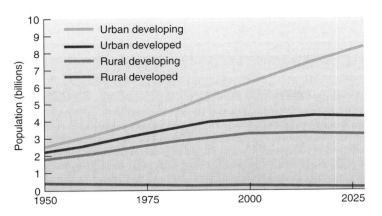

FIGURE 14.14   *Urban and rural growth in developed and developing countries.*
Source: Data from United Nations (UN) Population Division, *World Urbanization Prospects (The 1996 Revision)*, on diskette (UN, New York, 1996).

FIGURE 14.15   *Nine out of ten people in the world now have access to television. There may be only one TV per village, as in this scene from Nigeria, but it opens a window to the wider world. Most of what everyone watches is made in the United States and glorifies consumerism and materialism.*

year. This growth is occurring, as is most urban growth in the world, in a country with a sagging economy, an unstable government, and a high foreign debt load.

Causes of Urban Growth

Urban populations grow in two ways: by natural increase (more births than deaths) and by immigration. Natural increase is fueled by improved food supplies, better sanitation, and advances in medical care that reduce death rates and cause populations to grow both within cities and in the rural areas around them (see chapter 4). In Latin America and East Asia natural increase is responsible for two-thirds of urban population growth. In Africa and West Asia immigration is the largest source of urban growth. Immigration to cities can be caused both by **push factors** that force people out of the country and by **pull factors** that draw them into the city.

Immigration Push-and-Pull Factors

People migrate to cities for many reasons. In some areas the countryside is overpopulated and simply can't support more people. The "surplus" population is forced to migrate to cities in search of jobs, food, and housing. In some places economic forces or political, racial, or religious conflicts drive people out of their homes. Recent wars in Sierra Leone and the Congo, for example, depopulated the countryside and drove people into cities. The United Nations estimated that in 1992 at least 40 or 50 million people were displaced by political, economic, or social instability. Many of these refugees end up in the already overcrowded megacities of the developing world. Changes in agriculture also push people to cities. In the United States mechanization of agriculture made much farm labor obsolete early in the twentieth century. The same forces displace people from the land in other countries today. Where farmland is owned by a minority of wealthy landlords, as is the case in many developing countries, peasants are often ejected when new cash crops or cattle grazing become economically viable.

Many people also move to the city because of the opportunities there. Cities offer jobs, housing, entertainment, and freedom from the constraints of village traditions. Possibilities exist in the city for upward social mobility, prestige, and power not ordinarily available in the country. Cities support specialization in arts, crafts, and professions for which markets don't exist elsewhere.

Modern communications also draw people to cities by broadcasting images of luxury and opportunity. An estimated 90 percent of the people in Egypt, for instance, have access to a television set. The immediacy of television makes city life seem more familiar and attainable than ever before (fig. 14.15). We generally assume that poor people in the streets of teeming Third World cities have no place else to go, but in spite of what appear to be dismal conditions, life in the city may be preferable to what the country had to offer.

Government Policies

Government policies often favor urban over rural areas in ways that both push and pull people into the cities. Developing countries commonly spend most of their budgets on improving urban areas (especially around the capital city, where leaders live). This gives the major cities a virtual monopoly on new jobs, housing, education, and opportunities, all of which bring in rural people searching for a better life. Lima, for example, has only 20 percent of Peru's population, but has 50 percent of the national wealth, 60 percent of the manufacturing, 65 percent of the retail trade, 73 percent of the industrial wages, and 90 percent of all banking in the country. Similar statistics pertain to São Paulo, Mexico City, Manila, Cairo, Lagos, Bogotá, and a host of other cities.

Governments often manipulate exchange rates and food prices for the benefit of more politically powerful urban populations, but at the expense of rural people. Importing lower-priced food pleases city residents, but local farmers then find it uneconomical to grow crops. As a result, an increased number of people leave rural areas to become part of a large urban workforce, keeping wages down and industrial production high. Zambia, for instance, sets maize prices below the cost of local production to

INVESTIGATING Our Environment

Urban Ecology

Traditionally, most ecologists have studied pristine, natural ecosystems trying to understand how nature works without human disturbance. In recent years, however, we have come to recognize that cities are ecological systems too. If ecology is the relationship among organisms and their environment, what could be more ecological than studying the organism (humans) with the greatest impact on other species and our environment? As the human population becomes increasingly urbanized, perhaps the most important place to study human ecology is in cities. Rather than merely seeing humans as disturbing factors, some scientists are beginning to gather data on how urban ecosystems work.

Urban areas share many characteristics with natural ecosystems. Energy flows into and out of the city, and is degraded and dispersed as it is used to do work. Materials are used to create structures, some living and some nonliving. Species compete for shelter, food, habitat, and other resources. The most successful species flourish and proliferate; less fit ones dwindle and eventually vanish. Every day, a typical American city of a million people consumes roughly 1 million metric tons (roughly 2 billion lbs) of raw materials. This urban material flow includes 500,000 m^3 (1.3 million gal) of water, the energy equivalent of about 20,000 metric tons of fossil fuels, 13,000 metric tons of other minerals (including food packaging and construction materials), 12,000 metric tons of farm products, and 10,000 metric tons of wood and paper products.

To remove municipal waste from this city of a million residents, about 1,000 fully loaded garbage trucks travel every day to rural areas carrying the discarded remnants of this urban metabolism. In addition, the city's municipal wastewater treatment plants discharge somewhat more than 500,000 m^3 of water (that lost to evaporation is offset by precipitation and solids added to the wastewater stream). You may remember from chapter 2 that most natural ecosystems recycle materials. Recycling and reuse also occur in urban ecology, although rarely with the speed and efficiency of natural systems.

Urban biodiversity and wildlife also can be important. While many biologists have tended to assume that cities are devoid of life other than humans and our domesticated companion animals and plants, we are now beginning to appreciate that urban open space can be essential refuges for wild creatures as well. As cities grow, it's not uncommon for consolidated urban areas to be more than 100 miles (160 km) across. This would present an insurmountable barrier to many migrating animals if the city consists of nothing but concreate, glass, and metal. Urban parks, forests, corridors along rivers, and even ordinary yards, if managed properly, can provide a refuge for migrating wildlife and can sustain local populations that contribute significantly to overall biodiversity. Planting native vegetation and designing landscapes to provide maximum food and shelter is an important step in protecting and promoting urban biodiversity.

Long-term ecological research (LTER) sites are ecologically significant sites identified and funded by the National Science Foundation to allow long-term, in-depth research on ecological problems and questions. Although most of the LTER sites are in remote locations, where nature has minimal human interference, two new LTER sites were recently set up in Phoenix and Baltimore. Funded by the National Science Foundation for an initial period of six years at nearly a million dollars per year, but with an understanding that the research will go on much beyond that, these projects use modern technology to study every aspect of urban ecology. How does the city shape its own weather? What plants and animals live in the city, and where, and how? What ecological processes cycle materials and energy? How do human activities influence those processes?

One of the first things that scientists are discovering is that cities are not homogenous; like other ecosystems, they have shifting patches of contrasting conditions. Using computer models and sophisticated mapping techniques, ecologists can explore how patches change in space and time. Elsewhere, environmental scientists have investigated important problems of pollution and environmental health in cities. Where are toxic and hazardous materials generated, stored, and released in the city? How do they move around and where do they accumulate?

In Detroit, a group of students worked with experts to map data on more than 5,000 children with elevated blood lead levels. Not surprisingly, they found a correlation between low incomes, old housing, incidence of poisoning, and concentration of special-education students. Sometimes public awareness of a problem is one of the best outcomes of this research. Another large group of students in the Detroit area used geographic information systems (digital maps) together with chemical and biological analysis to prepare a detailed study of water quality and ecosystem health in the Rouge River. Perhaps students at your school could organize a similar study of the urban ecology in your area.

---

discourage farming and to maintain a large pool of workers for the mines. Keeping the currency exchange rate high stimulates export trade but makes it difficult for small farmers to buy the fuels, machinery, fertilizers, and seeds that they need. This depresses rural employment and rural income while stimulating the urban economy. The effect is to transfer wealth from the country to the city.

URBAN PROBLEMS IN DEVELOPING COUNTRIES

Cities in the developing world often lack infrastructure, organization, stability, and political will to supply adequate food, water, housing, jobs, and basic services for their residents. Much of the problem arises from unplanned and uncontrollable growth, with which city governments cannot hope to keep up. A first-time visitor to a supercity—particularly in a less-developed country—is often overwhelmed by the immense crush of pedestrians and vehicles of all sorts that clog the streets. Jakarta, for instance, is one of the most densely populated cities in the world (fig. 14.16). Traffic is chaotic almost all the time. People often spend three or four hours each way commuting to work from outlying areas.

Air and Water Pollution

The dense traffic (commonly old, poorly maintained vehicles), smoky factories, and use of wood or coal fires for cooking and heating often create a thick pall of air pollution in Third World

Principles of Environmental Science

www.mhhe.com/environmentalscience

FIGURE 14.16 *Motorized rickshaws, motor scooters, bicycles, street vendors, and pedestrians all vie for space on the crowded streets of Jakarta. The heat, noise, smells, and sights are overpowering. In spite of the difficulties of living here, people work hard and have hope for the future.*

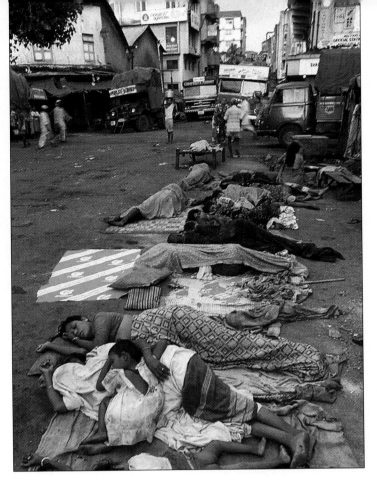

FIGURE 14.17 *In Mumbai, India, as many as half a million people sleep on the streets because they have no other place to live. Ten times as many live in crowded, dangerous slums and shantytowns throughout the city.*

cities. Lenient pollution laws, corrupt officials, inadequate testing equipment, ignorance about the sources and effects of pollution, and lack of funds to correct dangerous situations usually exacerbate the problem. What is its human toll? An estimated 60 percent of Calcutta's residents are thought to suffer from respiratory diseases linked to air pollution. Lung cancer mortality in Shanghai is reported to be four to seven times higher than rates in the countryside. Mexico City, which sits in a high mountain bowl with abundant sunshine, little rain, high traffic levels, and frequent air stagnation, has one of the highest levels of photochemical smog in the world (see chapter 9).

Few cities in developing countries can afford to build modern waste treatment systems for their rapidly growing populations. The World Bank estimates that only 35 percent of urban residents in developing countries have satisfactory sanitation services. The situation is especially desperate in Latin America, where only 2 percent of urban sewage receives any treatment. In Egypt, Cairo's sewer system was built about 50 years ago to serve a population of 2 million people. It is now being overwhelmed by more than 10 million people. Less than one-tenth of India's 3,000 towns and cities have even partial sewage systems and water treatment facilities. Some 150 million of India's urban residents lack access to sanitary sewer systems. In Colombia, the Bogotá River, 200 km (125 mi) downstream from Bogotá's 5 million residents, still has an average fecal bacteria count of 7.3 million cells per liter, more than 700,000 times the safe drinking level and 3,500 times higher than the safe limit for swimming.

Worldwide, clean water is more available to urban people than to rural people, but access to good water is a major problem. Available water is often unsafe, and diarrhea, dysentery, typhoid, and other waterborne diseases are common. Many rivers and streams in Third World countries are little more than open sewers, and yet they are all that poor people have for washing clothes, bathing, cooking, and—in the worst cases—drinking.

Housing

The United Nations estimates that at least 1 billion people—20 percent of the world's population—live in crowded, unsanitary slums of the central cities and in the vast shantytowns and squatter settlements that ring the outskirts of most Third World cities. Around 100 million people have no home at all. In Mumbai, India, for example, it is estimated that a half million people sleep on the streets, sidewalks, and traffic circles because they can find no other place to live (fig. 14.17).

Slums provide housing to large populations. These are generally legal but inadequate multifamily tenements or rooming houses, often converted from some other use. Families live crowded in small rooms with inadequate ventilation and sanitation. Often, the structures are rickety and unsafe.

Shantytowns, with shacks built of corrugated metal, discarded packing crates, brush, plastic sheets, and other scavenged materials, develop on the outskirts of most large Third World cities. Shantytowns are illegal, but they quickly fill in the unoccupied edges of town, where squatters (illegal residents) can build

FIGURE 14.18 *Shantytowns called* favelas *perch on hillsides above Rio de Janeiro.*

FIGURE 14.19 *In many older American cities, flight to the suburbs has left large areas of pollution and urban decay in the inner city.*

shelters close to jobs in the city. Many governments try to clean out illegal settlements by bulldozing the huts and sending riot police to drive out the settlers, but the people either move back in or relocate to another shantytown. Two-thirds of the population of Calcutta live in unplanned squatter settlements, and nearly half of the 20 million people in Mexico City live in uncontrolled, unauthorized shantytowns. Often, shantytowns occupy dangerous ground: In Bhopal, India, and Mexico City, for example, squatter settlements were built next to deadly industrial sites. In Rio de Janeiro, shantytowns called *favelas* perch on steep hillsides above the city (fig. 14.18). As desperate and inhumane as conditions are in these slums and shantytowns, many people do more than merely survive there. They keep themselves clean, raise families, educate their children, find jobs, and save a little money to send home to their parents.

Many of the worst urban environmental problems of the more developed countries have been substantially reduced in recent years. Improved sanitation and medical care have reduced or totally eliminated many of the communicable diseases that once afflicted urban residents. Air and water quality have improved dramatically as heavy industry such as steel smelting and chemical manufacturing have moved to developing countries. In consumer and information economies, workers no longer need to be concentrated in central cities. They can live and work in dispersed sites. Automobiles now make it possible for the working class to enjoy amenities such as single-family houses, yards, and access to recreation that once were available only to the elite.

In the United States, some of the most rapidly growing metropolitan areas like Phoenix, Arizona; Boulder, Colorado; Austin, Texas; and San Jose, California, are centers for high-technology companies located in landscaped suburban office parks. These cities often lack a recognizable downtown, being organized instead around low-density housing developments, national-chain shopping malls, and extensive freeway networks. For many high-tech companies, being located near industrial centers and shipping hubs is less important than a good climate, ready access to air travel, and amenities such as natural beauty and open space.

Old manufacturing cities such as Philadelphia and Detroit have suffered declining populations and infrastructure as their industrial base fades away (fig. 14.19). In some "rust belt" cities, up to half of inner-city housing and commercial property is abandoned and derelict. This presents an opportunity for redevelopment of brownfields (polluted industrial areas) and construction of new housing and urban parks, but financing these projects is difficult as tax revenues decline.

Urban Sprawl

While the move to suburbs and rural areas has brought many benefits to the average citizen, it also has caused numerous urban problems.

Principles of Environmental Science www.mhhe.com/environmentalscience

TABLE 14.6 Characteristics of Urban Sprawl

1. Unlimited outward extension
2. Low-density residential and commercial development
3. Leapfrog development that consumes farmland and natural areas
4. Fragmentation of power among many small units of government
5. Dominance of freeways and private automobiles
6. No centralized planning or control of land uses
7. Widespread strip-malls and "big-box" shopping centers
8. Great fiscal disparities among localities
9. Reliance on deteriorating older neighborhoods for low-income housing
10. Decaying city centers as new development occurs in previously rural areas

Source: PlannersWeb, Burlington, Vermont, 2001.

FIGURE 14.20 *Sprawl consumes natural areas, farmland, and open space. It wastes resources, results in traffic congestion, and requires large investments to replace schools, roads, parks, and shopping that is abandoned in the inner city.*

Cities that once were compact now spread over the landscape, consuming open space and wasting resources. This pattern of urban growth is known as **sprawl.**. While there is no universally accepted definition of the term, sprawl generally includes the characteristics outlined in table 14.6.

In most American metropolitan areas, the bulk of new housing is in large, tract developments that leapfrog out beyond the edge of the city in a search for inexpensive rural land with few restrictions on land use or building practices (fig. 14.20). The U.S. Department of Housing and Urban Development estimates that urban sprawl consumes some 200,000 ha (roughly 500,000 acres) of farmland each year. With planning authority divided among many small, local jurisdictions, most metropolitan areas have no way to regulate growth or provide for rational, efficient resource use. Small towns and township or county officials generally welcome this growth because it profits local landowners and businesses. Although the initial price of tract homes often is less than comparable urban property, there are external costs in the form of new roads, sewers, water mains, power lines, schools, shopping centers, and other infrastructure required by this low-density development. Ironically, people who move out to rural areas to escape from urban problems such as congestion, crime, and pollution often find that they have simply brought those problems with them. A neighborhood that seemed tranquil and remote when they first moved in, soon becomes just as crowded, noisy, and difficult as the city they left behind as more people join them in their rural retreat.

Because many Americans live far from where they work, shop, and recreate, they consider it essential to own a private automobile. The average U.S. driver spends about 443 hours per year behind a steering wheel. This means that for most people, the equivalent of one full 8-hour day per week is spent sitting in an

automobile. Building the roads, parking lots, filling stations, and other facilities needed for an automobile-centered society takes a vast amount of space and resources. In some metropolitan areas one-third of all land is devoted to the automobile. To make it easier for suburban residents to get from their homes to jobs and shopping, we provide an amazing network of freeways and highways. At a cost of several trillion dollars to build, the interstate highway system was designed to allow us to drive at high speeds from source to destination without ever having to stop.

As more and more drivers clog the highways, the reality is far different. In Los Angeles, for example, which has the worst congestion in the United States, the average speed in 1982 was 58 mph (93 km/hr) and the average driver spent less than 4 hours per year in traffic jams. In 2002, the average speed in Los Angeles was half that of 20 years earlier, and the typical driver spent 136 hours per year in bumper-to-bumper traffic. Although new automobiles are much more efficient and cleaner operating than those of a few decades ago, the fact that we drive so much farther today and spend so much more time idling in stalled traffic means that we burn more fuel and produce more pollution than ever before.

Altogether, traffic congestion costs the United States $78 billion per year in wasted time and fuel. Some people argue that the existence of traffic jams in cities shows that more freeways are needed. Often, however, building more traffic lanes simply encourages more people to drive farther than before. Rather than ease congestion and save fuel, more freeways can exacerbate the problem (fig. 14.21).

Smart Growth

Are there alternatives to unplanned sprawl and wasteful resource use? One option proposed by many urban planners is **smart growth**, which makes effective use of land resources and existing infrastructure by encouraging in-fill development that avoids

FIGURE 14.21 *Building freeways to reduce congestion is like trying to diet by loosening your belt.*

costly duplication of services and inefficient land use (table 14.7). Smart growth aims to provide a mix of land uses to create a variety of affordable housing choices and opportunities. It also attempts to provide a variety of transportation choices, including pedestrian friendly neighborhoods. This approach to planning also seeks to maintain a unique sense of place by respecting local cultural and natural features.

By making land-use planning open and democratic, smart growth makes urban expansion fair, predictable, and cost-effective. All stakeholders are encouraged to participate in creating a vision for the city and to collaborate rather than confront each other. Goals are established for staged and managed growth in urban transition areas with compact development patterns. This approach is not opposed to growth. It recognizes that the goal is not to block growth but to channel it to areas where it can be sustained over the long term. Smart growth strives to enhance access to equitable public and private resources for everyone and to promote the safety, livability, and revitalization of existing urban and rural communities.

Smart growth protects environmental quality. It tries to reduce traffic and to conserve farmlands, wetlands, and open space. As cities grow and transportation and communications enable more community interaction, the need for regional planning becomes greater and more pressing. Community and business leaders must make decisions based on a clear understanding of regional growth needs and how infrastructure can be built most efficiently and for the greatest good.

One of the best examples of successful urban land-use planning in the United States is Portland, Oregon, which has rigorously enforced a boundary on its outward expansion, requiring, instead, that development be focused on in-filling unused space within the city limits. Because of its many urban amenities, Portland is considered one of the best cities in America. Between 1970 and 1990 the Portland population grew by 50 percent, but its total land area grew only 2 percent. During this time, Portland property taxes decreased 29 percent and vehicle miles traveled increased only 2 percent. By contrast, Atlanta, which had similar population growth, experienced an explosion of urban sprawl that increased its land area three-fold, drove up property taxes 22 percent, and increased traffic miles by 17 percent. A result of this expanding traffic and increasing congestion was that Atlanta's air pollution increased by 5 percent, while Portland's, which has one of the best public transit systems in the nation, decreased by 86 percent.

Urban Sustainability in the Developed World

Rather than abandon the cultural history and infrastructure investment in existing cities, a group of architects and urban planners is attempting to redesign metropolitan areas to make them more appealing, efficient, and livable. European cities such as Stockholm, Sweden; Helsinki, Finland; Leichester, England; and Neerlands, the Netherlands, have a long history of innovative urban planning. In the United States Andres Duany, Elizabeth Plater-Zyberk, Peter Calthorpe, and Sym Van Der Ryn have been leaders in this movement. Sometimes called a neo-traditionalist approach, these designers attempt to recapture some of the best features of small towns and the best cities of the past. They are designing urban neighborhoods that integrate houses, offices, shops, and civic buildings. Ideally, no house should be more than a five-minute walk from a neighborhood center with a convenience store, a coffee shop, a bus stop, and other amenities. A mix of apartments, town houses, and detached houses in a variety of price ranges insures that neighborhoods

A MODEL SUSTAINABLE CITY

Chattanooga, Tennessee, is nestled in the Tennessee River valley, at the foot of the Cumberland Plateau. This medium-size industrial city provides an inspiring example of participatory city planning that has produced a clean urban environment, a vibrant economy, and strong social capital. Twenty-five years ago, the U.S. Department of Health, Education, and Welfare named Chattanooga America's dirtiest city. Factory smoke trapped between the mountains gave city air the highest particulate count in the nation—dirtier, even, than Los Angeles. Per capita tuberculosis rates were three times the national average. Chattanooga Creek was so polluted by toxic waste from coke ovens and steel mills that the Environmental Protection Agency designated 4 km (2.5 mi) of it a Superfund site.

Social conditions declined along with environmental quality. As industries moved out and the upper class fled to the suburbs, the city was left with abandoned factories, boarded-up buildings, a declining tax base, unemployment, crime, and despair. The worst conditions were concentrated in the poorest neighborhoods, producing further frustration and social division.

In 1984 the city council, led by David Crockett—the great nephew of the famous frontiersman—started a series of town meetings to get residents talking about the city and its future. More than 1,700 people brainstormed ways to achieve clean air and water, jobs, affordable housing, personal safety, a place to fish, and somewhere to walk along the river. By giving everyone a voice, the process built community spirit and a broad base of support. Suddenly, people with energy and good ideas from every part of the city got involved.

Public-private partnerships led to nearly $800 million in investments (two-thirds from the private sector) and created 8,000 new permanent and temporary jobs. The Chattanooga Neighborhood Enterprise Corporation invested $60 million in affordable housing. More than 1,000 first-time home buyers obtained low-interest loans, while other residents got home-improvement help. By 1990 Chattanooga was in compliance with national air quality standards. Public transport is widely used and efficient. The riverfront has been cleaned up. A 35-km (22-mi) long riverfront park with walking and bike trails has been built along both sides of the river through downtown. Fountains, flowers, shade trees, street musicians, and outdoor festivals now fill the park. Property values have shot up, and people are moving back to the center city to take advantage of the recreational and cultural opportunities.

Using principles of urban ecology, the city has created a livable environment, a strong sense of community, and a flourishing economy. As David Crockett says, it is "a city that is cleaner, greener, and safer; that values human and natural resources; and that provides an economic system that will keep our children here."

Chattanooga's waterfront has been cleaned up and has become an asset rather than a liability.

will include a diversity of ages and income levels (fig. 14.22). Some design principles of this movement include:

- Limit city size or organize cities in modules of 30,000 to 50,000 people—large enough to be a complete city but small enough to be a community.

- Maintain greenbelts in and around cities. These provide recreational space and promote efficient land use, as well as helping to ameliorate air and water pollution.

- Determine in advance where development will take place. This protects property values and prevents chaotic development.

FIGURE 14.22 *By clustering buildings together, Kentlands, Maryland, has maintained a high percentage of open space as public commons.*

Planning can also protect historical sites, agricultural resources, and ecological services of wetlands, clean rivers, and groundwater replenishment.

- Locate everyday shopping and services so people can meet daily needs with greater convenience, less stress, less automobile dependency, and less use of time and energy. This might be accomplished by encouraging small-scale commercial development in or close to residential areas.

- Encourage walking or the use of small, low-speed, energy-efficient vehicles (microcars, motorized tricycles, bicycles, etc.) for many local trips now performed in full-size automobiles. Creating special traffic lanes, reducing the number or size of parking spaces, or closing shopping streets to big cars might encourage such alternatives.

- Promote more diverse, flexible housing as an alternative to conventional, detached, single-family houses. "In-fill" building between existing houses saves energy, reduces land costs, and might help provide a variety of living arrangements. Allowing single-parent families or groups of unrelated adults to share housing and to use facilities cooperatively also provides alternatives to those not living in a traditional nuclear family.

- Make cities more self-sustainable by growing food locally, recycling wastes and water, using renewable energy sources, reducing noise and pollution, and creating a cleaner, safer environment. Encourage community gardening. Reclaimed inner-city space or a greenbelt of agricultural and forestland around the city provides food and open space, and also contributes valuable ecological services, such as purifying air, supplying clean water, and protecting wildlife habitat and recreation land.

- Invite public participation in decision making. Emphasize local history, culture, and environment to create a sense of community and identity. Create local networks in which residents take responsibility for crime prevention, fire protection, and home care of children and the elderly, sick, and disabled. Coordinate regional planning through metropolitan boards that cooperate with but do not supplant local governments.

- Plan cluster housing, or open-space zoning, which preserves at least half of a subdivision as natural areas, farmland, or other forms of open space. Studies have shown that people who move to the country don't necessarily want to live miles from the nearest neighbor; what most desire is long views across an interesting landscape and an opportunity to see wildlife. By carefully clustering houses on smaller lots, a conservation subdivision can provide the same number of buildable lots as a conventional subdivision and still preserve 50 to 70 percent of the land as open space (fig. 14.23, right). This not only reduces development costs (less distance to build roads, lay telephone lines, sewers, power cables, etc.) but also helps to foster a greater sense of community among new residents.

FIGURE 14.23 *Conventional subdivision* (left) *and an open-space plan* (right). *Although both plans provide 36 home sites, the conventional development allows for no open space. Cluster housing on smaller lots in the open-space design preserves at least half of the area as woods, prairie, wetlands, farms, or other conservation lands, while providing residents with more attractive vistas and recreational opportunities than a checkerboard development.*

- Preserve urban habitat: It can make a significant contribution toward saving biodiversity. In a groundbreaking series of habitat conservation plans triggered by the need to protect the endangered California gnatcatcher, some 85,000 ha (210,000 acres) of coastal scrub near San Diego was protected as open space within the rapidly expanding urban area. This is an area larger than Yosemite Valley, and it will benefit many other species, as well as humans.

Sustainable Development in the Third World

What can be done to improve conditions in Third World cities? Curitiba, Brazil (described at the beginning of this chapter) is an outstanding example of what can be done, even in relatively undeveloped countries, to improve transportation, protect central cities, and create a sense of civic pride. Other cities have far to go, however, to reach this standard. Among the immediate needs are housing, clean water, sanitation, food, education, health care, and basic transportation for their residents. The World Bank estimates that improved living conditions in urban households in the developing world could vastly increase health and productivity worldwide.

Some countries, recognizing the need to use vacant urban land, are redistributing unproductive land or closing their eyes to illegal land invasions. Indonesia, Peru, Tanzania, Zambia, and Pakistan have learned that squatter settlements make a valuable contribution to meeting national housing needs. Squatters' rights are being upheld in some cases, and such services as water, sewers, schools, and electricity are being provided to the settlements (fig. 14.24). Some countries intervene directly in land distribution and land prices. Tunisia, for instance, has a "rolling land bank" to buy and sell land. This strong and effective program controls urban land prices and reduces speculation and unproductive land ownership.

Many planners argue that social justice and sustainable economic development are answers to the urban problems discussed in this chapter. If people have the opportunity and money to buy better housing, adequate food, clean water, sanitation, and other things they need for a decent life, they will do so. Democracy, security, and improved economic conditions help in slowing population growth and reducing rural-to-city movement. An even more important measure of progress may be institution of a social welfare safety net guaranteeing that old or sick people will not be abandoned and alone.

FIGURE 14.24   *In this colonia on the outskirts of Mexico City, residents work with the government to bring in electricity, water, and sewers to shantytowns and squatter settlements. Like many developing countries, Mexico recognizes that helping people help themselves is the best way to improve urban living.*

Some countries have accomplished these goals even without industrialization and high incomes. Sri Lanka, for instance, has lessened the disparity between the core and periphery of the country. Giving all people equal access to food, shelter, education, and health care eliminates many incentives for interregional migration. Both population growth and city growth have been stabilized, even though the per capita income is only $800 per year. China has done something similar on a per capita income of around $300 per year.

Whether sustained, environmentally sound economic development is possible for a majority of the world's population remains one of the most important and most difficult questions in environmental science. The unequal relationship between the richer "northern" countries and their impoverished "southern" neighbors is a major part of this dilemma. Some people argue that the best hope for developing countries may be to "delink" themselves from the established international economic systems and develop direct south-south trade based on local self-sufficiency, regional cooperation, barter, and other forms of nontraditional exchange that are not biased in favor of the richer countries.

SUMMARY

Sustainable futures depend on how we use and conserve our resources and our environment. Our definitions of resources shape the ways we identify and use them, as well as the ways we theorize their future use. Ecological economics brings the insights of ecology into more traditional economic analysis. It calls for consideration of the role of natural capital and ecological services in planning and accounting. A resource is anything with potential use for creating wealth or giving satisfaction. Minerals and fossil fuels are examples of nonrenewable resources. Biological organisms and ecological processes are renewable and self-reproducing, but can be exhausted by overuse.

Questions about the scarcity of resources and their effects on economic development are important in determining what kind of society we have. Open-access systems encourage narrow self-interests and resource exploitation. In communal property resource systems, self-governing, locally based community management has successfully sustained natural resources for centuries in many cases.

Cost-benefit analysis can be a useful tool in environmental management. Ultimately, our aim should be to internalize costs that are now treated as externalities. GNP is used as a measure of economic progress, but new measures for human well-being and environmental health, such as the Human Development Index, are needed.

Green business and green design are efforts to reduce the environmental and social costs of the things we buy and use. These efforts are important contributors to sustainability, and they have proven profitable for businesses that use them.

Cities are where almost half of us live and where many of our resources are used. Cities are growing worldwide, especially in developing countries. Only Africa and South Asia remain predominantly rural, and cities are growing rapidly there as well. A variety of push-and-pull factors contribute to this urban growth. The consequences of urbanization in many developing areas include overstressed infrastructure, pollution, and crowding. Among the worst problems faced in these cities are traffic congestion, air pollution, inadequate or nonexistent sewers and waste disposal systems, water pollution, and housing shortages. Millions of people live in appalling slums and shantytowns, yet these people raise families, educate their children, learn new jobs and new ways of living, and have hope for the future.

In wealthier countries, problems of cities include urban decay and sprawl, and inefficiency and pollution associated with our reliance on cars and on low-density suburban development patterns. Planners propose a variety of strategies that may help make cities in both the developed and the developing world healthier, safer, and more environmentally sound and sustainable environments.

QUESTIONS FOR REVIEW

1. Identify the ways resources are defined by classical, neo-classical, and ecological economics.

2. Explain the difference between renewable and nonrenewable resources. Are nonrenewable resources always limiting factors in economic productivity?

3. Explain how scarcity and technological progress can extend resource availability and the carrying capacity of the environment.

4. Why does the marketplace sometimes fail to optimally allocate natural resource values?

5. What are some ways of accounting for environmental resources and including them in cost-benefit analysis?

6. Worldwide, how many people now live in cities, and how many live in rural areas?

7. What changes in urbanization are predicted to occur in the next 50 years, and where will those changes occur?

8. Where is most urban growth going on? What are the environmental and social costs of urbanization?

9. Why are urban areas in U.S. cities decaying? What can be done about this?

10. Describe some aspects of a "sustainable city."

THINKING SCIENTIFICALLY

1. When the ecologist warns that we are using up irreplaceable natural resources and the economist rejoins that ingenuity and enterprise will find substitutes for most resources, what underlying premises and definitions shape their arguments?

2. What would be the effect on the developing countries of the world if we were to change to a steady-state economic system? How could we achieve a just distribution of resource benefits, while still protecting environmental quality and future resource use?

3. If you were doing a cost-benefit study, how would you assign a value to the opportunity for good health or the existence of rare and endangered species in faraway places? Is there a danger or cost in simply saying some things are immeasurable and priceless and therefore off limits to discussion?

4. If natural capitalism or eco-efficiency has been so good for some entrepreneurs, why haven't all businesses moved in this direction?

5. What relationships can you draw between principles of thermodynamics and principles of ecological economics?

6. A city could be considered an ecosystem. Using what you learned in chapters 2 and 3, describe the structure and function of a city in ecological terms.

7. Look at the major urban area(s) in your state. Why were they built where they are? Are those features now a benefit or drawback?

8. Boulder, Colorado, has been a leader in controlling urban growth. One consequence is that housing costs have skyrocketed, and poor people have been driven out. If you lived in Boulder, would you vote for additional population limits? What do you think is an optimum city size?

9. This chapter presents a number of proposals for suburban redesign. Which of them would be appropriate or useful for your community? Try drawing up a plan for the ideal design of your neighborhood.

KEY TERMS

capital 322
communal resource
    management systems 325
cost-benefit analysis (CBA)
    325
discount rate 327
ecological economics 322
ecological services 322
externalizing costs 327
gross national product (GNP)
    322
internalizing costs 327
limits to growth 324
megacities 331

nonrenewable resources 322
open access system 325
pull factors 333
push factors 333
renewable resources 323
resources 320
sprawl 337
smart growth 337
steady-state economy 322
sustainable
    development 320
"The Tragedy of the
    Commons" 325
urbanization 331

Welcome to McGraw-Hill's Online Learning Center

Location: http://www.mhhe.com/environmentalscience

McGraw Hill

WEB EXERCISES

Calculating Your Ecological Footprint

One way of assessing the resource consumption patterns of communities and individuals is to calculate an ecological footprint. To learn about this concept and how it is done, go to www.rprogress.org/, the web page of Redefining Progress. You can find information here about individual consumption levels and those of whole nations as well as an interesting discussion of genuine progress indicators and real measures of satisfaction. For a quick calculation of your ecological footprint, go to www.earthday.net/footprint/info.asp. By answering 14 simple questions, you can find out how much land it takes to support your present lifestyle. Suggestions are given for average consumption rates of residents of Canada and the United States. At the end, you are asked how much space should be reserved for other species. This shows you how many earths it would take to support the whole human population at your level of consumption. For more detailed information on how ecological footprints are calculated, go to www.rprogress.org/programs/sustainability/ef/methods/calculating.html.

Land Use and Sustainability

Land is a resource that is at the center of many of our sustainability debates. How land should be used is a perennial policy and development problem. In the United States the Natural Resources Conservation Service (NRCS) conducts periodic surveys of land use and agricultural production. The NRCS website at www.nrcs.usda.gov/technical/land/ includes a large collection of mapped data.

1. Click on several themes in the index at the address above. Which topics would be most useful for providing information about sustainability

of land and water use in the United States? What information could you use from the different topics to support arguments about wise or foolish resource use decisions? Which of the resources mapped are renewable? Which are nonrenewable?

2. Urban growth into agricultural lands is a problem in many cities. Click on the link for "Urbanization" to find maps of acreage conversion to urban land use. For each map you can see an enlarged version by clicking on "Large GIF" just below the map image. Note the dates on the maps and the meaning of the colors. What do these maps tell about development rates in your area? How does growth impact sustainability in your area? Which regions are experiencing the most rapid urban expansion?

Cities and Sustainability

The New Colonist website at www.newcolonist.com/ is a website devoted to making cities more sustainable and more livable. Look at the list of current featured stories on the web page. Discuss the following questions with a group of your colleagues: What are the main themes? (transportation? food? housing? energy?) What important topics are missing? What are some of the recommendations for change in the stories on the main topics? Based on your understanding of sustainability, economics, and resource use, do you think the recommendations in these articles are practical? If so, what could you do to help pursue recommended changes? If not, how could conditions be changed to make the recommendations more practical?

You can never solve a problem with the same kind of thinking that created the problem in the first place.
–Albert Einstein

The Christmas Bird Count is an excellent example of citizen-science in action.

OBJECTIVES

After studying this chapter, you should be able to:

• understand how adaptive and precautionary principles can help us make decisions in "wicked" problems where scientific evidence is incomplete or contradictory.

• be aware of the goals and opportunities in environmental education and environmental careers.

• summarize the cycle by which policy is established.

• describe the path of a bill through the legislature.

• recognize the differences among civil, criminal, and administrative law.

• judge the effectiveness of litigation in environmental issues.

• consider the reasons that international treaties have or have not been successful.

• scrutinize collaborative, community-based planning methods.

• compare radical and mainline environmental groups and the tactics they employ to bring about social change.

Welcome to McGraw-Hill's Online Learning Center

Location: http://www.mhhe.com/environmentalscience

LEARNING ONLINE

Visit our webpage at www.mhhe.com/environmentalscience for data sources, further readings, additional case studies, current environmental news, and regional examples within the Online Learning Center to help you understand the material in this chapter. You'll also find active links to information pertaining to this chapter including:

The Center for International Law
Environmental Defense
NRDC: Natural Resources Defense Council
West Coast Environmental Law Website and Library

Citizen Science and the Christmas Bird Count

Every Christmas since 1900, teams of dedicated volunteers have counted and recorded all the birds they can find in a designated 24-km (15-mi) circle. This effort has become the largest, longest-running, citizen-science project in the world. For the 100th count, nearly 50,000 participants in about 1,800 teams observed 58 million birds belonging to 2,309 species. Although about 70 percent of the counts in 2000 were made in the United States or Canada, 650 teams in the Caribbean, Pacific Islands, and Central and South America also participated. Participants enter their bird counts on standardized data sheets, or submit their observations over the Internet. Compiled data can be viewed and investigated online, almost as soon as they are submitted.

The Christmas Bird Count was started in 1900 by Frank Chapman, the editor of *Bird-Lore* magazine and an officer in the newly formed Audubon Society. For years, hunters had gathered on Christmas Day for a competitive hunt, often killing hundreds of birds and mammals as teams tried to outshoot each other. Chapman suggested an alternative contest: to see which team could observe and identify the most birds, and the most species, in a day. The competition has grown and spread. In the 100th annual count, the winning team was in Monte Verde, Costa Rica, with an amazing 343 species tallied in a single day (Monte Verde's biodiversity is under threat, though: see opening story, chapter 6).

The tens of thousands of birdwatchers participating in the count gather vastly more information about the abundance and distribution of birds than biologists could gather themselves. Their data provide important information for scientific research on bird migrations, populations, and habitat change. Now that the entire record for a century of bird data is available on the BirdSource website (www.birdsource.org), both professional ornithologists and amateur birdwatchers can study the geographical distribution of a single species over time, or they could examine how all species vary at a single site through the years. Those concerned about changing climate can look for variation in long-term distribution of species. Climatologists can analyze the effects of weather patterns such as El Niño or La Niña on where birds occur. One of the most intriguing phenomena revealed by this continent-wide data collection is irruptive behavior: that is, appearance of massive numbers of a particular species in a given area in one year and then their move to other places in subsequent years following weather patterns, food availability, and other factors.

Following the success of the Christmas Bird Count, other citizen-science projects have been initiated. Project Feeder Watch enlists the efforts of schoolchildren and backyard birdwatchers to monitor birds. Farmers have monitored their own pasture and stream health and collaborated with students to explore how management can improve the land (see chapter 7, opening story). Volunteers are being enlisted to monitor water quality, climate and air quality, forest health, and many other environmental factors. Data collected from such widespread, coordinated projects provide an invaluable source of information for scientific research. How can you learn more about your local environment, and contribute to research, by participating in a citizen-science project? Contact your local Audubon chapter or your state's Department of Natural Resources to find out what you can do.

What does citizen science have to do with environmental policy? In a democratic society, well-informed citizens help shape policy. Scientifically literate individuals can help elected officials make sound decisions about developing and conserving resources. Citizen-science projects are one way individuals can learn more about their environment, and become more interested in their community. In this chapter, we'll investigate how public policy is made, and we'll look at the ways science and a well-educated public can help shape effective policies.

MAKING DECISIONS IN AN UNCERTAIN WORLD

Once you understand the environmental processes and problems we've seen in this book, what do you do next? One thing you can do is to put your knowledge to work helping society make good decisions and rules, or policies, that affect your environment and your community. But first it is important to know something about how policies are made, and how decisions can be formed in the face of uncertainty and competing interests.

Throughout this book, we've looked at difficult environmental problems. You have read that global carbon dioxide emissions are about four times what they were in 1950, and that about half the world's wetlands were lost in the twentieth century. You have learned that exposure to toxic and hazardous chemicals can cause birth defects, cancer, and chronic ailments.

More than 20 percent of the world's people lack access to adequate food, basic sanitation, and clean water (fig. 15.1). Both science and policy can help solve these problems. Table 15.1 summarizes some of the most important questions on which environmental scientists are currently working. In this chapter, we'll consider how answers provided by science can help establish public policy. We'll also look at ways you can get involved in this process.

In chapter 1 we defined science as a way of exploring and explaining the world around us. Ideally, science provides an orderly, methodical approach to collecting and analyzing information. The core of modern science is formulating hypotheses and then testing them in an objective, impartial, controlled manner. One of the most important things for you to understand about science is that hypotheses can be falsified—shown to be wrong—but they almost never can be shown to be unquestionably correct.

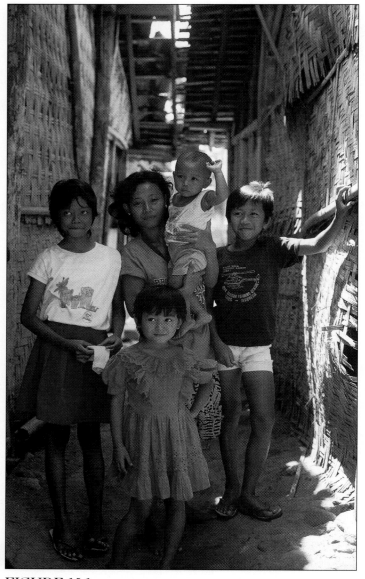

FIGURE 15.1 *More than 1 billion people—like this Indonesian family—live in acute poverty, lacking access to secure food supplies, safe drinking water, housing, education, health services, and decent jobs.*

TABLE 15.1 Some Important Current Questions in Environmental Science

1. What effects will global climate change have, and what can we do to prevent or mitigate adverse effects of these changes?

2. How and where can we best prevent habitat and biodiversity losses?

3. Where will freshwater shortages be most severe, and how can we conserve and extend our water supplies?

4. What are the most important sources of air and water pollution, and how can we reduce them?

5. How can we encourage energy conservation and a switch to renewable energy sources?

6. What are the effects of large-scale, chemical-intensive agriculture, and what are the barriers to sustainable, regenerative farming?

7. What can we do to encourage a demographic shift to balanced population growth?

8. How can we move toward an ecologically sustainable, socially just economy?

9. What health risks are associated with environmental factors, and how can we reduce those risks?

10. What are our ethical responsibilities toward nature, and how can we foster an ethic of environmental stewardship and sustainability?

Therefore, theories and explanations based on scientific data are always conditional. An overwhelming body of evidence currently may support a particular scientific interpretation, but we always have to be aware that the next investigation or experiment could provide evidence that completely overturns all our previous understanding of a particular topic. This may sound frustrating, but think of it as an ongoing process that gets us closer and closer to an understanding of the world around us.

Adaptive Management

How can we make decisions in the face of this scientific uncertainty? The answer is that our plans generally have to be conditional and contextual. Scientific information can help us understand environmental issues, but the policies we create based on this understanding will always depend on further study and more confirming evidence. An approach currently favored by many natural resource managers is called **adaptive management,** or "learning by doing." In adaptive management, policies are designed from the outset to use scientific principles to examine alternatives and assess outcomes. Rather than assume that what seems the best initial policy option will always remain so, adaptive management sets up scientific experiments to monitor how conditions are changing, and what effects our actions (or inactions) are having on both target and nontarget elements of the system. The goal of adaptive management is to enable us to live with the unexpected. It aims to yield understanding as much as to produce answers or solutions (table 15.2).

To give an example of adaptive management, suppose that a forest supervisor is debating whether clear-cutting or selective harvesting is the best option for forest harvesting. Using adaptive management principles, she or he would set up a series of carefully planned experiments to study the effects of projects as they are carried out. As information becomes available, management plans—and the monitoring programs that assess them—can constantly be adapted to changing conditions and interpretations. If the original hypothesis is confirmed, the management techniques may be applied on a wider scale or in some new situation. If unexpected results are discovered, new approaches can be tested. Another important aspect of adaptive design is that it tends to be broader and more holistic than the single-resource focus of previous management approaches. Ecological, social, and economic

TABLE 15.2 Institutional Conditions for Adaptive Management

1. There is a mandate to take action in the face of uncertainty.
2. Decision makers are aware they are experimenting anyway.
3. Decision makers care about improving outcomes over biological timescales.
4. Preservation of pristine environments is no longer an option, and human intervention cannot produce desired outcomes predictably.
5. Resources are sufficient to measure ecosystem-scale behavior.
6. Theory, models, and field methods are available to estimate and infer ecosystem-scale behavior.
7. Hypotheses can be formulated.
8. Organizational culture encourages learning from experience.
9. There is sufficient stability to measure long-term outcomes; institutional patience is essential.

Source: From Kai N. Lee, *Compass and Gyroscope*, 1993. Copyright © 1993 Island Press. Reprinted by permission of Alexander Hoyt & Associates.

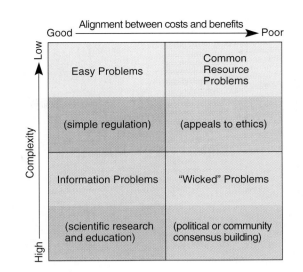

FIGURE 15.2 *The difficulty of environmental decision making increases as problems become more complex and the congruence between those who bear costs and enjoy benefits decreases. Solutions are shown in parentheses.*

impacts of our actions are considered equally important with the sheer availability or market cost of a commodity.

"Wicked" Problems

One reason why plans need to be conditional and why management should incorporate scientific experiments is that many environmental problems have no simple answer. Questions such as "What does ecosystem health mean?" or "How clean is clean?" don't have clear right or wrong answers. They depend on your worldview and how you define these terms. Different people come to different conclusions even if they share the same information.

Environmental scientists describe these issues as being **"wicked" problems,** not in the sense of having malicious intent, but rather as obstinate or intractable dilemmas. These problems often are nested within other sets of interlocking issues. The definition of both the problem and its solutions differ for various stakeholders. There are no value-free, objective answers for these puzzles, only choices that are better or worse, depending on your viewpoint. Wicked problems are important and have serious consequences. They also are complex, and those who bear the costs often are different from those who would benefit from proposed solutions (fig. 15.2). These problems usually can't be solved by simple regulations, more research, or appeals to ethics. Often, the best solution comes from community-based planning and consensus building. Inherent uncertainty gives these questions no clear end point. You cannot know when all possible solutions have been explored.

Recent advancements in understanding of how ecological systems work give us some insight into many wicked problems. Like biological organisms, social problems often change and evolve over time. Their history unfolds in complex ways, depending on chance interactions and unpredictable events. Like ecological systems, many environmental issues may never reach a stable

equilibrium. Each involves an assemblage of issues and actors that are unique in time and place. They can't be standardized. There are no good precedents from previous experience. The solutions are unique, and what may work today may not be applicable tomorrow.

The Precautionary Principle

The fact that many of the dilemmas we face are ambiguous and unclear doesn't mean that we can avoid decisions while we study the situation forever. In many cases if we wait until we have an absolutely certain solution, it may be too late to act. We need the best information available to make an informed decision, but sometimes, we need to take a courageous, but modest, stand that chooses among the best possible policy options. We need to recognize that science doesn't have all the answers. We may need to modify our position as we learn more.

Drawing on studies of ecological systems, many conservation biologists advocate a **precautionary principle** that says we should plan a margin of safety for error or surprises in natural systems. This approach says that when human health or the environment are threatened, precautionary measures should be taken even if some cause-and-effect relationships are not fully established scientifically. There are four basic tenets of precautionary action:

- People have a duty to take anticipatory steps to prevent harm. If you have a reasonable suspicion that something bad might happen, you have an obligation to try to stop it.
- The burden of proof of carelessness of a new technology, process, activity, or chemical lies with the proponents, not with the general public.
- Before using a new technology, process, or chemical, or starting a new activity, people have an obligation to examine a full range of alternatives, including the alternative of not using it.

FIGURE 15.3 *Outdoor activities are an essential part of environmental education, which strives to introduce environmental literacy to the general population and to encourage students to prepare for environmental careers.*

- Decisions applying the precautionary principle must be open, informed, and democratic, and must include the affected parties.

Opponents of this approach claim that it could prevent us from doing anything productive or innovative. What do you think? Is this principle just common sense or an invitation to decision paralysis?

ENVIRONMENTAL EDUCATION

We've defined environmental science in this book as a systematic study of our environment and our place in it. A relatively new field, environmental science is highly interdisciplinary; it integrates information from many fields of study to understand how the world works and how we should behave as environmental citizens. We also have described environmental science as mission oriented, implying that we all have a responsibility to get involved to do something about the problems we have created. What can each of us do to fulfill that responsibility?

The National Environmental Education Act sets two broad goals: (1) to improve understanding among the general public of natural and built environments and the relationships between humans and their environment, including global aspects of environmental problems, and (2) to encourage postsecondary students to pursue careers related to the environment. Specific objectives proposed to meet these goals include developing an awareness of

and appreciation for our natural and social/cultural environment, a knowledge of basic ecological concepts, an acquaintance with a broad range of current environmental issues, and experience in using investigative, critical-thinking, and problem-solving skills in solving environmental problems (fig. 15.3).

Outdoor activities and natural sciences are important components of this mission, but environmental topics such as responsible consumerism, waste disposal, and environmental ethics can and should be incorporated into reading, writing, arithmetic, and every other part of education. Table 15.3 presents one set of guidelines for an environmental education program.

Environmental Literacy

Speaking to the first broad goals of the National Environmental Education Act, former Environmental Protection Agency (EPA) administrator William K. Reilly called for broad **environmental literacy** in which every citizen is fluent in the principles of ecology and has a "working knowledge of the basic grammar and underlying syntax of environmental wisdom." Environmental literacy, according to Reilly, can help create a stewardship ethic—a sense of duty to care for and manage wisely our natural endowment and our productive resources for the long haul. "Environmental education," he said, "boils down to one profoundly important imperative: preparing ourselves for life in the next century. As it is now the twenty-first century, it will not be enough for a few specialists to know what is going on while the rest of us wander about in ignorance."

While you've made a great start toward learning about your environment by reading this book and taking a class in environmental science, we hope you will continue to read and expand your horizons after your class is finished. A large body of environmental literature can help you in this quest. Some of the most

TABLE 15.3 Outcomes from Environmental Education

1. *The natural context:* An environmentally educated person understands the scientific concepts and facts that underlie environmental issues and the interrelationships that shape nature.

2. *The social context:* An environmentally educated person understands how human society is influencing the environment, as well as the economic, legal, and political mechanisms that provide avenues for addressing issues and situations.

3. *The valuing context:* An environmentally educated person explores his or her values in relation to environmental issues; from an understanding of the natural and social contexts, the person decides whether to keep or change those values.

4. *The action context:* An environmentally educated person becomes involved in activities to improve, maintain, or restore natural resources and environmental quality for all.

Source: A Greenprint for Minnesota, Minnesota Office of Environmental Education, 1993.

TABLE 15.4 The Environmentalist's Bookshelf

What are some of the most influential and popular environmental books? In a survey[1] of environmental experts and leaders around the world on the best books on nature and the environment, the top ten were:

A Sand County Almanac by Aldo Leopold (100)[2]

Silent Spring by Rachel Carson (81)

State of the World by Lester Brown and the Worldwatch Institute (31)

The Population Bomb by Paul Ehrlich (28)

Walden by Henry David Thoreau (28)

Wilderness and the American Mind by Roderick Nash (21)

Small is Beautiful: Economics as if People Mattered by E. F. Schumacher (21)

Desert Solitaire: A Season in the Wilderness by Edward Abbey (20)

The Closing Circle: Nature, Man, and Technology by Barry Commoner (18)

The Limits to Growth: A Report for the Club of Rome's Project on the Predicament of Mankind by Donella H. Meadows et al. (17)

[1]Robert Merideth, 1992, G. K. Hall/Macmillan, Inc.

[2]Indicates number of votes for each book. Because the preponderance of respondents were from the United States (82 percent), American books are probably overrepresented.

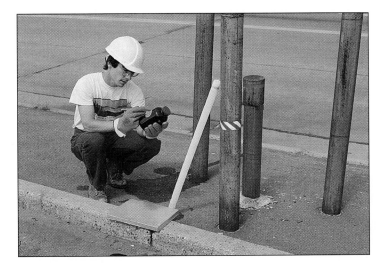

FIGURE 15.4 *Many interesting, well-paid jobs are opening up in environmental fields.*

influential environmental books of all time are listed in table 15.4. To this list, we would add some personal favorites: *The Singing Wilderness* by Sigurd F. Olson, *My First Summer in the Sierra* by John Muir, and *Remaking Society: Pathways to a Green Future* by Murray Bookchin.

Environmental Careers

The need for both environmental educators and environmental professionals opens up many job opportunities in environmental fields. The EPA estimates, for example, that 100,000 new professionals will be needed in the United States in the next five years alone to deal with hazardous-waste problems (fig. 15.4). Scientists are needed to understand the natural world and the effects of human activity on the environment. Lawyers and other specialists are central to developing government and industry policy, laws, and regulations to protect the environment. Engineers are in demand to develop technologies and products to clean up pollution and to prevent its production in the first place. Economists, geographers, and social scientists are key in evaluating the costs of pollution and resource depletion and in developing solutions that are socially, culturally, politically, and economically appropriate for different parts of the world. In addition, business will be looking for a new class of environmentally literate and responsible leaders who appreciate how products sold and services rendered affect our environment.

Trained people are essential in these professions at every level, from technical and clerical support staff to top managers.

Perhaps the biggest national demand over the next few years will be for environmental educators to help train an environmentally literate populace. We urgently need many more teachers at every level who are trained in environmental education.

Can environmental protection and resource conservation—a so-called green perspective—be a strategic advantage in business? Many companies think so. An increasing number are jumping on the environmental bandwagon, and most large corporations now have an environmental department. A few are beginning to explore integrated programs to design products and manufacturing processes that minimize environmental impacts. Often called "design for the environment," this approach is intended to avoid problems at the beginning, rather than deal with them later on a case-by-case basis. In the long run, executives believe this will save money and make their businesses more competitive in future markets. The alternative is to face increasing pollution control and waste disposal costs—now estimated to be more than $100 billion per year for all American businesses—as well as to be tied up in expensive litigation and administrative proceedings.

The market for pollution-control technology and know-how is also expected to be huge. Cleaning up the former East Germany alone is expected to cost some $200 billion. Many companies are positioning themselves to cash in on this enormous market. Germany and Japan appear to be ahead of America in the pollution control field because they have had more stringent laws for many years, giving them more experience in reducing effluents.

How can you prepare yourself to enter this market? The best bet is to get some technical training: environmental engineering, analytical chemistry, microbiology, ecology, limnology, groundwater hydrology, and computer science all have great potential. Currently, a chemical engineer with a graduate degree and some experience in an environmental field can practically name his or her salary. Some other very good possibilities are environmental law and business administration, both rapidly expanding fields.

For those who aren't inclined toward technical fields, there are still many opportunities for environmental careers. A good liberal arts education will help you develop skills such as communication, critical thinking, balance, vision, flexibility, and caring that should serve you well. Writers, editors, graphic arts professionals, and others who can communicate with the public have an important role to play in most environmental issues. Environmental attorneys, political scientists, anthropologists, and sociologists, among others, can help resolve environmental problems.

Citizen Science

While university classes often tend to be theoretical and abstract, many students are discovering they can also learn via hands-on experience and undergraduate research programs. Internships in agencies or environmental organizations are one way of doing this. Another is to get involved in organized **citizen science** projects in which ordinary people work with established scientists to answer real scientific questions. Community-based research was pioneered in the Netherlands, where several dozen research centers now study environmental issues, ranging from water quality in the Rhine River, cancer rates by geographic area, and substitutes for harmful organic solvents. In each project, students and neighborhood groups team with scientists and university personnel to collect data. Their results have been incorporated into official government policies.

Similar research opportunities exist in the United States and Canada. In addition to the Christmas Bird Count described at the beginning of this chapter, there are many other citizen-science projects in which you could participate. The Cornell University Bird Lab coordinates more than a dozen volunteer research projects dealing with birds, ranging from census of rare and endangered species, to the spread of exotic birds, to disease outbreaks. Look at their website at http://birds.cornell.edu/citsci/ to see if some might be of interest to you. Earthwatch offers a much smaller but more intense opportunity to take part in research. Every year hundreds of Earthwatch projects each field a team of a dozen or so volunteers who spend a week or two working on issues ranging from loon nesting behavior to archaeological digs. The Student Conservation Association places students in internships in which you can learn valuable job skills while also doing valuable resource work. And the American River Watch organizes teams of students to measure water quality. You might be able to get academic credit as well as helpful practical experience in one of these research experiences.

ENVIRONMENTAL POLICY

A policy is a rule or decision about how to act or deal with problems. On a personal, informal level, you might have a policy always to get your homework in on time. On a national, formal level, we have policies such as the Clean Air Act, a set of rules, agreed upon by a majority of the U.S. Congress, restricting air pollutants and setting fines for those who exceed legal limits. There are international policies, as well: the 1987 Montreal Protocol set

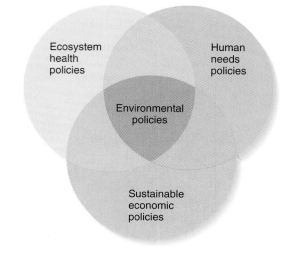

FIGURE 15.5 *The best environmental policies incorporate economic, ecological, and social/cultural considerations.*
Source: Modified from Ray Grizzle, *Bioscience*, vol. 44(4), April 1994.

agreed-upon limits to the production of ozone-depleting chemicals (chapter 9), and the Convention on International Trade in Endangered Species (CITES) is an international agreement to restrict trade in endangered species (chapter 5).

In this chapter, **environmental policy** will refer to official rules and regulations concerning the environment that are adopted, implemented, and enforced by some governmental agency, as well as to general public opinion about environmental issues. National policies are established through negotiation and compromise in a democratic society. Sometimes this wrangling can take decades. Theoretically, it allows all voices to be heard, and the resulting policy serves the interest of the majority (fig. 15.5).

Making policy isn't necessarily a clean process, though. In 2002 the U.S. Congress passed the Bipartisan Campaign Reform Act. This act established new rules limiting the amount of money that corporations, organizations, and individuals could donate to election campaigns. This law was passed in response to growing frustration among voters and elected officials over the power of money in politics. The unfair influence of money and power in government can be discouraging to voters. But sound policy for the public good can be formed—especially when the public is active and vocal in defending its interests. In the sections that follow, we will discuss the formation of rules and laws in general. Events don't always follow principles. The best way to make sure they do is to develop an informed and involved populace.

The Policy Cycle

How do policy issues and options make their way onto the stage of public debate? Problems are identified and acted upon in a **policy cycle** that acts to continually define and refine the public agenda (fig. 15.6). The first stage in this process is problem identification. Sometimes, the government identifies issues for groups that have no voice or don't recognize problems themselves. In other cases,

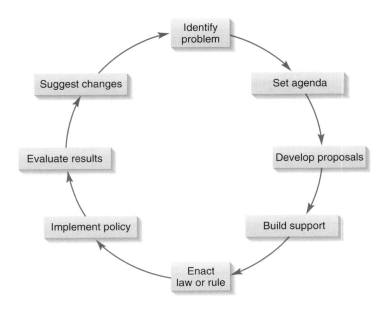

FIGURE 15.6   *The policy cycle.*

the public identifies a problem, such as loss of biodiversity or health effects of exposure to toxic waste, and demands redress by the government. In either case, proponents describe the issue—either privately or publicly—and characterize the risks and benefits of their preferred course of action.

Seizing the initiative in issue identification often allows a group to define terms, set the agenda, organize stakeholders, choose tactics, aggregate related issues, and legitimate (or delegitimate) issues and actors. It can be a great advantage to set the format or choose the location of a debate. Next, stakeholders develop proposals for preferred policy options, often in the form of legislative proposals or administrative rules. Proponents build support for their position through media campaigns, public education, and personal lobbying of decision makers. By following the legislative or administrative process through its many steps, interest groups ensure that their proposals finally get enacted into law or established as a rule or regulation.

The next step is implementation. Ideally, government agencies will faithfully carry out policy directives as they organize bureaucracies, provide services, and enforce rules and regulations, but often, it takes continued monitoring to make sure the system works as it should. Evaluating the results of policy decisions is as important as establishing them in the first place. Measuring impacts on target and nontarget populations shows us whether the intended goals, principles, and course of action are being attained. Finally, suggested changes or adjustments are considered that will make the policy fairer or more effective.

There are different routes by which this cycle is carried out. Special economic interest groups, such as industry associations, labor unions, or wealthy and powerful individuals, don't need (or often want) much public attention or support for their policy initiatives. They generally carry out the steps of issue identification, agenda setting, and proposal development privately because they

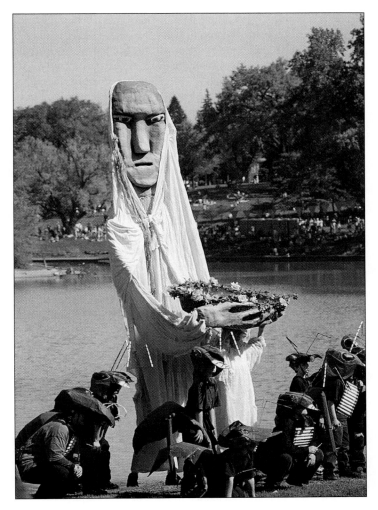

FIGURE 15.7   *Public protests and media events like this one are a way for public interest groups to influence policy.*

can influence legislative or administrative processes directly through their contacts with decision makers. Public interest groups, on the other hand, often lack direct access to corridors of power and need to rally broad general support to legitimate their proposals. An important method for getting their interests on the table is to attract media attention. Organizing a dramatic protest or media event can generate a lot of free publicity (fig. 15.7). Announcing some dire threat or sensational claim is a good way to gain attention. The problem is that it takes ever-increasing levels of hysteria and hyperbole to get yourself heard in the flood of shocking news with which we are bombarded every day. Ironically, many groups that bemoan the overload of rhetorical overstatement that engulfs us contribute to it to be heard above the din.

Political Decision Making

The policies we establish depend to a great extent on the system within which they operate. For many of us, the ideal political system is one that is open, honest, transparent, and reaches the best possible decisions to maximize benefits to everyone. In a pluralistic, democratic society, we aim to give everyone an equal voice in policy making. Ideally, many separate interests put forward their solutions to public problems that are discussed, debated, and evaluated fairly and equally. Facts and access are open to everyone. Policy choices are made democratically but compassionately; implementation is reasonable, fair, and productive. Unfortunately, this isn't always the way our system works. Our political system usually gives those with the most power and money the greatest voice in public affairs. Still, there have been altruistic movements in the past to ensure human rights and environmental protection that reveal an impulse for generosity and goodwill among the general public.

Another model for public decision making is rational choice and science-based management. In this utilitarian approach, no policy should have greater total costs than benefits. In choosing among policy alternatives, we should always prefer those with the greatest cumulative welfare and the least negative impacts. Professional administrators would weigh various options and make an objective, methodical decision that would bring maximum social gain. There are many reasons, however, that rational choice doesn't always work in the public arena.

- Many conflicting values and needs cannot be compared because they aren't comparable or we don't have perfect information.

- There are few generally agreed-upon broad societal goals, but rather, benefits to specific groups and individuals, many of which are in conflict.

- Policymakers generally aren't motivated to make decisions on the basis of societal goals, but rather to maximize their own rewards: power, status, money, or reelection.

- Large investments in existing programs and policies create "path dependence" and "sunken costs" that prevent policymakers from considering good alternatives foreclosed by previous decisions.

- Uncertainty about consequences of various policy options compels decision makers to stick as closely as possible to previous policies to reduce the likelihood of adverse, calamitous, unanticipated consequences.

- Policymakers, even if well meaning, don't have sufficient intelligence or adequate data or models to calculate accurate costs and benefits when large numbers of diverse political, social, economic, and cultural values are at stake.

- The segmented nature of policy making in large bureaucracies makes coordinating decision making difficult.

NEPA and EIS

The **National Environmental Policy Act (NEPA)** forms the cornerstone of U.S. environmental policy. Signed into law by President Nixon in 1970, NEPA does three important things: (1) it authorizes the Council on Environmental Quality (CEQ), the oversight board for general environmental conditions; (2) it directs federal agencies to take environmental consequences into account in decision making; and (3) it requires an Environmental Impact Statement (EIS) for every major federal project having a significant impact on the quality of the human environment. When NEPA was being debated, there were suggestions that it should be a constitutional amendment guaranteeing the right of a clean environment to everyone. Difficulties in defining what "clean" means, along with worries about how we could achieve this ambitious goal, limited NEPA to a statute with more limited, but still important, powers than a constitutional amendment. What do you think? Does everyone have an inherent right to a clean environment? How would you define "clean"?

The EIS process has proven to be one of the most powerful tools in the environmental arsenal. It requires more open and environmentally sensitive planning in both the agencies themselves and in private corporations seeking to do business with the government. An EIS can bring to light adverse aspects of a project that might otherwise remain hidden. It can provide valuable information about a proposal to opponents who can't afford to do their own research.

An EIS doesn't forbid environmentally destructive activities if they comply otherwise with relevant laws, but it demands that we admit what we're really doing. Once embarrassing information is revealed, however, few agencies will bulldoze ahead, ignoring public opinion.

What kinds of projects require an EIS? The activity must be federal and it must be major, with a significant impact on the human environment (fig. 15.8). Evaluations are always somewhat subjective as to whether specific activities meet these characteristics. Each case is unique and depends on context, geography, the balance of beneficial versus harmful effects, and whether any areas of cultural, scientific, or historical importance might be affected. To do a complete EIS for a project is usually time consuming and costly. The final document is often hundreds of pages long and generally takes six to nine months to prepare. Sometimes, just requesting an EIS is enough to sideline a questionable project. In other cases it gives adversaries time to rally public opposition and to research information with which to criticize what's being proposed. If agencies don't agree to prepare an EIS voluntarily, environmentalists can petition the courts to force them to do so.

Every EIS must contain the following elements: (1) purpose and need for the project, (2) alternatives to the proposed action (including taking no action), and (3) a statement of positive and negative environmental impacts of the proposed activities. In addition, an EIS should make clear the relationship between short-term resources and long-term productivity, as well as any irreversible commitment of resources resulting from project implementation.

ENVIRONMENTAL LAW

Laws are rules set by authority, society, or custom. Church laws, social mores, administrative regulations, and a variety of other codes of behavior can be considered laws if they are backed by

FIGURE 15.8 *Every major federal project in the United States must be preceded by an Environmental Impact Statement.*

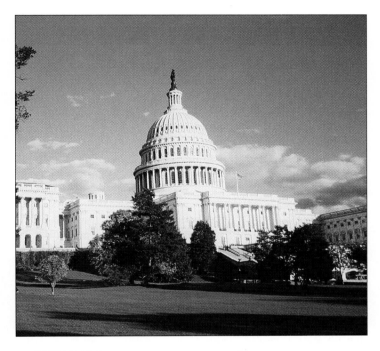

FIGURE 15.9 *Following the progress of bills as they pass through Congress is an important part of environmental action.*

APPLICATION:	Environment, Science, and Policy in Your Community

Take several minutes with your classmates and make a list of five current, environment-related policy issues in your community or state. (Hint: consider energy, economic developments, population changes, traffic congestion, water and air quality, or public health issues in the news.) Discuss which branch of government—legislative, executive, or judicial—you think should take the lead in addressing each issue. Why did you designate these authorities as you did?

Select one issue from your list, and discuss the following: What are some of the sources of uncertainty involved in the decision-making process? How could scientific research help identify or solve this policy issue? Where in the rule-making process could scientific data help inform decision makers?

some enforcement power. **Environmental law** includes official rules, decisions, and actions concerning environmental quality, natural resources, and ecological sustainability. In the United States a wide variety of environmental laws are promulgated at both the local and national level (fig. 15.9). Because every country has different legislative and legal processes, this chapter will focus primarily on the U.S. system in the interest of simplicity and space. Environmental law can be established or modified in each of the three branches of government: legislative, judicial, and executive. We describe these as statutory, case, and administrative law, respectively. Understanding how these systems work is an important step in becoming an environmentally literate person.

Statutory Law: The Legislative Branch

Establishing laws at either the state or federal level is one of the most important ways of protecting our environment. Many environmental groups spend a good deal of their time and resources trying to influence the legislative process. In this section we'll look at how that system works.

Federal laws (statutes) are enacted by Congress and must be signed by the president. They originate as legislative proposals called bills, which are usually drafted by the congressional staff, often in consultation with representatives of various interest groups. Thousands of bills are introduced every year in Congress. Some are very narrow, providing funds to build a specific section of road or to help a particular person, for instance. Others are extremely broad, perhaps overhauling the Social Security system or changing the entire tax code. Similarly, environmental legislation might deal with a very specific local problem or a national or international issue. Often, a number of competing bills on a single issue may be introduced as proponents from different sides attempt to incorporate their views into law. A bill may have a single sponsor if it is the pet project of a particular legislator, or it may have 100 or more co-authors if it is an issue of national importance.

A Convoluted Path

After introduction, each bill is referred to a committee or subcommittee with jurisdiction over the issue for hearings and debate. Most hearings take place in Washington, D.C., but if the bill is controversial or legislators want to attract publicity for themselves or the issue, they may conduct field hearings closer to the site of the controversy. The public often has an opportunity to give testimony at field hearings (fig. 15.10). Although it's not likely that you will change the opinions of many legislators, no matter how fervent or cogent your testimony, these events can be a good place to gain attention and educate the public about a topic. Hearings and debates also build a record of legislative intent that can be valuable in later interpretation and implementation of laws by courts and administrative agencies.

If a bill has sufficient support within the subcommittee, its language will be "marked up" or revised and modified to be more widely acceptable and to improve its chances of passage. At this stage several competing bills might be combined into a single compromise version and forwarded to the full committee for more hearings, debate, and a vote. If it fails in the full committee, the bill is sent back to the subcommittee for more work and further compromise. A bill that succeeds in the full committee is reported to the full House or Senate for a floor debate. Often, opponents of a bill will attempt to amend it during each of these stages to lessen its impact or to make it so unpalatable that even the original authors can no longer support it. Some amendments add completely unrelated material or even reverse the bill's intent. As bills move along this convoluted pathway, interested parties can follow their progress in the *Congressional Quarterly Weekly,* a publication both in print and online that tracks proposed legislation. Many environmental groups also maintain websites with up-to-date information on events in Congress.

By the time an issue has passed through both the House and Senate, the versions approved by the two bodies are likely to be different. They go then to conference committee to iron out any differences between them. After going back to the House and Senate for confirmation, the final bill goes to the president, who may either sign it into law or veto it. If the president vetoes the bill, it may still become law if two-thirds of the House and Senate vote to override the veto. If the president takes no action within 10 days of receiving a bill from Congress, the bill becomes law without his signature. One exception to this procedure is that, if Congress adjourns before the 10-day period elapses, the bill does not become law. The president, by doing nothing, is said to have exercised a "pocket veto."

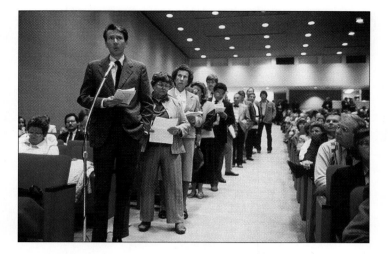

FIGURE 15.10 *Citizens line up to testify at a legislative hearing. By getting involved in the legislative process, you can be informed and have an impact on governmental policy.*

The Thomas website, maintained by Congress, also has current information about the progress of legislation. You can find out how your senators and representative voted on critical environmental issues by consulting The League of Conservation Voters, which ranks each member of Congress on their voting record. The Defenders of Wildlife maintains a daily email environmental news service called Greenwire that has up-to-date information about what's happening in Washington, D.C. Table 15.5 lists some of the most important recent federal environmental legislation.

Lobbying

Groups or individuals with an interest in pending legislation can often influence outcomes by **lobbying,** or using personal contacts, public pressure, and political action to persuade legislators to vote in their favor. Most major environmental organizations maintain offices in Washington, D.C., from which they monitor legislative and administrative programs and policies. Hundreds of professional environmental lobbyists and volunteer activists attend hearings, meet with legislators and agency personnel, draft proposed legislation and administrative rules, and work to shape the national environmental agenda.

In a survey of professional lobbyists, a majority agreed that personal contacts were the most effective way to influence decision makers. Your own senator or representative is more likely to be interested in your views than someone with whom you have no connection. Even if you aren't rich or illustrious or politically connected, you can often get a fair hearing from legislative staff—if not their bosses—if you have a persuasive case concerning an important topic.

Being involved in local election campaigns can greatly increase your access to legislators. Writing letters or making telephone calls also are highly effective ways to get your message across. All legislators now have email addresses, but letters and phone calls are usually taken more seriously. Getting media attention can sway the opinions of decision makers. Organizing protests,

Principles of Environmental Science

TABLE 15.5 Major U.S. Environmental Laws

LEGISLATION	PROVISIONS
National Environmental Policy Act of 1969	Declared national environmental policy, required Environmental Impact Statements, created Council on Environmental Quality.
Clean Air Act of 1970	Established national primary and secondary air quality standards. Required states to develop implementation plans. Major amendments in 1977 and 1990.
Clean Water Act of 1972	Set national water quality goals and created pollutant discharge permits. Major amendments in 1977 and 1996.
Federal Pesticides Control Act of 1972	Required registration of all pesticides in U.S. commerce. Major modifications in 1996.
Marine Protection Act of 1972	Regulated dumping of waste into oceans and coastal waters.
Coastal Zone Management Act of 1972	Provided funds for state planning and management of coastal areas.
Endangered Species Act of 1973	Protected threatened and endangered species. Directed FWS to prepare recovery plans.
Safe Drinking-Water Act of 1974	Set standards for safety of public drinking-water supplies and to safeguard groundwater. Major changes made in 1986 and 1996.
Toxic Substances Control Act of 1976	Authorized EPA to ban or regulate chemicals deemed a risk to health or the environment.
Federal Land Policy and Management Act of 1976	Charged the BLM with long-term management of public lands. Ended homesteading and most sales of public lands.
Resource Conservation and Recovery Act of 1976	Regulated hazardous-waste storage, treatment, transportation, and disposal. Major amendments in 1984.
National Forest Management Act of 1976	Gave statutory permanence to national forests. Directed USFS to manage forests for "multiple use."
Surface Mining Control and Reclamation Act of 1977	Limited strip-mining on farmland and steep slopes. Required restoration of land to original contours.
Alaska National Interest Lands Act of 1980	Protected 40 million ha (100 million acres) of parks, wilderness, and wildlife refuges.
Comprehensive Environmental Response, Compensation and Liabiity Act of 1980	Created $1.6 billion "Superfund" for emergency response, spill prevention, and site remediation for toxic wastes. Established liability for cleanup costs.
Superfund Amendments and Reauthorization Act of 1994	Increased Superfund to $8.5 billion. Shared responsibility for cleanup among potentially responsible parties. Emphasized remediation and public "right to know."

Source: N. Vig and M. Kraft, *Environmental Policy in the 1990s,* 3rd ed. Congressional Quarterly Press.

marches, demonstrations, or other kinds of public events can call attention to your issue (fig. 15.11). Public education campaigns, press conferences, TV ads, and a host of other activities can be helpful. Joining together with other like-minded groups can greatly increase your clout and ability to get things done. It's hard for a single individual or even a small group to have much impact, but if you can organize a mass movement, you may be very effective in influencing public policy.

Case Law: The Judicial Branch

Over the past 30 years, appeals to the judicial system have often been the most effective ways for seeking redress for environmental damage and forcing changes in how things are done. Activist judges and sympathetic juries in both federal and state court systems have been willing to take a stand where legislatures have been too timid or conservative to do so. Many groups spend a great deal of their time and energy bringing lawsuits that will shape environmental policy and law. The Environmental Defense Fund, for example, operates primarily in this arena.

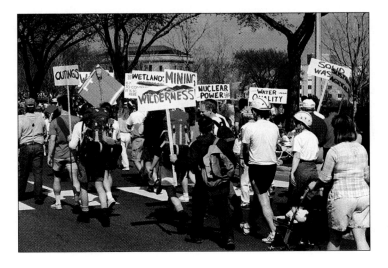

FIGURE 15.11 *Protests, marches, and public demonstrations can be an effective way to get your message out and to influence legislators.*

The judicial branch of government establishes environmental law by ruling on the constitutionality of statutes and interpreting their meaning. We describe the body of legal opinions built up by many court cases as **case law.** Often, legislation is written in vague and general terms so as to make it widely enough accepted to gain passage. Congress, especially in the environmental area, often leaves it to the courts to "fill in the gaps." When trying to interpret a law, the courts depend on the legislative record from hearings and debates to determine congressional intent. What was a particular statute meant to do by those who wrote and passed it?

Legal Standing

Before a trial can start, the litigants must satisfy certain threshold requirements. The first of these is **legal standing,** or whether the participants have a right to initiate an action. The main criterion for standing is a valid interest in the case. Plaintiffs must show that they are materially affected by the situation they petition the court to redress. This is an important point in environmental cases. Groups or individuals often want to sue a person or corporation for degrading the environment. Unless they can show that they personally suffer from the degradation, however, courts are likely to deny standing. In a landmark 1969 case, for example, the Sierra Club challenged a decision of the U.S. Forest Service and the Department of the Interior to lease public land in California to Walt Disney Enterprises for a ski resort. The land in question was a beautiful valley that cut into the southern boundary of Sequoia National Park (fig. 15.12). Building a road into the valley would have necessitated cutting through a grove of giant redwood trees within the park. The Sierra Club argued that it should be granted standing in the case to represent the trees, animals, rocks, and mountains that couldn't defend their own interests in court. After all, the club pointed out, corporations such as Walt Disney Enterprises are treated as persons and represented by attorneys in the courts. Why not grant trees the same rights? The case went all the way to the U.S. Supreme Court, which ruled that the Sierra Club failed to show that it or any of its members would be materially affected by the development. It turned out that the valley had too many avalanches to be safe for a resort. Still, the question of standing remains an important one.

Criminal Law

Criminal law derives from those federal and state statutes that prohibit wrongs against the state or society. Serious crimes, like murder or rape, that are punishable by long jail terms or heavy fines are called felonies. Lesser crimes, such as shoplifting or vandalism, that result in smaller fines or shorter sentences in a county or city jail are labeled misdemeanors. A criminal case is always initiated by a government prosecutor. Guilt or innocence of the defendant is determined by a jury of peers, but the sentence is imposed, often in consultation with the jury, by the judge (fig. 15.13). The judge is responsible for keeping order in the hearings and for determining points of law. The jury acts as a fact-finding body that weighs the truth and reliability of the witnesses and evidence.

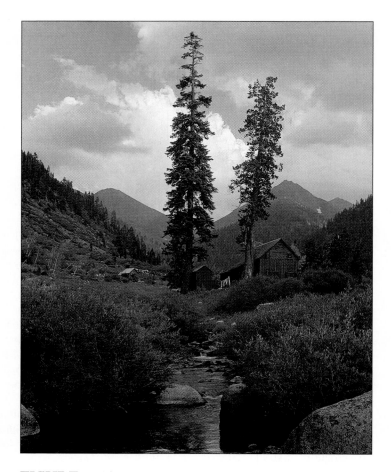

FIGURE 15.12 *Mineral King Valley at the southern border of Sequoia National Park was the focus of an important environmental law case in 1969. The Disney Corporation wanted to build a ski resort here, but the Sierra Club sued on behalf of the trees, rocks, and native wildlife to protect the valley.*

Violation of many environmental statutes constitutes a criminal offense. In 1975 the U.S. Supreme Court ruled that corporate officers can be held criminally liable for violations of environmental laws if they were grossly negligent, or the illegal actions can be considered willful and knowing violations. In 1982 the EPA created an Office for Criminal Investigation. Recently, as many as 100 criminal prosecutions per year have been brought to court for environmental crimes. In one of the toughest criminal sentences imposed so far, the president of a Colorado company was sentenced in 1999 to 14 years in prison for knowingly dumping chlorinated solvents that contaminated the water table. The company itself was fined $950,000 and put on probation for ten years. About the same time, two Connecticut men were convicted of three counts of abandoning 40 drums of hazardous waste in a parking lot in New Haven. The fact that they stripped off or painted over the identifying labels was taken as willful intent to commit a crime. They face up to seven years in prison and a $250,000 fine for each count.

FIGURE 15.13 *A trial court judge presides over trials, rules on motions, and interprets the law.*

Civil Law

Civil law is defined as a body of laws regulating relations between individuals or between individuals and corporations. Issues such as property rights and personal dignity and freedom are protected by civil law. The defendant in a civil case has a right to be tried by a jury, but in highly technical issues, this right often is waived and the case is heard only by the judge. Being found guilty of a civil offense can result in financial penalties but not jail time.

In contrast to a criminal case, where the burden of proof lies with the prosecution and defendants are considered innocent until proven guilty, civil cases can be decided on a "preponderance of evidence." This makes civil cases considerably easier to win than criminal cases when evidence is ambiguous. A number of mitigating factors also are taken into account in determining guilt and assigning penalties in civil cases. Guilt or innocence is based on whether the defendant could reasonably have anticipated and avoided the offense. A "good faith effort" to comply or solve the problem can be a factor. The compliance history is important. Is this a first-time offender or a habitual repeater? Finally, is there evidence of economic benefit to the perpetrator? That is, did the violator gain personally from the action? If so, it is more likely that willful intent was involved.

Civil cases can be brought in both state and federal court. In 1999 the Atlanta-based Colonial Pipeline Company was ordered to pay over $20 million in damages to local landowners and the state of South Carolina for spilling about 3.5 million liters (nearly a million gallons) of diesel fuel into the Reedy River near Simpsonville, South Carolina. And in 2000 the Koch Oil Company, one of the largest pipeline and refinery operators in the United States, agreed to pay $35 million in fines and penalties to state and federal authorities for negligence in more than 300 oil spills in Texas, Oklahoma, Kansas, Alabama, Louisiana, and Missouri over the previous decade. Koch also agreed to spend more than $1 billion on cleanup and improved operations.

Sometimes, the purpose of a civil suit is to seek an injunction or some other protection from the actions of an individual, a corporation, or a governmental agency. You might ask the courts, for example, to order the government to cease and desist from activities that are in violation of either the spirit or the letter of the law. This sort of civil action is heard only by a judge; no jury is present. Environmental groups have been very successful in asking courts to stop logging and mining operations, to enforce implementation of the Endangered Species Act, to require agencies to enforce air and water pollution laws, and a host of other efforts to protect the environment and conserve natural resources. Often, rather than sue a corporation directly for environmental damage, it is more effective to sue the government for not enforcing laws that would have prevented the damage. A big corporation with deep pockets may have the resources and incentive to tie up litigation for years with motions and countersuits. Federal or state agencies may be more inclined to agree that you are right and be willing to settle the matter quickly.

SLAPP Suits

Because defending a lawsuit is so expensive, the mere threat of litigation can be a chilling deterrent. Increasingly, environmental activists are being harassed with **Strategic Lawsuits against Political Participation (SLAPP).** Citizens who criticize businesses that pollute or government agencies that are derelict in their duty to protect the environment are often sued in retaliation. While most of these preemptive strikes are groundless and ultimately dismissed, defending yourself against them can be exorbitantly expensive and take up time that might have been spent working on the original issue. Public interest groups and individual activists—many of whom have little money to defend themselves—often are intimidated from taking on polluters. For example, a West Virginia farmer wrote an article about a coal company's pollution of the Buckhannon River. The company sued him for $200,000 for defamation. Similarly, citizen groups fighting a proposed incinerator in upstate New York were sued for $1.5 million by their own county governments. After a Texas woman called a nearby landfill a dump, her husband was named in a $5 million suit for failing to "control his wife." Of course, these suits also are expensive for the company or agency that initiates them, but they may be far cheaper than paying a fine or losing a big project.

Administrative Law: The Executive Branch

More than 100 federal agencies and thousands of state and local boards and commissions have environmental oversight. They usually have power to set rules, adjudicate disputes, and investigate misconduct. Federal agencies often delegate power to a matching state agency to decentralize authority. The enabling legislation to create each agency is called an "organic" act because it establishes a basic unit of governmental organization. In the federal government most executive agencies come under the jurisdiction of cabinet-level departments, such as Agriculture, Interior, or Justice (fig. 15.14).

Agency rule-making and standard-setting can be either formal or informal. In an informal case, notice and background for proposed rules are published in the *Federal Register.* All interested parties have opportunities to submit comments. This is often an important way for environmentally concerned citizens and public

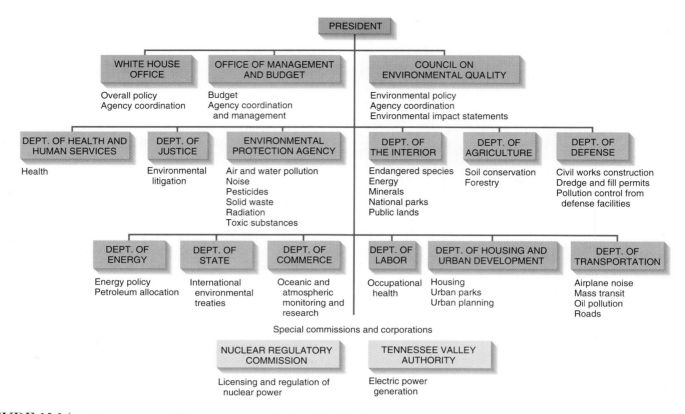

FIGURE 15.14 *Major agencies of the executive branch of the U.S. federal government with responsibility for resource management and environmental protection.*
Source: Data from U.S. General Accounting Office.

interest groups to have an impact on environmental policy. In formal rule-making, a public hearing is held, with witnesses and testimony much like a civil trial. Witnesses can be cross-examined. A complete transcript is made, and final findings are published in the public record. It is generally more difficult for individuals to intervene in a formal hearing, although sometimes, there is an opportunity to submit written comments. Rule-making is often a complex, highly technical process that is difficult for citizen groups to understand and monitor. The proceedings are usually less dramatic and colorful than criminal trials, and yet can be very important for environmental protection.

Regulatory Agencies

The EPA is the primary agency with responsibility for protecting environmental quality in the United States. Created in 1970 at the same time as NEPA, the EPA has more than 18,000 employees and ten regional offices. Often in conflict with Congress, other agencies of the executive branch, and environmental groups, the EPA has to balance many competing interests and conflicting opinions. Greatly influenced by politics, the agency changes dramatically, depending on which party is in power and what attitudes toward the environment prevail at any given time.

The Departments of the Interior and Agriculture are to natural resources what the EPA is to pollution. Interior is home to the National Park Service, which is responsible for more than 376

national parks, monuments, historic sites, and recreational areas. It also houses the Bureau of Land Management (BLM), which administers some 140 million ha (350 million acres) of land, mostly in the western United States. In addition, Interior is home to the U.S. Fish and Wildlife Service, which operates more than 500 national wildlife refuges and administers endangered species protection.

The Department of Agriculture is home to the U.S. Forest Service, which manages about 175 national forests and grasslands, totaling some 78 million ha (193 million acres). With 39,000 employees, the Forest Service is nearly twice as large as the EPA (fig. 15.15). The Department of Labor houses the Occupational Safety and Health Administration (OSHA), which oversees workplace safety. Research that forms the basis for OSHA standards is carried out by the National Institute for Occupational Safety and Health (NIOSH). In addition, several independent agencies that are not tied to any specific department also play a role in environmental protection and public health. The Consumer Products Safety Commission passes and enforces regulations to protect consumers, and the Food and Drug Administration is responsible for the purity and wholesomeness of food and drugs.

All of these agencies have a tendency to be "captured" by the industries they are supposed to be regulating. Many of the people with expertise to regulate specific areas came from the industry or sector of society that their agency oversees. Furthermore, the people with whom they work most closely and often develop

Principles of Environmental Science

www.mhhe.com/environmentalscience

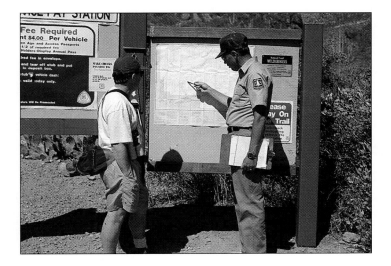

FIGURE 15.15 *A forest ranger explains wilderness rules to a hiker.*

friendships are those they are supposed to watch. And when they leave the agency to return to private life—as many do when the administration changes—they are likely to go back to the same industry or sector where their experience and expertise lies. The effect is often what's called a "revolving door," where workers move back and forth between industry and government. As a result, regulators often become overly sympathetic with and protective of the industry they should be overseeing.

INTERNATIONAL TREATIES AND CONVENTIONS

As recognition of the interconnections in our global environment has advanced, the willingness of nations to enter into protective **international treaties and conventions** has grown (table 15.6). The earliest of these conventions had no nations as participants;

they were negotiated entirely by panels of experts. Not only the number of parties taking part in these negotiations has grown, but the rate at which parties are signing on and the speed at which agreements take force also have increased rapidly (fig. 15.16). The Convention on International Trade in Endangered Species (CITES), for example, was not enforced until 14 years after ratification, but the Convention on Biological Diversity was enforceable after just one year and had 160 signatories only four years after introduction. Over the past 25 years, more than 170 treaties and conventions have been negotiated to protect our global environment. Designed to regulate activities ranging from intercontinental shipping of hazardous waste, to deforestation, overfishing, trade in endangered species, global warming, and wetland protection, these agreements theoretically cover almost every aspect of human impacts on the environment.

Unfortunately, many of these environmental treaties constitute little more than vague good intentions. Even though we often call them laws, there is no body that can legislate or enforce international environmental protection. The United Nations and a variety of regional organizations bring stakeholders together to negotiate solutions to a variety of problems, but the agreed-upon solutions generally rely on moral persuasion and public embarrassment for compliance. Most nations are unwilling to give up sovereignty. There is an international court, but it has no enforcement power. Nevertheless, there are creative ways to strengthen international environmental protection.

One of the principal problems with most international agreements is the tradition that they must be passed by unanimous consent. A single recalcitrant nation effectively has veto power over the wishes of the vast majority. For instance, more than 100 countries at the UN Conference on Environment and Development (UNCED), held in Rio de Janeiro in 1992, agreed to restrictions on the release of greenhouse gases. At the insistence of U.S. negotiators, however, the climate convention was reworded so that it only urged—but did not require—nations to stabilize their emissions.

TABLE 15.6 Some Major International Treaties and Conventions

DATE	TITLE	COVERS
1971	Convention on Wetlands of International Importance (Ramsar)	Protects wetlands, especially as waterfowl habitat
1972	Convention to Protect World Cultural and Natural Heritage	Protects cultural sites and natural resources
1973	Convention on International Trade in Endangered Species (CITES)	Restricts trade in endangered plants and animals
1979	Convention on Migratory Species (CMS)	Protects migratory species, especially birds
1982	United Nations Law of the Sea (UNCLOS)	Declares the oceans international commons
1985	Protocol on Substances that Deplete the Ozone Layer (Ozone)	Initiates phaseout of chloroflurocarbons
1989	Convention on the Transboundary Movements of Hazardous Waste (Basel)	Bans shipment of hazardous waste
1992	Convention on Biodiversity (CBD)	Protects biodiversity as national resources
1992	United Nations Framework Convention on Climate Change (UNFCCC)	Rolls back carbon dioxide production in industrialized countries
1994	Convention to Combat Desertification (CCD)	Provides assistance to fight desertification, especially in Africa

FIGURE 15.16   *Additions of participating parties to some major international environmental treaties. The thick portion of each line shows when the agreement went into effect (date in parentheses). See table 15.6 for complete treaty names.*
Source: United Nations Environment Programme from Global Environment Outlook–2000.

As a way of avoiding this problem, some treaties incorporate innovative voting mechanisms. When a consensus cannot be reached, they allow a qualified majority to add stronger measures in the form of amendments that do not need ratification. All members are legally bound to the whole document unless they expressly object. This approach was used in the Montreal Protocol, passed in 1987 to halt the destruction of stratospheric ozone by chlorofluorocarbons (CFCs). The agreement allowed a vote of two-thirds of the 140 participating nations to amend the protocol. Although initially, the protocol called for only a 50 percent reduction in CFC production, subsequent research showed that ozone was being depleted faster than previously thought (see chapter 9). The protocol was strengthened by amendment to an outright ban on CFC production, in spite of the objection of a few countries.

When strong accords with meaningful sanctions cannot be passed, sometimes the pressure of world opinion generated by revealing the sources of pollution can be effective. Activists can use this information to expose violators. For example, the environmental group Greenpeace discovered monitoring data in 1990 showing that Britain was disposing of coal ash in the North Sea. Although not explicitly forbidden by the Oslo Convention on ocean dumping, this evidence proved to be an embarrassment, and the practice was halted.

Trade sanctions can be an effective tool to compel compliance with international treaties. The Montreal Protocol, for example, bound signatory nations not to purchase CFCs or products made using them from countries that refused to ratify the treaty. Because many products employed CFCs in their manufacture, this stipulation proved to be very effective. On the other hand, trade agreements also can work against environmental protection. The World Trade Organization (WTO) was established to make international trade more fair and to encourage development. It has been used, however, to weaken national environmental laws. In 1990, after decades of public protests and boycotts, the United States banned the import of tuna caught using methods that killed thousands of dolphins each year. Shrimp caught with nets that kill endangered sea turtles were also banned. Mexico filed a complaint with the WTO,

contending that dolphin-safe tuna laws represented an illegal barrier to trade. Thailand, Malaysia, India, and Pakistan filed a similar suit over turtle-friendly shrimp laws. The WTO ordered the United States to allow the import of both tuna and shrimp from countries that allow fisheries to kill dolphins and turtles. Environmentalists point out that the WTO has never ruled against a corporate suit, in part because rules are decided by a committee composed mainly of appointed industry leaders, rather than by elected officials representing a spectrum of political and economic interests.

The North American Free Trade Agreement (NAFTA) has also been used to undermine national and local environmental laws. The Canadian company Methanex has sued to force California to overturn its ban on MTBE, a carcinogenic gasoline additive that has contaminated drinking water in thousands of cities and towns. The company seeks to recover nearly a billion dollars in lost MTBE sales and damages, as well as to force California to allow its continued use. In a similar NAFTA lawsuit, the Ethyl Corporation of the United States won a suit against Canada for banning the manganese-based gasoline additive, MMT, which is suspected to be a neurotoxin. Environmentalists are also concerned about a powerful international treaty currently being negotiated by the Organization of Economic Cooperation and Development, a group of 29 of the wealthiest countries in the world. The Multilateral Agreement on Investments, or MAI, may be more threatening to social justice and environmental protection than either the WTO or NAFTA (fig. 15.17).

DISPUTE RESOLUTION AND COMMUNITY-BASED PLANNING

The adversarial approach of our current legal system often fails to find good solutions for many complex environmental problems. Identifying and punishing an enemy for transgressions often seem more important to us than finding win/win compromises. Gridlocks occur in which conflicts between adversaries breed mutual suspicion and decision paralysis. The results can be continuing ecosystem deterioration, economic stagnation, and growing incivility and confrontation. The complexity of many environmental problems

FIGURE 15.17 *Trade agreements can weaken environmental protection as countries compete to market their resources. NAFTA, for example, encourages logging in British Columbia for U.S. markets.*

arises from the fact that they are not purely ecological, economic, or social, but a combination of all three. They require an understanding of the interrelations between nature and people. Are there ways to break these logjams and find creative solutions? In this section we will look at some new developments in mediation, dispute resolution, and alternative procedures for environmental decision making.

Arbitration and Mediation

Litigation and administrative challenges tend to be highly adversarial. You may win a battle but end up hardening the opposition to your ideas. Increasingly used to avoid the time, expense, and winner-take-all confrontation inherent in our legal system, arbitration and mediation encourage compromise and workable solutions with which everyone can live.

Arbitration is a formal process of dispute resolution somewhat like a trial. There are stringent rules of evidence and cross-examination of witnesses, and the process results in a legally binding decision. The arbitrator takes a more active role than a judge, however, and is not as constrained by precedent. The arbitrator is more interested in resolving the dispute than in strict application of the law. Arbitrators must have formal training and be certified by the Federal Mediation and Conciliation Service or the American Arbitration Association. Arbitration is usually an attractive prospect if you don't think you could win a formal lawsuit, but why would anyone agree to arbitration if they think they can win the whole enchilada in court? They might take this route just to avoid disagreeable surprises. Juries can be fickle. Furthermore, they might want to get an unpleasant process over with sooner rather than later. In addition, arbitration often is written into contracts so that the disputants have no choice in the matter.

There are disadvantages to arbitration. It doesn't create a legally binding precedent, something that often is the main motivation for a lawsuit. There is less opportunity to appeal if you

don't like the decision you get. There also is less protection from self-incrimination, false witnesses, or evidence you didn't expect. You don't generate nearly as much publicity because the proceedings and record are not public. For some litigants, the publicity generated by a trial is more valuable than the settlement itself. Finally, you are less likely to win the whole thing. Some sort of compromise is the most likely outcome.

Mediation is a process in which disputants are encouraged to sit down and talk to see if they can come up with a solution by themselves. The mediator makes no final decision but is simply a facilitator of communication. This process is especially useful in complex issues where there are multiple stakeholders with different interests, as is often the case in environmental controversies. There are both benefits and perils of mediation. It can be quicker and cheaper than court battles. It can lead to compromise and understanding that will lead to further cooperation, and that will solve problems faster than endless appeals. And it may find creative solutions that satisfy multiple parties and interests. On the other hand, no one can be forced to mediate or to do so in good faith. Rancorous participants can tie up the process in long, pointless arguments that only make others more angry. Ultimately, there can be a tyranny of the minority. A single person can veto an agreement that everyone else wants. Furthermore, mediation represses or denies certain irreconcilable structural conflicts, giving the impression of equality between disputants when none really exists. Unequal negotiating skills of the participants can lead to unfair outcomes and even more rancor and paranoia than before the mediation was attempted. As is the case with arbitration, mediation doesn't generate the publicity and complete victory that some groups may desire.

Collaborative Approaches to Community-Based Planning

Over the past several decades, natural resource managers have come to recognize the value of **community-based planning** that incorporates holistic, adaptive, pluralistic approaches. Involving all stakeholders and interest groups early in the planning process can help avoid the "train wrecks" in which adversaries become entrenched in nonnegotiable positions. Working with local communities can tap into traditional knowledge and gain acceptance for management plans that finally emerge from policy planning. This approach is especially important in nonlinear, nonequilibrium systems and "wicked" problems. Among the more important reasons to use collaborative approaches are:

- The way "wicked" problems are formulated depends on your worldview. Incorporating a variety of perspectives early in the process is more likely to lead to the development of acceptable solutions in the end.

- People have more commitment to plans they have helped develop. The first stage is therefore to identify those involved and to engage them in the process.

- There is truth in the old adage that "two heads are better than one." Involving multiple stakeholders and multiple sources of information enriches the process.

INVESTIGATING Our Environment

Reasoned Judgment and Scientific Uncertainty

Often, policymakers are called on to take action, even though scientific evidence is unclear or even contradictory. It often takes courage and foresight to take a position under such circumstances. You read in chapter 9 about the potential consequences if significant global warming were to occur. You have also read that data from computer models and other sources are equivocal, creating a degree of uncertainty about the timing and precise nature of the warming event.

Some people use this lack of absolute certainty to claim that we not undertake preventive changes until the uncertainty has been removed. They argue that to do so without absolute proof would be unscientific. They say that reducing greenhouse gas production would impose significant economic costs, their argument being that such actions are unwise because they could turn out to have been unnecessary. This position fails to acknowledge that some degree of uncertainty surrounds the vast majority of decisions and judgments society must make, including issues in science. By casting decision making as purely a process of separating fact from opinion, this position omits an important element of reasoned judgment.

The threat of significant, rapid, global climate change is but one of many issues confronting us in which factual evidence does not establish absolute certainty. Will your home be destroyed by fire? Will you develop lung cancer if you smoke cigarettes? Like global warming, available facts are insufficient to establish that any of these events will be visited upon someone with certainty. There is no doubt, however, that they are possible, or even probable.

In issues such as these, should we do nothing until certainty is established? The risk, of course, is that disaster may be unavoidable if we wait. Any serious effort to curb release of greenhouse gases, for example, will require time to implement. There will also be a lag time before such efforts can influence global climate. Mea-

sures undertaken only when scientific certainty is finally established may be too late to be of much help. An insurance company is unlikely to underwrite fire coverage on your house when the fire trucks are already on their way.

Or do we take preventive measures before an event is certain to occur, running the risk that our action may not have been needed? It certainly is true that your house may never burn down. And if it does not, the money spent for insurance premiums appears wasted.

In instances like these, we must use **reasoned judgment:** thoughtful decisions based on careful, logical examination of available evidence that recognize the potential for error or new information. To do otherwise is to paralyze actions, preventing timely intervention. Society has long used reasoned judgment to address issues of uncertainty similar to global warming. For most people, arriving at a reasoned judgment involves a thought process that is very

applicable to global warming. Two questions are evaluated:

1. What is the likelihood that the nasty event actually will occur?
2. How serious would it be if the nasty event occurs?

The more likely a bad event is to happen, or the more serious the effect of the event, the greater the advisability of taking preventive action. This concept is modeled in the graph in this box with some examples. Where on the graph would you plot global warming?

Thomas Lovejoy expresses the reasoned judgment of many ecologists when he argues against doing "total planet experiments that bet the biosphere, if there is even a small chance we may regret the result. After all, there is not even an experimental control planet to colonize if we lose at biosphere roulette."

The stakes seem high. Does reason require us to act or wait?

Society regularly takes preventive action to avoid or mitigate consequences of nasty events that may or may not happen.

- Community-based planning provides access to situation-specific information and experience that can often only be obtained by active involvement of local residents.

- Participation is an important management tool. Project-threatening resistance on the part of certain stakeholders can be minimized by inviting active cooperation of all stakeholders throughout the planning process.

- The knowledge and understanding needed by those who will carry out subsequent phases of a project can only be gained through active participation.

A good example of community-based planning can be seen in the Atlantic Coastal Action Programme (ACAP) in eastern Canada. The purpose of this project is to develop blueprints for the restoration and maintenance of environmentally degraded

FIGURE 15.18 *The Bay of Fundy is the site of an innovative community-based environmental planning process.*

harbors and estuaries in ways that are both biologically and socially sustainable (fig. 15.18). Officially established under Canada's Green Plan and supported by Environment Canada, this program created 13 community groups, some rural and some urban, with membership in each dominated by local residents. Federal and provincial government agencies are represented primarily as nonvoting observers and resource people. Each community group is provided with core funding for full-time staff who operate an office in the community and facilitate meetings. To cope with the complex biological and social problems affecting coastlines, ACAP is bringing together different stakeholders to create comprehensive plans for ecological, economic, and social sustainability. Through citizen monitoring and adaptive management, the community builds social capital (knowledge, cooperative spirit, trust, optimism, working relations), develops a sense of ownership in the planning process, and eliminates some of the fears and factional rivalry that often divide local groups, outsiders, and government agents.

CITIZEN PARTICIPATION

There are many things that individuals can do to live in a more sustainable manner. In nearly every chapter in this book, you can find a "What Can You Do?" box that lists some simple steps for living more lightly on the earth. Taking personal responsibility for our environmental impact can have many benefits. Recycling, buying "green" products, and other environmental actions set good examples for your friends and neighbors. They also strengthen your sense of involvement and commitment in valuable ways. There are limits, however, to how much we can do individually through our buying habits and personal actions to bring about the fundamental changes needed to save the earth. Green consumerism generally can do little about larger issues of global equity, chronic poverty, oppression, and the suffering of millions of people in the Third World. There is a danger that exclusive focus on such problems as whether to choose paper or plastic bags will divert our attention from the greater need to change basic institutions.

Environmentally concerned citizens have many opportunities to influence government policies. Get to know the positions of your state legislators, congressional representatives, and senators. Be active in party politics. Going to your district, county, and state political conventions gives you a voice in choosing candidates and establishing party platforms. Individuals can also contact legislators directly. You would probably be surprised at how little mail most legislators get from their constituents. Even on controversial issues, a representative might receive fewer than 100 personal letters or telephone calls. Since each congressional district includes about half a million people, the legislator tends to assume that each letter represents the views of 5,000 to 10,000 people. Your voice can have great impact. A single, well-written letter can make a difference in whether important legislation gets passed.

All legislators now have email access, but it isn't clear how many of them read Internet messages or how much attention they pay to them. Emails may fall in the category of form letters and petitions. Individually, they don't count for much, but if a decision maker gets a million of them, they probably will notice. If you don't have time to write, a telephone call to the local or Washington, D.C. office of your senator or representative can be effective. Follow the same steps in formulating your argument that you would use in a letter. You probably won't talk directly to the senator or representative, but your opinion will be registered by an aide. A well-organized, factual argument can be very persuasive and may change the position of your legislator.

Don't forget that many important decisions also are made at the local level. Your city planning and zoning board, county commissioners, state departments of natural resources or environmental protection, and a host of other agencies have power to establish many significant policies. These bodies are much more accessible to ordinary citizens. You might even run for a seat on a local board. What better way to ensure that local actions are sustainable than to become a policymaker yourself! Participating in practical environmental projects like litter cleanup or restoration projects can help build a sense of community, be educational, and also do good (fig. 15.19).

COLLECTIVE ACTION

Collective action multiplies individual power. You get encouragement and useful information from meeting regularly with others who share your interests. It's easy to get discouraged by the slow pace of change; having a support group helps maintain your enthusiasm. You should realize, however, that there is a broad spectrum of environmental and social action groups. Some will suit your particular interests, preferences, or beliefs more than others. In this section, we will look at some environmental organizations, as well as options for getting involved.

FIGURE 15.19 *Student volunteers plant native trees and bushes in a watershed protection project. Working together on a practical problem can be enjoyable and productive.*

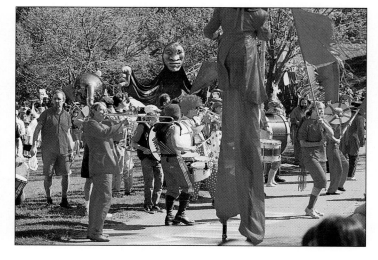

FIGURE 15.20 *Making a ruckus on behalf of environmental protection can attract attention to your cause.*

Student Environmental Groups

A number of organizations have been established to teach ecology and environmental ethics to elementary and secondary school students, as well as to get them involved in active projects to clean up their local community. Groups such as Kids Saving the Earth or Eco-Kids Corps are an important way to reach this vital audience. Family education results from these efforts as well. In a World Wildlife Fund survey, 63 percent of young people said they "lobby" their parents about recycling and buying environmentally responsible products.

Organizations for secondary and college students often are among our most active and effective groups for environmental change. The largest student environmental group in North America is the Student Environmental Action Coalition (SEAC). Formed in 1988 by students at the University of North Carolina at Chapel Hill, SEAC has grown rapidly to more than 30,000 members in some 500 campus environmental groups. SEAC is both an umbrella organization and grassroots network that functions as an information clearinghouse and a training center for student leaders. Member groups undertake a diverse spectrum of activities, ranging from politically neutral recycling promotion to confrontational protests of government or industrial projects (fig. 15.20). National conferences bring together thousands of activists who share tactics and inspiration while also having fun. If there isn't a group on your campus, why not look into organizing one?

Another important student organizing group is the network of Public Interest Research Groups (PIRGs) active on most campuses in the United States. While not focused exclusively on the environment, the PIRGs usually include environmental issues in their priorities for research. By becoming active, you could probably introduce environmental concerns to your local group if they are not already working on problems of importance to you.

One of the most important skills that you are likely to learn in SEAC or other groups committed to social change is how to organize. Organizing is a dynamic process in which you must constantly adapt to changing conditions. Some basic principles apply in most situations, however (table 15.7). Remember that you are not alone. Others share your concerns and want to work with you to bring about change; you just have to find them.

Mainline Environmental Organizations

Among the oldest, largest, and most influential environmental groups in the United States are the National Wildlife Federation, the World Wildlife Fund, the Audubon Society, the Sierra Club, the Izaak Walton League, Friends of the Earth, Greenpeace, Ducks Unlimited, the Natural Resources Defense Council, and The Wilderness Society. Sometimes known as the "group of 10," these organizations are criticized by radical environmentalists for their tendency to compromise and cooperate with the establishment. Although many of these groups were militant— even extremist— in their formative stages, they now tend to be more staid and conservative. Members are mostly passive and know little about the inner workings of the organization, joining as much for publications or social aspects as for their stands on environmental issues.

Still, these groups are powerful and important forces in environmental protection. Their mass membership, large professional staffs, and long history give them a degree of respectability and influence not found in newer, smaller groups. The Sierra Club, for instance, with about half a million members and chapters in almost every state, has a national staff of about 400, an annual budget over $20 million, and 20 full-time professional lobbyists in Washington, D.C. These national groups have become a potent force in Congress, especially when they band together to pass specific legislation, such as the Alaska National Interest Lands Act or the Clean Air Act.

In a survey that asked congressional staff and officials of government agencies to rate the effectiveness of groups that

TABLE 15.7 Organizing an Environmental Campaign

1. What do you want to change? Are your goals realistic, given the time and resources you have available?

2. What and who will be needed to get the job done? What resources do you have now, and how can you get more?

3. Who are the stakeholders in this issue? Who are your allies and constituents? How can you make contact with them?

4. How will your group make decisions and set priorities? Will you operate by consensus, majority vote, or informal agreement?

5. Have others already worked on this issue? What successes or failures did they have? Can you learn from their experience?

6. Who has the power to give you what you want or to solve the problem? Which individuals, organizations, corporations, or elected officials should your campaign target?

7. What tactics will be effective? Using the wrong tactics can alienate people and be worse than taking no action at all.

8. Are there social, cultural, or economic factors that should be recognized in this situation? Will the way you dress, talk, or behave offend or alienate your intended audience? Is it important to change your appearance or tactics to gain support?

9. How will you know when you have succeeded? How will you evaluate the possible outcomes?

10. What will you do when the battle is over? Is yours a single-issue organization, or will you want to maintain the interest, momentum, and network you have established?

Source: Based on material from "Grassroots Organizing for Everyone" by Claire Greensfelder and Mike Roselle from *Call to Action,* 1990 Sierra Book Club Books.

attempt to influence federal policy on pollution control, the top five were national environmental organizations. In spite of their large budgets and important connections, the American Petroleum Institute, the Chemical Manufacturers Association, and the Edison Electric Institute ranked far behind these environmental groups in terms of influence.

Although much of the focus of the big environmental groups is in Washington, Audubon, Sierra Club, and Izaak Walton have local chapters, outings, and conservation projects. This can be a good way to get involved. Go to some meetings, volunteer, offer to help. You may have to start out stuffing envelopes or some other unglamourous job, but if you persevere, you may have a chance to do something important and fun. It's a good way to learn and meet people.

Some environmental groups, such as the Environmental Defense Fund (EDF), The Nature Conservancy (TNC), the National Resources Defense Council (NRDC), and the Wilderness Society (WS), have limited contact with ordinary members except through their publications. They depend on a professional staff to carry out the goals of the organization through litigation (EDF and NRDC), land acquisition (TNC), or lobbying (WS). Although not often in the public eye, these groups can be very effective because of their unique focus. TNC buys land of high ecological value that is threatened by development. With more than $50 million in cash purchases and $44 million in donated

lands, TNC now owns some 900 parcels of land, the largest privately owned system of nature sanctuaries in the world. Altogether, it has saved more than 1 million ha (2.47 million acres) of land in its 40-year history (fig. 15.21).

Still, there often is bickering among groups. Members of Greenpeace and the Sierra Club angrily criticized TNC for taking "tainted money" from corporations such as Procter and Gamble (which tests products on animals), the Royal Caribbean Cruise Line (which was convicted twice in a year of dumping garbage in the ocean), and Gulf Oil, among others. TNC responds that it is trying to be part of the solution, rather than just criticizing these companies.

Radical Environmental Groups

A striking contrast to the mainline conservation organizations are the direct action groups, such as Earth First!, Sea Shepherd, and a few other groups that form either the "cutting edge" or the "radical fringe" of the environmental movement, depending on your outlook. Often associated with the deep ecology philosophy and bioregional ecological perspective, the strongest concerns of these militant environmentalists tend to be animal rights and protection of wild nature. Their main tactics are civil disobedience and attention-grabbing actions, such as guerrilla street theater, picketing, protest marches, road blockades, and other demonstrations. Many of these techniques are borrowed from the civil rights movement and Mahatma Gandhi's nonviolent civil disobedience. While often more innovative than the mainstream organizations, pioneering new issues and new approaches, the tactics of these groups can be controversial.

Members of Earth First! chained themselves to a giant tree-smashing bulldozer to prevent forest clearing in Texas and blockaded roads being built into wilderness areas. In one remarkable case, a 25-year-old activist named Julia "Butterfly" Hill spent more than two years sitting on a small, tarp-covered platform 55 m (180 ft) up in a redwood tree in northern California. She was

FIGURE 15.21 *The Nature Conservancy buys land, such as this short-grass prairie in Montana, to protect it from misuse and development.*

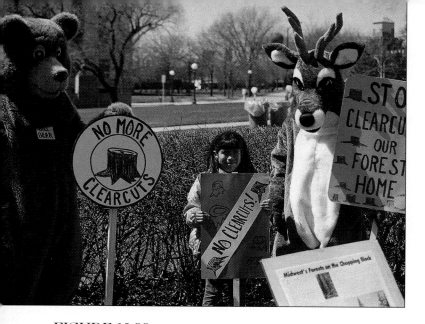

FIGURE 15.22 *Street theater can be a humorous, yet effective, way to convey a point in a nonthreatening manner. Confrontational tactics get attention, but they may alienate those who might be your allies and harden your opposition.*

protesting the clear-cutting of the Headwaters Forest by the Maxxam/Pacific Lumber Company. She came down from her perch in 1999 when her supporters raised $50,000 to buy the tree from Maxxam. A year later, however, vandals cut so deeply into the tree that it probably will die. Some protests are humorous and lighthearted, such as skits or street theater that get a point across in a nonthreatening way (fig. 15.22). What do you think? Are there circumstances in which it is appropriate or effective to use confrontation rather than cooperation and compromise to achieve your goals?

International Nongovernmental Organizations

The rise in international environmental organizations in recent years has been phenomenal. At the Stockholm Conference in 1972, only a handful of environmental groups attended, almost all from First World countries. Twenty years later, at the Rio Earth Summit, more than 30,000 individuals representing several thousand environmental groups, many from the Third World, held a global Ecoforum to debate issues and form alliances for a better world. We call these groups working for social change **nongovernmental organizations (NGOs).** They have become a powerful aspect of environmental protection.

Some NGOs are located primarily in the more highly developed countries of the North and work mainly on local issues. Others are headquartered in the North but focus their attention on the problems of developing countries in the South. Still others are truly global, with active groups in many different countries. A few are highly professional, combining private individuals with representatives of government agencies on quasi-governmental boards or standing committees with considerable power. Others are on the fringes of society, sometimes literally voices crying in the wilderness. Many work for political change, more specialize in gathering

and disseminating information, and some undertake direct action to protect a specific resource.

Public education and consciousness-raising using protest marches, demonstrations, civil disobedience, and other participatory public actions and media events are generally important tactics for these groups. Greenpeace, for instance, carries out well-publicized confrontations with whalers, seal hunters, toxic-waste dumpers, and others who threaten very specific and visible resources. Greenpeace may well be the largest environmental organization in the world, with some 2.5 million contributing members.

In contrast to these highly visible groups, others choose to work behind the scenes, but their impact may be equally important. Conservation International has been a leader in debt-for-nature swaps to protect areas particularly rich in biodiversity. It also has some interesting initiatives in economic development, seeking products made by local people that will provide income along with environmental protection (fig. 15.23).

The famous anthropologist Margaret Meade once wrote, "Never doubt that a small group of thoughtful, committed people can change the world; indeed it is the only thing that can." We hope that in this book you have found the inspiration and tools to make a positive change in our shared world.

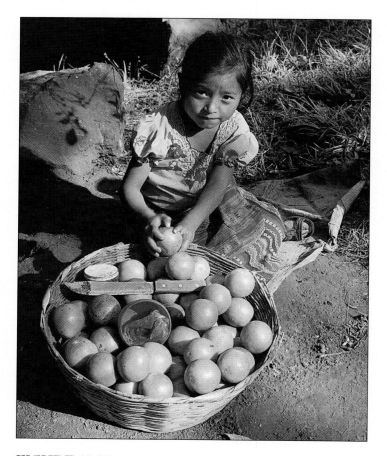

FIGURE 15.23 *International conservation groups often initiate economic development projects that provide local alternatives to natural resource destruction.*

SUMMARY

Throughout this book we've studied a variety of serious environmental problems. Science can help us understand why these problems occur and what we might do about them, but for many complex issues—sometimes called "wicked" problems because they have no single, clear answer—we can't wait until all possible evidence is gathered. We must make reasoned judgments based on the best available evidence. Learning from ecological systems, we see that some of the main goals for environmental policy and planning might be the precautionary principle and adaptive management.

Environmental education and literacy are important if we are to manage our natural resources for long-term sustainability. Citizen-science projects give hands-on, practical knowledge to participants and at the same time can contribute to solving real scientific problems. There are many opportunities for environmental careers in both natural and social sciences.

Although "policy" can have multiple meanings, in this chapter, environmental policy is taken to mean both public opinion as well as official rules and regulations concerning our environment. The policy cycle describes the steps by which problems are identified and defined, and solutions are proposed, debated, enacted into law, and monitored.

The National Environmental Policy Act (NEPA) forms the cornerstone of both environmental policy and law in the United States. One of its most important provisions is the requirement of Environmental Impact Statements (EIS) for all major federal projects and programs. Laws are rules established through legislation (statutes), judicial decisions (case law), or executive decisions (administrative law). Lobbying, litigation, and administrative interventions have been among environmentalists' most effective ways to shape policy and protect the environment. Individuals can play important roles in these processes, but often, it is more effective to work collectively. There are many different environmental organizations, and their philosophies and tactics range from quiet, behind-the-scene negotiations with decision makers to public protests and radical activism. Everyone should be able to find a group that fits their outlook.

Although many international treaties and conventions have been passed to protect our global environment, most are vague or toothless. Some innovative measures have been devised to compel compliance. Some alternatives to adversarial litigation include arbitration, mediation, and community-based planning. These techniques are useful in complex, unpredictable, multistakeholder, multivalue issues.

QUESTIONS FOR REVIEW

1. Describe adaptive management and the conditions necessary for its implementation.

2. What are "wicked" problems? Why are they difficult?

3. List two goals and four potential outcomes of environmental education.

4. What environmental career might you find rewarding and compatible with your interests?

5. What is the policy cycle, and how does it work?

6. Describe the three important provisions of NEPA (National Environmental Policy Act).

7. Discuss the differences and similarities among statutory, case, and administrative law.

8. List some of the major U.S. environmental legislation, as well as some important international environmental conventions and treaties.

9. Why have some international environmental treaties and conventions been effective, while most have not? Describe two such treaties.

10. What is collaborative, community-based planning?

THINKING SCIENTIFICALLY

1. Table 15.1 lists some important environmental questions. What would you add to this list? Which of these questions would you tackle first if you were head of a major foundation or research institute?

2. Think of an example of a "wicked" environmental problem. What makes it complex and difficult? What institutional changes could we implement to make this issue less wicked?

3. Which of the many environmental issues you have read about in this book are most interesting to you? Which laws in tables 15.5 or 15.6 pertain to those issues?

4. What fields of science help us understand or resolve the issues in question 3? What science classes could you take that would help you understand them or contribute to creative solutions? Which nonscience classes are also essential to understanding these issues?

5. As you read in chapter 1, scientists try to remain skeptical of evidence and rarely consider theories "proven." Why is this? How can politicians and lawyers make firm rules using scientific evidence when there is always uncertainty in scientific data?

6. Which is the most important step in the policy cycle (fig. 15.6)? If you were leader of a major environmental group, where would you put your efforts in establishing policy?

7. Do you believe that trees, wild animals, rocks, or mountains should have legal rights and standing in the courts? Why or why not? Are there partial rights or some other form of protection you would favor for nature?

8. It's sometimes difficult to determine whether a lawsuit is retaliatory or based on valid reason. How would you define a SLAPP suit and differentiate it from a legitimate case?

9. Under what conditions do you think mediation, arbitration, or collaborative planning could be more effective than litigation or confrontation? If you were planning a campaign to solve the problem you considered in question 2, which of these approaches would you use?

10. Try creating a list of arguments for and against an international body with power to enforce global environmental laws. Can you see a way to create a body that could satisfy both reasons for and against this power?

KEY TERMS

adaptive management 346
agency rule-making 357
arbitration 361
case law 356
citizen science 350
community-based planning 361
criminal law 356
environmental law 353
environmental literacy 348
environmental policy 350
federal laws (statutes) 353
international treaties and conventions 359
legal standing 356
lobbying 354
mediation 361
National Environmental Policy Act (NEPA) 352
nongovernmental organizations (NGOs) 366
policy cycle 350
precautionary principle 347
reasoned judgment 362
Strategic Lawsuits against Political Participation (SLAPP) 357
"wicked" problems 347

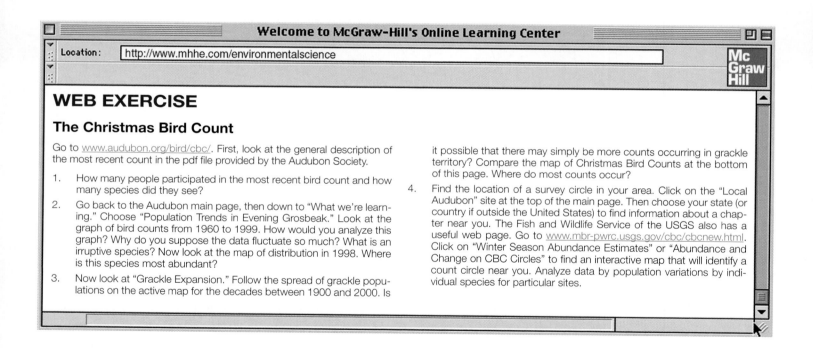

Welcome to McGraw-Hill's Online Learning Center

Location: http://www.mhhe.com/environmentalscience

WEB EXERCISE

The Christmas Bird Count

Go to www.audubon.org/bird/cbc/. First, look at the general description of the most recent count in the pdf file provided by the Audubon Society.

1. How many people participated in the most recent bird count and how many species did they see?

2. Go back to the Audubon main page, then down to "What we're learning." Choose "Population Trends in Evening Grosbeak." Look at the graph of bird counts from 1960 to 1999. How would you analyze this graph? Why do you suppose the data fluctuate so much? What is an irruptive species? Now look at the map of distribution in 1998. Where is this species most abundant?

3. Now look at "Grackle Expansion." Follow the spread of grackle populations on the active map for the decades between 1900 and 2000. Is

it possible that there may simply be more counts occurring in grackle territory? Compare the map of Christmas Bird Counts at the bottom of this page. Where do most counts occur?

4. Find the location of a survey circle in your area. Click on the "Local Audubon" site at the top of the main page. Then choose your state (or country if outside the United States) to find information about a chapter near you. The Fish and Wildlife Service of the USGS also has a useful web page. Go to www.mbr-pwrc.usgs.gov/cbc/cbcnew.html. Click on "Winter Season Abundance Estimates" or "Abundance and Change on CBC Circles" to find an interactive map that will identify a count circle near you. Analyze data by population variations by individual species for particular sites.

APPENDIXES

Political

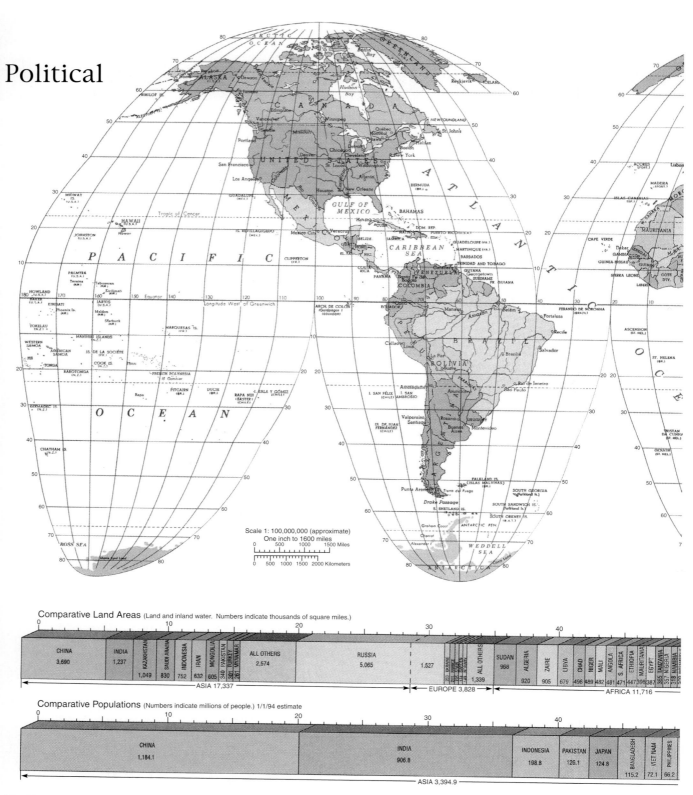

Scale 1: 100,000,000 (approximate)
One inch to 1600 miles

0   500   1000   1500 Miles

0   500 1000 1500 2000 Kilometers

Comparative Land Areas (Land and inland water. Numbers indicate thousands of square miles.)

CHINA	INDIA	KAZAKHSTAN	SAUDI ARABIA	INDONESIA	IRAN	MONGOLIA	PAKISTAN	TURKEY	MYANMAR	ALL OTHERS	RUSSIA		ALL OTHERS	SUDAN	ALGERIA	ZAIRE	LIBYA	CHAD	NIGER	MALI	ANGOLA	S. AFRICA	ETHIOPIA	MAURITANIA	EGYPT	TANZANIA	NIGERIA	NAMIBIA	MOZAMBIQUE
3,690	1,237	1,049	830	752	632	605	340	301	261	2,574	5,065	1,527	1,339	968	920	905	679	496	489	482	481	471	447	396	387	365	357	318	

←————— ASIA 17,337 —————→ ←— EUROPE 3,828 —→ ←———— AFRICA 11,716 ————→

Comparative Populations (Numbers indicate millions of people.) 1/1/94 estimate

CHINA	INDIA	INDONESIA	PAKISTAN	JAPAN	BANGLADESH	VIET NAM	PHILIPPINES
1,184.1	906.8	198.8	126.1	124.8	115.2	72.1	66.2

←———————————— ASIA 3,394.9 ————————————→

Principles of Environmental Science   www.mhhe.com/environmentalscience

Goode's Homolosine Equal Area Projection

WORLD TOTAL 57,900,000 square miles

ZAMBIA 291	ALL OTHERS 2,956	CANADA 3,850	UNITED STATES 3,787	GREENLAND 840	MEXICO 760	OTHER 292	BRAZIL 3,286	ARGENTINA 1,074	PERU 496	COLOMBIA 441	BOLIVIA 424	VENEZ. 352	CHILE 292	OTHER 518	AUSTRALIA 2,966	OTHER 318	ANTARCTICA 5,400

NORTH AMERICA 9,529 — SOUTH AMERICA 6,884 — OCEANIA 3,284 — ANTARCTICA 5,400

WORLD TOTAL 5,556,000,000 inhabitants

IRAN 63.9	TURKEY 61.5	THAILAND 59.0	S. KOREA 44.1	MYANMAR 43.6	ALL OTHERS 299.8	RUSSIA 120.8	28.6	GERMANY 80.9	UNITED KINGDOM 58.0	FRANCE 57.7	ITALY 56.7	UKRAINE 52.2	SPAIN 38.6	POLAND 38.5	ROMANIA 22.7	NETHER. 15.2	ALL OTHERS 158.9	NIGERIA 94.5	EGYPT 56.8	ETHIOPIA 54.2	S. AFRICA 42.4	ZAIRE 41.7	SUDAN 28.9	KENYA 28.3	MOROCCO 28.1	TANZANIA 27	ALGERIA	ALL OTHERS 254.7	UNITED STATES 259.4	MEXICO 90.9	CANADA 27.8	ALL OTHERS 66.5	BRAZIL 151.3	COLOMBIA 35.1	ARGENTINA 33.8	ALL OTHERS 84.4	OCEANIA 28.6

EUROPE 700.5 — AFRICA 683.8 — NORTH AMERICA 444.7 — S. AMERICA 304.5

Vegetation

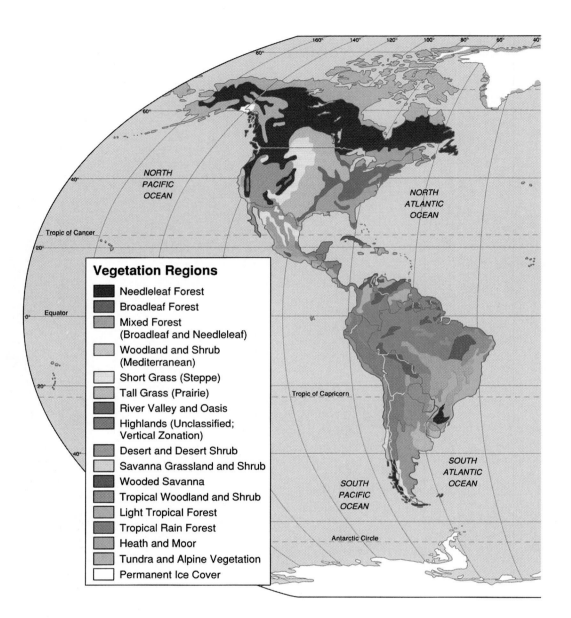

Vegetation Regions

- Needleleaf Forest
- Broadleaf Forest
- Mixed Forest (Broadleaf and Needleleaf)
- Woodland and Shrub (Mediterranean)
- Short Grass (Steppe)
- Tall Grass (Prairie)
- River Valley and Oasis
- Highlands (Unclassified; Vertical Zonation)
- Desert and Desert Shrub
- Savanna Grassland and Shrub
- Wooded Savanna
- Tropical Woodland and Shrub
- Light Tropical Forest
- Tropical Rain Forest
- Heath and Moor
- Tundra and Alpine Vegetation
- Permanent Ice Cover

Principles of Environmental Science   www.mhhe.com/environmentalscience

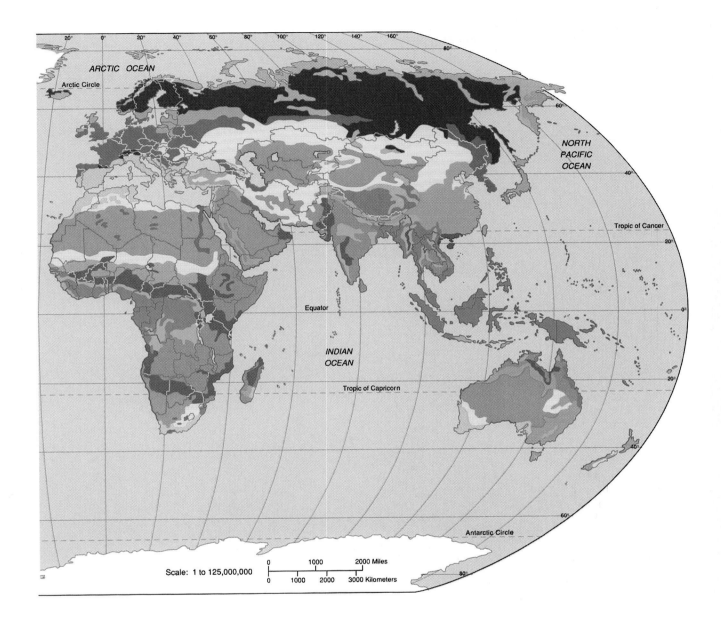

Vegetation is the most visible consequence of the distribution of temperature and precipitation. The global distribution of vegetation types and the global distribution of climate are closely related. But not all vegetation types are the consequence of temperature and precipitation or other climatic variables. Many types of vegetation, in many areas of the world, are the consequence of human activities, particularly the grazing of domesticated livestock, burning, and forest clearance.

World Population Density

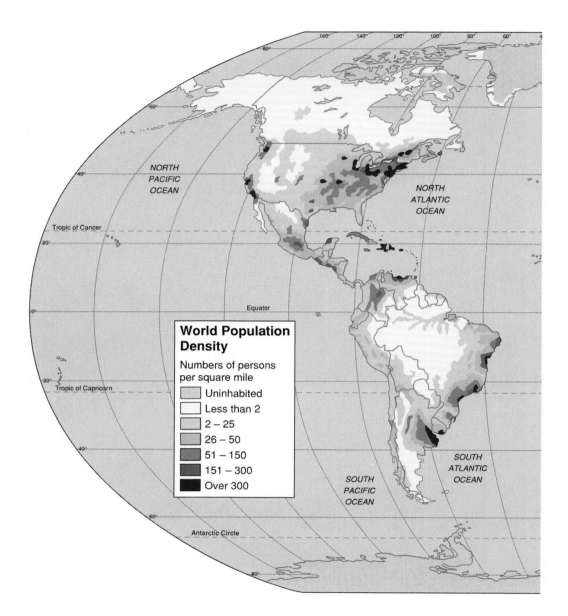

World Population Density

Numbers of persons per square mile

- Uninhabited
- Less than 2
- 2 – 25
- 26 – 50
- 51 – 150
- 151 – 300
- Over 300

No feature of human activity is more reflective of environmental conditions than where people live. In the areas of densest population, a mixture of natural and human factors have combined to allow maximum food production, maximum urbanization, and especially concentrated economic activity. Three such great concentrations appear on the map—East Asia, South Asia, and Europe—with a fourth lesser concentration in eastern North America (the "Megalopolis" region of the United States and Canada). One of these great population clusters—South Asia—is still growing rapidly and can be expected to become even more densely populated by the beginning of the twenty-first century. The other concentrations are likely to remain about as they are now. In Europe and North America, this is the result of

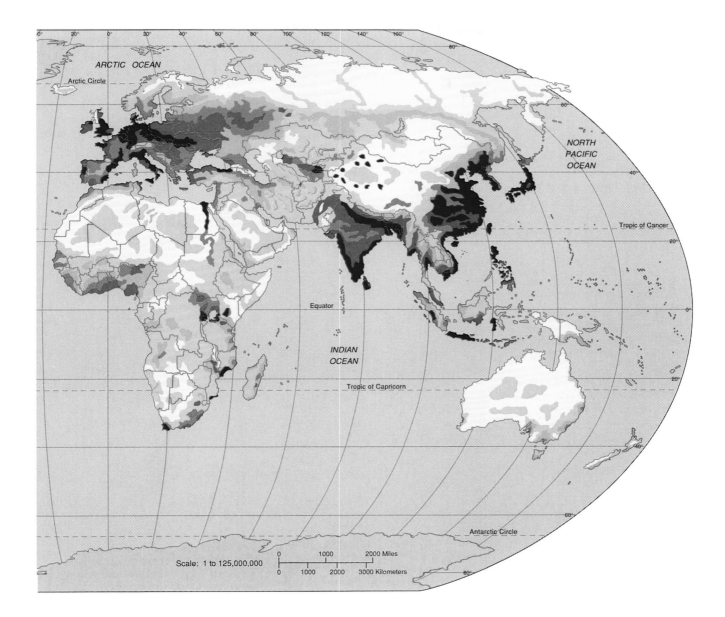

economic development that has caused population growth to level off during the last century. In East Asia, population has also begun to grow more slowly. In the case of Japan and the Koreas, this is the consequence of economic development; in the case of China, it is the consequence of government intervention in the form of strict family planning. The areas of future high density (in addition to those already existing) are likely to be in Middle and South America and Africa, where population growth rates are well above the world average. Population that is extremely dense or growing at an excessive rate when measured against a region's habitability is one of the greatest indicators of environmental deterioration.

Temperature Regions and Ocean Currents

Surface Temperature Regions

- Always cold: polar regions and high altitudes
- Cold winter and cool summer; always cool in tropical higher altitudes
- Cold winter and mild summer
- Cool winter and mild summer
- Hot summer and cold winter
- Hot summer and cool winter
- Hot summer and mild winter
- Always hot
- Always mild

Hot = above 68°F (20°C)
Mild = 50° − 68°F (10° − 20°C)
Cool = 32° − 50°F (0° − 10°C)
Cold = below 32°F (0°C)

→ Cool/cold current
→ Warm current

NORTH PACIFIC OCEAN
NORTH ATLANTIC OCEAN
SOUTH PACIFIC OCEAN
SOUTH ATLANTIC OCEAN

Tropic of Cancer
Equator
Tropic of Capricorn
Antarctic Circle

Along with precipitation, temperature is one of the two most important environmental variables, defining the climatic conditions so essential for the distribution of human activities and the human population. Ocean currents exert a significant influence over the climate of adjacent continents and are the most important mechanism for redistributing surplus heat from the equatorial region into middle and high latitudes.

APPENDIXES

GLOSSARY

A

acid precipitation Acidic rain, snow, or dry particles deposited from the air due to increased acids released by anthropogenic or natural resources.

acids Substances that release hydrogen atoms in water.

active solar system A mechanical system that uses moving substances to collect and transfer solar energy.

acute effects A sudden onset of symptoms or effects of exposure to some factor.

acute poverty Insufficient income or access to resources needed to provide the basic necessities for life such as food, shelter, sanitation, clean water, medical care, and education.

adaptation Physical changes that allow organisms to survive in a given environment.

adaptive management A management plan designed from the outset to "learn by doing," and to actively test hypotheses and adjust treatments as new information becomes available.

administrative law Executive orders, administrative rules and regulations, and enforcement decisions by administrative agencies and special administrative courts.

aerosols Minute particles or liquid droplets suspended in the air.

agency rule-making The formal process of establishing rules and standards by administrative agencies.

albedo A description of a surface's reflective properties.

allergens Substances that activate the immune system and cause an allergic response; may not be directly antigenic themselves but may make other materials antigenic.

ambient air The air immediately around us.

analytical thinking A way of systematic analysis that asks, "How can I break this problem down into its constituent parts?"

anemia Low levels of hemoglobin due to iron deficiency or lack of red blood cells.

anthropocentric Believing that humans hold a special place in nature; being centered primarily on humans and human affairs.

antigens Substances that stimulate the production of, and react with, specific antibodies.

aquifers Porous, water-bearing layers of sand, gravel, and rock below the earth's surface; reservoirs for groundwater.

arbitration A formal process of dispute resolution resulting in a legally binding decision that all parties must obey.

arithmetic growth A pattern of growth that increases at a constant amount per unit time, such as 1, 2, 3, 4 or 1, 3, 5, 7.

atmospheric deposition Sedimentation of solids, liquids, or gaseous materials from the air.

atom The smallest particle that exhibits the characteristics of an element.

atomic number The characteristic number of protons per atom of an element.

autotroph An organism that synthesizes food molecules from inorganic molecules by using an external energy source, such as light energy.

B

barrier islands Low, narrow, sandy islands that form offshore from a coastline.

bases Substances that readily bond with hydrogen ions in an aqueous solution.

Batesian mimicry Evolution by one species to resemble another species that is protected from predators by a venomous stinger, bad taste, or some other defensive adaptation.

benthos The bottom of a sea or lake.

bioaccumulation The selective absorption and concentration of molecules by cells.

biocentric preservation A philosophy that emphasizes the fundamental right of living organisms to exist and to pursue their own ends.

biocentrism The belief that all creatures have rights and values; being centered on nature rather than humans.

biochemical oxygen demand (BOD) A standard test for measuring the amount of dissolved oxygen utilized by aquatic microorganisms.

biodegradable plastics Plastics that can be decomposed by microorganisms.

biodiversity The genetic, species, and ecological diversity of the organisms in a given area.

biogeochemical cycles Movement of matter within or between ecosystems; caused by living organisms, geological forces, or chemical reactions. The cycling of nitrogen, carbon, sulfur, oxygen, phosphorus, and water are examples.

biogeographical area A region or ecosystem with characteristic biological, water, and land resources.

biological community The populations of plants, animals, and microorganisms living and interacting in a certain area at a given time.

biological controls Use of natural predators, pathogens, or competitors to regulate pest populations.

biomagnification Increase in concentration of certain stable chemicals (for example, heavy metals or fat-soluble pesticides) in successively higher trophic levels of a food chain or web.

biomass The accumulated biological material produced by living organisms.

biomass fuel Organic material produced by plants, animals, or microorganisms that can be burned directly as a heat source or converted into a gaseous or liquid fuel.

biomass pyramid A metaphor or diagram that explains the relationship between the amounts of biomass at different trophic levels.

biome A broad, regional type of ecosystem characterized by distinctive climate and soil conditions and a distinctive kind of biological community adapted to those conditions.

bioremediation Use of biological organisms to remove pollution or restore environmental quality.

biosphere The zone of air, land, and water at the surface of the earth that is occupied by organisms.

biosphere reserves World heritage sites identified by the IUCN as worthy for national park or wildlife refuge status because of high biological diversity or unique ecological features.

biota All organisms in a given area.

biotic potential The maximum reproductive rate of an organism, given unlimited resources and ideal environmental conditions. Compare with environmental resistance.

birth control Any method used to reduce births, including celibacy, delayed marriage, contraception; devices or medication that prevent implantation of fertilized zygotes, and induced abortions.

blind experiments A design in which researchers don't know which subjects were given experimental treatment until after data has been gathered and analyzed.

bog An area of waterlogged soil that tends to be peaty; fed mainly by precipitation; low productivity; some bogs are acidic.

boreal forest A broad band of mixed coniferous and deciduous trees that stretches across northern North America (and also Europe and Asia); its northernmost edge, the taiga, intergrades with the arctic tundra.

brownfields Abandoned or underused urban areas in which redevelopment is blocked by liability or financing issues related to toxic contamination.

C

cancer Invasive, out-of-control cell growth that results in malignant tumors.

capital Any form of wealth, resources, or knowledge available for use in the production of more wealth.

carbohydrate An organic compound consisting of a ring or chain of carbon atoms with hydrogen and oxygen attached; examples are sugars, starches, cellulose, and glycogen.

carbon cycle The circulation and reutilization of carbon atoms, especially via the processes of photosynthesis and respiration.

carbon management Projects to reduce carbon dioxide emissions from fossil fuel or to ameliorate their effects.

carbon monoxide (CO) Colorless, odorless, nonirritating but highly toxic gas produced by incomplete combustion of fuel, incineration of biomass or solid waste, or partially anaerobic decomposition of organic material.

carbon sink Places of carbon accumulation, such as in large forests (organic compounds) or ocean sediments (calcium carbonate).

carcinogens Substances that cause cancer.

carnivores Organisms that mainly prey upon animals.

carrying capacity The maximum number of individuals of any species that can be supported by a particular ecosystem on a long-term basis.

case law Precedents from both civil and criminal court cases.

cellular respiration The process in which a cell breaks down sugar or other organic compounds to release energy used for cellular work; may be anaerobic or aerobic, depending on the availability of oxygen.

chain reaction A self-sustaining reaction in which the fission of nuclei produces subatomic particles that cause the fission of other nuclei.

chaparral A biological community characterized by thick growth of thorny, evergreen shrubs typical of a Mediterranean climate.

chemical bond The force that holds molecules together.

chemical energy Potential energy stored in chemical bonds of molecules.

chlorofluorocarbons Chemical compounds with a carbon skeleton and one or more attached chlorine and fluorine atoms. Commonly used as refrigerants, solvents, fire retardants, and blowing agents.

chloroplasts Chlorophyll-containing organelles in eukaryotic organisms; sites of photosynthesis.

chronic effects Long-lasting results of exposure to a toxin; can be a permanent change caused by a single, acute exposure or a continuous, low-level exposure.

chronically undernourished Receiving less than 90 percent of the minimum caloric intake needed for normal growth and development, and a healthy, productive life.

citizen science Projects in which trained volunteers work with scientific researchers to answer real-world questions.

city A differentiated community with a sufficient population and resource base to allow residents to specialize in arts, crafts, services, and professional occupations.

civil law A body of laws regulating relations between individuals or between individuals and corporations concerning property rights, personal dignity and freedom, and personal injury.

classical economics Modern, western economic theories of the effects of resource scarcity, monetary policy, and competition on supply and demand of goods and services in the marketplace. This is the basis for the capitalist market system.

clear-cutting Cutting every tree in a given area, regardless of species or size; an appropriate harvest method for some species; can be destructive if not carefully controlled.

climate A description of the long-term pattern of weather in a particular area.

closed-canopy A forest where tree crowns spread over 20 percent of the ground; has the potential for commercial timber harvests.

cloud forests High mountain forests where temperatures are uniformly cool and fog or mist keeps vegetation wet all the time.

coevolution The process in which species exert selective pressure on each other and gradually evolve new features or behaviors as a result of those pressures.

cogeneration The simultaneous production of electricity and steam or hot water in the same plant.

cold front A moving boundary of cooler air displacing warmer air.

coliform bacteria Bacteria that live in the intestines (including the colon) of humans and other animals; used as a measure of the presence of feces in water or soil.

commensalism A symbiotic relationship in which one member is benefited and the second is neither harmed nor benefited.

common law the body of court decisions that constitute a working definition of individual rights and responsibilities where no formal statutes define these issues.

communal resource management systems Resources managed by a community for long-term sustainability.

community ecology The study of interactions of all populations living in the ecosystem of a given area.

community-based planning Involving community stake-holders in pluralistic, adaptive, inclusive, proactive planning.

competitive exclusion A theory that no two populations of different species will occupy the same niche and compete for exactly the same resources in the same habitat for very long.

complexity (ecological) The number of species at each trophic level and the number of trophic levels in a community.

composting The biological degradation of organic material under aerobic (oxygen-rich) conditions to produce compost, a nutrient-rich soil amendment and conditioner.

compound A molecule made up of two or more kinds of atoms held together by chemical bonds.

conifers Needle-bearing trees that produce seeds in cones.

conservation of matter In any chemical reaction, matter changes form; it is neither created nor destroyed.

conspicuous consumption A term coined by economist and social critic Thorstein Veblen to describe buying things we don't want or need to impress others.

consumer An organism that obtains energy and nutrients by feeding on other organisms or their remains. See also heterotroph.

consumption The fraction of withdrawn water that is lost in transmission or that is evaporated, absorbed, chemically transformed, or otherwise made unavailable for other purposes as a result of human use.

contour plowing Plowing along hill contours; reduces erosion.

control rods Neutron-absorbing material inserted into spaces between fuel assemblies in nuclear reactors to regulate fission reaction.

controlled studies Comparisons made between two populations that are identical (as far as possible) in every factor except the one being studied.

convection currents Rising or sinking air currents that stir the atmosphere and transport heat from one area to another. Convection currents also occur in water.

conventional (or criteria) pollutants The seven substances (sulfur dioxide, carbon monoxide, particulates, hydrocarbons, nitrogen oxides, photochemical oxidants, and lead) that make up the largest volume of air quality degradation; identified by the Clean Air Act as the most serious threat of all pollutants to human health and welfare.

convergent evolution Species evolve from different origins but under similar environmental conditions to have similar traits.

coral reefs Prominent oceanic features composed of hard, limy skeletons produced by coral animals; usually formed along edges of shallow, submerged ocean banks or along shelves in warm, shallow, tropical seas.

core The dense, intensely hot mass of molten metal, mostly iron and nickel, thousands of kilometers in diameter at the earth's center.

Coriolis effect The tendency for air above the earth to appear to be deflected to the right (in the northern hemisphere) or the left (in the South) because of the earth's rotation.

corridor A strip of natural habitat that connects two adjacent nature preserves to allow migration of organisms from one place to another.

cost-benefit analysis An evaluation of large-scale public projects by comparing the costs and benefits that accrue from them.

cover crops Plants, such as rye, alfalfa, or clover, that can be planted immediately after harvest to hold and protect the soil.

creative thinking Original, independent thinking that asks, "How might I approach this problem in new and inventive ways?"

criminal law A body of court decisions based on federal and state statutes concerning wrongs against persons or society.

criteria pollutants See conventional pollutants.

critical factor The single environmental factor closest to a tolerance limit for a given species at a given time. See limiting factors.

critical thinking An ability to evaluate information and opinions in a systematic, purposeful, efficient manner.

crude birthrate The number of births in a year divided by the midyear population.

crude death rate The number of deaths per thousand persons in a given year; also called crude mortality rate.

crust The cool, lightweight, outermost layer of the earth's surface that floats on the soft, pliable underlying layers; similar to the "skin" on a bowl of warm pudding.

cultural eutrophication An increase in biological productivity and ecosystem succession caused by human activities.

D

debt-for-nature swap Forgiveness of international debt in exchange for nature protection in developing countries.

deciduous Trees and shrubs that shed their leaves at the end of the growing season.

decline spiral A catastrophic deterioration of a species, community, or whole ecosystem; accelerates as functions are disrupted or lost in a downward cascade.

decomposers Fungi and bacteria that break complex organic material into smaller molecules.

deductive reasoning "Top down" reasoning in which we start with a general principle and derive a testable prediction about a specific case.

deep ecology A philosophy that calls for a profound shift in our attitudes and behavior based on rejection of anthropocentric attitudes; a belief in the sacredness of nature; and direct personal action to protect nature.

delta Fan-shaped sediment deposit found at the mouth of a river.

demanufacturing Disassembly of products so components can be reused or recycled.

demographic transition A pattern of falling death rates and birthrates in response to improved living conditions; could be reversed in deteriorating conditions.

demography The statistical study of human populations relating to growth rate, age structure, geographic distribution, etc., and their effects on social, economic, and environmental conditions.

dependency ratio The number of nonworking members compared to working members for a given population.

desalinization (or desalination) Removal of salt from water by distillation, freezing, or ultrafiltration.

desert A type of biome characterized by low moisture levels and infrequent and unpredictable precipitation. Daily and seasonal temperatures fluctuate widely.

desertification Denuding and degrading a once-fertile land, initiating a desert-producing cycle that feeds on itself and causes long-term changes in soil, climate, and biota of an area.

detritivore Organisms that consume organic litter, debris, and dung.

dew point The temperature at which condensation occurs for a given concentration of water vapor in the air.

dieback A sudden population decline; also called a population crash.

disability adjusted life years (DALYs) A health measure that assesses the total burden of disease by combining premature deaths and loss of a healthy life that result from illness or disability.

discharge The amount of water that passes a fixed point in a given amount of time; usually expressed as liters or cubic feet of water per second.

discount rate The amount we discount or reduce the value of a future payment. When you borrow money from the bank at 10 percent annual interest, you are in effect saying that having the money now is worth 10 percent more to you than having the same amount one year from now.

disease A deleterious change in the body's condition in response to destabilizing factors, such as nutrition, chemicals, or biological agents.

dissolved oxygen (DO) content Amount of oxygen dissolved in a given volume of water at a given temperature and atmospheric pressure; usually expressed in parts per million (ppm).

divergent evolution Separation of a species into new types.

diversity (species diversity, biological diversity) The number of species present in a community (species richness), as well as the relative abundance of each species.

DNA Deoxyribonucleic acid; the long, double-helix molecule in the nucleus of cells that contains the genetic code and directs the development and functioning of all cells.

double-blind design Neither the subject (participant) nor the experimenter knows which participants are receiving the experimental or the control treatments until after data have been gathered and analyzed.

drip irrigation Uses pipe or tubing perforated with very small holes to deliver water one drop at a time directly to the soil around each plant.

E

earth charter A set of principles for sustainable development, environmental protection, and social justice developed by a council appointed by the United Nations.

earthquakes Sudden, violent movement of the earth's crust.

ecocentric (ecologically centered) A philosophy that claims moral values and rights for both organisms and ecological systems and processes.

ecofeminism A pluralistic philosophy of respect for nature based on feminist philosophies of justice and egalitarianism.

ecological development A gradual process of environmental modification by organisms.

ecological economics Application of ecological insights to economic analysis; incorporating ecological principles and priorities into economic accounting systems.

ecological niche The functional role and position of a species in its ecosystem, including what resources it uses, how and when it uses the resources, and how it interacts with other species.

ecological services Processes or materials such as clean water, energy, climate regulation, and nutrient cycling provided by ecosystems.

ecological succession The process by which organisms gradually occupy a site, alter its ecological conditions, and are eventually replaced by other organisms.

ecology The scientific study of relationships between organisms and their environment. It is concerned with the life histories, distribution, and behavior of individual species as well as the structure and function of natural systems at the level of populations, communities, and ecosystems.

economic development A rise in real income per person; usually associated with new technology that increases productivity or resources.

economic growth An increase in the total wealth of a nation; if population grows faster than the economy, there may be real economic growth, but the share per person may decline.

ecosystem A specific biological community and its physical environment interacting in an exchange of matter and energy.

ecosystem management An integration of ecological, economic, and social goals in a unified systems approach to resource management.

ecosystem restoration To reinstate an entire community of organisms to as near its natural condition as possible.

ecotone A boundary between two types of ecological communities.

ecotourism A combination of adventure travel, cultural exploration, and nature appreciation in wild settings.

edge effects A change in species composition, physical conditions, or other ecological factors at the boundary between two ecosystems.

El Niño A climatic change marked by shifting of a large warm water pool from the western Pacific Ocean towards the east. Wind direction and precipitation patterns are changed over much of the Pacific and perhaps around the world.

electron A negatively charged subatomic particle that orbits around the nucleus of an atom.

element A substance that cannot be broken into simpler units by chemical means.

emergent disease A new disease or one that has been absent for at least 20 years.

emigration The movement of members from a population.

emission standards Regulations for restricting the amounts of air pollutants that can be released from specific point sources.

endangered species A species considered to be in imminent danger of extinction.

endemism A state in which species are restricted to a single region.

endocrine hormone disrupters Chemicals that interfere with the function of endocrine hormones such as estrogen, testosterone, thyroxine, adrenaline, or cortisone.

energy The capacity to do work, such as moving matter over a distance.

energy recovery Incineration of solid waste to produce useful energy.

environment The circumstances or conditions that surround an organism or group of organisms as well as the complex of social or cultural conditions that affect an individual or community.

environmental ethics A search for moral values and ethical principles in human relations with the natural world.

environmental impact statement (EIS) An analysis of the effects of any major program or project planned by a federal agency; required by provisions in the National Environmental Policy Act of 1970.

environmental indicators Organisms or physical factors that serve as a gauge for environmental changes. Indicator organisms, which cannot survive beyond certain environmental limits, are known as bioindicators.

environmental justice Fair access to a clean, healthy environment, regardless of class, race, or income level, or other status.

environmental law Legal rules, decisions, and actions concerning environmental quality, natural resources, and ecological sustainability.

environmental literacy A basic understanding of ecological principles and the ways society affects, or responds to, environmental conditions.

environmental policy The official rules or regulations concerning the environment adopted, implemented, and enforced by some governmental agency.

environmental racism Decisions that unfairly expose people to polluted or degraded environments on the basis of race.

environmental resistance All the limiting factors that tend to reduce population growth rates and set the maximum allowable population size or carrying capacity of an ecosystem.

environmental science The systematic, scientific study of our environment as well as our role in it.

enzymes Molecules, usually proteins or nucleic acids, that act as catalysts in biochemical reactions.

epidemiology The study of the distribution and causes of disease and injuries in human populations.

epiphyte A plant that grows on a substrate other than the soil, such as the surface of another organism.

equilibrium community Also called a disclimax community; a community subject to periodic disruptions, usually by fire, that prevent it from reaching a climax stage.

estuary A bay or drowned valley where a river empties into the sea.

eutrophic Rivers and lakes rich in organic material (eu = well; trophic = nourished).

evolution A theory that explains how random changes in genetic material and competition for scarce resources cause species to change gradually.

exhaustible resources Materials present in fixed amounts in the environment, especially the earth's geologic endowment: minerals, nonmineral resources, fossil fuels.

exotic organisms Alien species introduced by human agency into biological communities where they would not naturally occur.

exponential growth Growth at a constant rate of increase per unit of time; can be expressed as a constant fraction or exponent. See geometric growth.

externalizing costs Shifting expenses, monetary or otherwise, to someone other than the individuals or groups who use a resource.

extinction The irrevocable elimination of species; can be a normal process of the natural world as species out-compete or kill off others or as environmental conditions change.

extirpate To destroy totally; extinction caused by direct human action, such as hunting, trapping, etc.

F

family planning Controlling reproduction; planning the timing of birth and having only as many babies as are wanted and can be supported.

famines Acute food shortages characterized by large-scale loss of life, social disruption, and economic chaos.

fauna All of the animals present in a given region.

fecundity The physical ability to reproduce.

federal laws (statutes) Laws passed by the federal legislature and signed by the chief executive.

fen A wetland fed mainly by groundwater.

feral A domestic animal that has taken up a wild existence.

fertility The actual number of offspring produced through sexual reproduction; usually described in terms of number of offspring of females, since paternity can be difficult to determine.

fetal alcohol syndrome A tragic set of permanent physical and mental and behavioral birth defects that result when mothers drink alcohol during pregnancy.

fire-climax community An equilibrium community maintained by periodic fires; examples include grasslands, chaparral shrubland, and some pine forests.

first law of thermodynamics States that energy is conserved; that is, it is neither created nor destroyed under normal conditions.

First World The industrialized capitalist or market-economy countries of Western Europe, North America, Japan, Australia, and New Zealand.

flood An overflow of water onto land that normally is dry.

floodplains Low lands along riverbanks, lakes, and coastlines subjected to periodic inundation.

flora All of the plants present in a given region.

food chain A linked feeding series; in an ecosystem, the sequence of organisms through which energy and materials are transferred, in the form of food, from one trophic level to another.

food security The ability of individuals to obtain sufficient food on a day-to-day basis.

food web A complex, interlocking series of individual food chains in an ecosystem.

fossil fuels Petroleum, natural gas, and coal created by geological forces from organic wastes and dead bodies of formerly living biological organisms.

fragmentation Disruption of habitat into small, isolated fragments.

freshwater ecosystems Ecosystems in which the fresh (nonsalty) water of streams, rivers, ponds, or lakes plays a defining role.

fuel assembly A bundle of hollow metal rods containing uranium oxide pellets; used to fuel a nuclear reactor.

fuel cells Mechanical devices that use hydrogen or hydrogen-containing fuel such as methane to produce an electric current. Fuel cells are clean, quiet and highly efficient sources of electricity.

fugitive emissions Substances that enter the air without going through a smokestack, such as dust from soil erosion, strip mining, rock crushing, construction, and building demolition.

fungi Nonphotosynthetic, eukaryotic organisms with cell walls, filamentous bodies, and absorptive nutrition.

fungicide A chemical that kills fungi.

G

gamma rays Very short wavelength forms of the electromagnetic spectrum.

gap analysis A biogeographical technique of mapping biological diversity and endemic species to find gaps between protected areas that leave endangered habitats vulnerable to disruption.

gasohol A mixture of gasoline and ethanol.

gene A unit of heredity; a segment of DNA nucleus of the cell that contains information for the synthesis of a specific protein, such as an enzyme.

general fertility rate Crude birth rate multiplied by the percentage of reproductive age women.

genetic assimilation The disappearance of a species as its genes are diluted through crossbreeding with a closely related species.

genetic engineering Laboratory manipulation of genetic material using molecular biology.

genetically modified organisms (GMOs) Organisms created by combining natural or synthetic genes using the techniques of molecular biology.

geometric growth Growth that follows a geometric pattern of increase, such as 2, 4, 8, 16, etc. See exponential growth.

geothermal energy Energy drawn from the internal heat of the earth, either through geysers, fumaroles, hot springs, or other natural geothermal features, or through deep wells that pump heated groundwater.

global environmentalism Extension of modern environmental concerns to global issues.

grassland A biome dominated by grasses and associated herbaceous plants.

green plans Integrated national environmental plans for reducing pollution and resource consumption while achieving sustainable development and environmental restoration.

green political parties Political organizations based on environmental protection, participatory democracy, grassroots organization, and sustainable development.

green pricing Plans in which consumers can voluntarily pay premium prices for renewable energy.

green revolution Dramatically increased agricultural production brought about by "miracle" strains of grain; usually requires high inputs of water, plant nutrients, and pesticides.

greenhouse effect Trapping of heat by the earth's atmosphere, which is transparent to incoming visible light waves but absorbs outgoing long-wave infrared radiation.

greenhouse gas A gas that traps heat in the atmosphere.

gross domestic product (GDP) The total economic activity within national boundaries.

gross national product (GNP) The sum total of all goods and services produced in a national economy. Gross domestic product (GDP) is used to distinguish economic activity within a country from that of offshore corporations.

groundwater Water held in gravel deposits or porous rock below the earth's surface; does not include water or crystallization held by chemical bonds in rocks or moisture in upper soil layers.

gully erosion Removal of layers of soil, creating channels or ravines too large to be removed by normal tillage operations.

H

habitat The place or set of environmental conditions in which a particular organism lives.

habitat conservation plans Agreements under which property owners are allowed to harvest resources or develop land as long as habitat is conserved or replaced in ways that benefit resident endangered or threatened species in the long run. Some incidental "taking" or loss of endangered species is generally allowed in such plans.

half-life The time required for one-half of a sample to decay or change into some other form.

hazardous waste Any discarded material containing substances known to be toxic, mutagenic, carcinogenic, or teratogenic to humans or other life-forms; ignitable, corrosive, explosive, or highly reactive alone or with other materials.

health A state of physical and emotional well-being; the absence of disease or ailment.

heap-leach extraction A technique for separating gold from extremely low-grade ores. Crushed ore is piled in huge heaps and sprayed with a dilute alkaline-cyanide solution, which percolates through the pile to extract the gold, which is separated from the effluent in a processing plant. This process has a high potential for water pollution.

heat Total kinetic energy of atoms or molecules in a substance not associated with the bulk motion of the substance.

herbicide A chemical that kills plants.

herbivore An organism that eats only plants.

heterotroph An organism that is incapable of synthesizing its own food and, therefore, must feed upon organic compounds produced by other organisms.

high-level waste repository A place where intensely radioactive wastes can be buried and remain unexposed to groundwater and earthquakes for tens of thousands of years.

high-quality energy Intense, concentrated, and high-temperature energy that is considered high-quality because of its usefulness in carrying out work.

holistic science The study of entire, integrated systems rather than isolated parts. Often takes a descriptive or interpretive approach.

homeostasis Maintaining a dynamic, steady state in a living system through opposing, compensating adjustments.

humus Sticky, brown, insoluble residue from the bodies of dead plants and animals; gives soil its structure, coating mineral particles and holding them together; serves as a major source of plant nutrients.

hurricanes Large cyclonic oceanic storms with heavy rain and winds exceeding 119 km/hr (74 mph).

hypothesis A conditional explanation that can be verified or falsified by observation or experimentation.

I

igneous rocks Crystalline minerals solidified from molten magma from deep in the earth's interior; basalt, rhyolite, andesite, lava, and granite are examples.

inbreeding depression In a small population, an accumulation of harmful genetic traits (through random mutations and natural selection) that lowers viability and reproductive success of enough individuals to affect the whole population.

inductive reasoning "Bottom-up" in which we study specific examples and try to discover patterns and derive general explanations from collected observations.

industrial revolution Advances in science and technology that have given us power to understand and change our world.

infiltration The process of water percolation into the soil and pores and hollows of permeable rocks.

inherent value Ethical values or rights that exist as an intrinsic or essential characteristic of a particular thing or class of things simply by the fact of their existence.

inholdings Private lands within public parks, forests, or wildlife refuges.

insecticide A chemical that kills insects.

insolation Incoming solar radiation.

instrumental value Value or worth of objects that satisfy the needs and wants of moral agents. Objects that can be used as a means to some desirable end.

intangible resources Factors such as open space, beauty, serenity, wisdom, diversity, and satisfaction that cannot be grasped or contained. Ironically, these resources can be both infinite and exhaustible.

integrated pest management (IPM) An ecologically based pest-control strategy that relies on natural mortality factors, such as natural enemies, weather, cultural control methods, and carefully applied doses of pesticides.

Intergovernmental Panel on Climate Change (IPCC) A large group of scientists from many nations and a wide variety of fields assembled by the United Nations Environment Program and World Meteorological Organization to assess the current state of knowledge about climate change.

internalizing costs Planning so that those who reap the benefits of resource use also bear all the external costs.

international treaties and conventions Agreements between nations on important issues.

interspecific competition In a community, competition for resources between members of different species.

intraspecific competition In a community, competition for resources among members of the same species.

ionizing radiation High-energy electromagnetic radiation or energetic subatomic particles released by nuclear decay.

ionosphere The lower part of the thermosphere.

ions Electrically charged atoms that have gained or lost electrons.

irruptive growth See Malthusian growth.

island biogeography The study of rates of colonization and extinction of species on islands or other isolated areas based on size, shape, and distance from other inhabited regions.

isotopes Forms of a single element that differ in atomic mass due to a different number of neutrons in the nucleus.

J

J curve A growth curve that depicts exponential growth; called a J curve because of its shape.

jet streams Powerful winds or currents of air that circulate in shifting flows; similar to oceanic currents in extent and effect on climate.

joule A unit of energy. One joule is the energy expended in 1 second by a current of 1 amp flowing through a resistance of 1 ohm.

K

K-adapted Organisms whose population growth is regulated by internal (or intrinsic)

as well as external factors. Large animals such as whales and elephants, as well as top predators, generally fall in this category. They have relatively few offspring and often stabilize their population size near the carrying capacity of their environment.

keystone species A species whose impacts on its community or ecosystem are much larger and more influential than would be expected from mere abundance. This could be a top predator, a plant that shelters or feeds other organisms, or an organism that plays a critical ecological role.

kinetic energy Energy contained in moving objects such as a rock rolling down a hill, the wind blowing through the trees, or water flowing over a dam.

kwashiorkor A widespread human protein deficiency disease resulting from a starchy diet low in protein and essential amino acids.

Kyoto Protocol An international treaty adopted in Kyoto, Japan, in 1997, in which 160 nations agreed to roll back CO_2, methane, and nitrous oxide emissions to reduce the threat of global climate change.

L

landscape ecology The study of the reciprocal effects of spatial pattern on ecological processes.

landslides Mass wasting or mass movement of rock or soil downhill. Often triggered by seismic events or heavy rainfall.

LD50 A chemical dose lethal to 50 percent of a test population.

legal standing The right to take part in legal proceedings.

life expectancy The average age that a newborn infant can expect to attain in a particular time and place.

life span The longest period of life reached by a type of organism.

limiting factors Chemical or physical factors that limit the existence, growth, abundance, or distribution of an organism.

limits to growth A belief that the world has a fixed carrying capacity for humans

lobbying Using personal contacts, public pressure, or political action to persuade legislators to vote in a particular manner.

logical thinking A rational way of thought that asks, "How can orderly, deductive reasoning help me think clearly?"

logistic growth Growth rates regulated by internal and external factors that establish an equilibrium with environmental resources. See S curve.

longevity The length or duration of life; compare to survivorship.

low-quality energy Diffuse, dispersed energy at a low temperature that is difficult to gather and use for productive purposes.

LULUs Locally Unwanted Land Uses such as toxic waste dumps, incinerators, smelters, airports, freeways, and other sources of environmental, economic, or social degradation.

M

magma Molten rock from deep in the earth's interior; called lava when it spews from volcanic vents.

malnourishment A nutritional imbalance caused by lack of specific dietary components or inability to absorb or utilize essential nutrients.

Malthusian growth A population explosion followed by a population crash; also called irruptive growth.

Man and Biosphere (MAB) program A design for nature preserves that divides protected areas into zones with different purposes. A highly protected core is surrounded by a buffer zone and peripheral regions in which multiple-use resource harvesting is permitted.

mantle A hot, pliable layer of rock that surrounds the earth's core and underlies the cool, outer crust.

marasmus A widespread human protein deficiency disease caused by a diet low in calories and protein or imbalanced in essential amino acids.

marginal costs The cost to produce one additional unit of a good or service.

marsh Wetland without trees; in North America, this type of land is characterized by cattails and rushes.

mass burn Incineration of unsorted solid waste.

matter Anything that takes up space and has mass.

mean Average.

mediation An informal dispute resolution process in which parties try to reach agreement through discussion and compromise; often used as an alternative to resolving disputes through lawsuits.

Mediterranean climate areas Regions defined by warm, dry summers and cool, wet winters; these areas may have distinctive and endemic species.

megacity See megalopolis.

megalopolis Also known as a megacity or supercity; megalopolis indicates an urban area with more than 10 million inhabitants.

megawatt (MW) Unit of electrical power equal to one thousand kilowatts or one million watts.

mesosphere The atmospheric layer above the stratosphere and below the thermosphere; the middle layer; temperatures are usually very low.

metabolism All the energy and matter exchanges that occur within a living cell or organism; collectively, the life processes.

metamorphic rock Igneous and sedimentary rocks modified by heat, pressure, and chemical reactions.

methane hydrate Small bubbles or individual molecules of methane (natural gas) trapped in a crystalline matrix of frozen water.

midoceanic ridges Mountain ranges on the ocean floor where magma wells up through cracks and creates new crust.

Milankovitch cycles Periodic variations in tilt, eccentricity, and wobble in the earth's orbit; Milutin Milankovitch suggested these are responsible for cyclic weather changes.

mineral A naturally occurring, inorganic, crystalline solid with definite chemical composition, a specific internal crystal structure, and characteristic physical properties.

mitigation Repairing or rehabilitating a damaged ecosystem or compensating for damage by providing a substitute or replacement area.

modern environmentalism A fusion of conservation of natural resources and preservation of nature with concerns about pollution, environmental health, and social justice.

molecule A combination of two or more atoms.

monitored, retrievable storage Holding wastes in underground mines or secure surface facilities such as dry casks where they can be watched and repackaged, if necessary.

monoculture forestry Intensive planting of a single species; an efficient wood production approach, but one that encourages pests and disease infestations and conflicts with wildlife habitat or recreation uses.

monsoon A seasonal reversal of wind patterns caused by the different heating and cooling rates of the oceans and continents.

montane coniferous forests Coniferous forests of the mountains consisting of belts of different forest communities along an altitudinal gradient.

morals A set of ethical principles that guide our actions and relationships.

morbidity Illness or disease.

more-developed countries (MDC) Industrialized nations characterized by high per capita incomes, low birth and death rates, low population growth rates, and high levels of industrialization and urbanization.

mortality Death rate in a population, e.g., number of deaths per thousand people per year.

mulch Protective ground cover that protects the soil, saves water, and prevents weed growth; often straw, seaweed, leaves, or synthetic materials, such as heavy paper or plastic.

Müllerian (or Muellerian) mimicry Evolution of two species, both of which are unpalatable and, have poisonous stingers or some other defense mechanism, to resemble each other.

multiple use Many uses that occur simultaneously; used in forest management; limited to mutually compatible uses.

mutagens Agents, such as chemicals or radiation, that damage or alter genetic material (DNA) in cells.

mutation A change, either spontaneous or by external factors, in the genetic material of a cell; mutations in the gametes (sex cells) can be inherited by future generations of organisms.

mutualism A symbiotic relationship between individuals of two different species in which both species benefit from the association.

N

National Environmental Policy Act (NEPA) The law that established the Council on Environmental Quality and that requires environmental impact statements for all federal projects with significant environmental impacts.

natural increase Crude death rate subtracted from crude birth rate.

natural resources Goods and services supplied by the environment.

natural selection The mechanism for evolutionary change in which environmental pressures cause certain genetic combinations in a population to become more abundant; genetic combinations best adapted for present environmental conditions tend to become predominant.

negative feedbacks Factors that result from a process and, in turn, reduce that same process.

neo-classical economics A branch of economics that attempts to apply the principles of modern science to economic analysis in a mathematically rigorous, noncontextual, abstract, predictive manner.

neo-Malthusian A belief that the world is characterized by scarcity and competition in which too many people fight for too few resources. Named for Thomas Malthus, who predicted a dismal cycle of misery, vice, and starvation as a result of human overpopulation.

net energy yield Total useful energy produced during the lifetime of an entire energy system minus the energy used, lost, or wasted in making useful energy available.

neurotoxins Toxic substances, such as lead or mercury, that specifically poison nerve cells.

neutron A subatomic particle, found in the nucleus of the atom, that has no electromagnetic charge.

NIMBY Not-In-My-Back-Yard: the position of those opposed to LULUs.

nitrogen cycle The circulation and reutilization of nitrogen in both inorganic and organic phases.

nitrogen oxides Highly reactive gases formed when nitrogen in fuel or combustion air is heated to over 650°C (1,200°F) in the presence of oxygen or when bacteria in soil or water oxidize nitrogen-containing compounds.

nitrogen-fixing bacteria Bacteria that convert nitrogen from the atmosphere or soil solution into ammonia that can then be converted to plant nutrients by nitrite- and nitrate-forming bacteria.

noncriteria pollutants See unconventional air pollutants.

nongovernmental organizations (NGOs) A term referring collectively to pressure and research groups, advisory agencies, political parties, professional societies, and other groups concerned about environmental quality, resource use, and many other issues.

nonpoint sources Scattered, diffuse sources of pollutants, such as runoff from farm fields, golf courses, construction sites, etc.

nonrenewable resources Minerals, fossil fuels, and other materials present in essentially fixed amounts (within human time scales) in our environment.

nuclear fission The radioactive decay process in which isotopes split apart to create two smaller atoms.

nuclear fusion A process in which two smaller atomic nuclei fuse into one larger nucleus and release energy; the source of power in a hydrogen bomb.

nucleic acids Large organic molecules made of nucleotides that function in the transmission of hereditary traits, in protein synthesis, and in control of cellular activities.

nucleus The center of the atom; occupied by protons and neutrons. In cells, the organelle that contains the chromosomes (DNA).

numbers pyramid A diagram showing the relative population sizes at each trophic level in an ecosystem; usually corresponds to the biomass pyramid.

O

obese Pathologically overweight, having a body mass greater than 30 kg/m^2, or roughly 30 pounds above normal for an average person.

offset allowances A controversial component of air quality regulations that allows a polluter to avoid installation of control equipment on one source with an "offsetting" pollution reduction at another source.

oil shale A fine-grained sedimentary rock rich in solid organic material called kerogen. When heated, the kerogen liquefies to produce a fluid petroleum fuel.

old-growth forests Forests free from disturbance for long enough (generally 150 to 200 years) to have mature trees, physical conditions, species diversity, and other characteristics of equilibrium ecosystems.

oligotrophic Condition of rivers and lakes that have clear water and low biological productivity (oligo = little; trophic = nourished); are usually clear, cold, infertile headwater lakes and streams.

omnivore An organism that eats both plants and animals.

open access system A commonly held resource for which there are no management rules.

open canopy A forest where tree crowns cover less than 20 percent of the ground; also called woodland.

open system A system that exchanges energy and matter with its environment.

organic compounds Complex molecules organized around skeletons of carbon atoms arranged in rings or chains; includes biomolecules, molecules synthesized by living organisms.

overnutrition Receiving too many calories.

overshoot The extent to which a population exceeds the carrying capacity of its environment.

oxygen sag Oxygen decline downstream from a pollution source that introduces materials with high biological oxygen demands.

ozone A highly reactive molecule containing three oxygen atoms; a dangerous pollutant in ambient air. In the stratosphere, however, ozone forms an ultraviolet absorbing shield that protects us from mutagenic radiation.

P

paradigms Overarching models of the world that shape our worldviews and guide our interpretation of how things are.

parasite An organism that lives in or on another organism, deriving nourishment at the expense of its host, usually without killing it.

parsimony A principle that says where two equally plausible explanations for a phenomenon are possible, we should choose the simpler one (also known as Ockham's razor).

particulate material Atmospheric aerosols, such as dust, ash, soot, lint, smoke, pollen, spores, algal cells, and other suspended materials; originally applied only to solid particles but now extended to droplets of liquid.

parts per billion (ppb) Number of parts of a chemical found in one billion parts of a particular gas, liquid, or solid mixture.

parts per million (ppm) Number of parts of a chemical found in one million parts of a particular gas, liquid, or solid mixture.

parts per trillion (ppt) Number of parts of a chemical found in one trillion parts of a particular gas, liquid, or solid mixture.

passive heat absorption The use of natural materials or absorptive structures without moving parts to gather and hold heat; the simplest and oldest use of solar energy.

pastoralist Someone who lives by herding domestic animals.

patchiness Within a larger ecosystem, the presence of smaller areas that differ in some physical conditions and thus support somewhat different communities; promotes diversity in a system or area.

pathogen An organism that produces disease in a host organism, disease being an alteration of one or more metabolic functions in response to the presence of the organism.

peat Deposits of moist, acidic, semi-decayed organic matter.

perennial species Plants that grow for more than two years.

permafrost A permanently frozen layer of soil that underlies the arctic tundra.

permanent retrievable storage Placing waste storage containers in a secure location where they can be inspected periodically and retrieved, if necessary, for repacking or for transfer if a better means of disposal or reuse is developed.

persistent organic pollutant (POP) Chemical compounds that persist in the environment and retain biological activity for a long time.

pest Any organism that reduces the availability, quality, or value of a useful resource.

pest resurgence Rebound of pest populations due to acquired resistance to chemicals and nonspecific destruction of natural predators and competitors by broadscale pesticides.

pesticide Any chemical that kills, controls, drives away, or modifies the behavior of a pest.

pesticide treadmill A need for constantly increasing doses or new pesticides to prevent pest resurgence.

pH A value that indicates the acidity or alkalinity of a solution on a scale of 0 to 14, based on the proportion of H$^+$ ions present.

phosphorus cycle The movement of phosphorus atoms from rocks through the biosphere and hydrosphere and back to rocks.

photochemical oxidants Products of secondary atmospheric reactions. See smog.

photodegradable plastics Plastics that break down when exposed to sunlight or to a specific wavelength of light.

photosynthesis The biochemical process by which green plants and some bacteria capture light energy and use it to produce chemical bonds. Carbon dioxide and water are consumed while oxygen and simple sugars are produced.

photovoltaic cell An energy-conversion device that captures solar energy and directly converts it to electrical current.

phytoplankton Microscopic, free-floating, autotrophic organisms that function as producers in aquatic ecosystems.

pioneer species In primary succession on a terrestrial site, the plants, lichens, and microbes that first colonize the site.

plankton Primarily microscopic organisms that occupy the upper water layers in both freshwater and marine ecosystems.

plasma A hot, electrically neutral gas of ions and free electrons.

poaching Hunting wildlife illegally.

point sources Specific locations of highly concentrated pollution discharge, such as factories, power plants, sewage treatment plants, underground coal mines, and oil wells.

policy A societal plan to statement of intentions intended to accomplish some social or economic goal.

policy cycle The process by which problems are identified and acted upon in the public arena.

pollution To make foul, unclean, dirty; any physical, chemical, or biological change that adversely affects the health, survival, or activities of living organisms or that alters the environment in undesirable ways.

pollution charges Fees assessed per unit of pollution based on the "polluter pays" principle.

population All members of a species that live in the same area at the same time.

population crash A sudden population decline caused by predation, waste accumulation, or resource depletion; also called a dieback.

population explosion Growth of a population at exponential rates to a size that exceeds environmental carrying capacity; usually followed by a population crash.

population momentum A potential for increased population growth as young members reach reproductive age.

positive feedbacks Factors that result from a process and, in turn, increase that same process.

potential energy Stored energy that is latent but available for use. A rock poised at the top of a hill or water stored behind a dam are examples of potential energy.

power The rate of energy delivery; measured in horsepower or watts.

precautionary principle The rule that we should leave a margin of safety for unexpected developments. This principle implies that we should strive to prevent harm to human health and the environment even if risks are not fully understood.

precedent An act or decision that can be used as an example in dealing with subsequent similar situations.

predator An organism that feeds directly on other organisms in order to survive; live-feeders, such as herbivores and carnivores.

primary pollutants Chemicals released directly into the air in a harmful form.

primary producers Photosynthesizing organisms.

primary standards Regulations of the 1970 Clean Air Act; intended to protect human health.

primary succession Ecological succession that begins in an area where no biotic community previously existed.

primary treatment A process that removes solids from sewage before it is discharged or treated further.

principle of competitive exclusion A result of natural selection whereby two similar species in a community occupy different ecological niches, thereby reducing competition for food.

probability The likelihood that a situation, condition, or event will occur.

producer An organism that synthesizes food molecules from inorganic compounds by using an external energy source; most producers are photosynthetic.

productivity The amount of biomass (biological material) produced in a given area during a given period of time.

prokaryotic Cells that do not have a membrane-bounded nucleus or membrane-bounded organelles.

pronatalist pressures Influences that encourage people to have children.

prospective study A study in which experimental and control groups are identified before exposure to some factor. The groups are then monitored and compared for a specific time after the exposure to determine any effects the factor may have.

proteins Chains of amino acids linked by peptide bonds.

proton A positively charged subatomic particle found in the nucleus of an atom.

proven-in-place reserves Energy sources that have been thoroughly mapped and are likely to be economically recoverable with available technology.

pull factors (in urbanization) Conditions that draw people from the country into the city.

push factors (in urbanization) Conditions that force people out of the country and into the city.

R

r-adapted Organisms whose population growth is regulated mainly by external fac-

tors. They tend to have rapid reproduction and high mortality of offspring. Given optimum environmental conditions, they can grow exponentially. Many "weedy" or pioneer species fit in this category.

radioactive decay A change in the nuclei of radioactive isotopes that spontaneously emit high-energy electromagnetic radiation and/or subatomic particles while gradually changing into another isotope or different element.

rain shadow Dry area on the downwind side of a mountain.

rainforest A forest with high humidity, constant temperature, and abundant rainfall (generally over 380 cm [150 in] per year); can be tropical or temperate.

rangeland Grasslands and open woodlands suitable for livestock grazing.

rational choice Public decision making based on reason, logic, and science-based management.

reasoned judgment Thoughtful decisions based on careful, logical examination of available evidence.

recharge zone Area where water infiltrates into an aquifer.

reclamation Chemical, biological, or physical cleanup and reconstruction of severely contaminated or degraded sites to return them to something like their original topography and vegetation.

recycling Reprocessing of discarded materials into new, useful products; not the same as reuse of materials for their original purpose, but the terms are often used interchangeably.

red tide A population explosion or "bloom" of single-celled marine organisms called dinoflagellates. Billions of these cells can accumulate in protected bays where the toxins they contain can poison other marine life.

reduced tillage systems Farming methods that preserve soil and save energy and water through reduced cultivation; includes minimum till, conserve-till, and no-till systems.

reflective thinking A thoughtful, contemplative analysis that asks, "What does this all mean?"

reformer A device that strips hydrogen from fuels such as natural gas, methanol, ammonia, gasoline, or vegetable oil so they can be used in a fuel cell.

refuse-derived fuel Processing of solid waste to remove metal, glass, and other unburnable materials; organic residue is shredded, formed into pellets, and dried to make fuel for power plants.

regenerative farming Farming techniques and land stewardship that restore the health and productivity of the soil by rotating crops, planting ground cover, protecting the surface with crop residue, and reducing synthetic chemical inputs and mechanical compaction.

regulations Rules established by administrative agencies; regulations can be more important than statutory law in the day-to-day management of resources.

rehabilitation Rebuilding basic structure or function in an ecological system without necessarily achieving complete restoration to its original condition.

relative humidity At any given temperature, a comparison of the actual water content of the air with the amount of water that could be held at saturation.

relevé A rapid assessment of vegetation types and biodiversity in an area.

remediation Cleaning up chemical contaminants from a polluted area.

renewable resources Resources normally replaced or replenished by natural processes; resources not depleted by moderate use; examples include solar energy, biological resources such as forests and fisheries, biological organisms, and some biogeochemical cycles.

renewable water supplies Annual freshwater surface runoff plus annual infiltration into underground freshwater aquifers that are accessible for human use.

reproducibility Making an observation or obtaining a particular result consistently.

residence time The length of time a component, such as an individual water molecule, spends in a particular compartment or location before it moves on through a particular process or cycle.

resilience The ability of a community or ecosystem to recover from disturbances.

resistance (inertia) The ability of a community to resist being changed by potentially disruptive events.

resource In economic terms, anything with potential use in creating wealth or giving satisfaction.

resource partitioning In a biological community, various populations sharing environmental resources through specialization, thereby reducing direct competition. See also ecological niche.

restoration To bring something back to a former condition. Ecological restoration involves active manipulation of nature to re-create conditions that existed before human disturbance.

restoration ecology Seeks to repair or reconstruct ecosystems damaged by human actions.

retrospective study A study that looks back in history at a group of people (or other organisms) who suffer from some condition to try to identify something in their past life that the whole group shares in common but that is not found in the histories of a control group as near as possible to those being studied but who do not suffer from the same condition.

riders Amendments attached to bills in conference committee, often completely unrelated to the bill to which they are added.

rill erosion The removing of thin layers of soil as little rivulets of running water gather and cut small channels in the soil.

risk Probability that something undesirable will happen as a consequence of exposure to a hazard.

risk assessment Evaluation of the short-term and long-term risks associated with a particular activity or hazard; usually compared to benefits in a cost-benefit analysis.

rock A solid, cohesive, aggregate of one or more crystalline minerals.

rock cycle The process whereby rocks are broken down by chemical and physical forces; sediments are moved by wind, water, and gravity, sedimented and reformed into rock, and then crushed, folded, melted, and recrystallized into new forms.

rotational grazing Confining grazing animals in a small area for a short time to force them to eat weedy species as well as the more desirable grasses and forbes.

runoff The excess of precipitation over evaporation; the main source of surface water and, in broad terms, the water available for human use.

S

S curve A curve that depicts logistic growth; called an S curve because of its shape.

salinity Amount of dissolved salts (especially sodium chloride) in a given volume of water.

salinization A process in which mineral salts accumulate in the soil, killing plants; occurs when soils in dry climates are irrigated profusely.

saltwater intrusion Movement of saltwater into freshwater aquifers in coastal areas where groundwater is withdrawn faster than it is replenished.

sample Analyzing a small but representative portion of a population to estimate the characteristics of the entire class.

sanitary landfills A landfill in which garbage and municipal waste is buried every day under enough soil or fill to eliminate odors, vermin, and litter.

scientific method A systematic, precise, objective study of a problem. Generally this requires observation, hypothesis development and testing, data gathering, and interpretation.

scientific theory An explanation or idea accepted by a substantial number of scientists.

second law of thermodynamics States that, with each successive energy transfer or transformation in a system, less energy is available to do work.

secondary pollutants Chemicals modified to a hazardous form after entering the air or that are formed by chemical reactions as components of the air mix and interact.

secondary succession Succession on a site where an existing community has been disrupted.

secondary treatment Bacterial decomposition of suspended particulates and dissolved organic compounds that remain after primary sewage treatment.

secure landfill A solid waste disposal site lined and capped with an impermeable barrier to prevent leakage or leaching.

sedimentary rock Rock composed of accumulated, compacted mineral fragments, such as sand or clay; examples include shale, sandstone, breccia, and conglomerates.

sedimentation The deposition of organic materials or minerals by chemical, physical, or biological processes.

selective cutting Harvesting only mature trees of certain species and size; usually more expensive than clear-cutting, but less disruptive for wildlife and often better for forest regeneration.

selective pressure Limited resources or adverse environmental conditions that tend to favor certain adaptations in a population. Over many generations, this can lead to genetic change, or evolution.

sheet erosion Peeling off thin layers of soil from the land surface; accomplished primarily by wind and water.

shelterwood harvesting Mature trees are removed from the forest in a series of two or more cuts, leaving young trees and some mature trees as a seed source for future regeneration.

sick building syndrome A cluster of allergies and other illnesses caused by sensitivity to molds, synthetic chemicals, or other harmful compounds trapped in insufficiently ventilated buildings.

significant numbers Meaningful data that can be measured accurately and reproducibly.

sinkholes A large surface crater caused by the collapse of an underground channel or cavern; often triggered by groundwater withdrawal.

sludge Semisolid mixture of organic and inorganic materials that settles out of wastewater at a sewage treatment plant.

smart growth Efficient use of land resources and existing urban infrastructure that encourages in-fill development, provides a variety of affordable housing and transportation choices, and seeks to maintain a unique sense of place by respecting local cultural and natural features.

smelting Roasting ore to release metals from mineral compounds.

smog The term used to describe the combination of smoke and fog in the stagnant air of London; now often applied to photochemical pollution.

social justice Equitable access to resources and the benefits derived from them; a system that recognizes inalienable rights and adheres to what is fair, honest, and moral.

soil A complex mixture of weathered rock material, partially decomposed organic molecules, and a host of living organisms.

soil horizons Horizontal layers that reveal a soil's history, characteristics, and usefulness.

species All the organisms genetically similar enough to breed and produce live, fertile offspring in nature.

species diversity The number and relative abundance of species present in a community.

species recovery plan A plan for restoration of an endangered species through protection, habitat management, captive breeding, disease control, or other techniques that increase populations and encourage survival.

specific heat The amount of heat energy needed to change the temperature of a body. Water has a specific heat of 1, which is higher than most substances.

sprawl Unlimited, unplanned growth of urban areas that consumes open space and wastes resources.

stability In ecological terms, a dynamic equilibrium among the physical and biological factors in an ecosystem or a community; relative homeostasis.

stable runoff The fraction of water available year-round; usually more important than total runoff when determining human uses.

Standard Metropolitan Statistical Area (SMSA) An urbanized region with at least 100,000 inhabitants with strong economic and social ties to a central city of at least 50,000 people.

statistics Mathematical analysis of the collection, organization, and interpretation of numerical data.

statutory law Rules passed by a state or national legislature.

steady-state economy Characterized by low birth and death rates, use of renewable energy sources, recycling of materials, and emphasis on durability, efficiency, and stability.

stewardship A philosophy that holds that humans have a unique responsibility to manage, care for, and improve nature.

strategic lawsuits against public participation (SLAPP) Lawsuits that have no merit but are brought merely to intimidate and harass private citizens who act in the public interest.

strategic metals and minerals Materials a country cannot produce itself but that it uses for essential materials or processes.

stratosphere The zone in the atmosphere extending from the tropopause to about 50 km (30 mi) above the earth's surface; temperatures are stable or rise slightly with altitude; has very little water vapor but is rich in ozone.

stress Physical, chemical, or emotional factors that place a strain on an animal. Plants also experience physiological stress under adverse environmental conditions.

strip-cutting Harvesting trees in strips narrow enough to minimize edge effects and to allow natural regeneration of the forest.

strip-farming Planting different kinds of crops in alternating strips along land contours; when one crop is harvested, the other crop remains to protect the soil and prevent water from running straight down a hill.

strip-mining Extracting shallow mineral deposits (especially coal) by scraping off surface layers with giant, earth-moving equipment; creates a huge open-pit; an alternative to underground or deep open-pit mines.

subsidence Settling of the ground surface caused by the collapse of porous formations that result from withdrawal of large amounts of groundwater, oil, or other underground materials.

subsoil A layer of soil beneath the topsoil that has lower organic content and higher concentrations of fine mineral particles; often contains soluble compounds and clay particles carried down by percolating water.

sulfur cycle The chemical and physical reactions by which sulfur moves into or out of storage and through the environment.

sulfur dioxide A colorless, corrosive gas directly damaging to both plants and animals.

Superfund A fund established by Congress to pay for containment, cleanup, or remediation of abandoned toxic-waste sites. The fund is financed by fees paid by toxic-waste generators and by cost-recovery from cleanup projects.

surface mining Some minerals are also mined from surface pits. See strip-mining.

surface tension The tendency for a surface of water molecules to hold together, producing a surface that resists breaking.

survivorship The percentage of a population reaching a given age or the proportion of the maximum life span of the species reached by any individual.

sustainability Ecological, social, and economic systems that can last over the long term.

sustainable agriculture Ecologically sound, economically viable, socially just agricultural system. Stewardship, soil conservation, and integrated pest management are essential for sustainability.

sustainable development A real increase in well-being and standard of life for the average person that can be maintained over the long-term without degrading the environment or compromising the ability of future generations to meet their own needs.

sustained yield Utilization of a renewable resource at a rate that does not impair or damage its ability to be fully renewed on a long-term basis.

swamp Wetland with trees, such as the extensive swamp forests of the southern United States.

symbiosis The intimate living together of members of two different species; includes mutualism, commensalism, and, in some classifications, parasitism.

synergism When an injury caused by exposure to two environmental factors together is greater than the sum of exposure to each factor individually.

T

taiga The northernmost edge of the boreal forest, including species-poor woodland and peat deposits; intergrading with the arctic tundra.

tailings Mining waste left after mechanical or chemical separation of minerals from crushed ore.

taking Unconstitutional confiscation of private property.

tar sands Geologic deposits composed of sand and shale particles coated with bitumen, a viscous mixture of long-chain hydrocarbons.

tectonic plates Huge blocks of the earth's crust that slide around slowly, pulling apart to open new ocean basins or crashing ponderously into each other to create new, larger landmasses.

temperate rainforest The cool, dense, rainy forest of the northern Pacific coast; enshrouded in fog much of the time; dominated by large conifers.

temperature A measure of the speed of motion of a typical atom or molecule in a substance.

teratogens Chemicals or other factors that specifically cause abnormalities during embryonic growth and development.

terracing Shaping the land to create level shelves of earth to hold water and soil; requires extensive hand labor or expensive machinery, but it enables farmers to farm very steep hillsides.

tertiary treatment The removal of inorganic minerals and plant nutrients after primary and secondary treatment of sewage.

thermal pollution Artificially raising or lowering the temperature of a water body in a way that adversely affects the biota or water quality.

thermocline In water, a distinctive temperature transition zone that separates an upper layer that is mixed by the wind (the epilimnion) and a colder, deep layer that is not mixed (the hypolimnion).

thermodynamics A branch of physics that deals with transfers and conversions of energy.

thermosphere The highest atmospheric zone; a region of hot, dilute gases above the mesosphere extending out to about 1,600 km (1,000 mi) from the earth's surface.

Third World Less-developed countries that are not capitalistic and industrialized (First World) or centrally-planned socialist economies (Second World); not intended to be derogatory.

thorn scrub A dry, open woodland or shrubland characterized by sparse, spiny shrubs.

threatened species While still abundant in parts of its territorial range, this species has declined significantly in total numbers and may be on the verge of extinction in certain regions or localities.

tolerance limits See limiting factors.

topsoil The first true layer of soil; layer in which organic material is mixed with mineral particles; thickness ranges from a meter or more under virgin prairie to zero in some deserts.

total fertility rate The number of children born to an average woman in a population during her entire reproductive life.

total growth rate The net rate of population growth resulting from births, deaths, immigration, and emigration.

total maximum daily loads (TMDL) The amount of particular pollutant that a water body can receive from both point and nonpoint sources and still meet water quality standards.

toxic release inventory A program created by the Superfund Amendments and Reauthorization Act of 1984 that requires manufacturing facilities and waste handling and disposal sites to report annually on releases of more than 300 toxic materials. You can find out from the EPA whether any of these sites are in your neighborhood and what toxics they release.

toxins Poisonous chemicals that react with specific cellular components to kill cells or to alter growth or development in undesirable ways; often harmful, even in dilute concentrations.

tradable permits Pollution quotas or variances that can be bought or sold.

tragedy of the commons An inexorable process of degradation of communal resources due to selfish self-interest of "free riders" who use or destroy more than their fair share of common property. See open access system.

transpiration The evaporation of water from plant surfaces, especially through stomates.

trophic level Step in the movement of energy through an ecosystem; an organism's feeding status in an ecosystem.

tropical rainforest Forest near the equator in which rainfall is abundant—more than 200 cm (80 in) per year—and temperatures are warm to hot year-round.

tropical seasonal forest Semi-evergreen or partly deciduous forests tending toward open woodlands and grassy savannas dotted with scattered, drought-resistant trees.

tropopause The boundary between the troposphere and the stratosphere.

troposphere The layer of air nearest to the earth's surface; both temperature and pressure usually decrease with increasing altitude.

tundra Treeless arctic or alpine biome characterized by cold, dark winters, a short growing season, and potential for frost any month of the year; vegetation includes low-growing perennial plants, mosses, and lichens.

U

unconventional air pollutants Toxic or hazardous substances, such as asbestos, benzene, beryllium, mercury, polychlorinated biphenyls, and vinyl chloride, not listed in the original Clean Air Act because they were not released in large quantities; also called noncriteria pollutants.

urban renewal Programs to revitalize old, blighted sections of inner cities.

urbanization An increasing concentration of the population in cities and a transformation of land use to an urban pattern of organization.

utilitarian conservation A philosophy that resources should be used for the greatest good for the greatest number for the longest time.

V

values An estimation of the worth of things; a set of ethical beliefs and preferences that determine our sense of right and wrong.

vertical stratification The vertical distribution of specific subcommunities within a community.

visible light A portion of the electromagnetic spectrum that includes the wavelengths used for photosynthesis.

vitamins Organic molecules essential for life that we cannot make for ourselves; we must get them from our diet; they act as enzyme cofactors.

volatile organic compounds (VOCs) Organic chemicals that evaporate readily and exist as gases in the air.

volcano A vent in the earth's surface through which molten lava (magma), gases, and ash escape to create a mountain.

vulnerable species Naturally rare organisms or species whose numbers have been so reduced by human activities that they are susceptible to actions that could push them into threatened or endangered status.

W

warm front A long, wedge-shaped boundary caused when a warmer advancing air mass slides over neighboring cooler air parcels.

waste stream The steady flow of varied wastes, from domestic garbage and yard wastes to industrial, commercial, and construction refuse.

water cycle The recycling and reutilization of water on earth, including atmospheric, surface, and underground phases and biological and nonbiological components.

water table The top layer of the zone of saturation; undulates according to the surface topography and subsurface structure.

waterlogging Water saturation of soil that fills all air spaces and causes plant roots to die from lack of oxygen; a result of overirrigation.

watershed The land surface and groundwater aquifers drained by a particular river system.

weather Description of the physical conditions of the atmosphere (moisture, temperature, pressure, and wind).

weathering Changes in rocks brought about by exposure to air, water, changing temperatures, and reactive chemical agents.

wetlands Ecosystems of several types in which rooted vegetation is surrounded by standing water during part of the year. See also swamp, marsh, bog, fen.

wicked problems Problems with no simple right or wrong answer where there is no single, generally agreed on definition of or solution for the particular issue.

wilderness An area of undeveloped land affected primarily by the forces of nature; an area where humans are visitors who do not remain.

wildlife refuges Areas set aside to shelter, feed, and protect wildlife; due to political and economic pressures, refuges often allow hunting, trapping, mineral exploitation, and other activities that threaten wildlife.

wind farms Large numbers of windmills concentrated in a single area; usually owned by a utility or large-scale energy producer.

Wise Use Movement A coalition of ranchers, loggers, miners, industrialists, hunters, off-road vehicle users, land developers, and others who call for unrestricted access to natural resources and public lands.

withdrawal A description of the total amount of water taken from a lake, river, or aquifer.

woodland A forest where tree crowns cover less than 20 percent of the ground; also called open canopy forest.

work The application of force through a distance; requires energy input.

world conservation strategy A proposal for maintaining essential ecological processes, preserving genetic diversity, and ensuring that utilization of species and ecosystems is sustainable.

World Trade Organization (WTO) An association of 135 nations organized to regulate international trade.

X

X ray Very short wavelength in the electromagnetic spectrum; can penetrate soft tissue; although it is useful in medical diagnosis, it also damages tissue and causes mutations.

Z

zero population growth (ZPG) A condition in which births and immigration in a population just balance deaths and emigration.

zone of aeration Upper soil layers that hold both air and water.

zone of leaching The layer of soil just beneath the topsoil where water percolates, removing soluble nutrients that accumulate in the subsoil; may be very different in appearance and composition from the layers above and below it.

zone of saturation Lower soil layers where all spaces are filled with water.

CREDITS

LINE ART

Chapter 1

1.19: From United Nations (U.N.) Critical Trends: Global Change and Sustainable Development, 1997, p. 58: Growing Disparities in Incomes Among Regions.

Chapter 2

2.22: From Michael D. Moran, et al., *Environmental Science.* Copyright © 1991 McGraw-Hill Company, Inc. All Rights Reserved. Reprinted by permission.

Chapter 3

3.5: From Peter H. Raven and George B. Johnson, *Biology,* 4th edition. Copyright © 1996 McGraw-Hill Company, Inc. All Rights Reserved. Reprinted by permission; 3.26: From Verner, Jared et al. *Wildlife 2000.* © 1986. Reprinted by permission of The University of Wisconsin Press.

Chapter 4

4.2: Figure redrawn with permission from *Population Bulletin,* vol. 18, no. 1, 1985. Population Reference Bureau; 4.3: Figure redrawn by permission from *Population Bulletin,* vol. 18, no. 1, 1985. Population Reference Bureau; 4.4: From Thomas Merrick, with PRB staff, "World Population in Transition" *Population Bulletin,* vol. 41, no. 2 (reprinted 1991); 4.14: From *World Resources 1998-99,* Reprinted by permission of World Resources Institute.

Chapter 5

5.25: Reprinted with permission from J. Curtis in William L. Thomas (ed.), *Man's Role in Changing the Face of the Earth.* Copyright © 1956 by The University of Chicago Press. All Rights Reserved.

Chapter 6

6.6: From FAO Assessment Report, 2000. Used with permission.

Chapter 7

7.2: From *World Resources 1998-99,* World Resource Institute. Reprinted by permission; 7.3: From Hartwig de Haen, *Sofa* 2001. Used with permission; 7.9: Data from USDA as appeared in Peter H. Raven and George B. Johnson, *Understanding Biology,* 3rd edition. Copyright © 1995 McGraw-Hill Company, Inc. All Rights Reserved; 7.15: From FAO website: Global status of human-induces soil degradation. Used with permission; 7.16: *Source:* Data from www.fao.org; 7.24: From Information Systems for Biotechnology, Department of Biochemistry, Virginia Tech, www.isb.vt.edu. Used with permission.

Chapter 8

8.6: From Dr. Steven Morse, Rockefeller University in Time, September 12, 1994. Copyright © Time, Inc. Reprinted by permission.

Chapter 9

9.9a-c: From the Intergovernmental Panel on climate change (1990b) and Minister of Interior, Government of Canada, 1991. *The State of Canada's Environment,* Environmental Canada. Reproduced with the permission of the Minister of Public Works and Government Services Canada, 2000; 9.14: © Laurie Grace; 9.20: From *Government of Canada, 1991. The State of Canada's Environment.* Reproduced with permission of the Minister of Public Works and Government Services Canada, 2000; 9.28: Used with permission of General Motors Corporation.

Chapter 10

10.2: From Jerome Fellmann, et al., *Human Geography,* 4th edition. Copyright © 1995 McGraw-Hill Co. All Rights Reserved; 10.22: *Source:* From Environmental Protection Agency; 10.23: *Source:* From "Quality of the Nation's Waters," 1998, Environmental Protection Agency; 10.27: From *Student Atlas of Environmental Issues,* by John L. Allen, copyright © 1997 by The McGraw-Hill Companies, Inc. Reprinted by permission of McGraw-Hill/ Dushkin, a division of The McGraw-Hill Companies, Guilford, CT 06437.

Chapter 11

11.16: From Montgomery, *Environmental Geology* 6/e. Copyright © 2003 The McGraw-Hill Companies. All rights reserved.

Chapter 12

12.12: Courtesy of Northern States Power Co., Minneapolis, MN.

Chapter 14

14.6 and 14.7: Reprinted from *Beyond the Limits,* copyright © 1992 by Meadows, Meadows and Randers. With permission from Chelsea Green Publishing Co., White River Junction, Vermont; 14.23: From Randall Arenct, "Creating Open Space Networks" in *Environment & Development,* May/June 1996, American Planning Association, Chicago, IL. Reprinted by permission.

Chapter 15

15.2: From K. Sexton, et al., *Better Environment Decisions,* 1999. Copyright © 1999 Island Press. Reprinted by permission of Alexander Hoyt & Associates.

PHOTOS

Chapter 1

Opener: Courtesy Jennifer Stelten-Kivoja; 1.1: Courtesy Keene Engineering; 1.2: © Earth Images/Stone/Getty; 1.3: William Cunningham; 1.5: Courtesy Monumenti Musei E Gallerie Pontificie; 1.6: Courtesy Dr. David Mech; 1.8: © RJ Goldstein/

Visuals Unlimited; 1.10: Courtesy Dr. Mark Nelson; 1.11: Courtesy Grey Towers National Historic Landmark; 1.12: Courtesy of the Bancroft Library University of California, Berkeley; 1.13: © AP/Wide World Photo; 1.14: © Norbert Schiller/The Image Works; 1.15: NASA; 1.16: © Gary Milburn/Tom Stack & Associates; 1.17: William P. Cunningham; 1.18, 1.21: © The McGraw-Hill Companies, Inc./Barry W. Barker, photographer.

Chapter 2

Opener: William P. Cunningham; 2.1: © Carl Purcell/Photo Researchers, Inc.; 2.7: Courtesy Dr. Eldon Newcomb; 2.8: © G.I. Bernard/Animals Animals/Earth Scenes; 2.9: NASA/Goddard Space Flight Center; 2.21: © David Dennis/Tom Stack & Associates.

Chapter 3

Opener: © Art Wolfe/Stone/Getty; 3.1: Portrait of Charles Darwin,1840 by George Richmond (1809-96) Down House, Downe, Kent, UK/Bridgeman Art Library; 3.3: William P. Cunningham; 3.4: © Corbis/Volume 86; 3.6, 3.9: William P. Cunningham; 3.10: © Ray Coleman/Photo Researchers, Inc.; 3.11: D.P. Wilson/Photo Researchers, Inc.; 3.12, 3.13: William P. Cunningham; 3.14: © Michael Fogden/Animals Animals/Earth Scenes; 3.15a: © Edward S. Ross; 3.15b: © Edward S. Ross; 3.16: © Randy Morse/Tom Stack & Associates; 3.17: Courtesy National Museum of Natural History, © Smithsonian Institution; 3.24: © Fred Bavendam/Peter Arnold, Inc.; 3.25, 3.28, 3.29: William P. Cunningham; 3.30: © Gerard Lacz/Peter Arnold, Inc.

Chapter 4

Opener: © The McGraw-Hill Companies, Inc./Barry Barker, photographer; 4.1: Courtesy John Cunningham; 4.6, 4.10: William P. Cunningham; 4.15: © The McGraw Hill Companies, Inc./Bob Coyle, photographer.

Chapter 5

Opener: © Chad Ehlers/Stone/Getty; 5.1: Florida Gap Analysis Project/USGS; 5.5, 5.6, 5.7, 5.8, 5.9, 5.10, 5.11: William P. Cunningham; 5.12: © The McGraw-Hill Companies, Inc./Barry W. Barker, photographer; 5.13: William P. Cunningham; 5.16: © John D. Cunningham/Visuals Unlimited; 5.17: Courtesy Judy Fretwell, U.S. Geological Survey; 5.18: © Mike Bacon/Tom Stack & Associates; 5.19: © The McGraw-Hill Companies, Inc./Carlyn Iverson, photographer; 5.21: William P.

Cunningham; 5.22: Courtesy INBIO, San Jose, Costa Rica; 5.23: William P. Cunningham; 5.28: Courtesy Bell Museum/U. MN; 5.29: William P. Cunningham; 5.30: © Lynn Funkhouser/Peter Arnold, Inc.; 5.31(beetle): © Taina Litwak; 5.31(Kudzu vine): © David M. Dennis/Tom Stack & Associates; 5.32: © The McGraw-Hill Companies, Inc./Barry W. Barker, photographer.

Chapter 6

Opener: William P. Cunningham; 6.1: William P. Cunningham; 6.3: © Inga Spence/Tom Stack & Associates; 6.4: William P. Cunningham; 6.5, 6.7: © The McGraw-Hill Companies, Inc./Barry W. Barker, photographer; p.130: © Frans Lanting/Minden Pictures; 6.8: © The McGraw-Hill Companies, Inc./Steven P. Lynch; 6.9: © Greg Vaughn/Tom Stack & Associates; 6.10: © Gary Braasch/Stone/Getty; 6.11, 6.12: William P. Cunningham; 6.14: © Francois Gohier/Photo Researchers, Inc.; 6.15: William P. Cunningham; 6.16: © Corbis/Volume 46; 6.17,6.18: William P. Cunningham; 6.19: © American Heritage Center, University of Wyoming; p.138: © Gary Milburn/Tom Stack & Associates; 6.22: Courtesy R.O. Bierregaard; 6.23: © The McGraw-Hill Companies, Inc./Barry W. Barker, photographer; 6.25: William P. Cunningham; 6.26: © The McGraw-Hill Companies, Inc./Barry W. Barker, photographer.

Chapter 7

Opener: © PhotoDisc/Volume 16; 7.1: William P. Cunningham; 7.4: © Norbert Schiller/The Image Works; 7.5: Courtesy Thierry A. Brun; 7.7: © Lester Bergman/CORBIS; 7.8: © Scott Daniel Peterson/Gamma Liaison International; 7.10: © The McGraw-Hill Companies, Inc./Barry W. Barker, photographer; 7.11: © Corbis/Volume 102; 7.14: © Jacques Jangoux; 7.15: FAO; 7.17: © Larry Miller/Photo Researchers, Inc.; 7.18: William P. Cunningham; 7.19: © Corbis/Volume 13; 7.20: © Corbis/Volume 102; 7.21: William P. Cunningham; 7.23: © Michael Rosenfeld/Stone/Getty; 7.25: USDA; 7.26: © The McGraw-Hill Companies, Inc./Barry Barker, photographer; p.168: William P. Cunningham; 7.27: Courtesy of Dave Hansen, College of Agriculture Experiment Station, University of Minnesota; 7.28: © Tom Sweeny/Minneapolis Star Tribune; 7.29: William P. Cunningham.

Chapter 8

Opener: © Klaus Reisinger/BlackStar; 8.1: Getty Images/Taxi/Steve Kahn; 8.2: Courtesy of Stanley Erlandsen, University of Minnesota; 8.3: William P. Cunningham; 8.5: © The McGraw-Hill Companies, Inc./Diane R. Nelson; 8.11: © Corbis/Volume 5; 8.13: William P. Cunningham.

Chapter 9

Opener: © Corbis/Volume 107; 9.4: © Corbis/Volume 107; 9.8: © Vic Englebert/Photo Researchers, Inc.; 9.15: William P. Cunningham; 9.17: Courtesy Dr. Delbert Swanson; 9.18: © Corbis/Volume 160; 9.19: © Guang Hui Xie/Image Bank; 9.21: Courtesy Ronald Nagatani/National Centers for Environmental Prediction/NOAA; 9.24, 9.26: William P. Cunningham; 9.27: © John D. Cunningham/Visuals Unlimited.

Chapter 10

Opener: William P. Cunningham; 10.4: William P. Cunningham; 10.7: © Ray Ellis/Photo Researchers, Inc.; 10.9: © Corbis/Volume 102; 10.10: © Sylvan Whitwer/Visuals Unlimited; 10.11: © The McGraw-Hill Companies, Inc./Barry W. Barker, photographer; 10.13: © Corbis/Volume 16; 10.14: Courtesy Tim McCabe, Soil Conservation Service, USDA; 10.15: William P. Cunningham; 10.16: © Tom Stack/Tom Stack & Associates; 10.17: © Joe McDonald/Animals Animals/Earth Scenes; 10.19: William P. Cunningham; 10.20: © Lawrence Lowry/Photo Researchers, Inc.; 10.24: © Les Stone/Sygma/CORBIS; 10.26: © Frans Lanting/Photo Researchers, Inc.

Chapter 11

Opener: Department of Energy; 11.1: Department of Energy; 11.7: © Corbis/Volume 72; 11.10: © Bryan F. Peterson; 11.11: © AP Photo/Bob Bird; 11.13: © Joseph Nettis/Photo Researchers, Inc.; 11.14: © Reuter/Sankei Shimbun; 11.15: Courtesy Chris G. Newhall/U.S. Geological Survey; 11.17: © James Shaffer; 11.18: © The McGraw-Hill Companies, Inc., Bob Coyle, photographer; 11.19: © Al Seib/LA Times; 11.20: © Stephen Rose/Gamma Liaison International.

Chapter 12

Opener: William P. Cunningham; 12.7: © James P. Blair/National Geographic Image Collection; 12.9, 12.11: © Corbis/Volume 160; p. 284: Courtesy Toyota; 12.15: © Mike Zerby/2002 Star Tribune/ Minneapolis-St. Paul; 12.17: William P. Cunningham; 12.19: © The McGraw-Hill

Companies, Inc./Doug Sherman, photographer; 12.21: © Corbis/Volume 160; 12.23: William P. Cunningham; 12.24: Courtesy Dr. Douglas Pratt; 12.25: © Robert E. Ford/BPS/Terraphotographics; 12.27: © Corbis/Volume 13.

Chapter 13

Opener: © Ray Pfortner/Peter Arnold, Inc.; 13.3: © Fred McConnaughey/Photo Researchers, Inc.; p. 304: © Barbara Gauntt/The Clarion-Ledger, Jackson, MS; 13.6: William P. Cunningham; 13.10: © Reuters News Picture Service, Photo by Bobby Yip 2002; 13.11: Courtesy Urban Ore, Inc., Berkeley, CA; 13.13: © J. Cancalosi/Valan Photos; 13.16: © Michael Greenlar/The Images Works; 13.18: © Piet Van Lier.

Chapter 14

Opener: © Corbis/R-F Website; 14.4: © Corbis/Volume 6; 14.9: © The McGraw-Hill Companies, Inc./Barry W. Barker, photographer; 14.10: © Mark Luthringer; 14.15: © John Chiasson/Gamma Liaison International; 14.16: William P. Cunningham; 14.17: © Louis Psihoyos/Matrix International, Inc.; 14.18: © The McGraw-Hill Companies, Inc./Barry Barker, photographer; 14.19, 14.20, 14.21: William P. Cunningham; p. 339: © Bob Boyer/Shoot The Works; 14.22, 14.24: William P. Cunningham.

Chapter 15

Opener: William P. Cunningham; 15.1: William P. Cunningham; 15.3: © Robert Queen, Wisconsin Department of Natural Resources; 15.4, 15.7: William P. Cunningham; 15.8: © Corbis/Volume 160; 15.9: William P. Cunningham; 15.10: © Bob Daemmrich/The Image Works; 15.11: William P. Cunningham; 15.12: © Carr Clifton; 15.13: © PhotoDisc/Volume 25; 15.15, 15.17: William P. Cunningham; 15.18: © Phil Degginger/Animals Animals/Earth Scenes; 15.19, 15.20, 15.21, 15.22, 15.23: William P. Cunningham.

INDEX

B

Bacon, Francis, 5
bacteria
    bioaccumulation of heavy metals in, 182
    in bioremediation, 315
    black band disease from, 178
    coliform, 237–38, 240
    in converting biomass to methane gas, 292
    denitrifying, 43
    diarrhea and, 175
    in freshwater ecosystems, 105
    in municipal sewage treatment, 248
    nitrogen-fixing bacteria, 41, 42, 43
    in soil, 156
    sulfur, 45
bacterial dysentery, 237
Baikal, Lake (Russia), 228
baloney detection, 11
Baltic Sea, eutrophication in, 239
Banff National Park (Canada), 136
Bangladesh
    fuelwood consumption, 292
    Grammeen Bank network, 328–29
    monsoon rains and, 200–201
    population growth, annual, 80
barley, 154
barnacles
    critical factors and Pacific Northwest, 51
    predation and, 57
barrier islands, 269
basalt, 256, 258
bases, 30–31, 241
Bates, H.W., 59
Batesian mimicry, 59, 60
batteries, storage of electricity, 288
bauxite, as strategic mineral, 261
bays. See also individual bays
    water pollution in, 245
beach erosion, 269
beach pollution, 247
Beagle, 50
bean family, nitrogen-fixing bacteria and, 41,
    42, 43
bears
    brown (grizzly) (see brown bears)
    polar, 195, 206, 274
Beaufort Sea, oil reserves, 279
Becquerel, Alexandre-Edmond, 288
beech trees, in temperate deciduous
    forests, 102
bees, honeybees, communication, 7
beetles
    Asian long-horned, as introduced species in
        US, 117
    carabid, in soil, 156
Belize, crude death rates, 84
Beni Biosphere Reserve (Bolivia), 129
Benin, natural resource protection plans, 140
Bennett, Elena, 45
benthos, 105
benzene
    as air pollutant, 212, 213
    in indoor air pollution, 213
    at Superfund sites, 313

beryllium
    as air pollutant, 212
    in computers, 309
best practicable control technology (BPT), 251
Beyer, Peter, 149
Beyond the Limits, 324
Bhagirathi River (Nepal), 234
B horizon, 157
Bhutan
    biomass, energy from, 275
    energy consumption, per capita, 275
    fuelwood consumption, 292
    natural resource protection plans, 140
Big Cypress National Preserve (Florida), 137
bioaccumulation, 162, 182–83
biocentric preservation, 15–16
biochemical oxygen demand (BOD), 238, 240
biodegradable plastics, 310
biodiversity, 107–21
    aesthetic and cultural benefits, 110–11
    benefits of, 108–10
    commercial products, 114, 115–16
    defined, 107
    diseases, 117
    drugs and medicines, 108–10
    ecological benefits, 110
    exotic species introductions, 116–17
    extinction (see extinction)
    food, 108, 109
    fragmentation, 112–13
    genetic assimilation, 117
    habitat destruction, 111–12
    hot spots, 108, 109
    human-caused reductions in, 111–17
    hunting and fishing, 113–15, 118
    minimum viable populations, 119–20
    pollution, 117
    predator and pest control, 116
    private land and critical habitat, 120
    protecting, 118–21, 167, 168
    rapid forest assessment and, 130
    recovery plans, 119
    species, number of known living, 107–8
    and stability, 66
    threats to, 111–17
    urban, 334
biogenetics, international wildlife treaties, 121
biogeochemical cycling, 38, 40–45
biogeographical areas, 138–39
biological communities. See also ecosystems
    abundance, 65–66
    climax communities, 70–71
    complexity, 66–68
    defined, 36
    diversity, 65–66
    edges and boundaries, 68–69
    introduced species and community change,
        71–72
    productivity, 64–65
    resilience, 66, 67–68
    stability, 66–68
    structure, 68
    succession, 69–70, 71

biological communities—Cont.
    in transition, 69–72
biological communities as systems, 26
biological diversity, cultural diversity and, 22
biomagnification, 182–83
biomass, 291–93
    alcohol from, 293
    biogas, 292
    current use of, as energy source, 275
    defined, 36
    dung, 292–93
    forest, 125–26
    fuelwood, 291–92
    methane (see methane)
    productivity, 64–65
    pyramids, 38, 39
    in tropical rainforests, 104–5
biomes
    broad-leaved deciduous forests, 99, 102–3
    climate and, 98, 99, 100, 102
    conifer forests, 99, 101–2
    defined, 98
    deserts, 99, 101
    estuaries, 106
    evergreen forests, 102–3
    freshwater ecosystems, 105
    grasslands, 99, 100–101
    marine ecosystems, 106–7
    Mediterranean climate regions, 99, 103–4
    tropical moist forests, 99, 104–5
    tropical seasonal forests, 99, 105
    tundra, 99, 101
    wetlands, 105–6
    world distribution, 98, 99, 100
bioremediation, 250, 315
biosphere reserves, 140, 142–43
biotechnology
    pest-resistant crops, production of, 162
    rice, genetically modified, 149
biotic potential, 61
bioweapons, 174
Bipartisan Campaign Reform Act (2002), 350
birch trees
    in conifer forests, 102
    in temperate deciduous forests, 102
Bird-Lore, 345
birds. See also individual species
    avian malaria, 195
    Christmas Bird Count (Audubon Society),
        344, 345
    lead poisoning of, 117
    smuggling of live, 115
    toxins, collection and retention of, 182–83
    tundra as breeding ground, 56
    weedy species and, 56
    wetlands destruction, 106
birdwatching, 344, 345
birth control, 79, 91–92, 93
birthrates, 83–86
    birth reduction pressures, 87–88
    crude birth rate, 83
    in demographic transition, 88
    zero population growth, 83

chromium—*Cont.*
    at Superfund sites, 313
    as water pollutant, 244
chronic effects, toxins and, 188
chronic undernourishment, 150–51
chronic wasting disease (CWD), 178
cicadas, 156
circulation models, general, 205
circumpolar vortex, 215
citizen participation, 363
civil law, 357
clams, r-adapted reproduction in, 63
classical economics, 321–22
Classical Economics, 321
Clean Air Act (1963, 1970), 210–12, 220, 355
Clean Water Act (1972, 1985), 242, 250–51, 263, 355
clear-cutting, 70, 131
Clements, F. E., 71
climagraphs, 98, 100
climate
    and air pollution (*see* air pollution)
    biomes and, 98, 99, 100, 102
    change, 201–9
    defined, 195
    diversity and, 65–66
    El Niño/La Niña/Southern Oscillation, 202–3
    global warming (*see* global warming)
    human-caused global climate change (*see* global warming)
    international climate negotiations, 208
    Milankovitch cycles and, 202
    monsoons, 200–201
    and ocean currents, relationship between, 199–200
    satellite imagery, study of climate with, 203
    seasonal winds, 200–201
climax communities, 70–71
Clinton, William J. (president, US)
    listing of steelhead trout and salmon as endangered species, 131
    new parks and national monuments, 139, 140
    road building on federal lands, debate over, 131
closed-canopy forests, 125
closed systems (ecosystems), 36
cloud forests, 104, 124
cloud seeding, 232
clover, as cover crop, 167
Club of Rome, 324
cluster housing, 340
coal, 277–78
    air pollution from burning, 16, 219, 278
    carbon in, 40
    consumption, per capita, 275, 276
    creation of, 277
    in current US energy policy, 296
    in debris from mountaintop removal mining, 263
    deposits, location of, 278
    as economic resource, 261
    efficiency, 277
    mining, 241, 262–63, 278
    proven reserves, 277–78

coal—*Cont.*
    sulfur removal from, 219
    total resource, 277–78
Coastal Barrier Resources Act (1982), 270
Coastal Zone Management Act (1972), 355
cobalt, as strategic mineral, 261
Cobb, John, 326
cocoa, conserving biodiversity with shade-grown, 167, 168
cod, Atlantic, 115, 323
coevolution, 57, 59
coffee, conserving biodiversity with shade-grown, 167, 168
Coliseum (Italy), acid precipitation damage to, 218
collective action, 363–366
Colombia
    Nevado del Ruíz volcano (1985), 267
    population control in, 88
    sanitation in Bogota, 335
colonia, 341
Colorado
    dams, effect of, 234
    environmental damage from Summitville Mine, 264
    green pricing of electrical rates, 290
    Indian Peaks Wilderness, 123
    oil shale deposits, 280
    toxic waste site at Rocky Flats, 255
Colorado River, salinity, 234, 241
Columbia River, 119
commensalism, 58
commercial fishing
    biodiversity and, 113–15, 118
    endangered species and, 131
    overharvesting, 113, 115, 155
Committee on the Status of Endangered Wildlife in Canada (COSEWIC), 118
Commoner, Barry, 16
communal resource management systems, 325
communities, biological. *See* biological communities
community-based planning, 360–63
community-unit theory, 71
comparative advantage, theory of, 327–28
competition, 57–58
complexity, in biological communities, 66–68
composting, 308
compounds, molecular, 29
Comprehensive Environmental Response, Compensation, and Liability Act (CERCLA) (1980, 1995), 251, 312–13, 355
condoms, 91, 92
conductivity, 240
cone of depression, 232
confidence levels, 8–9
Congo, poverty in, 20–21
Congressional Quarterly Weekly, 354
conifer forests, 99, 101–2
conservation. *See also* environmentalism; nature preservation
    and economic development, 142, 143
    energy, 283–85, 310, 314

conservation—*Cont.*
    historical overview, 14–17
    soil, 148, 166
    US energy policy and, 296
    utilitarian, 15
    water, 235–36
    wildlife, 119
    world conservation strategy, 140–41
Conservation International, 129, 130, 366
conserv-till farming, 169
construction
    architectural design for environment, 329–30
    on beaches and barrier islands, 269–70
    cluster housing, 340
    earthquake proofing buildings, 266
    energy savings and, 283, 285
    in floodplains, 267, 268, 269–70
    open-space zoning, 340
consumerism, personally responsible, 330
Consumer Products Safety Commission (CPSC), 358
consumers, 36, 37, 38
consumption, water, 225, 230–31
contour plowing, 166–67
contraceptives, 91–92, 93
controlled studies, 6
control rods, in nuclear reactors, 281–82
convection, atmospheric, 198, 199
convection currents, 195
conventional pollutants, 210, 220, 221
Convention on Biodiversity (CBD) (1992), 359
Convention on International Trade in Endangered Species (CITES) (1973), 121, 350, 359
Convention on Migratory Species (CMS) (1979), 359
Convention on the Transboundary Movements of Hazardous Waste (1989), 359
Convention on Wetlands of International Importance (Ramsar) (1971), 359
Convention to Combat Desertification (CCD) (1994), 359
Convention to Protect World Cultural and Natural Heritage (1972), 359
convergent evolution, 54
Cook Islands, obesity and, 152
Co-Op America, 329
copper
    in computers, 309
    as metallic resource, 259, 260
    mining, 263
    new materials substituted for, 265
    processing, 263–64
    recycling, 265
Coral Reef National Monument (US Virgin Islands), 140
coral reefs, 107
    biodiversity and, 108
    ecosystems, 36
    preservation, 140
    threats to, 140
    tropical fish smuggling from, 115
corals, black-band disease, 178

Principles of Environmental Science

developing countries
    air pollution, 209–10, 213, 221, 334–35
    biomass, energy from, 275
    burden of illness in, 177
    chronic undernourishment and, 150–51
    cities in, 332–33
    ecotourism and, 142, 143
    environmental degradation, causes of, 17–19
    fuelwood, consumption, 127, 291–92
    housing, 335–36
    human health threats from air pollution in, 216
    indigenous peoples, 22
    indoor air pollution, 213
    influence of World Bank on, 328
    life expectancy, 84–85
    meat and milk products, consumption of, 154
    natural resource consumption and, 21
    population control in, 88–89
    population growth, 80, 82–83, 332–33
    pronatalist pressures, 86
    sewage disposal and water treatment in, 237, 244, 245
    surface water quality in, 244, 245
    sustainable development in Third World, 341
    urban growth, causes of, 333
    waste stream, composition of, 308
    water pollution, 334–35
    wilderness and, 144
    wildlife and wildlife products, as main source of, 115
    wood products and, 126
development
    architectural design for environment, 329–30
    brownfield redevelopment, 314
    conservation and economic, 142, 143
    earthquake proofing buildings, 266
    housing, in developing countries, 335–36, 341
    landslides and, 268
    sustainable (see sustainable development)
    urban (see urban development)
developmental stage, and toxins, 183
Diamond, Jared, 113
diamonds, 258
diaphragms, 91
diarrhea, 335
    death from, 175
    protozoans and, 175
dichlorodiphenyltrichloroethane. See DDT (dichlorodiphenyltrichloroethane)
dieldrin, 184
dimethylsulfide (DMS), 45, 46
dinoflagellates, 239
dinosaurs, 111
dioxins, 237, 241
    in incinerator ash, 306
    persistence, 184
    seal poisoning from, 117
diphtheria
    death from, 175
    vaccination campaigns for, 176
dirt bikes, 137
Disability-Adjusted Life Years (DALYs), 85, 176
discharge, stream, 228

disease, defined, 175
diseases. See also specific diseases
    biodiversity and, 117
    disease clusters, significance of, 185
    ecological diseases, emerging, 178
    emergent, 177–78
dispute resolution, 360–63
dissolved oxygen content (DO), 238, 240
distemper, 178
divergent evolution, 53–54
diversity, 65–66
DNA (deoxyribonucleic acid), 29
    effect of mutagens on, 180
    in genetic engineering, 164
dolphins
    dolphin-safe tuna laws, 360
    striped, death from toxic pollutants, 117
    Yangtze River, 225
Dombeck, Mike, 235
Dominican Republic, population control in, 88
dopamine, 185
double-blind design of experiments, 6
doubling times, for human populations, 77–78
drancunculiasis, 177
dredge spoil, 302
drip irrigation, 230, 236
drugs (pharmaceutical), biodiversity and, 108–10
dry cask storage of radioactive waste, 282–83
dry tropical forests, 99, 105
Duany, Andres, 338
Dubois, René, 17
ducks
    black, genetic assimilation and, 117
    lead poisoning, 117
    wood, restoration of, 118
Ducks Unlimited, 365
duckweed, 250
Dumping on Dixie, 304
dune buggies, 137
dung. See also manure
    energy from, 292–93
DuPont Corporation, eco-efficiency and, 329
Duracell Corporation, eco-efficiency and, 329
Dursban, 183
dust domes, 216
dysentery, 335

E

eagles, bald
    DDT and, 182
    Klamath Basin and, 233
    recovery of, 119
    as threatened species, 118
earth
    core, 256
    crust, 256
    habitability, 3, 4
    illustration, 3
    mantle, 256
    Milankovitch cycles, 202
Earth Day, 17
Earth First!, 365

Earth Observing System (EOS), 203
earthquakes, 265–66
    dams and, 234
    defined, 265
    Kobe, Japan (1995), 266
    New Madrid, Missouri (1812), 266
    at plate boundaries, 256–57, 266
    ring of fire, 256–57
    Yucca Mountain radioactive waste repository and, 29, 255, 285
Earth Summit (Brazil, 1992), 17, 93, 203, 208, 359, 366
Earthwatch, 350
Ebola, 174, 175, 177
Ecoforum (Earth Summit, 1992), 366
Eco-Kids Corps, 364
ecological development, 70
ecological diversity, 107
ecological economics, 322–24
ecological niche, 54–55, 56
ecological pyramids, 38, 39
ecological services, 322, 323, 326
ecological succession, 69–70, 71
ecology
    defined, 26
    long-term ecological research, 334
    sustainability, 21–22
    urban, 334
Ecology, 26
economic mineralogy, 259–62
    coal, 261
    defined, 259
    metals, 259–60
    nonmetal mineral resources, 260–61
    oil, 261
    strategic metals and minerals, 261–62
    uranium, 261
economics
    accounting, natural resource, 325–26
    capital, 322
    communal property and "Tragedy of the Commons," 325
    costs, internal and external, 327
    discount rates, 327
    ecological, 322–24
    green business (see green business)
    gross national product, 322, 326
    Index of Sustainable Economic Welfare, 326
    natural resource, 322
    neoclassical, 322
    nonmonetary resources, accounting for, 326–27
    resources (see natural resources; resources)
    steady-state economy, 322
economy
    eco-efficient economy, goals for, 329
    steady-state economy, 322
ecosystems. See also biological communities; *specific ecosystems*
    defined, 36
    ecosystem management policy, 132–33
    parks as, 138–39
ecotones, 69
ecotourism, 142, 143

hazardous waste—*Cont.*
  brownfields, 314
  defined, 311
  disposal, 305, 312–14
  exporting, 303–5
  landfills and, 303
  less hazardous substances, converting to, 314–15
  less waste, production of, 314
  liability and degree of purity required with cleanup, 314
  management, 314–16
  minority communities and, sites, 304
  nuclear waste (*see* nuclear waste)
  recycling of, 305
  retrievable storage, 315–16
  secure landfills, 316
  sites, EPA data on, 241
  Superfund sites, 312–14
  in waste stream, 301
Headwaters Forest (California), 366
health, defined, 175
Health, Education, and Welfare, US Department of, 339
Health, US Department of, 191
health hazards
  accepting risks, 189–90
  air pollution, 173, 216
  antibiotic resistance, 178–79
  assessing risks, 189, 190
  biological warfare and terrorism, 174
  bioweapons, 174
  diet, 181
  ecological diseases, emerging, 178
  emergent diseases and environmental change, 177–78
  endocrine hormone disrupters, 181
  infectious organisms, 175–76
  morbidity, 175, 176–77
  pesticide resistance, 178–79
  toxic chemicals, 179–81
  types of environmental, 175–82
heap-leach extraction, 263, 264
heart attacks, from air pollution, 216
heat
  defined, 27
  latent, 198, 199
  specific heat, water and, 27
heat islands, 216
heat of vaporization, 33–34
heavy metals
  bioaccumulation of, 182
  as neurotoxins, 180
helium, atmospheric, 195, 196
helium-3, 29
hemlock trees, 102
hemorrhagic fever, 174, 177
hepatitis, infectious, 237
heptachlor, 184
herbicides
  disease clusters and, 185
  genetic engineering and, 165
  persistence, 184, 185

herbivores, 37, 38
heroin, 185
hexachlorobenzene (HCB), 184
high responders, 163, 164
high-temperature solar energy, 287
Hill, Julia "Butterfly," 365–66
Himalayas (Asia)
  dams, 234
  heavy rainfall at foot of, 200, 226
  rising, 256
Hodgkins disease, 110
hogs, 155
  low-input farming and, 169
  mechanized farming and, 155, 169
Honda
  fuel cell automobiles, 290
  low-emission, hybrid gas-electric vehicles, 283, 284, 330
hormones, endocrine hormone disrupters, 181
Housing and Urban Development, US Department of, 337
Huang He (China)
  erosion and, 161
  and global climatic changes, 195
Huascarán National Park (Peru), 140
Hubbert, Stanley, 323
Hubbert curves, 323–24
Hudson Bay Company (Canada), 61, 62
Hudson River, mandated clean-up of polluted sediment, 2
Human Development Index (HDI), 326–27
human populations
  birthrates (*see* birthrates)
  birth reduction pressures, 87–88
  demographics (*see* demographics)
  doubling times, 77–78
  of earth, 17
  future of, 92–93
  global village, world as, 82
  growth, 76–78
  history, 77, 78
  limits to growth, 78–80
  Malthusian theories on, 78–79, 89
  Marxian theories on, 79
  more people, effect of, 79–80
  population density map, world, 374–75
  pronatalist pressures, 86–87
  rural to urban shift, 331–32
  technology, human population growth and, 79
  world population, 76, 80, 82–83
Humboldt Bay, 250
Hungary
  dangerously polluted air in, 216
  population growth, annual, 80
  water quality in, 244
hunting
  biodiversity and, 113–15, 118
  overharvesting, 113, 115
Hupa people (Native Americans), 233
Hurricane Andrew (1992), 194
Hurricane Floyd (1999), 155, 169
Hutchinson, G. Evelyn, 26, 54
hyacinths, water, 315

hydrocarbons
  in air pollution, 212, 278
  controls, 219–20
  nonmethane, in air pollution, 210, 211
  as organic compounds, 31, 32
  removal of, 219
  from wood burning, 291
hydrochlorofluorocarbons (HCFCs), 215
hydroelectricity, 275
hydrofluorocarbons, in Kyoto Protocol, 208
hydrogen, 28, 29
  in acids and bases, 30
  atmospheric, 195, 196
  atomic number, 29
  fuel cells, 290–91
  molecule, 29
  in organic compounds, 31, 32
  in photosynthesis, 35–36
hydrogen chloride, molecule, 29
hydrogen cyanide, excretion of, 186
hydrogen sulfide, 45
  odors from manure lagoons, 169
hydrologic cycle, 40, 227, 228
hydrolysis, 259
hydropower, 293–94
hyenas, 145
hypolimnion, 105
hypotheses, scientific, 7–8

I

Iceland
  geothermal energy at, 295
  wealth in, 20, 21
Idso, Sherwood, 207
igneous rocks, 258
Illinois, field testing of genetically engineered crops, 164
immigration, 331–33
  push-and-pull factors, 333
impala, 135, 145
inbreeding, in biodiversity, 120
incineration, 305–6, 314–15, 316
Index of Sustainable Economic Welfare (ISW), 326
India
  air pollution in, 209, 335
  Chipko Andean movement, 129
  DDT used against malaria, 178
  deforestation, 128
  dung, use of, 292
  forest protection, protests for, 129
  72 ft. of rain at Cherrapunji, 227
  immigration to Mumbai, 332
  iron deficiencies, 152
  life expectancy, 84–85
  methane generators, use of, 308
  monsoon rains and, 200
  population control, 76
  population growth, 76, 80
  poverty in, 20, 75
  respiratory diseases in Calcutta, 335
  sanitation in, 335

India—Cont.
    Sardar Sarovar Dam, 234
    slums and shantytowns in Mumbai, 335
    squatter settlements in Bhopal, 336
    Taj Mahal, acid precipitation damage to, 218
    turtle-friendly shrimp laws, WTO complaint
        over, 360
    waste-fed aquaculture system in Calcutta, 250
    water quality in, 245
Indian Ocean, long range transport of air
    pollutants to, 214
indigenous peoples, 22, 142–43. *See also
    individual tribes or peoples*
Indonesia
    acute poverty in, 346
    banning of CCA-treated wood, 185
    biodiversity of, 108, 109
    deforestation, 128
    immigration to Jakarta, 332
    indigenous peoples, treatment of, 22
    Krakatoa volcano (1863), 266
    old-growth forests in, 126
    population control in, 88
    poverty in, 20
    sea turtle egg hunters in, 140
    soil erosion in, 326
    squatter settlements, 341
    Tambora volcano (1815), 267
    terrace cultivation of rice, 167
    tropical fish smuggling, 115
    urbanization in, 334, 335
indoor air pollution, 213–14
inductive reasoning, 6–7
Indus River (India), and global climatic
    changes, 195
industry
    energy, uses of, 276
    energy savings and, 283
    waste, 300, 310
    water use, 241
infanticide, 91
infant mortality and women's rights, 90–91
infectious diseases
    death from, 175–76
    new, appearance of, 177–78
    outbreaks, map of, 177
    progress in eliminating, 176
infiltration, 228
influenza
    death from, 175
    epidemic, 177–78
infrared radiation, 34, 35
    in global warming, 204
    remote sensing of, 203
    in solar radiation, 196–98
inholdings, 137
inorganic pollutants, 238, 240–41
insecticides, genetic engineering and, 165
insects
    biodiversity and, 108
    in freshwater ecosystems, 105
    as infectious agents, 237
    predation in, 57

insects—Cont.
    tropical rainforests and, 104–5
    and tundras, 101
insight, in science, 6
insolation, 196
Instituto Nacional de Biodiversidad (INBIO)
    project (Costa Rico), 110
integrated pest management, 162
intensive agriculture, 248
Intergovernmental Panel on Climate Change
    (IPCC), 203–4
Interior, US Department of the, 356, 358
internalizing costs, 327
International Council on Mining and Metals, 263
International Geophysical Year, 203
International Institute for Aerospace Survey
    (Netherlands), 262–63
international nongovernmental organizations
    (NGOs), 366
International Rice Institute (Philippines), 163
International Soil Reference and Information
    Centre (Netherlands), 158
international trade, 327–28
international treaties and conventions, 359–60, 361
International Union for the Conservation of
    Nature and Natural Resources (IUCN),
    138–39, 140–41
International Whaling Commission, 114–15
Internet, 363
interspecific competition, 57
intraspecific competition, 57
intrauterine devices (IUDs), 91, 93
introduced species, and community change, 71–72
Inuit people, 214
iodine, 152, 212
ionic compounds, 29–30
ions, 28–29
Iowa
    Davenport flood (1993), 268
    genetically engineered crops, field testing
        of, 164
    Revolving Loan Fund for renewable
        energy, 290
Iran, family planning, 88, 89
iron
    in core of the earth, 256
    deficiencies, 152
    as metallic resource, 259, 260
    minimills, 265
    recycling, 265
iron disulfide (pyrite), 44
irrigation
    efficiency, 161, 230, 236
    global freshwater use for, data on, 161
    water consumption, 230
island biogeography, 112–13
isoprenes, 212
isotopes, 29–30
Israel
    mining groundwater, 229
    plastic mulch, use of, 167
    solar collectors, domestic hot water from, 287
    wealth in, 20

Italy
    Coliseum, acid precipitation damage to, 218
    fertility rate, 84
    population growth, annual, 80
    solar collectors, domestic hot water
        from, 287
    United Nations World Food Summit (Rome)
        (2002), 150
    Vesuvius, Mount, volcano (A.D. 79), 266
Izaak Walton League, 365

J

jade, 258
James Bay project (Canada), 234
Janzen, Dan, 129
Japan
    amorphous photovoltaic cells, 289
    anthropogenic carbon dioxide, production of,
        205, 206
    coral reefs, threatened, 140
    eco-efficiency and, 330
    energy consumption, per capita, 276
    geothermal energy at, 295
    Kobe earthquake (1995), 266
    megacities, 331
    metals, consumption of, 259–60
    population growth, annual, 80
    recycling, 307
    reforestation in, 127
    seafood consumption, 155
    waste-to-energy plants, 305
    water quality improvement, 244
    wealth in, 21
    wind energy, use of, 294
Java, tropical rainforests in, 104
J curve, 61, 62, 63
jellyfish
    damage to Black Sea, 117
    Leidy's comb jelly, 117
jet streams, 199, 200
jobs and the environment, 330
Johnson and Johnson, as green company, 329
Jordan, demographic transition in, 88
Joshua Tree National Park (California), 101
Joshua trees, 101
joules, 27, 275
Journal of Public Health, 20
judicial branch in environmental law, 355–57

K

K-adapted species, 63–64
Kampuchea (Cambodia), deforestation, 128
Kansas, wind farms, 294
kelp, keystone species and, 60
Kenya
    demographic transition in, 88
    population momentum in, 86
    Serengeti ecosystem, 145
kerogen, 280
Kesterson Wildlife Refuge (California),
    144–45, 241

O

pertussis
   death from, 175
   vaccination campaigns for, 176
Peru
   government policies and improvements in
      Lima, 333
   squatter settlements, 341
pesticides
   agricultural use of, 162–63
   alternatives, 162
   dangers from, 162
   dietary pesticide risk, 181
   disease clusters and, 185
   effect of, on biodiversity, 116
   persistence, 184–85
   reducing, in your food, 163
   resistance, 178–79
   runoff as water pollutant, 241
pests
   agricultural damage from, 162
   control of, genetic engineering and, 165
   integrated pest management, 162
PET (polyethylene terphthalate), 307, 308
pets, trade in wild species for, 114, 115
Pfiesteria piscicida, 239
pH, 217–18, 240
pheasants, edge effects and, 69
phenols, 212, 314
   at Superfund sites, 313
Philippines
   coral reefs, threatened, 140
   garbage problems in Manila, 302
   immigration to Manila, 332
   indigenous peoples, treatment of, 22
   Mayon volcano (1984), 266
   Pinatubo, Mount, volcano (1991), 205, 267
   rice terraces in Chico River Valley, 167
   scavenging in Manila, 310
   Smoky Mountain metropolitan dump in
      Manila, 302
   tropical fish smuggling, 115
   water quality in, 245
phosphates, 244
   aquatic pollution from, 161
   in freshwater ecosystems, 105
   sewage treatment and, 248
phosphogypsum, 305
phosphorus, 28, 45
   agricultural use of, 45
   imbalances, 45
   in organic compounds, 31, 32
   plants and, 161
phosphorus cycle, 43–44
photochemical oxidants, 210, 211, 212
   as air pollutants, 211
   plant losses from, 217
photochemical smog, 216
photodegradable plastics, 310
photosynthesis
   in aquatic ecosystems, 105
   biomass production and, 36
   energy capture and, 35–36, 286
   in marine ecosystems, 106–7

photosynthesis—*Cont.*
   in wetlands, 106
photovoltaic energy, 288–89
pH scale, 30–31
phytoplankton, 182, 208–9
phytoremediation, 315
pigeon, passenger, 113–14, 323
pigs, as introduced species, 72
Pimentel, David, 79
Pinchot, Gifford, 15, 16
Pindus National Park (Greece), 140
pine trees
   in conifer forests, 102
   lodgepole, succession and, 71
   Ponderosa, selective cutting of, 131
   white, tolerance limits and eastern, 52
pioneer species, 56, 70
placer mining, 262
plague, 174
plague (Black Death), 177
plankton
   in freshwater ecosystems, 105
   in marine ecosystems, 106
   predation and, 57
planning
   community-based planning, 360–63
   urban land-use planning, 320, 338–41
plants. *See also individual species*
   biomass (*see* biomass)
   in bioremediation, 250, 315
   biotechnology (*see* biotechnology)
   cells, 31, 33
   clustering and, 68
   competition and, 57–58
   edible wild plant species, 108
   extinction (*see* extinction)
   genetic engineering and (*see* genetic
      engineering)
   global warming, effects of, 206
   grazing and, 134–35
   in hydrologic cycle, 227, 228
   nitrogen cycle and, 41–43
   overharvesting, 115, 116
   pathology from air pollution, 216–17
   perennial species, reducing erosion with, 167
   photosynthesis (*see* photosynthesis)
   rapid forest assessment and, 130
   smuggling, 114, 115, 116
   taxonomy, 54
   threatened species of, 118, 134
   vegetation map, world, 372–73
   weedy species, 55–56
plasma atomic emission spectrometry, 240
plastics
   degradable, 310
   recycling, 306, 307
   reuse, 309–10
   waste reduction, 310
   water pollution and, 245–46, 247
Plater-Zyberk, Elizabeth, 338
platinum
   in fuel cells, 290–91
   as metallic resource, 260

platinum—*Cont.*
   recycling, 32, 265
   as strategic mineral, 261
Plato, 5
plutonium, 29, 212
pneumonia, death from, 175
poaching, 145
point source pollution, 236, 237
Poivre, Pierre, 15
Poland
   acid rain damage to aquatic ecosystems, 218
   cost of reducing particulate air pollution in
      Tarnobrzeg, 326
   dangerously polluted air in, 216
   environmentally linked diseases in, 173
   poor air quality in, 221
   water quality in, 244
polar-front jet streams, 199, 200
policy cycle, 350–51
poliomyelitis, 237
   death from, 175
   near eradication, 176
   vaccination campaigns for, 176
pollution
   air (*see* air pollution)
   biodiversity and, 117
   environmental racism and, 304
   odor pollution, 169
   water (*see* water pollution)
polybrominated diphenyl ethers (PBDE),
      persistence, 184
polychlorinated biphenyls (PCBs). *See* PCBs
      (polychlorinated biphenyls)
polycyclic aromatic compounds, 291
polymers, 265
polyvinyl alcohol, 288
ponds, as freshwater ecosystem, 105
poplar trees, 315
population dynamics, 60–64
   boom and bust cycles, 61–62
   community structure, 68
   complexity, 66–68
   edges and boundaries, 68–69
   growth, 60–61, 62
   K-adapted species, 63–64
   limiting factors, 62–63
   r-adapted species, 63–64
   resilience, 66, 67–68
   stability, 66–68
   stable population, growth to a, 62, 63
population momentum, 86
populations
   defined, 36
   human (*see* human populations)
Ports and Waterways Safety Act (1972), 251
positive crankcase ventilation (PCV), 219–20
potash, as nonmetallic resource, 260, 261
potassium
   in freshwater ecosystems, 105
   in organic compounds, 31, 32
   plants and, 161
potatoes, 154
potential energy, 27

Potrykus, Ingo, 149
poultry, 155
poverty
   acute, 19, 20
   burden of illness and, 177
   developed *versus* developing countries, 20–21
   effects of, 19–20
   food security and, 17
   life expectancy and, 84–85
   population growth and, 79
Powell, Lake, 234
power, defined, 275
power towers, 287
prairie dogs, extermination of, 116
prairies, 100, 133–34. *See also* rangelands
Precambrian time, 277
precautionary principle, 347–48
precipitation. *See also* rainfall
   aerosols and, 195
   in biome distribution, 98, 99, 100
   changes in, global warming and, 205–6
   convection currents and, 198, 199
   hydrologic cycle, 227
   infiltration of, 228
   mean annual, 226
   protecting forests to preserve rain, 124
predation, 56–57
predator control, biodiversity and, 116
pressurized water reactors (PWR), 282
primary producers, 36, 38
primary sewage treatment, 248–49
primary succession, 69–70
primates, emergent diseases and, 177, 178
Principe Island (Africa), 6
prions, 178
private land and critical habitat, 120
probability, 9, 10
processing, metal, 263–64
Proctor and Gamble, 365
producers, 36, 38
productivity, 36, 64–65
products, chemical, 30
progesterone, 91, 92
Project Feeder Watch, 345
pronatalist pressures, 86
propane, molecular structure, 32
prospective study of toxins, 187
proteins
   as organic compounds, 31, 32
   protein deficiencies, 152, 153
Protocol on Substances that Deplete the Ozone Layer (Ozone) (1985), 359
protons, 28–29
protozoans
   as cause of diarrhea, 175
   in soil, 156
proven-in-place reserves, 277–78
Prudhoe Bay (Alaska), 274
pseudoscience, 11
pseudoscorpions, 156
Public Interest Research Groups (PIRGs), 364
Public Utilities Regulatory Policies Act (1978), 295

Puerto Rico, field testing of genetically engineered crops, 164
pull factors, 333
pumped-hydro storage of electricity, 289
punctuated equilibrium, 50
pupfish, desert, tolerance limits and, 51
push factors, 333
PVC (polyvinyl chloride) plastics, 307
   persistence and, 184
pyrite, 44
pyroclastic flows, 266
Pythagoras, 5
pythons, albino, smuggling, 115

Q

Qapdo, 233
quality of life, 176
quartzite, 258
quetzal, 124

R

rabbits, jack, in chaparral, 103
raccoons, as weedy species, 56
r-adapted species, 63–64
radiation, sunlight and, 34–35
radical environmental groups, 365–66
radioactive isotopes, 29
radioactive waste. *See* nuclear waste
radio wavelengths, 34, 35
radon, 8
rainfall. *See also* precipitation
   convection currents and, 198, 199
   hydrologic cycle, 227
   monsoons, 200–201
   rain, process of, 198–99
   water availability and, 230
Rainier, Mount (Washington), 267
rain shadow, 226
ranching
   new approaches to, 135
   predator control, 116
   wolves and, 138
rangelands, 133–35
   management, 134
   ranching, new approaches to, 135
   in United States, 133, 134–35
rape seed, 165
rapid forest assessment, 130
Rathje, William, 300
rational choice, as model for public decision making, 352
rats
   as introduced species, 72
   kangaroo, in chaparral, 103
   as weedy species, 56
reactants, 30
reasoned judgement, 362
recharge zones, 228
recycling, 306–8
   battery recycling program in Minnesota, 306
   benefits, 307–8

recycling—*Cont.*
   challenges, 306–7
   defined, 306
   electronics recycling systems, 309
   hazardous waste, 305, 310, 314
   incentives for, creating, 308
   nonrenewable resources and, 323
   profitability and, 306–7
   study of, in Phoenix, 300
Red Sea, coral reefs threatened, 140
red tides, 239
reduced nicotinamide adenine dinucleoside phosphate (NADPH), 35
reduced tillage systems, 168–69
reduction, chemical, 30
redwood trees, giant, 356
Reedy River, pollution of, 357
reefs. *See* coral reefs
reflective thinking, 11, 13
reformer, in fuel cells, 290
refuse-derived fuel incinerator, 305
regenerative farming. *See* sustainable agriculture
regulatory agencies, 358–59
Reilly, William K., 348
reintroductions, of endangered species, 119
relativity, theory of, 6–7, 10
relevé, 130
remote sensing, 203
renewable energy, promoting, 289–90
renewable resources, 323
renewable water supplies, 229
reproducibility, science and, 6
reproduction, K-adapted and r-adapted species, 63–64
reptiles, smuggling, 115
reservoirs, 232
residence time, 227
resilience, in biological communities, 66, 67–68
resistance, antibiotic, 178–79
Resource Conservation and Recovery Act (RCRA) (1976), 251, 312, 313, 355
resource partitioning, 55, 56
resources
   in classical economics, 321–22
   communal property and "Tragedy of the Commons," 325
   communal resource management systems, 325
   cost-benefit analysis and natural resource accounting, 325–26
   costs, internal and external, 327
   defined, 320
   discount rates, 327
   in ecological economics, 322–24
   extraction (*see* mining)
   limits to growth, 324
   mineral (*see* economic mineralogy)
   natural (*see* natural resources)
   in neoclassical economics, 322
   nonmonetary resources, accounting for, 326–27
   nonrenewable, 322–23
   open access systems, 325
   renewable, 323

Principles of Environmental Science

tubal ligations, 91
tuberculosis
　　death from, 175
　　vaccination campaigns for, 176
tuff, 258
tularemia, 174
tuna
　　dolphin-safe tuna laws, 360
　　dolphins killed by use of nets to catch, 360
tundra, biome, 99, 101
Tuolumne River, 234
turkeys, wild, restoration of, 118
Turner, Frederick Jackson, 143
Turner, Ted, 135
turtles, Blandings, 13
Tuvalu, 207
Typhlodromus, 57
typhoid, 237, 335

U

Ukraine, Chernobyl nuclear accident (1986), 282
ultraviolet radiation
　　absorption by ozone, 196, 214, 215
　　remote sensing of, 203
　　sewage treatment and, 248–49
　　from sunlight, 34, 35
　　UV-B, 196
unconventional pollutants, 23
underground mining, 262, 263
ungulates, 135, 145
United Arab Emirates, wealth in, 20
United Kingdom
　　carbon dioxide emissions, progress in
　　　reducing, 208
　　wind farms, 294
United Nations
　　Conference on Environment and
　　　Development (Brazil, 1992), 17, 93,
　　　203, 208, 359, 366
　　Conference on Population and Development
　　　(Egypt, 1994), 76, 90, 91, 93
　　Conference on the Human Environment
　　　(Sweden, 1972), 17, 366
　　Development Programme (UNDP), 108,
　　　326–27
　　Educational, Scientific, and Cultural
　　　Organization (UNESCO), 143
　　Environment Programme (UNEP), 140, 203–4
　　Family Planning Fund, 93
　　Food and Agricultural Organization (FAO),
　　　125, 127–28, 150, 152, 153, 161
　　Framework Convention on Climate Change
　　　(1992), 203, 208, 359
　　freshwater shortages, data on, 231
　　global ecosystems, study on health of, 17
　　Human Development Index, 326–27
　　human population, data on, 80
　　instability, data on population movement due
　　　to, 333
　　lack of enforcement powers, 359
　　Law of the Sea (1982), 359
　　Man and Biosphere program (UNESCO), 143

United Nations—*Cont.*
　　polluted air, data on dangerously, 216
　　Population Division, 92
　　slums and shantytowns, data on, 335
　　Summit (South Africa, 2002), 231
　　and water shortages, 225
　　World Food Summit (Rome, 2002), 150
United States. *See also individual states*
　　acid precipitation, 217–18
　　air pollution, 216, 219, 220–21
　　anthropogenic carbon dioxide, production of,
　　　205, 206
　　antibiotic use, 179
　　automobile fuel efficiency, 283, 284
　　biological and chemical warfare programs, 174
　　chronic wasting disease in, 178
　　citizen science in, 345, 350
　　clear-cutting in, 131
　　consumption and waste generation in, 21
　　demographics, 80, 85, 86
　　downtowns, lack of recognizable, 336
　　eco-entrepreneurs, 329
　　energy, uses and consumption, 162, 275,
　　　276–77, 286–87
　　energy policy, 296
　　erosion rates, 160
　　fecal contamination infections in, 245
　　fuelwood, use of, 291
　　genetically engineered crops in, 164–65
　　geothermal energy at, 295
　　groundwater in, 231–32, 245
　　hazardous waste, 311, 313–14, 315
　　household water use, 235–36
　　introduced species, damage from, 116–17
　　landslides, 268, 269
　　life expectancy, 85
　　logging, controversies over, 130–32
　　mechanized farming in, 169
　　megacities in, 331
　　metals, consumption of, 259–60
　　microlending in, 329
　　mine fires, smoldering, 262–63
　　nuclear energy, use of, 275, 280–81
　　obesity, 152, 181
　　ocean dumping, 302
　　overnutrition in, 152
　　parks, 135, 136
　　petroleum resource, recoverable, 279
　　population shift away from older cities, 336
　　poverty, 20
　　radioactive waste in, 282–83
　　rangelands in, 133, 134–35
　　recycling, 264, 307, 309
　　rural to urban shift, 331–32
　　solar radiation, average daily, 286
　　strategic metals and ores, stockpiles of, 261
　　streams contaminated by mine drainage, 263
　　surplus material exchanges, 314
　　threatened plant species, 134
　　trash disposal costs in, 303
　　urban material flow, 334
　　water consumption and use, 225, 235–36
　　water policies, 234–35

United States—*Cont.*
　　water quality, 242, 243–44
　　wealth, 20, 21
　　wetlands in, 106, 112
　　wilderness areas, 144
　　wildfires, increase in western, 132
　　wildlife refuges, 144–45
　　wind energy, use of, 294–95
　　wood products, producer of, 126
Upper Klamath Lake, 233
Upper Missouri River Breaks National
　　Monument (Montana), 137
uranium
　　as air pollutant, 212
　　as economic resource, 261
　　mining, radioactive waste from, 282
uranium-235, 281
urban development, 330–34
　　developed world, urban sustainability in,
　　　338–41
　　government policies, 333–34
　　green office projects, 329
　　immigration push-and-pull factors, 333
　　land-use planning, 320, 338–41
　　planning, urban, 320
　　smart growth, 337–38
　　sustainable cities, 330–31
　　Third World, sustainable development
　　　in, 341
　　urban growth, 332–34
　　world's largest metropolitan regions, 331
urban ecology, 334
urbanization, defined, 331
urban problems
　　air pollution in urban climates, 215–16
　　in developing countries, 334–37
　　housing, 335–36, 341
　　urban sprawl, 336–37
urban sprawl, 336–37
urine
　　nitrogen in, 43
　　toxin accumulation in, 186
Utah
　　Bingham Canyon copper mine, 263
　　Glen Canyon Dam (Arizona/Utah), 232, 234
　　oil shale deposits, 280
utilitarian conservation, 15
Uzbekistan, land degradation, 158

V

vaginal rings, 92
vaginal sponges, 91
valuing context in environmental
　　education, 348
vanadium, in cars, 265
Vancouver Island (Canada), as temperate
　　rainforest, 103
Van Der Ryn, Sym, 338
Vanguard I, 288
vasectomies, 91
vegetables, dietary, 153, 154, 181
Venezuela, tar sands resources, 279

Vermont
  acid precipitation damage to Camel's Hump
      Mountain, 218
  cull wood fuel for power plant in
      Burlington, 291
  private energy purchased by utilities, 295
vertical stratification, in freshwater
      ecosystems, 105
Vietnam
  banning of CCA-treated wood, 185
  deforestation, 128
vinblastine, 110
vincristine, 110
vinyl chloride, 212
visibility reduction, air pollution and, 217
vitamin A, 181
  carcinogens and, 185
  deficiency, golden rice and, 149
  dietary deficiencies, 153
vitamin C, 181
vitamin deficiencies, 152, 153
vitamin E, 181, 185
volatile organic compounds (VOCs)
  as air pollutants, 210, 211, 212
  controls, 219–20
  from oil combustion, 279
volcanoes
  defined, 266
  Krakatoa (Indonesia) (1883), 266
  Mayon (Philippines) (1984), 266
  Mount St. Helens (Washington) (1980), 267
  Nevado del Ruíz (Colombia) (1985), 267
  Pelee, Mount, (Martinique) (1902), 267
  Pinatubo, Mount, (Philippines) (1991), 205, 267
  Tambora (Indonesia) (1815), 267
  tectonic plates and, 256–57
  Vesuvius, Mount, (Italy) (AD 79), 266
  Yucca Mountain radioactive waste repository
      and, 29, 255, 283
Von Frisch, Karl, 7
vulnerable species, 118
vultures, 145

W

Wallace, Alfred Russel, 52
walruses, 206
Ward, Barbara, 4, 17
Washington
  Mount St. Helens volcano (1980), 267
  northern spotted owls, controversy over
      protection of, 119
  toxic waste site at Hanford, 255
Washington, DC, acid precipitation damage to
      Washington Monument, 218
wasps, Batesian mimicry in, 59, 60
waste, 300–311
  agricultural, 300
  animal, 155
  composting, 308
  demanufacturing, 308–9
  disease-causing organisms from
      inadequately treated, 237

waste—Cont.
  disposal costs, 303
  energy from waste, 308
  environmental racism and, 304
  exporting, 303–5
  garbology, 300
  hazardous waste (see hazardous waste)
  incineration, 305–6, 314–15, 316
  industrial, 300
  landfills (see landfills)
  methane gas from, 292–93
  municipal, 300–301
  ocean dumping, 302
  open dumping, 301–2, 313–14
  producing less, 310–11
  recycling (see recycling)
  reuse, 309–10, 314
  waste stream, 301
waste-to-energy, 305
water. See also aquatic ecosystems
  agricultural consumption, 161, 230
  atmospheric, 229
  availability, 229–30
  compartments, major water, 227–29
  as compound, 29
  conservation, 235–36
  consumption, 225, 230–31
  dams (see dams; individual dams)
  domestic conservation, 235–36
  domestic consumption, 231
  erosion, 159–60
  as future environmental issue, 17
  hydrologic cycle, 40, 227, 228
  increase water supplies, efforts to, 232
  industrial consumption, 231
  lakes (see lakes)
  legislation in US, 250–51
  management, 235–36
  measurement, 226
  mining, 229, 232
  molecule, 29
  oceans (see oceans)
  phosphorus cycle and, 43, 44
  policy, 234–35
  price mechanisms, 234–35
  properties, 32–34
  renewable water supplies, 229
  residence time, 227
  resources, 225–26
  rivers (see streams)
  shortages, freshwater, 224, 225, 231–35
  specific heat and, 27
  streams (see streams)
  sulfur cycle and, 44–45
  testing, 240
  use, 225, 230–31, 235
  water-rich and water-poor countries, 229–30
Water Act (Canada, 1970), 242
water banks, 233
waterfowl. See also individual species
  Klamath Basin and, 233
  lead poisoning of, 117
  wetlands and, 106

waterlogging, 161
water pollution, 246–50
  acids and bases, 241
  from agricultural runoff, 161
  animal waste as, 155, 169
  biological pollution, 237
  Canada, water quality in, 242, 243–44
  categories of water pollutants, 237–38
  control, 247–50
  cultural eutrophication, 238–39
  in developing countries, 334–35
  drinking water, use of polluted water for, 231
  eutrophication, 238, 239
  fecal coliform bacteria, 237–38, 240
  as future environmental issue, 18
  groundwater pollution, 228, 231, 232, 245, 246
  Hudson River, mandated clean-up of, 2
  infectious agents, 237–40
  inorganic pollutants, 238, 240–41
  legislation in US, 250–51
  metals, 240–41
  from mining, 241, 262–63
  nonmetallic salts, 241
  nonpoint source pollution, 236–37,
      243–44, 248
  ocean pollution, 245–47
  from oil drilling, 279
  from ore processing, 263–64
  organic chemicals, 238, 241
  in other countries, 244–45
  oxygen demand, 237–38
  plant nutrients, 238, 239
  point source pollution, 236, 237
  from sediment, 238, 241–42
  sewage treatment (see sewage treatment)
  tar sands and, 280
  testing water, 240
  thermal, 238, 241, 242
  toxic tides, 239
  water quality today, 242–47
  watershed approaches, 242–43
  and wildlife refuges, 144–45
watershed, 235, 242–43
water table, 228
water vapor
  absorption of solar radiation, 196
  atmospheric, 195
  latent heat in, 198, 199
wave energy, 295–96
weather
  atmospheric pressure and, 198, 199
  convection currents and, 198, 199
  defined, 195
  El Niño/La Niña/Southern Oscillation, 202–3
  human-caused global climate change, 195
weathering, 259
  carbon released by, 41
  of spoil banks, 263
weedy species, 55–56
West Antarctic Ice Sheet, shrinking of, 206
Western Ghats (India), 200
West Virginia, debate over mountaintop
      removal mining, 263

Principles of Environmental Science   www.mhhe.com/environmentalscience